Birkhäuser

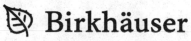

Gary F. Birkenmeier • Jae Keol Park •
S. Tariq Rizvi

Extensions of Rings and Modules

Birkhäuser

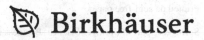

Gary F. Birkenmeier
Department of Mathematics
University of Louisiana at Lafayette
Lafayette, LA, USA

S. Tariq Rizvi
Department of Mathematics
The Ohio State University at Lima
Lima, OH, USA

Jae Keol Park
Department of Mathematics
Busan National University
Busan, South Korea

ISBN 978-1-4899-9714-2
ISBN 978-0-387-92716-9 (eBook)
DOI 10.1007/978-0-387-92716-9
Springer New York Heidelberg Dordrecht London

Mathematics Subject Classification (2010): 16-02, 16DXX, 16E50, 16E60, 16GXX, 16LXX, 16N20, 16N60, 16P60, 16P70, 16R20, 16S34, 16S35, 16S36, 16S50, 16S60, 16UXX, 16W10, 16W20, 16W22, 46H10, 46L05, 46L08, 46L35, 46L40, 46L45, 15A12, 15A21, 15A33, 65F35

Printed on acid-free paper

Springer is part of Springer Science+Business Media (www.springer.com)

Dedicated to Our Parents, Families, and Teachers

Preface

Since the discovery of the existence of the injective hull of an arbitrary module independently in 1952 by Shoda and in 1953 by Eckmann and Schopf, there have been numerous papers dedicated to the study and description of various types of hulls or "minimal" extensions of rings and modules satisfying some generalizations of injectivity or of related conditions. The study of these overrings, overmodules and extensions satisfying these conditions has been dealt with in detail in different papers. The question of when do certain properties transfer from any ring R to its many types of extensions such as matrix ring extensions, (skew-)group rings, polynomial ring extensions and Ore extensions, has also been of interest to many for a long time. It appears that the research work on various types of hulls and on the wide varieties of extensions is spread throughout the literature in disparate research papers. Thus, a book which presents (at least some part of) the state of the art on the subject and includes some of the most recent work done on these topics, is needed. That there has not been a comprehensive treatment of these topics, is one of the main reasons for us to write this research monograph.

Since the properties such hulls and extensions satisfy may be unlimited and thus cannot possibly be covered in one book, we wish to emphasize that *in this monograph we have focused mainly on hulls and extensions that satisfy certain conditions on direct summands*. We have however also made an attempt to provide a general theory for hulls belonging to arbitrary classes of rings (or modules) which can satisfy other properties.

Among other reasons, the need to present the results on the transference of certain algebraic properties to and from base rings and modules to various ring and module extensions belonging to specific classes, in a systematic way, has also been a motivation in writing this research monograph. To ensure some efficiency in the transfer of information between a ring or a module to its overrings or overmodules, respectively, we use the notion of a "minimal essential extension" with respect to belonging to a special class. We term this a "hull" (belonging to that particular class) and show that the "closeness" of such hulls to the base ring (or module) enriches the transfer of information from the base ring (or module) to such a hull. This will be shown as a useful tool in analyzing the structure of a ring (or of a module). Our desire is to present research work we have been involved in for over two decades as well as that done by others on the various topics of this monograph. We also wish to showcase the various applications of this research to Algebra and Functional Analysis.

Our view in writing this book is also to stimulate new and further research on the topics presented. A number of open research problems are listed at the end of the book to generate interest in research on these topics. It is our hope that the reader will find the material presented in an accessible and unified manner. The book is intended for research mathematicians in algebra and analysis and for advanced graduate students in mathematics. Each section includes exercises of varying degrees of difficulty for graduate students.

While we have attempted to make this monograph as self-contained as possible, it has been difficult in view of the limitations on the size of this book. To keep the book to a reasonable length, some proofs (including a few highly technical ones) have been omitted and appropriate references to research papers or books have been included. Some results have been included as exercises in various chapters, with proper references, for a motivated reader. We have also listed other references related to the material presented in the book. These, and the brief historical notes provided at the end of each chapter, should be useful for researchers interested in further investigations. There are many excellent papers which we regrettably could not include in this book in order to keep it within a moderate length.

We are very thankful to Gangyong Lee, Cosmin Roman, Henry E. Heatherly, and Toma Albu, for their numerous constructive comments, painstaking proof-reading, and suggestions for improvements in this book. The technical help provided by Cosmin Roman and Gangyong Lee, and corrections in several proofs they pointed out have been crucial during the preparation of this manuscript. There are many others who helped in proof-reading various parts of the book. These include Mohammad Ashraf, Xiaoxiang Zhang, Asma Ali, Shakir Ali, and Faiz Rizvi. We are thankful to them for their time and efforts. The errors that still may remain in the book, are our own fault.

There are others who have also played an important role, directly or indirectly, in the formation of this book. We are thankful to (late) E.H. Feller, Bruno J. Müller, Joe W. Fisher, T.Y. Lam, Barbara L. Osofsky, S.K. Jain, Efraim P. Armendariz, Sergio López-Permouth, Dinh Van Huynh, Pere Ara, Edmund R. Puczyłowski, and Mikhail Chebotar for their influence on our work, words of encouragement, advice and support.

The work on the book was supported by grants from The Ohio State University at Lima, Mathematics Research Institute, The Ohio State University, Columbus, Ohio, USA, Busan National University, Busan, South Korea, and the University of Louisiana at Lafayette, Lafayette, Louisiana, USA. We express our gratitude to these institutions for their support. S.T. Rizvi wishes to acknowledge the support of a 3-Year Stimulus Research Grant from the Mathematics Research Institute, Ohio State University, Columbus, USA, for his work in the final stages of this monograph. Jae Keol Park was supported by a 2-Year Research Grant of Pusan National University.

This work would not have been possible without the help and support of our families. We cannot ever fully express our heartfelt thanks to our respective wives, Betti, Insook, and Seema, for their unstinting support throughout the preparation of this work. Their continuous support, and that of our children, have been a source of strength for us for which we are truly very grateful.

We are grateful to Ann Kostant for the support and cooperation in the earlier stages of processing of this research monograph. The helpful cooperation and patience we received from Allen Mann and Mitch Moulton in the final processing of the book has been crucial. We greatly appreciate that and thank them for their personal interest and professional support. We are also thankful to other staff at Birkhäuser who have been helpful in our book project, including Tom Grasso.

Lafayette, Louisiana, USA
Busan, South Korea
Lima, Ohio, USA
February 12, 2013

Gary F. Birkenmeier
Jae Keol Park
S. Tariq Rizvi

Introduction

Among the major efforts in Ring Theory, one has been to find, for a given ring R, a "well behaved" overring S in the sense that S has better properties than R and such that some useful information can transfer between R and S. Alternatively, given a well behaved ring, to find conditions describing those subrings for which there is a fruitful inheritance of properties between the given ring and its subrings. A similar quest between a module and an overmodule has been pursued in Module Theory. These have been important topics of research and have been crucial in the development of Algebra—especially of Ring and Module Theory. Having yielded such important classes of rings and modules, as the rings and modules of quotients, injective hulls and right orders, this quest has been truly rewarding to ring and module theorists.

Another effort has been, to investigate when properties of a given ring R transfer to its various ring extensions and vice versa. A number of research papers have been published on investigations to address such questions. The ring extensions (for example, polynomial extensions, matrix ring extensions, triangular matrix ring extensions, group ring extensions, and skew group ring extensions, etc.) form important classes of rings and have been a focus of extensive research. While results on some particular types of ring extensions and the transfer of some limited algebraic properties have been included in a few existing research books or graduate texts, it appears that there is presently no research monograph covering the wide varieties of extensions and much of the recent work is spread in disparate research papers.

The focus of our book is related to the two quests mentioned above. As we mentioned in the preface, for a given ring R (or a given module M), we consider a "minimal essential extension" of R (or of M) with respect to belonging to a particular class. We call this a "hull" of R (or of M) belonging to that particular class and show that such hulls lie closer to the ring R (or to the module M) than its injective hull. This in turn allows for a better transfer of information between R (or M) and the hull of R (or of M) from these classes than between R (or M) and its injective hull. These hulls prove to be useful tools for the study of the structure of R (or of M).

While some of the techniques presented here can be applied in more general settings, our focus in this book is on certain properties of rings and modules related to their direct summands and direct sums. In 1940, R. Baer [35] introduced the notion of an injective module and showed that a module M_R is injective if and only if, whenever $M_R \leq N_R$, M_R is a direct summand of N_R. This generalization of a vector space is one of the cornerstones of Module Theory. The notion of injectivity and its generalizations have been a direction of extensive research. The need to study generalizations of injectivity arises from the fact that such classes of modules properly contain the class of injective modules while still enjoying some worthwhile advantages of injective modules. One such generalization that has been of interest for about three decades, is the notion of *extending* (or *CS*) *modules*, namely modules in which every submodule is essential in a direct summand. This notion was explicitly named for the case of rings (as CS-rings) by Chatters and Hajarnavis in 1977 [119] and was also studied earlier by Utumi [398]. It is easy to see that such a module is a common generalization of injective and semisimple modules. Analogously, a module M is called an *FI-extending module* if every *fully invariant* submodule of M is essential in a direct summand of M. These classes of modules and related notions will form an important focus of results in this book because of the interesting connections, as we will see later, to other topics of our study.

Among overrings of R, its right rings of quotients provide handy tools for studying R. However they become useless when R coincides with its maximal right ring of quotients such as when R is a right Kasch ring. To study overrings of such rings one can consider classes of rings that lie between R and its right injective hull $E(R_R)$. This motivates the notion of essential overring extensions that we present in Chap. 7 and further utilize in later chapters. We call an overring S of a ring R, a *right essential overring* of R if R_R is essential in S_R. This notion will prove to be a useful tool. The study of such extensions is also motivated by a result in Chap. 8 which shows that any right essential overring of a right FI-extending ring is right FI-extending. Therefore, all right essential overrings of a right FI-extending hull (if it exists) of a ring R are right FI-extending. A ring is called *quasi-Baer* if the left annihilator of an ideal is generated by an idempotent. Any right and left essential overring of a quasi-Baer ring is quasi-Baer. Also right essential overrings provide a natural setting for defining the notion of ring hulls in Chap. 8.

In 1936, Murray and von Neumann [311] developed the theory of von Neumann algebras (also called W^*-algebras) in an attempt to provide a rigorous mathematical model for quantum theory (see also [403–406], and [407]). Their theory was based on rings of operators on a Hilbert space. Rickart [353] in 1946 studied C^*-algebras (i.e., Banach $*$-algebras such that $\|xx^*\| = \|x\|^2$) which satisfy the condition that the right annihilator of every single element is generated by a projection (an idempotent e is called a projection if $e = e^*$). Rickart also showed that all von Neumann algebras satisfy this property (i.e., the right annihilator of any element is generated by a projection). These algebras werc later named *Rickart C^*-algebras* by Kaplansky.

Motivated by the work of Murray, von Neumann, and Rickart, Kaplansky in the 1950s showed that von Neumann algebras, in fact, satisfied a stronger annihilator condition, namely, that these are rings with identity in which the right annihilator of any nonempty subset is generated by an idempotent. He termed a ring with this property a *Baer ring* to honor R. Baer who had studied this condition in [36]. The Baer ring property is left-right symmetric. Kaplansky recognized that the notions of a Baer ring and a Baer $*$-ring provide a framework to study the algebraic properties of operator algebras and each is interesting in its own right. The theory of Baer rings, Baer $*$-rings, and AW^*-algebras (C^*-algebras which are Baer $*$-rings) have been studied in [246] and [45].

Maeda in 1960 [287] defined a Rickart ring. He called a ring *right* (*left*) *Rickart* if the right (left) annihilator of any single element is generated by an idempotent. It is clear that every Baer ring is right and left Rickart. The same year, Hattori [200] introduced the notion of a *right PP ring*, namely a ring in which every principal right ideal is projective. It was later discovered that a right Rickart ring is precisely the same as a right PP ring.

While the classes of Baer rings and right Rickart rings have many noteworthy properties, these are not closed under matrix ring or polynomial ring extensions. Finding additional conditions which allow for the transfer of the Baer and right Rickart properties to various types of ring extensions, has attracted interest.

A *quasi-Baer ring* (i.e., a ring for which the left annihilator of every ideal is generated by an idempotent) was defined by Clark in 1967 [128]. Similar to the Baer ring property, the quasi-Baer ring property is left-right symmetric. He also showed that any finite distributive lattice is isomorphic to a sublattice of the lattice of all ideals of an Artinian quasi-Baer ring. Remarkably, Pollingher and Zaks [347] in 1971 proved that unlike the class of Baer rings, the class of quasi-Baer rings is indeed closed under full and triangular matrix extensions. By [128] and [347], the quasi-Baer property is a Morita invariant property. It was shown in [77] that in contrast to the Baer property, the quasi-Baer property transfers, without any additional requirements, from a ring R to many types of its ring extensions (this will be a topic of our discussions). Therefore, at least a "weaker form" of the Baer property does transfer to several of the ring extensions of a Baer ring R.

Analogous to a right Rickart ring, a ring R is called *right principally quasi-Baer* (simply, *right p.q.-Baer*) if the right annihilator of any principal ideal is generated by an idempotent as a right ideal [78]. A generalization of the quasi-Baer notion, this property is also a Morita invariant and the class of right p.q.-Baer rings is closed under triangular matrix ring and polynomial ring extensions without any additional conditions on the base ring.

There are strong bonds between Baer and right extending rings. In particular, every regular right extending (e.g., regular right self-injective) ring is Baer. In 1980, Chatters and Khuri [121] provided an important characterization connecting the class of Baer rings to that of right extending rings. More specifically, they showed that a ring R is right nonsingular right extending if and only if R is Baer and right cononsingular. This useful link and its analogues provide another motivation for several topics that are included in this book. In a similar fashion, there are close links between the FI-extending property and the quasi-Baer property (not only for rings but also for modules). The algebraic properties we discussed in the preceding, will be of special interest in our discussions throughout the book. After the introductory first chapter, Chap. 2 is devoted to a discussion of the injective property and related notions. We shall discuss basic properties and results on (quasi-)Baer, Rickart, and p.q.-Baer rings in Chap. 3.

Module theoretic analogues of Baer and quasi-Baer rings using the endomorphism ring of a module were introduced in [357] in 2004 (see also [359, 360], and [361]). These will be the topics of our discussions in Chap. 4. Similar to the fact that every Baer ring is nonsingular, we will see that a Baer module also satisfies a weaker notion of nonsingularity of modules (called $\mathcal{K}$-nonsingularity) which depends on the endomorphism ring of the module. Strong connections between a Baer module and an extending module are demonstrated via an effective use of this weak nonsingularity and its dual notion. It is shown that an extending module which is $\mathcal{K}$-nonsingular is precisely a $\mathcal{K}$-cononsingular Baer module. This provides a useful module theoretic analogue of the Chatters-Khuri theorem. We shall use module theoretic methods to obtain conditions for the transfer of the Baer property to certain matrix ring extensions. The chapter also includes a section on some of the latest work on Rickart modules ([269, 270], and [271]). A Rickart module generalizes the notion of a Baer module analogous to the case of rings.

In Chap. 5, we consider generalized triangular matrix representations and discuss triangulating idempotents. A structure theorem for a quasi-Baer ring with a complete set of triangulating idempotents will be presented. The following well-known results are among the consequences of this structure theorem: Levy's decomposition theorem of semiprime right Goldie rings, Faith's characterization of semiprime right FPF rings with no infinite set of central orthogonal idempotents, Gordon and Small's characterization of piecewise domains, and Chatters' decomposition theorem of hereditary Noetherian rings.

Continuing the study of the transference of some properties to matrix ring extensions that were initiated in Chap. 3, we further consider the transference of various algebraic properties to other matrix, Ore, and group ring extensions, in Chap. 6.

Chapter 7 is mainly devoted to the study of right essential overring extensions of a ring and their ring structures. The techniques introduced here allow us to investigate overrings which are incomparable with the maximal ring of quotients of a ring. A ring R is said to be *right Osofsky compatible* if an injective hull $E(R_R)$ of R_R has a ring structure for which the ring multiplication extends the R-module scalar multiplication of $E(R_R)$ over R. We discuss and show when certain rings are (or are not) right Osofsky compatible.

Motivated by the results and examples of Chap. 7, in Chap. 8 we introduce the notion of a ring hull belonging to a particular class. This important concept provides a basis for a general theory of hulls. To facilitate our search for ring hulls from various classes of rings satisfying certain conditions related to idempotents, we determine exactly the set of ideals of R which are dense in ring direct summands of $Q(R)$, the maximal right ring of quotients of R. We note the ubiquity of these hulls by showing that every semiprime ring has a quasi-Baer ring hull and also has a right (and a left) principally quasi-Baer ring hull. As a consequence, a commutative semiprime ring has a Baer and a Rickart ring hull. Further, if R is a semiprime ring, then the quasi-Baer ring hull of R is explicitly described and it is proved that this quasi-Baer ring hull of R coincides with the FI-extending ring hull of R. It is shown that there is a fruitful transmission of information between rings and their quasi-Baer ring hulls. In the concluding part of the chapter, we discuss the notion of module hulls and show the existence of various module hulls including those which generalize injective hulls. Furthermore, using quasi-Baer ring hulls, we show that every finitely generated projective module over a semiprime ring has an FI-extending module hull. Continuing our discussions on hulls from Chap. 8, we next focus on ring hulls of ring extensions in Chap. 9. In particular, we obtain and describe ring hulls of monoid and matrix ring extensions belonging to the various classes of rings we discussed earlier.

Chapter 10 is devoted to applications of the results presented in earlier chapters to Ring Theory and Functional Analysis. In particular, among applications to the structure of rings of quotients, necessary and sufficient conditions are shown for a ring R such that $Q(R)$ can be decomposed into a direct product of indecomposable rings or a direct product of prime rings. Among applications to C^*-algebras, a C^*-algebra with only finitely many minimal prime ideals is characterized. It is shown that a unital C^*-algebra A is boundedly centrally closed if and only if A is a quasi-Baer ring. For a cardinal number $\aleph$ and a C^*-algebra A, we see that the extended centroid of A is $\mathbb{C}^{\aleph}$ (where $\mathbb{C}$ is the field of complex numbers) if and only if the local multiplier algebra $M_{\mathrm{loc}}(A)$ of A is a C^*-direct product of $\aleph$ prime C^*-algebras. A characterization of boundedly centrally closed intermediate C^*-algebras between a C^*-algebra and $M_{\mathrm{loc}}(A)$ is presented.

We have taken an opportunity in the book to improve and extend some of the results that appeared in our earlier papers. A few of the results in those papers have been corrected and some new unpublished ones have been included. We could not include all of the fascinating literature that is presently available, however we encourage the motivated reader to consult the papers cited in the references at the end of the book and other related papers for further research on the topics considered in this book. A number of open problems and questions are proposed for further investigations as well.

Contents

A Partial List of Symbols

$\mathbb{N}$	The set of positive integers
$\mathbb{Z}$	The ring of integers
$\mathbb{Z}_n (n > 1)$	The ring of integers modulo n
$\mathbb{Q}$	The field of rational numbers
$\mathbb{R}$	The field of real numbers
$\mathbb{C}$	The field of complex numbers
$\text{Mat}_n(R)$	The $n \times n$ matrix ring over R
$T_n(R)$	The $n \times n$ upper triangular matrix ring over R
$\mathbf{I}(R)$	The set of all idempotents of R
$\mathcal{B}(R)$	The set of all central idempotents of R
$\text{Cen}(R)$	The center of R
$\langle X \rangle_R$	The subring of R generated by $X \subseteq R$
$Z(M_R)$ or $Z(M)$	The singular submodule of M_R
$Z_2(M_R)$ or $Z_2(M)$	The second singular submodule of M_R
$\text{Soc}(M)$ or $\text{Soc}(M_R)$	The socle of M_R
$\text{Rad}(M)$ or $\text{Rad}(M_R)$	The Jacobson radical of M_R
$J(R)$	The Jacobson radical of R
$P(R)$	The prime radical of R
$\text{pd}(M_R)$	Projective dimension of M_R
$\text{r.gl.dim}(R)$	Right global dimension of a ring R
udim	Uniform dimension
$Q^r_{c\ell}(R)$	The classical right ring of quotients of R
$\mathbf{S}_\ell(R)$	The set of all left semicentral idempotents of R
$\mathbf{S}_r(R)$	The set of all right semicentral idempotents of R
$E(M_R)$ or $E(M)$	The injective hull of M_R
$\widetilde{E}(M_R)$ or $\widetilde{E}(M)$	The rational hull of M_R
$Q(R)$	The maximal right ring of quotients of R
$Q^\ell(R)$	The maximal left ring of quotients of R
$Q^m(R)$	The Martindale right ring of quotients of R
$Q^s(R)$	The symmetric ring of quotients of R
Mod-R	The category of right R-modules

$Q_{\mathfrak{K}}(R)$	The $\mathfrak{K}$ absolute ring hull of R
$Q_{\mathbf{qB}}(R)$	The quasi-Baer absolute ring hull of R
$\widehat{Q}_{\mathfrak{K}}(R)$	The $\mathfrak{K}$ absolute to $Q(R)$ ring hull of R
$\widehat{Q}_{\mathbf{qB}}(R)$	The quasi-Baer absolute to $Q(R)$ ring hull of R
$\mathrm{Tdim}(R)$	Triangulating dimension of R
$\mathrm{Spec}(R)$	The set of prime ideals of R
$\mathrm{MinSpec}(R)$	The set of minimal prime ideals of R
$\mathrm{CFM}_{\Gamma}(R)$	The $\Gamma \times \Gamma$ column finite matrix ring over R
$\mathrm{RFM}_{\Gamma}(R)$	The $\Gamma \times \Gamma$ row finite matrix ring over R
$\mathrm{CRFM}_{\Gamma}(R)$	The $\Gamma \times \Gamma$ column and row finite matrix ring over R
B	The class of Baer rings
eB	The class of right essentially Baer rings
qB	The class of quasi-Baer rings
eqB	The class of right essentially quasi-Baer rings
Con	The class of right continuous rings or modules
qCon	The class of right quasi-continuous rings or modules
SI	The class of right self-injective rings
E	The class of right extending rings or modules
FI	The class of right FI-extending rings or modules
SFI	The class of right strongly FI-extending rings or modules
C_∞	The infinite cyclic group
C_n	The cyclic group of order n ($n > 1$)
jdim	Johnson dimension
PWD	Piecewise domain
PWP ring	Piecewise prime ring
D-E class	Definition 8.1.5
$\mathfrak{D}_{\mathfrak{K}}(R)$	Definition 8.1.5
$\delta_{\mathfrak{C}}(R)$ and $\delta_{\mathfrak{C}}(R)(1)$	cf. Definition 8.2.8
$R(\mathfrak{C}, S)$ and $R(\mathfrak{C}, \rho, S)$	Definition 8.2.8
IC	Idempotent closure class
$Q_b(A)$	The bounded symmetric algebra of quotients
$M_{\mathrm{loc}}(A)$	The local multiplier algebra

Chapter 1
Preliminaries and Basic Results

We begin with basic notions, definitions, results, terminology, and notations used in the book. While we recommend standard graduate text books such as [8, 259], or [262] for more details on this material, we include some preliminary material in this chapter for the convenience of the reader.

1.1 Basic Notions and Definitions

Most of the material in this section is standard and is intended to provide background information.

1.1.1 All our rings are assumed to have identity unless indicated otherwise. Modules are assumed to be unitary. For a ring R, we use 1_R or $\mathbf{1}_R$ to denote the identity of R. In using subscripts, when the context is clear, the subscript may be omitted. A subring S of a ring R with 1_R means a subring $S \subseteq R$ with $1_R \in S$. Ideals without the adjective left or right mean two-sided ideals. When R is a ring, $I \trianglelefteq R$ denotes that I is an ideal of R. A Noetherian (resp., Artinian) ring means both a right and left Noetherian (resp., a right and left Artinian) ring.

1.1.2 $\text{Mat}_n(R)$ and $T_n(R)$ denote the $n \times n$ matrix ring and the $n \times n$ upper triangular matrix ring over a ring R, respectively. The notation $[r_{ij}]$ stands for the matrix whose (i, j)-position is r_{ij}.

Further, $\mathbf{I}(R)$ and $\mathcal{B}(R)$ are used for the set of idempotents and the set of central idempotents of R, respectively. The center of R is denoted by $\text{Cen}(R)$. For $X \subseteq R$, $\langle X \rangle_R$ denotes the subring of R generated by X.

The letters $\mathbb{Z}$, $\mathbb{Z}_n$, $\mathbb{Q}$, $\mathbb{R}$, and $\mathbb{C}$ are used for the ring of integers, the ring of integers modulo n (n is an integer greater than 1), the field of rational numbers, the field of real numbers, and the field of complex numbers, respectively. For given subsets A and B of a set X, $A \setminus B = \{a \in A \mid a \notin B\}$, and we use $|X|$ to denote the cardinal number for X.

G.F. Birkenmeier et al., *Extensions of Rings and Modules*,
DOI 10.1007/978-0-387-92716-9_1,

1.1.3 For right R-modules M_R and N_R, we use $\mathrm{Hom}(M_R, N_R)$, $\mathrm{Hom}_R(M, N)$, or $\mathrm{Hom}(M, N)$ to denote the set of all R-homomorphisms from M_R to N_R. Likewise, $\mathrm{End}(M_R)$, $\mathrm{End}_R(M)$, or $\mathrm{End}(M)$ denotes the endomorphism ring of an R-module M. For an R-homomorphism $f \in \mathrm{Hom}(M, N)$, $\mathrm{Image}(f)$ and $\mathrm{Ker}(f)$ denote the image and the kernel of f, respectively. Further, for $X \subseteq M$, $f(X)$ or fX is used for the image of X under f. For a submodule V of M, $f|_V$ means the restriction of f to V. Also, "homomorphism" is used for "R-homomorphism". When the context is clear, the subscript R may be omitted.

A submodule W of a module V is said to be a *fully invariant* submodule of V if $f(W) \subseteq W$ for each $f \in \mathrm{End}(V)$. We use $W \trianglelefteq V$ to denote that W is a fully invariant submodule of V. When R is a ring, the fully invariant submodules of R_R are precisely the ideals of R.

A submodule N of a module M is said to be *essential* (or *large*) in M if $N \cap U \neq 0$ for every nonzero submodule U of M. For a submodule V of a module M, Zorn's lemma guarantees the existence of a submodule W of M such that W is a maximal essential extension of V in M. In this case, there is no proper essential extension of W in M. A submodule N of M is said to be *closed* in M if N does not have a proper essential extension in M. A maximal essential extension of N in M is called a *closure* of N in M. A right ideal I of a ring R is called *closed* if I_R is closed in R_R. A closed left ideal is defined similarly.

For a module M, $N \leq M$ and $L \leq^{\mathrm{ess}} M$ denote that N is a submodule of M and L is an essential submodule of M, respectively. When $N \leq M$, $N \leq^{\oplus} M$ indicates that N is a direct summand of M.

Let S and R be rings. Then we use ${}_SM_R$ to denote that M is a left S-right R-bimodule, or simply an (S, R)-bimodule. The notion ${}_SN_R \leq {}_SM_R$ means that N is an (S, R)-subbimodule of M. If an R-module M is a direct sum $M = U \oplus V$ of R-modules U and V, the R-homomorphism $f : M \to U$ defined by $f(u + v) = u$, for any $u \in U$ and $v \in V$, is called the *canonical projection* from M onto U.

1.1.4 Let M be a right R-module. Then the *annihilator of* M *in* R is the set $r_R(M) = \{a \in R \mid Ma = 0\}$. It is easy to see that $r_R(M) \trianglelefteq R$. In general, if $\emptyset \neq U \subseteq M$, $r_R(U) = \{a \in R \mid Ua = 0\}$ is a right ideal of R. Similarly, for a left R-module M, we denote $\ell_R(M) = \{a \in R \mid aM = 0\}$, and for any $\emptyset \neq V \subseteq M$, we let $\ell_R(V) = \{a \in R \mid aV = 0\}$. Then $\ell_R(M) \trianglelefteq R$ and $\ell_R(V)$ is a left ideal of R.

A right R-module M_R (resp., a left R-module ${}_RM$) is called *faithful* if $r_R(M) = 0$ (resp., $\ell_R(M) = 0$). If M is a right R-module and $\emptyset \neq B \subseteq R$, then the *annihilator of* B *in* M is the set $\ell_M(B) = \{m \in M \mid mB = 0\}$. Similarly, for a left R-module M and $\emptyset \neq D \subseteq R$, the *annihilator of* D *in* M is $r_M(D) = \{m \in M \mid Dm = 0\}$.

1.1.5 When $M = R$ and $\emptyset \neq X \subseteq R$, $r_R(X) = \{a \in R \mid Xa = 0\}$ and $\ell_R(X) = \{a \in R \mid aX = 0\}$, which are called the *right annihilator of* X *in* R and the *left annihilator of* X *in* R, respectively.

1.1.6 For a module M and an index set Λ, let $M^{(\Lambda)}$ and M^{Λ} denote the direct sum and the direct product of $|\Lambda|$ copies of M, respectively. When Λ is finite with $|\Lambda| = n$, then direct product and direct sum coincide and we use $M^{(n)}$ to denote it.

1.1.7 (Modular Law) Let M be a module and V, W be submodules of M. Then, for any N such that $V \leq N \leq M$, $N \cap (V + W) = V + (N \cap W)$. In particular, if $M = V + W$ and $V \leq N \leq M$, then $N = V + (N \cap W)$.

1.1.8 The submodule $Z(M_R) = \{m \in M \mid r_R(m) \leq^{\text{ess}} R_R\}$, of a right R-module M, is called the *singular submodule* of M. The module M_R is said to be *nonsingular* if $Z(M_R) = 0$. The submodule $Z_2(M_R)$ of M_R is defined by the condition $Z_2(M_R)/Z(M_R) = Z(M/Z(M_R)_R)$. It is known that $Z_2(M_R)$ is the unique closure of $Z(M_R)$. The submodule $Z_2(M_R)$ is called the *second singular submodule* of M. It is easy to see that each of $Z(M)$ and $Z_2(M)$ is a fully invariant submodule of M. The singular submodule and the second singular submodule for a left module are defined similarly. A ring R is called *right* (resp., *left*) *nonsingular* if $Z(R_R) = 0$ (resp., $Z(_RR) = 0$). It can be verified that if $N \leq^{\text{ess}} M$, then $Z(M/N) = M/N$.

1.1.9 For a module M, the *socle* of M is the sum of all simple submodules of M and is denoted by $\text{Soc}(M)$. It is well known that $\text{Soc}(M)$ is the intersection of all essential submodules of M. A module M is said to be *semisimple* if M is a sum of simple modules. Therefore, M is semisimple if and only if $M = \text{Soc}(M)$. When $M = R_R$, then $\text{Soc}(R_R)$ is called the *right socle* of R. The left socle $\text{Soc}(_RR)$ of R is defined similarly. When $\text{Soc}(R_R) = \text{Soc}(_RR)$, we write $\text{Soc}(R)$ for $\text{Soc}(R_R)$ or $\text{Soc}(_RR)$.

We use $\text{Rad}(M)$ to denote the intersection of all maximal submodules of M, which is called the (*Jacobson*) *radical* of M. The *Jacobson radical* $J(R)$ of a ring R is the intersection of all maximal right ideals of R. It is also the intersection of all maximal left ideals of R.

An element a of a ring R is called *right* (resp., *left*) *quasi-regular* if $1 - a$ is right (resp., left) invertible in R. An element a of a ring R is called *quasi-regular* if $1 - a$ is invertible in R. A one-sided ideal I of R is called (resp., *right*, *left*) *quasi-regular* if every element of I is (resp., *right*, *left*) quasi-regular. A right ideal I is right quasi-regular if and only if I is quasi-regular (see [8, Proposition 15.2]). The Jacobson radical $J(R)$ of R is the sum of all quasi-regular ideals of R. Furthermore,

$$J(R) = \{r \in R \mid ra \text{ is quasi-regular for all } a \in R\}.$$

1.1.10 Let R be a ring and I an ideal of R. We say that each idempotent of R/I *lifts to an idempotent* of R if for each idempotent $\alpha \in R/I$, there is an idempotent $e \in R$ such that $\alpha = e + I$. If I is a nil ideal of a ring R, then every idempotent of the ring R/I lifts to an idempotent of R. Further, let $V \subseteq J(R)$ be an ideal of R such that each idempotent of R/V lifts to an idempotent of R. Then for any countable (possibly finite) set of orthogonal idempotents $\{\alpha_1, \alpha_2, \dots\}$ in R/V, there exists a set of orthogonal idempotents $\{e_1, e_2, \dots\}$ of R such that $e_i + V = \alpha_i$ for all i (see [259, Proposition 21.25]).

1.1.11 An ideal P of a ring R is said to be a *prime ideal* if $AB \subseteq P$ with $A, B \trianglelefteq R$ implies $A \subseteq P$ or $B \subseteq P$, or equivalently, for all $a, b \in R$, $aRb \subseteq P$ implies $a \in P$

or $b \in P$. A ring R is said to be a *prime ring* if 0 is a prime ideal. The intersection of all prime ideals of a ring R is called the *prime radical* of R and is denoted by $P(R)$. A ring R is said to be a *semiprime ring* if $P(R) = 0$. Thus, a ring R is semiprime if and only if for any $a \in R$, $aRa = 0$ implies $a = 0$ if and only if R has no nonzero nilpotent ideal.

1.1.12 A ring is said to be *reduced* if it contains no nonzero nilpotent elements. Reduced rings are semiprime. It is known that a commutative ring R is semiprime if and only if R is reduced if and only if R is nonsingular. Also if a ring R is reduced, then $ab = 0$ with $a, b \in R$ implies $aRb = 0$. Thus, for a reduced ring R, $eR(1-e) = 0$ and $(1-e)Re = 0$ for $e^2 = e \in R$. Hence, all idempotents are central in a reduced ring. So every reduced ring is an Abelian ring (a ring is called *Abelian* if every idempotent is central).

A ring R is called *regular* (in the sense of von Neumann) if for any $x \in R$ there exists $y \in R$ such that $x = xyx$. It is well known that a ring R is regular if and only if every finitely generated right ideal is generated by an idempotent. A ring R is said to be *strongly regular* if R is regular and reduced. We remark that R is a strongly regular ring if and only if R is an Abelian regular ring.

1.1.13 A ring R is said to be *right hereditary* if every right ideal of R is projective as a right R-module. A ring R is called *right semihereditary* if every finitely generated right ideal of R is projective. A left hereditary ring and a left semihereditary ring are defined similarly. A ring R is called *(semi)hereditary* if it is both right and left (semi)hereditary.

The *projective dimension* $\text{pd}(M_R)$ of a module M_R is defined to be the shortest length n of an exact sequence

$$0 \to P_n \to P_{n-1} \to \cdots \to P_0 \to M \to 0$$

of modules, where $P_n, P_{n-1}, \ldots, P_0$ are projective. If no such sequence exists, then we write $\text{pd}(M_R) = \infty$. The *right global dimension* $\text{r.gl.dim}(R)$ of a ring R is $\sup\{\text{pd}(M) \mid M \text{ is a right } R\text{-module}\}$. For a ring R, it is well known that $\text{r.gl.dim}(R) = 0$ if and only if R is a semisimple Artinian ring; and $\text{r.gl.dim}(R) \leq 1$ if and only if R is right hereditary.

1.1.14 A ring R is called *semilocal* if $R/J(R)$ is semisimple Artinian. A semilocal ring is called *semiperfect* if each idempotent of $R/J(R)$ lifts to an idempotent of R; a semilocal ring R is said to be *semiprimary* if $J(R)$ is nilpotent. A ring R is called *right perfect* if every right R-module has a projective cover. A left perfect ring is defined similarly. By a result of Bass [38], a ring R is right perfect if and only if R satisfies DCC on principal left ideals (see [8, Theorem 28.4]). If R is a right perfect ring, then $\text{Soc}({}_RR) \leq^{\text{ess}} {}_RR$ (see [259, Theorem 23.20]). Also, if R is a right (or left) perfect ring, then $J(R) = P(R)$ (see [259, Proposition 23.15]).

1.1.15 A nonzero module M is called *uniform* if every nonzero submodule of M is essential in M. We say that a module M has *uniform dimension* n (written udim$(M) = n$) if there is an essential submodule V of M which is a direct sum of n uniform submodules. On the other hand, if no such integer n exists, we write udim$(M) = \infty$. A ring R is called *right Goldie* if R has ACC on right annihilators and udim(R_R) is finite. A left Goldie ring is defined similarly.

1.1.16 Let R be a ring. A subset $\{e_{ij} \mid 1 \leq i, j \leq n\}$ of R is called a set of *matrix units* if $\sum_{i=1}^{n} e_{ii} = 1$ and $e_{ij}e_{k\ell} = \delta_{jk}e_{i\ell}$, where δ_{jk} is the Kronecker delta. The set $A = \{a \in R \mid ae_{ij} = e_{ij}a$ for all $i, j, 1 \leq i, j \leq n\}$ forms a ring. Then $R \cong \text{Mat}_n(A)$ and $A \cong e_{11}Re_{11}$ (see [221, Proposition 6, p. 52]).

1.1.17 An overring T of a ring R is called the *classical right ring of quotients* of R if (i) every nonzero-divisor of R is invertible in T; (ii) every $x \in T$ is of the form $x = as^{-1}$, where $a, s \in R$ and s is a nonzero-divisor. We write $T = Q^r_{c\ell}(R)$. For a ring R, $Q^r_{c\ell}(R)$ exists if and only if R satisfies the *right Ore condition*, that is, for a and s in R with s a nonzero-divisor, there exist b and t in R with t a nonzero-divisor such that $at = sb$ (see [382, Proposition 1.6, p. 52]). A ring R is said to be *right Ore* if R satisfies the right Ore condition. A domain with the right Ore condition is called a *right Ore domain*. The classical left ring of quotients of a ring, the left Ore condition, a left Ore ring, and a left Ore domain are defined similarly. If R is a right Ore domain, then $Q^r_{c\ell}(R)$ is a division ring.

1.2 Idempotents-Some Basic Results

Idempotents play an important role in the structure theory of rings and modules as they generate the direct summands in any ring direct sum decomposition and provide an analogous decomposition for any module. We include some basic results and facts related to idempotents.

Definition 1.2.1 Let R be a ring. An idempotent $e \in R$ is called *left* (resp., *right*) *semicentral* if $ae = eae$ (resp., $ea = eae$) for all $a \in R$.

We use $\mathbf{S}_\ell(R)$ (resp., $\mathbf{S}_r(R)$) to denote the set of all left (resp., right) semicentral idempotents of R.

Proposition 1.2.2 *Let $e^2 = e \in R$. Then the following are equivalent.*

(i) $e \in \mathbf{S}_\ell(R)$.
(ii) *eR is an ideal of R.*
(iii) $1 - e \in \mathbf{S}_r(R)$.
(iv) *$R(1 - e)$ is an ideal of R.*
(v) $(1 - e)Re = 0$.

Proof The proof is routine from Definition 1.2.1. □

Example 1.2.3 (i) Say $R = T_2(A)$, where A is a ring. Let $e_{ij} \in R$ be the matrix with 1 in the (i, j)-position and 0 elsewhere. Then for any $a \in A$, $e_{11} + ae_{12} \in \mathbf{S}_\ell(R)$ and $ae_{12} + e_{22} \in \mathbf{S}_r(R)$.

(ii) Let R be a ring and $e^2 = e \in R$. If Re (resp., eR) contains no nonzero nilpotent elements, then $e \in \mathbf{S}_\ell(R)$ (resp., $e \in \mathbf{S}_r(R)$).

Proposition 1.2.4 (i) *Let $e, f \in \mathbf{S}_\ell(R)$. Then $e + f - ef \in \mathbf{S}_\ell(R)$ and $ef \in \mathbf{S}_\ell(R)$. Further, $eR + fR = (e + f - ef)R$ and $eR \cap fR = efR$.*

(ii) *Let $e, f \in \mathbf{S}_r(R)$. Then $e + f - ef \in \mathbf{S}_r(R)$ and $ef \in \mathbf{S}_r(R)$. Further, $Re + Rf = R(e + f - ef)$ and $Re \cap Rf = Ref$.*

Proof The proof is straightforward. □

For a more comprehensive list of results on semicentral idempotents, see [79] and [202].

Proposition 1.2.5 *Let R be a ring. Then $\{eR \mid e \in \mathbf{S}_\ell(R)\}$ is a distributive sublattice of the lattice of ideals of R. More generally, if $\{e_\lambda\}_{\lambda \in \Lambda} \subseteq \mathbf{S}_\ell(R)$ and $f \in \mathbf{S}_\ell(R)$, then $fR \cap (\sum_{\lambda \in \Lambda} e_\lambda R) = \sum_{\lambda \in \Lambda} (fR \cap e_\lambda R)$.*

Proof By Proposition 1.2.4, $\{eR \mid e \in \mathbf{S}_\ell(R)\}$ forms a sublattice of the lattice of all ideals of R. Say e_1, e_2, and f are in $\mathbf{S}_\ell(R)$. Put $e = e_1 + e_2 - e_1e_2$. Then by Proposition 1.2.4, we observe that $fR \cap (e_1R + e_2R) = fR \cap eR = feR$ and that $(fR \cap e_1R) + (fR \cap e_2R) = fe_1R + fe_2R = feR$. □

Proposition 1.2.6 *Let R be a ring. Then:*

(i) $\mathcal{B}(R) = \mathbf{S}_\ell(R) \cap \mathbf{S}_r(R)$.

(ii) *If R is semiprime, then $\mathbf{S}_\ell(R) = \mathbf{S}_r(R) = \mathcal{B}(R)$.*

Proof The proof is routine. □

We now consider semicentral idempotents of polynomial rings. It can be easily seen that $\mathcal{B}(R) = \mathcal{B}(R[x])$ by routine computation. But the next example shows that, in general, $\mathbf{S}_\ell(R) \neq \mathbf{S}_\ell(R[x])$.

Example 1.2.7 Let $R = T_2(A)$, where A is a ring. Say e_{ij} is the matrix in R with 1 in the (i, j)-position and 0 elsewhere. Then $e = e_{11} + e_{12}x \in \mathbf{S}_\ell(R[x])$, however $e \notin R$.

In spite of Example 1.2.7, we have the following result which describes semicentral idempotents of $R[x]$ (see [71] for the proof).

Theorem 1.2.8 *Let $e(x)^2 = e(x) \in R[x]$. If $e(x) \in \mathbf{S}_\ell(R[x])$, then $e_0 \in \mathbf{S}_\ell(R)$ and $e(x)R[x] = e_0R[x]$, where e_0 is the constant term of $e(x)$.*

The notations $R[x,x^{-1}]$ and $R[[x,x^{-1}]]$ stand for the Laurent polynomial ring and the Laurent formal power series ring over a ring R, respectively. For more details on these rings, see [259]. Motivated by Theorem 1.2.8, the next interesting result describes semicentral idempotents of $R[x,x^{-1}]$ and $R[[x,x^{-1}]]$. See [77] for the proof.

Theorem 1.2.9 *Assume that $T = R[x,x^{-1}]$ or $T = R[[x,x^{-1}]]$ for a ring R, and say $e(x)^2 = e(x) \in T$ with e_0 the constant term. If $e(x) \in \mathbf{S}_\ell(T)$, then $e_0 \in \mathbf{S}_\ell(R)$ and $e(x)T = e_0T$.*

Definition 1.2.10 A nonzero idempotent e of a ring R is called *left* (resp., *right*) *semicentral reduced* if $\mathbf{S}_\ell(eRe) = \{0, e\}$ (resp., $\mathbf{S}_r(eRe) = \{0, e\}$). A ring R is called *left* (resp., *right*) *semicentral reduced* if 1 is left (resp., right) semicentral reduced.

Proposition 1.2.11 (i) *A ring R is left semicentral reduced if and only if R is right semicentral reduced.*

(ii) *A nonzero idempotent e of a ring R is left semicentral reduced if and only if e is right semicentral reduced.*

Proof The proof follows easily from Proposition 1.2.2. □

In view of Proposition 1.2.11, we say that a nonzero idempotent e is *semicentral reduced* if it is left (or right) semicentral reduced. Let $R = T_2(F)$, where F is a field. Let $e_{ij} \in R$ be the matrix with 1 in the (i, j)-position and 0 elsewhere. Then e_{11} and e_{22} are semicentral reduced idempotents.

Definition 1.2.12 A ring is said to be *orthogonally finite* if there is no set of infinitely many orthogonal idempotents.

Right Noetherian rings, rings with finite right uniform dimension, and semilocal rings are orthogonally finite.

Proposition 1.2.13 *The following are equivalent for a ring R.*

(i) *R is orthogonally finite.*
(ii) *$\{eR \mid e^2 = e \in R\}$ has ACC.*
(iii) *$\{eR \mid e^2 = e \in R\}$ has DCC.*
(iv) *$\{Re \mid e^2 = e \in R\}$ has ACC.*
(v) *$\{Re \mid e^2 = e \in R\}$ has DCC.*

Proof See [262, Proposition 6.59] for the proof. □

Definition 1.2.14 A nonzero idempotent e of a ring R is called *primitive* if it cannot be written as a sum of two nonzero orthogonal idempotents. A ring R is said to have a *complete set of primitive idempotents* if there is a set of nonzero orthogonal idempotents $\{e_1, e_2, \ldots, e_n\}$ for which each e_i is primitive, and $e_1 + \cdots + e_n = 1$.

It is easy to see that a nonzero idempotent $e \in R$ is primitive if and only if eR_R is indecomposable if and only if ${}_RRe$ is indecomposable (see [259, Proposition 21.8]). Thus, R has a complete set of primitive idempotents if and only if R_R (resp., ${}_RR$) is a finite direct sum of nonzero indecomposable submodules of R_R (resp., ${}_RR$). The proof of the next result is straightforward.

Proposition 1.2.15 *If R is orthogonally finite, then R has a complete set of primitive idempotents.*

The converse of Proposition 1.2.15 does not hold true. Shepherdson in [368] has shown that there is a domain D with $a, b \in R := \text{Mat}_2(D)$ satisfying $ab = 1$ and $ba \neq 1$. Let e_{ij} be the matrix in R with 1 in (i, j)-position and 0 elsewhere. Then $\{e_{11}, e_{22}\}$ is a complete set of primitive idempotents. But R is not orthogonally finite as $b^n(1 - ba)a^n$ for all positive integers n, form an infinite set of orthogonal idempotents (see [120, pp. 112–113]).

A ring R is called an *I-ring* (also called a *Zorn ring* by Kaplansky [246, p. 19]) if every nonnil right ideal of R contains a nonzero idempotent (see [221, Definition 1, p. 210]). We see that R is an I-ring if and only if every nonnil left ideal contains a nonzero idempotent.

Theorem 1.2.16 *Assume that R is an I-ring. Then either R contains an infinite number of orthogonal idempotents, or else R is semilocal.*

Proof Let R be an orthogonally finite I-ring. We prove that R is semilocal. As R is an I-ring, $J(R)$ is nil. We note that $A := R/J(R)$ is orthogonally finite as R is orthogonally finite, and any countable set of orthogonal idempotents of A lifts to a countable set of orthogonal idempotents of R (see 1.1.10). Also A is an I-ring. By Proposition 1.2.13, we can choose an idempotent e_1 of A such that e_1A is minimal in the set $\{hA \mid 0 \neq h^2 = h \in A\}$. As every nonzero right ideal of A contains a nonzero idempotent, the right ideal e_1A is a nonzero minimal right ideal of A. Thus, Ae_1A_A is a semisimple A-module.

We claim that Ae_1A is a simple Artinian ring. Put $f_1 = e_1$. If $r_{Ae_1A}(f_1) = 0$, then the map $\varphi : Ae_1A_A \to Ae_1A_A$ defined by $\varphi(x) = f_1x$ for $x \in Ae_1A$ is a monomorphism. Therefore, $Ae_1A_A \cong \varphi(Ae_1A)_A \leq^{\oplus} Ae_1A_A$. Thus, we have that $Ae_1A = \varphi(Ae_1A) \oplus N$ with $N_A \leq Ae_1A_A$. Since φ is a monomorphism, $\varphi^2(Ae_1A) \cap \varphi(N) = 0$, so $\varphi(Ae_1A) = \varphi^2(Ae_1A) \oplus \varphi(N)$. On the other hand, note that $\varphi(Ae_1A) = e_1Ae_1A$, so $\varphi^2(Ae_1A) = \varphi(Ae_1A)$. So we have that $\varphi(N) = 0$ and hence $N = 0$. Thus, $Ae_1A = \varphi(Ae_1A)$, so $f_1Ae_1A = Ae_1A$. Note that $f_1A = f_1f_1A = f_1e_1A \subseteq f_1Ae_1A$, so $f_1A = f_1Ae_1A = Ae_1A$. Because $f_1A \trianglelefteq A$ and A is semiprime, $f_1 \in \mathcal{B}(A)$ by Propositions 1.2.2 and 1.2.6(ii). So $Ae_1A = f_1Af_1 \cong \text{End}(f_1A_A)$ is a division ring as $f_1A_A = e_1A_A$ is simple (note that $f_1 = e_1$).

If $r_{Ae_1A}(f_1) \neq 0$, then choose $e_2^2 = e_2 \in r_{Ae_1A}(f_1)$ such that e_2A_A is a nonzero minimal right ideal of A as Ae_1A_A is semisimple and $r_{Ae_1A}(f_1)_A \leq Ae_1A_A$. Since $e_2A_A \leq Ae_1A_A$, $e_2A_A \cong e_1A_A$. Now $f_1e_2 = 0$, so $f_1 + e_2 - e_2f_1 \in Ae_1A$ is an idempotent and $f_1A+e_2A=(f_1+e_2-e_2f_1)A$. Put $f_2=f_1+e_2-e_2f_1$. Because $f_1e_2=0$, $f_1A\cap e_2A=0$ and so $f_2A=f_1A+e_2A=f_1A\oplus e_2A=e_1A\oplus e_2A$. Further, $f_1A\subsetneq f_2A$.

If $r_{Ae_1A}(f_2)=0$, then $f_2Ae_1A=Ae_1A$ as in the preceding argument. As $e_2\in Ae_1A$, $f_2A=e_1A\oplus e_2A\subseteq Ae_1A$. Hence, $f_2A=f_2f_2A\subseteq f_2Ae_1A$, so $f_2A=f_2Ae_1A=Ae_1A$. Thus, $f_2\in\mathcal{B}(A)$ by Propositions 1.2.2 and 1.2.6(ii). Hence $f_2A=f_2Af_2\cong\operatorname{End}(f_2A_A)=\operatorname{End}(e_1A_A\oplus e_2A_A)\cong\operatorname{Mat}_2(D)$, where $D=e_1Ae_1$ is a division ring because $e_1A_A\cong e_2A_A$ and e_1A_A is simple. Therefore, the ring f_2A is a simple Artinian ring.

Next, suppose that $r_{Ae_1A}(f_2)\neq 0$. As in the previous step, we can choose $e_3^2=e_3\in r_{Ae_1A}(f_2)$ such that e_3A_A is a simple A-module. Since e_3A_A is simple and $e_3A_A\leq Ae_1A_A$, $e_3A_A\cong e_1A_A$. We note that $f_2e_3=0$. Put $f_3=f_2+e_3-e_3f_2$. Then $f_3^2=f_3\in Ae_1A$, $f_2A\cap e_3A=0$, and $f_2A\oplus e_3A=f_2A+e_3A=f_3A$. Thus

$$f_3A=e_1A\oplus e_2A\oplus e_3A \quad \text{such that} \quad e_1A_A\cong e_2A_A\cong e_3A_A.$$

So we get $f_1A\subsetneq f_2A\subsetneq f_3A$, and so on.

As A is orthogonally finite, this process will be finished within finite steps by Proposition 1.2.13, thereby there exists a smallest positive integer n satisfying $r_{Ae_1A}(f_n)=0$ and $Ae_1A=f_nA=e_1A\oplus e_2A\oplus\cdots\oplus e_nA$ with f_n a central idempotent in A. Put $f=f_n$. Then $fA\cong\operatorname{Mat}_n(D)$, where $D\cong e_1Ae_1$. Therefore, $A=fA\oplus(1-f)A$ (ring direct sum) with $Ae_1A=fA$ a simple Artinian ring. We may continue in this fashion to split off simple Artinian ring direct summands. If the process does not terminate, we get an infinite number of orthogonal idempotents, a contradiction. Thus, A is a semisimple Artinian ring, so R is semilocal. □

A ring R is called *π-regular* if for each $a\in R$ there exist a positive integer n (depending on a) and $x\in R$ such that $a^n=a^nxa^n$. It is well known that π-regular rings are I-rings (see [221, Proposition 1(1), p. 210]).

On the other hand, as a generalization of rings with minimum condition, a ring R is called *right π-regular* if for each $a\in R$ there exist $x\in R$ and a positive integer n (depending on a) such that $a^n=a^{n+1}x$. The right π-regularity of R is equivalent to each descending chain of right ideals of the form $aR\supseteq a^2R\supseteq\dots$ terminating. A left π-regular ring is defined similarly.

A ring which is either left or right π-regular is called *strongly π-regular*. The next result exhibits a relationship between the strong π-regularity and the π-regularity of a ring.

Theorem 1.2.17 (i) *Every strongly π-regular ring is π-regular* (*see* [33]).
(ii) *A ring R is right π-regular if and only if R is left π-regular* (*see* [138]).

Theorem 1.2.18 *The following conditions for a ring R are equivalent.*

(i) *R is strongly π-regular.*
(ii) *Each prime factor ring of R is strongly π-regular.*

Proof (i)⇒(ii) is obvious. For (ii)⇒(i), we need to see that for each $a\in R$ there is a positive integer n with $a^nR=a^{n+1}R$. Assume on the contrary that there exists

$a \in R$ such that the descending chain $\{a^i R \mid i = 1, 2, \dots\}$ of right ideals does not terminate. By Zorn's lemma, choose an ideal I of R which is maximal with respect to the property that the descending chain $\{a^i R + I \mid i = 1, 2, \dots\}$ does not terminate. Clearly, I is not prime from (ii). By passing to R/I, we may assume that $I = 0$ and R is not a prime ring. Hence, there exist nonzero ideals J and K of R such that $JK = 0$. By the choice of I, there is a positive integer m with $a^m \equiv a^{m+1}x \equiv a^{m+r}x^r \,(\mathrm{mod}\, J)$ with $x \in R$ and $a^m \equiv a^{m+1}y \equiv a^{m+r}y^r \,(\mathrm{mod}\, K)$ with $y \in R$ for any r. Take $r = m + 1$, and from $(a^m - a^{m+r}x^r)(a^m - a^{m+r}y^r) = 0$, we deduce that $a^{2m} = a^{2m+1}z$ for some $z \in R$. This contradicts the choice of I. Thus, R is strongly π-regular. □

A ring R is said to have *bounded index* (*of nilpotency*) if there exists a positive integer n such that $x^n = 0$ whenever x is a nilpotent element of R. The least such positive integer is called *index of nilpotency* of R. Hence, a reduced ring is exactly a ring with index of nilpotency 1.

Say R is a ring not necessarily with identity. Then R is called a *nil* ring if every element of R is nilpotent. Next, R is said to be *nilpotent* if there exists a positive integer n such that $R^n = 0$. We say that R is *locally nilpotent* if the subring generated by every finite subset of R is nilpotent.

Theorem 1.2.19 *If R is a nil ring with bounded index (of nilpotency), then R is locally nilpotent.*

Proof See [139, Theorem 53]. □

Theorem 1.2.20 *Let R be a semiprime ring with index of nilpotency at most n. Then:*

(i) *For each $x \in R$, $r_R(x^n) = r_R(x^k)$ for any integer $k \geq n$.*
(ii) *R is right and left nonsingular.*

Proof (i) Choose $x \in R$. We put $X_1 = r_R(x)x, X_2 = r_R(x^2)x^2, \dots,$ and $X_n = r_R(x^n)x^n$. Then $y_i y_j = 0$ for $i \geq j$, where $y_i \in X_i$ and $y_j \in X_j$. We claim that $y_1 y_2 \cdots y_n = 0$ for $y_i \in X_i$, $1 \leq i \leq n$. Indeed, take $y_i \in X_i$ for $1 \leq i \leq n$ and put $z = y_1 + y_2 + \cdots + y_n$. Then $z^n = y_1 y_2 \cdots y_n$ and $z^{n+1} = 0$ since $y_i y_j = 0$ for $i \geq j$. As R has index of nilpotency at most n, $z^n = 0$ and so $y_1 y_2 \cdots y_n = 0$. Therefore, $r_R(x)[x r_R(x^2)] \cdots [x^{n-1} r_R(x^n)]x^n = 0$.

We observe that $x^n r_R(x^{n+1}) \subseteq r_R(x)$, $x^n r_R(x^{n+1}) \subseteq x r_R(x^2), \dots,$ and $x^n r_R(x^{n+1}) \subseteq x^{n-1} r_R(x^n)$. Thus, $[x^n r_R(x^{n+1})]^{n+1} = 0$. As R is semiprime, $x^n r_R(x^{n+1}) = 0$. Hence, $r_R(x^{n+1}) \subseteq r_R(x^n)$. Therefore, $r_R(x^n) = r_R(x^{n+1})$. Hence, we see that $r_R(x^n) = r_R(x^k)$ for each integer $k \geq n$.

(ii) Take $x \in Z(R_R)$. If $x^n \neq 0$, then there exists $0 \neq b \in r_R(x) \cap x^n R$ because $r_R(x) \leq^{\text{ess}} R_R$. Say $b = x^n r$ with $r \in R$. Then $xb = 0$ and hence $x^{n+1} r = 0$. As $r_R(x^{n+1}) = r_R(x^n)$ by part (i), $r \in r_R(x^n)$, and so $b = x^n r = 0$, a contradiction. Hence, $x^n = 0$ for each $x \in Z(R_R)$. By Theorem 1.2.19, $Z(R_R)$ is locally nilpotent.

Let F be a finite subset of $Z(R_R)$, and let S be the subring of $Z(R_R)$ generated by F. Then S is nilpotent. Thus, there exists a smallest positive integer k such that $S^k R$ is nilpotent. We claim that $k \leq n$. Assume on the contrary that $k > n$. Note that $I := RS^k R$ is nilpotent, so R/I has index of nilpotency at most n. Put $Y_i = S^{k-i} RS^i$ for $i = 1, 2, \ldots, n$. Then for $i \geq j$, $Y_i Y_j = S^{k-i} RS^{i-j} S^k RS^j \subseteq I$. Take

$$V_i = \{y + I \in R/I \mid y \in Y_i\}$$

for $i = 1, 2, \ldots, n$. Then $V_i V_j = 0$ for $i \geq j$. Since R/I has index of nilpotency at most n, $V_1 V_2 \cdots V_n = 0$ as in the proof of part (i). So $Y_1 Y_2 \cdots Y_n \subseteq I$. Thus we have that $(S^{k-1} RS)(S^{k-2} RS^2)(S^{k-3} RS^3) \cdots (S^{k-n} RS^n) \subseteq I$.

So $(S^{k-1} R)^n S^n \subseteq I$. Because $k > n$, $k - 1 \geq n$ and thus it follows that $(S^{k-1} R)^{n+1} = (S^{k-1} R)^n S^{k-1} R = (S^{k-1} R)^n S^n S^{(k-1)-n} R \subseteq I$. Hence, $S^{k-1} R$ is nilpotent, a contradiction to the choice of k. Thus $k \leq n$, so $S^n R$ is nilpotent. Since R is semiprime, $S^n R = 0$, and hence $S^n = 0$. Consequently $Z(R_R)^n = 0$, and hence $Z(R_R) = 0$. Similarly, $Z({}_R R) = 0$. □

1.3 Rational Extensions

We discuss dense submodules and rational extensions in this section. A brief description of maximal right rings of quotients is also included.

Definition 1.3.1 A right R-module E is said to be *injective* if it satisfies any one of the following equivalent conditions.

(i) For any right R-module V and any $W \leq V$, every R-homomorphism $W \to E$ can be extended to an R-homomorphism $V \to E$.
(ii) (Baer's Criterion) Every R-homomorphism from a right ideal I of R to E can be extended to an R-homomorphism from R to E.
(iii) For any right R-module U, every R-monomorphism $E \to U$ splits.

A ring R is called *right self-injective* if R_R is injective. A left self-injective ring is defined similarly. There exists a right self-injective ring which is not left self-injective. For example, let R be the endomorphism ring of an infinite dimensional right vector space over a division ring. Then R is right self-injective, but it is not left self-injective (see [262, Example 3.74B]).

Theorem 1.3.2 *Every module M has a minimal injective extension which is, equivalently, a maximal essential extension of M. This extension of M is unique up to isomorphism.*

Proof See [264, Sect. 4.2]. □

A minimal injective extension of a module M is called the *injective hull* of M. The injective hull of M is denoted by $E(M)$ or $E(M_R)$.

Definition 1.3.3 A ring R is called *quasi-Frobenius* (simply, *QF*) if R is right self-injective and right Artinian (right Noetherian).

Theorem 1.3.4 *Let R be a ring. Then the following are equivalent.*

(i) *R is a QF-ring.*
(ii) *R is right self-injective left Artinian.*
(iii) *R is left self-injective right Artinian.*
(iv) *R is left self-injective left Artinian.*
(v) *R is right self-injective left Noetherian.*
(vi) *R is left self-injective right Noetherian.*
(vii) *R is left self-injective left Noetherian.*
(viii) *R is right Noetherian with $r_R(\ell_R(I)) = I$ and $\ell_R(r_R(J)) = J$ for each right ideal I and each left ideal J of R, respectively.*

Proof See [262, Theorem 15.1]. □

Definition 1.3.5 A submodule N_R of M_R is called *dense* (or *rational*) in M_R if for any $x, y \in M$ with $y \neq 0$, there exists $r \in R$ such that $xr \in N$ and $yr \neq 0$. We denote this by $N_R \leq^{\text{den}} M_R$.

Every dense submodule of M is an essential submodule of M. But the converse is not true. For example, take $M = \mathbb{Z}/p^{n+1}\mathbb{Z}$ and $N = p\mathbb{Z}/p^{n+1}\mathbb{Z}$ as $\mathbb{Z}$-modules, where p is a prime integer and n is an integer such that $n \geq 1$. Then N is essential in M, but N is not dense in M. We will see from Proposition 1.3.14 that these two notions coincide if M is nonsingular.

To consider a rational hull (or rational completion) of a module and a maximal right ring of quotients of a ring, we start with the following well known result.

Proposition 1.3.6 *For modules $N \leq M$, the following are equivalent.*

(i) $N \leq^{\text{den}} M$.
(ii) $\text{Hom}(M/N, E(M)) = 0$.
(iii) *For any K with $N \leq K \leq M$,* $\text{Hom}(K/N, M) = 0$.

Proof See [262, Proposition 8.6]. □

Let M_R be a right R-module, and we let $S = \text{End}(E(M))$. Put

$$\widetilde{E}(M) = \{x \in E(M) \mid h(M) = 0 \text{ with } h \in S \text{ implies } h(x) = 0\}.$$

Then $M \leq \widetilde{E}(M) \leq E(M)$. The next two results are related to $\widetilde{E}(M)$.

Proposition 1.3.7 *Assume that $M \leq V \leq E(M)$. Then $M \leq^{\text{den}} V$ if and only if $V \leq \widetilde{E}(M)$.*

Proof Assume that $M \leq^{\text{den}} V$. Put $E = E(M)$. From the short exact sequence $0 \to M \to V \to V/M \to 0$, we have the short exact sequence

$$0 \to \text{Hom}(V/M, E) \to \text{Hom}(V, E) \to \text{Hom}(M, E) \to 0$$

by the injectivity of E. As $M \leq^{\text{den}} V$ and $E(V) = E$, $\text{Hom}(V/M, E) = 0$ from Proposition 1.3.6. So $\text{Hom}(V, E) \cong \text{Hom}(M, E)$. Take $h \in S := \text{End}(E)$ with $h(M) = 0$. Then $h(V) = 0$ as $\text{Hom}(V, E) \cong \text{Hom}(M, E)$. So $V \leq \widetilde{E}(M_R)$.

Conversely, let $M \leq V \leq \widetilde{E}(M)$. Then we obtain the short exact sequence

$$0 \to \text{Hom}(\widetilde{E}(M)/M, E) \to \text{Hom}(\widetilde{E}(M), E) \to \text{Hom}(M, E) \to 0$$

from the short exact sequence $0 \to M \to \widetilde{E}(M) \to \widetilde{E}(M)/M \to 0$.

Say $f \in \text{Hom}(M, E)$ and φ is an extension of f to E. If $f(M) = 0$, then $\varphi(\widetilde{E}(M)) = 0$ because $\varphi \in \text{End}(E)$. So $\text{Hom}(\widetilde{E}(M), E) \cong \text{Hom}(M, E)$, and hence $0 = \text{Hom}(\widetilde{E}(M)/M, E) = \text{Hom}(\widetilde{E}(M)/M, E(\widetilde{E}(M)))$. Therefore $M \leq^{\text{den}} \widetilde{E}(M)$ by Proposition 1.3.6. As $M \leq V \leq \widetilde{E}(M)$, $M \leq^{\text{den}} V$. □

Proposition 1.3.8 *Let M be a right R-module and $M \leq^{\text{den}} V$. Then there exists a unique monomorphism $g : V \to \widetilde{E}(M)$ extending the inclusion map $M \hookrightarrow \widetilde{E}(M)$.*

Proof As $M \leq^{\text{ess}} V$, the inclusion map $M \hookrightarrow E(M)$ extends to a monomorphism $g : V \to E(M)$. Clearly $M = g(M) \leq^{\text{den}} g(V)$ as $M \leq^{\text{den}} V$, and so $g(V) \leq \widetilde{E}(M)$ by Proposition 1.3.7. Suppose that $g_1, g_2 : V \to \widetilde{E}(M)$ both extend the inclusion map $M \hookrightarrow \widetilde{E}(M)$. As $M \leq^{\text{ess}} V$, g_1 and g_2 are monomorphisms. Consider the map $f : V \to \widetilde{E}(M)$ defined by $f(v) = g_1(v) - g_2(v)$ for $v \in V$. Say $\varphi \in \text{End}(E(M))$ such that $\varphi|_V = f$. Because $\varphi(M) = f(M) = 0$, $\varphi(\widetilde{E}(M)) = 0$ and therefore $\varphi(V) = f(V) = 0$. So $g_1(v) = g_2(v)$ for all $v \in V$. Thus $g_1 = g_2$. □

By Propositions 1.3.7 and 1.3.8, $\widetilde{E}(M_R)$ is the unique maximal rational extension of M_R. We call it the *rational hull* (or *rational completion*) of M_R.

Let R be a ring, $E = E_R = E(R_R)$ be an injective hull of R_R, and let $H = \text{End}(E_R)$, operating on the left of E_R. Furthermore, let $Q = \text{End}({}_HE)$, operating on the right of E. Thus $E = {}_HE_Q$ is an (H, Q)-bimodule.

Theorem 1.3.9 (i) *${}_HE$ is a cyclic H-module generated by 1_R.*

(ii) *Let $\varepsilon : Q \to E$ be defined by $\varepsilon(q) = 1_R \cdot q$, for $q \in Q$. Then ε is an R-isomorphism from Q onto $\widetilde{E}(R_R)$.*

Proof (i) Let $x \in E$. The R-homomorphism $R_R \to E_R$ sending 1_R to x can be extended to some $h \in \text{End}(E_R) = H$, so $x = h(1_R) \in H \cdot 1_R$.

(ii) Clearly, ε is an R-homomorphism. If $q \in \text{Ker}(\varepsilon)$, then $1_R \cdot q = 0$. By part (i), $0 = H \cdot (1_R \cdot q) = (H \cdot 1_R) \cdot q = E \cdot q$. Hence $q = 0$, so ε is one-to-one.

We claim that $\text{Image}(\varepsilon) = \widetilde{E}(R_R)$. For this, say $h \in H$ such that $h(R) = 0$. Then $h(1_R \cdot Q) = (h \cdot 1_R) \cdot Q = 0$. Hence $\text{Image}(\varepsilon) = 1_R \cdot Q \subseteq \widetilde{E}(R_R)$.

Next, let $y \in \widetilde{E}(R_R)$. Define

$$\phi : E = H \cdot 1_R \to E \quad \text{by} \quad \phi(h \cdot 1_R) = h(y) \quad \text{for} \quad h \in H.$$

If $h \cdot 1_R = 0$, then $h(1_R) = 0$ and so $h(R) = 0$. Thus $h(\widetilde{E}(R_R)) = 0$, $h(y) = 0$.

Therefore ϕ is well-defined and it is an H-homomorphism of ${}_H E$. Thus there is $q \in Q$ with $h(y) = \phi(h \cdot 1_R) = (h \cdot 1_R) \cdot q$ for all $h \in H$. When $h = 1$, we get that $y = 1_R \cdot q = \varepsilon(q) \in \text{Image}(\varepsilon)$. So $\text{Image}(\varepsilon) = \widetilde{E}(R_R)$. □

Definition 1.3.10 By using the R-isomorphism ε in Theorem 1.3.9, Q is identified with $\widetilde{E}(R_R)$. This then gives $\widetilde{E}(R_R)$ a ring structure extending its given right R-module structure. The ring Q is denoted by $Q(R)$ and is called the *maximal right ring of quotients* of R. Any intermediate ring between R and $Q(R)$ is called a *right ring of quotients* of R. The maximal left ring of quotients $Q^\ell(R)$ and a left ring of quotients of R are defined similarly.

Another description of $Q(R)$ will be provided in Theorem 1.3.13. For this, we need some preparations.

Proposition 1.3.11 (i) *Assume that D is a dense right ideal of R. Then for any $q \in Q(R)$, $q^{-1}D := \{r \in R \mid qr \in D\}$ is a dense right ideal of R.*

(ii) *If D is a dense right ideal of R, then $\ell_{Q(R)}(D) = 0$.*

(iii) *If D_1 and D_2 are dense right ideals of R, then so is $D_1 \cap D_2$.*

(iv) *For $D \trianglelefteq R$, $D_R \leq^{\text{den}} R_R$ if and only if $\ell_R(D) = 0$.*

Proof (i) Clearly, $q^{-1}D$ is a right ideal of R. Since $D_R \leq^{\text{den}} R_R \leq^{\text{den}} Q(R)_R$, $D_R \leq^{\text{den}} Q(R)_R$. Take $x, y \in R$ such that $y \neq 0$. There is $a \in R$ such that $(qx)a \in D$ and $ya \neq 0$. So $xa \in q^{-1}D$ and $ya \neq 0$. Hence, $(q^{-1}D)_R \leq^{\text{den}} R_R$.

(ii) If $xD = 0$ for some $0 \neq x \in Q(R)$, then there is $r \in R$ with $0 \neq xr \in R$. By part (i), $r^{-1}D_R \leq^{\text{den}} R_R$. So there is $a \in R$ such that $a = 1a \in r^{-1}D$ and $xra \neq 0$. Thus $0 \neq xra \in xr(r^{-1}D) \subseteq xD = 0$, a contradiction.

(iii) The proof is routine.

(iv) Say $D_R \leq^{\text{den}} R_R$. Then $\ell_R(D) = 0$ by part (ii). Conversely, take x, y in R with $y \neq 0$. Then $yD \neq 0$. There is $d \in D$ with $yd \neq 0$. As $xd \in D$, $D_R \leq^{\text{den}} R_R$. □

Proposition 1.3.12 *Assume that D is a dense right ideal of R and put $(R : D) = \{q \in Q(R) \mid qD \subseteq R\}$. Then $((R : D), +) \cong (\text{Hom}(D_R, R_R), +)$.*

Proof Define $\theta : (R : D) \to \text{Hom}(D_R, R_R)$ by $\theta(q)(d) = qd$ for $q \in (R : D)$ and $d \in D$. Then θ is additive. To show that θ is one-to-one, assume that $\theta(q) = 0$ with $q \in (R : D)$. Then $qD = 0$, so $q = 0$ by Proposition 1.3.11(ii). Hence, θ is one-to-one.

We show that θ is onto. Say $f \in \text{Hom}(D_R, R_R)$ and $g \in \text{End}(E_R)$ is an extension of f, where E_R is an injective hull of R_R. Then $f(d) = g(d) = g(1_R)d$ for $d \in D$.

To see that $g(1_R) \in Q(R)$, take $h \in \text{End}(E_R)$ such that $h(R) = 0$. If $h(g(1_R)) \neq 0$, then there exists $a \in R$ with $0 \neq h(g(1_R))a \in R$. Also

$$h(g(1_R))a(a^{-1}D) \subseteq h(g(1_R))D = h(g(D)) = h(f(D)) \subseteq h(R) = 0.$$

Since $a^{-1}D$ is a dense right ideal of R by Proposition 1.3.11(i), $h(g(1_R))a = 0$ from Proposition 1.3.11(ii), a contradiction. Whence $h(g(1_R)) = 0$, so it follows that $q := g(1_R) \in Q(R)$. Thereby $f(d) = qd$ for all $d \in D$. Thus $q \in (R : D)$ and $\theta(q) = f$. Therefore θ is onto. □

Our next theorem provides a construction for the maximal right ring of quotients of a ring R. For the existence of the maximal right ring of quotients of R, we need not assume that R has an identity; but we assume that $\ell_R(R) = 0$. This construction is due to Johnson [234] and Utumi [395].

Let $\mathfrak{D}$ be the set of all dense right ideals of R and let

$$\mathcal{Q} = \bigcup_{D \in \mathfrak{D}} \text{Hom}_R(D, R)/\sim,$$

where $\sim$ is an equivalence relation on $\bigcup_{D \in \mathfrak{D}} \text{Hom}_R(D, R)$ such that, for given $f \in \text{Hom}_R(D, R)$ and $g \in \text{Hom}_R(D', R)$, $f \sim g$ means that $f = g$ on $D \cap D'$. Let $[f]$ denote the equivalence class of f.

Say $f_1 \in \text{Hom}_R(D_1, R)$ and $f_2 \in \text{Hom}_R(D_2, R)$. We define

$$f_1 + f_2 \in \text{Hom}_R(D_1 \cap D_2, R) \text{ and } f_1 f_2 \in \text{Hom}_R(f_2^{-1}(D_1), R)$$

by $(f_1 + f_2)(d) = f_1(d) + f_2(d)$ for $d \in D_1 \cap D_2$; and $(f_1 f_2)(d) = f_1(f_2(d))$ for $d \in f_2^{-1}(D_1)$. Define $[f_1] + [f_2] = [f_1 + f_2]$ and $[f_1] \cdot [f_2] = [f_1 f_2]$.

Theorem 1.3.13 $Q(R) \cong (\mathcal{Q}, +, \cdot)$.

Proof The proof is a routine verification. Here we first show that $f_2^{-1}(D_1)$ is a dense right ideal of R. Thus, the description of multiplication makes sense, and gives an indication of the ring isomorphism between $Q(R)$ and $\mathcal{Q}$.

Say $x, y \in R$ with $y \neq 0$. As D_2 is a dense right ideal of R, there is $a \in R$ such that $xa \in D_2$ and $ya \neq 0$. Next, since D_1 is a dense right ideal of R, there is $b \in R$ with $f_2(xa)b \in D_1$ and $yab \neq 0$. So $xab \in f_2^{-1}(D_1)$ and $yab \neq 0$. Therefore, $f_2^{-1}(D_1)$ is a dense right ideal of R.

Let $q \in Q(R)$. Then $D := q^{-1}R$ is a dense right ideal of R by Proposition 1.3.11(i). Therefore $q \in (R : D)$, where q corresponds to $f \in \text{Hom}_R(D, R)$ such that $f(d) = qd$ for all $d \in D$ by Proposition 1.3.12. Say $q_i \in (R : D_i)$ correspond to $f_i \in \text{Hom}_R(D_i, R)$ for $i = 1, 2$. Then $q_1 = q_2$ if and only if $[f_1] = [f_2]$. In fact, if $q_1 = q_2$, then $f_1 = f_2$ on $D_1 \cap D_2$. Thus, $[f_1] = [f_2]$. Conversely, assume that $[f_1] = [f_2]$. Then $f_1(d) = f_2(d)$ for all $d \in D_1 \cap D_2$. Thus, $q_1(d) = q_2(d)$ for all $d \in D_1 \cap D_2$, so $(q_1 - q_2)(D_1 \cap D_2) = 0$. Thus, $q_1 = q_2$ by Proposition 1.3.11(ii) and (iii). Also we see that $q_1 + q_2$ and $q_1 q_2$ of $Q(R)$ correspond to $[f_1 + f_2]$ and $[f_1 f_2]$, respectively. □

Proposition 1.3.14 *Let M be a nonsingular right R-module and N a submodule of M. Then $N \leq^{\text{ess}} M$ if and only if $N \leq^{\text{den}} M$.*

Proof Assume that $N \leq^{\text{ess}} M$. Let $x, y \in M$ and $y \neq 0$. Then we see that $x^{-1}N := \{r \in R \mid xr \in N\}$ is an essential right ideal since $N \leq^{\text{ess}} M$. As M is nonsingular, $y \cdot x^{-1}N \neq 0$. Thus, $N \leq^{\text{den}} M$. The converse is obvious. □

Corollary 1.3.15 *Let a ring R be right nonsingular. Then $R_R \leq^{\text{den}} E(R_R)$, and so $Q(R) = \widetilde{E}(R_R) = E(R_R)$.*

Proof Because $Z(R_R) = 0$, $E(R_R)$ is also nonsingular. By Proposition 1.3.14, $R_R \leq^{\text{den}} E(R_R)$, so $Q(R) = \widetilde{E}(R_R) = E(R_R)$ from Proposition 1.3.7. □

If R is a semiprime ring and $I \trianglelefteq R$, then $\ell_R(I) = r_R(I)$. Thereby, in this case, we also use $\text{Ann}_R(I)$ to denote $\ell_R(I)$ or $r_R(I)$.

Proposition 1.3.16 *Assume that R is a semiprime ring and $I \trianglelefteq R$. Then the following are equivalent.*

(i) $\text{Ann}_R(I) = 0$.
(ii) $I_R \leq^{\text{den}} R_R$.
(iii) $I_R \leq^{\text{ess}} R_R$.
(iv) ${}_RI_R \leq^{\text{ess}} {}_RR_R$ *(i.e., $I \cap J \neq 0$ for any $0 \neq J \trianglelefteq R$).*

Proof The proof is straightforward. □

Definition 1.3.17 Assume that R is a semiprime ring and we let $\mathcal{F}$ be the set of all ideals I of R such that $\text{Ann}_R(I) = 0$. Observe that each $I \in \mathcal{F}$ is a dense right ideal of R by Proposition 1.3.11(iv) or Proposition 1.3.16.

Now we define $Q^m(R) = \{q \in Q(R) \mid qI \subseteq R \text{ for some } I \in \mathcal{F}\}$ and

$$Q^s(R) = \{q \in Q(R) \mid qI \subseteq R \text{ and } Jq \subseteq R \text{ for some } I, J \in \mathcal{F}\}.$$

If $I_1, I_2 \in \mathcal{F}$, then $I_1I_2, I_2I_1 \in \mathcal{F}$ and so $I_1 \cap I_2 \in \mathcal{F}$. Take $q_1, q_2 \in Q^m(R)$ such that $q_1I_1 \subseteq R$ and $q_2I_2 \subseteq R$, with $I_1, I_2 \in \mathcal{F}$. Then

$$q_1q_2(I_2I_1) \subseteq q_1RI_1 \subseteq q_1I_1 \subseteq R.$$

Thus $q_1q_2 \in Q^m(R)$. The other ring axioms for $Q^m(R)$ can be routinely verified, therefore $Q^m(R)$ is a subring of $Q(R)$.

Let $q_1, q_2 \in Q^s(R)$ with $q_1I_1 \subseteq R$, $J_1q_1 \subseteq R$ and $q_2I_2 \subseteq R$, $J_2q_2 \subseteq R$, where $I_1, I_2, J_1, J_2 \in \mathcal{F}$. Then as above, $q_1q_2(I_2I_1) \subseteq R$. Also

$$(J_2J_1)q_1q_2 \subseteq J_2(J_1q_1)q_2 \subseteq J_2Rq_2 \subseteq J_2q_2 \subseteq R.$$

Thus, $q_1q_2 \in Q^s(R)$. The other ring axioms for $Q^s(R)$ can be routinely checked. Thus $Q^s(R)$ is a subring of $Q(R)$ and $Q^s(R) \subseteq Q^m(R) \subseteq Q(R)$.

The ring $Q^m(R)$ is called the *Martindale right ring of quotients* of R, and the ring $Q^s(R)$ is known as the *symmetric ring of quotients* of R.

Let $N = \{q \in Q(R) \mid qR = Rq\}$. Then the ring RN is called the *normal closure* of R. The *central closure* of R is defined to be $R\mathrm{Cen}(Q(R))$. Then RN and $R\mathrm{Cen}(Q(R))$ are subrings of $Q^s(R)$ and $R\mathrm{Cen}(Q(R)) \subseteq RN$ (see [262, Theorem 14.30 and Corollary 14.31]).

A ring R is called *right Kasch* if every simple right R-module is embedded in R_R. The next result provides information about right Kasch rings.

Proposition 1.3.18 *Let R be a ring. Then the following are equivalent.*

(i) *R is right Kasch.*
(ii) *If M is a maximal right ideal of R, then $M = r_R(v)$ for some $v \in R$.*
(iii) *The left annihilator of any maximal right ideal is nonzero.*
(iv) *The only dense right ideal of R is R itself.*

In this case, $R = Q(R)$.

Proof See [262, Corollaries 8.28 and 13.24]. □

Example 1.3.19 There exists a ring R with $R = Q(R)$, but R is not right Kasch. Indeed, let R be the endomorphism ring of an infinite dimensional right vector space over a field. Then R is a prime, regular, and right self-injective ring, so $R = Q(R)$. Let $I = \{f \in R \mid f(V)$ is finite dimensional$\}$. Then I is a nonzero proper ideal of R and hence $\ell_R(I) = 0$ because R is prime. Thus, $I_R \leq^{\mathrm{den}} R_R$ by Proposition 1.3.16. So R is not right Kasch from Proposition 1.3.18 because $I \neq R$.

Historical Notes Semicentral idempotents in Definition 1.2.1 were initially defined and studied by Birkenmeier in [57]. Proposition 1.2.5 appears in [78]. Theorem 1.2.16 was obtained by Kaplansky [243], while Theorem 1.2.18 was shown by Fisher and Snider [170]. Theorem 1.2.19 is due to Levitzki in [278]. Theorem 1.2.20 was obtained by Hannah [195]. Theorem 1.3.9 is due to Lambek [263] (see also [264]). In [234], Johnson developed the notion of the right singular ideal and gave the construction in Theorem 1.3.13 for $Q(R)$ when $Z(R_R) = 0$. Utumi [395] introduced the notion of a dense right ideal and generalized Johnson's construction to a ring R for which $Z(R_R)$ is not necessarily zero. Most results of Sect. 1.3 can be found in [262] and [264]. See [317] for further materials on QF-rings. Additional references on related material include [22, 51, 134, 165–167, 174], and [346].

Chapter 2
Injectivity and Some of Its Generalizations

Recall that an injective module generalizes a vector space as well as a divisible Abelian group. This generalization is one of the cornerstones of Module theory. The fact that every module has a unique (up to isomorphism) injective hull makes this notion very useful. Among many basic properties, one which characterizes an injective module is that it is a direct summand of every overmodule of itself (see Definition 1.3.1). Moreover, the class of injective modules is closed under direct summands and direct products (hence finite direct sums).

In view of the above, it is natural to seek effective generalizations of injectivity in order to obtain classes of modules properly containing the class of injective modules which further enjoy some of the advantages of these modules. From the preceding properties, one can see that an injective module satisfies the condition that each submodule is essential in a direct summand. This is known as the (C_1) condition in the literature and is a common generalization of injective and semisimple modules. Modules with the (C_1) condition are called *extending modules* (also known as *CS-modules*).

In this chapter, we introduce various conditions which are related to extending or injective modules. It is known that direct summands of modules satisfying either of (C_1), (C_2) or (C_3) conditions, or of (quasi-)injective modules, or of (quasi-) continuous modules inherit these respective properties. On the other hand, these classes of modules are generally not closed under direct sums. One focus of this chapter is to discuss conditions which ensure that such classes of modules are closed under direct sums. Applications of these conditions, which include decomposition theorems, are also considered.

We shall also discuss FI-extending modules (i.e., modules for which every fully invariant submodule is essential in a direct summand). This provides a (natural) generalization of the extending property. An ongoing open problem is the precise characterization of when a direct sum of extending modules is extending. However, in Theorem 2.3.5, we shall see that an arbitrary direct sum of extending modules satisfies at least the extending property for fully invariant submodules *without any additional conditions*. Direct summands of FI-extending modules will also be considered. Later in Chaps. 3 and 4, we shall observe that there are strong connections

G.F. Birkenmeier et al., *Extensions of Rings and Modules*,
DOI 10.1007/978-0-387-92716-9_2,

between the (FI-)extending and the (quasi-)Baer properties for rings and for modules, respectively.

2.1 Quasi-continuous and Extending Modules

We introduce some generalizations of injectivity which are related to the extending property of modules. More specifically, we discuss the notions of quasi-injective, continuous, quasi-continuous, and extending modules. Examples, characterizations, and basic results for these modules are provided in this section. We begin with the following definition of relative injectivity.

Definition 2.1.1 Let M and N be modules.

(i) M is called *N-injective* if, for any $W \leq N$ and $f \in \mathrm{Hom}(W, M)$, there exists $\phi \in \mathrm{Hom}(N, M)$ such that $\phi|_W = f$.

(ii) M and N are said to be *relatively injective* if M is N-injective and N is M-injective.

If N is semisimple, then every module M is N-injective. A module M is injective if and only if it is N-injective for all modules N. By Baer's Criterion, a right R-module M_R is injective if and only if M_R is R_R-injective.

The next result is a useful criterion for the relative injectivity between two modules.

Theorem 2.1.2 *Let M and N be right R-modules. Then M is N-injective if and only if $\varphi(N) \subseteq M$ for every $\varphi \in \mathrm{Hom}(E(N), E(M))$.*

Proof Assume that M be N-injective. Take $\varphi \in \mathrm{Hom}(E(N), E(M))$. Now we let $W = \{w \in N \mid \varphi(w) \in M\}$. Then $\varphi|_W \in \mathrm{Hom}(W, M)$. As M is N-injective, there is $f \in \mathrm{Hom}(N, M)$, an extension of $\varphi|_W$. Suppose that $f(n_0) \neq \varphi(n_0)$ for some $n_0 \in N$. Then $0 \neq (f(n_0) - \varphi(n_0))r \in M$ for some $r \in R$ as $M \leq^{\text{ess}} E(M)$. We may note that $f(n_0)r \in M$. So $\varphi(n_0)r = \varphi(n_0 r) \in M$ and $n_0 r \in W$.

Thus $f(n_0 r) = \varphi(n_0 r)$, a contradiction. Hence $f(n) = \varphi(n)$ for all $n \in N$, and so $\varphi(N) = f(N) \subseteq M$.

Conversely, to see that M is N-injective, let $W \leq N$ and $\phi \in \mathrm{Hom}(W, M)$. There exists $\alpha \in \mathrm{Hom}(E(N), E(M))$ with $\alpha|_W = \phi$. Thus, M is N-injective. □

Corollary 2.1.3 *Assume that A and B are relatively injective modules. If $E(A) \cong E(B)$, then $A \cong B$.*

Proof Let $\varphi : E(A) \to E(B)$ be an isomorphism. As B is A-injective, we have that $\varphi(A) \subseteq B$ by Theorem 2.1.2. Similarly, $\varphi^{-1}(B) \subseteq A$. Therefore,

$$B = \varphi\varphi^{-1}(B) \subseteq \varphi(A) \subseteq B.$$

So $\varphi(A) = B$ and hence $A \cong B$. □

Theorem 2.1.4 *Let $M = M_1 \oplus M_2$ with M_1 and M_2 modules. Then M_1 is M_2-injective if and only if, for every $K \leq M$ with $M_1 \cap K = 0$, there exists $N \leq M$ such that $K \leq N$ and $M = M_1 \oplus N$.*

Proof See [145, Lemma 7.5] for the proof. □

Proposition 2.1.5 *Let N be a module and $\{M_i \mid i \in \Lambda\}$ be a set of modules. Then $\prod_{i\in\Lambda} M_i$ is N-injective if and only if M_i is N-injective for all $i \in \Lambda$.*

Proof Exercise. □

Proposition 2.1.6 *Let M and N be modules.*

(i) *Assume that M is N-injective and $K \leq N$. Then M is both K-injective and N/K-injective.*

(ii) *Let $N = \bigoplus_{i\in\Lambda} N_i$, a direct sum of modules N_i. Then M is N_i-injective for each $i \in \Lambda$ if and only if M is N-injective.*

Proof See [301, Propositions 1.3 and 1.5]. □

Theorem 2.1.7 *Assume that $M_1, M_2, \ldots, M_k$ and $N_1, N_2, \ldots, N_n$ are modules. Then M_i and N_j are relatively injective, for each i and j, if and only if $M_1 \oplus M_2 \oplus \cdots \oplus M_k$ and $N_1 \oplus N_2 \oplus \cdots \oplus N_n$ are relatively injective.*

Proof The proof follows from Propositions 2.1.5 and 2.1.6. □

Definition 2.1.8 A module M is called *quasi-injective* if M is M-injective.

Theorem 2.1.9 *A module M is quasi-injective if and only if $M \trianglelefteq E(M)$.*

Proof Theorem 2.1.2 and Definition 2.1.8 yield the result. □

Let M be a module and $N \leq M$. By Zorn's lemma, there is a submodule C of M which is maximal with respect to $N \cap C = 0$. Then $N \oplus C \leq^{\text{ess}} M$. In this case, C is called a *complement* of N in M.

Proposition 2.1.10 *Let M be a quasi-injective module. Then:*

(C_1) *Every submodule of M is essential in a direct summand of M.*
(C_2) *If $V \leq M$ and $V \cong N \leq^{\oplus} M$, then $V \leq^{\oplus} M$.*

Proof (C_1) Assume that $N \leq M$ and K a complement of N in M. Then we have that $N \oplus K \leq^{\text{ess}} M$. So $E(M) = E(N) \oplus E(K)$. Since $M \trianglelefteq E(M)$ from Theorem 2.1.9, $M = (E(N) \cap M) \oplus (E(K) \cap M)$ (see Exercise 2.1.37.2). Therefore, it follows that $N \leq^{\text{ess}} (E(N) \cap M) \leq^{\oplus} M$.

(C_2) Since M is M-injective, N is M-injective by Proposition 2.1.5. So V is M-injective, hence the identity map of V extends a homomorphism $f : M \to V$. Thus, $M = V \oplus \text{Ker}(f)$. □

Proposition 2.1.11 *A module M with* (C_2) *satisfies the following condition.*

(C_3) *If M_1 and M_2 are direct summands of M with $M_1 \cap M_2 = 0$, then $M_1 \oplus M_2$ is a direct summand of M.*

Proof Let M satisfy (C_2) condition. Say $M = M_1 \oplus V_1$ for some $V_1 \leq M$, and let $\pi : M = M_1 \oplus V_1 \to V_1$ be the canonical projection. Then $\text{Ker}(\pi|_{M_2}) = 0$ since $M_1 \cap M_2 = 0$. So $\pi|_{M_2} : M_2 \cong \pi(M_2)$.

From (C_2) condition, $\pi(M_2) \leq^{\oplus} M$. Let $M = \pi(M_2) \oplus K$ with $K \leq M$. Then $V_1 = \pi(M_2) \oplus (K \cap V_1)$ by the modular law as $\pi(M_2) \leq V_1$. Hence,

$$M = M_1 \oplus V_1 = M_1 \oplus \pi(M_2) \oplus (K \cap V_1).$$

Now $M_1 \oplus M_2 \cong M_1 \oplus \pi(M_2) \leq^{\oplus} M$, so $M_1 \oplus M_2 \leq^{\oplus} M$ by (C_2) condition. □

Quasi-injectivity is an important generalization of injectivity. We will see that conditions (C_1), (C_2), and (C_3) satisfied by quasi-injective modules are also interesting in their own right. The next definition provides three useful generalizations of (quasi-)injectivity. While we discuss some properties of these three notions in this chapter, the reader is referred to [145, 301], and [317] for more details.

Definition 2.1.12 Let M be a module.

(i) M is called *continuous* if it satisfies (C_1) and (C_2) conditions.
(ii) M is said to be *quasi-continuous* if it has (C_1) and (C_3) conditions.
(iii) M is called *extending* (or *CS*) if it satisfies (C_1) condition.

A ring R is said to be *right (quasi-)continuous* if R_R is (quasi-)continuous. A ring R is called *right extending* (or *right CS*) if R_R is extending. A left (quasi-)continuous ring and a left extending (or left CS) ring are defined similarly. From the preceding, the following implications hold true for modules:

injective ⇒ quasi-injective ⇒ continuous ⇒ quasi-continuous ⇒ extending,

while the reverse implications do not hold as illustrated in Example 2.1.14.

Lemma 2.1.13 *If R is a semiprime ring, then $I_R \leq^{\text{ess}} r_R(\ell_R(I))_R$ for any $I \trianglelefteq R$.*

Proof Exercise. □

Example 2.1.14 (i) Every injective module and every semisimple module are quasi-injective. There exists a simple module which is not injective (e.g., $\mathbb{Z}_p$ for any prime integer p as a $\mathbb{Z}$-module). Further, there is a quasi-injective module which is neither injective nor semisimple. Let $R = \mathbb{Z}$ and $M = \mathbb{Z}_{p^n}$, with p a prime integer and n an integer such that $n > 1$. Then $E(M) = \mathbb{Z}_{p^\infty}$, the Prüfer p-group, and $\text{End}(E(M)) = \widehat{\mathbb{Z}}_p$, the ring of p-adic integers. Thus, $f(M) \subseteq M$ for any $f \in \text{End}(E(M))$. So M is quasi-injective by Theorem 2.1.9. But M is neither injective nor semisimple (see [153, Example, p. 22]).

(ii) Let K be a field and F be a proper subfield of K. Set $K_n = K$ for all $n = 1, 2 \ldots$. Take $R = \{(a_n)_{n=1}^{\infty} \in \prod_{n=1}^{\infty} K_n \mid a_n \in F \text{ eventually}\}$, which is a subring of $\prod_{n=1}^{\infty} K_n$. Say $I \trianglelefteq R$. Then $r_R(I) = eR$ with $e^2 = e \in R$ (see also Example 3.1.7). From Lemma 2.1.13, $I_R \leq^{\text{ess}} r_R(\ell_R(I)) = (1-e)R_R$ as R is semiprime. So R_R is extending.

Further, since R is regular, R_R also satisfies (C_2) condition. Thus, R_R is continuous. As $E(R_R) = \prod_{n=1}^{\infty} K_n$, R_R is not injective, so R_R is not quasi-injective. For another example of a continuous module, which is not quasi-injective, see Example 2.1.36.

(iii) Let R be a right Ore domain which is not a division ring (e.g., the ring $\mathbb{Z}$ of integers). Then R_R is quasi-continuous. Take $0 \neq x \in R$ such that $xR \neq R$. Then $xR_R \cong R_R$, but xR_R is not a direct summand of R_R. Thus R_R is not continuous.

(iv) Let F be a field and $R = T_2(F)$. Then we see that R_R is extending. Let $e_{ij} \in R$ be the matrix with 1 in the (i, j)-position and 0 elsewhere. Put $e = e_{12} + e_{22}$ and $f = e_{22}$. Then $e^2 = e$ and $f^2 = f$. Note that $eR \cap fR = 0$. But $eR_R \oplus fR_R$ is not a direct summand of R_R. Thus, R_R is not quasi-continuous.

It can be checked that a submodule N of a module M is closed if and only if N is a complement of some submodule of M (Exercise 2.1.37.3).

Proposition 2.1.15 *Let M be a module. Then the following are equivalent.*

(i) *M is an extending module.*
(ii) *Every closed submodule of M is a direct summand of M.*
(iii) *Every complement submodule of M is a direct summand of M.*

Proof Exercise. □

Motivated by Proposition 2.1.15, an extending module is also called a *CS-module*. It is well known that every right R-module is injective if and only if R is semisimple Artinian (see [259, Theorem 2.9]). The next result has been obtained by Dung and Smith [144] (see also Vanaja and Purav [400]).

Theorem 2.1.16 *Let R be a ring. Then the following are equivalent.*

(i) *Every right R-module is extending.*
(ii) *R is Artinian serial and $J(R)^2 = 0$.*

Proof See [145, 13.5] for the proof. □

The following result is [155, Corollary 25.4.3].

Proposition 2.1.17 *If R is a semiprimary ring such that $R/J(R)^2$ is QF, then R is Artinian serial.*

Let a ring R be QF with $J(R)^2 = 0$. From Proposition 2.1.17, R is Artinian serial. So every right R-module is extending by Theorem 2.1.16. However, in the next example, there exists a QF-ring R such that $J(R)^3 = 0$ and R is serial, but not even every principal right ideal of R is extending.

Example 2.1.18 Let $R = \mathbb{Z}_3[S_3]$ be the group algebra of the symmetric group S_3 on three symbols $\{1, 2, 3\}$ over the field $\mathbb{Z}_3$. It is well known that R is QF. Denote $\sigma = (123)$ and $\tau = (12)$ in S_3. Then R has the following properties (see [72] for more details).

(i) $J(R)^3 = 0$.

(ii) Each of $(2 + \tau)R_R$ and $(2 + 2\tau)R_R$ has a unique composition series. So $R_R = (2+\tau)R_R \oplus (2+2\tau)R_R$ is serial. Similarly, ${}_RR = {}_RR(2+\tau) \oplus {}_RR(2+2\tau)$ is serial.

(iii) There exist a principal right ideal I of R and a right ideal J of R such that $J_R \leq^{\text{ess}} I_R \leq^{\text{ess}} R_R$ and J_R is extending. However, I_R is not extending. In fact, $I = (1 + \sigma + \tau)R$ is such a principal right ideal and J is the sum of all nonzero ideals of R contained in I.

Let M be a module. A (finitely generated) submodule of a factor module of M is called a (*finitely generated*) *subfactor* of M. A cyclic submodule of a factor module of M is called a *cyclic subfactor* of M.

Theorem 2.1.19 *Let M be a finitely generated module such that every finitely generated subfactor of M is extending. Then M is a finite direct sum of uniform modules.*

Proof From [145, 7.12], M satisfies ACC on direct summands. Thus, we have that $M = M_1 \oplus \cdots \oplus M_n$, a finite direct sum of indecomposable modules M_i. As M is extending, so is each M_i. Thus, each M_i is uniform. □

The next important result, due to Osofsky and Smith [333, Theorem 1], can also be found in [145, Corollary 7.13].

Theorem 2.1.20 *Let M be a cyclic module. Assume that all cyclic subfactors of M are extending. Then M is a finite direct sum of uniform modules.*

A module M is called a *V-module* if every simple module is M-injective, or equivalently, if any submodule of M is an intersection of maximal submodules.

Lemma 2.1.21 *Let M be a V-module such that every factor module of M has finite uniform dimension. Then M is Noetherian.*

Proof See [145, 16.14]. □

The next result appears in [145, Corollary 7.4].

Lemma 2.1.22 *Let M be a uniserial module with a unique composition series $M \supsetneq U \supsetneq V \supsetneq 0$. Then $M \oplus (U/V)$ is not extending.*

Let $\{S_\alpha\}$ be the socle series of M defined as follows:

$$S_1 = \mathrm{Soc}(M),\ S_\alpha/S_{\alpha-1} = \mathrm{Soc}(M/S_{\alpha-1})$$

and for a limit ordinal α, $S_\alpha = \cup_{\beta<\alpha} S_\beta$. If there exists the least ordinal α such that $M = S_\alpha$, then α is called *Loewy length* of M.

Let M be an R-module. Then an R-module N is called *M-generated* if there exists an epimorphism $M^{(\Lambda)} \to N$ for some index set Λ. For a module M, $\sigma[M]$ denotes the full subcategory of the category of R-modules whose objects are submodules of M-generated modules (see [412]).

For convenience, we say that a module M *satisfies* ($\star$) *if every finitely generated module in* $\sigma[M]$ *is extending.*

Theorem 2.1.23 *Let M be a finitely generated right R-module satisfying* ($\star$). *Then M is Noetherian.*

Proof Let $\{S_\alpha\}$ be the socle series of M. Put $S = \cup S_\alpha$. Then the module $H = M/S$ has zero socle. We first show that H is Noetherian. This is trivial if $H = 0$. Therefore, we assume that $H \neq 0$.

Observe that H is finitely generated. Say H/K is a factor module of H. By ($\star$), every finitely generated subfactor of H/K is extending (in particular H/K is also extending). From Theorem 2.1.19, H/K is a finite direct sum of uniform modules. Thus, every factor module of H is a finite direct sum of uniform modules. Hence, it is enough to show that H is a V-module, since we then apply Lemma 2.1.21 to obtain that H is Noetherian. So we may assume that H is uniform without loss of generality.

Let X be a simple right R-module. We claim that X is H-injective. First, assume that $X \notin \sigma[H]$. Then X is H-injective. Indeed, say $g \in \mathrm{Hom}(A, X)$, where $A \leq H$. Then there exists $\varphi \in \mathrm{Hom}(H, E(X))$ such that $\varphi|_A = g$. If $\varphi(H) \cap X \neq 0$, then $\varphi(H) \cap X = X$, since X is simple. So $X \leq \varphi(H) \in \sigma[H]$. Hence $X \in \sigma[H]$, a contradiction. Therefore, $\varphi(H) \cap X = 0$, so $\varphi(H) = 0$ as $X \leq^{\text{ess}} E(X)$. Thus $\varphi = 0$, so $g = 0$. Hence, X is H-injective.

Next, assume that $X \in \sigma[H]$. Then it is easy to see that $X \in \sigma[M]$. Since $\mathrm{Soc}(H) = 0$, $X \cap H = 0$. Also, $K := X \oplus H$ is finitely generated. Hence by ($\star$), K is extending since $K \in \sigma[M]$. Let $0 \neq V \leq H$ and $f \in \mathrm{Hom}(V, X)$. Put

$$U = \{a - f(a) \mid a \in V\} \leq K.$$

If $U = 0$, then $a = f(a)$ for all $a \in V$. Hence $V \leq X$, so $X = V \leq H$. But since $\mathrm{Soc}(H) = 0$, $X = 0$, a contradiction. Therefore, $U \neq 0$.

Since K is extending and $U \leq K$, U is essential in a direct summand U^* of K, say $K = U^* \oplus U_1$ for some submodule U_1 of K. To see that $U^* \cap X = 0$, let $a - f(a) = x \in X$, where $a \in V$. Then $a = f(a) + x \in V \cap X \subseteq H \cap X = 0$. Hence $x = a - f(a) = 0$, so $U \cap X = 0$. Therefore, $U^* \cap X = 0$ as $U \leq^{\text{ess}} U^*$.

We show that $X \leq U_1$. In fact, $\mathrm{Soc}(K) = \mathrm{Soc}(U^*) \oplus \mathrm{Soc}(U_1) = X$ as $K = X \oplus H$ and $\mathrm{Soc}(H) = 0$. Since X is simple, $X = xR$ for some $0 \neq x \in X$.

So $x \in \mathrm{Soc}(U^*) \oplus \mathrm{Soc}(U_1)$, thus $x = y + z$, where $y \in \mathrm{Soc}(U^*)$ and $z \in \mathrm{Soc}(U_1)$. Suppose $yR \neq 0$. Define $\theta : xR \to yR$ by $\theta(xr) = yr$. Then $\theta \neq 0$, so yR is simple and hence $yR \subseteq \mathrm{Soc}(K) = X$. As $yR \subseteq U^*$, $yR \subseteq X \cap U^* = 0$, a contradiction. So $x = z \in \mathrm{Soc}(U_1)$, and thus $X = xR \subseteq \mathrm{Soc}(U_1) \subseteq U_1$.

Note that $\mathrm{udim}(K) = 2$, $U^* \neq 0$, $U_1 \neq 0$, and $K = U^* \oplus U_1$. So we have that $\mathrm{udim}(U^*) = 1$ and $\mathrm{udim}(U^* \oplus X) = 2$. As $\mathrm{udim}(U_1) = 1$ and $X \leq U_1$, $X \leq^{\mathrm{ess}} U_1$. But X is closed in K, therefore $X = U_1$, and so

$$K = X \oplus H = U^* \oplus U_1 = U^* \oplus X.$$

Now let π be the canonical projection from $U^* \oplus X$ onto X. For $a \in V$,

$$a = (a - f(a)) + f(a) \in U^* \oplus X,$$

where $a - f(a) \in U \leq U^*$ and $f(a) \in X$. Thus $\pi|_H(a) = f(a)$ for $a \in V$, so $\pi|_H$ is an extension of f from H to X. This shows that X is H-injective. Therefore, H is a V-module. Consequently, by Lemma 2.1.21, H is Noetherian.

Assume that $S_3/S_2 \neq 0$ (where S_2 and S_3 are the second and the third socles of M, respectively). Then there is $0 \neq y + S_2 \in S_3/S_2$. Take $Y = yR$. Let C/B be a cyclic submodule of Y/B, where $B \leq C \leq Y$. Then

$$C/B \leq Y/B \leq M/B \in \sigma[M],$$

so $C/B \in \sigma[M]$. By ($\star$), C/B is extending. Thus from Theorem 2.1.20, we see that $Y = Y_1 \oplus \cdots \oplus Y_k$, where each Y_i is uniform. Hence one of the Y_i, say Y_1, has Loewy length 3.

Because Y is cyclic, so is Y_1 and hence $Y_1/\mathrm{Soc}(Y_1)$ is also cyclic. Say $D/\mathrm{Soc}(Y_1) \leq Y_1/\mathrm{Soc}(Y_1)$, and consider $(Y_1/\mathrm{Soc}(Y_1))/(D/\mathrm{Soc}(Y_1))(\cong Y_1/D)$, a factor module of $Y_1/\mathrm{Soc}(Y_1)$. Let $(B/\mathrm{Soc}(Y_1))/(D/\mathrm{Soc}(Y_1))(\cong B/D)$ be a cyclic submodule of $(Y_1/\mathrm{Soc}(Y_1))/(D/\mathrm{Soc}(Y_1))$. Then $B/D \leq M/D \in \sigma[M]$. As a consequence, $B/D \in \sigma[M]$.

By ($\star$), B/D is extending. So every cyclic subfactor of $Y_1/\mathrm{Soc}(Y_1)$ is extending. By Theorem 2.1.20, $Y_1/\mathrm{Soc}(Y_1) = T_1 \oplus \cdots \oplus T_m$, where each T_i is uniform and one of T_i, say T_1, has Loewy length 2. Thus, we have that $\mathrm{Soc}(T_1)$ is simple and $\mathrm{Soc}(T_1/\mathrm{Soc}(T_1)) = T_1/\mathrm{Soc}(T_1)$.

Say $T_0/\mathrm{Soc}(T_1)$ is a simple submodule of $T_1/\mathrm{Soc}(T_1)$. Then since $T_0 \leq T_1$ and T_1 is uniform, T_0 is uniform and $\mathrm{Soc}(T_0) = \mathrm{Soc}(T_1)$ is simple. So T_0 is a uniserial module with a unique composition series $T_0 \supsetneqq \mathrm{Soc}(T_1) \supsetneqq 0$. Let W be the inverse image of T_0 and W_1 be the inverse image of $\mathrm{Soc}(T_1)$ in Y_1, respectively. Then W is a uniserial module with a unique composition series $W \supsetneqq W_1 \supsetneqq \mathrm{Soc}(Y_1) \supsetneqq 0$.

As $T_0/\mathrm{Soc}(T_0)$ and $\mathrm{Soc}(T_1) = \mathrm{Soc}(T_0)$ are simple, T_0 is finitely generated. Next $\mathrm{Soc}(Y_1)$ is simple (note that Y_1 is uniform). Since $W/\mathrm{Soc}(Y_1) = T_0$, W is finitely generated. Therefore $W \oplus (W_1/\mathrm{Soc}(Y_1))$ is finitely generated and $W \oplus (W_1/\mathrm{Soc}(Y_1)) \in \sigma[M]$. By ($\star$), $W \oplus (W_1/\mathrm{Soc}(Y_1))$ is extending. But from Lemma 2.1.22, $W \oplus (W_1/\mathrm{Soc}(Y_1))$ is not an extending module, a contradiction. Hence $S_3 = S_2$, and so $S = \cup S_\alpha$ has Loewy length at most 2.

To see that udim(M) is finite by applying Theorem 2.1.19, say M/A is a factor module of M and L/A is a finitely generated submodule of M/A. We observe that $M/A \in \sigma[M]$, so $L/A \in \sigma[M]$. By ($\star$), L/A is extending. Hence, every finitely generated subfactor of M is extending. By Theorem 2.1.19, udim(M) is finite, and hence Soc(M) is semisimple Artinian. Similarly, udim(M/Soc(M)) is finite. So $S/\mathrm{Soc}(M) = \mathrm{Soc}(M/\mathrm{Soc}(M))$ is semisimple Artinian. Therefore, S is a module of finite composition length. Because $H = M/S$ is Noetherian by the preceding argument, M is Noetherian. □

If we put $M = R$ in Theorem 2.1.23 and we assume that every finitely generated R-module is extending, then the factor ring R/S has zero right socle and it is a right Noetherian, right V-ring (recall that a ring A is called a right V-ring if A_A is a V-module, so any right ideal of A is an intersection of maximal right ideals). Thus $J(R/S) = 0$ and hence $J(R) \subseteq S$, so $J(R)^3 = 0$.

Corollary 2.1.24 *Let R be a ring such that every finitely generated right R-module is extending. Then R is right Noetherian and $J(R)^3 = 0$.*

One of the reasons which makes the notion of quasi-continuous modules of interest, is their useful characterization in Theorem 2.1.25 given below. This result allows one to transfer from any given decomposition of the injective hull $E(M)$ of a quasi-continuous module M, to a similar decomposition for M. This fact is also helpful in the transference of properties between M and its injective hull $E(M)$.

Theorem 2.1.25 *The following are equivalent for a module M.*

(i) *M is quasi-continuous.*
(ii) *$M = X \oplus Y$ for any two submodules X and Y which are complements of each other.*
(iii) *$fM \subseteq M$ for every $f^2 = f \in \mathrm{End}(E(M))$.*
(iv) *$E(M) = \bigoplus_{i\in\Lambda} E_i$ implies $M = \bigoplus_{i\in\Lambda}(M \cap E_i)$.*
(v) *Any essential extension V of M with a decomposition $V = \bigoplus_{\alpha\in\Gamma} V_\alpha$ implies $M = \bigoplus_{\alpha\in\Gamma}(M \cap V_\alpha)$.*

Proof (i)⇒(ii) From (C_1) condition, $X \leq^\oplus M$ and $Y \leq^\oplus M$ by Proposition 2.1.15. Thus $X \oplus Y \leq^\oplus M$ by (C_3) condition. Therefore, $M = X \oplus Y$ since $X \oplus Y \leq^{\mathrm{ess}} M$.

(ii)⇒(iii) Take $f^2 = f \in \mathrm{End}(E(M))$. We put

$$A_1 = M \cap fE(M) \quad \text{and} \quad A_2 = M \cap (1-f)E(M).$$

Let B_1 be a complement of A_2 such that $A_1 \subseteq B_1$ and let B_2 be a complement of B_1 with $A_2 \subseteq B_2$. Then B_1 is a complement of B_2. By assumption, we have that $M = B_1 \oplus B_2$. Let $\pi : M \to B_1$ be the canonical projection. Now we show that $M \cap (f - \pi)M = 0$. Let $x, y \in M$ such that $y = (f - \pi)(x)$. Then

$$f(x) = y + \pi(x) \in M,$$

so $f(x) \in A_1$. Also $(1-f)(x) \in M$ and hence $(1-f)(x) \in A_2$. As

$$x = f(x) + (1-f)(x) \in A_1 \oplus A_2 \subseteq B_1 \oplus B_2,$$

$\pi(x) = f(x)$. So $y = 0$. Hence $M \cap (f-\pi)M = 0$. As $M \leq^{\text{ess}} E(M)$, we obtain that $(f-\pi)M = 0$ and so $fM = \pi M = B_1 \subseteq M$.

(iii)⇒(iv) Let $E(M) = \oplus_{i\in\Lambda} E_i$. Clearly, $\oplus_{i\in\Lambda}(M \cap E_i) \subseteq M$. Take $m \in M$. There is a finite subset F of Λ with $m \in \oplus_{i\in F} E_i$, $E(M) = (\oplus_{i\in F} E_i) \oplus V$, and $V \leq E(M)$. Let $f_i : E(M) \to E_i$ be the canonical projection for $i \in \Lambda$. By assumption, $f_i M \subseteq M$ for each i. Now $m = \sum_{i\in F} f_i(m) \in \oplus_{i\in F}(M \cap E_i)$ because $f_i(m) \in f_i M \subseteq M$. Hence $M \subseteq \oplus_{i\in\Lambda}(M \cap E_i)$. So $M = \oplus_{i\in\Lambda}(M \cap E_i)$.

(iv)⇒(i) Let $A \leq M$. Then $E(M) = E(A) \oplus E_2$ for some $E_2 \leq E(M)$. So $M = (M \cap E(A)) \oplus (M \cap E_2)$ by assumption. We note that $A \leq^{\text{ess}} M \cap E(A)$, so M has (C_1) condition.

To see that M has (C_3) condition, let $M_1 \leq^{\oplus} M$ and $M_2 \leq^{\oplus} M$ such that $M_1 \cap M_2 = 0$. Then $E(M) = E(M_1) \oplus E(M_2) \oplus E_3$ for some $E_3 \leq E(M)$. By assumption, $M = (M \cap E(M_1)) \oplus (M \cap E(M_2)) \oplus (M \cap E_3)$.

On the other hand, $M = M_1 \oplus W$ for some $W \leq M$. By the modular law, we get that $M \cap E(M_1) = M_1 \oplus (W \cap (M \cap E(M_1)))$ since $M_1 \leq M \cap E(M_1)$. Further, as $M_1 \leq^{\text{ess}} M \cap E(M_1)$, $M \cap E(M_1) = M_1$. Similarly, $M \cap E(M_2) = M_2$.

From $M = (M \cap E(M_1)) \oplus (M \cap E(M_2)) \oplus (M \cap E_3)$,

$$M = M_1 \oplus M_2 \oplus (M \cap E_3).$$

So M satisfies (C_3) condition. Thus M is quasi-continuous.

(iv)⇒(v) Let $M \leq^{\text{ess}} V$ and $V = \oplus_{\alpha\in\Gamma} V_\alpha$. Then $\oplus_{\alpha\in\Gamma}(M \cap V_\alpha) \subseteq M$. Next, let $m \in M$. Then there exists a finite subset, say F, of Γ such that $m \in \oplus_{\alpha\in F} V_\alpha$. Then $E(M) = E(V) = E(\oplus_{\alpha\in F} V_\alpha) \oplus E(\oplus_{\alpha\in\Gamma\setminus F} V_\alpha)$. By assumption, $M = (\oplus_{\alpha\in F}(M \cap E(V_\alpha))) \oplus (M \cap E(\oplus_{\alpha\in\Gamma\setminus F} V_\alpha))$. Comparing the representation of m in decompositions, $m \in \oplus_{\alpha\in F}(M \cap V_\alpha) \subseteq \oplus_{\alpha\in\Gamma}(M \cap V_\alpha)$. So $M \subseteq \oplus_{\alpha\in\Gamma}(M \cap V_\alpha)$. Hence, $M = \oplus_{\alpha\in\Gamma}(M \cap V_\alpha)$.

(v)⇒(iv) The proof is evident. □

Corollary 2.1.26 *Let M be a quasi-continuous module and $N \leq M$. Then $N \leq^{\text{ess}} M \cap E(N) \leq^{\oplus} M$.*

Proof The proof follows from Theorem 2.1.25(iv). □

For the case of extending modules, an analogue to the equivalence of parts (i) and (iii) of Theorem 2.1.25 is presented next.

Proposition 2.1.27 *A module M is extending if and only if for any given $e^2 = e \in \text{End}(E(M))$ there exists $f^2 = f \in \text{End}(M)$ with $eE(M) \cap M = fM$.*

Proof Assume that M is extending and let $eE(M) \cap M \leq^{\text{ess}} C \leq^{\oplus} M$ for some C. Then $eE(M) = E(C)$, so $C \leq eE(M) \cap M$. Hence, $C = eE(M) \cap M$ and thus

$eE(M) \cap M = fM$ for some $f^2 = f \in \text{End}(M)$. Conversely, say $N \leq M$. Then there exists $e^2 = e \in \text{End}(E(M))$ such that $N \leq^{\text{ess}} E(N) = eE(M)$. By hypothesis, $eE(M) \cap M = fM$ with $f^2 = f \in \text{End}(M)$. So $N \leq^{\text{ess}} fM$, and hence M is extending. □

Lemma 2.1.28 *Let M_R be a right R-module and $S = \text{End}_R(M)$. Then:*

(i) *$\Delta := \{s \in S \mid \text{Ker}(s)_R \leq^{\text{ess}} M_R\}$ is an ideal of S.*
(ii) *If $\{e_i \mid i \in \Lambda\}$ is a set of idempotents in S such that $\{e_i + \Delta \mid i \in \Lambda\}$ are orthogonal in the ring S/Δ, then $\sum_{i \in \Lambda} e_i M = \bigoplus_{i \in \Lambda} e_i M$.*

Proof (i) Let $f, g \in \Delta$ and $\varphi \in S$. Then $\text{Ker}(f) \leq^{\text{ess}} M$ and $\text{Ker}(g) \leq^{\text{ess}} M$. As $\text{Ker}(f) \cap \text{Ker}(g) \subseteq \text{Ker}(f - g)$ and $\text{Ker}(f) \subseteq \text{Ker}(\varphi f)$, $\text{Ker}(f - g) \leq^{\text{ess}} M$ and $\text{Ker}(\varphi f) \leq^{\text{ess}} M$. So $f - g \in \Delta$ and $\varphi f \in \Delta$.

Take $N = \{m \in M \mid \varphi(m) \in \text{Ker}(f)\}$. Then $N_R \leq^{\text{ess}} M_R$ and $N \subseteq \text{Ker}(f\varphi)$. Hence $f\varphi \in \Delta$.

(ii) It is enough to consider a finite family e_i, $1 \leq i \leq n$. For $j \neq k$, $e_j e_k \in \Delta$, and so $\text{Ker}(e_j e_k) \leq^{\text{ess}} M$. Note that a finite intersection of essential submodules of M is again essential in M. Thus, there exists $K \leq^{\text{ess}} M$ such that $e_j e_k K = 0$ for all $j \neq k$, $1 \leq j, k \leq n$. Hence, $\sum_{i=1}^n e_i K = \oplus_{i=1}^n e_i K$. But as $e_i K \leq^{\text{ess}} e_i M$, $\sum_{i=1}^n e_i M = \oplus_{i=1}^n e_i M$. So $\sum_{i \in \Lambda} e_i M = \oplus_{i \in \Lambda} e_i M$. □

Theorem 2.1.29 *Assume that M is a continuous right R-module and let $S = \text{End}_R(M)$. Then:*

(i) *$S/J(S)$ is regular and every idempotent of $S/J(S)$ lifts to an idempotent of S.*
(ii) *$J(S) = \Delta$.*
(iii) *$S/J(S)$ is a right continuous ring.*
(iv) *Further, if M is quasi-injective, then $S/J(S)$ is right self-injective.*

Proof (i) and (ii) By Lemma 2.1.28(i), $\Delta \trianglelefteq S$. We show that $\Delta \subseteq J(S)$. For this, take $s \in \Delta$. Then, we note that $\text{Ker}(s) \leq^{\text{ess}} M$. Thus, $\text{Ker}(1 - s) = 0$ because $\text{Ker}(s) \cap \text{Ker}(1 - s) = 0$. So $(1 - s)M \cong M$. By (C_2) condition, $(1 - s)M \leq^{\oplus} M$. As $\text{Ker}(s) \leq (1 - s)M$ and $\text{Ker}(s) \leq^{\text{ess}} M$, $(1 - s)M = M$ and so $1 - s$ is invertible. For any $t \in S$, $st \in \Delta$ by Lemma 2.1.28(i), so $1 - st$ is invertible by the preceding argument. Hence, st is quasi-regular for all $t \in S$, and thus $s \in J(S)$ (see 1.1.9). Therefore, $\Delta \subseteq J(S)$.

Next, we show that the ring S/Δ is regular. Take $\alpha \in S$. By (C_1) condition of M, $\text{Ker}(\alpha) \leq^{\text{ess}} P \leq^{\oplus} M$ with $P \leq M$. Say $M = P \oplus N$ for some N. Then $\alpha N \cong N \leq^{\oplus} M$, and so $\alpha N \leq^{\oplus} M$ from (C_2) condition. Write $M = \alpha N \oplus W$ for some $W \leq M$. Define a map $\beta : M = \alpha N \oplus W \to M$ by $\beta(\alpha n + w) = n$, where $n \in N$ and $w \in W$. Now we let $\pi : M = P \oplus N \to N$ be the canonical projection. Then $\beta\alpha\pi = \pi$. Take $\tau = \alpha\pi\beta$. So $\tau^2 = \tau$ since $\pi^2 = \pi$. Also $\text{Ker}(\alpha) \oplus N \subseteq \text{Ker}(\alpha - \alpha\pi\beta\alpha)$. Thus, $\alpha - \alpha\pi\beta\alpha \in \Delta$ since $\text{Ker}(\alpha) \oplus N \leq^{\text{ess}} M$. So $\alpha + \Delta = (\alpha + \Delta)(\pi\beta + \Delta)(\alpha + \Delta)$ in S/Δ, hence S/Δ is regular and thus $J(S) \subseteq \Delta$. Therefore $J(S) = \Delta$ and $S/J(S)$ is regular.

We claim that each idempotent of $S/J(S)$ lifts to an idempotent of S. Take $e \in S$ such that $e^2 - e \in \Delta$. Put $K = \mathrm{Ker}(e^2 - e)$. Then $K \leq^{\text{ess}} M$ as $e^2 - e \in \Delta$. Note that $eK \cap (1-e)K = 0$ and $K \subseteq eK \oplus (1-e)K$. So $eK \oplus (1-e)K \leq^{\text{ess}} M$. Thus, $E(M) = E(eK) \oplus E((1-e)K)$. Because M is continuous, Theorem 2.1.25 yields that $M = (M \cap E(eK)) \oplus (M \cap E((1-e)K))$.

Put $M_1 = M \cap E(eK)$ and $M_2 = M \cap E((1-e)K)$. Then $M = M_1 \oplus M_2$, $eK \leq^{\text{ess}} M_1$, and $(1-e)K \leq^{\text{ess}} M_2$. Let $h : M_1 \oplus M_2 \to M_1$ be the canonical projection. Then $(h-e)K \subseteq (h-e)eK + (h-e)(1-e)K$. Observe that $(h-e)eK = (e-e^2)K = 0$ because $eK \subseteq M_1 = hM$. Also we observe that $(h-e)(1-e)K \subseteq h(1-e)K + e(1-e)K \subseteq hM_2 = 0$. Thus, $(h-e)K = 0$ and so $h - e \in \Delta$. Hence $h + \Delta = e + \Delta$ with $h^2 = h \in S$. Since $J(S) = \Delta$, each idempotent of $S/J(S)$ lifts to an idempotent of S.

(iii) To show that $S/J(S)$ is right continuous, it is enough to prove that $S/J(S)$ satisfies (C_1) condition, since $S/J(S)$ is regular. Let $\overline{S} = S/\Delta$ and let $\overline{A}$ be a right ideal of $\overline{S}$. By Zorn's lemma, there exists a maximal direct sum $\oplus_{i\in\Lambda}\overline{e_i}\,\overline{S}$ of principal right ideals $\overline{e_i}\,\overline{S}$ of $\overline{S}$ which contained in $\overline{A}$. Since $\overline{S}$ is regular, we may assume that each $\overline{e_i}$ is an idempotent. By the maximality, $\oplus_{i\in\Lambda}\overline{e_i}\,\overline{S}$ is essential in $\overline{A}$. Since idempotents of $\overline{S}$ lift to idempotents of S, we may assume that all e_i are idempotents in S.

We claim that $\sum_{i\in\Lambda} e_i M = \oplus_{i\in\Lambda} e_i M$. For this, it is enough to consider a finite number of idempotents, say $e_1, \ldots, e_n$. Note that $\oplus_{i=1}^{n}\overline{e_i}\,\overline{S}_{\overline{S}} \leq^{\oplus} \overline{S}_{\overline{S}}$, since $\overline{S}$ is regular. Thus there exist orthogonal idempotents $\overline{g_i}$ in $\overline{S}$ such that $\overline{e_i}\,\overline{S} = \overline{g_i}\,\overline{S}$ for $i = 1, \ldots, n$. By Lemma 2.1.28(ii), $\sum_{i=1}^{n} g_i M = \oplus_{i=1}^{n} g_i M$. Also since $\overline{e_i}\,\overline{S} = \overline{g_i}\,\overline{S}$, $\overline{e_i} = \overline{g_i}\,\overline{e_i}$ and hence $e_i - g_i e_i \in \Delta$. So there is $K_i \leq^{\text{ess}} M$, for each $i = 1, \ldots, n$, such that $(e_i - g_i e_i)K_i = 0$. Thus, $e_i K_i \subseteq g_i M$. Since $\sum_{i=1}^{n} g_i M = \oplus_{i=1}^{n} g_i M$, also $\sum_{i=1}^{n} e_i K_i = \oplus_{i=1}^{n} e_i K_i$. Now $e_i K_i \leq^{\text{ess}} e_i M$ for each $i = 1, \ldots, n$ yields that $\sum_{i=1}^{n} e_i M = \oplus_{i=1}^{n} e_i M$. So it follows that $\sum_{i\in\Lambda} e_i M = \oplus_{i\in\Lambda} e_i M$. From (C_1) condition, there exists $e^2 = e \in S$ such that $\oplus_{i\in\Lambda} e_i M \leq^{\text{ess}} eM$.

We show that $\oplus_{i\in\Lambda}\overline{e_i}\,\overline{S}_{\overline{S}} \leq^{\text{ess}} \overline{e}\,\overline{S}_{\overline{S}}$. Let $(\oplus_{i\in\Lambda}\overline{e_i}\,\overline{S}) \cap \overline{f}\,\overline{S} = 0$ for some $\overline{f}$ in $\overline{e}\,\overline{S}$. We may assume that f is an idempotent in S as in the preceding argument. As $\oplus_{i\in\Lambda}\overline{e_i}\,\overline{S} + \overline{f}\,\overline{S} = \oplus_{i\in\Lambda}\overline{e_i}\,\overline{S} \oplus \overline{f}\,\overline{S}$, by the preceding proof, we have that $(\oplus_{i\in\Lambda} e_i M) + fM = (\oplus_{i\in\Lambda} e_i M) \oplus fM$. So $(\oplus_{i\in\Lambda} e_i M) \cap fM = 0$.

Since $\overline{f} \in \overline{e}\,\overline{S}$, $\overline{e}\,\overline{f} = \overline{f}$ and so $ef - f \in \Delta$. Hence, $(ef - f)V = 0$ for some $V \leq^{\text{ess}} M$. Thus $fV = efV \subseteq eM$. So $fV = 0$ because $\oplus_{i\in\Lambda} e_i M \leq^{\text{ess}} eM$. Thus $f \in \Delta$ and so $\overline{f} = 0$. Therefore, $\oplus_{i\in\Lambda}\overline{e_i}\,\overline{S} \leq^{\text{ess}} \overline{e}\,\overline{S}$.

Finally, we claim that $\overline{A}_{\overline{S}} \leq^{\text{ess}} \overline{e}\,\overline{S}_{\overline{S}}$. For this, first we show that $\overline{A} \subseteq \overline{e}\,\overline{S}$. Say $\overline{a} \in \overline{A}$. Since $(\oplus_{i\in\Lambda}\overline{e_i}\,\overline{S})_{\overline{S}} \leq^{\text{ess}} \overline{A}_{\overline{S}}$, it follows that $[\overline{a}\,\overline{S} \cap (\oplus_{i\in\Lambda}\overline{e_i}\,\overline{S})]_{\overline{S}}$ is essential in $\overline{a}\overline{S} \cap \overline{A} = \overline{a}\,\overline{S}$. As $\overline{a}\,\overline{S} \cap (\oplus_{i\in\Lambda}\overline{e_i}\,\overline{S}) \subseteq \overline{a}\,\overline{S} \cap \overline{e}\,\overline{S} \subseteq \overline{a}\,\overline{S}$, $\overline{a}\,\overline{S} \cap \overline{e}\,\overline{S}$ is essential in $\overline{a}\,\overline{S}$. Since $\overline{S}$ is regular, $\overline{a}\,\overline{S} \cap \overline{e}\,\overline{S}$ is generated by an idempotent. So $\overline{a}\,\overline{S} \cap \overline{e}\,\overline{S} = \overline{a}\,\overline{S}$, thus $\overline{a}\,\overline{S} \subseteq \overline{e}\,\overline{S}$. Thus, $\overline{a} \in \overline{e}\,\overline{S}$, and hence $\overline{A} \subseteq \overline{e}\,\overline{S}$. Since $\oplus_{i\in\Lambda}\overline{e_i}\,\overline{S} \subseteq \overline{A}$ and $\oplus_{i\in\Lambda}\overline{e_i}\,\overline{S}$ is essential in $\overline{e}\,\overline{S}$, $\overline{A}$ is essential in $\overline{e}\,\overline{S}$. Thus $\overline{S}_{\overline{S}}$ satisfies (C_1) condition, so $\overline{S}$ is right continuous.

(iv) Further, assume that M_R is quasi-injective. To prove that $\overline{S} = S/J(S)$ is right self-injective, let $\overline{A}$ be a right ideal of $\overline{S}$ and $\varphi : \overline{A}_{\overline{S}} \to \overline{S}_{\overline{S}}$ be a homomorphism. As in the proof of part (iii), there is a set $\{e_i \mid i \in \Lambda\}$ of idempotents in S

with $\oplus_{i\in\Lambda}\overline{e_i}\,\overline{S}$ essential in $\overline{A}$ and $\sum_{i\in\Lambda} e_iM = \oplus_{i\in\Lambda}e_iM$. Let $\varphi(\overline{e_i}) = \overline{x_i} \in \overline{S}$. Then $\varphi(\overline{e_i}) = \varphi(\overline{e_i}\,\overline{e_i}) = \varphi(\overline{e_i})\overline{e_i} = \overline{x_i}\,\overline{e_i}$. For each i, define

$$\phi_i : e_iM \to M \text{ by } \phi_i(e_im) = x_ie_im,$$

where $m \in M$ and let $\phi(\sum_i e_im_i) = \sum_i \phi_i(e_im_i) = \sum_i x_ie_im_i$. Since M is quasi-injective, there exists $h \in S$, an extension of ϕ. Then it follows that $(h - x_ie_i)e_i = he_i - x_ie_i = 0$. So $\overline{he_i} - \varphi(\overline{e_i}) = \overline{h}\overline{e_i} - \overline{x_i}\,\overline{e_i} = 0$ for all $i \in \Lambda$.

Take $\overline{a} \in \overline{A}$. As $\oplus_{i\in\Lambda}\overline{e_i}\,\overline{S}_{\overline{S}} \le^{\text{ess}} \overline{A}_{\overline{S}}$, there is an essential right ideal $\overline{K}$ of $\overline{S}$ with $\overline{a}\overline{K} \subseteq \oplus_{i\in\Lambda}\overline{e_i}\,\overline{S}$. Now $(\overline{h}\overline{a} - \varphi(\overline{a}))\overline{K} = 0$ since $\overline{h}\,\overline{e_i} - \varphi(\overline{e_i}) = 0$ for all $i \in \Lambda$ and $\overline{a}\,\overline{K} \subseteq \oplus_{i\in\Lambda}\overline{e_i}\,\overline{S}$. As $\overline{S}$ is regular, it is right nonsingular. Hence, $\overline{h}\,\overline{a} = \varphi(\overline{a})$ for all $\overline{a} \in \overline{A}$. By Baer's Criterion, $\overline{S}$ is right self-injective. □

A ring R is called *semiregular* if $R/J(R)$ is a regular ring and every idempotent of $R/J(R)$ lifts to an idempotent of R. In Theorem 2.1.29(i), we show that $S = \text{End}(M)$ is semiregular when M is a continuous right R-module (see [314] and [317] for more details on semiregular rings). Now Theorem 2.1.29 yields the next two useful results.

Corollary 2.1.30 *Assume that R is a right continuous (resp., right self-injective) ring. Then R is semiregular, $J(R) = Z(R_R)$, and $R/J(R)$ is right continuous (resp., right self-injective).*

Proof The proof follows immediately from Theorem 2.1.29. □

Theorem 2.1.31 *Let R be a ring. Then R is right nonsingular if and only if $Q(R)$ is right self-injective and regular.*

Proof Let R be right nonsingular. By Corollary 1.3.15, $Q(R) = E(R_R)$ and hence $Q(R)_R$ is injective. Put $Q = Q(R)$ and $E = E(R_R)$.

Observe that $Q = \text{End}({}_HE)$, where $H = \text{End}(E_R)$, and $E = {}_HE_Q$ is an (H, Q)-bimodule. Thus $h(x)q = h(xq)$ for $h \in \text{End}(E_R)$, $x \in E$, and $q \in Q$. Therefore, $\text{End}(E_R) \subseteq \text{End}(E_Q)$. Obviously, $\text{End}(E_Q) \subseteq \text{End}(E_R)$. Hence $\text{End}(E_Q) = \text{End}(E_R)$. As $E = Q$, $\text{End}(Q_Q) = \text{End}(Q_R)$.

From Theorem 2.1.29, $J(Q) = \Delta = \{q \in Q \mid \text{Ker}(q)_R \le^{\text{ess}} Q_R\}$ because Q_R is injective and $\text{End}(Q_R) \cong Q$. Note that Q_R is nonsingular since R_R is nonsingular. Say $q \in J(Q)$. Then $\text{Ker}(q)_R \le^{\text{ess}} Q_R$, so $(\text{Ker}(q) \cap R)_R \le^{\text{ess}} R_R$. Take $I = \text{Ker}(q) \cap R$. Then $qI = 0$ and $I_R \le^{\text{ess}} R_R$. Thus, $q \in Z(Q_R)$. Because $Z(Q_R) = 0$, $q = 0$ and so $J(Q) = 0$. Therefore, Q is right self-injective regular by Theorem 2.1.29.

Conversely, let Q be a right self-injective regular ring. Then $Z(Q_Q) = 0$. Let $q \in Z(Q_R)$. Then there exists $I_R \le^{\text{ess}} R_R$ with $qI = 0$. As $R_R \le^{\text{ess}} Q_R$, $I_R \le^{\text{ess}} Q_R$ and so $IQ_R \le^{\text{ess}} Q_R$. Thus, $IQ_Q \le^{\text{ess}} Q_Q$. As $q(IQ) = (qI)Q = 0$, $q \in Z(Q_Q)$, and so $q = 0$. Therefore, $Z(Q_R) = 0$, hence $Z(R_R) = 0$. □

By the proof of Theorem 2.1.31, we are motivated to consider the following.

Proposition 2.1.32 *Let R be a ring and T a right ring of quotients of R. Then* $\mathrm{End}(T_T) = \mathrm{End}(T_R)$.

Proof It is enough to see that $\mathrm{End}(T_R) \subseteq \mathrm{End}(T_T)$. Let $\varphi \in \mathrm{End}(T_R)$. Assume on the contrary that there are $q_1, q_2 \in T$ such that $\varphi(q_1q_2) - \varphi(q_1)q_2 \neq 0$. As R_R is dense in T_R, there is $r \in R$ with $q_2r \in R$ and $(\varphi(q_1q_2) - \varphi(q_1)q_2)r \neq 0$. Now

$$(\varphi(q_1q_2) - \varphi(q_1)q_2)r = \varphi(q_1q_2)r - \varphi(q_1)q_2r = \varphi(q_1q_2r) - \varphi(q_1q_2r) = 0,$$

as $q_2r \in R$, which is a contradiction. Thus, $\mathrm{End}(T_T) = \mathrm{End}(T_R)$. □

Similar to self-injectivity, none of the conditions, continuous, quasi-continuous, or extending is left-right symmetric for a ring. Examples 2.1.33 and 2.1.36 illustrate this lack of symmetry.

Example 2.1.33 Let R be a right Ore domain which is not left Ore (see [262, p. 308]). Then R_R is uniform. However, ${}_RR$ is not uniform. We note that R_R is quasi-continuous (hence R_R is extending), but ${}_RR$ is not extending (hence ${}_RR$ is not quasi-continuous).

Lemma 2.1.34 *If M is an extending module with ACC on essential submodules, then $M = K \oplus N$, where K is semisimple and N is Noetherian.*

Proof See [145, Corollary 18.6] for the proof. □

Theorem 2.1.35 *Let R be a right continuous ring with ACC on essential right ideals. Then R is right Artinian.*

Proof By Lemma 2.1.34, R is right Noetherian. Now $J(R) = Z(R_R)$ from Corollary 2.1.30. Let $x \in J(R)$. Take a maximal element $r_R(x^n)$ in the set $\{r_R(x^k) \mid k = 1, 2, \ldots\}$. Put $a = x^n$. Then we see that $aR \cap r_R(a) = 0$. Since $a \in Z(R_R)$, $aR = 0$ and so $a = 0$. Thus, $J(R)$ is nil. As every nil ideal of a right Noetherian ring is nilpotent, $J(R)$ is nilpotent. By Corollary 2.1.30, $R/J(R)$ is regular. So $R/J(R)$ is semisimple Artinian as R is right Noetherian. Hence R is semiprimary, and thus R is right Artinian. □

The next example (see [154, Example 7.11′.1]) exhibits a left continuous left Artinian ring with ACC on essential right ideals, which is not right continuous.

Example 2.1.36 Let $F = \mathbb{Q}(x_1, x_2, \ldots)$ be the field of fractions of the polynomial ring $\mathbb{Q}[x_1, x_2, \ldots]$ with indeterminates $x_1, x_2, \ldots$, and we take $K = \mathbb{Q}(x_1^2, x_2^2, \ldots)$. Then F is infinite dimensional over K. Define $f : F \to K$ by $f(x_i) = x_i^2$, and $f(q) = q$ for $i = 1, 2, \ldots$ and $q \in \mathbb{Q}$. Let

$$R = \left\{ \begin{bmatrix} a & b \\ 0 & f(a) \end{bmatrix} \mid a, b \in F \right\},$$

which is a subring of the ring $T_2(F)$. Then R has only three left ideals 0, $J(R)$, and R itself. Clearly R is left continuous. Now $\text{Soc}(R_R) = J(R)$ and $R/\text{Soc}(R_R) \cong F$. So R has ACC on essential right ideals since every essential right ideal of R contains $\text{Soc}(R_R)$. As F is infinite dimensional over K, R is not right Artinian. So R is not right continuous by Theorem 2.1.35.

We remark that R is also not left self-injective since otherwise R will be QF, forcing it to be Artinian on both sides (see Theorem 1.3.4).

Exercise 2.1.37

1. Prove Proposition 2.1.5.
2. Let $M = \oplus_{i\in\Lambda} M_i$ be a direct sum of modules M_i and $N \trianglelefteq M$. Show that $N = \oplus_{i\in\Lambda}(N \cap M_i)$.
3. Prove that a submodule N of a module M is closed in M if and only if N is a complement of a submodule of M.
4. Prove Lemma 2.1.13 and Proposition 2.1.15.
5. ([177, Goel and Jain]) By Goel and Jain, a module M is called *π-injective* if for every pair of submodules M_1, M_2 of M with $M_1 \cap M_2 = 0$, each canonical projection $\pi_i : M_1 \oplus M_2 \to M_i$, $i = 1, 2$ can be extended to an endomorphism of M. Show that M is π-injective if and only if M is quasi-continuous.
6. ([25, Armendariz and Park]) Suppose that R is a right self-injective ring and the ring $R/\text{Soc}(R_R)$ is orthogonally finite. Show that $\text{Soc}(R_R)$ is a finitely generated right R-module.
7. ([25, Armendariz and Park]) Prove that if R is a right self-injective ring and $R/\text{Soc}(R_R)$ has ACC on right annihilators, then R is semiprimary.
8. ([25, Armendariz and Park]) Prove that if R is a right self-injective ring and the ring $R/\text{Soc}(R_R)$ is right Goldie, then R is QF.
9. ([18, Ara and Park]) Show that if R is a right and left continuous ring and $R/\text{Soc}(R_R)$ is right Goldie, then R is QF.
10. ([83, Birkenmeier, Müller, and Rizvi]) Assume that R is a right quasi-continuous ring with $Z(R_R) = 0$. Show that R is semiprime.
11. ([54, Birkenmeier]) A module K is said to be *ker-injective* if given any monomorphism $f \in \text{Hom}(A, B)$ and any homomorphism $g \in \text{Hom}(A, K)$, there exists $h \in \text{Hom}(B, K)$ such that $\text{Ker}(hf) \subseteq \text{Ker}(g)$. Show that the following are equivalent for a module M.
    (i) M is injective.
    (ii) M is ker-injective and satisfies (C_2).
    (iii) M is ker-injective and for every monomorphism $\alpha \in \text{End}(M)$, there exists $\beta \in \text{End}(M)$ such that $\beta\alpha$ is an isomorphism.
12. ([53, 56, Birkenmeier]) A right ideal I of a ring R is called *densely nil* (simply, *DN*) if $I = 0$ or if for each $0 \neq x \in I$ there exists $r \in R$ such that $xr \neq 0$ but $(xr)^2 = 0$. Prove the following.
    (i) $Z(R_R)$ is DN.
    (ii) Assume that R is a right extending (resp., right continuous) ring. Then $R = A \oplus B$ (right ideal decomposition), where A is a reduced right quasi-continuous (resp., strongly regular continuous) ring, and B is DN and min-

imal among direct summands containing the set of all nilpotent elements of R.

2.2 Internal Quasi-continuous Hulls and Decompositions

In this section, we first discuss direct sums and decompositions of quasi-continuous modules. Using these results, we obtain the uniqueness of internal quasi-continuous hulls (up to isomorphism). Some applications of this result will be presented.

Let M be a module and $N \leq^{\oplus} M$. It is well known that if M is injective, quasi-injective, continuous, quasi-continuous, or extending, then so is N. A finite direct sum of injective modules is injective. However, as we mentioned earlier, none of the classes of quasi-injective, continuous, quasi-continuous, or extending modules is closed even under finite direct sums. The next example illustrates this fact.

Example 2.2.1 (i) It is well known that any direct sum of injective right R-modules is injective if and only if R is right Noetherian (see [262, Theorem 3.46]). Thereby, if a ring R is not right Noetherian, then there is a family of injective right R-modules whose direct sum is not injective.

(ii) We note that $\mathbb{Z}_p$ and $\mathbb{Z}_{p^3}$ are quasi-injective $\mathbb{Z}$-modules, where p is a prime integer (see Example 2.1.14(i)). But recall that $M = \mathbb{Z}_p \oplus \mathbb{Z}_{p^3}$ is not extending (see [145, p. 56]). Hence, the classes of quasi-injective, continuous, quasi-continuous, and extending modules are not closed under direct sums.

(iii) Take $R = T_2(\mathbb{Z})$ and let e_{ij} be the matrix in R with 1 in the (i, j)-position and 0 elsewhere. Then $R_R = e_{11}R_R \oplus e_{22}R_R$ is a direct sum of uniform (hence extending) modules. We may observe that $(e_{12} + 2e_{22})R_R$ is not essential in a direct summand of R_R, so R_R is not extending.

Proposition 2.2.2 *Let $M_1, M_2, \ldots, M_n$ be modules. Then the following are equivalent.*

(i) *$M_1 \oplus M_2 \oplus \cdots \oplus M_n$ is quasi-injective.*
(ii) *M_i and M_j are relatively injective for $i, j = 1, 2, \ldots, n$.*

Proof Theorem 2.1.7 immediately yields the result. □

Corollary 2.2.3 *Let M be a module and n a positive integer. Then $M^{(n)}$ is quasi-injective if and only if M is quasi-injective.*

In the next result, we see that a direct summand of a quasi-continuous module is always relatively injective to all other direct summands.

Lemma 2.2.4 *Let $M = \bigoplus_{i \in \Lambda} M_i$. If M is quasi-continuous, then each M_i is quasi-continuous and M_j-injective for all $j \neq i$.*

Proof Let $M = \oplus_{i\in\Lambda} M_i$ be quasi-continuous. Each M_i is quasi-continuous since $M_i \leq^{\oplus} M$. Let $j \neq i$, where $i, j \in \Lambda$. We show that M_i is M_j-injective. For this, put $V_j = \oplus_{k\neq j} M_k$. Then $E(\oplus_{i\in\Lambda} M_i) = E(M_j) \oplus E(V_j)$.

Say $g \in \mathrm{Hom}(E(M_j), E(M_i))$. Then $g \in \mathrm{Hom}(E(M_j), E(V_j))$. Consider an idempotent

$$e = \begin{bmatrix} 1 & 0 \\ g & 0 \end{bmatrix} \in \begin{bmatrix} \mathrm{End}(E(M_j)) & \mathrm{Hom}(E(V_j), E(M_j)) \\ \mathrm{Hom}(E(M_j), E(V_j)) & \mathrm{End}(E(V_j)) \end{bmatrix}$$
$$= \mathrm{End}(E(M_j) \oplus E(V_j))$$
$$= \mathrm{End}(E(M)).$$

By assumption $M = M_j \oplus V_j$ is quasi-continuous. Hence $e(M_j \oplus V_j) \subseteq M_j \oplus V_j$ from Theorem 2.1.25. Thus,

$$\begin{bmatrix} 1 & 0 \\ g & 0 \end{bmatrix} \begin{bmatrix} M_j \\ V_j \end{bmatrix} \subseteq \begin{bmatrix} M_j \\ V_j \end{bmatrix}.$$

Therefore, $g(M_j) \subseteq V_j$. Thus, $g(M_j) \subseteq V_j \cap E(M_i) \subseteq M \cap E(M_i)$. Now we put $K = \oplus_{k\neq i} M_k$. Because $M = M_i \oplus K$ and $M_i \leq M \cap E(M_i)$, by the modular law $M \cap E(M_i) = M_i \oplus (K \cap M \cap E(M_i)) = M_i$, so $g(M_j) \subseteq M_i$. Hence, M_i is M_j-injective by Theorem 2.1.2. □

Corollary 2.2.3 and Lemma 2.2.4 yield that: for a module M and an integer n such that $n > 1$, $M^{(n)}$ is quasi-injective if and only if $M^{(n)}$ is quasi-continuous. For the case of quasi-continuous modules, we have the following.

Theorem 2.2.5 *Let $M = \bigoplus_{i\in\Lambda} M_i$, and assume that $\bigoplus_{i\in\Lambda} E(M_i)$ is an injective right R-module (e.g., Λ is finite or R is right Noetherian). Then M is quasi-continuous if and only if each M_i is quasi-continuous and M_j-injective for all $j \neq i$.*

Proof The necessity follows from Lemma 2.2.4. For the sufficiency, let all M_i be quasi-continuous and M_j-injective for $j \neq i$. That M is quasi-continuous, will be established by Theorem 2.1.25 once we show that $eM \subseteq M$, for every $e^2 = e \in \mathrm{End}(E(M))$. Since $\oplus_{i\in\Lambda} E(M_i)$ is injective by assumption, $E(M) = \oplus_{i\in\Lambda} E(M_i)$. Thus, e can be written as a matrix $e = [e_{ik}]$, with $e_{ik} \in \mathrm{Hom}(E(M_k), E(M_i))$. By Theorem 2.1.2, the M_k-injectivity of M_i yields that $e_{ik}(M_k) \subseteq M_i$, for all $k \neq i$. Hence, it is enough to show that $e_{ii}(M_i) \subseteq M_i$ for all i.

Now $e^2 = e$ implies that $e_{ik} = \sum_j e_{ij} e_{jk}$. Write $\beta_i = \sum_{j\neq i} e_{ij} e_{ji}$. Then

$$e_{ii} - e_{ii}^2 = \beta_i : E(M_i) \to E(M_i).$$

Put $K_i = \mathrm{Ker}(\beta_i)$. So $(e_{ii} - e_{ii}^2)(K_i) = 0$. As $\beta_i e_{ii} = e_{ii}\beta_i$, $e_{ii}(K_i) \subseteq K_i$. Therefore $e_{ii}|_{K_i}$ is an idempotent in the ring $\mathrm{End}(K_i)$. Hence

$$K_i = X_i \oplus Y_i, \quad \text{where} \quad X_i = e_{ii}(K_i) \quad \text{and} \quad Y_i = \mathrm{Ker}(e_{ii}) \cap K_i.$$

Since $X_i \oplus Y_i = K_i \subseteq E(M_i)$, $E(M_i) = E(X_i) \oplus F_i$ with $Y_i \leq F_i \leq E(M_i)$. By Theorem 2.1.25, $M_i = (M_i \cap E(X_i)) \oplus (M_i \cap F_i)$ as M_i is quasi-continuous.

We show that $\mathrm{Ker}(\beta_i|_{E(X_i)}) = X_i$ and $\mathrm{Ker}(\beta_i|_{F_i}) = Y_i$. Indeed,

$$\begin{aligned}\mathrm{Ker}(\beta_i|_{E(X_i)}) &= \mathrm{Ker}(\beta_i) \cap E(X_i) = K_i \cap E(X_i)\\ &= (X_i \oplus Y_i) \cap E(X_i) = X_i,\end{aligned}$$

because $K_i = \mathrm{Ker}(\beta_i)$ and $Y_i \cap E(X_i) = 0$. Similarly, we can prove that $\mathrm{Ker}(\beta_i|_{F_i}) = \mathrm{Ker}(\beta_i) \cap F_i = K_i \cap F_i = (X_i \oplus Y_i) \cap F_i = Y_i$ since $Y_i \leq F_i$ and $X_i \cap F_i = 0$.

We note that the map $g : E(X_i)/X_i \to E(M_i)$ defined by $g(t + X_i) = \beta_i(t)$, with $t \in E(X_i)$, is a monomorphism since $\mathrm{Ker}(\beta_i|_{E(X_i)}) = X_i$. Also, the map $f : E(X_i)/X_i \to E(M_i)$ defined by $f(t + X_i) = (1 - e_{ii})(t)$, with $t \in E(X_i)$, is a homomorphism, as $X_i = e_{ii}(K_i) \subseteq \mathrm{Ker}((1 - e_{ii})|_{E(X_i)})$. Since $E(M_i)$ is injective and g is a monomorphism, there is $\varphi : E(M_i) \to E(M_i)$ with $f = \varphi g$. Hence $(\varphi\beta_i)|_{E(X_i)} = (1 - e_{ii})|_{E(X_i)}$. This implies that

$$\begin{aligned}(1 - e_{ii})(M_i \cap E(X_i)) &= \varphi\beta_i(M_i \cap E(X_i))\\ &= \sum_{j\neq i}(\varphi e_{ij})e_{ji}(M_i \cap E(X_i))\\ &\subseteq M_i,\end{aligned}$$

as $(\varphi e_{ij})e_{ji}(M_i \cap E(X_i)) \subseteq (\varphi e_{ij})e_{ji}(M_i) \subseteq (\varphi e_{ij})(M_j) \subseteq M_i$, by the relative injectivity of M_i and M_j (Theorem 2.1.2). So $e_{ii}(M_i \cap E(X_i)) \subseteq M_i$.

To show that $e_{ii}(M_i \cap F_i) \subseteq M_i \cap F_i$, we define $h : F_i/Y_i \to E(M_i)$ by $h(t + Y_i) = \beta_i(t)$ for $t \in F_i$. Then since $Y_i = \mathrm{Ker}(\beta_i|_{F_i})$, h is a monomorphism. Consider the map $k : F_i/Y_i \to E(M_i)$ defined by $k(t + Y_i) = e_{ii}(t)$. Then k is a homomorphism as $Y_i = \mathrm{Ker}(e_{ii}) \cap K_i$. Hence, there is $\psi \in \mathrm{End}(E(M_i))$ with $k = \psi h$. So $e_{ii}(M_i \cap F_i) = \psi\beta_i(M_i \cap F_i) = \sum_{j\neq i}(\psi e_{ij})e_{ji}(M_i \cap F_i) \subseteq M_i$ by the relative injectivity of M_i and M_j.

Consequently, $e_{ii}(M_i) \subseteq M_i$ since $M_i = (M_i \cap E(X_i)) \oplus (M_i \cap F_i)$, and the proof is completed. □

A module is said to be *directly finite* if it is not isomorphic to a proper direct summand of itself. A module is called *purely infinite* if it is isomorphic to the direct sum of two copies of itself. Recall that a ring R is called directly finite if $xy = 1$ implies that $yx = 1$ for $x, y \in R$. We remark that a module M is directly finite if and only if the endomorphism ring $\mathrm{End}(M)$ is directly finite (see [183, Lemma 5.1]). If a module M is not directly finite, then M contains an infinite direct sum of nonzero isomorphic submodules (see [301, Lemma 1.26]). When M is an injective module, we have the next result.

Proposition 2.2.6 *The following are equivalent for an injective module M.*

(i) *M is not directly finite.*

(ii) *M contains an infinite direct sum of nonzero isomorphic submodules.*
(iii) *M has a nonzero direct summand B such that* $B \cong B \oplus B$.

Proof See [183, Proposition 5.7]. □

Lemma 2.2.7 *Let M be a directly finite injective module.*
(i) *If* $M = A_1 \oplus B_1 = A_2 \oplus B_2$ *with* $A_1 \cong A_2$, *then* $B_1 \cong B_2$.
(ii) *If* $M \oplus X \cong M \oplus Y$ *(where X and Y are modules), then* $X \cong Y$.

Proof See [301, Theorem 1.21, Propositions 1.23 and 1.28]. □

Refer to Exercise 2.2.19.5 for an extension of Lemma 2.2.7(ii) to the class of directly finite continuous modules. A module N is said to be *subisomorphic* to a module M if N is isomorphic to a submodule of M. Theorem 2.2.9 provides an algebraic proof of a result of Goodearl [181] (proved in a categorical way). This also shows "uniqueness" of the decomposition. We shall see later that this result is extended to a similar decomposition of a quasi-continuous module. For the proof of Theorem 2.2.9, we begin with an auxiliary observation.

Lemma 2.2.8 *Assume that A is a submodule of C and E(A) is directly finite, and assume that C is subisomorphic to an injective module I. Then every monomorphism* $f : A \to I$ *extends to a monomorphism* $C \to I$.

Proof Let $g : C \to I$ be a monomorphism. Then f and g extend to monomorphisms $\varphi : E(A) \to I$ and $\gamma : E(C) \to I$, respectively. Because $E(C) = E(A) \oplus X$ for some $X \le E(C)$, $I = \gamma(E(C)) \oplus Y = \gamma(E(A)) \oplus \gamma(X) \oplus Y$ with $Y \le I$. Also, we get that $I = \varphi(E(A)) \oplus Z$ for some $Z \le I$.

Note that $\varphi(E(A)) \cong E(A) \cong \gamma(E(A))$ is directly finite injective, and

$$\varphi(E(A)) \oplus Z = \gamma(E(A)) \oplus \gamma(X) \oplus Y \cong \varphi(E(A)) \oplus \gamma(X) \oplus Y.$$

Hence from Lemma 2.2.7(ii), $Z \cong \gamma(X) \oplus Y$. Therefore there exists a monomorphism $\mu : X \to Z$, and $\varphi \oplus \mu : E(C) = E(A) \oplus X \to \varphi(E(A)) \oplus Z = I$ is a monomorphism, whose restriction $(\varphi \oplus \mu)|_C$ extends f. □

Theorem 2.2.9 *Every injective module E has a direct sum decomposition,* $E = U \oplus V$ *with U directly finite, V purely infinite, and no nonzero isomorphic direct summands (or submodules) between U and V. If* $E = U_1 \oplus V_1 = U_2 \oplus V_2$ *are two such decompositions, then* $E = U_1 \oplus V_2$ *holds too, and consequently* $U_1 \cong U_2$ *and* $V_1 \cong V_2$.

Proof Step 1. Consider the set of triples (V, φ', φ''), where $V \le E$ and φ', φ'' are monomorphisms of V into itself such that $V = \varphi'(V) \oplus \varphi''(V)$. We order such triples: $(V, \varphi', \varphi'') \le (W, \psi', \psi'')$ if $V \subseteq W$ and $\varphi' = \psi'|_V$, $\varphi'' = \psi''|_V$. By Zorn's lemma, there is a maximal triple (V, φ', φ'').

We note that $\varphi'(V) \cong V \cong \varphi''(V)$ and $V = \varphi'(V) \oplus \varphi''(V)$. Thus V is purely infinite. Also V is injective because φ' and φ'' extend to isomorphisms

$$\phi' : E(V) \to E(\varphi'(V)),\ \phi'' : E(V) \to E(\varphi''(V))$$

of the injective hulls, and so $(V, \varphi', \varphi'') \le (E(V), \phi', \phi'')$. Hence $V = E(V)$.

The injectivity of V implies that $E = U \oplus V$ for some $U \le E$. We show that U is directly finite. In fact, if U is not directly finite, then by Proposition 2.2.6, there are nonzero submodules A' and A'' of U such that $A' \oplus A'' \le U$ with isomorphisms $\alpha' : A' \oplus A'' \to A'$ and $\alpha'' : A' \oplus A'' \to A''$. Thus we obtain

$$(V, \varphi', \varphi'') < (V \oplus A' \oplus A'', \varphi' \oplus \alpha', \varphi'' \oplus \alpha''),$$

a contradiction to the maximality of (V, φ', φ''). Therefore, U is directly finite.

Step 2. We study now a fixed but arbitrary decomposition $E = U \oplus V$, with U directly finite and V purely infinite. The set of all pairs (A, f), where $A \le U$ and a monomorphism $f : A \to V$, ordered by restriction, allows again the application of Zorn's lemma. So there is a maximal pair (A, f). Since f extends to a monomorphism $E(A) \to V$, $A = E(A)$ and so A is injective. Put $U = U' \oplus A$, where $U' \le U$. Thus A and U' are directly finite.

Let $V' = A \oplus V$. We claim that $V' \cong V$ (consequently, V' is purely infinite). Indeed, $V \cong V \oplus V$ and $V = X \oplus f(A) \cong X \oplus A$ for some $X \le V$, yield that $V \cong X \oplus A \oplus V$. Whence $V = X_1 \oplus A_1 \oplus V_1$, where $X_1, A_1, V_1 \le V$ such that $X_1 \cong X$, $A_1 \cong A$, and $V_1 \cong V$.

Iterating this procedure, $V = (X_1 \oplus A_1) \oplus \cdots \oplus (X_n \oplus A_n) \oplus V_n$, where $X_i \cong X$, $A_i \cong A$, and $V_n \cong V$. Therefore $\oplus_{i=1}^{\infty} A_i \subseteq V$, thus it follows that $V = E(\oplus_{i=1}^{\infty} A_i) \oplus Y$ for some $Y \le V$. So we have that

$$\begin{aligned} V &= E(\oplus_{i=1}^{\infty} A_i) \oplus Y = A_1 \oplus E(\oplus_{i=2}^{\infty} A_i) \oplus Y \\ &\cong A \oplus E(\oplus_{i=1}^{\infty} A_i) \oplus Y = A \oplus V = V'. \end{aligned}$$

So far, we have obtained a new decomposition, $E = U' \oplus V'$, again with U' directly finite and V' purely infinite. We claim now that it enjoys the additional property that U' and V' have no nonzero isomorphic submodules.

To this end, we consider a submodule B of U' which is subisomorphic to V'. Then $B \oplus A$ is subisomorphic to V, via $B \oplus A \to V' \oplus A \cong V \oplus A \cong V$. Note that A is directly finite since it is a direct summand of U. We can apply Lemma 2.2.8, with $C = B \oplus A$ and $I = V$, and we get a monomorphic extension $h : B \oplus A \to V$ of f. Consequently, $(A, f) \le (B \oplus A, h)$ holds. The maximality of (A, f) implies that $B = 0$.

Step 3. We turn now to the uniqueness statement. Thus, we are given two decompositions $E = U_1 \oplus V_1 = U_2 \oplus V_2$ with U_i directly finite, V_i purely infinite, and no nonzero isomorphic direct summands between U_i and V_i $(i = 1, 2)$.

The immediate goal is to show that U_1 and V_2 have no nonzero isomorphic direct summands either. We claim that for any nonzero injective module H which is subisomorphic to both U_1 and V_2, there exists a positive integer n such that $H^{(n)}$ is subisomorphic to U_1, but $H^{(n+1)}$ is not.

Assume on the contrary that $H^{(n)}$ is subisomorphic to U_1 for all n. Using induction on n, we show that $U_1 = X_n \oplus H_n \oplus \cdots \oplus H_1$ with $H_i \cong H$.

For $n = 1$, this is true by assumption. If it holds for n, then we have that $U_1 \cong X_n \oplus H^{(n)}$; but we also have that $U_1 \cong Y \oplus H^{(n+1)}$. Since H, being isomorphic to a direct summand of U_1, is directly finite and injective, $X_n \cong Y \oplus H$ by Lemma 2.2.7(ii). Thus $X_n = X_{n+1} \oplus H_{n+1}$ with $X_{n+1} \cong Y$ and $H_{n+1} \cong H$. So we deduce that U_1 contains $\oplus_{i=1}^{\infty} H_i$, a contradiction to the fact that U_1 is directly finite (see Proposition 2.2.6).

If U_1 and V_2 have nonzero isomorphic direct summands, then by our claim we can find a nonzero injective module A, which is subisomorphic to both U_1 and V_2, but such that $A \oplus A$ is not subisomorphic to U_1. We obtain that $U_1 = A_1 \oplus B$ with $A_1 \cong A$, so $E = U_1 \oplus V_1 = A_1 \oplus B \oplus V_1$ and there exists a monomorphism $A \oplus A \to V_2 \oplus V_2 \cong V_2 \to E$. Thus, $E = A_2 \oplus A_3 \oplus C$ such that $A_2 \cong A_3 \cong A$ and $C \leq E$. Therefore,

$$E = A_1 \oplus B \oplus V_1 \cong A \oplus B \oplus V_1 \text{ and } E = A_2 \oplus A_3 \oplus C \cong A \oplus A_3 \oplus C.$$

Because A is directly finite and injective, $B \oplus V_1 \cong A_3 \oplus C$ by Lemma 2.2.7(ii). As a consequence, $B \oplus V_1 = A_4 \oplus C'$, where $A_4, C' \leq B \oplus V_1$ with $A_4 \cong A_3$ and $C' \cong C$.

Let $\pi : B \oplus V_1 \to V_1$ be the canonical projection. Then we show that $B \cap A_4 = \text{Ker}(\pi|_{A_4}) \leq^{\text{ess}} A_4$. For this, let $X \leq A_4$ with $X \cap \text{Ker}(\pi|_{A_4}) = 0$. Then X is subisomorphic to V_1 via π. Also we note that X is subisomorphic to U_1 via $X \subseteq A_4 \cong A \rightarrowtail U_1$. So $X = 0$ by assumption on the decomposition $E = U_1 \oplus V_1$. Since $B \cap A_4 \leq^{\text{ess}} A_4$, $A_4 = E(B \cap A_4) \leq^{\oplus} B$. So $B = A_4 \oplus D$ for some $D \leq B$. Hence, $U_1 = A_1 \oplus B = A_1 \oplus A_4 \oplus D$ has the submodule $A_1 \oplus A_4$ which is isomorphic to $A \oplus A$, contrary to the choice of A.

We showed that U_1 and V_2 have no nonzero isomorphic direct summands. Thus $U_1 \cap V_2 = 0$ as the injective hulls of $U_1 \cap V_2$ in U_1 and in V_2 are isomorphic direct summands. So $E = U_1 \oplus V_2 \oplus F$ for some submodule F of E. From $E = U_1 \oplus V_1 = U_1 \oplus V_2 \oplus F$, it follows that $V_1 \cong V_2 \oplus F$. Also from $E = U_1 \oplus V_2 \oplus F = U_2 \oplus V_2$, we see that $U_2 \cong U_1 \oplus F$. This shows that F yields isomorphic direct summands of U_2 and V_1. Now $F = 0$ as U_2 and V_1 cannot have nonzero isomorphic direct summands by symmetry of our preceding arguments. So $E = U_1 \oplus V_2$. □

In Theorem 2.2.14, we shall show that this result also holds true for the larger class of quasi-continuous modules instead of injective modules.

Lemma 2.2.10 *The following holds for a quasi-continuous module M.*

(i) *M is purely infinite if and only if $E(M)$ is purely infinite.*
(ii) *M is directly finite if and only if $E(M)$ is directly finite.*

Proof (i) Let M be purely infinite. Then $M \cong M \oplus M$, thus we have that $E(M) \cong E(M) \oplus E(M)$. So $E(M)$ is purely infinite.

Conversely, let $E(M)$ be purely infinite. Then $E(M) \cong E(M) \oplus E(M)$. Thus $E(M) = E_1 \oplus E_2$, with $E_1 \cong E(M)$ and $E_2 \cong E(M)$. Since M is quasi-continuous, $M = (M \cap E_1) \oplus (M \cap E_2)$ by Theorem 2.1.25. Next, we set $M_1 = M \cap E_1$ and $M_2 = M \cap E_2$. From Lemma 2.2.4, M_1 and M_2 are relatively injective. We observe that $E_1 = E(M_1)$ and $E_2 = E(M_2)$. Hence $E(M_1) \cong E(M_2)$, so $M_1 \cong M_2$ by Corollary 2.1.3, and $M_1 \oplus M_2$ is quasi-injective by Proposition 2.2.2. So M_1 and M_2 are quasi-injective.

From Theorem 2.1.7, $M_1 \oplus M_2$ and M_1 are relatively injective. Further, note that $E(M_1 \oplus M_2) = E(M) \cong E_1 = E(M_1)$. By Corollary 2.1.3, $M_1 \oplus M_2 \cong M_1$. Since $M_1 \cong M_2$, $M_1 \cong M_1 \oplus M_2 \cong M_1 \oplus M_1$. Similarly, $M_2 \cong M_2 \oplus M_2$. So we get that $M \cong M \oplus M$. Thus, M is purely infinite.

(ii) Let M be directly finite, and assume on the contrary that, $E(M)$ is not directly finite. By Proposition 2.2.6, there is $0 \neq B \leq^{\oplus} E(M)$ such that $B \cong B \oplus B$. Say $E(M) = B \oplus Y$ for some $Y \leq E(M)$. Because M is quasi-continuous, by Theorem 2.1.25 $M = (M \cap B) \oplus (M \cap Y)$. Therefore, $M \cap B$ is quasi-continuous from Lemma 2.2.4.

Also because $M \cap B \leq^{\text{ess}} B$ and B is injective, $B = E(M \cap B)$ and therefore $E(M \cap B) \cong E(M \cap B) \oplus E(M \cap B)$ from $B \cong B \oplus B$. Thus $E(M \cap B)$ is purely infinite. Hence, $M \cap B$ is also purely infinite by part (i) since $M \cap B$ is quasi-continuous. Note that $M = (M \cap B) \oplus (M \cap Y)$ with $0 \neq M \cap B$ purely infinite. Since M is directly finite, so is $M \cap B$. This is absurd as $M \cap B$ is purely infinite.

Conversely, let $E(M)$ be directly finite. If M is not directly finite, then therefore $\oplus_{i=1}^{\infty} A_i \leq M$, where for each i, $0 \neq A_i \leq M$ and $A_i \cong A$ for some A. Hence, $\oplus_{i=1}^{\infty} A_i \leq E(M)$. By Proposition 2.2.6, $E(M)$ is not directly finite, a contradiction. □

Definition 2.2.11 Given a quasi-continuous module M and an arbitrary submodule A of M, there is a direct summand P of M containing A as an essential submodule (take $P = M \cap E(A)$) by Corollary 2.1.26. The overmodule P is called an *internal quasi-continuous hull* of the submodule A of M.

The next theorem shows that the internal quasi-continuous hull of A is unique up to isomorphism.

Theorem 2.2.12 *Let M be a quasi-continuous module, and for $i = 1, 2$, assume that $A_i \leq^{\text{ess}} P_i \leq^{\oplus} M$ $(i = 1, 2)$. If $A_1 \cong A_2$, then $P_1 \cong P_2$.*

Proof Put $D = A_1 \cap A_2$, and we let X_i be a complement of D in A_i, for $i = 1, 2$. Then $D \oplus X_i \leq^{\text{ess}} A_i$. Hence $E_i \oplus E(X_i) = E(A_i) = E(P_i)$, where E_i denotes an injective hull of D in $E(A_i)$ for $i = 1, 2$. We note that $E_1 \cong E_2$. Also $X_1 \cap X_2 = 0$ because $X_1 \cap X_2 \subseteq A_1 \cap A_2 \cap X_2 = D \cap X_2 = 0$.

Write $M = P_i \oplus Q_i$ and let $E_i = U_i \oplus V_i$ be a decomposition according to Theorem 2.2.9, for $i = 1, 2$. We obtain that $E(M) = E(P_i) \oplus E(Q_i)$ and

$E(P_i) = E_i \oplus E(X_i) = U_i \oplus V_i \oplus E(X_i)$. Because $E(M) = E(P_i) \oplus E(Q_i)$, we have that $M = (M \cap E(P_i)) \oplus (M \cap E(Q_i))$ by Theorem 2.1.25. Since $M = P_i \oplus Q_i$, $M \cap E(P_i) = P_i$ and $M \cap E(Q_i) = Q_i$ by using the modular law. Hence, $E(P_i) = E(M \cap E(P_i)) = U_i \oplus V_i \oplus E(X_i)$. Note that $P_i = M \cap E(P_i)$ is quasi-continuous by Lemma 2.2.4. Thus,

$$M \cap E(P_i) = (M \cap E(P_i) \cap U_i) \oplus (M \cap E(P_i) \cap V_i) \oplus (M \cap E(P_i) \cap E(X_i)).$$

We see that $M \cap E(P_i) \cap U_i = M \cap U_i$ and $M \cap E(P_i) \cap V_i = M \cap V_i$ because $U_i \subseteq E(P_i)$ and $V_i \subseteq E(P_i)$. Observe that $M \cap E(P_i) \cap E(X_i) = M \cap E(X_i)$ since $E(X_i) \subseteq E(P_i)$. Hence, for $i = 1, 2$,

$$P_i = M \cap E(P_i) = (M \cap U_i) \oplus (M \cap V_i) \oplus (M \cap E(X_i)).$$

Let Y be a complement of D in M. Then $D \oplus Y \leq^{\text{ess}} M$. As $D \leq^{\text{ess}} E_1 \leq E(M)$ and $D \cap Y = 0$, $E_1 \cap Y = 0$ and $E_1 \oplus Y \leq^{\text{ess}} E(M)$. Thus $E(M) = E_1 \oplus E(Y)$. Similarly, $E(M) = E_2 \oplus E(Y)$. Hence, there is an isomorphism $\sigma : E_1 \to E_2$, which is determined by $\sigma(e_1) = e_2$ if and only if $e_1 - e_2 \in E(Y)$.

Because $E(M) = E_i \oplus E(Y)$ and M is quasi-continuous, it follows that $M = (M \cap E_1) \oplus (M \cap E(Y)) = (M \cap E_2) \oplus (M \cap E(Y))$ by Theorem 2.1.25. So there is an isomorphism $\sigma' : M \cap E_1 \to M \cap E_2$, determined by $\sigma'(m_1) = m_2$ if and only if $m_1 - m_2 \in M \cap E(Y)$, if and only if $m_1 - m_2 \in E(Y)$. Thus, $\sigma' = \sigma|_{M \cap E_1}$, so $\sigma(M \cap E_1) = M \cap E_2$.

From $E_i = U_i \oplus V_i$ and $\sigma(E_1) = E_2$, we obtain the two decompositions $E_2 = U_2 \oplus V_2 = \sigma(U_1) \oplus \sigma(V_1)$. The uniqueness part of Theorem 2.2.9 implies that $E_2 = U_2 \oplus \sigma(V_1) = \sigma(U_1) \oplus V_2$.

On the other hand, from $E(M) = E_2 \oplus E(Y)$ and the quasi-continuity of M, $M = (M \cap E_2) \oplus (M \cap E(Y))$ by Theorem 2.1.25. Thus $C := M \cap E_2$ is quasi-continuous from Lemma 2.2.4. Also, $E(C) = E(M \cap E_2) = E_2 = U_2 \oplus V_2$ since $(M \cap E_2) \leq^{\text{ess}} E_2$. As $U_2 \subseteq E_2$, $C \cap U_2 = M \cap E_2 \cap U_2 = M \cap U_2$. Also since $V_2 \subseteq E_2$, $C \cap V_2 = M \cap V_2$.

Observe that $C = (C \cap U_2) \oplus (C \cap V_2) = (C \cap \sigma(U_1)) \oplus (C \cap V_2)$ by Theorem 2.1.25 since $E(C) = U_2 \oplus V_2 = \sigma(U_1) \oplus V_2$. So $C \cap U_2 \cong C \cap \sigma(U_1)$. Thus,

$$\begin{aligned} M \cap U_2 = C \cap U_2 \cong C \cap \sigma(U_1) &= (M \cap E_2) \cap \sigma(U_1) \\ &= \sigma(M \cap E_1) \cap \sigma(U_1) = \sigma(M \cap E_1 \cap U_1) \\ &= \sigma(M \cap U_1) \cong M \cap U_1. \end{aligned}$$

Similarly, $M \cap V_2 \cong M \cap V_1$. The given isomorphism $A_1 \cong A_2$ yields that

$$U_1 \oplus V_1 \oplus E(X_1) = E(A_1) \cong E(A_2) = U_2 \oplus V_2 \oplus E(X_2).$$

As $U_2 \cong \sigma(U_1)$ from the fact that $E_2 = U_2 \oplus V_2 = \sigma(U_1) \oplus V_2$, $U_2 \cong \sigma(U_1) \cong U_1$ and hence

$$U_1 \oplus V_1 \oplus E(X_1) \cong U_2 \oplus V_2 \oplus E(X_2) \cong U_1 \oplus V_2 \oplus E(X_2).$$

Now U_1 is directly finite and injective by Theorem 2.2.9. So by Lemma 2.2.7(ii), $V_1 \oplus E(X_1) \cong V_2 \oplus E(X_2)$. Also we observe that

$$(M \cap V_i) \leq^{\text{ess}} V_i \text{ and } M \cap E(X_i) \leq^{\text{ess}} E(X_i).$$

Thus $V_i = E(M \cap V_i)$ and $E(X_i) = E(M \cap E(X_i))$ for $i = 1, 2$.

So, from $V_1 \oplus E(X_1) \cong V_2 \oplus E(X_2)$,

$$E(M \cap V_1) \oplus E(M \cap E(X_1)) \cong E(M \cap V_2) \oplus E(M \cap E(X_2)).$$

Thus, $E[(M \cap V_1) \oplus (M \cap E(X_1))] \cong E[(M \cap V_2) \oplus (M \cap E(X_2))]$. We see that $E(M) = E(P_i) \oplus E(Q_i) = E_i \oplus E(X_i) \oplus E(Q_i) = U_i \oplus V_i \oplus E(X_i) \oplus E(Q_i)$ as $M = P_i \oplus Q_i$ for $i = 1, 2$. Therefore,

$$M = (M \cap U_i) \oplus (M \cap V_i) \oplus (M \cap E(X_i)) \oplus (M \cap E(Q_i))$$

for $i = 1, 2$ by Theorem 2.1.25 since M is quasi-continuous. Thus, from Lemma 2.2.4, $(M \cap V_i) \oplus (M \cap E(X_i))$ is quasi-continuous for $i = 1, 2$.

We claim that $(M \cap V_1) \oplus (M \cap E(X_1))$ and $(M \cap V_2) \oplus (M \cap E(X_2))$ are relatively injective. Because V_1 is purely infinite and $V_1 = E(M \cap V_1)$, Lemma 2.2.10(i) yields that $M \cap V_1$ is purely infinite (as $M \cap V_1$ is quasi-continuous by Lemma 2.2.4), so $(M \cap V_1) \oplus (M \cap V_1) \cong M \cap V_1$. As $M \cap V_2 \cong M \cap V_1$,

$$(M \cap V_1) \oplus (M \cap V_2) \cong (M \cap V_1) \oplus (M \cap V_1) \cong M \cap V_1 \leq^{\oplus} M.$$

Hence, $(M \cap V_1) \oplus (M \cap V_2)$ is quasi-continuous as $M \cap V_1$ is quasi-continuous. By Lemma 2.2.4, $M \cap V_1$ and $M \cap V_2$ are relatively injective.

Note that $(M \cap V_1) \oplus (M \cap E(X_2)) \cong (M \cap V_2) \oplus (M \cap E(X_2)) \leq^{\oplus} M$. The quasi-continuity of M implies that of $(M \cap V_1) \oplus (M \cap E(X_2))$ by Lemma 2.2.4. Again from Lemma 2.2.4, $M \cap V_1$ and $M \cap E(X_2)$ are relatively injective. Similarly, $(M \cap V_2) \oplus (M \cap E(X_1)) \cong (M \cap V_1) \oplus (M \cap E(X_1)) \leq^{\oplus} M$ and the quasi-continuity of M imply that $M \cap V_2$ and $M \cap E(X_1)$ are relatively injective.

Recall that $X_1 \cap X_2 = 0$. Hence, $E(M) = E(X_1) \oplus E(X_2) \oplus F$ for some $F \leq E(M)$. By Theorem 2.1.25, $M = (M \cap E(X_1)) \oplus (M \cap E(X_2)) \oplus (M \cap F)$. Therefore, $(M \cap E(X_1)) \oplus (M \cap E(X_2))$ is quasi-continuous by Lemma 2.2.4. Hence, $M \cap E(X_1)$ and $M \cap E(X_2)$ are relatively injective from Lemma 2.2.4. Consequently, $(M \cap V_1) \oplus (M \cap E(X_1))$ and $(M \cap V_2) \oplus (M \cap E(X_2))$ are relatively injective by Theorem 2.1.7.

Since $E[(M \cap V_1) \oplus (M \cap E(X_1))] \cong E[(M \cap V_2) \oplus (M \cap E(X_2))]$, Corollary 2.1.3 yields that $(M \cap V_1) \oplus (M \cap E(X_1)) \cong (M \cap V_2) \oplus (M \cap E(X_2))$. Consequently,

$$\begin{aligned} P_1 &= (M \cap U_1) \oplus (M \cap V_1) \oplus (M \cap E(X_1)) \\ &\cong (M \cap U_2) \oplus (M \cap V_2) \oplus (M \cap E(X_2)) \\ &= P_2, \end{aligned}$$

because $M \cap U_1 \cong M \cap U_2$, which completes the proof. □

In the following series of results, we apply Theorem 2.2.12 to provide a decomposition of a quasi-continuous module into a directly finite direct summand and a purely infinite direct summand. Moreover, we obtain an analogue of Theorem 2.2.5 for continuous modules (see also Theorem 2.2.13 and Exercise 2.2.19.5 for other consequences of Theorem 2.2.12).

It may be worth noting that the isomorphism between P_1 and P_2 in Theorem 2.2.12 is not an extension of the isomorphism between A_1 and A_2 (see [308]). If A and B are direct summands of a quasi-continuous module M such that $A \cap B = 0$ and $E(A) \cong E(B)$, then A and B are relatively injective by Goel and Jain [177], so $A \cong B$ from Corollary 2.1.3. However, if $A \cap B \neq 0$, then this conclusion becomes difficult. Theorem 2.2.12 allows us to prove a powerful result, Theorem 2.2.13, showing that if A and B are direct summands of a quasi-continuous module M and $E(A) \cong E(B)$, then $A \cong B$ even when $A \cap B \neq 0$.

Theorem 2.2.13 *Let A and B be direct summands of a quasi-continuous module M. If $E(A) \cong E(B)$, then $A \cong B$.*

Proof Let $\varphi : E(A) \to E(B)$ be an isomorphism. Put

$$A_1 = \varphi^{-1}(B) \cap A \text{ and } A_2 = \varphi(A) \cap B.$$

Then $\varphi|_{A_1} : A_1 \to A_2$ is an isomorphism. Since $A \leq^{\text{ess}} E(A)$ and $B \leq^{\text{ess}} E(B)$,

$$\varphi(A) \leq^{\text{ess}} \varphi(E(A)) = E(B)$$

and so $\varphi(A) \cap B \leq^{\text{ess}} B$. Hence $A_2 \leq^{\text{ess}} B$. Similarly, $A_1 \leq^{\text{ess}} A$. By Theorem 2.2.12, $A \cong B$ because $A \leq^{\oplus} M$ and $B \leq^{\oplus} M$. □

Theorem 2.2.14 *Every quasi-continuous module M has a direct sum decomposition, $M = U \oplus V$ with U directly finite, and V purely infinite such that U and V have no nonzero isomorphic direct summands (or submodules). If $M = U_1 \oplus V_1 = U_2 \oplus V_2$ are two such decompositions, then $M = U_1 \oplus V_2$ holds too, and consequently $U_1 \cong U_2$ and $V_1 \cong V_2$.*

Proof Let M be a quasi-continuous module. Note that if $M = A \oplus B$ is any decomposition, such that A and B have no nonzero isomorphic direct summands, then they have no nonzero isomorphic submodules either. Indeed, say X and Y are isomorphic submodules of A and B, respectively. Note that A and B are quasi-continuous from Lemma 2.2.4. So $X \leq^{\text{ess}} P$ and $Y \leq^{\text{ess}} Q$ with $P \leq^{\oplus} A$ and $Q \leq^{\oplus} B$ by (C_1) condition. Therefore, $X \leq^{\text{ess}} P \leq^{\oplus} M$ and $Y \leq^{\text{ess}} Q \leq^{\oplus} M$. By Theorem 2.2.12, $P \cong Q$. Thus, $X = Y = 0$ as $P = Q = 0$. We apply Theorem 2.2.9 to $E(M)$ and obtain $E(M) = A \oplus B$, where A is directly finite and B is purely infinite. So $M = (M \cap A) \oplus (M \cap B)$ by Theorem 2.1.25. As $A = E(M \cap A)$

and $B = E(M \cap B)$, $M \cap A$ is directly finite and $M \cap B$ is purely infinite from Lemma 2.2.10.

Put $U = M \cap A$ and $V = M \cap B$. Then $M = U \oplus V$, where U is directly finite and V is purely infinite. Clearly, U and V cannot have nonzero isomorphic submodules since A and B do not have nonzero isomorphic submodules.

For uniqueness, let $M = U_1 \oplus V_1 = U_2 \oplus V_2$ be two such decompositions. Then $E(M) = E(U_i) \oplus E(V_i)$ with $E(U_i)$ directly finite and $E(V_i)$ purely infinite, for $i = 1, 2$, by Lemma 2.2.10.

If X and Y are isomorphic direct summands of $E(U_i)$ and $E(V_i)$, respectively, then from Theorem 2.1.25 $X \cap U_i$ and $Y \cap V_i$ are direct summands of U_i and V_i respectively, by the quasi-continuity of U_i and V_i. Hence, $X \cap U_i$ and $Y \cap V_i$ are direct summands of M. Since $E(X \cap U_i) = X \cong Y = E(Y \cap V_i)$, Theorem 2.2.13 yields that $X \cap U_i \cong Y \cap V_i$. So $X \cap U_i = Y \cap V_i = 0$ and hence $X = Y = 0$. Therefore the uniqueness statement from Theorem 2.2.9 gives

$$E(M) = E(U_1) \oplus E(V_2), \text{ and hence } M = (M \cap E(U_1)) \oplus (M \cap E(V_2)).$$

Consequently, we obtain $M \cap E(U_1) = U_1$ and $M \cap E(V_2) = V_2$ by the modular law because $M = U_1 \oplus V_1$ and $M = U_2 \oplus V_2$. Therefore, $M = U_1 \oplus V_2$. □

A homomorphism $f : V \to W$ is said to be *essential* if $f(V) \leq^{\text{ess}} W$.

Lemma 2.2.15 *Let M be a quasi-continuous module. Then the following are equivalent.*

(i) *M is continuous.*
(ii) *Every essential monomorphism $M \to M$ is an isomorphism.*
(iii) *No direct summand of M is isomorphic to a proper essential submodule of itself.*

Proof (i)⇒(ii) Say $f : M \to M$ is an essential monomorphism. Then we have that $M \cong f(M) \leq^{\text{ess}} M$. As $M \leq^{\oplus} M$, $f(M) \leq^{\oplus} M$ by (C_2) condition. Hence, $f(M) = M$.

(ii)⇒(iii) Assume that $P \cong A \leq^{\text{ess}} P \leq^{\oplus} M$. Then there exists $Q \leq M$ such that $M = P \oplus Q \cong A \oplus Q \leq^{\text{ess}} P \oplus Q = M$. By hypothesis, this is an isomorphism, and $A = P$.

(iii)⇒(i) To show that M is continuous, we only need to prove that M satisfies (C_2) condition because M is quasi-continuous. Assume that $A \leq M$ and $A \cong B \leq^{\oplus} M$. By ($C_1$) condition, $A \leq^{\text{ess}} P \leq^{\oplus} M$ for some P.

Since $B \leq^{\text{ess}} B \leq^{\oplus} M$ and $A \cong B$, we have that $P \cong B$ by Theorem 2.2.12. Thus $P \cong A \leq^{\text{ess}} P \leq^{\oplus} M$. Hence $A = P$ by assumption. So M has (C_2) condition, thus M is continuous. □

The next result is obtained as an application of Theorems 2.2.5 and 2.2.12.

Theorem 2.2.16 *Let $M = \bigoplus_{i \in \Lambda} M_i$, and assume that $\bigoplus_{i \in \Lambda} E(M_i)$ is an injective right R-module (e.g., Λ is finite or R is right Noetherian). Then M is continuous if and only if all M_i are continuous and M_j-injective for all $j \neq i$.*

Proof Let $M = \oplus_{i\in\Lambda} M_i$ be continuous. Then all M_i are continuous. Further, all M_i are M_j-injective for all $j \neq i$ by Lemma 2.2.4.

Conversely, let all M_i be continuous and M_j-injective for all $j \neq i$. Then $\oplus_{i\in\Lambda} M_i$ is quasi-continuous by Theorem 2.2.5. From Lemma 2.2.15, it suffices to show that each essential monomorphism $f : M \to M$ is onto. Since $M \cong f(M) = \oplus_{i\in\Lambda} f(M_i) \subseteq M$ and M is quasi-continuous, there is a direct summand P_i of M such that $f(M_i) \leq^{\text{ess}} P_i \leq^{\oplus} M$ for each i by (C_1) condition. Because $M_i \leq^{\text{ess}} M_i \leq^{\oplus} M$, Theorem 2.2.12 yields that $M_i \cong P_i$. Hence the essential monomorphism $M_i \cong f(M_i) \leq^{\text{ess}} P_i \cong M_i$, becomes an isomorphism from Lemma 2.2.15 because M_i is continuous. Therefore $f(M_i) = P_i \leq^{\oplus} M$, so $M \cap E(f(M_i)) = f(M_i)$ by the modular law.

Since $f(M_i) \cong M_i$, it follows that $E(f(M_i)) \cong E(M_i)$. Thus, $\oplus_{i\in\Lambda} E(f(M_i))$ is injective by assumption, and it is therefore a direct summand of $E(M)$. So $E(M) = [\oplus_{i\in\Lambda} E(f(M_i))] \oplus N$ for some $N \leq E(M)$.

As M is quasi-continuous, $M = [\oplus_{i\in\Lambda}(M \cap E(f(M_i)))] \oplus (M \cap N)$ by Theorem 2.1.25. So $M = [\oplus_{i\in\Lambda} f(M_i)] \oplus (M \cap N) = f(M) \oplus (M \cap N)$ because $M \cap E(f(M_i)) = f(M_i)$ for each i. But $f(M) \leq^{\text{ess}} M$, so $M \cap N = 0$. Therefore $M = f(M)$. By Lemma 2.2.15, M is continuous. □

We remark that there are weaker finiteness conditions which yield necessary and sufficient conditions for arbitrary direct sums of (quasi-)continuous modules to be (quasi-)continuous. We refer the reader to [301] and [324] for more details.

Proposition 2.2.17 *Let $M = M_1 \oplus M_2$ with M_1 and M_2 extending modules. Then M is extending if and only if every closed submodule K of M with $K \cap M_1 = 0$ or $K \cap M_2 = 0$ is a direct summand of M.*

Proof See [145, Lemma 7.9] for the proof. □

Theorem 2.2.18 *Let $M = M_1 \oplus \cdots \oplus M_n$ such that M_i is M_j-injective for any $i \neq j$. Then M is extending if and only if each M_i is extending.*

Proof The necessity is clear because every direct summand of an extending module is extending. Conversely, let each M_i be extending. First say $n = 2$. Let $K \leq M$ be closed in M and suppose that $M_1 \cap K = 0$ or $M_2 \cap K = 0$. We may assume that $M_1 \cap K = 0$. By Theorem 2.1.4, $M = M_1 \oplus N$ for some N such that $K \leq N \leq M$. Thus $N \cong M_2$, so N is extending. As $K \leq N \leq M$, K is closed in N. So $K \leq^{\oplus} N$ by Proposition 2.1.15. Say $N = K \oplus W$ for some $W \leq N$. Then $M = M_1 \oplus K \oplus W$, so $K \leq^{\oplus} M$. Thus, M is extending by Proposition 2.2.17.

Next, say $M = M_1 \oplus \cdots \oplus M_n$ with $n > 2$ such that M_i is M_j-injective for all $i \neq j$. By induction, $M_1 \oplus \cdots \oplus M_{n-1}$ is extending. Further, $M_1 \oplus \cdots \oplus M_{n-1}$ and M_n are relatively injective by Theorem 2.1.7. So $M_1 \oplus \cdots \oplus M_n$ is extending by the proof for the case when $n = 2$. □

Exercise 2.2.19

1. ([392, Tercan] and [121, Chatters and Khuri]) Let R be a right Ore domain. Prove the following.
   (i) R is a right extending ring.
   (ii) If $(R \oplus R)_R$ is an extending module, then R is a left Ore domain.
   (Hint: see [145, Corollary 12.9].)
2. ([69, Birkenmeier, Kim, and Park]) Let M be an extending module. Assume that the lattice of submodules of M is a distributive lattice. Show that every submodule of M is extending.
3. ([69, Birkenmeier, Kim, and Park]) Let R be a right extending ring and M a cyclic right R-module. Show that the following are equivalent.
   (i) M is nonsingular.
   (ii) Every cyclic submodule of M is projective and extending.
   (iii) Every cyclic submodule of M is projective.
4. ([54, Birkenmeier]) A module V is said to be *cancellative* if $V \oplus X \cong V \oplus Y$ (X and Y are modules), then $X \cong Y$ (cf. Lemma 2.2.7). Assume that M is a non-cancellative injective module. Prove that exactly one of the following holds.
   (i) $M = A \oplus B$, where A is cancellative, and B is semisimple and non-cancellative.
   (ii) $M = E(K)$, where K is ker-injective (see Exercise 2.1.37.11).
5. ([308, Müller and Rizvi]) Show that every directly finite continuous module is cancellative. Give an example of a directly finite quasi-continuous module which is not cancellative.

2.3 FI-Extending Property

The notion of an FI-extending module generalizes that of an extending module by requiring that only *every fully invariant* submodule is essential in a direct summand rather than *every* submodule. In Theorem 2.3.5, we show that any direct sum of FI-extending modules is FI-extending without any additional requirements. Thus, while a direct sum of extending modules may not be extending, it does satisfy the extending property for all its fully invariant submodules (which include many well-known submodules of any given module). Similar to the close connections that exist between the extending property and the Baer property (see Sect. 3.3), there are also close connections between the FI-extending property and the quasi-Baer property for rings (see Sect. 3.2).

Definition 2.3.1 A module M is called *FI-extending* if every fully invariant submodule of M is essential in a direct summand of M.

The complete sublattice of fully invariant submodules of the lattice of submodules of a module M is both extensive and contains many important submodules such as $\mathrm{Soc}(M)$, $\mathrm{Rad}(M)$, $Z(M)$, in fact $\rho(M)$ for any preradical ρ (see [382] for more details on preradicals). Moreover, for each ideal J of a ring R, $\ell_M(J)$ and MJ

are fully invariant. The FI-extending property, introduced in this section, effectively generalizes the extending property by targeting only the fully invariant submodules of M to be essential in direct summands. Thereby the FI-extending property ensures that the aforementioned preradicals are "essentially split-off".

Because the fully invariant submodules of a ring R are precisely all ideals of R, R_R is FI-extending if every ideal of R is right essential in a direct summand of R_R. Such a ring is called *right FI-extending*. A left FI-extending ring is defined similarly. A ring is called *FI-extending* if it is both right and left FI-extending.

Proposition 2.3.2 *Let M be a module. Then the following are equivalent.*

(i) *M is FI-extending.*
(ii) *For $N \trianglelefteq M$, there is $e^2 = e \in \text{End}(E(M))$ such that $N \leq^{\text{ess}} eE(M)$ and $eM \leq M$.*
(iii) *Each $N \trianglelefteq M$ has a complement which is a direct summand of M.*

Proof (i)⇒(ii) Assume that $N \trianglelefteq M$. Then there is $f^2 = f \in \text{End}(M)$ such that $N \leq^{\text{ess}} fM$. Let $e : E(M) \to E(fM)$ be the canonical projection. Then we see that $N \leq^{\text{ess}} eE(M)$ and $eM = fM \leq M$.

(ii)⇒(iii) Let $N \trianglelefteq M$. Then there exists $e^2 = e \in \text{End}(E(M))$ such that $N \leq^{\text{ess}} eE(M)$ and $eM \leq M$. Take $g = (1-e)|_M$. Then $g^2 = g \in \text{End}(M)$. We show that gM is a complement of N. For this, first note that $gM \cap N = 0$ as $gM = (1-e)M$. Say $K \leq M$ such that $gM = (1-e)M \leq K$ and $K \cap N = 0$.

From $M = (1-e)M \oplus eM$, $K = (1-e)M \oplus (K \cap eM)$ by the modular law. As $K \cap N = 0$ and $N \leq^{\text{ess}} eE(M)$, $K \cap eE(M) = 0$ and so $K \cap eM = 0$. Thus, we get that $K = (1-e)M$, then $K = gM$. Therefore gM is a complement of N.

(iii)⇒(i) Say $N \trianglelefteq M$. There exists $h^2 = h \in \text{End}(M)$ so that hM is a complement of N. As $N \trianglelefteq M$, $hN \leq N \cap hM = 0$. Hence, $N = (1-h)N$.

To show that M is FI-extending, we claim that $N \leq^{\text{ess}} (1-h)M$. For this, assume that $K \leq (1-h)M$ such that $N \cap K = 0$. Then note that $hM \cap K = 0$. Take $hm + k = n \in (hM \oplus K) \cap N$ with $m \in M$, $k \in K$, and $n \in N$. Then $(1-h)hm + (1-h)k = (1-h)n$, so $k = n \in K \cap N$ because $K \leq (1-h)M$ and $N = (1-h)N$. Now as $K \cap N = 0$, $k = n = 0$. Thus, $(hM \oplus K) \cap N = 0$. Since hM is a complement of N, $hM \oplus K = hM$ and so $K = 0$. Therefore, $N \leq^{\text{ess}} (1-h)M$. Hence, M is FI-extending. □

The following are some basic facts about fully invariant submodules.

Proposition 2.3.3 *The following hold true for a right R-module M.*

(i) *If $N_i \trianglelefteq M$ for $i \in \Lambda$, then $\bigcap_{i \in \Lambda} N_i \trianglelefteq M$ and $\sum_{i \in \Lambda} N_i \trianglelefteq M$.*
(ii) *If $W \trianglelefteq V$ and $V \trianglelefteq M$, then $W \trianglelefteq M$.*
(iii) *If $e^2 = e \in \text{End}(M)$, then $eM \trianglelefteq M$ if and only if $e \in \mathbf{S}_\ell(\text{End}(M))$.*
(iv) *If $e^2 = e \in \text{End}(M)$ and $V \trianglelefteq M$ with $V \leq^{\text{ess}} eM$, then $eM + Z(M) \trianglelefteq M$.*
(v) *Let $M = M_1 \oplus \cdots \oplus M_n$, where $M_i \trianglelefteq M$ for $i = 1, \ldots, n$. If $N \leq^{\oplus} M$, then $N = (M_1 \cap N) \oplus \cdots \oplus (M_n \cap N)$.*

Proof The proof of parts (i)–(iii) is routine.

(iv) Let $S = \text{End}(M)$. First, we show that $(1-e)SeM \subseteq Z(M)$. For this, say $m \in M$. Then there exists $K_R \leq^{\text{ess}} R_R$ such that $emK \subseteq V$ because $V \leq^{\text{ess}} eM$. Now $(1-e)SemK \subseteq (1-e)M \cap V = 0$, thus $(1-e)Sem \subseteq Z(M)$. Hence, it follows that $(1-e)SeM \subseteq Z(M)$. Therefore,

$$S(eM + Z(M)) = (eS + (1-e)S)(eM + Z(M)) \subseteq eM + Z(M).$$

So $eM + Z(M) \trianglelefteq M$.

(v) Assume that $M = N \oplus V$ with $V \leq M$. Then $M_i = (M_i \cap N) \oplus (M_i \cap V)$ (see Exercise 2.1.37.2) because $M_i \trianglelefteq M$, for $i = 1, 2, \ldots, n$. Therefore, we have that $M = \oplus_{i=1}^n M_i = [\oplus_{i=1}^n (M_i \cap N)] \oplus [\oplus_{i=1}^n (M_i \cap V)]$. From the modular law, $N = [\oplus_{i=1}^n (M_i \cap N)] \oplus [(\oplus_{i=1}^n (M_i \cap V)) \cap N]$. Because $V \cap N = 0$, it follows that $N = \oplus_{i=1}^n (M_i \cap N)$. □

Proposition 2.3.4 *Any fully invariant submodule of an FI-extending module is FI-extending.*

Proof Let M be an FI-extending module and $N \trianglelefteq M$. Take $V \trianglelefteq N$. By Proposition 2.3.3(ii), $V \trianglelefteq M$. Therefore $V \leq^{\text{ess}} D$ for some $D \leq^{\oplus} M$. Say $M = D \oplus W$, where $W \leq M$. So $N = (D \cap N) \oplus (W \cap N)$ (see Exercise 2.1.37.2). Thus, N is FI-extending because $V \leq^{\text{ess}} (D \cap N) \leq^{\oplus} N$. □

The next result shows that a direct sum of FI-extending modules inherits the property without any additional requirements.

Theorem 2.3.5 *Any direct sum of FI-extending modules is FI-extending.*

Proof Let $M = \oplus_{i \in \Lambda} M_i$ with each M_i an FI-extending module. Take $V \trianglelefteq M$. Then $V = \oplus_{i \in \Lambda} (V \cap M_i)$ (see Exercise 2.1.37.2), and $V \cap M_i \trianglelefteq M_i$ for each $i \in \Lambda$. As M_i is FI-extending, there is $D_i \leq^{\oplus} M_i$ with $(V \cap M_i) \leq^{\text{ess}} D_i$ for every $i \in \Lambda$. So $V = \oplus_{i \in \Lambda} (V \cap M_i) \leq^{\text{ess}} \oplus_{i \in \Lambda} D_i \leq^{\oplus} \oplus_{i \in \Lambda} M_i$. Therefore M is FI-extending. □

Corollary 2.3.6 *Let M be a direct sum of extending (e.g., uniform) modules. Then M is FI-extending.*

By Corollary 2.3.6, we see that while a direct sum of extending modules may not be extending in general (see Example 2.2.1(ii) and (iii)), it *has to be always FI-extending* without any additional conditions.

Corollary 2.3.7 *Let G be an Abelian group. If G satisfies any one of the following conditions, then G is an FI-extending Abelian group.*

(i) *G is finitely generated.*
(ii) *G is of bounded order* (*i.e.*, *$nG = 0$ for some positive integer n*).
(iii) *G is divisible.*

Proof (i) and (ii) follow from Corollary 2.3.6 as G is a direct sum of uniform $\mathbb{Z}$-modules in each case. (iii) is obvious as being divisible, G is injective. □

In contrast to Theorem 2.3.5, a direct product of FI-extending modules is not FI-extending as shown in the following example.

Example 2.3.8 Let $M = \prod_p \mathbb{Z}/p\mathbb{Z}$, where p varies through all prime integers. Then the torsion subgroup $t(M)$ of M is fully invariant and closed. But $t(M)$ is not a direct summand of M (see [362, Theorem 9.2]). Hence, the $\mathbb{Z}$-module M is not FI-extending.

It is presently an open problem to determine if a direct summand of an FI-extending module is always FI-extending (see [356]). The following three results show instances where this inheritance does occur.

Proposition 2.3.9 *If a module $M = B \oplus C$ is FI-extending and $B \trianglelefteq M$, then both B and C are FI-extending.*

Proof From Proposition 2.3.4, B is FI-extending. To prove that C is FI-extending, let $D \trianglelefteq C$. As $B \trianglelefteq M$, $\mathrm{Hom}(B, C) = 0$. Thus $B \oplus D \trianglelefteq M$, so there is $N \leq^{\oplus} M$ with $B \oplus D \leq^{\mathrm{ess}} N$. Say $M = N \oplus W$ for some $W \leq M$. From $M = B \oplus C$, $N = B \oplus (N \cap C)$ by the modular law. Since $B \oplus D \leq N$ and $D \leq C$, $D \leq N \cap C$. Further, $D \leq^{\mathrm{ess}} N \cap C$ as $B \oplus D \leq^{\mathrm{ess}} N$. Now $M = N \oplus W = (N \cap C) \oplus B \oplus W$. Thus, $C = (N \cap C) \oplus ((B \oplus W) \cap C)$ by the modular law. So $D \leq^{\mathrm{ess}} (N \cap C) \leq^{\oplus} C$. Therefore, C is FI-extending. □

Proposition 2.3.10 *Let M be a module. Then M is FI-extending if and only if $M = Z_2(M) \oplus N$, where $Z_2(M)$ and N are FI-extending.*

Proof Assume that M is FI-extending. Since $Z_2(M) \trianglelefteq M$ and $Z_2(M)$ is closed in M, $M = Z_2(M) \oplus N$ for some $N \leq M$. By Proposition 2.3.9, $Z_2(M)$ and N are FI-extending. The converse follows from Theorem 2.3.5. □

Proposition 2.3.11 *The following hold true for a ring R and $e \in \mathbf{S}_\ell(R)$.*

(i) *R_R is FI-extending if and only if eR_R and $(1-e)R_R$ are FI-extending.*
(ii) *If R is right FI-extending, then so is $(1-e)R(1-e)$.*

Proof (i) We see that $R_R = eR_R \oplus (1-e)R_R$ and $eR_R \trianglelefteq R_R$ since $e \in \mathbf{S}_\ell(R)$. Hence, Theorem 2.3.5 and Proposition 2.3.9 yield the desired result.

(ii) Let $W \trianglelefteq (1-e)R(1-e)$. Then $W_R \trianglelefteq (1-e)R_R$ because $1-e \in \mathbf{S}_r(R)$ and $\mathrm{End}((1-e)R_R) \cong (1-e)R(1-e)$. By part (i), since $(1-e)R_R$ is FI-extending, there exists $g^2 = g \in (1-e)R(1-e)$ such that $W_R \leq^{\mathrm{ess}} g(1-e)R_R$. We see that $W_{(1-e)R(1-e)} \leq^{\mathrm{ess}} g(1-e)R(1-e)_{(1-e)R(1-e)}$ as $1-e \in \mathbf{S}_r(R)$. Further,

$$g(1-e)R(1-e)_{(1-e)R(1-e)} \leq^{\oplus} (1-e)R(1-e)_{(1-e)R(1-e)}.$$

Hence $(1-e)R(1-e)$ is right FI-extending. □

Theorem 2.3.12 *Let R be a right FI-extending ring. Then $\text{Mat}_n(R)$ is a right FI-extending ring for all positive integer n.*

Proof Let $K \trianglelefteq \text{Mat}_n(R)$. Then $K = \text{Mat}_n(I)$ for some $I \trianglelefteq R$. As R is right FI-extending, there exists $e^2 = e \in R$ such that $I_R \leq^{\text{ess}} eR_R$. This yields that as a right ideal of $\text{Mat}_n(R)$, K is essential in a direct summand $(e\mathbf{1})\text{Mat}_n(R)$ of $\text{Mat}_n(R)$, where $\mathbf{1}$ is the identity matrix of $\text{Mat}_n(R)$. Therefore $\text{Mat}_n(R)$ is right FI-extending. □

Motivated by Theorem 2.3.12, the right FI-extending property for matrix rings will be discussed in Chap. 6, where matrix ring extensions will be dealt with in detail (e.g., see Theorem 6.1.17).

Example 2.3.13 Let R be a commutative domain. Then $\text{Mat}_n(R)$ is right FI-extending by Theorem 2.3.12. However, if R is not semihereditary and $n > 1$, then $\text{Mat}_n(R)$ is neither right nor left extending (see Theorem 6.1.4).

Example 2.3.14 Let $R = \begin{bmatrix} \mathbb{Z}_2 & \mathbb{Z}_2 \\ 0 & \mathbb{Z} \end{bmatrix}$. By computation, we see that the ring R is right FI-extending, but it is not left FI-extending (see Theorem 5.6.10 and Corollary 5.6.11 related to this example).

By Example 2.3.14, the FI-extending property for rings is not left-right symmetric. This motivates us to consider rings which are FI-extending on both sides.

Theorem 2.3.15 *Let R be a ring and $A \trianglelefteq R$. Assume that R is right and left FI-extending, $A \cap \ell_R(A) = 0$, and $A \cap r_R(A) = 0$. Then there exists $c \in \mathcal{B}(R)$ such that $A_R \leq^{\text{ess}} cR_R$, ${}_RA \leq^{\text{ess}} {}_RRc$, and $\ell_R(A) = r_R(A) = (1-c)R$.*

Proof There is $e^2 = e \in R$ such that $A_R \leq^{\text{ess}} eR_R$. Say $0 \neq y \in eR(1-e)$. Then there is $s \in R$ with $0 \neq ys \in A$. But $ysA \subseteq eR(1-e)A = 0$. Therefore $ys \in A \cap \ell_R(A) = 0$, a contradiction. So $eR(1-e) = 0$, thus $e \in \mathbf{S}_r(R)$ by Proposition 1.2.2. Similarly, there is $f \in \mathbf{S}_\ell(R)$ such that ${}_RA \leq^{\text{ess}} {}_RRf$.

To show that $r_R(A) = (1-e)R$, we notice that $A(1-e) \subseteq eR(1-e) = 0$. Thus $(1-e)R \subseteq r_R(A)$. Next, if $e r_R(A) \neq 0$, then take $0 \neq ex \in e r_R(A)$ with $x \in r_R(A)$. Since $A_R \leq^{\text{ess}} eR_R$, there exists $r \in R$ satisfying $0 \neq exr \in A$. Observe that $Aexr \subseteq Axr = 0$ since $x \in r_R(A)$, so $exr \in r_R(A) \cap A = 0$. Thus $exr = 0$, it is absurd. So $e r_R(A) = 0$, hence $r_R(A) \subseteq (1-e)R$. Therefore, $r_R(A) = (1-e)R$. Similarly, $\ell_R(A) = R(1-f)$.

As $A \cap \ell_R(A) = 0$ by assumption, $A\ell_R(A) = 0$, therefore $\ell_R(A) \subseteq r_R(A)$. From $A \cap r_R(A) = 0$, $r_R(A) \subseteq \ell_R(A)$. Thus, $\ell_R(A) = r_R(A)$, so $(1-e)R = R(1-f)$. Hence, $e = f \in \mathbf{S}_\ell(R) \cap \mathbf{S}_r(R)$, so $e \in \mathcal{B}(R)$ by Proposition 1.2.6(i). Take $c = e$. Then c is the desired idempotent. □

Let R be a ring. An ideal K of R is said to be a *regular ideal* if for each $x \in K$ there exists $y \in K$ satisfying $x = xyx$. Put

$$\mathcal{M}(R) = \{x \in R \mid RxR \text{ is a regular ideal of} R\}.$$

Then $\mathcal{M}(R)$ is a regular ideal of R containing all regular ideals of R (and is a Kurosh-Amitsur radical, see [183, Proposition 1.5] and [176]). We use φ to denote an assignment on the class of all rings such that $\varphi(R)$ is an ideal of R. A ring R is called *φ-regular* if $R/\varphi(R)$ is regular and $\varphi(R) \cap \mathcal{M}(R) = 0$. A ring R is said to be *J-regular* if $R/J(R)$ is regular (automatically $J(R) \cap \mathcal{M}(R) = 0$). A right (or left) ideal K of a ring R is said to be *ideal essential* in R if K has nonzero intersection with every nonzero ideal of R. In this case, R is called an *ideal essential extension* of K.

Theorem 2.3.16 *Let a ring R be right FI-extending.*
(i) *If R is left FI-extending and φ-regular, then $R = \mathcal{M}(R) \oplus B$ (ring direct sum) and B is an ideal essential extension of $\varphi(R)$.*
(ii) *If R is left continuous, then $R = \mathcal{M}(R) \oplus B$ (ring direct sum), where B is an ideal essential extension of $J(R)$.*

Proof (i) We note that $\mathcal{M}(R)$ is a semiprime ring since it is regular. Thus, we have that $\mathcal{M}(R) \cap r_R(\mathcal{M}(R)) = 0$ because $\mathcal{M}(R) \cap r_R(\mathcal{M}(R)) \trianglelefteq \mathcal{M}(R)$ and $[\mathcal{M} \cap r_R(\mathcal{M}(R))]^2 = 0$. Similarly, $\mathcal{M}(R) \cap \ell_R(\mathcal{M}(R)) = 0$. By Theorem 2.3.15, there is $c \in \mathcal{B}(R)$ such that $\mathcal{M}(R)_R \leq^{\text{ess}} cR_R$. As $\mathcal{M}(R) \cap \varphi(R) = 0$, $cR \cap \varphi(R) = 0$ and so $\varphi(R)c = 0$. Whence $\varphi(R) \subseteq (1-c)R$, thus $R/(1-c)R$ is a ring homomorphic image of the regular ring $R/\varphi(R)$. So $R/(1-c)R \cong cR$ is a regular ring. Thus cR is a regular ideal, and hence $\mathcal{M}(R) = cR$. Therefore, $R = \mathcal{M}(R) \oplus B$, where $B = (1-c)R$.

Let $I \trianglelefteq B$ with $\varphi(R) \cap I = 0$. Take $a \in I$. Then there exists $r \in R$ satisfying $a - ara \in \varphi(R)$ as $R/\varphi(R)$ is regular. Thus, $a - ara = 0$ since $a - ara \in I$. Hence, $a(rar)a = (ara)ra = ara = a$ and $rar \in I$. So I is a regular ideal of B (hence I is a regular ideal of R). Thus $I \subseteq \mathcal{M}(R) \cap B = 0$, so B is an ideal essential extension of $\varphi(R)$.

(ii) As R is left continuous, R is J-regular by Corollary 2.1.30. Therefore, $R = \mathcal{M}(R) \oplus B$ and B is an ideal essential extension of $J(R)$ by part (i). □

Let R be the ring of Example 2.1.36. Then R is right FI-extending and left continuous, but R is not right continuous. Thereby, Theorem 2.3.16 is a proper generalization of the result of Faith [158] for two-sided continuous rings.

Upon examining the FI-extending property, it is natural to ask when a module has the property that every fully invariant submodule is essential in a fully invariant direct summand. This question motivates the next definition.

Definition 2.3.17 A module M is said to be *strongly FI-extending* if every fully invariant submodule of M is essential in a fully invariant direct summand of M. A ring

R is called *right strongly FI-extending* if R_R is strongly FI-extending. A left strongly FI-extending ring is defined similarly. A ring R is called *strongly FI-extending* if R is right and left strongly FI-extending.

While every strongly FI-extending module is FI-extending, there exists an FI-extending module which is not strongly FI-extending. By Theorem 2.3.5, the module $M = \mathbb{Z} \oplus \mathbb{Z}_p$ is FI-extending $\mathbb{Z}$-module for any prime integer p. However, M is not strongly FI-extending by [80, Theorem 7.1] (see also Exercise 2.3.34.7).

It should be noted that unlike the FI-extending property, the strongly FI-extending property does not generalize injective modules. In Example 2.3.18, a right self-injective ring which is not right strongly FI-extending is provided. Thus, the right strongly FI-extending property does not belong to a hierarchy of generalizations of injectivity. However, the strongly FI-extending property coincides with the FI-extending property for nonsingular modules (Theorem 2.3.27). Also, the strongly FI-extending modules will be helpful in our study of FI-extending module hulls (see Sect. 8.4).

Example 2.3.18 Let $R = \mathbb{Z}_3[S_3]$, the group algebra of the symmetric group S_3 on three symbols $\{1, 2, 3\}$ over the field $\mathbb{Z}_3$ (see Example 2.1.18). Then R is a QF-ring. Let $\sigma = (123)$ and $\tau = (12)$ in S_3. Put $e = 2 + \tau \in R$. Then $e^2 = e$ and hence $eR = \{a + b\sigma + c\sigma^2 + 2a\tau + 2c\sigma\tau + 2b\sigma^2\tau \mid a, b, c \in \mathbb{Z}_3\}$.

Let $\omega(\mathbb{Z}_3[N]) = \{a + b\sigma + c\sigma^2 \mid a + b + c = 0,\ a, b, c \in \mathbb{Z}_3\}$, the augmentation ideal of $\mathbb{Z}_3[N]$, where $N = \{1, \sigma, \sigma^2\}$. Then $J(R) = \omega(\mathbb{Z}_3[N])R$ by [341, Exercise 8, p. 106]. We see that

$$\mathrm{Soc}(R_R) = \{a(1 + \sigma + \sigma^2) + b(1 + \sigma + \sigma^2)\tau \mid a, b \in \mathbb{Z}_3\}$$

since $\mathrm{Soc}(R_R) = \ell_R(J(R)) = \ell_R(\omega(\mathbb{Z}_3[N]))$. Therefore,

$$\mathrm{Soc}(eR_R) = eR \cap \mathrm{Soc}(R_R) = \{a(1+\sigma+\sigma^2) + 2a(1+\sigma+\sigma^2)\tau \mid a \in \mathbb{Z}_3\}.$$

We observe that $\mathrm{Soc}(eR_R) \trianglelefteq R$. Also $\mathrm{Soc}(eR_R) \leq^{\mathrm{ess}} eR_R$. By direct computation, R is semicentral reduced (Exercise 2.3.34.2). If R is right strongly FI-extending, then there is $f \in \mathbf{S}_\ell(R)$ with $\mathrm{Soc}(eR_R) \leq^{\mathrm{ess}} fR_R$. Thus, $f = 1$. Hence $\mathrm{Soc}(eR_R) \leq^{\mathrm{ess}} R_R$, a contradiction because $\mathrm{Soc}(eR_R) \leq^{\mathrm{ess}} eR_R$. Thus, R is not right strongly FI-extending.

Generally, if R is an indecomposable QF-ring with $0 \neq I \trianglelefteq R$ such that I_R is not essential in R_R, then R is neither right nor left strongly FI-extending by [262, Exercise 16, p. 421]. As mentioned earlier, at present it is unknown if a direct summand of an FI-extending module is FI-extending. However, the strongly FI-extending property of a module is inherited by its direct summands.

Theorem 2.3.19 *Every direct summand of a strongly FI-extending module is strongly FI-extending.*

Proof Let M_R be a strongly FI-extending module and $N \leq^{\oplus} M$. We let $S = \text{End}(M)$. Then $N = eM$ with $e^2 = e \in S$. Let $V \trianglelefteq N$. Then $SV \trianglelefteq M$. Since M is strongly FI-extending, there is $f \in \mathbf{S}_\ell(S)$ such that $SV \leq^{\text{ess}} fM$. Obviously, $V \subseteq SV \cap eM$. As $V \trianglelefteq eM$ and $eSe = \text{End}(eM)$, $eSeV \subseteq V$. So

$$SV \cap eM = eSV \cap eM = eSeV \cap eM \subseteq V \cap eM = V.$$

Hence it follows that $V = SV \cap eM \leq^{\text{ess}} fM \cap eM$. Furthermore, $(ef)^2 = ef$ and $efM \subseteq eM \cap fM$ because $f \in \mathbf{S}_\ell(S)$.

Let $y \in eM \cap fM$. Then there are $m, m' \in M$ with $y = em = fm'$. Thus, we get that $y = ey = efm' \in efM$, so $efM = eM \cap fM$ holds. Hence, we have that $V \leq^{\text{ess}} eM \cap fM = efM$. From $M = efM \oplus (1 - ef)M$ and $efM \leq eM$, by the modular law $eM = efM \oplus (eM \cap (1 - ef)M)$.

To see that $efM \trianglelefteq eM$, note first that $\text{End}(eM) = eSe$. As $f \in \mathbf{S}_\ell(S)$, we see that $(eSe)efM = ef(Se)fM \subseteq efM$, so $V \leq^{\text{ess}} efM \trianglelefteq eM = N$ and $efM \leq^{\oplus} eM$. Therefore, N is strongly FI-extending. □

Unlike the FI-extending modules, a direct sum of strongly FI-extending modules is not, in general, strongly FI-extending.

Example 2.3.20 Let $R = \mathbb{Z}_3[S_3]$, the group algebra as in Example 2.3.18. Say V_R is a nonzero proper direct summand of R_R. The vector space dimension of $\text{Soc}(R_R)$ over $\mathbb{Z}_3$ is 2. Thus $\text{Soc}(V)$ is a one dimensional vector spaces over $\mathbb{Z}_3$. Hence V_R is uniform, and so it is strongly FI-extending. Thus, every proper direct summand of R_R is strongly FI-extending. But R_R itself is not strongly FI-extending from Example 2.3.18.

In spite of Example 2.3.20, certain direct sums of strongly FI-extending modules are strongly FI-extending.

Theorem 2.3.21 *Let $M = \bigoplus_{i \in \Lambda} M_i$ and let $M_i \trianglelefteq M$ for each $i \in \Lambda$. If each M_i is strongly FI-extending, then M is strongly FI-extending.*

Proof Assume that each M_i is strongly FI-extending. Write $M_i = e_i M$, where $e_i^2 = e_i \in S := \text{End}(M)$. Say $V \trianglelefteq M$. Then $V = \oplus_{i \in \Lambda}(V \cap M_i) = \oplus_{i \in \Lambda} e_i V$ (see Exercise 2.1.37.2). Observe that $e_i V \trianglelefteq e_i M = M_i$, for each $i \in \Lambda$, because $(e_i S e_i)(e_i V) = e_i(Se_i V) \subseteq e_i V$ and $V \trianglelefteq M$. Thus there is $W_i \leq^{\oplus} M_i$, where $W_i \trianglelefteq M_i$ and $e_i V \leq^{\text{ess}} W_i$. So $V = \oplus_{i \in \Lambda} e_i V \leq^{\text{ess}} \oplus_{i \in \Lambda} W_i$. As $W_i \leq^{\oplus} M_i$ for each $i \in \Lambda$, $\oplus_{i \in \Lambda} W_i \leq^{\oplus} \oplus_{i \in \Lambda} M_i = M$. Also $W_i \trianglelefteq M_i$ and $M_i \trianglelefteq M$ for each $i \in \Lambda$, by Proposition 2.3.3(ii) $W_i \trianglelefteq M$ for each $i \in \Lambda$. So $\oplus_{i \in \Lambda} W_i \trianglelefteq M$ from Proposition 2.3.3(i). Thus, M is strongly FI-extending. □

Lemma 2.3.22 *Let $N \leq^{\text{ess}} eM$ with $e \in \mathbf{S}_\ell(\text{End}(M))$. If $N \leq^{\text{ess}} fM$ with $f^2 = f \in \text{End}(M)$, then $eM = fM$.*

Proof We observe that $eM \cap fM = feM$ and $(fe)^2 = fe \in \mathrm{End}(M)$ because $e \in \mathbf{S}_\ell(\mathrm{End}(M))$. Since $N \leq^{\mathrm{ess}} feM$, $feM = eM$ and $feM = fM$ from the modular law. Thus $eM = fM$. □

Theorem 2.3.23 *Let $M = \bigoplus_{i \in \Lambda} M_i$, where $M_i \cong M_j$, and M_i is strongly FI-extending for all $i, j \in \Lambda$. Then M is strongly FI-extending.*

Proof Assume that $N \trianglelefteq M$. Then $N = \oplus_{i \in \Lambda}(N \cap M_i)$, where $N \cap M_i \trianglelefteq M_i$. Since M_i is strongly FI-extending, $M_i = e_i M_i \oplus (1 - e_i)M_i$, where $e_i \in \mathbf{S}_\ell(\mathrm{End}(M_i))$ and $N \cap M_i \leq^{\mathrm{ess}} e_i M_i$. Set σ_{ji} to be the the isomorphism from M_i to M_j.

As $N \trianglelefteq M$, $\sigma_{ji}(N \cap M_i) \subseteq N \cap M_j$ and $\sigma_{ji}^{-1}(N \cap M_j) \subseteq N \cap M_i$. Whence

$$\sigma_{ji}^{-1}\sigma_{ji}(N \cap M_i) \subseteq \sigma_{ji}^{-1}(N \cap M_j) \subseteq N \cap M_i,$$

so $N \cap M_i \subseteq \sigma_{ji}^{-1}(N \cap M_j) \subseteq N \cap M_i$, thus $\sigma_{ji}^{-1}(N \cap M_j) = N \cap M_i$. Therefore $\sigma_{ji}(N \cap M_i) = N \cap M_j$.

Note that $N \cap M_j = \sigma_{ji}(N \cap M_i) \leq^{\mathrm{ess}} \sigma_{ji}(e_i M_i) \leq^{\oplus} \sigma_{ji}(M_i) = M_j$ because $N \cap M_i \leq^{\mathrm{ess}} e_i M_i \leq^{\oplus} M_i$. From Lemma 2.3.22, $\sigma_{ji}(e_i M_i) = e_j M_j$ since $e_j \in \mathbf{S}_\ell(\mathrm{End}(M_j))$, $N \cap M_j \leq^{\mathrm{ess}} e_j M_j$, and $N \cap M_j \leq^{\mathrm{ess}} \sigma_{ji}(e_i M_i) \leq^{\oplus} M_j$. Because $N \leq^{\mathrm{ess}} \oplus_{i \in \Lambda} e_i M_i$ and $\oplus_{i \in \Lambda} e_i M_i \leq^{\oplus} M$, to complete the proof it suffices to show that $\oplus_{i \in \Lambda} e_i M_i \trianglelefteq M$.

Let $h \in \mathrm{End}(M)$, and let $x \in \oplus_{i \in \Lambda} e_i M_i$. Without loss of generality, we assume that $x = e_i m_i$ for some $i \in \Lambda$. So $h(x) = h(e_i m_i) = \sum_{j \in J} m'_j$ for a finite subset J of Λ. To prove that $h(e_i m_i) \in \oplus_{i \in \Lambda} e_i M_i$, we consider, without loss of generality, $\pi_j h(e_i m_i) = m'_j$ (where $\pi_k : M \to M_k$, $k \in \Lambda$, are the canonical projections), and show that $m'_j \in e_j M_j$. Then $\pi_j h(e_i m_i) = \pi_j h \pi_i(e_i m_i)$, hence $\sigma_{ji}^{-1}(\pi_j h \pi_i)(e_i m_i) = \sigma_{ji}^{-1}(m'_j)$. Note that $(\sigma_{ji}^{-1}\pi_j h \pi_i)|_{M_i} \in \mathrm{End}_R(M_i)$.

We see that $(e_i \sigma_{ji}^{-1}\pi_j h \pi_i)(e_i m_i) = \sigma_{ji}^{-1}(\pi_j h \pi_i)(e_i m_i) = \sigma_{ji}^{-1}(m'_j)$, where $e_i(\sigma_{ji}^{-1}\pi_j h \pi_i)(e_i m_i) \in e_i M_i$, because $e_i \in \mathbf{S}_\ell(\mathrm{End}_R(M_i))$. Hence, it follows that $\sigma_{ji}(e_i \sigma_{ji}^{-1}\pi_j h \pi_i)(e_i m_i) = m'_j \in \sigma_{ji}(e_i M_i) = e_j M_j$. Thus, $\oplus_{i \in \Lambda} e_i M_i$ is a fully invariant direct summand of M. So M is strongly FI-extending. □

Corollary 2.3.24 *Let R be a right strongly FI-extending ring. Then every projective right R-module is strongly FI-extending.*

Proof The proof follows from Theorems 2.3.23 and 2.3.19. □

Lemma 2.3.25 *Assume that P is a generator in the category* Mod-R *of right R-modules. Let $S = \mathrm{End}(P)$ and let A, B be right ideals of S. Then $(A \cap B)P = AP \cap BP$.*

Proof See [229, Theorem 1.3]. □

Theorem 2.3.26 *The right strongly FI-extending property is a Morita invariant property.*

Proof Assume that R is a right strongly FI-extending ring. Let P_R be a progenerator in the category Mod-R of right R-modules. By Corollary 2.3.24, P_R is strongly FI-extending. Let $S = \mathrm{End}(P)$ and $I \trianglelefteq S$. Then $IP \trianglelefteq P$. As P is strongly FI-extending, there is $e \in \mathbf{S}_\ell(S)$ with $IP \leq^{\mathrm{ess}} eP = eSP$. We show that $I_S \leq^{\mathrm{ess}} eS_S$. For $0 \neq es \in eS$ with $s \in S$, assume on the contrary that $I \cap esS = 0$. Then $0 = (I \cap esS)P = IP \cap esP$ by Lemma 2.3.25. But since $0 \neq esP \subseteq eP$ and $IP \leq^{\mathrm{ess}} eP$, a contradiction. Thus, $I_S \leq^{\mathrm{ess}} eS_S$ and $eS \trianglelefteq S$. Hence, S is right strongly FI-extending. So the right strongly FI-extending property is Morita invariant. □

Theorem 2.3.27 *A nonsingular module M is FI-extending if and only if M is strongly FI-extending.*

Proof Let M be FI-extending. Take $N \trianglelefteq M$. There is $e^2 = e \in \mathrm{End}_R(M)$ with $N \leq^{\mathrm{ess}} eM$. From Proposition 2.3.3(iv), $eM \trianglelefteq M$ as $Z(M) = 0$. So M is strongly FI-extending. The converse is obvious. □

Remark 2.3.28 We remark that Theorem 2.3.27 can be extended as follows: A $\mathcal{K}$-nonsingular module M is FI-extending if and only if M is strongly FI-extending. (See Definition 4.1.3 for definition of $\mathcal{K}$-nonsingular modules.)

Let R be a right nonsingular ring. By Theorem 2.1.25 and Corollary 1.3.15, R is right quasi-continuous if and only if every idempotent of $Q(R)$ lies in R. The following result is a result along similar lines.

Theorem 2.3.29 *Let R be a right nonsingular ring. Then the following are equivalent.*

(i) *R is right FI-extending.*
(ii) *For every $e^2 = e \in Q(R)$ with $Re = eRe$, there exists $f \in \mathbf{S}_\ell(R)$ such that $eQ(R) = fQ(R)$.*

Proof (i)⇒(ii) Assume that R is right FI-extending. Take $e^2 = e \in Q(R)$ such that $Re = eRe$. Then $R \cap eQ(R) \trianglelefteq R$. By Theorem 2.3.27, there is f in $\mathbf{S}_\ell(R)$ such that $(R \cap eQ(R))_R \leq^{\mathrm{ess}} fR_R$. As $(R \cap eQ(R))_R \leq^{\mathrm{ess}} eQ(R)_R$, it follows that

$$(R \cap eQ(R))_R \leq^{\mathrm{ess}} (fQ(R) \cap eQ(R))_R, \text{ and } (fQ(R) \cap eQ(R))_R = feQ(R)_R$$

since $fe = efe$. Thus $feQ(R)_R = eQ(R)_R$ as $feQ(R)_R \leq^{\mathrm{ess}} eQ(R)_R$ and fe is an idempotent. Also $feQ(R)_R = fQ(R)_R$. Hence $eQ(R) = fQ(R)$.

(ii)⇒(i) Let $A \trianglelefteq R$. As $Z(R_R) = 0$, $Q(R) = E(R_R)$ by Corollary 1.3.15. From Proposition 2.1.32, $\mathrm{End}_R(E(R_R)) = \mathrm{End}_R(Q(R)) \cong Q(R)$, so there is

$e^2 = e \in Q(R)$ with $A_R \leq^{\text{ess}} eQ(R)_R$. We take $I = \{r \in R \mid er \in A\}$. Then $I_R \leq^{\text{ess}} R_R$ and $eI \subseteq A$. Thus $(1-e)ReI \subseteq (1-e)RA \subseteq (1-e)A = 0$.

We observe that $Z(Q(R)_R) = 0$ and so $(1-e)Re = 0$ because $I_R \leq^{\text{ess}} R_R$. Thus $Re = eRe$ holds. Therefore, $eQ(R) = fQ(R)$ for some $f \in \mathbf{S}_\ell(R)$ by assumption. Hence, $A_R \leq^{\text{ess}} fQ(R)_R$, so $A_R \leq^{\text{ess}} fR_R$. Therefore, R is right FI-extending. □

Two modules are said to be *orthogonal* if they have no nonzero isomorphic submodules (this concept was used in Sect. 2.2, for example, Theorems 2.2.9 and 2.2.14). For a class $\mathfrak{A}$ of modules, $\mathfrak{A}^\perp$ denotes the class of modules orthogonal to all members of $\mathfrak{A}$. Classes $\mathfrak{A}$ and $\mathfrak{B}$ form an *orthogonal pair* if $\mathfrak{A}^\perp = \mathfrak{B}$ and $\mathfrak{B}^\perp = \mathfrak{A}$. To obtain a ring direct sum decomposition of a right nonsingular right FI-extending ring whose summands are from a given orthogonal pair (see Theorem 2.3.31), we need the next lemma, which is of interest also in its own right.

Lemma 2.3.30 *Let $Z(R_R) = 0$. Given an orthogonal pair $\mathfrak{A}$ and $\mathfrak{B}$ of classes of right R-modules, there exist ideals A and B of R, such that A and B are maximal among the right ideals of R contained in $\mathfrak{A}$ and $\mathfrak{B}$, respectively, and $(A \oplus B)_R \leq^{\text{ess}} R_R$.*

Proof Let $\mathfrak{A}$ and $\mathfrak{B}$ be an orthogonal pair. We use Zorn's lemma to obtain a right ideal A which is maximal among the right ideals (of R) in $\mathfrak{A}$ which is orthogonal to all members of $\mathfrak{B}$, and a right ideal B maximal among the right ideals (of R) in $\mathfrak{B}$ which is orthogonal to all members of $\mathfrak{A}$.

Obviously, $A \cap B = 0$. We show that $(A \oplus B)_R \leq^{\text{ess}} R_R$. If not, then there exists a nonzero right ideal I of R such that $(A \oplus B) \cap I = 0$.

Case 1. I has a nonzero submodule J in $\mathfrak{A}$. We show that $A \oplus J \in \mathfrak{A}$. For this, assume on the contrary that $A \oplus J$ is not orthogonal to some $0 \neq C \in \mathfrak{B}$. Then there exists $0 \neq v = a + y \in A \oplus J$ with $a \in A$ and $y \in J$ such that vR is subisomorphic to C. Define $f : vR \to aR$ by $f(vr) = ar$, where $r \in R$. If $\text{Ker}(f) = 0$, then $vR_R \cong aR_R$, and so aR_R is nonzero and it is subisomorphic to C. But since $aR_R \leq A_R$, it is absurd. If $\text{Ker}(f) \neq 0$, then

$$\text{Ker}(f) = \{vr \in vR \mid ar = 0, r \in R\} = \{yr \mid r \in R\},$$

thus $0 \neq \text{Ker}(f)_R \leq J_R$. Since vR is subisomorphic to C, so is $\text{Ker}(f)$. However, $\text{Ker}(f) \subseteq J$, it is also absurd as $J \in \mathfrak{A}$. Thus $A \oplus J \in \mathfrak{A}$. By the maximality of A, this cannot happen.

Case 2. I does not have any nonzero submodule J in $\mathfrak{A}$. First, we show that $B \oplus I \in \mathfrak{B}$. Assume on the contrary that $B \oplus I$ is not orthogonal to some $0 \neq D \in \mathfrak{A}$. Then there exists $0 \neq w = b + x \in B \oplus I$ with $b \in B$ and $x \in I$ such that wR_R is subisomorphic to D. Define $g : wR \to bR$ by $g(wr) = br$, where $r \in R$. If $\text{Ker}(g) = 0$, then $wR_R \cong bR_R$ and so bR_R is subisomorphic to $D \in \mathfrak{A}$, which is impossible.

If $\text{Ker}(g) \neq 0$, then $\text{Ker}(g) = \{xr \mid r \in R\}$ is subisomorphic to D because $\text{Ker}(g)_R \leq wR_R$. So we have that $0 \neq \text{Ker}(g)_R \leq I_R$ and it is subisomorphic to

$D \in \mathfrak{A}$. Say $D_1 \leq D$ such that $\operatorname{Ker}(g) \cong D_1$. We note that $D_1 \in \mathfrak{A}$ because $D \in \mathfrak{A}$ and $D_1 \leq D$. Thus, $0 \neq \operatorname{Ker}(g) \in \mathfrak{A}$ and $\operatorname{Ker}(g)_R \leq I_R$, which contradicts the hypothesis. Hence, $B \oplus I$ is orthogonal to $\mathfrak{A}$, so $B \oplus I \in \mathfrak{A}^{\perp} = \mathfrak{B}$. This is impossible by the maximality of B. So this case also cannot happen. Consequently, by Cases 1 and 2, $(A \oplus B)_R \leq^{\text{ess}} R_R$.

We claim that A and B are ideals of R. Since R is right nonsingular, we get that $Q(R) = E(R_R)$ from Corollary 1.3.15. So $Q := Q(R) = E(A_R) \oplus E(B_R)$.

Put $E(A_R) = eQ$ and $E(B_R) = (1-e)Q$ for some $e^2 = e \in Q$. Let $t \in Q$. Define a map $f : (1-e)Q \to eQ$ such that $f((1-e)q) = et(1-e)q$ for $q \in Q$.

For the claim, we show that $\operatorname{Ker}(f)_R \leq^{\text{ess}} (1-e)Q_R$. If not, then there exists $0 \neq C_R \leq (1-e)Q_R$ with $\operatorname{Ker}(f) \cap C = 0$. Put $V = C \cap B$. Then $V \neq 0$ because $B_R \leq^{\text{ess}} (1-e)Q_R$. Since $\operatorname{Ker}(f) \cap V = 0$, the restriction f_0 of f to V is an isomorphism to $f_0(V)_R \leq eQ_R$. We let $A_0 = f_0(V) \cap A$ and $V_0 = f_0^{-1}(A_0)$. Then A_0 is a nonzero submodule of A_R and is isomorphic to $V_{0R} \leq B_R$, a contradiction. Therefore $\operatorname{Ker}(f)_R \leq^{\text{ess}} (1-e)Q_R$, and hence $\operatorname{Ker}(f)_Q \leq^{\text{ess}} (1-e)Q_Q$.

From $[(1-e)Q/\operatorname{Ker}(f)]_Q \cong et(1-e)Q_Q$, $Z(et(1-e)Q_Q) = et(1-e)Q$ (see 1.1.8) for all $t \in Q$. Note that $Z(Q_Q) = 0$ since Q is regular from Theorem 2.1.31. Thus, $et(1-e)Q = Z(et(1-e)Q_Q) = 0$ for all $t \in Q$. So $eQ(1-e) = 0$, so e is central by Propositions 1.2.2 and 1.2.6(ii). Because $A_R \leq^{\text{ess}} (eQ \cap R)_R$, $eQ \cap R \in \mathfrak{A}$ and so maximality of A yields that $A = eQ \cap R$. Similarly, $B = (1-e)Q \cap R$. Thus A and B are ideals of R as e is central. □

Theorem 2.3.31 *Let R be a right FI-extending ring with $Z(R_R) = 0$. Then any orthogonal pair $\mathfrak{A}$ and $\mathfrak{B}$ of classes of R-modules yields a ring direct sum decomposition $R = R_1 \oplus R_2$, where R_1 and R_2 are maximal among the right ideals of R in $\mathfrak{A}$ and $\mathfrak{B}$, respectively.*

Proof We let $Q = Q(R)$. From Lemma 2.3.30, $(R_1 \oplus R_2)_R \leq^{\text{ess}} R_R$, where $R_1 = eQ \cap R$ and $R_2 = (1-e)Q \cap R$ for some $e \in \mathcal{B}(Q)$. Because R is right FI-extending, $R_{1R} \leq^{\text{ess}} fR_R$, where $f^2 = f \in R$. Also $R_{1R} \leq^{\text{ess}} eQ_R$. Since $R_{1R} \leq^{\text{ess}} fR_R \leq^{\text{ess}} fQ_R$ and $e \in \mathcal{B}(Q)$, it follows that $e = f \in R$. Therefore, $R_1 = eR$, $R_2 = (1-e)R$, and $R = R_1 \oplus R_2$. □

The next example shows that Theorem 2.3.31 fails if R is not right FI-extending in the hypothesis (the example also illustrates Lemma 2.3.30).

Example 2.3.32 Let F be any field and let $F_i = F$, $i \in \Lambda$, where Λ is infinite. Define $R = \oplus_{i \in \Lambda} F_i + F\,1$, which is an F-subalgebra of $\prod_{i \in \Lambda} F_i$, where 1 is the identity of $\prod_{i \in \Lambda} F_i$. Then R is a regular ring. Let $\Lambda = \Lambda_1 \cup \Lambda_2$ be a nontrivial disjoint union. Then $A = \oplus_{i \in \Lambda_1} F_i$ and $B = \oplus_{i \in \Lambda_2} F_i$ are ideals of R.

Let $\mathfrak{A} = \{A\}^{\perp\perp}$ and $\mathfrak{B} = \{A\}^{\perp}$. We see that $(A \oplus B)_R \leq^{\text{ess}} R_R$ and $A \oplus B \neq R$. Note that A_R and B_R are closed in R_R. Also $A \in \mathfrak{A}$ and $B \in \mathfrak{B}$.

To show that A and B are maximal among ideals of R in $\mathfrak{A}$ and $\mathfrak{B}$, respectively, say $A \leq C$ and $C \in \mathfrak{A}$. Assume on the contrary that $A \neq C$. Then A_R is not essential in C_R (as A_R is closed in R_R), so there exists $0 \neq V \leq C$ such that $A \cap V = 0$. Since

$(A \oplus B)_R \leq^{\text{ess}} R_R$, $(A \oplus B) \cap V \neq 0$. We note that $A \oplus B$ is a semisimple R-module, so there is $0 \neq v \in (A \oplus B) \cap V$ such that vR is a simple R-module. Say $v = a + b$ with $a \in A$ and $b \in B$.

If $a \neq 0$ and $b \neq 0$, then $vR \cong aR$ by the R-homomorphism corresponding vr to ar for $r \in R$. Also $vR \cong bR$. Hence, $aR \cong bR$. But since $aR \leq A$ and $bR \leq B$, this cannot happen. If $a \neq 0$ and $b = 0$, then $vR = aR \leq A$, which contradicts $A \cap V = 0$. If $a = 0$ and $b \neq 0$, then $vR = bR \leq B$. Since $vR \leq C$, $C \notin \mathfrak{A}$, a contradiction. Therefore $A = C$, so A is maximal among ideals of R in $\mathfrak{A}$. Similarly, B is maximal among ideals of R in $\mathfrak{B}$.

Remark 2.3.33 ($\mathcal{G}$-extending and C_{11}-Modules) Let M be a module. Consider two relations ω and β on submodules of M: Let $L, N \leq M$. Then we say that $L \, \omega \, N$ if $L \leq^{\text{ess}} L + N$ and $N \leq^{\text{ess}} L + N$. We say that $L \, \beta \, N$ if $L \cap N \leq^{\text{ess}} L$ and $L \cap N \leq^{\text{ess}} N$. The relation β has been initially defined (as the relation ρ) and studied in [374]. It is an equivalence relation and was introduced in [178] for right ideals of a ring. Note that if $L \, \omega \, N$, then $L \, \beta \, N$.

We routinely see that a module M is extending if and only if for each $N \leq M$, there is $D \leq^{\oplus} M$ such that $N \, \omega \, D$. Motivated by this fact and the use of the equivalence relation β by Goldie [178], another interesting generalization of extending modules is defined in [1]. A module M is called *$\mathcal{G}$-extending* (i.e., Goldie extending) if for each $N \leq M$, there exists $D \leq^{\oplus} M$ such that $N \, \beta \, D$ (see [1]). Thus an extending module is a $\mathcal{G}$-extending module. When M is a UC-module (UC for unique closure) (e.g., M is nonsingular), M is $\mathcal{G}$-extending if and only if M is extending. Every canonical cogenerator is $\mathcal{G}$-extending; but, in general, it is not extending [1].

In [2], $\mathcal{G}$-extending modules over a Dedekind domain and those over a commutative PID are characterized which extend the characterization of $\mathcal{G}$-extending Abelian groups [1]. These characterizations involve the condition that pure submodules are direct summands. From these results, it is shown in [2] that every finitely generated module over a Dedekind domain is $\mathcal{G}$-extending (this result also appears in [3] with a different proof). Also, in [2], it is proved that the class of $\mathcal{G}$-extending modules over a commutative PID is closed under direct summands, and the class of $\mathcal{G}$-extending torsion modules over a Dedekind domain is closed under finite direct sums.

Furthermore in [3], $\mathcal{G}$-extending 2×2 generalized triangular matrix rings and $T_n(R)$ $(n > 1)$ over a ring R are characterized. From these characterizations, it follows that $T_n(R)$ $(n > 1)$ over a right self-injective ring R is right $\mathcal{G}$-extending, but not, in general, right extending.

Another class of modules generalizing the class of extending modules is the class of C_{11}-modules. The investigation of these modules was posed as Open Problem 9 in [301]. A module is called a *C_{11}-module* if every submodule has a complement which is a direct summand (see [376]). Every $\mathcal{G}$-extending module is a C_{11}-module and every C_{11}-module is an FI-extending module, the reverse implications do not hold. In [376], it is shown that a direct sum of C_{11}-modules is a C_{11}-module. Thereby any direct sum of extending modules is a C_{11}-module. Further work on C_{11}-modules appears in [67].

Exercise 2.3.34

1. Prove Proposition 2.3.3(i), (ii), and (iii).
2. Let $R = \mathbb{Z}[S_3]$ as in Example 2.3.18. Show that R is semicentral reduced.
3. ([83, Birkenmeier, Müller, and Rizvi]) Let a ring R be right and left FI-extending. Prove that every ideal, which is semiprime (as a ring), is right and left essential in a ring direct summand.
4. ([83, Birkenmeier, Müller, and Rizvi]) Show that a right and left FI-extending ring is a direct sum of a reduced ring and a ring in which every nonzero ideal contains a nonzero nilpotent element.
5. ([84, Birkenmeier, Park, and Rizvi]) Assume that R is a right strongly FI-extending ring. Prove that if R is right (semi-)hereditary, then every (finitely generated) submodule of a projective right R-module is strongly FI-extending.
6. ([84, Birkenmeier, Park, and Rizvi]) Let M_R be a free R-module and let $S = \text{End}(M)$. Prove that if M_R is FI-extending or strongly FI-extending, then so is S_S (cf. Theorem 6.1.17).
7. ([80, Birkenmeier, Călugăreanu, Fuchs, and Goeters]) Show that an Abelian group A is strongly FI-extending if and only if $A = B \oplus C \oplus D$, where
   (1) B is a direct sum of p-groups each of which is the direct sum of cyclic groups of the same order;
   (2) C is a torsion-free FI-extending group; and
   (3) D is a divisible group
   such that if B has a nontrivial p-component, then C is p-divisible.

Historical Notes Quasi-injective modules were defined by Johnson and Wong in [238]. Theorem 2.1.23 was shown by Huynh, Rizvi, and Yousif in [219]. Theorem 2.1.25 comprises results of Jeremy [230], Goel and Jain [177], and Oshiro and Rizvi [324]. Theorem 2.1.29(ii) and the fact that $S/J(S)$ is regular were first proved for injective modules by Utumi [396], and were extended to quasi-injective modules by Faith and Utumi [161].

For an injective module M, it was proved that if $\Delta = 0$, then S is right self-injective by Johnson and Wong [238]. This result was generalized by Osofsky [330] showing that if M is quasi-injective, then S/Δ is right self-injective and orthogonal idempotents of S/Δ lift to orthogonal idempotents of S. For a right continuous ring R, Utumi [398] proved that $J(R) = \Delta$ and $R/J(R)$ is right continuous. This result was generalized by Mohamed and Bouhy [300] to continuous modules. The endomorphism ring of a quasi-continuous module has been studied by Jeremy [230]. Theorem 2.1.35 is due to Jain, López-Permouth, and Rizvi [223].

Theorem 2.2.5 is due to Müller and Rizvi [308]. For quasi-continuity of a direct sum of modules, see also [301]. Refer to [410] for Lemma 2.2.7. Regarding further developments on Lemma 2.2.7(ii), it was shown that cancellation property also holds for the larger class of directly finite quasi-injective modules by Fuchs [173] and Birkenmeier [52] (see also [389] for Lemma 2.2.7(i)). This result was further extended to the class of directly finite continuous modules by Müller and Rizvi [308] as described in Exercise 2.2.19.5. It was also proved by Müller and Rizvi [308, Examples (1), p. 206] that the result cannot be further weakened to the class of directly finite quasi-continuous modules.

It was shown in [184] that any nonsingular injective module has a (unique) direct sum decomposition into a directly finite and a purely infinite part. The existence part of the decomposition in Theorem 2.2.9 for arbitrary injective modules was shown by Goodearl [181] using categorical techniques. Uniqueness (up to isomorphism) of the decomposition for an arbitrary injective module was shown together with direct proof (without categorical methods) by Müller and Rizvi in [308] and has been included here. A (relatively) short and direct proof of Theorem 2.2.9, due to Müller and Rizvi [308] has been provided. Internal quasi-continuous hulls were defined by Müller and Rizvi [308]. Theorem 2.2.12 is due to Müller and Rizvi [308] which shows that, for quasi-continuous modules, the isomorphism type of the internal quasi-continuous hull of any submodule is determined by the isomorphism type of the submodules. Theorem 2.2.13, Theorem 2.2.14, Lemma 2.2.15, and Theorem 2.2.16 appear in [308]. Theorem 2.2.18 is [145, Proposition 7.10].

A number of mathematicians have contributed greatly to the development of the theory of extending and (quasi-)continuous modules. These include B.J. Müller, M. Harada, K. Oshiro, S.H. Mohamed, P.F. Smith, B.L. Osofsky, D.V. Huynh, R. Wisbauer and many others who we could not list here. For further results and materials on (quasi-)continuous and extending modules can be found, for example, in [35, 37, 44, 144, 145, 149, 198, 199, 211, 213–216, 219, 220, 223, 224, 227, 228, 230, 241, 242, 284, 324, 338, 355, 375, 376], and [377], etc. Some other papers on injectivity or its generalizations have also been included in references.

Definition 2.3.1 is from [83], where FI-extending modules were defined. FI-extending Abelian groups have been initially defined and investigated by Birkenmeier, Călugăreanu, Fuchs, and Goeters in [80]. The equivalence of (i) and (ii) of Proposition 2.3.2 is in [83]. Proposition 2.3.3(i)–(iv) appears in [83], while Lemma 2.3.3(v) is [361, Lemma 1.3.18]. Its proof is due to Gangyong Lee. Results 2.3.4–2.3.8 and 2.3.10 appear in [83]. Proposition 2.3.9 is in [80]. Proposition 2.3.11(ii) is in [59]. Theorem 2.3.12 is in [83]. Example 2.3.14 is taken from [85]. Theorems 2.3.15 and 2.3.16 are due to Birkenmeier, Müller, and Rizvi in [83] (see also [60, Theorem 3.9 and Corollary 3.10]).

Strongly FI-extending modules were defined and studied in [84] and [85]. Example 2.3.18 appears in [72] and [84]. Results 2.3.19, 2.3.21, 2.3.23, 2.3.24, 2.3.26, 2.3.27, and 2.3.29 appear in [84]. Lemma 2.3.30, Theorem 2.3.31, and Example 2.3.32 appear in [83]. Exercise 2.3.34.4 generalizes a result on quasi-continuous rings in [230].

There has been an extensive work done on injectivity and its generalizations. While we cannot cite all such references, some related references include [5, 6, 21, 31, 41, 106, 126, 146, 150, 152, 207, 217, 218, 222, 225, 237, 306, 309, 310, 315, 318, 326, 328, 331, 390, 399], and [408].

Chapter 3
Baer, Rickart, and Quasi-Baer Rings

The notion of a Baer ring was introduced by Kaplansky in 1955 [245] (Kaplansky's book [246] was published in 1968). A ring is called Baer if the right (left) annihilator of every nonempty subset of R is generated by an idempotent, as a right (left) ideal. Kaplansky and Berberian were instrumental in developing the theory of Rickart and Baer rings ([246] and [45]). In 1960, Maeda [287] defined Rickart rings in an arbitrary setting. A ring is called right Rickart if the right annihilator of any single element is generated by an idempotent. In 1960, Hattori [200] introduced the notion of a right PP ring (i.e., a ring with the property that every principal right ideal is projective). It was later shown that right PP rings are precisely right Rickart rings.

A ring for which the left annihilator of every ideal is generated by an idempotent was termed a quasi-Baer ring by Clark in 1967 [128]. He also showed that any finite distributive lattice is isomorphic to a certain sublattice of the lattice of all ideals of an Artinian quasi-Baer ring. Analogous to a right Rickart ring, a ring R is called right principally quasi-Baer (simply, right p.q.-Baer) if the right annihilator of any principal ideal is generated by an idempotent as a right ideal. Principally right quasi-Baer rings were initially defined and studied in [78]. The concept of a right p.q.-Baer ring generalizes those of a quasi-Baer ring and a biregular ring.

In this chapter, we begin with basic properties and results on these classes of rings which will be instrumental in developing the subject of our study in later chapters. It will be shown that the Baer and the Rickart properties of rings do not transfer to the rings of matrices or to the polynomial ring extensions, while the quasi-Baer and the principally quasi-Baer properties of rings do so. In particular, for a commutative domain R, $\mathrm{Mat}_n(R)$ is Baer (Rickart) for every positive integer n if and only if R is a Prüfer domain. We shall compare and contrast the notions of Baer and Rickart rings in Sect. 3.1 and the notions of quasi-Baer and principally quasi-Baer rings in Sect. 3.2, respectively. Also in Sect. 3.2, we shall observe that there are close connections between the FI-extending and the quasi-Baer properties for rings.

A result of Chatters and Khuri shows that there are strong bonds between the extending and the Baer properties of rings (Theorem 3.3.1). We shall also see some instances where the two notions coincide.

G.F. Birkenmeier et al., *Extensions of Rings and Modules*,

DOI 10.1007/978-0-387-92716-9_3,

One of the motivations for the study of the quasi-Baer and principally quasi-Baer rings is the fact that they behave better with respect to various extensions than the Baer and Rickart rings. For example, as we shall see in this chapter, each of the quasi-Baer and the principally quasi-Baer properties is Morita invariant. This useful behavior will be applied in later chapters.

The results on the transference (or the lack of transference) of these properties to matrix and polynomial ring extensions included here are intended to motivate further investigations on when these properties transfer to various extensions. A detailed treatment of this topic will be included in Chaps. 5 and 6.

3.1 Baer and Rickart Rings

The focus of our discussion in this section is mainly on the properties of Baer and Rickart rings. We shall observe similarities and contrasts between these two classes of rings. While most of the basic material can be found in [45, 246], and [47], we introduce several new results. These results will also provide motivation for the study of quasi-Baer rings and principally quasi-Baer rings treated in Sect. 3.2.

Proposition 3.1.1 *Let R be a ring not necessarily with identity. Then any two of the following conditions imply the third condition.*

(i) *For each $\emptyset \neq X \subseteq R$, there exists $e^2 = e \in R$ such that $r_R(X) = eR$.*
(ii) *For each $\emptyset \neq Y \subseteq R$, there exists $f^2 = f \in R$ such that $\ell_R(Y) = Rf$.*
(iii) *R has an identity.*

Proof Assume that (i) and (ii) hold. Note that $r_R(0) = R$. By (i), $R = eR$ with $e^2 = e \in R$. Thus, e is a left identity. Similarly (ii) implies that R has a right identity. Thus R has an identity and hence (iii) holds.

Suppose that (ii) and (iii) hold. Take $\emptyset \neq X \subseteq R$. First, we observe that $r_R(X) = r_R(\ell_R(r_R(X)))$. Say $A = \ell_R(r_R(X))$. By (ii), $A = Rf$ for some $f^2 = f \in R$. Hence, $r_R(X) = r_R(A) = (1 - f)R$ by (iii) as R has an identity. Put $e = 1 - f$. Then $e^2 = e \in R$ and $r_R(X) = eR$. So (ii) and (iii) imply (i). Similarly, (i) and (iii) imply (ii). □

Proposition 3.1.1 suggests the following definition.

Definition 3.1.2 A ring R is called *Baer* if R satisfies any two (hence, all three) of the conditions of Proposition 3.1.1.

By Proposition 3.1.1, any Baer ring always has an identity, and the condition for a ring to be Baer is left-right symmetric. The next result, essentially due to the work of Baer [36], has been an important model for defining a Baer ring by Kaplansky.

Theorem 3.1.3 *The endomorphism ring of a semisimple module is Baer.*

Proof Assume that M_R is a semisimple R-module and let $S = \text{End}_R(M)$. Take $\emptyset \neq X \subseteq S$ and put $X = \{\varphi_\alpha\}_{\alpha \in \Lambda}$. Let $U = \sum_{\alpha \in \Lambda} \varphi_\alpha(M)$. Then $M = U \oplus W$ for some $W_R \leq M_R$. Let e be the canonical projection from M onto W. We claim that $\ell_S(X) = Se$. For this, note that $e\varphi_\alpha(M) = 0$ for each $\alpha \in \Lambda$. So $e\varphi_\alpha = 0$ for every $\alpha \in \Lambda$. Thus, $e \in \ell_S(X)$, so $Se \subseteq \ell_S(X)$.

Next, let $g \in \ell_S(X)$. Then $g\varphi_\alpha = 0$ for each $\alpha \in \Lambda$. Take $m \in M$. Then $m = u + w$ with $u \in U$ and $w \in W$. Hence, $g(m) = g(u) + g(w) = g(w) = ge(m)$, so $(g - ge)(m) = 0$ for each $m \in M$. Thus, $g = ge \in Se$. Hence, $\ell_S(X) \subseteq Se$. Therefore, $\ell_S(X) = Se$ and $e^2 = e \in S$. So S is a Baer ring. □

Example 3.1.4 (i) The endomorphism ring of a vector space over a field is Baer (see Theorem 3.1.3).

(ii) A ring R is a domain if and only if R is Baer and 1 is a primitive idempotent.

(iii) Every von Neumann algebra is a Baer ring.

(iv) Every orthogonally finite right semihereditary ring is Baer (see Theorem 3.1.25). In particular, every right Noetherian right hereditary ring is a Baer ring.

(v) Any right extending right nonsingular ring is Baer (see Theorem 3.3.1). In particular, any right self-injective regular ring is Baer.

Proposition 3.1.5 (i) *Let $\{R_i \mid i \in \Lambda\}$ be a set of rings and $R = \prod_{i \in \Lambda} R_i$. Then R is Baer if and only if each R_i is Baer.*

(ii) *If R is a subring of a Baer ring S and R contains all idempotents of S, then R is a Baer ring.*

Proof (i) The proof is straightforward.

(ii) Let $\emptyset \neq X \subseteq R$. Then we see that $r_S(X) = eS$, where $e^2 = e \in S$. Hence, $r_R(X) = eS \cap R = eR$. Therefore, R is a Baer ring. □

A subring R of $\prod_{i \in \Lambda} S_i$ is called a *subdirect product* of rings $S_i, i \in \Lambda$, if $S_i \cong R/K_i$, where each K_i is an ideal of R and $\cap_{i \in \Lambda} K_i = 0$. Note that any semiprime ring is a subdirect product of its prime factor rings.

Example 3.1.6 In contrast to Proposition 3.1.5(i), a subdirect product of Baer rings need not be Baer, in general. Let $R = \mathbb{Z}[C_2]$ be the group ring, where $C_2 = \{1, g\}$ is the group of order 2. Then R has only trivial idempotents. Because $(1 + g)(1 - g) = 0$, R is not Baer (see also Example 6.3.11). But as R is commutative semiprime, it is a subdirect product of commutative domains.

The next example illustrates how Proposition 3.1.5 can be used to check when a subring of a Baer ring will also be Baer.

Example 3.1.7 Let A be a domain and B a subring of A. Set $A_n = A$ for every $n = 1, 2 \dots$. Take $R = \{(a_n)_{n=1}^{\infty} \in \prod_{n=1}^{\infty} A_n \mid a_n \in B \text{ eventually}\}$, a subring of $\prod_{n=1}^{\infty} A_n$. From Proposition 3.1.5(i), $\prod_{n=1}^{\infty} A_n$ is a Baer ring. Also, R contains all idempotents of $\prod_{n=1}^{\infty} A_n$, so R is a Baer ring by Proposition 3.1.5(ii).

Theorem 3.1.8 *A ring R is Baer if and only if eRe is Baer for every idempotent e of R.*

Proof Assume that R is a Baer ring and let $\emptyset \neq X \subseteq eRe$. Then $r_R(X) = fR$ with $f^2 = f \in R$. We show that $r_{eRe}(X) = r_R(X) \cap eRe = fR \cap eRe = efeRe$. Indeed, $X(efeRe) = XfeRe = 0$ because $X \subseteq eRe$ and $Xf = 0$. Hence $efeRe \subseteq r_R(X) = fR$. So $efeRe \subseteq fR \cap eRe$. Next, take $y \in fR \cap eRe$. Then $y = fy$ and $y = eye$, and so $y = eye = efye = efeye \in efeRe$. Whence $r_{eRe}(X) = fR \cap eRe = efeRe$. As $X(ef - fe) = 0$, $ef - fe \in r_R(X) = fR$. Thus, $ef - fe = f(ef - fe) = fef - fe$, so $ef = fef$. Hence $(efe)^2 = efe \in eRe$. Therefore, the ring eRe is a Baer ring. The converse is clear by taking $e = 1$. □

The next example of a ring R shows that even when eRe is Baer for all nonidentity idempotents e, R itself may not be Baer.

Example 3.1.9 Let $R = T_2(\mathbb{Z})$. Then $eRe \cong \mathbb{Z}$ for any nontrivial idempotent e, is a domain and hence eRe is Baer, but R is not Baer.

Remark 3.1.10 Rizvi and Roman [357] extended the notion of a Baer ring to a Baer module by using the endomorphism ring of a module. Theorem 3.1.8 served as one of the motivations for initial results on Baer modules. We shall discuss Baer modules and their applications in Chap. 4. Among other results, Theorem 3.1.8 can also be proved using module theoretic methods. (Indeed, if a ring R is Baer, then eR_R is a Baer module for each $e^2 = e \in R$ by Theorem 4.1.22. Hence, the ring $eRe = \text{End}(eR_R)$ is Baer by Theorem 4.2.8.)

Theorem 3.1.11 *Let R be a Baer ring with only countably many idempotents. Then R is orthogonally finite. Additionally, if R is regular, then R is semisimple Artinian.*

Proof Assume on the contrary that R has an infinite set $E = \{e_n\}_{n=1}^{\infty}$ of nonzero orthogonal idempotents. Let U and V be nonempty distinct subsets of E. As R is a Baer ring, $r_R(U) = f_u R$ and $r_R(V) = f_v R$ for some idempotents f_u and f_v of R. We claim that $f_u \neq f_v$. For this, suppose that $f_u = f_v$. Then $r_R(U) = r_R(V) = f_u R$. Now there exists $e_i \in V$ and $e_i \notin U$. Observe that $e_i \in r_R(U) = f_u R$, so $e_i = f_u e_i$. Since $e_i \in V$ and $r_R(V) = f_u R$, $e_i f_u = 0$ and hence $e_i = e_i^2 = e_i f_u e_i = 0$, a contradiction. So $f_u \neq f_v$. Thus R has uncountably many idempotents, also a contradiction. Hence, R is orthogonally finite. Additionally, assume that R is regular. Since R is orthogonally finite, R has a complete set of primitive idempotents by Proposition 1.2.15. Thus R is semisimple Artinian. □

Corollary 3.1.12 *Let R be a regular Baer ring with only countably many idempotents. If R is an algebra over an uncountable field, then R is a finite direct sum of division rings.*

Proof By Theorem 3.1.11, R is semisimple Artinian, so $R = \oplus_{\ell=1}^{k} \text{Mat}_{n_\ell}(D_\ell)$ for some positive integers n_ℓ and division rings D_ℓ, $1 \leq \ell \leq k$. Suppose that $n_1 > 1$. Let e_{ij} be the matrix in $\text{Mat}_{n_1}(D_1)$ with 1 in the (i, j)-position and 0 elsewhere. As D_1 is uncountable, $\{e_{11} + xe_{12} \mid x \in D_1\}$ is an uncountable set of idempotents in R, a contradiction. So $n_1 = 1$. Similarly, each $n_i = 1$. Hence, $R = D_1 \oplus \cdots \oplus D_k$. □

A structural property for a π-regular Baer ring with only countably many idempotents will appear in Corollary 5.4.18. The next result, due to Rangaswamy [350], follows from Theorem 3.1.11.

Corollary 3.1.13 *A countable regular Baer ring is semisimple Artinian.*

We shall later see that every right nonsingular right extending ring is Baer (Theorem 3.3.1). Therefore, Corollary 3.1.13 yields that every countable regular right extending ring is semisimple Artinian. The next example shows that being Baer and being regular are independent notions for a ring.

Example 3.1.14 (i) Let R be a domain which is not a division ring. Then R is Baer, but not regular.

(ii) Assume that F is a field and take $F_n = F$, for $n = 1, 2, \ldots$. Consider $R = \{(a_n)_{n=1}^{\infty} \in \prod_{n=1}^{\infty} F_n \mid a_n$ is constant eventually$\}$, which is a subring of $\prod_{n=1}^{\infty} F_n$. Take $x_1 = (1, 0, 0, \ldots)$, $x_3 = (0, 0, 1, 0, \ldots)$, and so on. We let $X = \{x_1, x_3, \ldots\}$. Then there is no $e^2 = e \in R$ such that $r_R(X) = eR$. Thus R is not Baer. But observe that R is regular.

Definition 3.1.15 A ring R is called *right Rickart* if the right annihilator of any element is generated by an idempotent. A left Rickart ring is defined similarly. A ring which is right and left Rickart is called *Rickart*.

One can easily see that the class of right Rickart rings is a generalization of the classes of Baer rings and regular rings. Similar to the case of Baer rings, a ring R is a domain if and only if R is right Rickart and 1 is a primitive idempotent (cf. Example 3.1.4(ii)). The next example provides a Rickart ring that is neither Baer nor regular.

Example 3.1.16 Let $A_n = \mathbb{Z}$ for $n = 1, 2, \ldots$, and put

$$R = \{(a_n)_{n=1}^{\infty} \in \prod_{n=1}^{\infty} A_n \mid a_n \text{ is constant eventually}\},$$

a subring of $\prod_{n=1}^{\infty} A_n$. Then R is a Rickart ring. Indeed, say $\alpha = (a_n)_{n=1}^{\infty} \in R$. Let $e = (e_n)_{n=1}^{\infty}$, where $e_n = 1$ if $a_n = 0$ and $e_n = 0$ if $a_n \neq 0$. Since a_n is constant eventually, so is e_n. We see that $e^2 = e \in R$ and $r_R(\alpha) = eR$. But R is neither Baer (similar to the case of Example 3.1.14(ii)) nor regular.

Proposition 3.1.17 *A ring R is right Rickart if and only if aR is a projective right R-module for each $a \in R$.*

Proof Assume that R is a right Rickart ring and let $a \in R$. Then $r_R(a) = eR$ with $e^2 = e \in R$. Thus, $(1-e)R \cong R/r_R(a) \cong aR$ as right R-modules, so aR_R is projective. Conversely, suppose that every principal right ideal of R is projective. Take $x \in R$. Then the homomorphism $\theta : R_R \to xR_R$, defined by $\theta(r) = xr$, splits because xR_R is projective. So $\text{Ker}(\theta) = r_R(x) = fR$ for some $f^2 = f \in R$. Thus, R is right Rickart. □

Because of Proposition 3.1.17, right (left) Rickart rings are also called *right* (*left*) *PP* rings. A ring R is called *PP* if it is both right and left PP.

Proposition 3.1.18 *Every right Rickart ring is right nonsingular. Therefore, every Baer ring is right and left nonsingular.*

Proof Let R be a right Rickart ring and take $a \in Z(R_R)$. Then $r_R(a) = eR$ with $e^2 = e \in R$ and $eR_R \leq^{\text{ess}} R_R$. Thus $e = 1$, so $r_R(a) = R$. Hence $a = 0$, therefore $Z(R_R) = 0$. □

The following example presents a right and left nonsingular ring which is neither right nor left Rickart.

Example 3.1.19 The ring $R = T_2(\mathbb{Z})$ is right and left nonsingular. Let $e_{ij} \in R$ be the matrix with 1 in the (i, j)-position and 0 elsewhere. Then $r_R(2e_{11} + e_{12})$ is not generated by an idempotent, so R is not right Rickart. It can also be checked that R is not left Rickart.

Recall from 1.1.13 that a ring R is called right (semi)hereditary if every (finitely generated) right ideal of R is projective. A left (semi)hereditary ring is defined similarly. Right semihereditary rings are right Rickart. The hereditary, semihereditary, and Rickart notions need not be left-right symmetric as shown in the next example.

Example 3.1.20 There is a ring R such that R is left hereditary (hence left Rickart), but R is not right Rickart. Take $F_n = \mathbb{Z}_2$ for $n = 1, 2, \ldots$.

We let $A = \{(a_n)_{n=1}^{\infty} \in \prod_{n=1}^{\infty} F_n \mid a_n \text{ is constant eventually}\}$, which is a subring of $\prod_{n=1}^{\infty} F_n$, and let $I = \{(a_n)_{n=1}^{\infty} \in \prod_{n=1}^{\infty} F_n \mid a_n = 0 \text{ eventually}\}$, which is an ideal of A.

Put $S = A/I$ and

$$R = \begin{bmatrix} S & S \\ 0 & A \end{bmatrix}.$$

Then R is left hereditary (see [120, Example 8.2]). Hence, R is a left Rickart ring. Say $\alpha = \begin{bmatrix} 0 & \mathbf{1}_S \\ 0 & 0 \end{bmatrix} \in R$, where $\mathbf{1}_S$ is the identity of S. Then $r_R(\alpha)$ is not generated by an idempotent of R. Therefore, R is not right Rickart.

Proposition 3.1.21 *Let $\{R_i \mid i \in \Lambda\}$ be a set of rings and $R = \prod_{i\in\Lambda} R_i$. Then R is right Rickart if and only if each R_i is right Rickart.*

Proof The proof is routine. □

In contrast to Proposition 3.1.21, Example 3.1.6 also shows that a subdirect product of right Rickart rings need not be right Rickart. Similar to the case of Baer rings in Theorem 3.1.8, we obtain the following for Rickart rings.

Theorem 3.1.22 (i) *If R is a right Rickart ring, then eRe is a right Rickart ring for every $e^2 = e \in R$.*

(ii) *The center of a right Rickart ring is a Rickart ring.*

Proof (i) Assume that R is a right Rickart ring and $e \in R$ is an idempotent. Similar to the proof of Theorem 3.1.8, eRe is a right Rickart ring.

(ii) Let R be a right Rickart ring and put $C = \text{Cen}(R)$, the center of R. Say $a \in C$. Then $r_R(a) = eR$ with $e^2 = e \in R$. Since $a \in C$, $r_R(a) = eR$ is an ideal, hence $e \in \mathbf{S}_\ell(R)$ from Proposition 1.2.2.

Say $z \in R$. Then $r_R(ez - ze) = fR$ with $f^2 = f \in R$. Since $(ez - ze)a = 0$, $a \in fR$ and thus $a = fa$. Hence $(1 - f)a = a(1 - f) = 0$, so $1 - f \in r_R(a) = eR$. Therefore, $(1 - e)(1 - f) = 0$. Because $e \in \mathbf{S}_\ell(R)$, $ze = eze$ and we have that $(ez - ze)(1 - f) = (ez - eze)(1 - f) = ez(1 - e)(1 - f) = 0$. Therefore, we see that $1 - f \in r_R(ez - ze) = fR$, thus $1 - f = 0$, and so $f = 1$. Hence, $r_R(ez - ze) = fR = R$, therefore $ez - ze = 0$ and $e \in C$. From $r_R(a) = eR$, $r_C(a) = eC$ with $e \in C$. Consequently, C is a Rickart ring. □

The module theoretic methods also yield an alternate proof of Theorem 3.1.22(i) (see Proposition 4.5.4(i) and (v)). The next result provides conditions on a Rickart ring which ensure that the ring is Baer.

Theorem 3.1.23 *Let R be a ring and let $\mathcal{R} = \{eR \mid e^2 = e \in R\}$. Then the following are equivalent.*

(i) *R is a Baer ring.*

(ii) *R is Rickart and $\mathcal{R}$ is a complete lattice under inclusion.*

Proof (i)⇒(ii) Assume that R is a Baer ring. Then R is a Rickart ring.

Consider a subset $\{e_i R \mid i \in \Lambda\}$ of $\mathcal{R}$. As R is Baer, $r_R(\ell_R(\sum e_i R)) = hR$ for some $h^2 = h \in R$. Note that $e_i R \subseteq hR$ for each i. Next, let $g^2 = g \in R$ such that $e_i R \subseteq gR$ for all i. Then $\sum e_i R \subseteq gR$, so $hR = r_R(\ell_R(\sum e_i R)) \subseteq gR$. Hence $hR = \sup\{e_i R \mid i \in \Lambda\}$. As $\mathcal{R}$ is a partially ordered set under inclusion, $\mathcal{R}$ is a complete lattice under inclusion by [382, Proposition 1.2, p. 64].

(ii)⇒(i) Let $\emptyset \neq X = \{x_i \mid i \in \Lambda\} \subseteq R$. As R is Rickart, there exists $e_i^2 = e_i \in R$ with $r_R(x_i) = e_i R$. Then $r_R(X) = \cap_{i \in \Lambda} r_R(x_i) = \cap_{i \in \Lambda} e_i R$. Since $\mathcal{R}$ is complete, there exists $e^2 = e \in R$ such that $eR = \inf\{e_i R \mid i \in \Lambda\}$.

We show that $eR = \cap_{i \in \Lambda} e_i R$. As $eR \subseteq e_i R$ for each i, $eR \subseteq \cap_{i \in \Lambda} e_i R$. Next, take $x \in \cap_{i \in \Lambda} e_i R$. Then $\ell_R(x) = Rh$ for some $h^2 = h$ since R is Rickart. Therefore, $r_R(\ell_R(x)) = (1-h)R \in \mathcal{R}$. As $x \in e_i R$, $r_R(\ell_R(x)) \subseteq r_R(\ell_R(e_i R)) = e_i R$. So $r_R(\ell_R(x)) \subseteq e_i R$ for each $i \in \Lambda$. As $r_R(\ell_R(x)) \in \mathcal{R}$ and $eR = \inf\{e_i R \mid i \in \Lambda\}$, we get $r_R(\ell_R(x)) \subseteq eR$.

Therefore $x \in r_R(\ell_R(x)) \subseteq eR$, and hence $\cap_{i \in \Lambda} e_i R \subseteq eR$. Consequently, we have that $eR = \cap_{i \in \Lambda} e_i R = r_R(X)$. Whence R is a Baer ring. □

We remark that the proof (ii)⇒(i) of Theorem 3.1.23 shows that R_R satisfies the SSIP (see Definition 4.1.20 and Theorem 4.1.21 for the SSIP). Completeness under inclusion of the lattice of principal right ideals for a regular ring yields that it is Baer as shown the following corollary.

Corollary 3.1.24 *Let R be a regular ring. Then the following are equivalent.*

(i) *R is Baer.*
(ii) *The set of principal right ideals of R forms a complete lattice under inclusion.*

Proof The result follows immediately from Theorem 3.1.23. □

The next result (due to Small [372]) shows that when a ring is orthogonally finite, the notion of a right Rickart ring coincides with that of a Baer ring.

Theorem 3.1.25 *Any orthogonally finite right Rickart ring is Baer.*

Proof Let R be an orthogonally finite right Rickart ring. Say L is a nonzero left annihilator, write $L = \ell_R(X)$ with $\emptyset \neq X \subseteq R$. Take $0 \neq s \in \ell_R(X)$. Then $r_R(\ell_R(X)) \subseteq r_R(s) = gR$ for some $g^2 = g \in R$. Hence, we have that

$$R(1-g) = \ell_R(gR) = \ell_R(r_R(s)) \subseteq \ell_R(r_R(\ell_R(X))) = \ell_R(X) = L.$$

If $g = 1$, then $r_R(s) = gR = R$, so $s = 0$, a contradiction. Hence, L contains a nonzero idempotent $1-g$.

As R is orthogonally finite, we can choose an idempotent $e \in L$ with $\ell_R(e)$ minimal in $\{\ell_R(h) \mid h^2 = h \in L\}$ by Proposition 1.2.13. We claim that $\ell_R(e) \cap L = 0$. Assume on the contrary that $\ell_R(e) \cap L \neq 0$. As $\ell_R(e) \cap L$ is a left annihilator, $\ell_R(e) \cap L$ contains $f^2 = f \neq 0$ from the preceding argument. Let $b = e + f - ef$.

Then $b \in L$ and $b^2 = b$ since $fe = 0$. Note that $bR = eR + fR$, so $eR \subseteq bR$ and hence $\ell_R(b) \subseteq \ell_R(e)$. But, $fe = 0$ and $fb = f \neq 0$. Thus $\ell_R(b) \subsetneq \ell_R(e)$ and $b \in L$, a contradiction to the choice of e. So $\ell_R(e) \cap L = 0$. If $x \in L$, then $(x - xe)e = 0$ and $x - xe \in L$. Hence $x - xe \in \ell_R(e) \cap L = 0$, so $x = xe \in Re$. Thus $L = Re$. So R is Baer. □

For the case of right p.q.-Baer rings, it will be shown that an orthogonally finite right p.q.-Baer ring is quasi-Baer (see Propositions 5.2.13 and 5.4.5). Using the endomorphism ring of a module, the notion of a Rickart module will be introduced and discussed in Sect. 4.5. Theorem 3.1.25 will be extended to Theorem 4.5.13 in a general module theoretical setting to Rickart modules.

In the next theorem, we consider an orthogonally finite right Rickart (hence Baer by Theorem 3.1.25) ring which is an I-ring.

Theorem 3.1.26 *Let R be an orthogonally finite right Rickart ring. Then the following are equivalent.*

(i) *R is an I-ring.*
(ii) *R is a semiprimary ring.*
(iii) *R is a strongly π-regular ring.*
(iv) *R is a π-regular ring.*

Proof (i)⇒(ii) Let R be an I-ring. By Theorem 1.2.16, R is semilocal because R is orthogonally finite. To show that R is semiprimary, we need to prove that $J(R)$ is nilpotent. Write $R = \sum_{i=1}^{n} e_i R$, where $\{e_1, e_2, \dots, e_n\}$ is a complete set of primitive idempotents (recall that R has a complete set of primitive idempotents from Proposition 1.2.15 since R is orthogonally finite). Hence $J(R) = e_1 J(R) + \cdots + e_n J(R)$.

We claim that each $e_i J(R)$ is nilpotent. Assume on the contrary that there exists some $e_k J(R)$ which is not nilpotent. Then $(e_k J(R))^2 \neq 0$, hence $e_k x e_k \neq 0$ for some $x \in J(R)$. Consider the map $\theta : e_k R \to e_k R$ defined by $\theta(e_k r) = e_k x e_k r$. As R is right Rickart, Image$(\theta) = e_k x e_k R$ is a projective right R-module (see Proposition 3.1.17). Hence, Ker(θ) is a direct summand of $e_k R_R$. Since $e_k R_R$ is indecomposable, Ker$(\theta) = 0$ or Ker$(\theta) = e_k R$. But $e_k x e_k \neq 0$, so $\theta \neq 0$. Hence, Ker$(\theta) = 0$. Note that $J(R)$ is nil (as R is an I-ring) and $e_k x \in J(R)$. So there is a positive integer m such that $(e_k x)^m = 0$ and $(e_k x)^{m-1} \neq 0$. But $(e_k x)^{m-1} \in$ Ker$(\theta) = 0$, a contradiction. Hence each $e_i J(R)$ is nilpotent. Thus, we see that $e_1 J(R) + e_2 J(R)$ is nilpotent by direct computation. Since e_1 and e_2 are orthogonal,

$$e_1 J(R) + e_2 J(R) = (e_1 + e_2) J(R),$$

and hence $e_1 J(R) + e_2 J(R) + e_3 J(R) = (e_1 + e_2) J(R) + e_3 J(R)$ is nilpotent by the same method, since $(e_1 + e_2) J(R)$ and $e_3 J(R)$ are nilpotent. Inductively, $J(R) = e_1 J(R) + \cdots + e_n J(R)$ is nilpotent.

(ii)⇒(iii) Let R be semiprimary. Say P is a prime ideal of R. Then R/P is simple Artinian, so R/P is strongly π-regular for each prime ideal P of R. Therefore, R is strongly π-regular by Theorem 1.2.18.

(iii)⇒(iv) It is obvious from Theorem 1.2.17.

(iv)⇒(i) From [221, Proposition 1(1), p. 210], every π-regular ring is an I-ring. □

The following example provides a Baer (hence right Rickart) ring which is an I-ring, but is not π-regular. Thereby, the condition that "R be orthogonally finite" in Theorem 3.1.26 is not superfluous.

Example 3.1.27 Let D be a commutative domain which is not a field, and let F be the field of fractions of D. Put $F_n = F$ for $n = 1, 2, \ldots$. Now consider $R = \{(a_n)_{n=1}^{\infty} \in \prod_{n=1}^{\infty} F_n \mid a_n \in D \text{ eventually}\}$, a subring of $\prod_{n=1}^{\infty} F_n$. Then $\prod_{n=1}^{\infty} F_n$ is Baer. Also R contains all idempotents of $\prod_{n=1}^{\infty} F_n$. Hence, R is a Baer ring by Proposition 3.1.5(ii). So R is a right Rickart ring.

Next to see that R is an I-ring, let V be a nonnil (right) ideal of R. Then there is $0 \neq x \in V$ with a nonzero k-th coordinate, say x_k for some k. Let $y \in \prod_{n=1}^{\infty} F_n$ with the k-th coordinate x_k^{-1} and 0 for all other coordinates. Then $y \in R$ and $xy \in V$ is a nonzero idempotent. Hence, R is an I-ring. Finally, let $0 \neq a \in D$ which is not invertible in D, and put $\alpha = (a, a, \ldots) \in R$. If R is π-regular, then there exist a positive integer n and an element $\beta \in R$ such that $\alpha^n = \alpha^n \beta \alpha^n$. Hence, there is $b \in D$ so that $a^n = a^n b a^n$. Thus, a is invertible in D, a contradiction. Whence R is not π-regular.

From Example 3.1.19, we have already seen that, the Baer and Rickart properties of a ring R are not inherited by triangular matrix rings over R (note that $\mathbb{Z}$ is a Baer ring while $T_2(\mathbb{Z})$ is not even right or left Rickart). Similarly, the Baer and Rickart properties of a ring R are not inherited by matrix rings and by polynomial rings over R, as shown by the next example.

Example 3.1.28 Let $S = \mathrm{Mat}_2(\mathbb{Z}[x])$. Then obviously $\mathbb{Z}[x]$ is Baer. But S is not Baer. Say $\alpha = \begin{bmatrix} 2 & 0 \\ x & 0 \end{bmatrix} \in S$. P.M. Cohn (see [239]) has shown that $\ell_S(\alpha)$ is not generated by an idempotent in S. Thus, the Baer ring property cannot transfer to the matrix ring from the base ring. Further, the right (left) Rickart property of a ring also cannot transfer to the matrix ring from the base ring. The ring $\mathrm{Mat}_2(\mathbb{Z})$ is Baer. But, $S = \mathrm{Mat}_2(\mathbb{Z})[x] = \mathrm{Mat}_2(\mathbb{Z}[x])$ is not Baer (also neither right nor left Rickart by Theorem 3.1.25). So in general the Baer ring property and the right (left) Rickart ring property of the base ring cannot transfer to the polynomial ring extension.

The following result shows that for $\mathrm{Mat}_n(R)$ to be right Rickart, we need stronger conditions on the base ring R than just being right Rickart.

Theorem 3.1.29 *A ring R is right semihereditary if and only if* $\mathrm{Mat}_n(R)$ *is right Rickart for all positive integers n.*

Proof If R is right semihereditary, then so is $\mathrm{Mat}_n(R)$ for every positive integer n because the right semihereditary property is Morita invariant.

Conversely, let $I = a_1R + \cdots + a_nR$ with $a_i \in R$. Put $\alpha = [c_{ij}] \in \mathrm{Mat}_n(R)$ where $c_{1i} = a_i$ for $i = 1, \ldots, n$ and all other entries of α are 0. Then $\alpha\mathrm{Mat}_n(R)$ is a projective right $\mathrm{Mat}_n(R)$-module by Proposition 3.1.17. So $\alpha\mathrm{Mat}_n(R)$ is a projective right R-module because $\mathrm{Mat}_n(R)_R$ is a free right R-module. As right R-modules, $\alpha\mathrm{Mat}_n(R)$ is isomorphic to $I^{(n)}$. So I_R is projective. Hence, R is right semihereditary. □

The topic of the transference of the Baer, Rickart, and other related properties to various matrix and polynomial ring extensions will be dealt with further in Chap. 6.

Exercise 3.1.30

1. Prove that the central idempotents of a Baer ring form a complete Boolean algebra.
2. ([148, Endo]) Let a ring R be Abelian. Prove that R is right Rickart if and only if R is left Rickart.
3. Show that a ring R is right Rickart if and only if for any nonempty finite set F of R there exists $e^2 = e \in R$ with $r_R(F) = eR$ (see also Proposition 4.5.4(iv)).
4. A ring R is called *compressible* if $\mathrm{Cen}(eRe) = e\mathrm{Cen}(R)$ for each idempotent $e \in R$. Not every Baer ring is compressible. Let
$$R = \begin{bmatrix} \mathbb{C} & \mathbb{H} \\ 0 & \mathbb{H} \end{bmatrix},$$
where $\mathbb{H}$ is the division ring of real quaternions. Prove that R is a Baer ring, but not compressible. (This example is due to Armendariz [46].)
5. ([231, Jeremy]) Show that a right self-injective regular ring is compressible.
6. ([27, Armendariz, Koo, and Park]) Let $F[G]$ be a semiprime group algebra of a group G over a field F. Prove that $\mathrm{Mat}_n(F[G])$ is compressible for every positive integer n.
7. ([59, Birkenmeier]) Let R be a ring and $N_{\mathbf{I}(R)}(R) = \cup_{e \in \mathbf{I}(R)} eR(1-e)$, where $\mathbf{I}(R)$ is the set of all idempotents of R. Assume that R is right Rickart. Show that the subring (not necessarily with identity) generated by $N_{\mathbf{I}(R)}(R)$ is the ideal generated by the set of all nilpotent elements of R.

3.2 Quasi-Baer and Principally Quasi-Baer Rings

As shown in Example 3.1.28, the Baer ring property and the Rickart ring property do not transfer from a ring R to two of its important ring extensions, namely, the matrix rings and the polynomial rings over R. Thus, neither the Baer ring property nor the Rickart ring property is Morita invariant. The difficulties in these cases motivate the need to study classes of rings for which such transfers can take place easily—even under somewhat weaker conditions. This brings us to the notions of quasi-Baer (resp., principally quasi-Baer) rings where one studies a "generalized" Baer property in which the annihilators of ideals (resp., principal ideals) instead of nonempty subsets of the rings are generated by idempotents.

In this section, we introduce quasi-Baer rings and their basic properties. Two remarkable results show that the quasi-Baer ring property can transfer to matrix rings (Theorem 3.2.12) and to polynomial ring extensions (as shown in Theorem 6.2.4) *without any additional requirements*. It is shown that the quasi-Baer ring property is Morita invariant. These results stimulate further investigations on the transference of the quasi-Baer property for various other types of ring extensions which will be presented in Chap. 6. We will see that there are strong connections between the quasi-Baer and the FI-extending properties for rings.

Right principally quasi-Baer rings generalize the class of quasi-Baer rings analogous to the way right Rickart rings generalize the class of Baer rings. The right principally quasi-Baer property unifies the quasi-Baer property and the biregular property of a ring into one concept. It is shown that this property is also Morita invariant among other included results. We shall see that the right principally quasi-Baer property and the right Rickart property are independent (Examples 3.2.28 and 3.2.31). Connections between quasi-Baer rings and biregular rings are presented.

Among applications, the results on quasi-Baer rings will be used to establish the existence of the quasi-Baer ring hull of a semiprime ring in Chap. 8. We shall then use the quasi-Baer ring hulls to investigate boundedly centrally closed C^*-algebras and extended centroids of C^*-algebras in Chap. 10.

Let S be a semigroup with zero, F a field, and let $\varphi : S \times S \to F$ satisfy the following.

(i) $\varphi(s,t) = 0$ if and only if $st = 0$.
(ii) $\varphi(r,st)\varphi(s,t) = \varphi(rs,t)\varphi(r,s)$ whenever $rst \neq 0$.

Further, let $F_\varphi[S]$ denote the vector space of all formal finite linear combinations $\sum \alpha_i s_i$, where $\alpha_i \in F$, $s_i \in S$, and $s_i \neq 0$ for each i. Define a multiplication by $s \cdot t = \varphi(s,t)st$ for s and t nonzero elements of S, and extend this multiplication linearly to all of $F_\varphi[S]$. Then we see that (ii) above is exactly what is required to ensure the associativity of $F_\varphi[S]$. In this case, $F_\varphi[S]$ is called a *twisted semigroup algebra* of S over F. If S has no zero, $F_\varphi[S]$ is defined similarly with the obvious modification.

Say n is a positive integer and let $MU(n)$ denote the full semigroup of matrix units $\{e_{ij} \mid 1 \leq i, j \leq n\} \cup \{0\}$, where $e_{hi}e_{jk} = \delta_{ij}e_{hk}$ and δ_{ij} is the Kronecker delta. By a *matrix units semigroup* is meant a subsemigroup of $MU(n)$ which contains $e_{11}, \ldots, e_{nn}$.

In the next result, due to Clark [128], a finite dimensional algebra over an algebraically closed field which is a twisted semigroup algebra is characterized.

Theorem 3.2.1 *Let R be a finite dimensional algebra over an algebraically closed field F. Then the following are equivalent.*

(i) $R \cong F_\varphi[S]$ *for some matrix units semigroup S.*

(ii) *The left annihilator of every ideal of R is generated by an idempotent and R has a finite ideal lattice.*

Furthermore, every finite distributive lattice is isomorphic to a certain sublattice of the lattice of all ideals of an Artinian ring satisfying condition (ii) of Theorem 3.2.1 as follows.

Theorem 3.2.2 *Let L be a finite distributive lattice. Then there exists an Artinian ring R such that:*

(i) *the left annihilator of any ideal of R is generated by an idempotent;*
(ii) *the lattice L is isomorphic to the sublattice $\{\ell_R(I) \mid {}_RI \trianglelefteq {}_RR\}$ of the lattice of all ideals of R.*

Therefore, the condition (ii) of Theorem 3.2.1 and the condition (i) of Theorem 3.2.2 motivate the following definition.

Definition 3.2.3 A ring R is called *quasi-Baer* if the left annihilator of every ideal of R is generated by an idempotent of R.

Proposition 3.2.4 *Let R be a ring. Then the following are equivalent.*

(i) *R is a quasi-Baer ring.*
(ii) *For each $I \trianglelefteq R$, there exists $e^2 = e \in R$ such that $r_R(I) = eR$.*

Proof The proof is routine (see also Exercise 3.2.44.1). □

From Proposition 3.2.4, the quasi-Baer condition is left-right symmetric. If R is a quasi-Baer ring and $I \trianglelefteq R$, then $r_R(I) = eR$ and $\ell_R(I) = Rf$ for some $e^2 = e \in R$ and $f^2 = f \in R$, respectively. We note that $eR \trianglelefteq R$ and $Rf \trianglelefteq R$, thereby $e \in \mathbf{S}_\ell(R)$ and $f \in \mathbf{S}_r(R)$ (see Proposition 1.2.2).

In the next result, prime rings are described in terms of quasi-Baer rings.

Proposition 3.2.5 *A ring R is prime if and only if R is quasi-Baer and semicentral reduced.*

Proof Assume that R is a prime ring. Say $e \in \mathbf{S}_\ell(R)$. Then $(1-e)Re = 0$ by Proposition 1.2.2. Thus $e = 1$ or $e = 0$, and so R is semicentral reduced. Take $I \trianglelefteq R$. If $I = 0$, then $r_R(I) = R$. If $I \neq 0$, then $r_R(I) = 0$ as R is prime. Thus R is quasi-Baer. Conversely, suppose that R is quasi-Baer and semicentral reduced. Say $I \trianglelefteq R$ and $J \trianglelefteq R$ with $IJ = 0$. Then $J \subseteq r_R(I) = fR$ for some $f^2 = f \in R$. As $fR \trianglelefteq R$, $f \in \mathbf{S}_\ell(R)$ from Proposition 1.2.2. Thus $f = 0$ or $f = 1$. If $f = 0$, then $J = 0$. If $f = 1$, then $I = 0$. So R is prime. □

Some examples of quasi-Baer rings are provided in the following.

Example 3.2.6 (i) Every Baer ring is a quasi-Baer ring.

(ii) If R is a quasi-Baer ring, then $\text{Mat}_n(R)$ is a quasi-Baer ring for every positive integer n. When R is a commutative domain, $\text{Mat}_n(R)$ is Baer for every positive integer n if and only if R is semihereditary (see Theorems 3.2.12 and 6.1.4).

(iii) If a ring R is quasi-Baer, then $T_n(R)$ is quasi-Baer for every positive integer n. When R is a commutative domain, $T_n(R)$ is Baer for every positive integer n if and only if R is a field (see Theorems 5.6.7 and 5.6.2).

(iv) The endomorphism ring of a projective (hence a free) module over a quasi-Baer ring is quasi-Baer (see Theorem 4.6.19).

(v) If a ring R is quasi-Baer, then $R[x]$ is quasi-Baer (see Theorem 6.2.4).

(vi) Any group algebra $F[G]$ of a polycyclic-by-finite group G over a field F with characteristic zero is quasi-Baer (see Corollary 6.3.4).

(vii) Every semiprime right FPF ring is quasi-Baer [157, p. 168] (a ring R is called right *FPF* if every faithful finitely generated right R-module generates the category Mod-R of right R-modules).

(viii) The local multiplier algebra $M_{\text{loc}}(A)$ of a C^*-algebra A is a quasi-Baer ring (see Theorem 10.3.10).

(ix) Any unital boundedly centrally closed C^*-algebra is a quasi-Baer ring (see Theorem 10.3.20).

By Proposition 3.1.18, a Baer ring is right and left nonsingular. Hence any prime ring which is not right nonsingular is quasi-Baer but not Baer. Such examples appear in [107, 267], and [329]. In Example 3.2.7(ii), there is a simple ring (hence quasi-Baer) which is not Baer.

Example 3.2.7 (i) Let $A = \mathbb{Z}_2\{x, y_0, y_1, y_2, \dots\}$ be the free algebra over $\mathbb{Z}_2$. A word w in A is a finite product of generators $w = x^{i_1} y_{j_2} x^{i_2} \cdots y_{j_n} x^{i_n}$, where $i_k, j_k \geq 0$, $n \geq 1$, and $x^0 = 1$. The length of the word w is defined by $\ell(w) = \sum_{k=1}^{n} i_k$, and the maximum subscript of w is defined by $m(w)$, where $m(w)$ is the largest subscript of y in w if $n \geq 2$, and $m(w) = 0$ if $n = 1$.

Let I be the ideal of A generated by all words w such that $m(w) > 0$ and $\ell(w) > m(w)k(m(w))$, where $k(m(w))$ is the number of times $y_{m(w)}$ appears in w. Put $R = A/I$. It is shown in [329] that R is a prime ring, but $Z(R_R) \neq 0$.

(ii) By Zalesskii and Neroslavskii, there is a simple Noetherian ring R with only trivial idempotents 0 and 1, which is not a domain. Clearly R is quasi-Baer. But R is not Baer (see [120, Example 14.17] and [182]).

Proposition 3.2.8 *Let $\{R_i \mid i \in \Lambda\}$ be a set of rings and $R = \prod_{i \in \Lambda} R_i$. Then R is quasi-Baer if and only if each R_i is quasi-Baer.*

Proof The proof is straightforward. □

A subdirect product of quasi-Baer rings is not quasi-Baer from Example 3.1.6. We remark that a quasi-Baer analogue of Proposition 3.1.5(ii) is something like: If R is a subring of a quasi-Baer ring S and R contains all left and all right semicentral idempotents of S, then R is quasi-Baer. This statement is not true as evident by the following example.

Example 3.2.9 For a field F, let

$$R = \begin{bmatrix} F\mathbf{1} & \text{Mat}_2(F) & \text{Mat}_2(F) \\ 0 & F\mathbf{1} & \text{Mat}_2(F) \\ 0 & 0 & F\mathbf{1} \end{bmatrix}$$

be a subring of $S := T_3(\text{Mat}_2(F))$, where $\mathbf{1}$ is the identity matrix in $\text{Mat}_2(F)$. The ring S is quasi-Baer (see Theorem 5.6.7). Furthermore, $\mathbf{S}_\ell(S) \subseteq R$, and $\mathbf{S}_r(S) \subseteq R$. So R contains all left semicentral and all right semicentral idempotents of S. But R is not quasi-Baer by direct computation (or Corollary 5.4.2). Thus, a quasi-Baer analogue of Proposition 3.1.5(ii) does not hold.

Theorem 3.2.10 *If R is a quasi-Baer ring, then eRe is a quasi-Baer ring for each $e^2 = e \in R$.*

Proof Let R be a quasi-Baer ring and $I \trianglelefteq eRe$. Then $\ell_R(RI) = Rf$ for some $f^2 = f \in R$. We first show that $\ell_{eRe}(I) = e\ell_R(RI)e$. Let $x \in \ell_{eRe}(I)$. Then $xRI = xeReI \subseteq xI = 0$. Thus, $x \in e\ell_R(RI)e$. Next, if $y \in e\ell_R(RI)e$, then $y = eue$ with $u \in \ell_R(RI)$ and so $yI = eueI \subseteq euRI = 0$. Hence, $y \in \ell_{eRe}(I)$. This shows that $\ell_{eRe}(I) = e\ell_R(RI)e = eRfe$. Because $Rf \trianglelefteq R$, $f \in \mathbf{S}_r(R)$ by Proposition 1.2.2, hence $(efe)^2 = efe$. Let $g = efe$. Then $eReg = eRe(efe) \subseteq eRfe$. Further, $eRfe = eRfefe = eRfe(efe) \subseteq eReg$. Therefore, $eReg = eRfe$. So $\ell_{eRe}(I) = eRfe = eReg = eRe(efe)$. Hence, eRe is quasi-Baer. □

As one of the motivations for the study of the quasi-Baer property, our next result shows that the quasi-Baer property is Morita invariant.

Theorem 3.2.11 *The endomorphism ring of a finitely generated projective module over a quasi-Baer ring is a quasi-Baer ring. In particular, the quasi-Baer ring property is Morita invariant.*

Proof Let P be a finitely generated projective right R-module over a quasi-Baer ring R. Then there exist a positive integer n and $e^2 = e \in \text{Mat}_n(R)$ such that $\text{End}(P) \cong e\text{Mat}_n(R)e$. Take $A \trianglelefteq \text{Mat}_n(R)$. Then $A = \text{Mat}_n(I)$ for some $I \trianglelefteq R$, and thus $r_R(I) = fR$ for some $f^2 = f \in R$. Say $\mathbf{1}$ is the identity matrix of $\text{Mat}_n(R)$. Then we see that $r_{\text{Mat}_n(R)}(A) = (f\mathbf{1})\text{Mat}_n(R)$ and $(f\mathbf{1})^2 = f\mathbf{1} \in \text{Mat}_n(R)$. So $\text{Mat}_n(R)$ is a quasi-Baer ring.

As $\text{End}(P) \cong e\text{Mat}_n(R)e$, $\text{End}(P)$ is quasi-Baer by Theorem 3.2.10. In particular, the quasi-Baer ring property is Morita invariant. □

Motivated by Theorems 3.2.10 and 3.2.11, the notion of a quasi-Baer module using the endomorphism ring of the module will be introduced and studied in Sect. 4.6.

As an application of this module theoretic notion, we shall see in Theorem 4.6.19 that every projective module over a quasi-Baer ring is a quasi-Baer module and its endomorphism ring is a quasi-Baer ring.

In contrast to Example 3.1.28 (cf. Theorems 6.1.3 and 6.1.4), the quasi-Baer property transfers from a ring R to its $n \times n$ matrix ring over R and vice versa without any additional requirements.

Theorem 3.2.12 *The following are equivalent for a ring R.*

(i) *R is a quasi-Baer ring.*
(ii) $\mathrm{Mat}_n(R)$ *is a quasi-Baer ring for every positive integer n.*
(iii) $\mathrm{Mat}_k(R)$ *is a quasi-Baer ring for some integer $k > 1$.*
(iv) $\mathrm{Mat}_2(R)$ *is a quasi-Baer ring.*

Proof (i)⇒(ii) Theorem 3.2.11 yields the implication.

(ii)⇒(iii) It is obvious.

(iii)⇒(iv) Let e_{ij} be the matrix in $\mathrm{Mat}_k(R)$ with 1 in the (i, j)-position and 0 elsewhere. Put $f = e_{11} + e_{22}$. Then $f^2 = f \in \mathrm{Mat}_k(R)$. Further, we get that $\mathrm{Mat}_2(R) \cong f\mathrm{Mat}_k(R)f$. By Theorem 3.2.10, $\mathrm{Mat}_2(R)$ is quasi-Baer.

(iv)⇒(i) Let $e \in \mathrm{Mat}_2(R)$ be the matrix with 1 in (1, 1)-position and 0 elsewhere. Then $e^2 = e$ and $e\mathrm{Mat}_2(R)e \cong R$. From Theorem 3.2.10, R is quasi-Baer. □

Theorem 3.2.13 *The center of a quasi-Baer ring is a Baer ring.*

Proof Let R be a quasi-Baer ring. Put $C = \mathrm{Cen}(R)$. Take $\emptyset \neq Y \subseteq C$. Then there exists $e \in \mathbf{S}_\ell(R)$ such that $r_R(Y) = r_R(YR) = eR$. We observe that

$$r_R(Y) = \ell_R(Y) = \ell_R(RY) = Rf \quad \text{for some} \quad f \in \mathbf{S}_r(R).$$

Because $eR = Rf$, $e = f \in \mathbf{S}_\ell(R) \cap \mathbf{S}_r(R) = \mathcal{B}(R)$ by Proposition 1.2.6(i). Therefore $e \in C$, so $r_C(Y) = r_R(Y) \cap C = eR \cap C = eC$. Hence $\mathrm{Cen}(R)$ is Baer. □

The following example provides a reduced ring whose center is a Baer ring, but the ring itself is not quasi-Baer.

Example 3.2.14 Let K be a field, x, y, z not necessarily commuting indeterminates and take $R = K[x, y, z]$ subject to $xy = xz = zx = yx = 0$ and $yz \neq zy$. Then R is a reduced ring with $\mathrm{Cen}(R) = K[x]$ and $K[x]$ is a Baer ring. However, $r_R(y) = Rx = K[x]x$, so $r_R(y)$ contains no nonzero idempotent. Thus, R is not a Baer ring. Thereby, R cannot be a quasi-Baer ring since otherwise R will be Baer because R is reduced.

Theorem 3.2.13 and Example 3.2.14 motivate us to discuss conditions under which a ring R is quasi-Baer if $\mathrm{Cen}(R)$ is Baer.

Definition 3.2.15 A ring R (not necessarily with identity) is said to satisfy a *polynomial identity* or simply, *PI*, if there is a polynomial $f(x_1, \ldots, x_n)$ with integer coefficients and not necessarily commuting indeterminates $x_1, \ldots, x_n$ such that the coefficient of one of the monomials in $f(x_1, \ldots, x_n)$ of maximal degree is 1 and $f(a_1, \ldots, a_n) = 0$ for all $a_1, \ldots, a_n$ in R.

The next theorem due to Rowen shows that ideals of a semiprime PI-ring R have a close connection with the center of R (see [364] or [366]).

Theorem 3.2.16 *Let R be a semiprime PI-ring. Then $I \cap \text{Cen}(R) \neq 0$ for any nonzero ideal I of R.*

Let R be a ring and S a subring of R. Then R is said to be an *ideal intrinsic extension* of S if $I \cap S \neq 0$ for any $0 \neq I \trianglelefteq R$. If R is a semiprime PI-ring with identity, then by Theorem 3.2.16 R is an ideal intrinsic extension of $\text{Cen}(R)$. See [30] for more details on ideal intrinsic extensions.

Theorem 3.2.17 (i) *If a ring R is ideal intrinsic over* $\text{Cen}(R)$, *then R is quasi-Baer if and only if* $\text{Cen}(R)$ *is Baer.*

(ii) *If R is a semiprime PI-ring with identity, then R is quasi-Baer if and only if* $\text{Cen}(R)$ *is Baer.*

(iii) *If R is a reduced PI-ring with identity, then R is Baer if and only if* $\text{Cen}(R)$ *is Baer.*

Proof (i) We only need to prove the sufficiency in view of Theorem 3.2.13. So assume that $C := \text{Cen}(R)$ is Baer. Note that the commutative ring C is semiprime because C is nonsingular from Proposition 3.1.18. Let $I \trianglelefteq R$. If $I = 0$, then $r_R(I) = R$, so we are done. Next, suppose that $I \neq 0$. Then $I \cap C \neq 0$, and $r_C(I \cap C) = eC$ for some $e^2 = e \in C$. If $Ie \neq 0$, then $0 \neq Ie \cap C \subseteq (I \cap C) \cap r_C(I \cap C)$. As C is semiprime, $(I \cap C) \cap r_C(I \cap C) = 0$, hence $Ie \cap C = 0$, a contradiction. Thus, $Ie = 0$, so $eR \subseteq r_R(I)$.

By the modular law, $r_R(I) = eR \oplus [(1-e)R \cap r_R(I)]$. To show that $r_R(I) = eR$, put $A = (1-e)R \cap r_R(I) \trianglelefteq R$. If $A \neq 0$, then $A \cap C \neq 0$. Say $0 \neq z \in A \cap C$. Then $(I \cap C)z = 0$ as $A \subseteq r_R(I)$. Therefore, we have that $z \in r_C(I \cap C) = eC \subseteq eR$, so $z \in eR \cap A = 0$, a contradiction. Whence $A = 0$, thus $r_R(I) = eR$. So R is quasi-Baer.

(ii) The proof follows immediately from Theorem 3.2.16 and part (i).

(iii) We observe that any reduced quasi-Baer ring is Baer (see also Exercise 3.2.44.10). So the proof follows from part (ii). □

We remark that there exists a PI-ring R with identity such that $\text{Cen}(R)$ is Baer, but R is not quasi-Baer (see Example 3.2.9 and Theorem 3.2.17(ii)).

For an algebra R over a commutative ring C, we can form the enveloping algebra $R^e = R \otimes_C R^o$, where R^o denotes the algebra opposite to R. We see that R has a structure as a left R^e-module induced by $(x \otimes y)r = xry$ for $x, r \in R$ and $y \in R^o$.

The algebra R is called *separable* if R is a projective left R^e-module (see [137] for more details on separable algebras). In contrast to Example 3.2.14, we obtain the next result for a separable algebra.

Corollary 3.2.18 *Any separable algebra with its center Baer is quasi-Baer.*

Proof Since R is an ideal intrinsic extension of $\text{Cen}(R)$ (see [137, Corollary 3.7, p. 54 and Theorem 3.8, p. 55]), the result follows immediately from Theorem 3.2.17(i). □

Theorem 3.2.19 *Let R be a prime PI-ring and let $F = Q(\text{Cen}(R))$. Then $Q(R) = \text{Mat}_n(D)$ for a positive integer n and a division ring D. Further, $\text{Cen}(D) = F$ and D is finite dimensional over the field F.*

Proof See [348] and [366, Theorems 1.5.16 and 1.7.9]. □

The next example illustrates Theorem 3.2.17 and Corollary 3.2.18.

Example 3.2.20 (i) There exists a semiprime ring R which is not PI, but R is ideal intrinsic over $\text{Cen}(R)$ and $\text{Cen}(R)$ is Baer, hence by Theorem 3.2.17(i) R is quasi-Baer. Take $R = \mathbb{Z}\{x, y\}/A$, where $\mathbb{Z}\{x, y\}$ is the free algebra over $\mathbb{Z}$ and A is the ideal of $\mathbb{Z}\{x, y\}$ generated by $yx - xy - 1$. Then R is a domain. By direct computation, we see that R is intrinsic over $\text{Cen}(R) = \mathbb{Z}$.

We note that $W := \mathbb{Q}\{x, y\}/K$, where K is the ideal of $\mathbb{Q}\{x, y\}$ generated by $yx - xy - 1$, is the first Weyl algebra over $\mathbb{Q}$. Then $Q(R) = Q(W)$ is a division ring and it is infinite dimensional over its center $\mathbb{Q}$. So R is not a PI-ring by Theorem 3.2.19.

(ii) There is a semiprime PI-ring R with $\text{Cen}(R)$ Baer, but R is not Baer. Let $R = \text{Mat}_n(\mathbb{Z}[x])$ $(n > 1)$. Then R is semiprime PI. Further, R is a separable algebra over $\mathbb{Z}[x]$. Observe that $\text{Cen}(R) \cong \mathbb{Z}[x]$ is Baer, and by Theorem 3.2.12 R is quasi-Baer, but R is not Baer (see Example 3.1.28).

A ring R is called *biregular* if for each $x \in R$ there is a central idempotent $e \in R$ with $RxR = eR$.

Example 3.2.21 The following rings are biregular: (i) Boolean rings; (ii) simple rings; (iii) reduced π-regular rings [221, Proposition 1(3), p. 210]; (iv) right self-injective regular PI-rings; and (v) the maximal right ring of quotients of a semiprime PI-ring.

Theorem 3.2.22 *Let R be a biregular ring. Then:*

(i) $J(R) = 0$.
(ii) *The concepts of prime, right primitive, and maximal ideals coincide.*

(iii) *R is a subdirect product of simple rings.*
(iv) *Every ideal is the intersection of the maximal ideals containing it.*

Proof Exercise. □

We see that a biregular ring is an ideal intrinsic extension of its center. In fact, let I be a nonzero ideal of a biregular ring R. Then for $0 \neq a \in I$, $RaR = eR$ for some $e \in \mathcal{B}(R)$, so $0 \neq e \in I \cap \text{Cen}(R)$. Hence, if R is a biregular ring, then R is quasi-Baer if and only if Cen(R) is Baer by Theorem 3.2.17(i).

Recall that a ring is regular if and only if every principal right ideal is generated by an idempotent. Thus, a biregular ring is an ideal analogue of a regular ring. But the following example shows that the two notions are independent.

Example 3.2.23 (i) Let R be the endomorphism ring of an infinite dimensional right vector space V over a field (see Example 1.3.19). Then R is regular. Further, R is prime. Thus if R is biregular, then R is simple by Theorem 3.2.22(ii). Consider $I = \{f \in R \mid f(V) \text{ is finite dimensional}\}$. Then I is a proper nonzero ideal of R, a contradiction. Hence, R is not biregular.

(ii) Let R be the first Weyl algebra over a field F with characteristic zero. That is, $R = F\{x, y\}/A$, where $F\{x, y\}$ is the free algebra over F and A is the ideal of $F\{x, y\}$ generated by $yx - xy - 1$. Then R is a simple domain, hence R is biregular. But R is not regular because R is not a division ring.

Being a quasi-Baer ring and being a biregular ring are independent notions for a ring. The ring in Example 3.1.14(ii) is biregular but it is not quasi-Baer, while the ring of Example 3.2.23(i) is quasi-Baer but it is not biregular.

Definition 3.2.24 A ring R is called *right principally quasi-Baer* (simply, *right p.q.-Baer*) if the right annihilator of any principal ideal is generated by an idempotent as a right ideal. A left principally quasi-Baer (simply, left p.q.-Baer) ring is defined similarly. Rings which are right and left principally quasi-Baer are called *principally quasi-Baer* (simply, *p.q.-Baer*).

We note that the class of right p.q.-Baer rings is a generalization of the classes of quasi-Baer rings and biregular rings. In the next result, prime rings are described in terms of right p.q.-Baer rings (cf. Proposition 3.2.5).

Proposition 3.2.25 *A ring R is a prime ring if and only if R is a right p.q.-Baer ring and semicentral reduced.*

Proof The proof is similar to that of Proposition 3.2.5. □

The ring R in Example 3.1.16 is p.q.-Baer, but R is neither quasi-Baer nor biregular. Another similar example will given in Example 3.2.31.

Proposition 3.2.26 *A ring R is right p.q.-Baer if and only if for each finitely generated ideal I of R, there exists $e^2 = e \in R$ such that $r_R(I) = eR$.*

Proof Assume that R is a right p.q.-Baer ring and let $I = \sum_{i=1}^n Ra_iR$. Then $r_R(I) = \cap_{i=1}^n r_R(Ra_iR) = \cap_{i=1}^n e_iR$, where each $r_R(Ra_iR) = e_iR$ and $e_i^2 = e_i$. So each $e_i \in \mathbf{S}_\ell(R)$. From Proposition 1.2.4(i), $r_R(I) = \cap_{i=1}^n e_iR = eR$ for some $e \in \mathbf{S}_\ell(R)$. The converse is obvious. □

The following is an effective criterion for a ring to be regular.

Proposition 3.2.27 *Let $e_1, \ldots, e_n$ be orthogonal idempotents in a ring R such that $e_1 + \cdots + e_n = 1$. Then R is regular if and only if for each $x \in e_iRe_j$ there exists $y \in e_jRe_i$ such that $x = xyx$.*

Proof See [183, Lemma 1.6]. □

The ring $S = \text{Mat}_2(\mathbb{Z}[x])$ is quasi-Baer (hence p.q.-Baer) by Theorem 3.2.12. But S is neither right nor left Rickart as in Example 3.1.28. We note that a domain which is not a division ring is p.q.-Baer, but it is not regular. In the next example, there is a regular ring (hence a Rickart ring) which is neither right nor left p.q.-Baer. Thereby the notions of being right (or left) p.q.-Baer and being right (or left) Rickart are independent.

Example 3.2.28 The ring R described below is a regular ring (hence a Rickart ring) that is neither right nor left p.q.-Baer. For a field F, take $F_n = F$ for $n = 1, 2, \ldots$, and let

$$R = \begin{bmatrix} \prod_{n=1}^\infty F_n & \oplus_{n=1}^\infty F_n \\ \oplus_{n=1}^\infty F_n & \langle \oplus_{n=1}^\infty F_n, 1\rangle \end{bmatrix},$$

a subring of the 2×2 matrix ring over the ring $\prod_{n=1}^\infty F_n$, where $\langle \oplus_{n=1}^\infty F_n, 1\rangle$ is the F-subalgebra of $\prod_{n=1}^\infty F_n$ generated by $\oplus_{n=1}^\infty F_n$ and $1 = 1_{\prod_{n=1}^\infty F_n}$.

Take $e_1 = \begin{bmatrix} 1 & 0 \\ 0 & 0 \end{bmatrix} \in R$ and $e_2 = \begin{bmatrix} 0 & 0 \\ 0 & 1 \end{bmatrix} \in R$. Then for each $x \in e_iRe_j$, there exists $y \in e_jRe_i$ such that $x = xyx$, where $i, j = 1, 2$. Thus by Proposition 3.2.27, R is a regular ring. Obviously, R is a PI-ring.

Let $a = (a_n)_{n=1}^\infty \in \prod_{n=1}^\infty F_n$ such that $a_n = 1_{F_n}$ if n is odd and $a_n = 0$ if n is even, and let $\alpha = \begin{bmatrix} a & 0 \\ 0 & 0 \end{bmatrix} \in R$. Assume that there is $e^2 = e \in R$ such that $r_R(\alpha R) = eR$. Then $e \in \mathbf{S}_\ell(R)$. Since R is semiprime, $e \in \mathcal{B}(R)$ by Proposition 1.2.6(ii). But this is impossible. Whence R is not right p.q.-Baer. Similarly, R is not left p.q.-Baer.

From Example 3.2.21(iv), a right self-injective regular PI-ring is biregular. However, Example 3.2.28 shows that a regular PI-ring need not be biregular. For further illustration of Definition 3.2.24, we present a left p.q.-Baer ring which is not right p.q.-Baer. Thus, similar to the Rickart property, the p.q.-Baer property is not left-right symmetric (unlike the Baer and the quasi-Baer properties).

Example 3.2.29 Let R be the ring as in Example 3.1.20. Then R is left p.q.-Baer by computation. Let α be the element of R as in Example 3.1.20. Then $r_R(\alpha R)$ is not generated by an idempotent. So R is not right p.q.-Baer.

Proposition 3.2.30 *Let $\{R_i \mid i \in \Lambda\}$ be a family of rings and $R = \prod_{i\in\Lambda} R_i$. Then R is right p.q.-Baer if and only if each R_i is right p.q.-Baer.*

Proof The proof is routine. □

In contrast to Proposition 3.2.30, Example 3.1.6 also shows that a subdirect product of right p.q.-Baer rings need not be right p.q.-Baer. In the next example, we provide a right p.q.-Baer ring which is neither quasi-Baer nor right Rickart. Moreover, it is also not biregular.

Example 3.2.31 Let $S = A \oplus R$, where A is a prime ring which is not right nonsingular (see Example 3.2.7(i)) and R is the ring in Example 3.1.14(ii). Hence A is not right Rickart by Proposition 3.1.18. But A is right p.q.-Baer by Proposition 3.2.25. Observe that the ring R is not quasi-Baer, but R is p.q.-Baer. From Propositions 3.1.21, 3.2.8, and 3.2.30, S is the desired example. Further, S is not biregular, since otherwise so is A and hence A is simple by Theorem 3.2.22(ii), a contradiction.

The following result shows that in a quasi-Baer ring, the lattice of principal right ideals, each generated by a left semicentral idempotent, is complete. This is analogous to Theorem 3.1.23 for the case of Baer rings.

Theorem 3.2.32 *Let R be a ring and let $\mathcal{LS} = \{eR \mid e \in \mathbf{S}_\ell(R)\}$. Then the following are equivalent.*

(i) *R is a quasi-Baer ring.*
(ii) *R is p.q.-Baer and $\mathcal{LS}$ is a complete lattice under inclusion.*

Proof (i)⇒(ii) Assume that R is a quasi-Baer ring. Then R is a p.q.-Baer ring. Consider a subset $\{e_i R \mid i \in \Lambda\}$ of $\mathcal{LS}$. As $\sum e_i R$ is an ideal and R is quasi-Baer, $r_R(\ell_R(\sum e_i R)) = fR$ for some $f \in \mathbf{S}_\ell(R)$. We can easily check that $fR = \sup\{e_i R \mid i \in \Lambda\}$. As $\mathcal{LS}$ is a partially ordered set under inclusion, $\mathcal{LS}$ is a complete lattice under inclusion by [382, Proposition 1.2, p. 64].

(ii)⇒(i) Let $I \trianglelefteq R$ and write $I = \sum_{i\in\Lambda} Rx_i R$. Because R is p.q.-Baer, there is $e_i \in \mathbf{S}_\ell(R)$ for each i with $r_R(I) = \cap_{i\in\Lambda} r_R(Rx_i R) = \cap_{i\in\Lambda} e_i R$. By completeness, there exists $e \in \mathbf{S}_\ell(R)$ such that $eR = \inf\{e_i R \mid i \in \Lambda\}$. Thus, $eR \subseteq \cap_{i\in\Lambda} e_i R$. Now let $y \in \cap_{i\in\Lambda} e_i R$. Then $\ell_R(Ry) = Rh$ for some $h \in \mathbf{S}_r(R)$ since R is p.q.-Baer. As $y \in e_i R$ and $e_i R \trianglelefteq R$, $Ry \subseteq e_i R$. Thus, we have that

$$(1-h)R = r_R(\ell_R(Ry)) \subseteq r_R(\ell_R(e_i R)) = e_i R$$

for each i.

Note that $1-h \in \mathbf{S}_\ell(R)$ by Proposition 1.2.2. Thus $(1-h)R \subseteq eR$, so $y \in Ry \subseteq r_R(\ell_R(Ry)) = (1-h)R \subseteq eR$. Hence, $\cap_{i\in\Lambda} e_i R \subseteq eR$. Therefore, $eR = \cap_{i\in\Lambda} e_i R$, and so $r_R(I) = eR$. Hence, R is quasi-Baer. □

By Proposition 1.2.5, $\mathcal{LS}$ is a sublattice of the lattice of ideals of a ring R. So one may expect that $\mathcal{LS}$ is a complete sublattice of the lattice of ideals of R if R quasi-Baer. But this does not hold. For example, let F be a field and $F_n = F$ for $n = 1, 2, \ldots$. Put $R = \prod_{i=1}^{\infty} F_n$. Then R is Baer (hence quasi-Baer) by Proposition 3.1.5(i).

Consider $e_1 = (1, 0, 0, \ldots)$, $e_2 = (0, 1, 0, 0, \ldots)$, and so on. Now we let $f \in \mathbf{S}_\ell(R)$ such that $fR = \sup\{e_i R \mid i = 1, 2, \ldots\}$. As $\sum_{i=1}^{\infty} e_i R \subseteq fR$, we see that $f = 1$ and so $fR = R$. Hence, $\sup\{e_i R \mid i = 1, 2, \ldots\} = R$. Note that $\sum_{i=1}^{\infty} e_i R = \oplus_{i=1}^{\infty} F_n$. So there is no $e \in \mathbf{S}_\ell(R)$ such that $\sum_{i=1}^{\infty} e_i R = eR$. Therefore, $\mathcal{LS}$ cannot be a complete sublattice of the lattice of ideals of R.

The next corollary shows that the completeness of the lattice of principal ideals for a biregular ring yields that it is quasi-Baer.

Corollary 3.2.33 *Assume that R is a biregular ring. Then the following are equivalent.*

(i) *R is quasi-Baer.*
(ii) *The set of principal ideals of R forms a complete lattice under inclusion.*

Proof Theorem 3.2.32 yields the result. □

Similar to the case of quasi-Baer rings in Theorems 3.2.10 and 3.2.13, we obtain the following for right p.q.-Baer rings.

Theorem 3.2.34 *Let R be a right p.q.-Baer ring. Then:*

(i) *eRe is a right p.q.-Baer ring for each $e^2 = e \in R$.*
(ii) *The center* $\mathrm{Cen}(R)$ *is a Rickart ring.*

Proof (i) Let R be a right p.q.-Baer ring. Say $x \in eRe$. Then it follows that $r_{eRe}(xeRe) = r_{eRe}((exe)(eRe)) = r_R((exe)R) \cap eRe$. As R is right p.q.-Baer, $r_R((exe)R) = fR$ with $f \in \mathbf{S}_\ell(R)$. So $r_{eRe}((exe)(eRe)) = fR \cap eRe$. Because $f \in \mathbf{S}_\ell(R)$, efe is an idempotent. Also, $fR \cap eRe = (efe)(eRe)$, as a consequence $r_{eRe}(xeRe) = (efe)(eRe)$. Therefore, eRe is right p.q.-Baer.

(ii) Assume that R is a right p.q.-Baer ring and let $C = \mathrm{Cen}(R)$. Take $a \in C$. Then $\ell_R(a) = \ell_R(Ra) = r_R(aR) = eR$ with $e \in \mathbf{S}_\ell(R)$ from Proposition 1.2.2. Note that $\ell_R(Ra) = \ell_R[r_R(\ell_R(Ra))] = \ell_R(r_R(eR))$. Say $r_R(eR) = fR$ with $f \in \mathbf{S}_\ell(R)$. Therefore $1 - f \in \mathbf{S}_r(R)$ by Proposition 1.2.2. Hence, we have that

$$eR = \ell_R(Ra) = \ell_R(r_R(eR)) = \ell_R(fR) = R(1-f).$$

So $e = 1 - f$. Because $e \in \mathbf{S}_\ell(R)$ and $1 - f \in \mathbf{S}_r(R)$, $e = 1 - f \in \mathcal{B}(R)$ by Proposition 1.2.6(i). Thus $r_C(a) = r_R(a) \cap C = eR \cap C = eC$. So $\mathrm{Cen}(R)$ is Rickart. □

In Theorem 3.2.11, we showed that the quasi-Baer ring property is Morita invariant. The next result proves that the right p.q.-Baer ring property is also a Morita invariant property.

Theorem 3.2.35 *The endomorphism ring of a finitely generated projective module over a right p.q.-Baer ring is a right p.q.-Baer ring. In particular, the right p.q.-Baer ring property is Morita invariant.*

Proof Let n be a positive integer and take $\alpha \in \mathrm{Mat}_n(R)$. Then we see that $\mathrm{Mat}_n(R)\alpha\mathrm{Mat}_n(R) = \mathrm{Mat}_n(K)$ for a finitely generated ideal K of R. Since R is right p.q.-Baer, there exists $e^2 = e \in R$ with $r_R(K) = eR$ by Proposition 3.2.26. We see that $r_{\mathrm{Mat}_n(R)}(\mathrm{Mat}_n(R)\alpha\mathrm{Mat}_n(R)) = (e\mathbf{1})\mathrm{Mat}_n(R)$, where $\mathbf{1}$ is the identity matrix of $\mathrm{Mat}_n(R)$. So $\mathrm{Mat}_n(R)$ is right p.q.-Baer.

Let P be a finitely generated projective right R-module. Then there exist a positive integer n and $f^2 = f \in \mathrm{Mat}_n(R)$ such that $\mathrm{End}(P) \cong f\mathrm{Mat}_n(R)f$. As $\mathrm{Mat}_n(R)$ is right p.q.-Baer, $\mathrm{End}(P)$ is right p.q.-Baer by Theorem 3.2.34(i). In particular, the right p.q.-Baer ring property is Morita invariant. □

Theorem 3.2.36 *The following are equivalent for a ring R.*

(i) *R is a right p.q.-Baer ring.*
(ii) $\mathrm{Mat}_n(R)$ *is a right p.q.-Baer ring for every positive integer n.*
(iii) $\mathrm{Mat}_k(R)$ *is a right p.q.-Baer ring for some integer $k > 1$.*
(iv) $\mathrm{Mat}_2(R)$ *is a right p.q.-Baer ring.*

Proof By Theorems 3.2.34 and 3.2.35, we see that the proof is similar to that of Theorem 3.2.12. □

We next present some strong connections between the quasi-Baer and the FI-extending properties for rings. The following result shows that, for a semiprime ring, the right FI-extending, left FI-extending, right strongly FI-extending, left strongly FI-extending, and the quasi-Baer properties coincide.

Theorem 3.2.37 *Assume that R is a semiprime ring. Then the following are equivalent.*

(i) *R is quasi-Baer.*
(ii) *If $I \trianglelefteq R$, then $I_R \leq^{\mathrm{ess}} eR_R$ for some $e \in \mathcal{B}(R)$.*
(iii) *R is right FI-extending.*
(iv) *R is left FI-extending.*
(v) *R is right strongly FI-extending.*
(vi) *R is left strongly FI-extending.*

(vii) *If* $I \trianglelefteq R$, *then* $r_R(I)_R \leq^{\text{ess}} eR_R$ *for some* $e^2 = e \in R$.

Proof (i)⇒(ii) Let $I \trianglelefteq R$. By Lemma 2.1.13, $I_R \leq^{\text{ess}} r_R(\ell_R(I))_R$. Since R is semiprime quasi-Baer, $r_R(\ell_R(I)) = eR$ for some $e \in \mathcal{B}(R)$ from Proposition 1.2.6(ii), so $I_R \leq^{\text{ess}} eR_R$. (ii)⇒(iii) is obvious.

For (iii)⇒(i), say $I \trianglelefteq R$. Then $I_R \leq^{\text{ess}} eR_R$ for some $e^2 = e \in R$. So we get that $R(1-e) \subseteq \ell_R(I)$. By the modular law, $\ell_R(I) = R(1-e) \oplus (\ell_R(I) \cap Re)$. We put $J = \ell_R(I) \cap Re$. Then $Je = J$. If $J \neq 0$, then $J^2 = JeJe \neq 0$ since R is semiprime. Hence, $eJ \neq 0$. Take $0 \neq ex \in eJ \subseteq eR \cap \ell_R(I)$. As $I_R \leq^{\text{ess}} eR_R$, it follows that $0 \neq exr \in I$ with $r \in R$. So $0 \neq exr \in I \cap \ell_R(I)$, a contradiction as R is semiprime. Thus $\ell_R(I) = R(1-e)$, hence R is a quasi-Baer ring.

(ii)⇔(v) follows from Proposition 1.2.6(ii) since R is semiprime.

(i)⇔(vii) If R is quasi-Baer, then (vii) follows from (ii). For (vii)⇒(i), let $I \trianglelefteq R$. Using the proof of (iii)⇒(i) by replacing I with $r_R(I)$, we obtain that $\ell_R(r_R(I)) = R(1-e)$, so $r_R(I) = r_R(\ell_R(r_R(I))) = eR$. Thus, R is quasi-Baer.

The equivalences (i)⇔(iv)⇔(vi) can be proved similarly. □

The condition (vii) of Theorem 3.2.37 is considered in Definition 8.1.1. When R is a right nonsingular ring, the next result demonstrates the connection between right FI-extending property and the quasi-Baer property.

Theorem 3.2.38 *Let* R *be a right nonsingular ring. Then* R_R *is FI-extending if and only if* R *is quasi-Baer and* $A_R \leq^{\text{ess}} r_R(\ell_R(A))_R$ *for every* $A \trianglelefteq R$.

Proof Let R be right FI-extending and $A \trianglelefteq R$. Then $A_R \leq^{\text{ess}} eR_R$ for some $e^2 = e \in R$. Obviously, $\ell_R(eR) \subseteq \ell_R(A)$. Take $x \in \ell_R(A)$. To show that $x \in \ell_R(eR)$, say $a \in R$. We let $V = \{r \in R \mid ear \in A\}$. Then we can check that $V_R \leq^{\text{ess}} R_R$. Now $xeaV = 0$ and $Z(R_R) = 0$. Therefore, $xea = 0$ and hence $x \in \ell_R(eR)$. Thus, $\ell_R(A) = \ell_R(eR) = R(1-e)$. So R is quasi-Baer.

Also, $A_R \leq^{\text{ess}} eR_R = r_R(\ell_R(eR)) = r_R(\ell_R(A))$. The converse is obvious. □

In Theorem 4.6.12, a complete characterization of FI-extending modules in terms of quasi-Baer modules will be provided. By Theorem 3.2.38, every right nonsingular FI-extending ring is a quasi-Baer ring. However, there is a right nonsingular quasi-Baer ring which is not right FI-extending in the next example.

Example 3.2.39 Let F be a field and

$$R = \left\{ \begin{bmatrix} a & 0 & x \\ 0 & a & y \\ 0 & 0 & b \end{bmatrix} \mid a, b, x, y \in F \right\} \cong \begin{bmatrix} F & F \oplus F \\ 0 & F \end{bmatrix}.$$

Then $Q(R) = \text{Mat}_3(F)$, so R is right nonsingular. By calculation R is quasi-Baer (or see Theorem 5.6.5). Also by direct computation, R is not right FI-extending (or see Corollary 5.6.11).

By Theorems 2.3.27, 3.2.37, and 3.2.38, one may wonder if there is a right strongly FI-extending ring that is neither semiprime, nor quasi-Baer, nor right nonsingular. The next example provides a class of such rings.

Example 3.2.40 Let p be a prime integer and let $R = T_2(\mathbb{Z}_{p^n})$, where n is a integer such that $n \geq 2$. Then R satisfies the following properties.

(i) R is not semiprime.
(ii) R is not right nonsingular.
(iii) R is not quasi-Baer (see Theorem 5.6.7).
(iv) R is right strongly FI-extending (see Theorem 5.6.18).

We say that a ring R is *right principally FI-extending* (resp., *right finitely generated FI-extending*) if every principal ideal (resp., finitely generated ideal) of R is essential in a direct summand of R_R.

Proposition 3.2.41 *Assume that R is a semiprime ring. Then the following are equivalent.*

(i) *R is right p.q.-Baer.*
(ii) *If I is a finitely generated ideal of R, then $I_R \leq^{\text{ess}} eR_R$ for some $e^2 = e \in \mathcal{B}(R)$.*
(iii) *R is right finitely generated FI-extending.*
(iv) *R is right principally FI-extending.*
(v) *R is left p.q.-Baer.*

Proof (i)⇒(ii) Say I is a finitely generated ideal. Then $I_R \leq^{\text{ess}} r_R(\ell_R(I))_R$ by Lemma 2.1.13. Because R is semiprime, $\ell_R(I) = r_R(I)$. From Proposition 3.2.26, $\ell_R(I) = r_R(I) = fR$ for some $f \in \mathbf{S}_\ell(R)$ since R is right p.q.-Baer. By Proposition 1.2.6(ii), $f \in \mathcal{B}(R)$. Thus $r_R(\ell_R(I)) = (1 - f)R$. Take $e = 1 - f$. Then $e \in \mathcal{B}(R)$ and $I_R \leq^{\text{ess}} eR_R$.

(ii)⇒(iii)⇒(iv) These implications are clear.

(iv)⇒(i) Let I be a principal ideal of R. By hypothesis, $I_R \leq^{\text{ess}} eR_R$ with $e^2 = e \in R$. Now $\ell_R(I) = R(1 - e)$ similar to the proof of (iii)⇒(i) in Theorem 3.2.37. Since $R(1 - e) \trianglelefteq R$, $1 - e \in \mathcal{B}(R)$ by Propositions 1.2.2 and 1.2.6(ii). As R is semiprime, $r_R(I) = \ell_R(I) = (1 - e)R$. So R is right p.q.-Baer.

(i)⇔(v) The proof follows from the fact that for a semiprime ring, left and right annihilators of ideals coincide and from Proposition 1.2.6(ii), semicentral idempotents are central. □

Homomorphic images of quasi-Baer rings are not quasi-Baer, in general (e.g., $\mathbb{Z}/4\mathbb{Z}$). If R is a quasi-Baer ring, then there exists $e \in \mathbf{S}_\ell(R)$ such that $P(R) \leq^{\text{ess}} eR_R$ and $(1 - e)R = (1 - e)R(1 - e)$ is a semiprime quasi-Baer ring (see Theorem 3.2.10 and Exercise 3.2.44.8). Moreover, each semiprime ring has a right ring of quotients which is quasi-Baer (see Theorem 8.3.17). Thus it is natural to ask: *If R is a quasi-Baer ring, must $R/P(R)$ be quasi-Baer?* In the following example, we provide a negative answer to this question in general. Nevertheless, in

Theorem 3.2.43 we exhibit a positive answer by considering the nilpotency of the prime radical.

Example 3.2.42 Let A be a prime ring and S the upper triangular (column finite) $\aleph_0 \times \aleph_0$ matrix ring over A. We let R be the ring of $[s_{ij}] \in S$ such that $s_{ij} = 0$ for all but finitely many $i \neq j$ and s_{nn} is constant eventually. Then:

(i) R is quasi-Baer.
(ii) $R/P(R)$ is not quasi-Baer.

Say I is a nonzero right ideal of R with $I = \sum_{k\in\Lambda} u_k R$ for an index set Λ. Let $u_k = [{}^k a_{ij}] \in R$ for each $k \in \Lambda$. Since A is prime, it is quasi-Baer.

We see that there exist $e_i \in S_\ell(A)$, $i = 1, 2, \dots$ such that

$$r_A(\sum_{k\in\Lambda} {}^k a_{11}A) = e_1 A,\ r_A(\sum_{k\in\Lambda} {}^k a_{11}A + {}^k a_{12}A + {}^k a_{22}A) = e_2 A,$$

$$r_A(\sum_{k\in\Lambda} {}^k a_{11}A + {}^k a_{12}A + {}^k a_{13}A + {}^k a_{22}A + {}^k a_{23}A + {}^k a_{33}A) = e_3 A,$$

and so on. Since A is a prime ring, $\mathbf{S}_\ell(A) = \mathcal{B}(A) = \{0, 1\}$. So $e_i = 0$ or 1 for each i. Also $e_j A \subseteq e_i A$ when $j \geq i$. If $e_i = 0$, then $e_j = 0$ for any $j \geq i$.

Let g be the diagonal matrix with e_i in the (i, i)-position for each i. Since e_i is eventually constant to 0 or 1, and $g^2 = g \in R$. We show that $r_R(I) = gR$ so that R is quasi-Baer. Indeed, $Ig = 0$, so $gR \subseteq r_R(I)$. Next, let $[z_{ij}] \in r_R(I)$. Then $I[z_{ij}] = 0$. Thus $z_{1m} \in r_R(\sum_{k\in\Lambda} {}^k a_{11}A) = e_1 A$ for all $m = 1, 2, \dots$. Let $\alpha_{1m} \in A$ for $m = 1, 2, \dots$ such that $z_{1m} = e_1\alpha_{1m}$ and $\alpha_{1m} = 0$ when $z_{1m} = 0$. We see that $z_{2m} \in r_R(\sum_{k\in\Lambda} {}^k a_{11}A + {}^k a_{12}A + {}^k a_{22}A) = e_2 A$, where $m = 2, 3, \dots$. Let $\alpha_{2m} \in A$ for $m = 2, 3, \dots$ such that $z_{2m} = e_2\alpha_{2m}$ and $\alpha_{2m} = 0$ when $z_{2m} = 0$. Next

$$z_{3m} \in r_A(\sum_{k\in\Lambda} {}^k a_{11}A + {}^k a_{12}A + {}^k a_{13}A + {}^k a_{22}A + {}^k a_{23}A + {}^k a_{33}A) = e_3 A,$$

for $m = 3, 4, \dots$. Let $\alpha_{3m} \in A$ for $m = 3, 4, \dots$ such that $z_{3m} = e_2\alpha_{3m}$ and $\alpha_{3m} = 0$ when $z_{3m} = 0$, and so on. Then we see that $[e_i\alpha_{ij}] \in R$ as $[z_{ij}] \in R$. Moreover, $[z_{ij}] = g[\alpha_{ij}] \in gR$. Thus $r_R(I) \subseteq gR$, so $r_R(I) = gR$. Hence R is quasi-Baer.

For i, j with $i \leq j$ and $a \in A$, let ae_{ij} be the matrix in R with a in the (i, j)-position and 0 elsewhere. Then we see that $ae_{ij}Rae_{ij} = 0$ for $i < j$. As a consequence, $ae_{ij} \in P(R)$ for any $a \in A$ and i, j with $i < j$. Therefore, $P(R)$ is the set of all matrices in R with zero diagonal.

Note that $R/P(R) \cong \{(s_n) \in \prod_{n=1}^{\infty} A \mid s_n \text{ is constant eventually}\}$. As in Example 3.1.14(ii), the ring $R/P(R)$ is not quasi-Baer.

In spite of Example 3.2.42, we obtain the following result when the prime radical is nilpotent.

Theorem 3.2.43 *Let R be a quasi-Baer (resp., Baer) ring such that $P(R)$ is nilpotent. Then the ring $R/P(R)$ is quasi-Baer (resp., Baer).*

Proof We may assume that $P(R)^n = 0$ and $P(R)^{n-1} \neq 0$ for some integer $n \geq 2$. We give a proof by induction on n.

First, assume that $n = 2$. Since R is quasi-Baer, there exists $e \in \mathbf{S}_\ell(R)$ such that $r_R(P(R)) = eR$. Thus $P(R)e = 0$, so $P(R) \subseteq R(1-e)$. Since $P(R)^2 = 0$, $P(R) \subseteq r_R(P(R)) = eR$. So $P(R) \subseteq eR \cap R(1-e) = eR(1-e)$. Note that $(1-e)Re = 0$ because $e \in \mathbf{S}_\ell(R)$. Thus,

$$R = \begin{bmatrix} eRe & eR(1-e) \\ 0 & (1-e)R(1-e) \end{bmatrix}.$$

Now $[eR(1-e)]^2 = 0$ and $eR(1-e) \trianglelefteq R$, so $eR(1-e) \subseteq P(R)$. Thus $P(R) = eR(1-e)$, so $R/P(R) = R/eR(1-e) \cong eRe \oplus (1-e)R(1-e)$. Since R is quasi-Baer, eRe and $(1-e)R(1-e)$ are quasi-Baer by Theorem 3.2.10. Thus, $R/P(R)$ is quasi-Baer by Proposition 3.2.8.

Assume that $S/P(S)$ is quasi-Baer for a given quasi-Baer ring S, whenever $P(S)^m = 0$ for $m < n$. We let $r_R(P(R)^k) = f_k R$, where $f_k \in S_\ell(R)$ and $k = 1, \ldots, n-1$. By the modular law, $f_{k+1}R = f_k R \oplus [(1-f_k)R \cap f_{k+1}R]$, for $k = 1, \ldots, n-2$.

As $f_k, f_{k+1} \in \mathbf{S}_\ell(R)$, we have that $(1-f_k)R \cap f_{k+1}R = (1-f_k)f_{k+1}R$ and $(1-f_k)f_{k+1}$ is an idempotent. So $f_{k+1}R = f_k R \oplus (1-f_k)f_{k+1}R$. Put

$$e_1 = f_1, e_2 = (1-f_1)f_2, \ldots, e_{n-1} = (1-f_{n-2})f_{n-1}, \text{ and } e_n = 1 - f_{n-1}.$$

Therefore $f_1 R = e_1 R$, $f_2 R = e_1 R \oplus e_2 R, \ldots,$ and $f_{n-1}R = e_1 R \oplus \cdots \oplus e_{n-1}R$. Also $R = e_1 R \oplus e_2 R \oplus \cdots \oplus e_n R$.

Note that $f_{n-1} \in \mathbf{S}_\ell(R)$, hence $(1-f_{n-1})Rf_{n-1} = 0$ by Proposition 1.2.2. From $0 = (1-f_{n-1})Rf_{n-1}R = e_n R(e_1 R \oplus \cdots \oplus e_{n-1}R)$, $e_n Re_i = 0$, $i = 1, \ldots, n-1$. So $e_{n-1}Rf_{n-2}R \subseteq e_{n-1}R \cap f_{n-2}R$ as $e_{n-1}Rf_{n-2}R = f_{n-2}e_{n-1}Rf_{n-2}R$.

Now $e_{n-1}R \cap f_{n-2}R = e_{n-1}R \cap (e_1 R \oplus \cdots \oplus e_{n-2}R) = 0$, thus we have that $e_{n-1}R(e_1 R \oplus \cdots \oplus e_{n-2}R) = e_{n-1}Rf_{n-2}R = 0$. Hence $e_{n-1}Re_i = 0$ for $i = 1, 2, \ldots, n-2$.

Continue this process to obtain $e_j Re_i = 0$ for $i < j$. Thus,

$$R = \begin{bmatrix} e_1 Re_1 & e_1 Re_2 & \cdots & e_1 Re_n \\ 0 & e_2 Re_2 & \cdots & e_2 Re_n \\ \vdots & \vdots & \ddots & \vdots \\ 0 & 0 & \cdots & e_n Re_n \end{bmatrix}.$$

Therefore, $R/P(R) \cong \oplus_{i=1}^n [e_i Re_i / P(e_i Re_i)]$.

In general, for $e^2 = e \in R$, $eP(R)e$ is a semiprime ideal of eRe, and thus $P(eRe) \subseteq eP(R)e$. Because $P(R)e_1 = 0$, $P(e_1 Re_1) \subseteq e_1 P(R)e_1 = 0$, and so $e_1 Re_1$ is a semiprime ring. Also, for $k = 2, \ldots, n-1$,

$$[e_k P(R)e_k]^{n-1} \subseteq e_k P(R)^{n-1} e_k = e_k P(R)^{n-1-k} P(R)^k e_k = 0$$

since $r_R(P(R)^k) = e_1R \oplus \cdots \oplus e_kR$. Note that $P(R)^{n-1} \subseteq r_R(P(R)) = e_1R$, so $P(R)^{n-1} = e_1P(R)^{n-1}$. Since $P(e_nRe_n) \subseteq e_nP(R)e_n$ and $e_nRe_1 = 0$, it follows that $[P(e_nRe_n)]^{n-1} \subseteq e_nP(R)^{n-1}e_n = e_ne_1P(R)^{n-1}e_n = 0$. Hence,

$$R/P(R) \cong e_1Re_1 \oplus [e_2Re_2/P(e_2Re_2)] \oplus \cdots \oplus [e_nRe_n/P(e_nRe_n)]$$

and $[P(e_kRe_k)]^{n-1} = 0$ for $k = 2, \ldots, n$. Since R is quasi-Baer, e_kRe_k is quasi-Baer for each k by Theorem 3.2.10. Thus by induction $e_kRe_k/P(e_kRe_k)$ is quasi-Baer for $k = 2, \ldots, n$. By Theorem 3.2.10, e_1Re_1 is quasi-Baer as R is quasi-Baer. Therefore, $R/P(R)$ is quasi-Baer by Proposition 3.2.8.

From Theorem 3.1.8, if R is Baer, then eRe is also Baer for any $e^2 = e \in R$. Thus by the same argument for the case of quasi-Baer rings, we see that $R/P(R)$ is Baer if R is Baer. □

Exercise 3.2.44

1. Let R be a ring not necessarily with identity. Show that any two of the following conditions imply the third condition.
   (i) For each $I \trianglelefteq R$, there exists $e^2 = e \in R$ such that $r_R(I) = eR$.
   (ii) For each $J \trianglelefteq R$, there exists $f^2 = f \in R$ such that $\ell_R(J) = Rf$.
   (iii) R has an identity.
2. ([73, Birkenmeier, Kim, and Park]) Prove that any semiprimary quasi-Baer ring is right nonsingular.
3. ([128, Clark]) Let e be a primitive idempotent in an Artinian quasi-Baer ring. Show that eRe is a division ring.
4. Prove Theorem 3.2.22.
5. ([56, Birkenmeier]) Let R be a right essentially quasi-Baer ring (see Definition 8.1.1). Prove the following.
   (i) There is a unique smallest idempotent generated right ideal B of R containing all nilpotent elements of R (B is called the minimal direct summand containing the nilpotent elements, MDSN).
   (ii) B is an ideal of R such that $RNR_R \leq^{\text{ess}} B_R$, where N is the set of nilpotent elements of R.
   (iii) $R = A \oplus B$ (right ideal direct sum), where A is a right ideal that is a reduced Baer ring.
6. Assume that R is a semiprime quasi-Baer ring.
   (i) Show that for each $I \trianglelefteq R$, there is $e \in \mathcal{B}(R)$ such that $I_R \leq^{\text{ess}} eR_R$ and $R/I = (eR/I) \oplus ((1-e)R + I)/I$ (ring direct sum). Further, R/I is a quasi-Baer ring (resp., right FI-extending ring) if and only if eR/I is a quasi-Baer ring (resp., right FI-extending ring).
   (ii) Prove that every homomorphic image of R is a quasi-Baer ring (resp., right FI-extending ring) if and only if R/I is a quasi-Baer ring (resp., right FI-extending ring), for all $I \trianglelefteq R$ with $I_R \leq^{\text{ess}} R_R$.
7. Show that if R is a quasi-Baer ring with I a nilpotent ideal, then I_R is not essential in R_R.

8. ([73, Birkenmeier, Kim, and Park]) Prove that a ring R is quasi-Baer if and only if for each $I \trianglelefteq R$, there exists $e \in \mathbf{S}_\ell(R)$ such that $I \subseteq eR$ and $\ell_R(I) \cap eR = eR(1-e)$. In this case, $(I + eR(1-e))_R \leq^{\text{ess}} eR_R$ and $eR(1-e) \trianglelefteq R$. If $I = P(R)$, then $P(R)_R \leq^{\text{ess}} eR_R$.
9. ([78, Birkenmeier, Kim, and Park]) Show that a ring R is right p.q.-Baer if and only if whenever I is a principal ideal of R there exists $e \in \mathbf{S}_r(R)$ such that $I \subseteq Re$ and $r_R(I) \cap Re = (1-e)Re$.
10. Let R be a ring. Prove the following.
    (i) R is Abelian and Baer if and only if R is reduced and quasi-Baer.
    (ii) R is Abelian and Rickart if and only if R is reduced and p.q.-Baer.
11. ([192, Han, Hirano, and Kim]) Let R be a semiprime ring. Show that R is a quasi-Baer ring if and only if R is extending as a left R^e-module, where $R^e = R \otimes_{\mathbb{Z}} R^o$.

3.3 Extending versus Baer Rings

The Chatters-Khuri theorem below states that a ring R is right extending and right nonsingular if and only if R is Baer and right cononsingular. This result indicates an interrelationship between the extending property and the Baer property of rings. As an application of the Chatters-Khuri theorem, it is shown that a right nonsingular ring R with $Q(R) = Q^\ell(R)$ is right extending if and only if R is left extending if and only if R is Baer. Thus a semiprime PI-ring R is right extending if and only if R is left extending if and only if R is Baer. Further close connections between extending rings and Baer rings will be discussed in Sects. 5.6 and 6.1.

A ring R is called *right cononsingular* if $\ell_R(I) = 0$ with $I_R \leq R_R$ implies $I_R \leq^{\text{ess}} R_R$.

Theorem 3.3.1 *The following are equivalent for a ring R.*

(i) *R is right nonsingular and right extending.*
(ii) *R is Baer and right cononsingular.*

Proof (i)$\Rightarrow$(ii) Take $\emptyset \neq X \subseteq R$ and let $I = XR$. By (C_1) condition, there is $e^2 = e \in R$ with $I_R \leq^{\text{ess}} eR_R$. Thus, $R(1-e) = \ell_R(eR) = \ell_R(I)$ as in the proof of Theorem 3.2.38. So $\ell_R(X) = \ell_R(I) = R(1-e)$. Therefore, R is Baer.

To show that R is right cononsingular, let $A_R \leq R_R$ with $\ell_R(A) = 0$. Since R is right extending, $A_R \leq^{\text{ess}} gR_R$ with $g^2 = g \in R$. Thus $(1-g)A = 0$, so $1-g = 0$. Hence $g = 1$ and $A_R \leq^{\text{ess}} R_R$. Whence R is right cononsingular.

(ii)$\Rightarrow$(i) Since R is Baer, $Z(R_R) = 0$ by Proposition 3.1.18. To show that R is right extending, let $V_R \leq R_R$. Then $\ell_R(V) = Rf$ for some $f^2 = f \in R$. Therefore $V \subseteq r_R(\ell_R(V)) = r_R(Rf) = (1-f)R$.

If V_R is not essential in $(1-f)R_R$, then there exists $0 \neq W_R \leq (1-f)R_R$ such that $V \cap W = 0$. Let K be a complement of W in R with $V \subseteq K$. Then

$$K \cap W = 0 \text{ and } (K \oplus W)_R \leq^{\text{ess}} R_R.$$

Since K_R is not essential in R_R and R is right cononsingular, $\ell_R(K) \neq 0$. Take

$$0 \neq s \in \ell_R(K) \subseteq \ell_R(V) = Rf.$$

Then $s(1-f)=0$, so $sW = s(1-f)W = 0$ because $W \subseteq (1-f)R$. Thus, we have that $s(K \oplus W) = 0$. As $(K \oplus W)_R \leq^{\text{ess}} R_R$ and $Z(R_R)=0$, $s=0$, a contradiction. Thus, $V_R \leq^{\text{ess}} (1-f)R_R$, so R is right extending. □

Motivated by Theorem 3.3.1, in Chap. 4 this result will be extended to a general module theoretic setting by the introduction of the notion of Baer modules (see Theorem 4.1.15). Recall that $Q^{\ell}(R)$ denotes the maximal left ring of quotients of R.

Proposition 3.3.2 *Assume that R is a right nonsingular ring such that $Q(R) = Q^{\ell}(R)$. Then R is right cononsingular.*

Proof Let A be a closed right ideal of R. We claim that $A = r_R(\ell_R(A))$. Indeed, if $A \neq r_R(\ell_R(A))$, then $r_R(\ell_R(A))$ is not an essential extension of A as A is closed in R. Thus, there is $0 \neq C_R \leq r_R(\ell_R(A))_R$ with $A \cap C = 0$. Let D_R be a complement of C_R in R_R with $A \subseteq D$. Then D is a closed right ideal of R (see Exercise 2.1.37.3). By Corollary 1.3.15 and Proposition 2.1.32, there is $e^2 = e \in Q(R)$ with $DQ(R)_R \leq^{\text{ess}} eQ(R)_R$.

For $D_R \leq^{\text{ess}} DQ(R)_R$, say $0 \neq \alpha = d_1x_1 + d_2x_2 + \cdots + d_nx_n \in DQ(R)$, where $d_i \in D$ and $x_i \in Q(R)$, $i = 1, 2, \ldots, n$. As $R_R \leq^{\text{den}} Q(R)_R$, there is $r_1 \in R$ such that $x_1r_1 \in R$ and $\alpha r_1 = \sum_{i=1}^{n} d_ix_ir_1 \neq 0$. Again there is $r_2 \in R$ with $x_2r_1r_2 \in R$ and $\alpha r_1r_2 = \sum_{i=1}^{n} d_ix_ir_1r_2 \neq 0$. Continuing this procedure, we get $r = r_1r_2 \cdots r_n \in R$ such that $x_1r, x_2r, \ldots, x_nr \in R$ and $\alpha r \neq 0$. So $\alpha r \in D$. Thus, $D_R \leq^{\text{ess}} DQ(R)_R$. Since $DQ(R)_R \leq^{\text{ess}} eQ(R)_R$ and $D_R \leq^{\text{ess}} DQ(R)_R$, $D_R \leq^{\text{ess}} (eQ(R) \cap R)_R$. So $D = eQ(R) \cap R$ because D is a closed right ideal of R. Note that $C \neq 0$, so $D \neq R$. Thus $e \neq 1$, hence $Q(R)(1-e) \neq 0$. Therefore, $Q(R)(1-e) \cap R \neq 0$ because $Q(R) = Q^{\ell}(R)$. Now take $0 \neq a \in Q(R)(1-e) \cap R$. Then $aD = 0$, and hence $aA = 0$. Therefore, $a \in \ell_R(A) = \ell_R(r_R(\ell_R(A)))$, so $ar_R(\ell_R(A)) = 0$ and $aC = 0$. Hence, $a(C \oplus D) = 0$. Since $(C \oplus D)_R \leq^{\text{ess}} R_R$ and $Z(R_R) = 0$, we have that $a = 0$, a contradiction. So $A = r_R(\ell_R(A))$.

To show that R is right cononsingular, say $I_R \leq R_R$ such that $\ell_R(I) = 0$. Let K_R be a closure of I_R in R_R. Because K is a closed right ideal of R, $K = r_R(\ell_R(K))$ by the preceding argument. Now as $I \subseteq K$ and $\ell_R(I) = 0$, $\ell_R(K) = 0$ and thus $K = r_R(\ell_R(K)) = R$. Whence $I_R \leq^{\text{ess}} R_R$, and therefore R is right cononsingular. □

The next result provides another connection between the Baer property and the extending property of rings.

Corollary 3.3.3 *Let R be a right nonsingular ring such that $Q(R) = Q^{\ell}(R)$. Then the following are equivalent.*

(i) *R is right extending.*
(ii) *R is left extending.*
(iii) *R is a Baer ring.*

Proof For (i)⇔(iii), let R be right extending. Since R is right nonsingular, R is Baer by Theorem 3.3.1. Conversely, let R be a Baer ring. By Proposition 3.3.2, R is right cononsingular. So R is right extending by Theorem 3.3.1.

Next, for (ii)⇔(iii), let R be left extending. Since $Q(R) = Q^{\ell}(R)$ is regular by Theorem 2.1.31, $Q^{\ell}(R)$ is left nonsingular. As in the proof of Theorem 2.1.31, we see that R is left nonsingular. So R is Baer by the left-sided version of Theorem 3.3.1. Conversely, if R is Baer, then R is left nonsingular by Proposition 3.1.18. From the left-sided version of Proposition 3.3.2, R is left cononsingular. Therefore, R is left extending by the left-sided version of Theorem 3.3.1. □

Theorem 3.3.4 *Let R be a semiprime PI-ring. Then:*
(i) *R is right and left nonsingular.*
(ii) $Q(R) = Q^{s}(R) = Q^{\ell}(R)$.
(iii) $Q(R)$ *is a self-injective regular PI-ring.*

Proof Let R be a semiprime PI-ring. Then R is right and left nonsingular by Fisher [168]. Also $Q(R) = Q^{\ell}(R)$ by Martindale [292] and Rowen [365], and $Q(R)$ is a PI-ring by Martindale [292]. Further, $Q(R) = Q^{s}(R)$ by Armendariz, Birkenmeier, and Park [30]. □

Corollary 3.3.5 *Let R be a semiprime PI-ring. Then R is right extending if and only if R is left extending if and only if R is Baer.*

Proof The proof follows from Corollary 3.3.3 and Theorem 3.3.4. □

Exercise 3.3.6

1. ([69, Birkenmeier, Kim, and Park]) Prove that the following are equivalent for a ring R.
   (i) R is right extending and right nonsingular.
   (ii) R is right extending and Baer.
   (iii) R is right extending and right Rickart.
   (iv) R is right nonsingular and every principal right ideal of R is extending.

Historical Notes Proposition 3.1.1, Definition 3.1.2, Proposition 3.1.5, and Theorem 3.1.8 are in [246]. Theorem 3.1.11 was obtained by Kim and Park [253]. Example 3.1.20 is due to Chase [115]. It has been treated in detail in [73, 78], and [120]. Proposition 3.1.21 and Theorem 3.1.22 are in [287]. Also Theorem 3.1.22(ii) appears in [48]. Theorem 3.1.23 is due to Maeda [287]. Theorem 3.1.26 and Example 3.1.27 appear in [76]. Theorem 3.1.29 was shown by Small [372] and its proof is taken from that of [262, Proposition 7.63]. Theorem 3.1.29 also appears in [277]. Exercises 3.1.30.1 and 3.1.30.4 are in [47]. Exercise 3.1.30.4 appears also in [46]. Bergman [49] constructed an example of a compressible ring R for which $\mathrm{Mat}_2(R)$ is not compressible. See also [23, 46, 47], and [112] for compressible rings.

Propositions 3.2.5 and 3.2.8 are in [78]. Example 3.2.7(i) is from [329]. Example 3.2.9 is in [82]. Theorem 3.2.10 was shown in [128]. Theorems 3.2.11 is essentially due to Clark [128], and Pollingher and Zaks [347]. Theorem 3.2.12 is in [347]. Theorem 3.2.13 is taken from [73]. By Kaplansky [246], the center of a Baer ring is a Baer ring. Example 3.2.14 is due to Armendariz [20]. Theorem 3.2.17(i) and (ii), Corollary 3.2.18, and Example 3.2.20 appear in [73]. Theorem 3.2.17(iii) appears in [20].

Stone [385] showed that a Boolean ring with identity is isomorphic to the ring of all continuous functions from a zero-dimensional compact space to the field $\mathbb{Z}_2$ (with discrete topology). Arens and Kaplansky [19] defined biregular rings and extended several results on Boolean rings to biregular rings. Example 3.2.21(iv) is due to Armendariz and Steinberg [26], while Example 3.2.21(v) follows from [26] and [292]. Theorem 3.2.22 is in [19].

Right principally quasi-Baer rings were defined and studied by Birkenmeier, Kim, and Park in [78]. Proposition 3.2.25 appears in [78] and [70], while Proposition 3.2.26 is in [78]. Examples 3.2.28, 3.2.29, 3.2.31, Proposition 3.2.30 are in [78]. Theorem 3.2.32 is a corrected version of Theorem 3.3 in [78]. Corollary 3.2.33 appears in [73]. In Lemma 3.3 [73] and Lemma 3.2 [78], "complete sublattice of the lattice of ideals" should be replaced by "complete lattice under inclusion". Results 3.2.34–3.2.36 are in [78]. Theorems 3.2.37 and 3.2.38 appear in [83] (see also [58]). Example 3.2.39 is in [89], while Example 3.2.40 is taken from [85] (also see [84]). Right principally FI-extending rings and right finitely generated FI-extending rings are defined in [101]. Proposition 3.2.41 appears in [78]. Example 3.2.42 and Theorem 3.2.43 are in [103].

Theorem 3.3.1 appears in [121]. Proposition 3.3.2 is taken from [382, Proposition 4.7, p. 251]. For further material on Baer and Rickart rings, see [132, 148, 151, 200, 210, 282, 287, 379, 380], and [381]. For results on Baer and quasi-Baer near-rings see [61, 62], and [63]. Additional references include [24, 51, 127, 160, 283, 291, 316, 349], and [417].

Chapter 4
Baer, Quasi-Baer Modules, and Their Applications

In this chapter, the Baer and the quasi-Baer properties for arbitrary modules are introduced and studied. The definition of a Baer module using its endomorphism ring is proposed in Sect. 4.1. It was shown in Proposition 3.1.18 that every Baer ring is nonsingular. We shall see that a Baer module also satisfies a weaker nonsingularity of modules (called $\mathcal{K}$-nonsingularity) which depends on the endomorphism ring of the module. Strong connections between a Baer module and an extending module will be exhibited via this weak nonsingularity and a dual notion. We shall see that an extending module which is $\mathcal{K}$-nonsingular is precisely a $\mathcal{K}$-cononsingular Baer module. This provides a module theoretic analogue of the Chatters-Khuri theorem for rings (Theorem 3.3.1).

Direct summands of Baer and quasi-Baer modules respectively inherit these properties. This provides us a rich source of examples of Baer and quasi-Baer modules. For example, one can readily see that for any Baer ring R and an idempotent $e \in R$, the right R-module eR_R is always a Baer module. Among other results, a Baer module is characterized in terms of the strong summand intersection property. The connections between a (quasi-)Baer module and its endomorphism ring will be discussed. Characterizations of classes of rings via the Baer property of certain classes of free modules over them, will be presented. These results are applied later to matrix ring extensions in Chap. 6.

We shall apply the results of this chapter to present a type theory for Baer modules. An application also yields a type theory for ($\mathcal{K}$-)nonsingular extending (continuous) modules which, in particular, improves the type theory for nonsingular injective modules provided by Goodearl and Boyle [184]. Analogous to right Rickart rings, we shall include the notion of Rickart modules as another application of the theory of Baer modules. Some recent results on Rickart modules are included.

Similar to the case of Baer modules, close links between quasi-Baer modules and FI-extending modules are established and a characterization connecting the two notions is shown. The concepts of FI-$\mathcal{K}$-nonsingularity and FI-$\mathcal{K}$-cononsingularity are introduced to obtain this.

G.F. Birkenmeier et al., *Extensions of Rings and Modules*,
DOI 10.1007/978-0-387-92716-9_4,

4.1 Baer Modules

We begin this section with the definition of a Baer module M_R via its endomorphism ring $S = \text{End}(M_R)$ in contrast to defining this notion in terms of the base ring R. The use of the endomorphism ring instead of the base ring R appears to offer a more natural generalization of Baer rings in the module theoretic setting (see Definition 4.1.1 and the comment after Definition 4.1.1).

Properties of Baer modules are investigated and examples are provided in this section. Similar to the ring theoretic concepts of nonsingularity and cononsingularity, the $\mathcal{K}$-nonsingularity and the $\mathcal{K}$-cononsingularity respectively, are introduced for modules. These conditions provide a characterization of a $\mathcal{K}$-cononsingular Baer module as an extending module which is $\mathcal{K}$-nonsingular, generalizing the Chatters-Khuri theorem to the module theoretic setting.

Throughout this section, we let $S = \text{End}(M_R)$.

Definition 4.1.1 A right R-module M is called a *Baer module* if $\ell_S(N) = Se$ with $e^2 = e \in S$ for all $N \leq M$. A left R-module which is Baer is defined similarly.

In [274, Definition 2.1], Lee and Zhou also called a module M *Baer* if, for any nonempty subset X of M, $r_R(X) = eR$ with $e^2 = e \in R$. But our Definition 4.1.1 is distinct from their definition. In fact, any semisimple module is a Baer module by Definition 4.1.1, but it may not be a Baer module in the sense of Lee and Zhou [274] (for example, $\mathbb{Z}_p$ as a $\mathbb{Z}$-module, where p is a prime integer, is a Baer module in our sense).

From Proposition 3.1.1, the Baer property of rings is a left-right symmetric property. The next result shows a module theoretic analogue of this fact.

Proposition 4.1.2 *The following are equivalent for a right R-module M.*

(i) *M is a Baer module.*
(ii) *For any left ideal I of S, $r_M(I) = fM$ with $f^2 = f \in S$.*

Proof (i)⇒(ii) Let I be a left ideal of S. Then $r_M(I) \leq M$, so $\ell_S(r_M(I)) = Se$ with $e^2 = e \in S$. Hence, $r_M(I) = r_M(\ell_S(r_M(I))) = r_M(Se) = (1-e)M$. Put $f = 1-e$. Then $r_M(I) = fM$ and $f^2 = f \in S$.

(ii)⇒(i) Let $N \leq M$. Then $\ell_S(N)$ is a left ideal of S. So $r_M(\ell_S(N)) = fM$ with $f^2 = f \in S$. So $\ell_S(N) = \ell_S(r_M(\ell_S(N))) = \ell_S(fM) = S(1-f)$. Let $e = 1-f$. Then $\ell_S(N) = Se$ and $e^2 = e \in S$. □

All Baer rings viewed as right modules over themselves are Baer modules, and as was noted before all semisimple modules are obviously Baer modules. Several other examples will follow from our results later (for example, if R is any Baer ring, then eR_R is a Baer module, where $e^2 = e \in R$). Next, we introduce the following concept of nonsingularity, in utilizing the endomorphism ring of a module.

Definition 4.1.3 Let M be a module. Then M is called *$\mathcal{K}$-nonsingular* if, for $\phi \in S$, $r_M(\phi) = \text{Ker}(\phi) \le^{\text{ess}} M$ implies $\phi = 0$. We observe that M is $\mathcal{K}$-nonsingular if and only if $\Delta = 0$ (see Lemma 2.1.28 for Δ).

Example 4.1.4 Any semisimple module is $\mathcal{K}$-nonsingular. In general, any Baer module is $\mathcal{K}$-nonsingular (see Lemma 4.1.18).

A module M is called *polyform* (also called *non-M-singular*) if every essential submodule of M is a dense submodule.

Proposition 4.1.5 *Every polyform module is $\mathcal{K}$-nonsingular.*

Proof Let M be a polyform right R-module. Say $\phi \in S = \text{End}(M)$ such that $\text{Ker}(\phi) \le^{\text{ess}} M$. If $\phi(M) \neq 0$, then there is $x \in M$ such that $\phi(x) \neq 0$. Since $\text{Ker}(\phi) \le^{\text{den}} M$, there exists $r \in R$ such that $xr \in \text{Ker}(\phi)$ and $\phi(x)r \neq 0$, a contradiction. Thus $\phi(M) = 0$, so $\phi = 0$. □

Corollary 4.1.6 *Every nonsingular module is $\mathcal{K}$-nonsingular.*

Proof The proof follows from the fact that every nonsingular module is polyform by Proposition 1.3.14. □

While the nonsingularity of a module M provides the uniqueness of closures in M (i.e., M is a UC-module), the $\mathcal{K}$-nonsingularity provides the uniqueness of closures which are direct summands of M.

Theorem 4.1.7 *Assume that M is a $\mathcal{K}$-nonsingular module, and let $N \le M$. If $N \le^{\text{ess}} N_i \le^{\oplus} M$, for $i = 1, 2$, then $N_1 = N_2$.*

Proof Write $M = N_1 \oplus V_1$ and $M = N_2 \oplus V_2$ with $V_1, V_2 \le M$. Consider $(1 - \pi_1)\pi_2$, where π_i is the canonical projection of M onto N_i, $i = 1, 2$. Then $((1 - \pi_1)\pi_2)N = (1 - \pi_1)(\pi_2 N) = (1 - \pi_1)(\pi_1 N) = 0$, since $N \subseteq N_1 \cap N_2$. Also $((1 - \pi_1)\pi_2)V_2 = (1 - \pi_1)(\pi_2 V_2) = 0$. Hence $N \oplus V_2 \subseteq \text{Ker}((1 - \pi_1)\pi_2)$, but $N \oplus V_2 \le^{\text{ess}} N_2 \oplus V_2 = M$. Thus, $\text{Ker}((1 - \pi_1)\pi_2) \le^{\text{ess}} M$, so $(1 - \pi_1)\pi_2 = 0$ because M is $\mathcal{K}$-nonsingular. Hence $\pi_2 = \pi_1\pi_2$. Similarly, $\pi_1 = \pi_2\pi_1$. We see that $N_2 = \pi_2(M) = \pi_1\pi_2(M) = \pi_1(N_2) \subseteq \pi_1(M) = N_1$. Similarly, $N_1 \subseteq N_2$. Therefore, $N_1 = N_2$. □

Definition 4.1.8 Let M and N be modules. Then M is called *$\mathcal{K}$-nonsingular relative to N* if for $\varphi \in \text{Hom}(M, N)$, $\text{Ker}(\varphi) \le^{\text{ess}} M$ implies $\varphi = 0$.

Theorem 4.1.9 *Let $\{M_i\}_{i \in \Lambda}$ be a set of modules. Then $\bigoplus_{i \in \Lambda} M_i$ is $\mathcal{K}$-nonsingular if and only if M_i is $\mathcal{K}$-nonsingular relative to M_j for all $i, j \in \Lambda$.*

Proof Let $M = \oplus_{i\in\Lambda} M_i$ be $\mathcal{K}$-nonsingular. Say $\varphi \in \mathrm{Hom}(M_\alpha, M_\beta)$, where $\alpha, \beta \in \Lambda$, such that $\mathrm{Ker}(\varphi) \leq^{\mathrm{ess}} M_\alpha$. Define $\psi \in \mathrm{End}(M)$ by $\psi(x) = \varphi(x)$ for $x \in M_\alpha$ and $\psi(x) = 0$ for $x \in \oplus_{i\in\Lambda, i\neq\alpha} M_i$. Now we see that $\mathrm{Ker}(\psi)$ is $\mathrm{Ker}(\varphi) \oplus (\oplus_{i\in\Lambda, i\neq\alpha} M_i)$, so $\mathrm{Ker}(\varphi) \leq^{\mathrm{ess}} M$. As M is $\mathcal{K}$-nonsingular, $\psi = 0$ and hence $\varphi = 0$.

Conversely, suppose that M_i is $\mathcal{K}$-nonsingular relative to M_j for all $i, j \in \Lambda$. Say $\phi \in \mathrm{End}(M)$ and $\mathrm{Ker}(\phi) \leq^{\mathrm{ess}} M$. Let π_j be the canonical projection of M onto M_j. As $\mathrm{Ker}(\phi) \cap M_i \leq^{\mathrm{ess}} M_i$ and $\mathrm{Ker}(\phi) \cap M_i \leq \mathrm{Ker}(\pi_j \phi|_{M_i})$ for each $i \in \Lambda$, $\mathrm{Ker}(\pi_j \phi|_{M_i}) \leq^{\mathrm{ess}} M_i$ for each $i \in \Lambda$. Because M_i is $\mathcal{K}$-nonsingular relative to M_j, $\pi_j \phi|_{M_i} = 0$ for all $j \in \Lambda$. Hence $\phi|_{M_i} = 0$ for all $i \in \Lambda$, and so $\phi = 0$. Therefore, M is $\mathcal{K}$-nonsingular. □

The next example shows that the $\mathcal{K}$-nonsingularity of modules is a proper generalization of the concepts of polyform and nonsingularity; in particular, the converse of Proposition 4.1.5 and Corollary 4.1.6 do not hold true.

Example 4.1.10 (i) The $\mathbb{Z}$-module $\mathbb{Z}_p$, where p is a prime integer, is $\mathcal{K}$-nonsingular. However, the module $\mathbb{Z}_p$ is not nonsingular.

(ii) Let $M = \mathbb{Q} \oplus \mathbb{Z}_2$. Then $\mathbb{Q} \trianglelefteq M$ and $\mathbb{Z}_2 \trianglelefteq M$ as $\mathrm{Hom}(\mathbb{Q}, \mathbb{Z}_2) = 0$ and $\mathrm{Hom}(\mathbb{Z}_2, \mathbb{Q}) = 0$. From Theorem 4.1.9, M is a $\mathcal{K}$-nonsingular $\mathbb{Z}$-module because $\mathbb{Q}$ is $\mathcal{K}$-nonsingular (in fact it is nonsingular) and $\mathbb{Z}_2$ is $\mathcal{K}$-nonsingular. But $0 \neq (0, \overline{1}) \in Z(M_{\mathbb{Z}})$, so $Z(M_{\mathbb{Z}}) \neq 0$. Consider $(0, \overline{1}), (1/2, \overline{0}) \in M$. Then there is no $a \in \mathbb{Z}$ with $(1/2, \overline{0})a \in \mathbb{Z} \oplus \mathbb{Z}_2$ and $(0, \overline{1})a \neq 0$. So $\mathbb{Z} \oplus \mathbb{Z}_2$ is not dense in M. However, $\mathbb{Z} \oplus \mathbb{Z}_2$ is essential in M. Thus, M is neither nonsingular nor polyform.

However, when the module $M = R$, the three concepts coincide.

Proposition 4.1.11 *Let R be a ring. Then R_R is $\mathcal{K}$-nonsingular if and only if R_R is nonsingular if and only if R_R is polyform.*

Proof The proof easily follows from Proposition 1.3.14 and the fact that $\mathrm{End}(R_R)$ consists of left multiplications by elements of R. Further, assume that R_R is polyform. Say $x \in Z(R_R)$. Then $xI = 0$ for some $I_R \leq^{\mathrm{ess}} R_R$. Thus $I_R \leq^{\mathrm{den}} R_R$. Therefore $x = 0$ from Proposition 1.3.11(ii), so R is right nonsingular. □

Proposition 4.1.12 *The following are equivalent for a right R-module M.*

(i) *M is $\mathcal{K}$-nonsingular.*
(ii) *For each left ideal I of S, $r_M(I) \leq^{\mathrm{ess}} eM$ with $e^2 = e \in S$ implies $I \cap Se = 0$.*
(iii) *For every left ideal J of S, $r_M(J) \leq^{\mathrm{ess}} M$ implies $J = 0$.*

Proof For (i)⇒(ii), let I be a left ideal of S such that $r_M(I) \leq^{\mathrm{ess}} eM$ with $e^2 = e \in S$. Then $r_M(I) \oplus (1-e)M \leq^{\mathrm{ess}} M$ and $r_M(I) \oplus (1-e)M \subseteq r_M(I \cap Se)$. So $r_M(I \cap Se) \leq^{\mathrm{ess}} M$, and hence $I \cap Se = 0$ by $\mathcal{K}$-nonsingularity of M.

(ii)⇒(iii) is evident. For (iii)⇒(i), take $\phi \in S$ with $\mathrm{Ker}(\phi) \leq^{\mathrm{ess}} M$. Then $\mathrm{Ker}(\phi) = r_M(S\phi)$. Hence $S\phi = 0$, so $\phi = 0$. Thus, M is $\mathcal{K}$-nonsingular. □

We introduce a module theoretic version for cononsingularity as follows.

Definition 4.1.13 A right R-module M is said to be $\mathcal{K}$-*cononsingular* if, for $N \leq M$, $\ell_S(N) = 0$ implies $N \leq^{\text{ess}} M$.

Proposition 4.1.14 *A right R-module M is $\mathcal{K}$-cononsingular if and only if, for $N \leq M$, $r_M(\ell_S(N)) \leq^{\oplus} M$ implies $N \leq^{\text{ess}} r_M(\ell_S(N))$.*

Proof Let M be $\mathcal{K}$-cononsingular and $N_R \leq M_R$. If $r_M(\ell_S(N)) = eM$ for some $e^2 = e \in S$, then $\ell_S(N) = \ell_S(r_M(\ell_S(N))) = \ell_S(eM) = S(1-e)$. Therefore, we obtain $\ell_S(N \oplus (1-e)M) = S(1-e) \cap Se = 0$. From $\mathcal{K}$-cononsingularity of M, it follows that $N \oplus (1-e)M \leq^{\text{ess}} M$. Thus, $N \leq^{\text{ess}} eM = r_M(\ell_S(N))$.

Conversely, let $V \leq M$ such that $\ell_S(V) = 0$. Then $r_M(\ell_S(V)) = M$, so $V \leq^{\text{ess}} M$ by assumption. Therefore, M is $\mathcal{K}$-cononsingular. □

Now we are ready to extend Theorem 3.3.1 for arbitrary modules in the following theorem.

Theorem 4.1.15 *A module M is extending and $\mathcal{K}$-nonsingular if and only if M is Baer and $\mathcal{K}$-cononsingular.*

The proof of Theorem 4.1.15 is comprised of the next four lemmas, each of which is of interest in its own right. These results also provide us with a good source of examples.

Lemma 4.1.16 *Each extending module is $\mathcal{K}$-cononsingular.*

Proof Let M be an extending right R-module. Take $N \leq M$ with $\ell_S(N) = 0$. Because M is extending, $N \leq^{\text{ess}} eM$ for some $e^2 = e \in S$. Thus it follows that $S(1-e) = \ell_S(eM) \subseteq \ell_S(N)$. Hence $1 - e = 0$, and so $N \leq^{\text{ess}} M$. □

Lemma 4.1.17 *Any $\mathcal{K}$-nonsingular extending module is a Baer module.*

Proof Let M be a $\mathcal{K}$-nonsingular extending right R-module. Say $N \leq M$. Then there is $e^2 = e \in S$ with $N \leq^{\text{ess}} eM$. So $S(1-e) = \ell_S(eM) \subseteq \ell_S(N)$.

Say $\phi \in \ell_S(N)$. Then $\phi eN = 0$ as $N \subseteq eM$. So $\phi e(N \oplus (1-e)M) = 0$. Now since $N \oplus (1-e)M \leq^{\text{ess}} M$, $\phi e = 0$ from the $\mathcal{K}$-nonsingularity of M. Hence, we obtain $\phi = \phi(1-e) \in S(1-e)$, so $\ell_S(N) = S(1-e)$. Thus, M is Baer. □

Lemma 4.1.18 *Every Baer module is $\mathcal{K}$-nonsingular.*

Proof Let M be a Baer right R-module. Say $\phi \in S$ with $\text{Ker}(\phi) \leq^{\text{ess}} M$. Since M is Baer, $\text{Ker}(\phi) = r_M(S\phi) = fM$ for some $f^2 = f \in S$ by Proposition 4.1.2. Thus $\text{Ker}(\phi) = M$, and so $\phi = 0$. Hence, M is $\mathcal{K}$-nonsingular. □

Lemma 4.1.19 *Any $\mathcal{K}$-cononsingular Baer module is extending.*

Proof Let a right R-module M be $\mathcal{K}$-cononsingular and Baer. By Lemma 4.1.18, M is $\mathcal{K}$-nonsingular. To show that M is extending, say $N \leq M$. Then $\ell_S(N) = Sf$ with $f^2 = f \in S$. So $N \subseteq r_M(\ell_S(N)) = (1-f)M$. Assume on the contrary that N is not essential in $(1-f)M$. Then there is $0 \neq P \leq (1-f)M$ with $N \cap P = 0$. Let K be a complement of P in M such that $N \subseteq K$. Note that $\ell_S(K) \neq 0$ by $\mathcal{K}$-cononsingularity of M as K is not essential in M. Take $0 \neq s \in \ell_S(K) \subseteq \ell_S(N) = Sf$, so $s(1-f) = 0$. Hence $s(1-f)M = 0$, and thus $sP = 0$. So $s(K \oplus P) = 0$. But $K \oplus P \leq^{\text{ess}} M$. By $\mathcal{K}$-nonsingularity of M, $s = 0$, a contradiction. Thus, we obtain $N \leq^{\text{ess}} (1-f)M$, so M is extending. □

Definition 4.1.20 A module M is said to have the *summand intersection property* (*SIP*) if the intersection of any two direct summands of M is a direct summand. A module M is said to have the *strong summand intersection property* (*SSIP*) if the intersection of any family of direct summands of M is a direct summand.

Every free module over a commutative PID has the SIP (see [247, Exercise 51(b), p. 49]). If R is a regular ring, which is not a Baer ring, then R satisfies the SIP, but R does not have the SSIP. A characterization of Baer modules via the SSIP is provided as follows.

Theorem 4.1.21 *Let M be a right R-module. Then M is Baer if and only if M has the SSIP and* $\text{Ker}(\phi) \leq^{\oplus} M$ *for any* $\phi \in S$.

Proof Let M be Baer and let $\{e_i M\}$ be a set of direct summands of M, where each $e_i^2 = e_i \in S$. Then $\cap_i e_i M = r_M(\sum_i S(1-e_i))$. Since M is Baer, we get that $r_M(\sum_i S(1-e_i)) \leq^{\oplus} M$ by Proposition 4.1.2. Thus $\cap_i e_i M \leq^{\oplus} M$, so M has the SSIP. Next, $\text{Ker}(\phi) = r_M(S\phi) = eM$ for some $e^2 = e \in S$ from Proposition 4.1.2. Hence $\text{Ker}(\phi) \leq^{\oplus} M$.

Conversely, assume that I is a left ideal of S. Then $\text{Ker}(f) \leq^{\oplus} M$ for each f in I, by assumption. Also we have that $r_M(I) = \cap_{f \in I} \text{Ker}(f) \leq^{\oplus} M$ by the SSIP. Therefore M is Baer. □

Theorem 4.1.22 *Every direct summand of a Baer module is a Baer module.*

Proof Let M be a Baer right R-module and $N \leq^{\oplus} M$. Say $M = N \oplus P$ for some $P \leq M$. By Theorem 4.1.21, M satisfies the SSIP. Thus, N also has the SSIP by using the modular law. Let $H = \text{End}(N)$. Then for $h \in H$, there exists $\phi \in S$ defined as $\phi = h \oplus 0$, where 0 is the zero map of P. Since $\phi \in S$, $\text{Ker}(\phi) \leq^{\oplus} M$ from Theorem 4.1.21. Thus, $\text{Ker}(h) \oplus P \leq^{\oplus} M$.

Say $M = \text{Ker}(h) \oplus P \oplus V$ with $V \leq M$. Then $N = \text{Ker}(h) \oplus (N \cap (V \oplus P))$ by the modular law since $\text{Ker}(h) \leq N$. So $\text{Ker}(h) \leq^{\oplus} N$. Hence, N is a Baer module by Theorem 4.1.21. □

Example 4.1.23 Let R be a Baer ring and $e^2 = e \in R$. Then by Theorem 4.1.22, the R-module eR_R is Baer.

As an application of the previous results, we obtain the next characterization of a certain class of Baer modules over a commutative PID.

Theorem 4.1.24 *Let $M = \bigoplus_{\alpha \in \Lambda} M_\alpha$ be a direct sum of cyclic R-modules M_α, where Λ is a countable set, over a commutative PID R. Then the following are equivalent.*

(i) *M is a Baer module.*
(ii) *M is either semisimple or torsion-free.*

Proof (i)⇒(ii) Let M be a Baer module. Say $M = t(M) \oplus f(M)$, where $t(M)$ is the torsion submodule of M and $f(M)$ is the torsion-free submodule of M. As each M_α is cyclic, note that $f(M_\alpha) \cong R$ or $t(M_\alpha) = \oplus_{P_i \in \mathcal{P}_\alpha} R/{P_i}^{n_i}$, where $\mathcal{P}_\alpha$ is a finite set of nonzero prime ideals P_i and n_i is a positive integer (depending on P_i) for each $P_i \in \mathcal{P}_\alpha$.

Suppose that $f(M) \neq 0$ and $t(M) \neq 0$. Then there are M_α and M_β such that $M_\alpha \cong R$ and $M_\beta \cong R/{P_0}^{n_0} \oplus (\oplus_{P_i \in \mathcal{P}} R/{P_i}^{n_i})$, where P_0 and each P_i are nonzero prime ideals of R such that $n_0 \geq 1$.

Let $\varphi : R \to R/{P_0}^{n_0}$ be the homomorphism defined by $\varphi(x) = x + {P_0}^{n_0}$ for $x \in R$. Then $\text{Ker}(\varphi) = {P_0}^{n_0}$ is not a direct summand of R_R. Say $\eta : M_\alpha \to R$ is the given isomorphism and $\mu : R/{P_0}^{n_0} \to M_\beta$ is the given monomorphism. Put $\phi = \mu\varphi\eta$. Then $\text{Ker}(\phi)$ is not a direct summand of M_α.

Write $M = M_\alpha \oplus V$ for some $V \leq M$. Define $h : M = M_\alpha \oplus V \to M$ by $h(x + v) = \phi(x)$ for $x \in M_\alpha$ and $v \in V$. Then $\text{Ker}(h) = \text{Ker}(\phi) \oplus V$. Since M is Baer, $M = \text{Ker}(h) \oplus U$ for some $U \leq M$ by Theorem 4.1.21. Then $M = \text{Ker}(\phi) \oplus (V \oplus U)$. By the modular law, $M_\alpha = \text{Ker}(\phi) \oplus (M_\alpha \cap (V \oplus U))$, a contradiction. Therefore, $f(M) = 0$ or $t(M) = 0$.

Assume that $f(M) = 0$. Then $t(M) = M = \oplus_{\alpha \in \Lambda} M_\alpha$. We note that each $M_\alpha \cong \oplus_{P_i \in \mathcal{P}_\alpha} R/{P_i}^{n_i}$, where each P_i is a nonzero prime ideal of R and each n_i is a positive integer. So every $R/{P_i}^{n_i}$ is a Baer module by Theorem 4.1.22.

Suppose that there exists P_i such that $n_i > 1$. Note that we can choose $a \in {P_i}^{n_i - 1} \setminus {P_i}^{n_i}$. Define $g : R/{P_i}^{n_i} \to R/{P_i}^{n_i}$ by $g(x + {P_i}^{n_i}) = ax + {P_i}^{n_i}$. Then $g(1 + {P_i}^{n_i}) = a + {P_i}^{n_i} \neq 0$. So $g \neq 0$. Further, $0 \neq a + {P_i}^{n_i} \in \text{Ker}(g)$ because $n_i > 1$. Since $\text{Ker}(g) \leq^{\text{ess}} R/{P_i}^{n_i}$ and $\text{Ker}(g) \neq R/P_i^{n_i}$, $\text{Ker}(g)$ is not a direct summand of $R/{P_i}^{n_i}$. Thus $R/{P_i}^{n_i}$ is not a Baer module by Theorem 4.1.21, a contradiction. So each $n_i = 1$. Therefore $M = t(M) = \oplus_{P_i \in \mathcal{P}} R/P_i$, where $\mathcal{P}$ is a set of nonzero prime ideals of R (possibly in multiple instance). Thus M is semisimple because every nonzero prime ideal of a commutative PID is maximal.

Next, if $t(M) = 0$, then M is torsion-free.

(ii)⇒(i) If M is semisimple, then M is trivially a Baer module. Next, assume that M is torsion-free. Then M is a free module of countable rank over R. As R is a commutative PID, M has the SSIP (see [247, Exercise 51(c), p. 49]). Let $\varphi \in \text{End}_R(M)$. Since R is a commutative PID, R is a hereditary ring. So $\varphi(M)$ is projective as M is free and $\varphi(M) \leq M$. Thus, $\text{Ker}(\varphi)$ is a direct summand of M. Hence, M is a Baer module by Theorem 4.1.21. □

All finitely generated Baer modules over a commutative PID can be characterized as follows.

Corollary 4.1.25 *Let M be a finitely generated module over a commutative PID. Then M is a Baer module if and only if either M is semisimple Artinian or M is torsion-free.*

Theorem 4.1.26 *A module M is an indecomposable Baer module if and only if any nonzero endomorphism of M is a monomorphism.*

Proof Let M be indecomposable and Baer. Take $0 \neq \phi \in S$. By Proposition 4.1.2, $\text{Ker}(\phi) = \text{Ker}(S\phi) \leq^{\oplus} M$. Thus, $\text{Ker}(\phi) = 0$ or $\text{Ker}(\phi) = M$, and so $\text{Ker}(\phi) = 0$. Hence, ϕ is a monomorphism.

Conversely, assume on the contrary that $M = M_1 \oplus M_2$ with $M_1 \neq 0$ and $M_2 \neq 0$. Take $\phi = \pi_1$, the canonical projection of M onto M_1. Then $\phi \neq 0$ and $\text{Ker}\ (\phi) = M_2 \neq 0$, a contradiction. Thus, M is indecomposable. Next, let I be a left ideal of S. Then $r_M(I) = 0$ (if $I \neq 0$) or $r_M(I) = M$ (if $I = 0$). Thus, M is Baer. □

Corollary 4.1.27 *Every indecomposable Baer module is Hopfian* (*i.e., every epimorphism is an isomorphism*).

Proof Let M be an indecomposable Baer module and $f \in S$ an epimorphism. Then $f \neq 0$, so $\text{Ker}(f) = 0$ by Theorem 4.1.26. So f is an isomorphism. □

Exercise 4.1.28

1. ([359, Rizvi and Roman]) Let M be a $\mathcal{K}$-nonsingular module. Show that $M^{(\Lambda)}$ for any index set Λ and any direct summand of M are $\mathcal{K}$-nonsingular.
2. ([357, Rizvi and Roman]) Let M be a $\mathcal{K}$-nonsingular module, $V \trianglelefteq M$, and $V \leq^{\text{ess}} N \leq^{\oplus} M$. Prove that $N \trianglelefteq M$.
3. ([359, Rizvi and Roman]) Assume that M is a module and $S = \text{End}(M)$. Let $Z^{\mathcal{K}}(M) = \sum\{\varphi(M) \mid \varphi \in S \text{ and } \text{Ker}(\varphi) \leq^{\text{ess}} M\}$, which is called the *$\mathcal{K}$-singular submodule* of M. Show that a module M is $\mathcal{K}$-nonsingular if and only if $Z^{\mathcal{K}}(M) = 0$.
4. ([359, Rizvi and Roman]) Let M be a module. Prove the following.
   (i) $Z^{\mathcal{K}}(M) \trianglelefteq M$ and $Z^{\mathcal{K}}(M) \subseteq Z(M)$.
   (ii) If $E(M)$ is $\mathcal{K}$-nonsingular, then M is $\mathcal{K}$-nonsingular.
5. ([357, Rizvi and Roman]) Let M be a $\mathcal{K}$-nonsingular module. Prove that M is FI-extending if and only if M is strongly FI-extending (cf. Theorem 2.3.27).

4.2 Direct Sums and Endomorphism Rings of Baer Modules

We begin with showing the connections between the Baer property of a module and that of its endomorphism ring. A characterization of a Baer module via its endomorphism ring is established. We shall see that a direct sum of Baer modules may not be Baer, in general. Some results on when a direct sum of Baer modules is Baer are obtained. We start with the definition of retractable modules and their generalizations.

Definition 4.2.1 Let M be a right R-module and $S = \mathrm{End}(M)$. Then M is said to be *retractable* if, for every $0 \neq N \leq M$, there exists $0 \neq \phi \in S$ such that $\phi(M) \subseteq N$ (i.e., $\mathrm{Hom}(M, N) \neq 0$).

Examples of retractable modules include free modules, generators, and semisimple modules. For a full characterization of a Baer module via its endomorphism ring, a more general form of retractability (which we shall see that every Baer module satisfies) is defined as follows.

Definition 4.2.2 Let M be a right R-module and $S = \mathrm{End}(M)$. Then M is called *quasi-retractable* if $\mathrm{Hom}(M, r_M(I)) \neq 0$ for every left ideal I of S with $r_M(I) \neq 0$.

We remark that a module M_R is quasi-retractable if and only if $r_S(I) \neq 0$ for every left ideal I of S with $r_M(I) \neq 0$. The following fact shows that the concept of quasi-retractability is a generalization of retractability.

Proposition 4.2.3 *Every retractable module is quasi-retractable.*

Proof Assume that M_R is a retractable module. Take a left ideal I of S such that $r_M(I) \neq 0$. By retractability, there is $\phi \in S$ with $0 \neq \phi(M) \subseteq r_M(I)$. As a consequence, $0 \neq \phi \in \mathrm{Hom}(M, r_M(I))$. Therefore, M_R is quasi-retractable. □

The next example exhibits a module which is quasi-retractable but not retractable, showing that the class of retractable modules is a proper subclass of the class of quasi-retractable modules.

Example 4.2.4 Let K be a field. Put

$$R = \begin{bmatrix} K & K \\ 0 & K \end{bmatrix} \text{ and } e = \begin{bmatrix} 1 & 0 \\ 0 & 0 \end{bmatrix} \in R.$$

Consider the module $M = eR$. Observe that $S = \mathrm{End}(M_R) \cong eRe \cong K$, which is a field. Let I be a left ideal of S such that $r_M(I) \neq 0$. Then $I = 0$ and so $r_M(I) = M$. Hence, $\mathrm{Hom}(M_R, r_M(I)) = \mathrm{End}(M_R) \cong K \neq 0$. Thus, M_R is quasi-retractable. But M_R is not retractable, since the endomorphism ring of M_R, which is isomorphic to K, consists of isomorphisms and the zero endomorphism. On the other hand, as M_R is not simple, retractability of M_R implies that there exist nonzero endomorphisms of M_R which are not onto (note that M_R is extending and nonsingular, hence M_R is Baer by Theorem 4.1.15).

The following result shows that the property of retractability passes to arbitrary direct sums of retractable modules.

Proposition 4.2.5 *Let $\{M_i\}_{i\in\Lambda}$ be a set of retractable right R-modules. Then $\bigoplus_{i\in\Lambda} M_i$ is retractable.*

Proof Note that retractability of a right R-module M is equivalent to the fact that for each $0 \neq x \in M$ there is $0 \neq \phi \in \mathrm{End}(M)$ with $\phi(M) \subseteq xR$. Let $0 \neq x \in \oplus_{i\in\Lambda} M_i$. Then there is a finite subset $F \subseteq \Lambda$ such that $x \in \oplus_{i\in F} M_i$. So $xR \subseteq \oplus_{i\in F} M_i$. Hence, it suffices to show that any finite direct sum of retractable modules is retractable.

Let M_1 and M_2 be retractable and $0 \neq N \leq M_1 \oplus M_2$. If $N \cap M_1 \neq 0$, then we are done by the retractability of M_1. Let $\pi_2 : M_1 \oplus M_2 \to M_2$ be the canonical projection. If $N \cap M_1 = 0$, then $\pi_2(N) \cong N$, so $0 \neq \pi_2(N) \subseteq M_2$. As M_2 is retractable, $\mathrm{Hom}(M_2, \pi_2(N)) \neq 0$. Hence $\mathrm{Hom}(M_2, N) \neq 0$, so $\mathrm{Hom}(M_1 \oplus M_2, N) \neq 0$. Thus, $M_1 \oplus M_2$ is retractable. Similarly, any finite direct sum of retractable modules is retractable. □

A direct summand of a retractable module may not be retractable, as the following example demonstrates.

Example 4.2.6 Let M be a right R-module which is not retractable. Take $P = R \oplus M$. Then the module P_R is retractable. Indeed, let $0 \neq N \leq P$ and take $0 \neq x \in N$. Define a homomorphism f from P to N by sending $1 \in R$ to x and mapping elements from M to 0. Then $0 \neq f \in \mathrm{Hom}(P, N)$.

Proposition 4.2.7 *A module M is retractable if and only if any direct sum of copies of M is retractable.*

Proof The necessity follows from Proposition 4.2.5. For the sufficiency, let $M^{(\Lambda)}$ be retractable for a set Λ. Take $0 \neq N \leq M$, and view M as one of the direct summand of $M^{(\Lambda)}$. Therefore, $N \leq M \leq M^{(\Lambda)}$. Hence, there exists $0 \neq \phi \in \mathrm{End}(M^{(\Lambda)})$ with $\phi(M^{(\Lambda)}) \subseteq N$. As $\phi \neq 0$, there is $k \in \Lambda$ such that $\phi\,\iota_k \neq 0$, where ι_k is the canonical injection of the k-th coordinate in $M^{(\Lambda)}$. So $0 \neq \phi\,\iota_k \in \mathrm{Hom}(M, N)$, thus M is retractable. □

Next, we provide a characterization of a module M whose endomorphism ring is a Baer ring.

Theorem 4.2.8 *A module M is a Baer module if and only if $S = \mathrm{End}(M)$ is a Baer ring and M is quasi-retractable.*

Proof Assume that M_R is Baer. Let I be a left ideal of S. Then $r_M(I) = eM$ with $e^2 = e \in S$ by Proposition 4.1.2. We show that $r_S(I) = eS$. Indeed, as $IeM = 0$ from $r_M(I) = eM$, $Ie = 0$ and so $IeS = 0$. Thus, $eS \subseteq r_S(I)$. Next, say $\phi \in r_S(I)$. Since $I\phi = 0$, $I\phi(M) = 0$ and so $\phi(M) \subseteq r_M(I) = eM$. Hence, $\phi(m) = e\phi(m)$ for each $m \in M$, so $\phi = e\phi$. Thus, $r_S(I) \subseteq eS$. Therefore $r_S(I) = eS$. Thus, S is a Baer ring. To see that M_R is quasi-retractable, suppose that $r_M(J) \neq 0$, where J is a left ideal of S. Then $r_M(J) = fM$ with $0 \neq f^2 = f \in S$ from Proposition 4.1.2, as M is a Baer module. Thus $0 \neq f \in \text{Hom}(M, r_M(J))$, so M is quasi-retractable.

Conversely, let S be a Baer ring and M_R be quasi-retractable. Say I is a left ideal of S. As S is a Baer ring, $r_S(I) = eS$ for some $e^2 = e \in S$. So

$$I \subseteq \ell_S(r_S(I)) = S(1-e),$$

hence $eM \subseteq r_M(I)$. Next, let $m \in r_M(I)$. Then $0 = Im = I(em + (1-e)m)$. Thus, we obtain $I(1-e)m = 0$ as $Ie = 0$. Put $J = I + Se$. Then

$$r_S(J) = r_S(I) \cap r_S(Se) = eS \cap (1-e)S = 0.$$

Thus, $r_M(J) = 0$ because M is quasi-retractable.

Now $J(1-e)m = 0$ since $I(1-e)m = 0$. Hence, $(1-e)m \in r_M(J)$. Consequently, $m = em$ and thus $r_M(I) \subseteq eM$. So $r_M(I) = eM$. Therefore, M is a Baer module by Proposition 4.1.2. □

The following example shows that there is a module M whose endomorphism ring S is Baer, but M is not a Baer module. Thus, quasi-retractability of M is required for it to be Baer.

Example 4.2.9 Let $M = \mathbb{Z}_{p^\infty}$ (p a prime integer), considered as a $\mathbb{Z}$-module. Then $S = \text{End}(M) = \widehat{\mathbb{Z}}_p$, the ring of p-adic integers. Hence, S is a Baer ring. But M is not a Baer module.

Corollary 4.2.10 *A module M is an indecomposable Baer module if and only if $S = \text{End}(M)$ is a domain and M is quasi-retractable.*

Proof Let M be indecomposable Baer. By Theorem 4.2.8 S is a Baer ring. Say $x, y \in S$ with $xy = 0$. If $x \neq 0$, then $r_S(x) = 0$ as S is a Baer ring with trivial idempotents, so $y = 0$. Thus, S is a domain. From Theorem 4.2.8, M is quasi-retractable. Conversely, if S is a domain and M is quasi-retractable, then S is a Baer ring, so M is a Baer module by Theorem 4.2.8. Since 0 and 1 are only idempotents in S, M is indecomposable. □

The next corollary exhibits a relationship between certain Baer modules and extending modules when the base ring is a semiprime PI-ring.

Corollary 4.2.11 *Let R be a semiprime PI-ring and n a positive integer. Then $R_R^{(n)}$ is a Baer module if and only if $R_R^{(n)}$ is an extending module.*

Proof Note that $R_R^{(n)}$ is quasi-retractable. By Theorem 4.2.8, $R_R^{(n)}$ is a Baer module if and only if $\text{Mat}_n(R)$ is a Baer ring. The rest of the proof follows from Corollary 3.3.5 (see also Exercise 6.1.18.1). □

In general, a direct sum of Baer modules is not a Baer module, as the next examples show.

Example 4.2.12 (i) Note that $\mathbb{Z}$ and $\mathbb{Z}_2$ are Baer $\mathbb{Z}$-modules, but $\mathbb{Z} \oplus \mathbb{Z}_2$ is not a Baer $\mathbb{Z}$-module by Corollary 4.1.25.

(ii) Even a (finite) direct sum of copies of a Baer module may not be Baer. Let $M = \mathbb{Z}[x]$. Then $M \oplus M$ is not Baer as a $\mathbb{Z}[x]$-module (see Example 3.1.28 and Theorem 4.2.8).

Definition 4.2.13 Let M and N be R-modules. We say that M is *N-Rickart* if $\text{Ker}(\phi) \leq^{\oplus} M$ for each $\phi \in \text{Hom}(M, N)$. Modules M and N are said to be *relatively Rickart* if M is N-Rickart and N is M-Rickart.

Proposition 4.2.14 *Let M and N be modules. If M is N-Rickart, then U is V-Rickart for any $U \leq^{\oplus} M$ and $V \leq N$.*

Proof There exists $e^2 = e \in \text{End}(M)$ with $U = eM$. Let $\psi \in \text{Hom}(eM, V)$. Define $\phi \in \text{Hom}(M, N)$ by $\phi(m) = \psi(em)$ for $m \in M$. Since M is N-Rickart, $\text{Ker}(\phi) = \text{Ker}(\psi) \oplus (1-e)M \leq^{\oplus} M$. Say $M = \text{Ker}(\psi) \oplus (1-e)M \oplus K$ for some $K \leq M$. Therefore $eM = \text{Ker}(\psi) \oplus [eM \cap ((1-e)M \oplus K)]$ by the modular law because $\text{Ker}(\psi) \leq eM$. So $\text{Ker}(\psi) \leq^{\oplus} eM$, and thus eM is V-Rickart. □

Proposition 4.2.15 *Let $\{M_\lambda\}_{\lambda \in \Lambda}$ be a set of modules. If $\bigoplus_{\lambda \in \Lambda} M_\lambda$ is a Baer module, then for all $i, j \in \Lambda$, M_i and M_j are relatively Rickart.*

Proof Put $M = \oplus_{\lambda \in \Lambda} M_\lambda$. Then M is M-Rickart by Theorem 4.1.21. From Proposition 4.2.14, for all $i, j \in \Lambda$, M_i and M_j are relatively Rickart. □

Lemma 4.2.16 *Let $\{M_i\}_{1 \leq i \leq n}$ be a finite set of modules such that M_i is M_j-injective for all $i < j$, and let N be a module. Then $\bigoplus_{i=1}^n M_i$ is N-Rickart if and only if M_i is N-Rickart for all $i = 1, \ldots, n$.*

Proof The necessity follows from Proposition 4.2.14. Conversely, suppose that M_i is N-Rickart for all i. We show that $\oplus_{i=1}^n M_i$ is N-Rickart by induction on n. Start with $n = 2$. Suppose that M_i is N-Rickart for $i = 1, 2$ and M_1 is M_2-injective. Say $\varphi \in \text{Hom}(M_1 \oplus M_2, N)$. Let $\varphi_1 = \varphi|_{M_1}$ and $\varphi_2 = \varphi|_{M_2}$.

We claim that $\text{Ker}(\varphi) \leq^{\oplus} M_1 \oplus M_2$. For the claim, set $K = \text{Ker}(\varphi)$. As M_i is N-Rickart, there exists V_i such that $M_i = \text{Ker}(\varphi_i) \oplus V_i$ for $i = 1, 2$, and hence $M = \text{Ker}(\varphi_1) \oplus \text{Ker}(\varphi_2) \oplus V_1 \oplus V_2$. Since $\text{Ker}(\varphi_1) \oplus \text{Ker}(\varphi_2) \leq K$, by the modular law $K = \text{Ker}(\varphi_1) \oplus \text{Ker}(\varphi_2) \oplus (K \cap (V_1 \oplus V_2))$. Also, since $V_i \leq^{\oplus} M_i$, V_i is N-Rickart for $i = 1, 2$ by Proposition 4.2.14.

Now to see that $V_1 \cap K = \text{Ker}(\varphi|_{V_1}) = 0$, we take $v_1 \in V_1 \cap K$. Then it follows that $0 = \varphi(v_1) = \varphi_1(v_1)$, so $v_1 \in \text{Ker}(\varphi_1) \cap V_1 = 0$. Hence, $V_1 \cap K = 0$. So

$$V_1 \cap [K \cap (V_1 \oplus V_2)] = V_1 \cap K = \text{Ker}(\varphi|_{V_1}) = 0.$$

Since V_1 is V_2-injective from Propositions 2.1.5 and 2.1.6, there is $L \leq V_1 \oplus V_2$ such that $K \cap (V_1 \oplus V_2) \leq L$ and $V_1 \oplus V_2 = V_1 \oplus L$ from Theorem 2.1.4.

We see that $\text{Ker}(\varphi|_L) = K \cap L \subseteq K \cap (V_1 \oplus V_2) \subseteq L$, so we get

$$K \cap L \subseteq K \cap (V_1 \oplus V_2) \subseteq K \cap L.$$

Thus, $\text{Ker}(\varphi|_L) = K \cap L = K \cap (V_1 \oplus V_2)$.

Now $\varphi|_L \in \text{Hom}(L, N)$ and $L \cong V_2$, thus L is N-Rickart because V_2 is N-Rickart. So $K \cap (V_1 \oplus V_2) = \text{Ker}(\varphi|_L) \leq^{\oplus} L$. Say $L = [K \cap (V_1 \oplus V_2)] \oplus U$ for some submodule $U \leq L$. Then $V_1 \oplus V_2 = V_1 \oplus L = V_1 \oplus [K \cap (V_1 \oplus V_2)] \oplus U$. Therefore,

$$M = M_1 \oplus M_2 = \text{Ker}(\varphi_1) \oplus \text{Ker}(\varphi_2) \oplus V_1 \oplus V_2 = K \oplus V_1 \oplus U,$$

as $K = \text{Ker}(\varphi_1) \oplus \text{Ker}(\varphi_2) \oplus (K \cap (V_1 \oplus V_2))$. Hence, $M_1 \oplus M_2$ is N-Rickart.

Put $W_n := \oplus_{i=1}^{n-1} M_i$. From Proposition 2.1.5, W_n is M_n-injective. By induction hypothesis, W_n is N-Rickart. Because M_n is N-Rickart, $W_n \oplus M_n$ is N-Rickart by similar arguments as in the preceding case for $n = 2$. □

In the next result, necessary conditions are provided for a direct sum of Baer modules to be Baer (cf. Example 4.2.12). Also, under the additional requirement of relative injectivity for each M_i, we obtain the converse for Proposition 4.2.15.

Theorem 4.2.17 *Let $\{M_i\}_{1 \leq i \leq n}$ be a finite set of Baer modules. Assume that, for any $i \neq j$, M_i and M_j are relatively Rickart, and M_i is M_j-injective for any $i < j$. Then $\bigoplus_{i=1}^n M_i$ is a Baer module.*

Proof The result will be shown by induction n. We start with $n = 2$.

Step 1. Assume that $n = 2$. Let $\{\phi_j\}_{j \in \Lambda}$ be a set of endomorphisms of $M_1 \oplus M_2$. We claim that

$$K := \cap_{j \in \Lambda} \text{Ker}(\phi_j) \leq^{\oplus} M_1 \oplus M_2.$$

Let π_1, π_2 be the canonical projections of $M_1 \oplus M_2$ onto M_1 and M_2, respectively, and ι_1, ι_2 be the canonical injections. Then we see that

$$\text{Ker}(\phi_j) \cap M_1 = \text{Ker}(\pi_1 \phi_j \iota_1) \cap \text{Ker}(\pi_2 \phi_j \iota_1).$$

Note that $\text{Ker}(\pi_1 \phi_j \iota_1) \leq^{\oplus} M_1$, since M_1 is Baer. Also as M_1 and M_2 are relatively Rickart, we have that $\text{Ker}(\pi_2 \phi_j \iota_1) \leq^{\oplus} M_1$. Note that M_1 has the SSIP by Theorem 4.1.21. Therefore,

$$\text{Ker}(\phi_j) \cap M_1 = \text{Ker}(\pi_1 \phi_j \iota_1) \cap \text{Ker}(\pi_2 \phi_j \iota_1) \leq^{\oplus} M_1.$$

Further, $K \cap M_1 = [\cap_{j \in \Lambda} \text{Ker}(\phi_j)] \cap M_1 = \cap_{j \in \Lambda} (\text{Ker}(\phi_j) \cap M_1) \leq^{\oplus} M_1$ by the SSIP of M_1. Similarly, $K \cap M_2 \leq^{\oplus} M_2$.

Say $M_1 = (K \cap M_1) \oplus U_1$ with some $U_1 \leq M_1$ and $M_2 = (K \cap M_2) \oplus U_2$ with some $U_2 \leq M_2$. So $M_1 \oplus M_2 = (K \cap M_1) \oplus (K \cap M_2) \oplus U_1 \oplus U_2$. By the modular law, we obtain that $K = (K \cap M_1) \oplus (K \cap M_2) \oplus (K \cap (U_1 \oplus U_2))$. Now, we set $W = K \cap (U_1 \oplus U_2)$. Then

$$K = (K \cap M_1) \oplus (K \cap M_2) \oplus W$$

and $W \cap U_1 = K \cap (U_1 \oplus U_2) \cap U_1 = K \cap U_1 = (K \cap M_1) \cap U_1 = 0$ as $U_1 \subseteq M_1$. Similarly, $W \cap U_2 = 0$.

It suffices to show that $W \leq^{\oplus} U_1 \oplus U_2$ for the claim. By assumption M_1 is M_2-injective. So U_1 is U_2-injective by Propositions 2.1.5 and 2.1.6. Since $W \leq U_1 \oplus U_2$

and $U_1 \cap W = 0$, there is $V_2 \leq U_1 \oplus U_2$ with $U_1 \oplus U_2 = U_1 \oplus V_2$ and $W \leq V_2$ by Theorem 2.1.4. Hence, $U_2 \cong V_2$.

For $j \in \Lambda$, we put $\psi_j = \phi_j|_{U_1 \oplus U_2} : U_1 \oplus U_2 \to M_1 \oplus M_2$. Then clearly $\text{Ker}(\psi_j) = \text{Ker}(\phi_j) \cap (U_1 \oplus U_2)$, therefore

$$\cap_{j\in\Lambda}\text{Ker}(\psi_j) = [\cap_{j\in\Lambda}\text{Ker}(\phi_j)] \cap (U_1 \oplus U_2) = K \cap (U_1 \oplus U_2) = W.$$

Next, we observe that

$$\begin{aligned} M_1 \oplus M_2 &= (K \cap M_1) \oplus (K \cap M_2) \oplus U_1 \oplus U_2 \\ &= (K \cap M_1) \oplus (K \cap M_2) \oplus U_1 \oplus V_2. \end{aligned}$$

Let q_1, q_2, q_3, and q_4 be the canonical projections from $M_1 \oplus M_2$ onto $K \cap M_1$, $K \cap M_2$, U_1, and V_2, respectively. Also let $\kappa : V_2 \to U_1 \oplus V_2 = U_1 \oplus U_2$ be the canonical injection.

Note that $x \in W$ if and only if $\psi_j\kappa(x) = 0$ for all $j \in \Lambda$ if and only if $q_1\psi_j\kappa(x) = 0$, $q_2\psi_j\kappa(x) = 0$, $q_3\psi_j\kappa(x) = 0$, and $q_4\psi_j\kappa(x) = 0$ for all $j \in \Lambda$, because $W \subseteq V_2$. Therefore,

$$W = \cap_{j\in\Lambda}[\text{Ker}(q_1\psi_j\kappa) \cap \text{Ker}(q_2\psi_j\kappa) \cap \text{Ker}(q_3\psi_j\kappa) \cap \text{Ker}(q_4\psi_j\kappa)].$$

Because M_2 is M_1-Rickart and $M_2 = (K \cap M_2) \oplus U_2$, U_2 is $K \cap M_1$-Rickart by Proposition 4.2.14.

Further, as $V_2 \cong U_2$, V_2 is $K \cap M_1$-Rickart. So $\text{Ker}(q_1\psi_j\kappa) \leq^{\oplus} V_2$. Note that $M_1 = (K \cap M_1) \oplus U_1$, $M_2 = (K \cap M_2) \oplus U_2$, and M_1, M_2 are relatively Rickart. So Proposition 4.2.14 yields that U_2 is U_1-Rickart. Hence V_2 is U_1-Rickart as $V_2 \cong U_2$, and so $\text{Ker}(q_3\psi_j\kappa) \leq^{\oplus} V_2$.

Since $M_2 = (K \cap M_2) \oplus U_2$ is a Baer module, U_2 is $K \cap M_2$-Rickart by Proposition 4.2.15. Thus, V_2 is $K \cap M_2$-Rickart as $V_2 \cong U_2$. So we have that $\text{Ker}(q_2\psi_j\kappa) \leq^{\oplus} V_2$. As U_2 is a Baer module by Theorem 4.1.22, U_2 is U_2-Rickart from Theorem 4.1.21, so V_2 is V_2-Rickart. Hence, $\text{Ker}(q_4\psi_j\kappa) \leq^{\oplus} V_2$. Since U_2 is a Baer module and $V_2 \cong U_2$, V_2 is a Baer module. Hence, V_2 has the SSIP by Theorem 4.1.21. So $W \leq^{\oplus} V_2 \leq^{\oplus} U_1 \oplus U_2$. Thus $K \leq^{\oplus} M_1 \oplus M_2$.

Let I be a left ideal of $S = \text{End}(M_1 \oplus M_2)$. Say $I = \sum_{i\in\Lambda} S\phi_i$ with $\phi_i \in S$ and $i \in \Lambda$. Then $r_{M_1\oplus M_2}(I) = \cap_{i\in\Lambda}\text{Ker}(\phi_i) \leq^{\oplus} M_1 \oplus M_2$ by the preceding argument. So $r_{M_1\oplus M_2}(I) = f(M_1 \oplus M_2)$ for some $f^2 = f \in S$. Thus, $M_1 \oplus M_2$ is a Baer module by Proposition 4.1.2.

Step 2. Assume that $\oplus_{i=1}^{n-1} M_i$ is a Baer module. We claim that $\oplus_{i=1}^{n-1} M_i$ and M_n are relatively Rickart. First, to show that M_n is $(\oplus_{i=1}^{n-1} M_i)$-Rickart, we assume that $f : M_n \to \oplus_{i=1}^{n-1} M_i$ is a homomorphism. Let π_i be the canonical projections from $\oplus_{i=1}^{n-1} M_i$ onto M_i, for i, $1 \leq i \leq n-1$. Then $\text{Ker}(f) = \cap_{i=1}^{n-1}\text{Ker}(\pi_i f)$. Because M_n is M_i-Rickart for i, where $1 \leq i \leq n-1$, $\text{Ker}(\pi_i f)$ is a direct summand of M_n for each i, $1 \leq i \leq n-1$. Thus, $\text{Ker}(f) \leq^{\oplus} M_n$ from the SSIP because M_n is a Baer module (see Theorem 4.1.21). So M_n is $(\oplus_{i=1}^{n-1} M_i)$-Rickart.

Next, by Lemma 4.2.16, $\oplus_{i=1}^{n-1} M_i$ is M_n-Rickart. We note that $\oplus_{i=1}^{n-1} M_i$ is M_n-injective by Proposition 2.1.5. Hence, $\oplus_{i=1}^{n} M_i$ is a Baer module by the argument in Step 1. □

Theorem 4.2.18 *Let V be a nonsingular injective module and M be a nonsingular extending module. Then $V \oplus M$ is a Baer module.*

Proof By Theorem 4.1.15, $V \oplus E(M)$ is a Baer module because $V \oplus E(M)$ is a nonsingular injective module. First, to show that V is M-Rickart, let $f \in \mathrm{Hom}(V, M) \subseteq \mathrm{Hom}(V, E(M))$. Consider $g \in \mathrm{End}(V \oplus E(M))$ defined by $g(x + y) = f(x)$, for $x \in V$ and $y \in E(M)$. Then $\mathrm{Ker}(g) = \mathrm{Ker}(f) \oplus E(M)$. As $V \oplus E(M)$ is Baer, $V \oplus E(M) = \mathrm{Ker}(g) \oplus U$ for some $U \leq V \oplus E(M)$ by Theorem 4.1.21. So $V \oplus E(M) = \mathrm{Ker}(f) \oplus E(M) \oplus U$. By the modular law,

$$V = \mathrm{Ker}(f) \oplus (V \cap (E(M) \oplus U)).$$

So $\mathrm{Ker}(f) \leq^{\oplus} V$, thus V is M-Rickart. As $V \oplus E(M)$ is Baer, V is $E(M)$-Rickart and $E(M)$ is V-Rickart by Proposition 4.2.15.

Next to prove that M is V-Rickart, let $\varphi \in \mathrm{Hom}(M, V)$. Then there exists $\phi \in \mathrm{Hom}(E(M), V)$ an extension of φ. So $\mathrm{Ker}(\varphi) = \mathrm{Ker}(\phi) \cap M \leq^{\mathrm{ess}} \mathrm{Ker}(\phi)$. Because $E(M)$ is V-Rickart, $\mathrm{Ker}(\phi) \leq^{\oplus} E(M)$. Also since M is extending, $\mathrm{Ker}(\varphi) \leq^{\mathrm{ess}} L$ for some $L \leq^{\oplus} M$. Thus $\mathrm{Ker}(\varphi) \leq^{\mathrm{ess}} E(L) \leq^{\oplus} E(M)$. Therefore $\mathrm{Ker}(\phi)$ and $E(L)$ are closures of $\mathrm{Ker}(\varphi)$ in $E(M)$.

Since $E(M)$ is nonsingular, $\mathrm{Ker}(\varphi)$ has a unique closure in $E(M)$. Thus, $\mathrm{Ker}(\phi) = E(L)$, so $L \leq \mathrm{Ker}(\phi) \cap M = \mathrm{Ker}(\varphi)$. Hence, $\mathrm{Ker}(\varphi) = L \leq^{\oplus} M$. So M is V-Rickart. Note that V is M-injective. From Theorem 4.1.15, M and V are Baer modules. Thus, $V \oplus M$ is a Baer module by Theorem 4.2.17. □

Remark 4.2.19 The preceding proof of Theorem 4.2.18 via Baer module theory was provided by G. Lee and C. Roman using ideas from [271]. In [1], Proposition 1.8(ii) and Corollary 3.3(i) obtained earlier by Akalan, Birkenmeier, and Tercan imply that $V \oplus M$ is extending for a nonsingular injective module V and a nonsingular extending module M. An application of Theorem 4.1.15 now yields that $V \oplus M$ is Baer.

The study of rings R for which a certain class of R-modules is Baer is of natural interest. We see that R is semisimple Artinian if and only if *every* R-module is Baer.

Theorem 4.2.20 *The following are equivalent for a ring R.*

(i) *Every injective right R-module is Baer.*
(ii) *Every right R-module is Baer.*
(iii) *R is semisimple Artinian.*

Proof To show (i)⇒(iii), we let $B = E(M) \oplus E(E(M)/M)$, where M is a right R-module. Since B is injective, B is a Baer module by hypothesis. Let

$$\varphi : E(M) \to E(E(M)/M) \text{ be defined by } \varphi(x) = x + M$$

for $x \in E(M)$. By Proposition 4.2.15, $E(M)$ and $E(E(M)/M)$ are relatively Rickart. Thus, $\mathrm{Ker}(\varphi) = M \leq^{\oplus} E(M)$, so $M = E(M)$. Hence, every right R-module is injective. So R is semisimple Artinian.

(iii)⇒(ii)⇒(i) are obvious. □

Exercise 4.2.21

1. ([251, Khuri]) Let M_R be a nonsingular and retractable module. Show that $\text{End}(M)$ is a right extending ring if and only if M_R is an extending module.
2. ([360, Rizvi and Roman]) Assume that $M = \oplus_{i \in \Lambda} M_i$, where each M_i is an indecomposable Baer module and relatively Rickart to M_j, for every $i, j \in \Lambda$. Let $N \leq^{\oplus} M$. Prove that for each $i \in \Lambda$ either $N \cap M_i = M_i$ or $N \cap M_i = 0$.
3. ([359, Rizvi and Roman]) Let M_R be a $\mathcal{K}$-nonsingular continuous module. Show that $S = \text{End}(M)$ is regular and right continuous.
4. ([359, Rizvi and Roman]) Assume that M is a module such that $\text{End}(M)$ is regular. Show that M is $\mathcal{K}$-nonsingular.
5. ([359, Rizvi and Roman]) Let M be a retractable $\mathcal{K}$-nonsingular right R-module. Prove that $\text{End}(M)$ is right nonsingular.
6. ([359, Rizvi and Roman]) Assume that R is a ring. Show that the following are equivalent.
   (i) Every right R-module is $\mathcal{K}$-nonsingular.
   (ii) Every injective right R-module is $\mathcal{K}$-nonsingular.
   (iii) R is semisimple Artinian.
7. ([359, Rizvi and Roman]) Assume that M is an extending module and $S = \text{End}(M)$ is a regular ring. Prove that M is a Baer module.
8. ([359, Rizvi and Roman]) Let M_R be a Baer module with only countably many direct summands. Show that M_R is semisimple Artinian if any one of the following conditions holds.
   (i) M_R is retractable and $\text{End}(M)$ is a regular ring.
   (ii) Every cyclic submodule of M_R is a direct summand of M_R.
   (iii) For each $m \in M$ there is $f \in \text{Hom}_R(M, R)$ with $m = mf(m)$ (i.e., M_R is a regular module in the sense of Zelmanowitz [420]).

4.3 Applications to Free Modules

Results shown in previous sections are applied in this section to characterize a ring R for which every (finitely generated) free R-module is a Baer module. It will be seen that every free right R-module is a Baer module if and only if R is a semiprimary hereditary ring. Also it will be shown that if R is an n-fir, then $R_R^{(n)}$ is a Baer module. This result yields an example of a module M such that $M^{(n)}$ is Baer, but $M^{(n+1)}$ is not Baer, for $n > 1$. The utility of the theory of Baer modules will be seen as the results of this section are applied to consider the Baer property of various matrix ring extensions in Sect. 6.1.

An R-module M is said to be *finitely presented* if there exists a short exact sequence of R-modules $0 \to K \to R^{(n)} \to M \to 0$, where n is a positive integer and K is a finitely generated R-module. The following result is due to Chase [114, Theorem 3.3] which characterizes when the direct product of projective modules is projective.

Theorem 4.3.1 *For a ring R, the following are equivalent.*

(i) *The direct product of any family of projective right R-modules is projective.*
(ii) *The direct product of any family of copies of R_R is projective.*
(iii) *The ring R is right perfect, and every finitely generated left ideal is finitely presented.*

A ring satisfying the condition "every finitely generated left ideal is finitely presented" in Theorem 4.3.1(iii) is called a *left coherent* ring. A right coherent ring is defined similarly.

Proposition 4.3.2 *Let R be a semiprimary ring. Then R is right hereditary if and only if R is left hereditary.*

Proof See [262, Corollary 5.71]. □

The following result provides a characterization of rings R for which every free R-module is Baer.

Theorem 4.3.3 *The following are equivalent for a ring R.*

(i) *Every free right R-module is a Baer module.*
(ii) *Every projective right R-module is a Baer module.*
(iii) *R is a semiprimary hereditary (Baer) ring.*

Proof (i)⇔(ii) The equivalence follows from Theorem 4.1.22.

(i)⇒(iii) Let I be a right ideal of R. Then there exist a free R-module F_R and an epimorphism $f : F_R \to I_R$. Thus, f can be viewed as an endomorphism of F_R. As F_R is Baer, $\operatorname{Ker}(f) \leq^{\oplus} F$ by Theorem 4.1.21, so I_R is isomorphic to a direct summand of F_R. Thus, I_R is projective, so R is right hereditary.

To show that R is semiprimary, we prove first that R is right perfect. For this, let M be an arbitrary direct product of copies of R_R, say $M = R^J$. There is a set A such that we can construct a short exact sequence of right R-modules

$$0 \to K \to R^{(A)} \xrightarrow{\varphi} M \to 0.$$

Then $K = \cap_{j\in J} \operatorname{Ker}(\pi_j\varphi)$, where π_j is the canonical projection from M onto its j-th component. But each $\pi_j\varphi$ can be viewed as an endomorphism of $R^{(A)}$, so $\operatorname{Ker}(\pi_j\varphi) \leq^{\oplus} R^{(A)}$ from Theorem 4.1.21 as $R^{(A)}$ is a Baer module. Further, $R^{(A)}$ has the SSIP by Theorem 4.1.21. So $K = \cap_{j\in J} \operatorname{Ker}(\pi_j\varphi) \leq^{\oplus} R^{(A)}$. Thus M is isomorphic to a direct summand of $R^{(A)}$, hence M is projective. By Theorem 4.3.1, R is right perfect. So $R/J(R)$ is semisimple Artinian and thus R is orthogonally finite. From 1.1.14, R satisfies DCC on principal left ideals. Therefore R is left π-regular (i.e., R is strongly π-regular). Moreover, since R is a Baer ring, R is semiprimary by Theorem 3.1.26.

(iii)⇒(i) Let F be a free right R-module. Since R is right hereditary, $\operatorname{Image}(\varphi)$ is projective for each $\varphi \in \operatorname{End}(F_R)$. Therefore φ splits, and so $\operatorname{Ker}(\varphi) \leq^{\oplus} F$. We only need to show that F satisfies the SSIP to obtain that F is a Baer module by

Theorem 4.1.21. As R is semiprimary and right hereditary, R is left hereditary from Proposition 4.3.2. Let I be a finitely generated left ideal of R. Then ${}_RI$ is projective since R is (left) hereditary. So there is a positive integer k such that the short exact sequence

$$0 \to {}_RK \to {}_RR^{(k)} \xrightarrow{g} {}_RI \to 0$$

splits, where $K = \mathrm{Ker}(g)$. Hence ${}_RK$ is a direct summand of ${}_RR^{(k)}$, so ${}_RK$ is finitely generated. Thus, every finitely generated left ideal of R is finitely presented. Using Theorem 4.3.1, the direct product of any family of projective right R-modules is projective since R is semiprimary.

To show that F satisfies the SSIP, let $\{V_j\}_{j\in\Lambda}$ be an arbitrary family of direct summands of F. Say $W_j \leq^{\oplus} F$ such that $F = V_j \oplus W_j$ for each $j \in \Lambda$. Now we define $\Pi\pi_j : F \to \prod_{j\in\Lambda} W_j$ by $(\Pi\pi_j)(x) = (\pi_j(x)) \in \prod_{j\in\Lambda} W_j$, where π_j is the canonical projection from $F = V_j \oplus W_j$ to W_j. Since $\prod_{j\in\Lambda} W_j$ is a direct product of projective modules, it is projective. Thus the submodule $\mathrm{Image}(\Pi\pi_j)$, of the module $\prod_{j\in\Lambda} W_j$, is projective as R is right hereditary. So the short exact sequence,

$$0 \to \cap_{j\in\Lambda} V_j \to F \xrightarrow{\Pi\pi_j} \mathrm{Image}(\Pi\pi_j) \to 0$$

splits, and hence $\cap_{j\in\Lambda} V_j \leq^{\oplus} F$. Thus, F satisfies the SSIP. From Theorem 4.1.21, F is a Baer module. □

In Example 4.2.12(ii), we observed that even a (finite) direct sum of copies of a Baer module M may not Baer. The next result provides a characterization for this to happen for an arbitrary direct sum when M is finitely generated and retractable. For a ring A and a nonempty ordered set Γ, $\mathrm{CFM}_\Gamma(A)$ denotes the ring of $\Gamma \times \Gamma$ column finite matrix ring over A.

Corollary 4.3.4 *Let M be a finitely generated retractable right R-module. Then an arbitrary direct sum of copies of M is a Baer module if and only if $S = \mathrm{End}(M)$ is semiprimary hereditary.*

Proof We note that, for a finitely generated module M and $S = \mathrm{End}(M)$, $\mathrm{End}(M^{(J)}) \cong \mathrm{CFM}_J(S)$, where J is an arbitrary nonempty set. So, if $M^{(J)}$ is a Baer module, its endomorphism ring is a Baer ring by Theorem 4.2.8. Thus $\mathrm{CFM}_J(S)$ is a Baer ring. So $S_S^{(J)}$ is a Baer module by Theorem 4.2.8 since $S_S^{(J)}$ is retractable. Hence, S is semiprimary hereditary by Theorem 4.3.3.

Conversely, for an arbitrary set J, $S_S^{(J)}$ is a Baer module by Theorem 4.3.3. So $\mathrm{CFM}_J(S)$ is a Baer ring by Theorem 4.2.8. Thus, $\mathrm{End}(M^{(J)})$ is a Baer ring because $\mathrm{End}(M^{(J)}) \cong \mathrm{CFM}_J(S)$. Since $M^{(J)}$ is also retractable from Proposition 4.2.5, $M^{(J)}$ is a Baer module by Theorem 4.2.8. □

A module M_R is called *torsionless* if it can be embedded in a direct product of copies of R_R. The following result characterizes a ring R for which every finitely generated free right R-module is a Baer module.

Theorem 4.3.5 *The following are equivalent for a ring R.*

(i) *Every finitely generated free right R-module is a Baer module.*
(ii) *Every finitely generated projective right R-module is a Baer module.*
(iii) *Every finitely generated torsionless right R-module is projective.*
(iv) *Every finitely generated torsionless left R-module is projective.*
(v) *$Mat_n(R)$ is a Baer ring for every positive integer n.*

Proof (i)⇔(ii) The equivalence is a direct consequence of Theorem 4.1.22.

(i)⇒(iii) Let M be a finitely generated torsionless right R-module. Then there exist sets A and B with A finite, and $\varphi \in \mathrm{Hom}(R_R^{(A)}, R_R^B)$ where $M = \mathrm{Image}(\varphi) \subseteq R^B$. In the short exact sequence of right R-modules

$$0 \to \mathrm{Ker}(\varphi) \to R^{(A)} \xrightarrow{\varphi} M \to 0,$$

$\mathrm{Ker}(\varphi) = \cap_{b\in B}\mathrm{Ker}(\pi_b\varphi)$, where π_b is the canonical projection of R^B onto its b-th component. But each $\pi_b\varphi$ can be considered as an endomorphism of $R^{(A)}$.

Because $R_R^{(A)}$ is a Baer module, by Theorem 4.1.21 $\mathrm{Ker}(\pi_b\varphi) \le^{\oplus} R_R^{(A)}$ and $\mathrm{Ker}(\varphi) = \cap_{b\in B}\mathrm{Ker}(\pi_b\varphi) \le^{\oplus} R_R^{(A)}$. Therefore, M is isomorphic to a direct summand of $R_R^{(A)}$, so M is projective.

(iii)⇒(i) Let A be a finite set. Take $\varphi \in \mathrm{End}(R_R^{(A)})$. Then $\mathrm{Image}(\varphi)$ is a finitely generated torsionless R-module. Thus $\mathrm{Image}(\varphi)$ is projective by assumption, so the short exact sequence

$$0 \to \mathrm{Ker}(\varphi) \to R^{(A)} \xrightarrow{\varphi} \mathrm{Image}(\varphi) \to 0$$

splits. Hence, $\mathrm{Ker}(\varphi) \le^{\oplus} R_R^{(A)}$. To show that $R_R^{(A)}$ is a Baer module, by Theorem 4.1.21, we only need to prove that $R_R^{(A)}$ satisfies the SSIP. For this, let $\{V_b\}_{b\in B}$ be a family of direct summands of $R^{(A)}$, where B is an index set. Say $R^{(A)} = V_b \oplus W_b$ for some submodule W_b for each $b \in B$. Let $\varphi_b : R^{(A)} \to W_b$ be the canonical projection, where $b \in B$. Then $V_b = \mathrm{Ker}(\varphi_b)$ for $b \in B$.

Now we define $\Pi_{b\in B}\varphi_b : R^{(A)} \to (R^{(A)})^B$ by $(\Pi_{b\in B}\varphi_b)(x) = (\varphi_b(x))_{b\in B}$ for $x \in R^{(A)}$. From the following short exact sequence

$$0 \to \mathrm{Ker}(\Pi_{b\in B}\varphi_b) \to R^{(A)} \xrightarrow{\Pi_{b\in B}\varphi_b} \mathrm{Image}(\Pi_{b\in B}\varphi_b) \to 0,$$

we see that $\mathrm{Image}(\Pi_{b\in B}\varphi_b)$ is a finitely generated torsionless R-module as $R_R^{(A)}$ is finitely generated, hence it is projective. So $\mathrm{Ker}(\Pi_{b\in B}\varphi_b) \le^{\oplus} R_R^{(A)}$. Note that $\cap_{b\in B}\mathrm{Ker}(\varphi_b) = \mathrm{Ker}(\Pi_{b\in B}\varphi_b) \le^{\oplus} R_R^{(A)}$. Thus $\cap_{b\in B} V_b \le^{\oplus} R_R^{(A)}$, and hence $R_R^{(A)}$ has the SSIP. So $R_R^{(A)}$ is a Baer module by Theorem 4.1.21.

(i)⇔(v) It follows from Theorem 4.2.8 since any free right R-module is retractable.

(iv)⇒(v) Assume that (iv) holds. From the left-sided version of (i)⇔(iii), every finitely generated free left R-module is a Baer module. Thus ${}_RR^{(n)}$ is a Baer module for every positive integer n, and hence $\mathrm{End}({}_RR^{(n)}) \cong \mathrm{Mat}_n(R)$ is a Baer ring by Theorem 4.2.8.

(v)⇒(iv) Assume that $\mathrm{Mat}_n(R)$ is a Baer ring for every positive integer n. Then by Theorem 4.2.8, ${}_RR^{(n)}$ is a Baer module for every positive integer n. Thus, we obtain (iv) from the left-sided version of (i)⇒(iii). □

We will discuss C^*-algebras and AW^*-algebras in Sect. 10.3, but it is worthwhile to mention that every AW^*-algebra is a Baer ring. Hilbert C^*-modules are introduced in [17] and [105]. For the convenience of the reader, we include a definition for the next consequence of Theorem 4.3.5.

Let A be a C^*-algebra. A right A-module M is called a (right) *Hilbert A-module* [17, p. 38] (also called a (right) *C^*-module* in [105, Definition 8.1.1]) if there is an A-valued inner product $\langle\ ,\ \rangle$ satisfying the following properties for all x, y, z in M, a in A, and α, β in $\mathbb{C}$:

(i) $\langle x, \alpha y + \beta z\rangle = \alpha\langle x, y\rangle + \beta\langle x, z\rangle$;
(ii) $\langle x, ya\rangle = \langle x, y\rangle a$;
(iii) $\langle y, x\rangle = \langle x, y\rangle^*$;
(iv) $\langle x, x\rangle \geq 0$ and $\langle x, x\rangle = 0$ if and only if $x = 0$;
(v) M is complete with respect to the norm given by $||x||^2 = ||\langle x, x\rangle||$.

Corollary 4.3.6 *Any finitely generated right Hilbert A-module over an AW^*-algebra A is a Baer module.*

Proof Say M is a finitely generated right Hilbert A-module over an AW^*-algebra A. Then M is a finitely generated projective right A-module (see [105, p. 352(a)]). Since A is an AW^*-algebra, $\mathrm{Mat}_n(A)$ is an AW^*-algebra for all positive integers n (see [45, Corollary 62.1]). Hence, $\mathrm{Mat}_n(A)$ is a Baer ring. By Theorem 4.3.5, every finitely generated projective right A-module is a Baer module. Therefore, M is a Baer module. □

For a positive integer n, recall that an n-generated module means a module which is generated by n elements. A ring R is said to be *right n-hereditary* if every n-generated right ideal of R is projective. Thus, a ring R is right semihereditary if and only if it is right n-hereditary for all positive integers n. Given a fixed positive integer n, we obtain the following characterization for every n-generated free R-module to be Baer.

Corollary 4.3.7 *Let R be a ring and n a positive integer. Then the following are equivalent.*

(i) *Every n-generated free right R-module is a Baer module.*
(ii) *Every n-generated projective right R-module is a Baer module.*
(iii) *Every n-generated torsionless right R-module is projective (therefore R is right n-hereditary).*
(iv) $\mathrm{Mat}_n(R)$ *is a Baer ring.*

Proof The proof follows the same outline as in Theorem 4.3.5, where we replace "finite" with "n elements". □

Corollary 4.3.8 *Let R be a ring. Then R is a Baer ring if and only if every cyclic torsionless right R-module is projective.*

Proof It is an immediate consequence of Corollary 4.3.7. □

Definition 4.3.9 A ring R is said to be an *n-fir* if any right ideal of R generated by at most n elements is free of unique rank.

We note that if R is an n-fir, then R is a domain. Indeed, take $a \in R$. Then $aR \cong R/r_R(a)$. As R is an n-fir, aR_R is free. Now $R = r_R(a) \oplus V$ as right R-modules, where $V_R \cong aR_R$, shows that $r_R(a)$ is a homomorphic image of R, hence $r_R(a)$ is principal. So $r_R(a)$ is free. By the uniqueness of the rank of R_R, either $r_R(a) = 0$ or $aR = 0$. Thus, R is a domain (see [130] for more details on n-firs).

Recall that a ring R is said to have *IBN* (*invariant basis number*) if each finitely generated free right R-module has a unique rank. Thus, a ring has IBN if and only if $R_R^{(m)} \cong R_R^{(n)}$ with m, n positive integers implies that $m = n$. It is well-known that every right Ore domain has IBN. See [342, p. 18] or [262, p. 3] for rings with IBN.

We note that if $M \oplus M$ is an injective module, a quasi-injective module, a continuous module, or a quasi-continuous module, respectively, then so is $M \oplus M \oplus M$ (see Corollary 2.2.3, Theorems 2.2.5, and 2.2.16). In contrast, if $M \oplus M$ is a Baer module, it does not imply that $M \oplus M \oplus M$ is also Baer. To see this, we begin with the next lemma.

Lemma 4.3.10 *Let R be an n-fir. Then:*

(i) *Any submodule of a free right R-module generated by at most n elements is free.*
(ii) *If $R_R^{(m)} \cong R_R^{(k)}$ with $1 \le m \le n$ and k a positive integer, then $m = k$.*

Proof (i) Let F be a free right R-module and M be a submodule of F generated by at most n elements. Then M is a submodule of a free right R-module with finite rank. So we may assume that $F = R_R^{(\ell)}$ for some positive integer ℓ since M can only involve finitely many components of F.

We use induction on ℓ. If $\ell = 1$, then M is free by Definition 4.3.9. Assume that every submodule of $R_R^{(\ell-1)}$ generated by at most n elements is free. Let $\pi : R^{(\ell)} = R^{(\ell-1)} \oplus R \to R$ be the canonical projection on the last component R of $R^{(\ell)}$. Then $\pi(M)$ is generated by at most n elements, so $\pi(M)$ is free since R is an n-fir. Hence, the short exact sequence

$$0 \to \mathrm{Ker}(\pi|_M) \to M \to \pi(M) \to 0$$

splits, so $M \cong \mathrm{Ker}(\pi|_M) \oplus \pi(M)$. Also $\mathrm{Ker}(\pi|_M) \le \mathrm{Ker}(\pi) = R^{(\ell-1)}$. By induction hypothesis, $\mathrm{Ker}(\pi|_M)$ is a free module. So M is a free module.

(ii) As was noted before, R is a domain. If R is right Ore, then R has IBN. Therefore, any finitely generated free right R-module has a unique rank.

Next, assume that R is not right Ore. Then there are two nonzero elements $a, b \in R$ with $aR \cap bR = 0$. Hence

$$bR + abR + a^2bR + \cdots = bR \oplus abR \oplus a^2bR \oplus \cdots,$$

and $a^ibR_R \cong R_R$ for each i. Let $1 \leq m \leq n$ and assume that $R_R^{(m)} \cong R_R^{(k)}$. We note that $R_R^{(m)} \cong bR_R \oplus abR_R \oplus \cdots \oplus a^{m-1}bR_R$.

Because $1 \leq m \leq n$ and $bR_R \oplus abR_R \oplus \cdots \oplus a^{m-1}bR_R \cong R^{(k)}$ by assumption, we see that $m = k$ as R is an n-fir and $bR_R \oplus abR_R \oplus \cdots \oplus a^{m-1}bR_R$ is a right ideal of R generated by m elements. □

Theorem 4.3.11 *Let R be an n-fir. Then $R_R^{(n)}$ is a Baer module. Hence,* $\text{Mat}_n(R)$ *is a Baer ring.*

Proof Since R is an n-fir, it is a domain. Thus, R is a Baer ring and so R_R is a Baer module. We prove the result by induction on n.

Assume that R is a 2-fir. We claim that $R_R^{(2)}$ is a Baer module. First, we show that $R_R^{(2)}$ satisfies the SSIP. For this, take a set of nonzero proper direct summands $\{N_i\}_{i \in \Lambda}$ of $R_R^{(2)}$, and fix an index $i_0 \in \Lambda$. Then $R_R^{(2)} = N_{i_0} \oplus V$ for some $V \leq R_R^{(2)}$. By Lemma 4.3.10(i), N_{i_0} and V are free. As N_{i_0} and V must have lower rank than that of $R_R^{(2)}$ (otherwise conflicting with the uniqueness of rank of $R_R^{(2)}$), we obtain that $N_{i_0} \cong R_R \cong V$. Thus, N_{i_0} and V are Baer modules. Since R_R is a Baer module, N_{i_0} and V are relatively Rickart by Proposition 4.2.15.

By the same argument, $N_i \cong R_R$ and so, for each $i \in \Lambda$, $N_i \cong N_{i_0}$. For $i \in \Lambda$, say $R_R^{(2)} = N_i \oplus V_i$ for some V_i. Then similarly N_i and V_i are relatively Rickart. Hence, N_{i_0} and V_i are relatively Rickart for each $i \in \Lambda$.

Let $\pi_i : N_i \oplus V_i \to V_i$ be the canonical projection for each $i \in \Lambda$. Then $\text{Ker}(\pi_i|_{N_{i_0}}) = N_{i_0} \cap N_i \leq^{\oplus} N_{i_0}$ for each $i \in \Lambda$ because N_{i_0} is V_i-Rickart. Since N_{i_0} is a Baer module, it has the SSIP from Theorem 4.1.21. Thus, we obtain that $\cap_{i \in \Lambda} N_i = \cap_{i \in \Lambda}(N_{i_0} \cap N_i) \leq^{\oplus} N_{i_0} \leq^{\oplus} R_R^{(2)}$. So $R_R^{(2)}$ has the SSIP.

The image of every endomorphism of $R_R^{(2)}$ is free from Lemma 4.3.10(i). So the kernel of every endomorphism of $R_R^{(2)}$ is a direct summand of $R_R^{(2)}$. By Theorem 4.1.21, $R_R^{(2)}$ is a Baer module.

We suppose that if a ring T is an $(n-1)$-fir, then $T_T^{(n-1)}$ is a Baer module. Let R be an n-fir. To prove that $R_R^{(n)}$ is a Baer module, take $\varphi \in \text{End}(R^{(n)})$. Then $\text{Image}(\varphi)$ is generated by n elements. Thus, $\text{Image}(\varphi)$ is a free R-module from Lemma 4.3.10(i). Hence, the short exact sequence

$$0 \to \text{Ker}(\varphi) \to R_R^{(n)} \to \text{Image}(\varphi) \to 0$$

splits. Therefore, $\text{Ker}(\varphi) \leq^{\oplus} R_R^{(n)}$. Thus to show that $R_R^{(n)}$ is a Baer module, we only need to prove that $R_R^{(n)}$ has the SSIP by Theorem 4.1.21. Let $\{N_i\}_{i \in \Lambda}$ be a set of nonzero proper direct summands of $R_R^{(n)}$. Select one particular direct summand $N_{i_0} \in \{N_i\}_{i\in\Lambda}$. Then $R_R^{(n)} = N_{i_0} \oplus V$ for some $V \leq R_R^{(n)}$. By Lemma 4.3.10(i), $N_{i_0} \cong R_R^{(k)}$ and $V \cong R_R^{(s)}$. Thus, $R_R^{(n)} \cong R_R^{(k)} \oplus R_R^{(s)} = R_R^{(k+s)}$. Thus, $n = k+s$ from Lemma 4.3.10(ii). So $1 \leq k \leq n-1$.

Note that R is also a k-fir, so $R_R^{(k)}$ is a Baer module by induction hypothesis, and N_{i_0} is a Baer module since $N_{i_0} \cong R_R^{(k)}$ with $1 \leq k \leq n-1$. For each $i \in \Lambda$, say $R_R^{(n)} = N_i \oplus V_i$ for some $V_i \leq R_R^{(n)}$. Then V_i is free by Lemma 4.3.10(i). Let $\varphi_i \in \mathrm{Hom}_R(N_{i_0}, V_i)$. Then $\varphi_i(N_{i_0})$ is a free R-module by Lemma 4.3.10(i). So $\mathrm{Ker}(\varphi_i) \leq^{\oplus} N_{i_0}$. Thus, N_{i_0} is V_i-Rickart for $i \in \Lambda$.

Let $\pi_i : N_i \oplus V_i \to V_i$ be the canonical projection for each $i \in \Lambda$. Then $\mathrm{Ker}(\pi_i|_{N_{i_0}}) = N_{i_0} \cap N_i \leq^{\oplus} N_{i_0}$ for each $i \in \Lambda$ because N_{i_0} is V_i-Rickart. Since N_{i_0} is a Baer module, it has the SSIP from Theorem 4.1.21. As a consequence, $\cap_{i\in\Lambda} N_i = \cap_{i\in\Lambda}(N_{i_0} \cap N_i) \leq^{\oplus} N_{i_0} \leq^{\oplus} R_R^{(n)}$. So $R_R^{(n)}$ has the SSIP. By Theorem 4.1.21, $R_R^{(n)}$ is a Baer module. □

Example 4.3.13, due to Cohn and Jøndrup, shows that for each positive integer n, there exists a module M such that $M^{(n)}$ is a Baer module, but $M^{(n+1)}$ is not Baer (see [239]). First, a result from the same paper is the following.

Theorem 4.3.12 *For every integer $n \geq 1$, there exists a ring R such that any n-generated left ideal is flat, while there exists a nonflat $(n+1)$-generated left ideal. We can choose R to be an n-fir.*

Proof See [239, Theorem 2.3]. □

Example 4.3.13 Let n be a positive integer and let R be the K-algebra (K is a field) on the $2(n+1)$ generators x_i, y_i, $i = 1, \ldots, n+1$, with the defining relation $x_1y_1 + \cdots + x_ny_n + x_{n+1}y_{n+1} = 0$.

It is shown that R is an n-fir [239, Example]. Hence $R_R^{(n)}$ is a Baer module by Theorem 4.3.11. However, not every $(n+1)$-generated left ideal of R is flat [239, Example]. Thus R is not left $(n+1)$-hereditary, so ${}_RR^{(n+1)}$ is not a Baer module by the left-sided version of Corollary 4.3.7. Whence $\mathrm{Mat}_{n+1}(R)$ is not a Baer ring from Theorem 4.2.8. So $R_R^{(n+1)}$ is not a Baer module again by Theorem 4.2.8.

Exercise 4.3.14

1. ([360, Rizvi and Roman]) Let M_R be a retractable Baer right R-module and let $S = \mathrm{End}(M)$. Prove that any finite direct sum of copies of M is a Baer module if and only if S is left semihereditary and right Π-coherent. (A ring R is called *right Π-coherent* if every finitely generated torsionless right R-module is finitely presented. A left Π-coherent ring is defined similarly.)
2. ([360, Rizvi and Roman]) Let M_R be a $\mathcal{K}$-nonsingular right R-module and $S = \mathrm{End}(M)$. Show that the following hold true.

(i) If $M^{(n)}$ is extending, where n is a positive integer, then every n-generated torsionless right S-module is projective. Therefore, S is a right n-hereditary ring.
(ii) If $M^{(n)}$ is extending for every positive integer n, then S is a right semihereditary and left Π-coherent ring.
(iii) If M is finitely generated and $M^{(\Lambda)}$ is extending for every index set Λ, then S is a semiprimary hereditary ring.

3. Prove Theorem 4.3.12.

4.4 Applications to Type Theory

A useful type theory was provided by Kaplansky [246] for Baer rings who classified Baer rings into five types and showed that every Baer ring can be uniquely decomposed as a ring direct sum of these five types (Theorem 4.4.7). As a generalization of this theory, Goodearl and Boyle [184] established a type theory for nonsingular injective modules following similar lines as the one done by Kaplansky for Baer rings. One benefit of such a decomposition into types is that one can then study each type separately in a more effective manner and obtain a better understanding of the structure.

We apply $\mathcal{K}$-nonsingularity to various generalizations of injective modules, such as the extending modules or the (quasi-)continuous modules to extend and generalize a number of results from the type theory of Kaplansky [246], and of Goodearl and Boyle [184]. In particular, we weaken the hypotheses for the existing type theory for nonsingular injective modules in two ways: Firstly, we replace the nonsingular condition in the hypothesis by the more general $\mathcal{K}$-nonsingular condition, and secondly, we replace the class of injective modules by the larger and more general class of extending (or continuous) modules. Internal characterizations for type I, II, and type III $\mathcal{K}$-nonsingular continuous modules are obtained similar to the case of nonsingular injective modules.

A module M is called *Abelian* if the endomorphism ring of M is an Abelian ring (i.e., every idempotent is central). A characterization of Abelian modules on the lines of Goodearl and Boyle [184] is provided. Our proof, however, is for arbitrary modules instead of nonsingular injective ones.

Theorem 4.4.1 *The following are equivalent for a module M.*

(i) *M is Abelian.*
(ii) *Every direct summand of M is fully invariant.*
(iii) *Isomorphic direct summands of M are equal.*
(iv) *If N_1 and N_2 are direct summands of M and $N_1 \cap N_2 = 0$, then* $\mathrm{Hom}(N_1, N_2) = 0$.

Proof (i)⇒(ii) Let $N \leq^{\oplus} M$. Then $N = eM$ for some $e^2 = e \in S$, where $S = \mathrm{End}(M)$. Since M is Abelian, e is central and so $\varphi(eM) = e(\varphi M) \subseteq eM$ for each $\varphi \in S$. Thus $N = eM \trianglelefteq M$.

(ii)⇒(iii) Let $N_1, N_2 \leq^{\oplus} M$ and $f : N_1 \to N_2$ be an isomorphism. Put $M = N_1 \oplus V$ with $V \leq M$. As $N_2 \trianglelefteq M$, $N_2 = (N_2 \cap N_1) \oplus (N_2 \cap V)$ (Exercise 2.1.37.2). We claim that $N_2 \cap V = 0$. Note that $N_2 \cap V \leq^{\oplus} N_2$, so $f^{-1}(N_2 \cap V) \leq^{\oplus} N_1$. Hence, $N_1 = f^{-1}(N_2 \cap V) \oplus W$ for some $W \leq N_1$.

Define $g : N_1 \to V$ which is equal to f on $f^{-1}(N_2 \cap V)$ and zero on W. Then $g(N_1) = N_2 \cap V$. Extend g to $h \in S = \text{End}(M)$ by $h(x) = g(x)$ for $x \in N_1$ and $h(x) = 0$ for $x \in V$.

As $N_1 \trianglelefteq M$ by hypothesis, $g(N_1) = h(N_1) \subseteq N_1$. So $N_2 \cap V = g(N_1) \subseteq N_1$, hence $N_2 \cap V \subseteq N_1 \cap N_2$. Thus, $N_2 \cap V = (N_2 \cap V) \cap (N_1 \cap N_2) = 0$. Therefore, we have that $N_2 = (N_2 \cap N_1) \oplus (N_2 \cap V) = N_2 \cap N_1$. Hence $N_2 \subseteq N_1$. Similarly, $N_1 \subseteq N_2$. Therefore, $N_1 = N_2$.

(iii)⇒(iv) Let $N_1, N_2 \leq^{\oplus} M$ with $N_1 \cap N_2 = 0$. Assume on the contrary that $\text{Hom}(N_1, N_2) \neq 0$. Say $0 \neq \varphi \in \text{Hom}(N_1, N_2)$. Let $M = N_1 \oplus V$ with $V \leq M$, and let $\pi : M \to V$ be the canonical projection. If $\pi\varphi(N_1) = 0$, then we obtain that $0 \neq \varphi(N_1) \subseteq N_2 \cap N_1 = 0$, a contradiction. Hence, $\pi\varphi(N_1) \neq 0$.

Let $P = \{n + \pi\varphi(n) \mid n \in N_1\} \leq M$. Then $P \cap V = 0$ and $M = P + V$. Thus, $M = P \oplus V$. Since $M = N_1 \oplus V$, $N_1 \cong P$ and thus $N_1 = P$ by hypothesis. For $n \in N_1$, $n + \pi\varphi(n) \in N_1$ and so $\pi\varphi(n) \in N_1 \cap V = 0$. Thus $\pi\varphi(n) = 0$ for any $n \in N_1$, so $\pi\varphi(N_1) = 0$, a contradiction. Hence, $\text{Hom}(N_1, N_2) = 0$.

(iv)⇒(i) Take $e^2 = e \in S = \text{End}(M)$. Let $N_1 = eM$ and $N_2 = (1 - e)M$. Then $(1 - e)Se = \text{Hom}(N_1, N_2) = 0$. Also $eS(1 - e) = \text{Hom}(N_2, N_1) = 0$. Consequently, we get $e \in \mathbf{S}_{\ell}(S) \cap \mathbf{S}_r(S) = \mathcal{B}(S)$ by Propositions 1.2.2 and 1.2.6(i), so e is central. Therefore, M is Abelian. □

Proposition 4.4.2 *Let M be an Abelian Baer module. If $N \leq^{\oplus} M$, then N is an Abelian Baer module.*

Proof Let $V \leq^{\oplus} N$. As M is Abelian, $V \trianglelefteq M$ by Theorem 4.4.1. Note that any endomorphism of N is extended to an endomorphism of M. So $V \trianglelefteq N$, hence from Theorem 4.4.1, N is Abelian. By Theorem 4.1.22, N is Baer. □

Definition 4.4.3 An idempotent endomorphism e of M is said to be *directly finite* if eM is a directly finite module.

We note that a module M is directly finite if and only if $\text{End}(M)$ is directly finite. Thus for $e^2 = e \in \text{End}(M)$, e is directly finite if and only if $\text{End}(eM)$ is a directly finite ring. We remark that Lemma 4.4.4(ii) has been used in Sect. 2.2 implicitly.

Lemma 4.4.4 (i) *Every Abelian module is directly finite.*

(ii) *Any direct summand of a directly finite module is directly finite.*

Proof (i) Say M is an Abelian module and $S = \text{End}(M)$. Let $x, y \in S$ with $xy = 1$. Then $(yx)^2 = yx$. Since M is Abelian, yx is central in S, and so

$$1 = (xy)(xy) = x(yx)y = (yx)(xy) = yx.$$

Hence, M is directly finite.

(ii) The proof is routine. □

Let R be a Baer ring and $x \in R$. Take $U = \{u_i \mid i \in \Lambda\}$, the set of all central idempotents u_i of R satisfying $xR \subseteq u_i R$. Since R is a Baer ring,

$$\cap_{i \in \Lambda} u_i R = r_R(\{1 - u_i\}_{i \in \Lambda}) = eR$$

for some $e^2 = e \in R$. Further, $r_R(\{1 - u_i\}_{i \in \Lambda}) = \ell_R(\{1 - u_i\}_{i \in \Lambda}) = Rf$ for some $f^2 = f \in R$. We note that $e \in \mathbf{S}_\ell(R)$ and $f \in \mathbf{S}_r(R)$.

Because $eR = Rf$, $e = f \in \mathbf{S}_\ell(R) \cap \mathbf{S}_r(R) = \mathcal{B}(R)$ from Proposition 1.2.6(i). If $xR \subseteq wR$ with $w^2 = w \in \mathcal{B}(R)$, then $eR \subseteq wR$. So e is the smallest central idempotent in R satisfying $xR \subseteq eR$ (i.e., $x = ex$). We write $e = C(x)$ and e is called the *central cover* of x.

Definition 4.4.5 (i) A nonzero idempotent e in a Baer ring is called *faithful* if $C(e) = 1$. Equivalently, e is faithful if 0 is the only central idempotent orthogonal to e.

(ii) An idempotent f of a ring R is said to be *Abelian* if the ring fRf is Abelian.

(iii) An idempotent g of a ring R is called *directly finite* if the ring gRg is directly finite.

Next, we include the description of the various types which occur in the decomposition theory of Baer rings, from Kaplansky [246].

Definition 4.4.6 A Baer ring is said to be of *type* I if it has a faithful Abelian idempotent. A Baer ring is said to be of *type* II if it has a faithful directly finite idempotent, but no nonzero Abelian idempotents. A Baer ring is said to be of *type* III if it has no nonzero directly finite idempotents. A Baer ring is called *purely infinite* if it has no nonzero central directly finite idempotents.

A Baer ring of *type* I_f means a Baer ring of type I which is directly finite, while a Baer ring of *type* I_∞ means a Baer ring of type I which is purely infinite. A Baer ring of type II which is directly finite is called a Baer ring of *type* II_f. Also a Baer ring of type II which is purely infinite is called a Baer ring of *type* II_∞.

In [246], Kaplansky proved that a Baer ring decomposes uniquely into a ring direct sum of these five components, as described in the following result.

Theorem 4.4.7 *A Baer ring decomposes uniquely into a ring direct sum of Baer rings of types* I_f; I_∞; II_f; II_∞; *and* III.

We now define the five types of $\mathcal{K}$-nonsingular extending modules in terms of the types of their endomorphism rings.

Definition 4.4.8 We call a $\mathcal{K}$-nonsingular extending module M of *type* T if the endomorphism ring, $S = \mathrm{End}(M)$ is of type T, where $\mathrm{T} \in \{\mathrm{I}_f, \mathrm{I}_\infty, \mathrm{II}_f, \mathrm{II}_\infty, \mathrm{III}\}$.

Note that this definition is meaningful, since the endomorphism ring of a $\mathcal{K}$-nonsingular extending module is Baer by Theorems 4.1.15 and 4.2.8. At the same time, this type theory applies, respectively, to the class of $\mathcal{K}$-nonsingular quasi-continuous modules and to that of $\mathcal{K}$-nonsingular continuous modules. The types given in Definition 4.4.8 coincide with those given in [184] for the case when M is nonsingular and injective. Now we come to the main theorem of the section, which provides a unique decomposition of a $\mathcal{K}$-nonsingular extending module into the five types listed.

Theorem 4.4.9 *Any $\mathcal{K}$-nonsingular extending module decomposes uniquely into a direct sum of fully invariant direct summands of types* I_f; I_∞; II_f; II_∞; *and* III.

Proof Let M be a $\mathcal{K}$-nonsingular extending module. Then $S = \mathrm{End}(M)$ is a Baer ring by Theorems 4.1.15 and 4.2.8. So S decomposes uniquely, as a ring direct sum, into $S = e_{\mathrm{I}_f}S \oplus e_{\mathrm{I}_\infty}S \oplus e_{\mathrm{II}_f}S \oplus e_{\mathrm{II}_\infty}S \oplus e_{\mathrm{III}}S$, where e_{T} is a central idempotent in S and $e_{\mathrm{T}}S$ is of type T with T in $\{\mathrm{I}_f, \mathrm{I}_\infty, \mathrm{II}_f, \mathrm{II}_\infty, \mathrm{III}\}$. Since this is a ring direct sum decomposition,

$$M = e_{\mathrm{I}_f}M \oplus e_{\mathrm{I}_\infty}M \oplus e_{\mathrm{II}_f}M \oplus e_{\mathrm{II}_\infty}M \oplus e_{\mathrm{III}}M$$

has the property that each of five direct summands is a fully invariant direct summand of M (each idempotent occurring is central). Because (C_1) condition is inherited by direct summands, each of five direct summands is $\mathcal{K}$-nonsingular extending (see Exercise 4.1.28.1). The endomorphism ring of $e_{\mathrm{I}_f}M$ is $e_{\mathrm{I}_f}Se_{\mathrm{I}_f} = e_{\mathrm{I}_f}S$, so $e_{\mathrm{I}_f}M$ is of type I_f; the endomorphism ring of $e_{\mathrm{I}_\infty}M$ is $e_{\mathrm{I}_\infty}Se_{\mathrm{I}_\infty} = e_{\mathrm{I}_\infty}S$, hence $e_{\mathrm{I}_\infty}M$ is of type I_∞. The proof follows similarly for the remaining direct summands.

To prove uniqueness, say $M = g_{\mathrm{I}_f}M \oplus g_{\mathrm{I}_\infty}M \oplus g_{\mathrm{II}_f}M \oplus g_{\mathrm{II}_\infty}M \oplus g_{\mathrm{III}}M$ is another decomposition with each direct summand fully invariant, and each direct summand of, respectively, type I_f; I_∞; II_f; II_∞; and III. Then we see that $S = g_{\mathrm{I}_f}S \oplus g_{\mathrm{I}_\infty}S \oplus g_{\mathrm{II}_f}S \oplus g_{\mathrm{II}_\infty}S \oplus g_{\mathrm{III}}S$ is a ring direct sum decomposition. Because the type decomposition for Baer rings is unique, $e_{\mathrm{I}_f}S = g_{\mathrm{I}_f}S$ and therefore $e_{\mathrm{I}_f}M = g_{\mathrm{I}_f}M$. Similar equalities hold for the remaining four types. □

For a $\mathcal{K}$-nonsingular extending module M, we put $M_{\mathrm{I}} = e_{\mathrm{I}_f}M \oplus e_{\mathrm{I}_\infty}M$, $M_{\mathrm{II}} = e_{\mathrm{II}_f}M \oplus e_{\mathrm{II}_\infty}M$, and $M_{\mathrm{III}} = e_{\mathrm{III}}M$. Then M_{I} is of type I, M_{II} is of type II, and M_{III} is of type III. Further, $M = M_{\mathrm{I}} \oplus M_{\mathrm{II}} \oplus M_{\mathrm{III}}$ is the unique decomposition into direct summands of types I, II, and III. These direct summands are fully invariant. Recall that modules U and V are said to be orthogonal if U and V have no nonzero isomorphic submodules.

Corollary 4.4.10 *Every $\mathcal{K}$-nonsingular (quasi-)continuous module decomposes uniquely into a direct sum of orthogonal direct summands of types* I_f; I_∞; II_f; II_∞; *and* III.

Proof Let M be a $\mathcal{K}$-nonsingular (quasi-)continuous module. Then from Theorem 4.4.9 and Lemma 2.2.4, $M = M_{\mathrm{I}_f} \oplus M_{\mathrm{I}_\infty} \oplus M_{\mathrm{II}_f} \oplus M_{\mathrm{II}_\infty} \oplus M_{\mathrm{III}}$, such that M_{T} a $\mathcal{K}$-nonsingular (quasi-)continuous module of type T, where T is in $\{\mathrm{I}_f, \mathrm{I}_\infty, \mathrm{II}_f, \mathrm{II}_\infty, \mathrm{III}\}$. Further, each M_{T} is fully invariant in M. Since M is (quasi-)continuous, M_{T} and $M_{\mathrm{T}'}$ are relatively injective for $\mathrm{T} \neq \mathrm{T}'$ by Lemma 2.2.4. Thus if there exist $0 \neq N \leq M_{\mathrm{T}}$ and a monomorphism $\varphi : N \to M_{\mathrm{T}'}$, we extend φ to $0 \neq \overline{\varphi} : M_{\mathrm{T}} \to M_{\mathrm{T}'}$. Thus, $0 \neq \overline{\varphi}(M_T) \subseteq M_{T'}$. This is absurd because $M_{\mathrm{T}} \trianglelefteq M$ and $M_{\mathrm{T}} \cap M_{\mathrm{T}'} = 0$. Thus M_{T} and $M_{\mathrm{T}'}$ are orthogonal for $\mathrm{T} \neq \mathrm{T}'$, where T and T′ are in $\{\mathrm{I}_f, \mathrm{I}_\infty, \mathrm{II}_f, \mathrm{II}_\infty, \mathrm{III}\}$. □

The next example shows that the type decomposition in Theorem 4.4.9 is a proper generalization of Goodearl-Boyle type decomposition for nonsingular injective modules. Also, it demonstrates that we cannot always use the Goodearl-Boyle type decomposition to obtain our general type decomposition.

Example 4.4.11 Let p be a prime integer. Then the $\mathbb{Z}$-module $M = \mathbb{Z}_p$ is a $\mathcal{K}$-nonsingular continuous module of type I (since it is indecomposable). But, we note that $E(M) = \mathbb{Z}_{p^\infty}$ is a singular injective module, for which the type theory developed by Goodearl and Boyle does not apply.

Furthermore, Example 4.4.11 illustrates that the type decomposition for a $\mathcal{K}$-nonsingular (quasi-)continuous module M cannot, in general, be obtained simply by going up to the injective hull $E(M)$, decomposing it into the type decomposition $E(M) = E_{\mathrm{I}} \oplus E_{\mathrm{II}} \oplus E_{\mathrm{III}}$, where E_{T} is of type T with $\mathrm{T} \in \{\mathrm{I, II, III}\}$ and using (quasi-)continuity of M and Theorem 2.1.25 to express

$$M = (M \cap E_{\mathrm{I}}) \oplus (M \cap E_{\mathrm{II}}) \oplus (M \cap E_{\mathrm{III}}).$$

Goodearl and Boyle [184] provide a number of characterizations for nonsingular injective modules of various types. These characterizations are extended as follows. Some of our proofs are based on the ideas behind the proofs of Goodearl and Boyle.

Lemma 4.4.12 *Assume that R is a semiprime, right extending, and right nonsingular ring. Then $e^2 = e \in R$ is a faithful idempotent of R if and only if $fRe \neq 0$ for any $0 \neq f^2 = f \in R$.*

Proof Note that R is a Baer ring by Theorem 3.3.1. Assume that e is faithful. As R is semiprime and right extending, R is right strongly FI-extending by Theorem 3.2.37. So there is $h \in \mathbf{S}_\ell(R)$ with $ReR_R \leq^{\mathrm{ess}} hR_R$. By Proposition 1.2.6(ii), h is central as R is semiprime. Since h is central, $he = e$, and e is faithful, it follows that $h = 1$ and so $ReR_R \leq^{\mathrm{ess}} R_R$. If $0 \neq f^2 = f \in R$, then $fReR \neq 0$ because R is right nonsingular. So $fRe \neq 0$.

Conversely, let $g \in \mathcal{B}(R)$ with $ge = e$. Then $(1-g)Re = R(1-g)e = 0$. So we obtain $1 - g = 0$, so $g = 1$ and hence e is faithful. □

Theorem 4.4.13 *Let M be a $\mathcal{K}$-nonsingular continuous module. Then the following are equivalent.*

(i) *M is of type* I.
(ii) *Every nonzero direct summand of M contains a nonzero Abelian direct summand.*
(iii) *The sum of all Abelian direct summands of M forms an essential submodule of M.*

Proof (i)⇒(ii) Since M is of type I, there is a faithful Abelian idempotent $e \in S = \text{End}(M)$. Let $0 \neq N \leq^{\oplus} M$. Then $N = fM$ with $0 \neq f^2 = f \in S$.

We see that $\Delta = \{g \in S \mid \text{Ker}(g) \leq^{\text{ess}} M\} = 0$ by $\mathcal{K}$-nonsingularity of M. From Theorem 2.1.29, S is right continuous and regular. In particular, S is right extending, right nonsingular, and semiprime. By Lemma 4.4.12, $fSe \neq 0$. Let $s \in S$ such that $fse \neq 0$. So $0 \neq fseM \subseteq fM$. As M is Baer from Theorem 4.1.15, $\text{Ker}(fs) \leq^{\oplus} M$ by Theorem 4.1.21. Also we have that $\text{Ker}(fs|_{eM}) = eM \cap \text{Ker}(fs) \leq^{\oplus} M$ by the SSIP from Theorem 4.1.21.

By the modular law, $\text{Ker}(fs|_{eM}) \leq^{\oplus} eM$. Say $eM = \text{Ker}(fs|_{eM}) \oplus V$ for some $V \leq eM$. Then $fseM \cong V \leq^{\oplus} eM$. Hence, $fseM \leq^{\oplus} M$ from (C_2) condition of M. Again by the modular law, $N = fM = fseM \oplus U$ for some $U \leq M$. By Proposition 4.4.2, $fseM$ is Abelian since $fseM \cong V \leq^{\oplus} eM$ and eM is Abelian. Thus N has a nonzero Abelian direct summand $fseM$.

(ii)⇒(iii) Assume on the contrary that the sum of all Abelian direct summands of M is not essential in M. Then, by (C_1) condition, the sum of all Abelian direct summands is essential in a direct summand, say N of M. Put $M = N \oplus U$ for some $U \leq M$. But, if $U \neq 0$, then U contains a nonzero Abelian direct summand, so $N \cap U \neq 0$, a contradiction. Thus the sum of Abelian direct summands of M is essential in M.

(iii)⇒(i) By Theorem 4.4.9, $M = M_{\text{I}} \oplus M_{\text{II}} \oplus M_{\text{III}}$. Let N be an Abelian direct summand of M. Because M_{I}, M_{II}, and M_{III} are fully invariant submodules of M, $N = (N \cap M_{\text{I}}) \oplus (N \cap M_{\text{II}}) \oplus (N \cap M_{\text{III}})$ from Proposition 2.3.3(v). Since direct summands of an Abelian module are Abelian by Proposition 4.4.2, and since M_{II} and M_{III} do not contain nonzero Abelian direct summands, $N = N \cap M_{\text{I}}$ and so $N \subseteq M_{\text{I}}$. Thus, the sum of all Abelian direct summands is a submodule of M_{I}. By (iii), M_{I} is essential in M. Therefore, $M = M_{\text{I}}$ because $M_{\text{I}} \leq^{\oplus} M$. □

Theorem 4.4.14 *Assume that M is a $\mathcal{K}$-nonsingular continuous module. Then the following are equivalent.*

(i) *M is of type* II.
(ii) *Every nonzero direct summand of M contains a nonzero directly finite direct summand, but M has no nonzero Abelian direct summands.*
(iii) *The sum of all directly finite direct summands of M is an essential submodule of M, but M has no nonzero Abelian direct summands.*

Proof (i)⇒(ii) Since M is of type II, there exists $e^2 = e \in S := \text{End}(M)$ which is a faithful directly finite idempotent, but S has no nonzero Abelian idempotents. So M has no nonzero Abelian direct summands. Say $N = fM$ for some $0 \neq f^2 = f \in S$. We note that S is right continuous and regular from Theorem 2.1.29 because $\Delta = \{g \in S \mid \text{Ker}(g)_R \leq^{\text{ess}} M_R\} = 0$ from $\mathcal{K}$-nonsingularity of M. By Lemma 4.4.12, $fSe \neq 0$.

Take $s \in S$ such that $fse \neq 0$. Then $0 \neq fseM \subseteq fM$. Using the arguments in the proof of Theorem 4.4.13, (i)⇒(ii), $fseM \leq^{\oplus} fM$ and $fseM$ is isomorphic to a direct summand of eM. As eM is directly finite, $fseM$ is also directly finite from Lemma 4.4.4(ii). Thus, we see that $fseM$ is a nonzero directly finite direct summand of $fM = N$.

(ii)⇒(iii) By (C_1) condition of M, the sum of all directly finite direct summands is essential in a direct summand N of M. Let $M = N \oplus V$ for some $V \leq M$. But, if $V \neq 0$, then by hypothesis, V contains a nonzero directly finite direct summand, a contradiction. Thus $V = 0$, so $N = M$. Hence, the sum of all directly finite direct summands of M is essential in M.

(iii)⇒(i) By Theorem 4.4.9, $M = M_{\mathrm{I}} \oplus M_{\mathrm{II}} \oplus M_{\mathrm{III}}$. Let N be a directly finite direct summand of M. Then $N = (N \cap M_{\mathrm{I}}) \oplus (N \cap M_{\mathrm{II}}) \oplus (N \cap M_{\mathrm{III}})$ by Proposition 2.3.3(v) because M_{I}, M_{II}, and M_{III} are fully invariant. Recall from Lemma 4.4.4(ii) that direct summands of a directly finite module are directly finite. Hence $N \cap M_{\mathrm{I}}$, $N \cap M_{\mathrm{II}}$, and $N \cap M_{\mathrm{III}}$ are directly finite. Because S has no nonzero Abelian idempotents, $M_{\mathrm{I}} = 0$. Thus $N = (N \cap M_{\mathrm{II}}) \oplus (N \cap M_{\mathrm{III}})$ and hence $M = (N \cap M_{\mathrm{II}}) \oplus (N \cap M_{\mathrm{III}}) \oplus K$ for some $K \leq M$. By the modular law, $M_{\mathrm{III}} = (N \cap M_{\mathrm{III}}) \oplus V$ for some $V \leq M$. So $N \cap M_{\mathrm{III}} = 0$ as M_{III} does not contain nonzero directly finite direct summands. Thus $N = N \cap M_{\mathrm{II}}$, so $N \subseteq M_{\mathrm{II}}$. Hence, all directly finite direct summands of M are contained in M_{II}. By hypothesis, M_{II} is essential in M. So $M = M_{\mathrm{II}}$ as $M_{\mathrm{II}} \leq^{\oplus} M$. □

Theorem 4.4.15 *Let M be a $\mathcal{K}$-nonsingular continuous module, and assume that M has no nonzero Abelian direct summands. Then for each positive integer n, there exists $M_n \leq^{\oplus} M$ such that $M_n^{(n)} \cong M$.*

Proof We begin the proof by showing that for any $0 \neq N \leq^{\oplus} M$ and for each positive integer n, there is $0 \neq N' \leq^{\oplus} N$ with $N' \cong P^{(n)}$ for some $P \leq M$. For this, say $0 \neq N \leq^{\oplus} M$. Since N is not Abelian, there is $e^2 = e \in B := \mathrm{End}(N)$, which is not central. Hence, either $(1-e)Be \neq 0$ or $eB(1-e) \neq 0$. Without loss of generality, let $(1-e)Be \neq 0$.

Put $K_1 = eN$ and $K_2 = (1-e)N$. Then $\mathrm{Hom}(K_1, K_2) = (1-e)Be \neq 0$. Take $0 \neq \varphi \in \mathrm{Hom}(K_1, K_2)$. Say $M = K_1 \oplus Y$ with $Y \leq M$. Define $\phi \in \mathrm{End}_R(M)$ such that $\phi|_{K_1} = \varphi$ and $\phi|_Y = 0$. Then $\mathrm{Ker}(\phi) = \mathrm{Ker}(\varphi) \oplus Y$.

Since M is $\mathcal{K}$-nonsingular and continuous, M is Baer from Theorem 4.1.15. So $\mathrm{Ker}(\phi) \leq^{\oplus} M$ by Theorem 4.1.21. Thus, $M = \mathrm{Ker}(\varphi) \oplus Y \oplus U$ with $U \leq M$. By the modular law, $K_1 = \mathrm{Ker}(\varphi) \oplus (K_1 \cap (Y \oplus U))$.

Put $V = K_1 \cap (Y \oplus U)$. As $K_1 = \mathrm{Ker}(\varphi) \oplus V$, $V \cong \varphi(K_1)$. We note that, since $K_1 = eN \leq^{\oplus} N \leq^{\oplus} M$, $V \leq^{\oplus} K_1 \leq^{\oplus} M$ and hence $\varphi(K_1) \leq^{\oplus} M$ by (C_2) condition of M. So by the modular law, $\varphi(K_1) \leq^{\oplus} K_2$ as $\varphi(K_1) \leq K_2$.

Let $N_1 = V \oplus \varphi(K_1)$ and $P_1 = \varphi(K_1)$. As $V \leq^{\oplus} eN$ and $P_1 \leq^{\oplus} (1-e)N$, $N_1 \leq^{\oplus} N$. Note that $N_1 = V \oplus \varphi(K_1) \cong \varphi(K_1) \oplus \varphi(K_1) = P_1 \oplus P_1$. By the same argument, there exist $N_2 \leq^{\oplus} P_1$ and $P_2 \leq^{\oplus} M$ such that $N_2 \cong P_2 \oplus P_2$. So $P_2 \oplus P_2 \cong N_2 \leq^{\oplus} P_1$, and hence

$$P_2^{(4)} \cong N_2^{(2)} \leq^{\oplus} P_1^{(2)} \cong N_1.$$

Again by the same argument, there exist $N_3 \leq^{\oplus} P_2$ and $P_3 \leq^{\oplus} M$ such that $N_3 \cong P_3 \oplus P_3$. Thus

$$P_3^{(8)} \cong N_3^{(4)} \leq^{\oplus} P_2^{(4)} \cong N_2^{(2)} \leq^{\oplus} P_1^{(2)} \cong N_1.$$

Iterating this process, for each positive integer k, there exists $0 \neq P_k \leq^{\oplus} M$ such that $P_k^{(2^k)}$ is isomorphic to a direct summand of N_1. We fix a positive integer n and choose a positive integer k so that $2^k \geq n$. Then $P_k^{(n)}$ is isomorphic to a direct summand, say N' of N_1. So $0 \neq N' \leq^{\oplus} N_1 \leq^{\oplus} N$ and $N' \cong P_k^{(n)}$.

Let $\mathcal{V}$ be the family of all nonempty sets of independent direct summands V of M such that $V \cong P_V^{(n)}$ for some $P_V \leq^{\oplus} M$. Then by Zorn's lemma, there exists a maximal element, say $\mathcal{C}_n$, in $\mathcal{V}$. By (C_1) condition, there exists $W_1 \leq^{\oplus} M$ such that $\oplus_{V \in \mathcal{C}_n} V \leq^{\text{ess}} W_1$. Say $M = W_1 \oplus W_2$. If $W_2 \neq 0$, then there exists $0 \neq L \leq^{\oplus} W_2$ so that $L \cong X^{(n)}$ for some $X \leq M$. This contradicts the maximality of the chain $\mathcal{C}_n$. Thus, $W_2 = 0$, so $\oplus_{V \in \mathcal{C}_n} V \leq^{\text{ess}} M$. Hence,

$$E(M) = E\left(\oplus_{V \in \mathcal{C}_n} V\right) \cong E\left(\oplus_{V \in \mathcal{C}_n,\, V \cong P_V^{(n)}} P_V^{(n)}\right)$$
$$= \left(E\left(\oplus_{V \in \mathcal{C}_n,\, V \cong P_V^{(n)}} P_V\right)\right)^{(n)} = E^{(n)},$$

where $E = E\left(\oplus_{V \in \mathcal{C}_n,\, V \cong P_V^{(n)}} P_V\right)$. We obtain that $E(M) = \oplus_{i=1}^n E_i$, $E_i \cong E$ for i with $1 \leq i \leq n$. Hence by Theorem 2.1.25, $M = \oplus_{i=1}^n (M \cap E_i)$. Moreover, $M \cap E_i$ and $M \cap E_j$ are relatively injective for $i \neq j$ from Lemma 2.2.4. Also $E(M \cap E_i) = E_i \cong E_j = E(M \cap E_j)$. Therefore by Proposition 2.1.3, $M \cap E_i \cong M \cap E_j$ for $i \neq j$, $1 \leq i, j \leq n$. Let $M_n = M \cap E_1$. Then $M_n \leq^{\oplus} M$ and $M \cong M_n^{(n)}$. □

Corollary 4.4.16 *Let M be a $\mathcal{K}$-nonsingular continuous module of type* II *or type* III. *Then $M \cong M_2^{(2)}$ for some $M_2 \leq M$. Hence, M is quasi-injective.*

Proof Because type II or type III $\mathcal{K}$-nonsingular continuous modules do not contain nonzero Abelian direct summands, $M \cong M_2^{(2)}$ with $M_2 \leq M$ by Theorem 4.4.15. Since M is continuous, M_2 is M_2-injective by Lemma 2.2.4, and hence M_2 is quasi-injective. Therefore, $M \cong M_2^{(2)}$ is quasi-injective by Corollary 2.2.3. □

We remark that, in view of Corollary 4.4.16, any $\mathcal{K}$-nonsingular continuous module is quasi-injective if and only if its type I component is quasi-injective.

Theorem 4.4.17 *Let M be a $\mathcal{K}$-nonsingular quasi-continuous module. Then M is of type* III *if and only if $N \cong N \oplus N$ for any $N \leq^{\oplus} M$. In this case, M is quasi-injective.*

Proof Assume that M is of type III. It is enough to prove the statements for M since a direct summand of a $\mathcal{K}$-nonsingular quasi-continuous module is $\mathcal{K}$-nonsingular quasi-continuous, and a nonzero direct summand of a type III module is of type III, by definition. Since no nonzero direct summands of M are directly finite, no nonzero direct summands of $E(M)$ are directly finite by Lemma 2.2.10(ii) and Theorem 2.1.25.

Let $\mathcal{V}$ be the family of all nonempty sets of independent direct summands V of M satisfying $V \cong V \oplus V$. By Zorn's lemma, there exists a maximal element, say $\mathcal{C}$ in

$\mathcal{V}$. As $E(M)$ is injective, there exists E_1 such that $\oplus_{V\in\mathcal{C}}V \leq^{\text{ess}} E_1 \leq^{\oplus} E(M)$. Let $E(M) = E_1 \oplus E_2$ for some $E_2 \leq E(M)$. If E_2 contains a nonzero direct summand which is not directly finite, then by Proposition 2.2.6 there exists a nonzero direct summand W of E_2 satisfying $W \cong W \oplus W$. But this contradicts the maximality of $\mathcal{C}$. So all direct summands of E_2 are directly finite. But since $E(M)$ does not have nonzero directly finite direct summand, $E_2 = 0$. Thus

$$E(M) = E\left(\oplus_{V\in\mathcal{C}}V\right) \cong E\left(\oplus_{V\in\mathcal{C}}(V \oplus V)\right)$$
$$= E\left(\oplus_{V\in\mathcal{C}}V\right) \oplus E\left(\oplus_{V\in\mathcal{C}}V\right).$$

So $E(M) \cong E(M) \oplus E(M)$. From this fact, we have that $E(M) = A \oplus B$, where $A, B \leq E(M)$, $A \cong E(M)$, and $B \cong E(M)$. By Theorem 2.1.25,

$$M = (M \cap A) \oplus (M \cap B)$$

since M is quasi-continuous. Note that $E(M \cap A) = A$ and $E(M \cap B) = B$. Thus it follows that $E(M \cap A) = A \cong E(M)$ and $E(M \cap B) = B \cong E(M)$. By Theorem 2.2.13, $M \cap A \cong M$ and $M \cap B \cong M$. So $M = (M \cap A) \oplus (M \cap B) \cong M \oplus M$.

Conversely, assume that $N \cong N \oplus N$ for any $0 \neq N \leq^{\oplus} M$. Then N is not directly finite by Proposition 2.2.6. Thus, M is of type III by definition. Further, M is quasi-injective because $M \cong M \oplus M$ by assumption. □

By Theorem 4.4.9, any $\mathcal{K}$-nonsingular continuous module decomposes into a direct sum of $\mathcal{K}$-nonsingular continuous modules, of type I, II, and III, respectively. In this case, however, the type II and type III continuous direct summands are in fact quasi-injective by Corollary 4.4.16.

Exercise 4.4.18

1. ([359, Rizvi and Roman]) Let $M = M_1 \oplus M_2$ be a $\mathcal{K}$-nonsingular continuous module, with M_1 and M_2 indecomposable modules. Then either $M_1 \cong M_2$ (in which case M_1 and M_2 are quasi-injective), or both M_1 and M_2 are orthogonal.
2. ([359, Rizvi and Roman]) Let M be a $\mathcal{K}$-nonsingular extending module such that $S = \text{End}(M)$ is right and left extending, and semiprime. Prove that the following conditions are equivalent.
   (i) M is of type I.
   (ii) Every nonzero direct summand of M contains a nonzero Abelian submodule which is isomorphic to a direct summand of M.
   (iii) The sum of all Abelian submodules which are isomorphic to direct summands of M is an essential submodule of M.
3. ([359, Rizvi and Roman]) Let M be a $\mathcal{K}$-nonsingular extending module, for which $S = \text{End}(M)$ is right and left extending, and semiprime. Show that the following are equivalent.
   (i) M is of type II.
   (ii) Every nonzero direct summand of M contains a nonzero directly finite submodule which is isomorphic to a direct summand of M, and S does not have nonzero Abelian idempotents.

(iii) The sum of all directly finite submodules which are isomorphic to direct summands of M forms an essential submodule of M, and S does not have nonzero Abelian idempotents.

4.5 Rickart Modules

In this section, we briefly discuss Rickart modules as a generalization of right Rickart rings studied in Chap. 3. This can be considered as an application of the results of Sect. 4.2. Basic definitions and properties related to Rickart modules will be presented here. For further information, we refer the reader to [268–271], and [273].

Definition 4.5.1 Let M be a right R-module and $S = \text{End}(M)$. Then M is called a *Rickart module* if for each $\phi \in S$, $r_M(\phi) = \text{Ker}(\phi) = eM$ for some $e^2 = e \in S$.

From Definition 4.2.13, we see that a module M is Rickart if and only if M is M-Rickart. The concept of Rickart modules also appears in Theorem 4.1.21. Indeed, Theorem 4.1.21 says that a module M is a Baer module if and only if M is a Rickart module with the SSIP. In particular, an indecomposable Rickart module is always a Baer module. Some examples of Rickart modules include the following.

Example 4.5.2 (i) Let R be a ring. Then R_R is a Rickart module if R is a right Rickart ring.

(ii) Every semisimple module is a Rickart module.

(iii) Every Baer module is a Rickart module. Hence, every nonsingular injective (or extending) module is Rickart (see Theorem 4.1.15).

(iv) Every projective right R-module over a right hereditary ring R is a Rickart module (see Theorem 4.5.8).

(v) The free $\mathbb{Z}$-module $\mathbb{Z}^{(\Lambda)}$, for any nonempty index set Λ, is Rickart, while $\mathbb{Z}^{(\Lambda)}$ is not a Baer $\mathbb{Z}$-module if Λ is uncountable (see Remark 4.5.10). In particular, $\mathbb{Z}^{(\mathbb{N})}$ ($\cong \mathbb{Z}[x]$) is a Rickart (and Baer) $\mathbb{Z}$-module, while $\mathbb{Z}^{(\mathbb{R})}$ is a Rickart but not a Baer $\mathbb{Z}$-module. In general, if R is a right hereditary ring which is not Baer, then every free R-module is Rickart but not Baer (Proposition 4.5.11).

(vi) $\mathbb{Z}_{p^\infty}$ is injective, while $\mathbb{Z}_4$ is quasi-injective as $\mathbb{Z}$-modules. However, neither of these is a Rickart $\mathbb{Z}$-module.

The Rickart property does not always transfer from a module to its submodules or conversely as the next example illustrates.

Example 4.5.3 Every submodule of a Rickart module is not Rickart, in general. Let $M = \mathbb{Q} \oplus \mathbb{Z}_2$, which is a $\mathbb{Z}$-module (see Example 4.1.10(ii)). Then $\text{Hom}(\mathbb{Q}, \mathbb{Z}_2) = 0$ and $\text{Hom}(\mathbb{Z}_2, \mathbb{Q}) = 0$. Further, $\mathbb{Q}$ and $\mathbb{Z}_2$ are Baer $\mathbb{Z}$-modules. Hence we see that M is Baer (see Theorem 4.2.17), so it is Rickart. However, the submodule $N := \mathbb{Z} \oplus \mathbb{Z}_2$

is not a Rickart $\mathbb{Z}$-module, even though $\mathbb{Z}$ and $\mathbb{Z}_2$ are both Rickart $\mathbb{Z}$-modules. In fact, the map $(m, \overline{n}) \to (0, \overline{m})$ has the kernel $2\mathbb{Z} \oplus \mathbb{Z}_2$, which is not a direct summand of N.

Recall that a module is said to have the SIP if the intersection of any two direct summands is a direct summand (see Definition 4.1.20). A module M is said to satisfy (D_2) *condition* if, for any $N \leq M$ with $M/N \cong K \leq^{\oplus} M$, we have $N \leq^{\oplus} M$. For (D_2) condition in detail, see [301].

Proposition 4.5.4 *Let M_R be a Rickart right R-module with $S = \text{End}(M)$. Then:*

(i) *Every direct summand of M is a Rickart module.*
(ii) *M is $\mathcal{K}$-nonsingular.*
(iii) *M satisfies (D_2) condition.*
(iv) *M has the SIP.*
(v) *S is a right Rickart ring.*

Proof (i) Let $N = eM$ for some $e^2 = e \in S$ and $\psi \in \text{End}(eM)$. Then we have that $\text{Ker}(\psi e) = \text{Ker}(\psi) \oplus (1-e)M$. Since $\psi e \in S$, $\text{Ker}(\psi e) \leq^{\oplus} M$ and hence $\text{Ker}(\psi) \leq^{\oplus} M$. Thus, $\text{Ker}(\psi) \leq^{\oplus} N$ by the modular law, so N is Rickart.

(ii) Assume that $\text{Ker}(\phi) \leq^{\text{ess}} M$, where $\phi \in S$. Then $\text{Ker}(\phi) = M$ because $\text{Ker}(\phi) \leq^{\oplus} M$. So $\phi = 0$.

(iii) Suppose that $N \leq M$ such that $M/N \cong K \leq^{\oplus} M$ and $M = K \oplus U$. We let $\varphi : M \to M/N$ be the natural homomorphism and $f : M/N \to K$ be the given isomorphism. Then $f\varphi \in \text{End}(M)$ and $\text{Ker}(f\varphi) = N$. Since M is Rickart, $N \leq^{\oplus} M$.

(iv) Let $L = eM$ and $N = fM$ for some nonzero idempotents $e, f \in S$. We claim that $\text{Ker}((1-f)e) = [eM \cap \text{Ker}(1-f)] \oplus (1-e)M$. For this claim, we first take $x \in \text{Ker}((1-f)e)$. Then $(1-f)(ex) = 0$ so $ex \in eM \cap \text{Ker}(1-f)$. Thus

$$x = ex + (1-e)x \in [eM \cap \text{Ker}(1-f)] \oplus (1-e)M.$$

The other inclusion is obvious. Since M is Rickart, $\text{Ker}((1-f)e) \leq^{\oplus} M$. Therefore, we get $L \cap N = eM \cap \text{Ker}(1-f)$ is a direct summand of M.

(v) Let $0 \neq \phi \in S$. Then $r_M(\phi) = eM$ for some $e^2 = e \in S$. As $\phi eM = 0$, $\phi e = 0$. So $e \in r_S(\phi)$. Thus, $eS \subseteq r_S(\phi)$. For $\psi \in r_S(\phi)$, $\phi\psi = 0$. As a consequence, we obtain $\psi M \subseteq r_M(\phi) = eM$, so $\psi = e\psi \in eS$ and hence $r_S(\phi) \subseteq eS$.

Thus, $eS = r_S(\phi)$. Therefore, S is a right Rickart ring. □

In reference to the endomorphism ring of a module, it is easy to see that the converse of Proposition 4.5.4(v) does not hold true. In Example 4.2.9, the $\mathbb{Z}$-module $\mathbb{Z}_{p^\infty}$ (p is a prime integer) is not Rickart, while $\text{End}(\mathbb{Z}_{p^\infty})$ is a domain (hence a Baer ring). By Proposition 4.5.4(iv), a ring R is right Rickart if and only if for any nonempty finite subset F of R, $r_R(F) = eR$ for some $e^2 = e \in R$. The following example illustrates that this cannot be improved to the case of countable subsets of R.

Example 4.5.5 Let $R = \{(a_n)_{n=1}^{\infty} \in \prod_{n=1}^{\infty} F_n \mid a_n \text{ is constant eventually}\}$, which is a subring of $\prod_{n=1}^{\infty} F_n$, where $F_n = \mathbb{Z}_2$ for $n = 1, 2, \ldots$. Then R is Rickart. If for any countable subset X of R, $r_R(X) = eR$ for some $e^2 = e \in R$, then R is a Baer ring because R itself is a countable infinite set. This is absurd since R is not Baer as seen in Example 3.1.14(ii).

A module M_R is called *k-local-retractable* (i.e., kernel-local-retractable) if $r_M(\phi) = r_S(\phi)(M)$ for any $\phi \in S$.

Theorem 4.5.6 *Let M be a right R-module and $S = \mathrm{End}(M)$. Then the following are equivalent.*

(i) *M is a Rickart module.*
(ii) *M satisfies (D_2) condition, and $\phi(M)$ is isomorphic to a direct summand of M for any $\phi \in S$.*
(iii) *S is a right Rickart ring and M is k-local-retractable.*

Proof (i)⇒(ii) From Proposition 4.5.4(iii), M satisfies (D_2) condition. For $\phi \in S$, $\mathrm{Ker}(\phi) \leq^{\oplus} M$ and so $M = \mathrm{Ker}(\phi) \oplus V$ for some $V \leq M$. Thus, we have that $\phi(M) \cong V \leq^{\oplus} M$.

(ii)⇒(i) Assume that $\varphi \in S$. Then $\varphi(M) \cong M/\mathrm{Ker}(\varphi)$ and by assumption $\varphi(M) \cong N \leq^{\oplus} M$ for some N. By (D_2) condition, $\mathrm{Ker}(\varphi) \leq^{\oplus} M$.

(i)⇒(iii) From Proposition 4.5.4(v), S is a right Rickart ring. For $\phi \in S$,

$$r_M(\phi) = eM \quad \text{with} \quad e^2 = e \in S,$$

and so $r_S(\phi) = eS$. Therefore $r_S(\phi)(M) = eM = r_M(\phi)$. Hence M is k-local-retractable.

(iii)⇒(i) Let $0 \neq \phi \in S$. Then $r_S(\phi) = eS$ for some $e^2 = e \in S$. Because M is k-local-retractable, $r_M(\phi) = r_S(\phi)(M) = eM \leq^{\oplus} M$. □

In the following two results, we characterize the class of semisimple Artinian rings and that of right hereditary rings via Rickart modules.

Theorem 4.5.7 *The following are equivalent for a ring R.*

(i) *Every right R-module is a Rickart module.*
(ii) *Every extending right R-module is a Rickart module.*
(iii) *Every injective right R-module is a Baer module.*
(iv) *R is semisimple Artinian.*

Proof (i)⇒(ii) It is evident.

(ii)⇒(iii) Let M be an injective right R-module. Then M is a Rickart module by assumption, so M is $\mathcal{K}$-nonsingular by Proposition 4.5.4(ii). Thus, M is a Baer module from Theorem 4.1.15.

(iii)⇒(iv) The proof follows immediately from Theorem 4.2.20.

(iv)⇒(i) The proof is obvious. □

Theorem 4.5.8 *The following are equivalent for a ring R.*

(i) *Every projective right R-module is a Rickart module.*

(ii) *Every free right R-module is a Rickart module.*
(iii) *R is right hereditary.*

Proof (i)⇒(ii) It is obvious.

(ii)⇒(iii) Let I be a right ideal of R. Then there is a set Λ and an epimorphism $\varphi : R_R^{(\Lambda)} \to I_R$. We can view φ as an endomorphism of $R_R^{(\Lambda)}$. By assumption, as $R_R^{(\Lambda)}$ is Rickart, $\text{Ker}(\varphi)_R \leq^{\oplus} R_R^{(\Lambda)}$, so I_R is isomorphic to a direct summand of $R_R^{(\Lambda)}$. Thus I_R is projective, so R is right hereditary.

(iii)⇒(i) Let M be any projective right R-module and $0 \neq \phi \in \text{End}(M)$. As R is right hereditary, $\phi(M)$ is projective. Hence $\text{Ker}(\phi) \leq^{\oplus} M$, so M is a Rickart module. □

As a consequence of Theorem 4.5.8, we obtain the following example.

Example 4.5.9 Let R be any right hereditary ring. Then $R^{(\mathbb{N})} (\cong R[x])$ and $R^{(\mathbb{R})}$ are Rickart R-modules.

Remark 4.5.10 Note that every free module F of countable rank over a commutative PID R has the SSIP (see [247, Exercise 51(c), p. 49]). Also, by Theorem 4.5.8, F is a Rickart R-module. Theorem 4.1.21 yields that F is Baer. In particular, $\mathbb{Z}^{(\mathbb{N})}$ is a Baer $\mathbb{Z}$-module. On the other hand, $\mathbb{Z}^{(\mathbb{R})}$ is a Rickart $\mathbb{Z}$-module which is not a Baer $\mathbb{Z}$-module because it does not satisfy the SSIP (see [411, Remark, p. 32]).

To distinguish between Rickart and Baer modules, our next result provides a rich source of examples where the two notions differ. Recall that the ring R in Example 3.1.20 is left hereditary, which is not a Baer ring. A right hereditary ring which is not Baer can be similarly constructed.

Proposition 4.5.11 *Let R be a right hereditary ring which is not a Baer ring. Then every free right R-module is Rickart but not Baer.*

Proof For a right hereditary ring R, every free right R-module is a Rickart module by Theorem 4.5.8. Also, any free right R-module can not be Baer from Theorem 4.1.22 since R_R is a direct summand which is not Baer. □

Lemma 4.5.12 *Let M be a Rickart right R-module and $S = \text{End}_R(M)$. If $\ell_S(X) \neq 0$, where $\emptyset \neq X \subseteq M$, then $\ell_S(X)$ contains a nonzero idempotent.*

Proof Say $I = \ell_S(X)$ with $\emptyset \neq X \subseteq M$. Let $I \neq 0$. Take $0 \neq \varphi \in I$, and let $N = r_M(I)$. As M is a Rickart module, $r_M(\varphi) = eM$ with $e^2 = e \in S$. Now $N = r_M(I) \subseteq r_M(\varphi) = eM$, and so $(1 - e)N = 0$. Therefore we have that $1 - e \in \ell_S(N) = \ell_S(r_M(I)) = \ell_S(r_M(\ell_S(X))) = \ell_S(X)$, and $1 - e \neq 0$ because $\varphi \neq 0$. □

As mentioned earlier, every indecomposable Rickart module is a Baer module. The next result extends this fact and it is a module theoretic analogue of Theorem 3.1.25.

Theorem 4.5.13 *Let M be a Rickart right R-module. If $S = \mathrm{End}(M)$ is orthogonally finite, then M is a Baer module.*

Proof Let $N \leq M$ and let $I = \ell_S(N)$. To show that M is a Baer module, we may assume that I is a nonzero proper left ideal of S. From Lemma 4.5.12, I contains a nontrivial idempotent of S. From Proposition 1.2.13, S has DCC on direct summand left ideals, and so among all nonzero idempotents in I, we can choose $0 \neq e \in I$ such that $S(1-e)$ is minimal.

We claim that $I \cap \ell_S(eM) = 0$. On the contrary, let $I \cap \ell_S(eM) \neq 0$. Then

$$0 \neq I \cap \ell_S(eM) = \ell_S(N) \cap \ell_S(eM) = \ell_S(N + eM).$$

By Lemma 4.5.12, there exists $0 \neq f^2 = f \in I \cap \ell_S(eM)$. Because $fe = 0$, we can check that $g := e + f - ef$ is an idempotent in I.

Also $ge = e$ as $fe = 0$. Thus $g \neq 0$ and $eM \subseteq gM$. So $\ell_S(gM) \subseteq \ell_S(eM)$. Thus $S(1-g) \subseteq S(1-e)$. As $0 \neq g \in I$, $S(1-g) = S(1-e)$. So $\ell_S(gM) = \ell_S(eM)$. But $\ell_S(gM) \subsetneq \ell_S(eM)$ because $fe = 0$ and $fg = f \neq 0$. This is a contradiction. So $I \cap \ell_S(eM) = 0$.

Finally, we show that $I = Se$. Because $e \in I$, $Se \subseteq I$. Take $\varphi \in I$. Then, we get that $\varphi(1-e) = \varphi - \varphi e \in I$. Thus $\varphi(1-e) \in I \cap \ell_S(eM) = 0$, so $\varphi = \varphi e \in Se$. Thus, $I \subseteq Se$. Hence $I = Se$, therefore M is a Baer module. □

Theorem 4.5.14 *The following are equivalent for a module M_R.*

(i) *M is a Rickart module with (C_2) condition.*
(ii) *$S = \mathrm{End}(M)$ is a regular ring.*
(iii) *For each $\phi \in S = \mathrm{End}(M)$, $\mathrm{Ker}(\phi) \leq^{\oplus} M$ and $\mathrm{Image}(\phi) \leq^{\oplus} M$.*

Proof (i)⇒(ii) Suppose that M is a Rickart module with (C_2) condition. We take $0 \neq \phi \in S$. Because M is Rickart, $M = \mathrm{Ker}(\phi) \oplus N$ for some $N \leq M$. Since $\phi N \cong N \leq^{\oplus} M$, $\phi N \leq^{\oplus} M$ by (C_2) condition. Also, there exists $0 \neq \psi \in S$ such that $(\psi\phi)|_N$ is the identity map of N. Then we have that $(\phi - \phi\psi\phi)(M) = (\phi - \phi\psi\phi)(\mathrm{Ker}(\phi) \oplus N) = (\phi - \phi\psi\phi)(N) = 0$. Hence $\phi - \phi\psi\phi = 0$, so S is a regular ring.

(ii)⇒(iii) Say $\phi \in S$. Since S is regular, $S\phi = Se$ and $\phi S = fS$ for some idempotents e and f in S. Thus, $\mathrm{Ker}(\phi) = r_M(Se) = (1-e)M \leq^{\oplus} M$ consequently $\phi(M) = \phi S(M) = fS(M) = fM \leq^{\oplus} M$.

(iii)⇒(i) By hypothesis, it suffices to show that M satisfies (C_2) condition. Let N be a submodule of M with an isomorphism $f : N \to eM$, where $e^2 = e \in S$. Consider $\varphi \in \mathrm{End}(M)$ such that $\varphi(em + (1-e)m) = f^{-1}(em)$. Then $N = \varphi(M) \leq^{\oplus} M$ by hypothesis. So M satisfies (C_2) condition. □

Theorem 4.5.15 *Let $\{M_i\}_{1 \leq i \leq n}$ be a finite set of modules such that M_i is M_j-injective for all $i < j$. Then $M = \bigoplus_{i=1}^n M_i$ is a Rickart module if and only if M_i is M_j-Rickart for all i and j.*

Proof The necessity follows from Proposition 4.2.14. Conversely, assume that M_i is M_j-Rickart for all i, j. We show that each M_k is M-Rickart. As M_k is M_k-Rickart, M_k is a Rickart module and so M_k has the SIP by Proposition 4.5.4(iv). Let $\varphi \in \mathrm{Hom}(M_k, M)$ and π_i be the canonical projection from M onto M_i. So $\mathrm{Ker}(\varphi) = \cap_{i=1}^{n} \mathrm{Ker}(\pi_i \varphi) \leq^{\oplus} M_k$ because $\mathrm{Ker}(\pi_i \varphi) \leq^{\oplus} M_k$ for each i and M_k has the SIP. So each M_k is M-Rickart. As M_i is M_j-injective by for all $i < j$ by hypothesis, $M = \oplus_{i=1}^{n} M_i$ is M-Rickart by Lemma 4.2.16. Thus, M is a Rickart module. □

Exercise 4.5.16

1. ([269, Lee, Rizvi, and Roman]) Show that a module M is k-local retractable if and only if for any $\varphi \in \mathrm{End}(M)$ and any $0 \neq m \in r_M(\varphi)$, there exists ψ_m in $\mathrm{Hom}(M, r_M(\varphi))$ such that $m \in \psi_m(M)$.
2. ([269, Lee, Rizvi, and Roman]) Prove that every free module is k-local-retractable.
3. ([269, Lee, Rizvi, and Roman]) Show that the following are equivalent for a module M.
   (i) M is a Rickart module.
   (ii) For any $N \leq M$, every direct summand L of M is N-Rickart.
   (iii) For every pair of direct summands L and N of M and for every ϕ in $\mathrm{Hom}(M, N)$, we have that $\mathrm{Ker}(\phi|_L) \leq^{\oplus} L$.
4. ([271, Lee, Rizvi, and Roman]) Let M be a Rickart module with (C_2) condition. Prove that any finite direct sum of copies of M is a Rickart module.
5. ([271, Lee, Rizvi, and Roman]) Let M be an indecomposable Rickart module with a nonzero maximal submodule N. Prove that M and M/N are Baer modules, but $M \oplus (M/N)$ is not a Rickart module.
6. ([269, Lee, Rizvi, and Roman]) Let M be an arbitrary direct sum of cyclic modules over a Dedekind domain (i.e., commutative hereditary domain). Show that M is a Rickart module if and only if M is either semisimple or torsion-free.

4.6 Quasi-Baer Modules

The concept of a quasi-Baer module is introduced and studied. It is shown that the endomorphism ring of a quasi-Baer module is a quasi-Baer ring. Connections between quasi-Baer modules and FI-extending modules are established, as both of these concepts depend on the behavior of fully invariant submodules. The concepts of FI-$\mathcal{K}$-nonsingularity and FI-$\mathcal{K}$-cononsingularity are introduced. These are used to provide a complete characterization for a quasi-Baer module which is FI-$\mathcal{K}$-cononsingular. It is shown that an arbitrary direct sum of mutually subisomorphic quasi-Baer modules is quasi-Baer and that a direct summand of a quasi-Baer module is quasi-Baer. Consequently, every projective module over a quasi-Baer ring is quasi-Baer.

Definition 4.6.1 A right R-module M is called a *quasi-Baer module* if for each $N \trianglelefteq M$, $\ell_S(N) = Se$ with $e^2 = e \in S$ where $S = \mathrm{End}(M)$.

Example 4.6.2 (i) All semisimple modules are quasi-Baer.

(ii) All Baer and quasi-Baer rings are quasi-Baer modules, viewed as modules over themselves.

(iii) Every Baer module is a quasi-Baer module.

(iv) Any free R-module over a quasi-Baer ring R which is not a Baer ring (e.g., $R = T_2(\mathbb{Z})$), is a quasi-Baer module, but it is not a Baer module (see Theorems 4.6.19 and 4.1.22).

We remark that in [274, Definition 2.1] Lee and Zhou defined a module M to be *quasi-Baer* if, for any submodule N of M, $r_R(N) = eR$ for some $e^2 = e \in R$. But our Definition 4.6.1 is distinct from their definition. In particular, semisimple modules are not quasi-Baer in the sense of their definition in general unlike Example 4.6.2(i).

Proposition 4.6.3 *The following are equivalent for a module M.*

(i) *M is a quasi-Baer module.*
(ii) *For any $I \trianglelefteq S$, $r_M(I) = fM$ with $f^2 = f \in S$, where $S = \mathrm{End}(M)$.*

Proof For (i)⇒(ii), say $I \trianglelefteq S$. Then $r_M(I) \trianglelefteq M$. Thus $\ell_S(r_M(I)) = Se$ for some $e^2 = e \in S$. So $r_M(I) = r_M(\ell_S(r_M(I))) = (1-e)M$. Take $f = 1-e$. Then $r_M(I) = fM$. Next, for (ii)⇒(i), let $N \trianglelefteq M$. Then $\ell_S(N) \trianglelefteq S$. Thus $r_M(\ell_S(N)) = fM$ with $f^2 = f \in S$. So $\ell_S(N) = \ell_S(r_M(\ell_S(N))) = S(1-f)$. Therefore M is a quasi-Baer module. □

Similar to the case of Baer modules, quasi-Baer modules also satisfy a weak nonsingularity condition. This condition will allow us to exhibit close links between quasi-Baer modules and FI-extending modules analogous to Theorem 4.1.15.

Definition 4.6.4 A right R-module M is called *FI-$\mathcal{K}$-nonsingular* if for any $I \trianglelefteq S$, $r_M(I) \leq^{\mathrm{ess}} eM$ with $e^2 = e \in S$ implies $Ie = 0$.

Definition 4.6.5 A right R-module M is said to be *FI-$\mathcal{K}$-cononsingular* if for any $N \trianglelefteq M$, $r_M(\ell_S(N)) \leq^{\oplus} M$ implies $N \leq^{\mathrm{ess}} r_M(\ell_S(N))$.

By Propositions 4.1.12 and 4.1.14, if a module M is $\mathcal{K}$-nonsingular (resp., $\mathcal{K}$-cononsingular), then M is FI-$\mathcal{K}$-nonsingular (resp., FI-$\mathcal{K}$-cononsingular).

Proposition 4.6.6 *Let M be a right R-module and $S = \mathrm{End}(M)$. Then:*

(i) *M is FI-$\mathcal{K}$-nonsingular if and only if for any $I \trianglelefteq S$ with $r_M(I) \leq^{\mathrm{ess}} eM$, where $e^2 = e \in S$, implies $r_M(I) = eM$.*
(ii) *M is FI-$\mathcal{K}$-cononsingular if and only if for every $N \trianglelefteq M$ with $N \leq^{\oplus} M$, and $V \trianglelefteq N$ such that $\varphi(V) \neq 0$ whenever $0 \neq \varphi \in \mathrm{End}(N)$, we get that $V \leq^{\mathrm{ess}} N$.*

Proof (i) Let M be FI-$\mathcal{K}$-nonsingular. Take $I \trianglelefteq S$ with $r_M(I) \leq^{\mathrm{ess}} eM$ for some $e^2 = e \in S$. Then $Ie = 0$ by assumption, and so $IeM = 0$. Therefore $eM \subseteq r_M(I)$, thus $r_M(I) = eM$. Conversely, let $J \trianglelefteq S$ with $r_M(J) \leq^{\text{ess}} fM$ for some $f^2 = f$ in S. Then $r_M(J) = fM$ by assumption, hence $JfM = 0$ and so $Jf = 0$. Therefore M is FI-$\mathcal{K}$-nonsingular.

(ii) Let M be FI-$\mathcal{K}$-cononsingular. Say $N \trianglelefteq M$ with $N \leq^{\oplus} M$, and $V \trianglelefteq N$ with $\varphi(V) \neq 0$ whenever $0 \neq \varphi \in \text{End}_R(N)$. Put $N = fM$ with $f^2 = f \in S$. Then $f \in \mathbf{S}_\ell(S)$ as $N \trianglelefteq M$. By Proposition 2.3.3(ii), $V \trianglelefteq M$. Note that $\ell_S(N) \subseteq \ell_S(V)$ and $\ell_S(N) = S(1-f)$. From $S = S(1-f) \oplus Sf = \ell_S(N) \oplus Sf$ and the modular law, $\ell_S(V) = S(1-f) \oplus (\ell_S(V) \cap Sf)$. Thus it follows that

$$r_M(\ell_S(V)) = fM \cap r_M(\ell_S(V) \cap Sf) = N \cap r_M(\ell_S(V) \cap Sf).$$

Next, we claim that $N \subseteq r_M(\ell_S(V) \cap Sf)$. For this, take $\nu \in \ell_S(V) \cap Sf$. Then $\nu(V) = 0$ and $\nu \in Sf = fSf$ as $f \in \mathbf{S}_\ell(S)$. So, $\nu|_N \in \text{End}_R(fM) = \text{End}_R(N)$. By assumption, $\nu(N) = 0$ as $\nu(V) = 0$. Hence, $N \subseteq r_M(\ell_S(V) \cap Sf)$. Now $r_M(\ell_S(V)) = N \cap r_M(\ell_S(V) \cap Sf) = N \leq^{\oplus} M$. Since M is FI-$\mathcal{K}$-cononsingular and $V \trianglelefteq M$, $V \leq^{\text{ess}} r_M(\ell_S(V)) = N$.

Conversely, to show that M is FI-$\mathcal{K}$-cononsingular, let $V \trianglelefteq M$ such that $r_M(\ell_S(V)) \leq^{\oplus} M$. Say $r_M(\ell_S(V)) = eM$ for some $e^2 = e \in S$. Then we see that $\ell_S(V) = \ell_S(r_M(\ell_S(V))) = \ell_S(eM) = S(1-e)$. Now $S(1-e) \trianglelefteq S$ as $V \trianglelefteq M$. Thus $1 - e \in \mathbf{S}_r(S)$, hence $e \in \mathbf{S}_\ell(S)$ by Proposition 1.2.2, so $r_M(\ell_S(V)) = eM \trianglelefteq M$. As $V \trianglelefteq M$ and $V \leq eM \trianglelefteq M$, $V \trianglelefteq eM$.

Say $\varphi \in \text{End}(r_M(\ell_S(V)))$ such that $\varphi(V) = 0$. We may observe that $\text{End}(r_M(\ell_S(V))) = \text{End}(eM) = eSe$, so $\varphi = ege$ for some $g \in S$. Because $\varphi(V) = (ege)(V) = 0$, $ege \in \ell_S(V) = S(1-e)$, hence $ege = 0$. So $\varphi = 0$. By assumption, $V \leq^{\text{ess}} r_M(\ell_S(V))$. Whence M is FI-$\mathcal{K}$-cononsingular. □

Example 4.6.7 (i) If a ring R is semiprime, then R_R is FI-$\mathcal{K}$-nonsingular. In fact, say $I \trianglelefteq R$ such that $r_R(I) \leq^{\text{ess}} eR_R$ with $e^2 = e \in R$. Assume on the contrary that $Ie \neq 0$. Since R is semiprime and $I \trianglelefteq R$, $eI \neq 0$ and so there is $x \in I$ with $0 \neq ex \in eI \subseteq eR$. Hence, there exists $r \in R$ with $0 \neq exr \in r_R(I)$ because $r_R(I) \leq^{\text{ess}} eR_R$. Thus $0 \neq exr \in r_R(I) \cap I = 0$, a contradiction. Hence, $Ie = 0$. Therefore, R_R is FI-$\mathcal{K}$-nonsingular.

(ii) There exists a ring R such that R_R is FI-$\mathcal{K}$-nonsingular, but R_R is not $\mathcal{K}$-nonsingular. In fact, by Example 3.2.7(i), there is a prime ring R with $Z(R_R) \neq 0$. Then R_R is not $\mathcal{K}$-nonsingular by Proposition 4.1.11. But note that, R_R is FI-$\mathcal{K}$-nonsingular from part (i).

(iii) There exists a ring R such that R_R is FI-$\mathcal{K}$-cononsingular, but R_R is not $\mathcal{K}$-cononsingular. For example, take

$$R = \begin{bmatrix} \mathbb{C} & \mathbb{C} \\ 0 & \mathbb{R} \end{bmatrix}.$$

Then R is a Baer ring since $T_2(\mathbb{C})$ is a Baer ring and R contains all idempotents of $T_2(\mathbb{C})$ (see Proposition 3.1.5(ii) and Theorem 5.6.2). Furthermore, R right FI-extending ring (see Corollary 5.6.11). Let $\alpha \in R$ be the matrix with $\mathbf{i}$ in the $(1, 2)$-position and 0 elsewhere, where $\mathbf{i}$ is the imaginary unit. There is no idempotent $e \in R$ with $\alpha R_R \leq^{\text{ess}} eR_R$. Thus, R is not right extending. Then R_R is not $\mathcal{K}$-cononsingular by Theorem 4.1.15. But R_R is FI-$\mathcal{K}$-cononsingular (see Lemma 4.6.8). In general, any module which is Baer and FI-extending, but not extending has the property that it is FI-$\mathcal{K}$-cononsingular but not $\mathcal{K}$-cononsingular.

We need the following series of lemmas to show the analogue of Theorem 4.1.15 for the quasi-Baer case in the module theoretic setting.

Lemma 4.6.8 *Every FI-extending module is FI-$\mathcal{K}$-cononsingular.*

Proof Let M be an FI-extending right R-module. We take $N \trianglelefteq M$ such that $N \leq^{\oplus} M$. Then from Proposition 2.3.4, N is FI-extending. Take $V \trianglelefteq N$ and suppose that $\varphi(V) \neq 0$ whenever $0 \neq \varphi \in \text{End}(N)$. By the FI-extending property of N, $V \leq^{\text{ess}} W \leq^{\oplus} N$ for some W. Let $N = W \oplus Y$ with $Y \leq N$, and $\pi : N \to Y$ be the canonical projection. Then $\pi \in \text{End}(N)$ and $\pi(V) = 0$. Thus, $\pi(N) = 0$ by assumption. So $Y = 0$ and $N = W$. Hence, $V \leq^{\text{ess}} N$. Therefore, M is FI-$\mathcal{K}$-cononsingular by Proposition 4.6.6(ii). □

Lemma 4.6.9 *Any FI-extending and FI-$\mathcal{K}$-nonsingular module is a quasi-Baer module.*

Proof Assume that M is a right R-module which is FI-extending and FI-$\mathcal{K}$-nonsingular. Let $I \trianglelefteq S$. Then $r_M(I) \trianglelefteq M$. By the FI-extending property of M, there is $e^2 = e \in S$ such that $r_M(I) \leq^{\text{ess}} eM$. As M is FI-$\mathcal{K}$-nonsingular, $r_M(I) = eM$ from Proposition 4.6.6(i). Thus, $r_M(I) \leq^{\oplus} M$. So M is a quasi-Baer module by Proposition 4.6.3. □

Lemma 4.6.10 *If M is a quasi-Baer module, then M is FI-$\mathcal{K}$-nonsingular.*

Proof Let $I \trianglelefteq S$ with $r_M(I) \leq^{\text{ess}} eM$ for some $e^2 = e \in S$. As M is quasi-Baer, $r_M(I) = fM$, where $f^2 = f \in S$. So $fM \leq^{\text{ess}} eM$. By the modular law, $eM = fM \oplus ((1 - f)M \cap eM)$. Thus $eM = fM$, hence $r_M(I) = eM$. By Proposition 4.6.6(i), M is FI-$\mathcal{K}$-nonsingular. □

Lemma 4.6.11 *If a right R-module M is FI-$\mathcal{K}$-cononsingular and quasi-Baer, then M is FI-extending.*

Proof Let $N \trianglelefteq M$. As M is quasi-Baer, $\ell_S(N) = Se$ for some $e^2 = e \in S$. Therefore $N \subseteq r_M(\ell_S(N)) = r_M(Se) = (1 - e)M$. Because M is FI-$\mathcal{K}$-cononsingular, we obtain $N \leq^{\text{ess}} (1 - e)M$. Thus, M is FI-extending. □

In the proof of Lemma 4.6.11, because $N \trianglelefteq M$, $\ell_S(N) = Se \trianglelefteq S$. Thus $e \in \mathbf{S}_r(S)$, and therefore $1 - e \in \mathbf{S}_\ell(S)$ by Proposition 1.2.2. Hence we have that $r_M(\ell_S(N)) = (1 - e)M \trianglelefteq M$. So M is, in fact, strongly FI-extending. As a consequence of the preceding lemmas, we obtain a full characterization of an FI-extending

FI-$\mathcal{K}$-nonsingular module in the following result. Similar to Theorem 4.1.15, the FI-extending property of modules is characterized in terms of the quasi-Baer property of modules as follows.

Theorem 4.6.12 *A module M is FI-extending and FI-$\mathcal{K}$-nonsingular if and only if M is quasi-Baer and FI-$\mathcal{K}$-cononsingular.*

Lemma 4.6.13 *Let M be a module with $M = M_1 \oplus M_2$, and let $V \trianglelefteq M_1$. Then there exists $W \trianglelefteq M_2$ such that $V \oplus W \trianglelefteq M$.*

Proof Take $W = \sum\{\varphi(V) \mid \varphi \in \text{Hom}(M_1, M_2)\}$. Then we can check that $W \trianglelefteq M_2$ and $V \oplus W \trianglelefteq M_1 \oplus M_2$. □

Theorem 4.6.14 *Any direct summand of a quasi-Baer module is a quasi-Baer module.*

Proof Let M be a quasi-Baer right R-module. Say $N \leq^{\oplus} M$. Then $N = eM$ for some $e^2 = e \in S = \text{End}(M)$. To show that N is quasi-Baer, let $V \trianglelefteq N$. Using Lemma 4.6.13, there is $W \trianglelefteq (1 - e)M$ with $V \oplus W \trianglelefteq M$. As M is quasi-Baer, $I := \ell_S(V \oplus W) = Sf$ with $f^2 = f \in S$. By Proposition 1.2.2, $f \in \mathbf{S}_r(S)$ since $I \trianglelefteq S$. We notice that $\text{End}(N) = eSe$ and $eIe \subseteq I = Sf$. Thus

$$eIe = eSfe = eSfef = eSfefe = (eSfe)(efe) \subseteq (eSe)(efe).$$

Also, $(eSe)(efe) \subseteq eSfe = eIe$. So $eIe = (eSe)(efe)$ and $(efe)^2 = efe$.

We show that $eIe = \ell_{eSe}(V)$. For this, first note that $(eIe)V = 0$, because $(eIe)V \subseteq IV = 0$. Therefore $eIe \subseteq \ell_{eSe}(V)$. Next, take $ese \in \ell_{eSe}(V)$. Then $eseV = 0$. Also $eseW \subseteq ese(1 - e)M = 0$, so $ese \in \ell_S(V \oplus W) = I$. As a consequence, $\ell_{eSe}(V) = eIe = (eSe)(efe)$. So N is a quasi-Baer module. □

Theorem 4.6.15 *Let $\{M_i\}_{i\in\Lambda}$ be a set of quasi-Baer modules. If M_i is subisomorphic to (i.e., isomorphic to a submodule of) M_j for all $i, j \in \Lambda$ with $i \neq j$. Then $M = \bigoplus_{i\in\Lambda} M_i$ is quasi-Baer.*

Proof Let $S = \text{End}(M)$. Write $\varphi \in S$ in matrix form $[\varphi_{hk}]_{(h,k)\in\Lambda\times\Lambda}$ with $\varphi_{hk} \in \text{Hom}(M_k, M_h)$. We need to show that, for $I \trianglelefteq S$, $r_M(I) \leq^{\oplus} M$. Since $I \trianglelefteq S$, $r_M(I) \trianglelefteq M$. Hence $r_M(I) = \oplus_{i\in\Lambda}(r_M(I) \cap M_i)$ (Exercise 2.1.37.2). For $i, j \in \Lambda$, let

$$I_{ij} = \{\alpha \in \text{Hom}(M_j, M_i) \mid \text{there is } \varphi = [\varphi_{hk}] \in I \text{ with } \alpha = \varphi_{ij}\}.$$

We claim that $r_M(I) \cap M_i = r_{M_i}(I_{ii})$ for each $i \in \Lambda$. For this, we first easily see that $r_M(I) \cap M_i \subseteq r_{M_i}(I_{ii})$. Next, let $x \in r_{M_i}(I_{ii})$. Take $\varphi = [\varphi_{hk}] \in I$. Then $\varphi_{ii}(x) = 0$. For $j \neq i$, let $\mu_{ij} : M_j \to M_i$ be a monomorphism, and let $\mu \in S$ with μ_{ij} in the (i, j)-position and 0 elsewhere. Then the (i, i)-position of $\mu\varphi \in I$ is $\mu_{ij}\varphi_{ji}$. So $\mu_{ij}\varphi_{ji} \in I_{ii}$ and $\mu_{ij}\varphi_{ji}(x) = 0$ since $I_{ii}x = 0$. As μ_{ij} is a monomorphism, $\varphi_{ji}(x) = 0$ for each $j \neq i$. So $\varphi(x) = [\varphi_{hk}](x) = 0$. Thus $Ix = 0$

and so $x \in r_M(I)$. Hence, $r_M(I) \cap M_i = r_{M_i}(I_{ii})$. Now we see that $r_M(I) \cap M_i = r_{M_i}(I_{ii}) \leq^{\oplus} M_i$ because each M_i is quasi-Baer and $I_{ii} \trianglelefteq \mathrm{End}_R(M_i)$. Hence,

$$r_M(I) = \oplus_{i \in \Lambda}(r_M(I) \cap M_i) \leq^{\oplus} \oplus_{i \in \Lambda} M_i.$$

Therefore, by Proposition 4.6.3 $\oplus_{i \in \Lambda} M_i$ is a quasi-Baer module. □

Theorem 4.6.16 *Let M be a quasi-Baer module. Then $S = \mathrm{End}(M)$ is a quasi-Baer ring.*

Proof Let $I \trianglelefteq S$. Since M is a quasi-Baer module, $r_M(I) = eM$ for some $e^2 = e \in S$ from Proposition 4.6.3. As in the proof of Theorem 4.2.8, we can show that $r_S(I) = eS$. Thus, S is a quasi-Baer ring. □

Corollary 4.6.17 *Let M be an FI-extending FI-$\mathcal{K}$-nonsingular module. Then $S = \mathrm{End}(M)$ is a quasi-Baer ring.*

Proof By Theorem 4.6.12, M is a quasi-Baer module. Hence, S is a quasi-Baer ring from Theorem 4.6.16. □

Example 4.6.18 The converse of Theorem 4.6.16 is not true. Let $M = \mathbb{Z}_{p^\infty}$, a $\mathbb{Z}$-module. Then S is a quasi-Baer ring, but M is not a quasi-Baer module.

Theorem 4.6.19 *Every projective module over a quasi-Baer ring is a quasi-Baer module and its endomorphism ring is a quasi-Baer ring.*

Proof Any free module over a quasi-Baer ring is a quasi-Baer module by Theorem 4.6.15. So Theorems 4.6.14 and 4.6.16 yield the result. □

For a ring R and a nonempty ordered set Γ, we use $\mathrm{RFM}_\Gamma(R)$ to denote the ring of $\Gamma \times \Gamma$ row finite matrices over R. Thus $\mathrm{RFM}_\Gamma(R) \cong \mathrm{End}({}_R R^{(\Gamma)})$, while $\mathrm{CFM}_\Gamma(R) \cong \mathrm{End}(R_R^{(\Gamma)})$.

Corollary 4.6.20 *The following are equivalent for a ring R.*

(i) *R is quasi-Baer.*
(ii) *$\mathrm{CFM}_\Gamma(R)$ is a quasi-Baer ring for any nonempty ordered set Γ.*
(iii) *$\mathrm{RFM}_\Gamma(R)$ is a quasi-Baer ring for any nonempty ordered set Γ.*

Proof It is a consequence of Theorems 4.6.19 and 3.2.10. □

Exercise 4.6.21

1. ([357, Rizvi and Roman]) Assume that M is a retractable module. Show that M is a quasi-Baer module if and only if $\mathrm{End}(M)$ is a quasi-Baer ring.

2. ([357, Rizvi and Roman]) Let M_1 and M_2 be quasi-Baer modules. Assume that $\varphi(x) = 0$, for any $\varphi \in \text{Hom}(M_i, M_j)$, implies $x = 0$, where $i \neq j$, $1 \leq i, j \leq 2$. Prove that $M_1 \oplus M_2$ is quasi-Baer.
3. Let $M = \oplus_{i=1}^n M_i$, where $M_i \trianglelefteq M$ for each i. Show that M is a quasi-Baer module if and only if each M_i is a quasi-Baer module.

Historical Notes The notions of Baer and quasi-Baer modules were introduced using the endomorphism ring of a module, by Rizvi and Roman [357]. Definition 4.1.1, Proposition 4.1.2, Definition 4.1.3, Proposition 4.1.12, Definition 4.1.13, and Proposition 4.1.14 are taken from [357], while Proposition 4.1.5, Theorem 4.1.7, Definition 4.1.8, Theorem 4.1.9, and Example 4.1.10 are from [359]. Theorem 4.1.15 is due to Rizvi and Roman [357]. The SIP and the SSIP for modules have been considered and studied in [247, Exercise 51, p. 49], [277, Satz 11], [411], and [201]. Theorems 4.1.21, 4.1.22, and 4.1.26 are due to Rizvi and Roman [357]. Theorem 4.1.24 is from [269].

Definition 4.2.2 was introduced in [360] and Proposition 4.2.3, Proposition 4.2.5, Theorem 4.2.8 are taken from that same paper. Example 4.2.4 is based on examples introduced in [250]. Proposition 4.2.14 is due to Lee, Rizvi, and Roman and is from [271]. Lemma 4.2.16 appears in [271]. Theorem 4.2.17 is due to Rizvi and Roman [360] modified by their work with Lee [271]. The proof of Theorem 4.2.18 using Baer module theory was provided by G. Lee and C. Roman. Theorem 4.2.20 is obtained by Rizvi and Roman [359].

Theorem 4.3.3, Corollary 4.3.4, Theorem 4.3.5, and Corollary 4.3.7 are due to Rizvi and Roman [360]. Corollary 4.3.8 appears in [121]. Theorem 4.3.5 is related to Satz 11 in [277]. By Lenzing [277], a ring R is called a B-ring if $\text{Mat}_n(R)$ is a Baer ring for every positive integer n. In Mao [289], the Baer property of the endomorphism ring of a module is described by using envelopes of modules. It was shown by Stephenson and Tsukerman [383] that the endomorphism ring of every free right R-module is a Baer ring if and only if R is a semiprimary hereditary ring.

Theorem 4.3.5 generalizes Theorem 2.2 in [159], which states that, for a regular ring R, every finitely generated torsionless right R-module embeds in a free right R-module (FGTF property) if and only if $\text{Mat}_n(R)$ is a Baer ring for every positive integer n. Theorem 4.3.5 in fact establishes that even in the absence of regularity of R, every finitely generated torsionless right R-module is projective if and only if $\text{Mat}_n(R)$ is Baer for every positive integer n. Lemma 4.3.10 can be found in [130]. Theorem 4.3.11 appears in [360]. Example 4.3.13 is taken from [239] and [360].

Most of the results and examples in Sect. 4.4 are due to Rizvi and Roman and taken from [359]. By Rizvi [355], commutative rings for which every continuous module is quasi-injective were studied and several classes of rings for which this holds true were identified. The problem remains open for arbitrary rings. Thus, in the presence of $\mathcal{K}$-nonsingularity, Corollary 4.4.16 reduces this problem to determining, when is a continuous module of type I, quasi-injective.

The module theoretic analogue of Rickart rings which we have discussed in Sect. 4.5 is due to recent works by Lee, Rizvi, and Roman in [271] and [269] (see also [270, 273], and [268]). The results of Sect. 4.5 are taken from [271] and [269]. We remark that (ii)⇔(iii) in Theorem 4.5.14 was known and also proved by Ware ([409, Corollary 3.2]). See [272] for a survey article on direct sums of Rickart modules. Most of the results on quasi-Baer modules in Sect. 4.6 are due to Rizvi and Roman and taken from [357]. For further results on FI-extending modules and quasi-Baer modules, see [357] and [361]. Work on topics related to this chapter can also be found in [175, 204, 252, 275, 276, 290], and [358].

Chapter 5
Triangular Matrix Representations and Triangular Matrix Extensions

A ring R is said to have a *generalized triangular matrix representation* if R is ring isomorphic to a generalized triangular matrix ring

$$\begin{bmatrix} R_1 & R_{12} & \cdots & R_{1n} \\ 0 & R_2 & \cdots & R_{2n} \\ \vdots & \vdots & \ddots & \vdots \\ 0 & 0 & \cdots & R_n \end{bmatrix},$$

where each R_i is a ring and R_{ij} is an (R_i, R_j)-bimodule for $i < j$, and the matrices obey the usual rules for matrix addition and multiplication. Generalized triangular matrix representations provide an effective tool in the investigation of the structures of a wide range of rings. In this chapter, these representations, in an abstract setting, are discussed by introducing the concept of a *set of left triangulating idempotents*.

The importance and applicability of the concept of a generalized triangular matrix representation can be seen from: (1) for any right R-module M, the generalized triangular matrix ring

$$\begin{bmatrix} S & M \\ 0 & R \end{bmatrix},$$

where $S = \text{End}(M)$, completely encodes the algebraic information of M into a single ring; (2) a ring R is ring isomorphic to

$$\begin{bmatrix} R_1 & R_{12} \\ 0 & R_2 \end{bmatrix},$$

where $R_1 \neq 0$ and $R_2 \neq 0$ if and only if there exists $e \in \mathbf{S}_\ell(R)$ with $e \neq 0$ and $e \neq 1$. From (2), we see that there is a natural connection between quasi-Baer rings and modules and generalized triangular matrix representation, since the "e" in Proposition 3.2.4(ii) is in $\mathbf{S}_\ell(R)$ and the "f" in Proposition 4.6.3(ii) is in $\mathbf{S}_\ell(\text{End}(M))$.

In a manner somewhat analogous to determining a matrix ring by a set of matrix units (see 1.1.16), a generalized triangular matrix ring is determined by a set

G.F. Birkenmeier et al., *Extensions of Rings and Modules*,
DOI 10.1007/978-0-387-92716-9_5,

of left (or right) triangulating idempotents. The existence of a set of left triangulating idempotents does not depend on any specific conditions on a ring (e.g., $\{1\}$ is a set of left triangulating idempotents); however, if the ring satisfies a mild finiteness condition, then such a set can be refined to a certain set of left triangulating idempotents in which each diagonal ring R_i has no nontrivial generalized triangular matrix representation. When this occurs, the generalized triangular matrix representation is said to be *complete*.

Complete triangular matrix representations and left triangulating idempotents are applied to get a structure theorem for a certain class of quasi-Baer rings (see Theorem 5.4.12). A number of well known results follow as consequences of this structure theorem. These include Levy's decomposition theorem of semiprime right Goldie rings, Faith's characterization of semiprime right FPF rings with no infinite set of central orthogonal idempotents, Gordon and Small's characterization of piecewise domains, and Chatters' decomposition theorem of hereditary Noetherian rings.

Further, a sheaf representation of quasi-Baer rings is studied as another application of our results of this chapter. Also the Baer, the quasi-Baer, the FI-extending, and the strongly FI-extending properties of (generalized) triangular matrix rings are discussed. Most results of Sects. 5.1, 5.2, and 5.3 are applicable to an algebra over a commutative ring.

5.1 Triangulating Idempotents

In this section, some basic properties of triangulating idempotents are discussed. Then a result showing the connection between triangulating idempotents and generalized triangular matrix rings is presented.

Definition 5.1.1 Let R be a ring. An ordered set $\{b_1, \ldots, b_n\}$ of nonzero distinct idempotents in R is called a *set of left triangulating idempotents* of R if the following conditions hold:

(i) $1 = b_1 + \cdots + b_n$;
(ii) $b_1 \in \mathbf{S}_\ell(R)$;
(iii) $b_{k+1} \in \mathbf{S}_\ell(c_k R c_k)$, where $c_k = 1 - (b_1 + \cdots + b_k)$, for $1 \le k \le n-1$.

Similarly, we define a *set of right triangulating idempotents* of R by using part (i) in the preceding, $b_1 \in \mathbf{S}_r(R)$, and $b_{k+1} \in \mathbf{S}_r(c_k R c_k)$. By condition (iii) of Definition 5.1.1, a set of left (right) triangulating idempotents is a set of orthogonal idempotents.

Definition 5.1.2 A set $\{b_1, \ldots, b_n\}$ of left (right) triangulating idempotents of R is said to be *complete* if each b_i is semicentral reduced.

Theorem 5.1.3 *Let $\{b_1, \ldots, b_n\}$ be an ordered set of nonzero idempotents of R. Then the following are equivalent.*

(i) $\{b_1, \dots, b_n\}$ *is a set of left triangulating idempotents.*
(ii) $b_1 + \cdots + b_n = 1$ *and* $b_j R b_i = 0$, *for all* $i < j \le n$.

Proof (i)⇒(ii) By definition, $b_1 + \cdots + b_n = 1$. As $b_2 \in (1 - b_1)R(1 - b_1)$ and $b_1 \in \mathbf{S}_\ell(R)$, $b_2 b_1 = 0$ and $b_2 R b_1 = b_2 b_1 R b_1 = 0$. Similarly we obtain $b_j R b_1 = 0$, for all $j > 1$. By assumption $b_2 \in \mathbf{S}_\ell((1 - b_1)R(1 - b_1))$ and $\{b_1, \dots, b_n\}$ is orthogonal, thus for $j > 2$,

$$\begin{aligned} b_j R b_2 &= b_j R(1 - b_1)b_2 = b_j(b_1 R + (1 - b_1)R)(1 - b_1)b_2 \\ &= b_j(1 - b_1)R(1 - b_1)b_2 = b_j b_2(1 - b_1)R(1 - b_1)b_2 \\ &= 0. \end{aligned}$$

Continue the process, using $(1 - b_1 - b_2)R(1 - b_1 - b_2)$ in the next step, and so on, to get $b_j R b_i = 0$ for all $i < j \le n$.

(ii)⇒(i) Note that $(1 - b_1)R b_1 = (b_2 + \cdots + b_n)R b_1 = 0$. So $b_1 \in \mathbf{S}_\ell(R)$ by Proposition 1.2.2. Now $b_2 \in (1 - b_1)R(1 - b_1)$ as $b_2(1 - b_1) = b_2 - b_2 b_1 = b_2$ and $(1 - b_1)b_2 = b_2$. Also $(1 - b_1 - b_2)(1 - b_1) = b_3 + b_4 + \cdots + b_n$. Therefore $(1 - b_1 - b_2)[(1 - b_1)R(1 - b_1)]b_2 = \sum_{i=3}^n b_i R(1 - b_1)b_2 = \sum_{i=3}^n b_i R b_2 = 0$. So $b_2 \in \mathbf{S}_\ell((1 - b_1)R(1 - b_1))$ by Proposition 1.2.2. Continuing this process yields the desired result. □

Theorem 5.1.4 *R has a (resp., complete) set of left triangulating idempotents if and only if R has a (resp., complete) generalized triangular matrix representation.*

Proof Let $\{b_1, \dots, b_n\}$ be a set of left triangulating idempotents of R. Using Theorem 5.1.3 and a routine argument shows that the map

$$\theta : R \to \begin{bmatrix} b_1 R b_1 & b_1 R b_2 & \cdots & b_1 R b_n \\ 0 & b_2 R b_2 & \cdots & b_2 R b_n \\ \vdots & \vdots & \ddots & \vdots \\ 0 & 0 & \cdots & b_n R b_n \end{bmatrix}$$

defined by $\theta(r) = [b_i r b_j]$ is a ring isomorphism, where $[b_i r b_j]$ is the matrix whose (i, j)-position is $b_i r b_j$. Conversely, assume that

$$\phi : R \to \begin{bmatrix} R_1 & R_{12} & \cdots & R_{1n} \\ 0 & R_2 & \cdots & R_{2n} \\ \vdots & \vdots & \ddots & \vdots \\ 0 & 0 & \cdots & R_n \end{bmatrix}$$

is a ring isomorphism. Then $\{\phi^{-1}(e_{11}), \dots, \phi^{-1}(e_{nn})\}$ is a set of left triangulating idempotents of R by a routine calculation, where e_{ii} is the matrix with 1_{R_i} in the (i, i)-position and 0 elsewhere. □

Lemma 5.1.5 (i) $\mathbf{S}_\ell(eRe) \subseteq \mathbf{S}_\ell(R)$ *for* $e \in \mathbf{S}_\ell(R)$.

(ii) $f\mathbf{S}_\ell(R)f \subseteq \mathbf{S}_\ell(fRf)$ *for* $f^2 = f \in R$.

(iii) *Let* $e \in \mathbf{S}_\ell(R)$. *If* f *is a primitive idempotent of* R *such that* $efe \neq 0$, *then* efe *is a primitive idempotent in* eRe *and* $fef = f$.

Proof (i) For $g \in \mathbf{S}_\ell(eRe)$, $gRg = geReg = eReg = Rg$. So $g \in \mathbf{S}_\ell(R)$.

(ii) Let $g \in \mathbf{S}_\ell(R)$ and $r \in R$. Then $(fgf)(frf)(fgf) = (ff)(frf)(fgf)$. Thus $(fgf)(frf)(fgf) = (frf)(fgf)$. So $fgf \in \mathbf{S}_\ell(fRf)$.

(iii) Note that $0 \neq efe = fe = fefe$, so $fef \neq 0$ and $(fef)^2 = fef$. As f is primitive, $fef = f$. To show that efe is a primitive idempotent of eRe, we note that $(efe)(efe) = e(fef)e = efe$. Let $0 \neq h^2 = h \in (efe)(eRe)(efe)$. Since $e \in \mathbf{S}_\ell(R)$, $he = h$, $fh = h$, so $hf = fhf$, and thus $(hf)(hf) = hf$. As $hf = 0$ implies that $h = hefe = hfe = 0$, hf is a nonzero idempotent in fRf. Thus, $hf = f$ since f is a primitive idempotent. Note that $(fe)^2 = fe$ and $h \in (efe)(eRe)(efe)$, so $h = hefe = hfe = fe = efe$. Thus, efe is a primitive idempotent eRe. □

Lemma 5.1.6 (i) *If* h *is a ring homomorphism from a ring* R *to a ring* A, *then* $h(\mathbf{S}_\ell(R)) \subseteq \mathbf{S}_\ell(h(R))$.

(ii) *Assume that* $e \in \mathbf{S}_\ell(R) \cup \mathbf{S}_r(R)$ *and* $f \in \mathbf{S}_\ell(eRe) \cup \mathbf{S}_r(eRe)$. *Then the map* $h : R \to fRf$, *defined by* $h(r) = frf$ *for* $r \in R$, *is a ring epimorphism.*

Proof (i) The proof is routine.

(ii) Say $x, y \in R$. Since $e \in \mathbf{S}_\ell(R) \cup \mathbf{S}_r(R)$ and $f \in \mathbf{S}_\ell(eRe) \cup \mathbf{S}_r(eRe)$,

$$fxyf = fexyef = fexeyef = fexefeyef = fxfyf.$$

Therefore, $h(xy) = h(x)h(y)$. □

Proposition 5.1.7 *Let* $\{b_1, \dots, b_n\}$ *be a set of left triangulating idempotents of* R. *Then:*

(i) $c_k \in \mathbf{S}_r(R)$, $k = 1, \dots, n-1$, *where* $c_k = 1 - (b_1 + \cdots + b_k)$.
(ii) $b_1 + \cdots + b_k \in \mathbf{S}_\ell(R)$, $k = 1, \dots, n$.
(iii) *The map* $h_j : R \to b_jRb_j$, *defined by* $h_j(r) = b_jrb_j$ *for all* $r \in R$, *is a ring epimorphism.*

Proof (i) Recall that $b_1 \in \mathbf{S}_\ell(R)$ implies $c_1 = 1 - b_1 \in \mathbf{S}_r(R)$ by Proposition 1.2.2. As $b_2 \in \mathbf{S}_\ell(c_1Rc_1)$, $c_2 = 1 - b_1 - b_2 \in \mathbf{S}_r(c_1Rc_1)$ by Proposition 1.2.2. Therefore $c_2 \in \mathbf{S}_r(R)$ by the right-sided version of Lemma 5.1.5(i). Using this procedure, an induction proof completes the argument.

(ii) It is a direct consequence of part (i) and Proposition 1.2.2.

(iii) Put $e = c_k$ and $f = b_{k+1}$. By part (i), $e \in \mathbf{S}_r(R)$, so $f \in \mathbf{S}_\ell(eRe)$. From Lemma 5.1.6(ii), the map $r \to frf$ is a ring epimorphism. □

Corollary 5.1.8 *The ordered set* $\{b_1, \dots, b_n\}$ *is a (complete) set of left triangulating idempotents of* R *if and only if the ordered set* $\{b_n, \dots, b_1\}$ *is a (complete) set of right triangulating idempotents.*

Proof Let $\{b_1, \ldots, b_n\}$ be a set of left triangulating idempotents of R. Then by Proposition 5.1.7(i), $1 - (b_1 + \cdots + b_{n-1}) = b_n \in \mathbf{S}_r(R)$. We next show that $b_{n-1} \in \mathbf{S}_r((1-b_n)R(1-b_n))$. For this, first it can be checked that $\{b_1, \ldots, b_{n-1}\}$ is a set of left triangulating idempotents of $(1-b_n)R(1-b_n)$ and $1-b_n$ is the identity of $(1-b_n)R(1-b_n)$. By Proposition 5.1.7(ii), $b_1 + \cdots + b_{n-2} \in \mathbf{S}_\ell(R)$, and hence $b_1 + \cdots + b_{n-2} \in \mathbf{S}_\ell((1-b_n)R(1-b_n))$. Therefore by Proposition 1.2.2,

$$(1-b_n) - (b_1 + b_2 + \cdots + b_{n-2}) = b_{n-1} \in \mathbf{S}_r((1-b_n)R(1-b_n))$$

and so on. By this argument, the ordered set $\{b_n, \ldots, b_1\}$ is a set of right triangulating idempotents. Also, if $\{b_1, \ldots, b_n\}$ is complete, then so is $\{b_n, \ldots, b_1\}$.

The converse is proved similarly. Further, completeness is left-right symmetric since $\mathbf{S}_\ell(b_i R b_i) = \{0, b_i\}$ if and only if $\mathbf{S}_r(b_i R b_i) = \{0, b_i\}$ (see Proposition 1.2.11). □

Exercise 5.1.9

1. Let R be a subdirectly irreducible ring (i.e., the intersection of all nonzero ideals of R is nonzero) and $\{b_1, \ldots, b_n\}$ a set of left triangulating idempotents. Prove the following.
   (i) For each $i \neq 1$ there exists $j < i$ such that $b_j R b_i \neq 0$.
   (ii) For each $i \neq n$ there exists $j > i$ such that $b_i R b_j \neq 0$.
   (iii) The heart of R (i.e., the intersection of all nonzero ideals of R) is contained in $b_1 R b_n$.
2. Let $\{b_1, \ldots, b_n\}$ be a set of left triangulating idempotents of a ring R. Prove the following.
   (i) $b_i \in \mathbf{S}_\ell(R)$ if and only if $b_j R b_i = 0$ for all $j < i$.
   (ii) $b_i \in \mathbf{S}_r(R)$ if and only if $b_i R b_j = 0$ for all $j > i$.

5.2 Generalized Triangular Matrix Representations

Rings with a complete generalized triangular matrix representation will be characterized. Then the uniqueness of a complete set of triangulating idempotents will be discussed. We shall see that if a ring R satisfies some mild finiteness conditions, then R has a generalized triangular matrix representation with semicentral reduced rings on the diagonal which satisfy the same finiteness condition as R. Thereby reducing the study of such rings to those which are semicentral reduced. Further, it will be shown that the condition of having a complete set of left triangulating idempotents is strictly between that of having a complete set of primitive idempotents and that of having a complete set of centrally primitive idempotents.

Lemma 5.2.1 *Let $0 \neq f^2 = f \in R$. If $fR = eR$ for every $0 \neq e \in \mathbf{S}_\ell(fRf)$, then f is semicentral reduced.*

Proof Let $0 \neq e \in \mathbf{S}_\ell(fRf)$. Then since $fR = eR$, $f = ex$ for some $x \in R$, and so $e = ef = eex = ex = f$. Thus, f is semicentral reduced. □

Lemma 5.2.2 (i) *A ring R has DCC on $\{bR \mid b \in \mathbf{S}_\ell(R)\}$ if and only if R has ACC on $\{Rc \mid c \in \mathbf{S}_r(R)\}$.*

(ii) *A ring R has ACC on $\{bR \mid b \in \mathbf{S}_\ell(R)\}$ if and only if R has DCC on $\{Rc \mid c \in \mathbf{S}_r(R)\}$.*

(iii) *If a ring R has DCC on $\{Rc \mid c \in \mathbf{S}_r(R)\}$, then R has DCC on $\{cR \mid c \in \mathbf{S}_r(R)\}$.*

Proof (i) Assume that R has DCC on $\{bR \mid b \in \mathbf{S}_\ell(R)\}$. Consider a chain $Rc_1 \subseteq Rc_2 \subseteq \dots$, where $c_i \in \mathbf{S}_r(R)$. Then $(1 - c_1)R \supseteq (1 - c_2)R \supseteq \dots$ with $1 - c_i \in \mathbf{S}_\ell(R)$ (see Proposition 1.2.2). This descending chain becomes stationary, say with $(1 - c_n)R = (1 - c_{n+j})R$ for each $j \geq 1$. Then we have that $\ell_R((1 - c_n)R) = \ell_R((1 - c_{n+j})R)$ for each $j > 1$. Thus, $Rc_n = Rc_{n+j}$ for each $j > 1$. The converse is proved similarly.

(ii) The proof is similar to that of part (i).

(iii) Assume that R has DCC on $\{Rc \mid c \in \mathbf{S}_r(R)\}$. Let $c_1R \supseteq c_2R \supseteq \dots$ be a descending chain with $c_i \in \mathbf{S}_r(R)$. Then $c_{i+1} = c_ic_{i+1}$. So it follows that $c_{i+1}c_i = c_ic_{i+1}c_i = c_ic_{i+1} = c_{i+1}$ because $c_i \in \mathbf{S}_r(R)$. Therefore $Rc_i \supseteq Rc_{i+1}$ for each i. Thus we have a descending chain $Rc_1 \supseteq Rc_2 \supseteq \dots$, so there is n with $Rc_n = Rc_{n+1} = \dots$. Therefore, $(1 - c_n)R = (1 - c_{n+1})R$. Hence, we obtain that $(1 - c_n)Rc_n = (1 - c_{n+1})Rc_n = (1 - c_{n+1})Rc_{n+1}$.

We observe that $Rc_n = c_nRc_n + (1 - c_n)Rc_n = c_nR + (1 - c_n)Rc_n$ and

$$Rc_{n+1} = c_{n+1}Rc_{n+1} + (1 - c_{n+1})Rc_{n+1} = c_{n+1}R + (1 - c_n)Rc_n$$

because $c_n, c_{n+1} \in \mathbf{S}_r(R)$ and $(1 - c_n)Rc_n = (1 - c_{n+1})Rc_{n+1}$. Therefore, we have that $c_nR + (1 - c_n)Rc_n = c_{n+1}R + (1 - c_n)Rc_n$ as $Rc_n = Rc_{n+1}$.

To show that $c_nR = c_{n+1}R$, it suffices to check that $c_nR \subseteq c_{n+1}R$ because $c_{n+1}R \subseteq c_nR$. Now $c_n = c_{n+1}y + \alpha$, where $y \in R$ and $\alpha \in (1 - c_n)Rc_n$, as $c_nR + (1 - c_n)Rc_n = c_{n+1}R + (1 - c_n)Rc_n$. Since $c_n\alpha = 0$ and $c_{n+1} = c_nc_{n+1}$ from $c_{n+1}R \subseteq c_nR$, $c_n = c_n^2 = c_nc_{n+1}y + c_n\alpha = c_{n+1}y \in c_{n+1}R$. Therefore $c_nR \subseteq c_{n+1}R$, and hence $c_nR = c_{n+1}R = \dots$. We conclude that R satisfies DCC on $\{cR \mid c \in \mathbf{S}_r(R)\}$. □

Lemma 5.2.3 *Let $e \in \mathbf{S}_r(R)$. If R has DCC on $\{bR \mid b \in \mathbf{S}_\ell(R)\}$, then eRe has DCC on $\{d(eRe) \mid d \in \mathbf{S}_\ell(eRe)\}$.*

Proof First, we show that $\{(eRe)c \mid c \in \mathbf{S}_r(eRe)\}$ has ACC. For this, assume that $(eRe)c_1 \subseteq (eRe)c_2 \subseteq \dots$ is an ascending chain, where $c_i \in \mathbf{S}_r(eRe)$ for $i = 1, 2, \dots$. By the right-sided version of Lemma 5.1.5(i), each $c_i \in \mathbf{S}_r(R)$. Note that $ec_ie \in (eRe)ec_ie \subseteq (eRe)ec_{i+1}e$.

So there exists $x \in eRe$ such that $ec_ie = xec_{i+1}e$. Thus,

$$\begin{aligned}(1 - e)Rc_i &= (1 - e)Rec_ie = (1 - e)Rxec_{i+1}e \\ &\subseteq (1 - e)Rec_{i+1}e = (1 - e)Rc_{i+1}.\end{aligned}$$

Therefore, for each i,

$$Rc_i = eRc_i + (1 - e)Rc_i = (eRe)ec_ie + (1 - e)Rc_i$$

$$\subseteq (eRe)ec_{i+1}e + (1-e)Rc_{i+1} = eRc_{i+1} + (1-e)Rc_{i+1}$$
$$= Rc_{i+1}.$$

By assumption and Lemma 5.2.2(i), $Rc_n = Rc_{n+1} = \dots$ for some n as each c_i is in $\mathbf{S}_r(R)$. Therefore, $eRc_n = eRc_{n+1} = \dots$, so $(eRe)c_n = (eRe)c_{n+1} = \dots$. From Lemma 5.2.2(i), eRe has DCC on $\{d(eRe) \mid d \in \mathbf{S}_\ell(eRe)\}$. □

Lemma 5.2.4 *Let $\{b_1, \dots, b_n\}$ be a complete set of left triangulating idempotents of R. If $e \in \mathbf{S}_\ell(R)$, then $eR = \bigoplus_i b_i R$, where the sum runs over a subset of $\{1, \dots, n\}$. Thus, $|\{eR \mid e \in \mathbf{S}_\ell(R)\}| \le 2^n$.*

Proof Assume that $0 \neq e \in \mathbf{S}_\ell(R)$. Consider i such that $b_i e \neq 0$. We show that $b_i eR = b_i R$. For this, note that $b_i e b_i e = b_i e \neq 0$, so $b_i e b_i \neq 0$. From Lemma 5.1.5(ii), $b_i \mathbf{S}_\ell(R) b_i \subseteq \mathbf{S}_\ell(b_i R b_i)$. Hence $b_i e b_i \in \mathbf{S}_\ell(b_i R b_i)$, but by hypothesis $\mathbf{S}_\ell(b_i R b_i) = \{0, b_i\}$. So $b_i e b_i = b_i$. Also $b_i R = b_i e b_i R \subseteq b_i eR \subseteq b_i R$, and thus $b_i eR = b_i R$. Recall that b_i are orthogonal. Hence, $b_i e b_j e = b_i b_j e = 0$ yields that $b_1 e, \dots, b_n e$ are orthogonal idempotents. Let $I = \{i \mid 1 \le i \le n \text{ and } b_i e \neq 0\}$. Then $eR = \oplus_{i \in I} b_i eR = \oplus_{i \in I} b_i R$. □

The next result characterizes rings with a complete generalized triangular matrix representation.

Theorem 5.2.5 *The following are equivalent for a ring R.*

(i) *R has a complete set of left triangulating idempotents.*
(ii) *$\{bR \mid b \in \mathbf{S}_\ell(R)\}$ is a finite set.*
(iii) *$\{bR \mid b \in \mathbf{S}_\ell(R)\}$ satisfies ACC and DCC.*
(iv) *$\{bR \mid b \in \mathbf{S}_\ell(R)\}$ and $\{Rc \mid c \in \mathbf{S}_r(R)\}$ satisfy ACC.*
(v) *$\{bR \mid b \in \mathbf{S}_\ell(R)\}$ and $\{Rc \mid c \in \mathbf{S}_r(R)\}$ satisfy DCC.*
(vi) *$\{bR \mid b \in \mathbf{S}_\ell(R)\}$ and $\{cR \mid c \in \mathbf{S}_r(R)\}$ satisfy DCC.*
(vii) *R has a complete set of right triangulating idempotents.*
(viii) *R has a complete generalized triangular matrix representation.*

Proof Lemma 5.2.4 yields (i)⇒(ii), and (ii)⇒(iii) is trivial. From Lemma 5.2.2, (iii)⇒(iv)⇒(v)⇒(vi) follows immediately.

We show that (vi)⇒(i). If $\mathbf{S}_\ell(R) = \{0, 1\}$, then we are finished. Otherwise take e_1 to be a nontrivial element of $\mathbf{S}_\ell(R)$.

If e_1 is not semicentral reduced, then there exists $0 \neq e_2 \in \mathbf{S}_\ell(e_1 R e_1)$ such that $e_1 R \neq e_2 R$ by Lemma 5.2.1, and so $e_1 R \supsetneq e_2 R$. From Lemma 5.1.5(i), $e_2 \in \mathbf{S}_\ell(R)$. If e_2 is not semicentral reduced, then by Lemmas 5.2.1 and 5.1.5(i) again there exists $0 \neq e_3 \in \mathbf{S}_\ell(e_2 R e_2) \subseteq \mathbf{S}_\ell(R)$ such that $e_2 R \neq e_3 R$. So we have that $e_2 R \supsetneq e_3 R$. This process should be stopped within a finite steps. Thus, we obtain a semicentral reduced idempotent $e_n \in \mathbf{S}_\ell(R)$ for some positive integer n because $\{eR \mid e \in \mathbf{S}_\ell(R)\}$ has DCC.

Starting a new process, let $b_1 = e_n$. Then $\mathbf{S}_\ell(b_1Rb_1) = \{0, b_1\}$. From Proposition 1.2.2, $1 - b_1 \in \mathbf{S}_r(R)$. If $1 - b_1$ is semicentral reduced, then we see that $\{b_1, 1 - b_1\}$ is a complete set of left triangulating idempotents.

Otherwise, we consider $R_1 = (1 - b_1)R(1 - b_1)$. Note that by Lemma 5.2.3, R_1 has DCC on $\{dR_1 \mid d \in \mathbf{S}_\ell(R_1)\}$. By a similar argument to that used to get b_1, we obtain $b_2 \in \mathbf{S}_\ell(R_1)$ such that $\mathbf{S}_\ell(b_2R_1b_2) = \{0, b_2\}$.

As $1 - b_1$ is the identity of R_1 and $b_2 \in R_1$, it follows that $b_2R_1b_2 = b_2Rb_2$, so $\mathbf{S}_\ell(b_2Rb_2) = \{0, b_2\}$. Also, $(1 - b_1) - b_2 \in \mathbf{S}_r(R_1)$. The right-sided version of Lemma 5.1.5(i) yields that $\mathbf{S}_r(R_1) \subseteq \mathbf{S}_r(R)$. Therefore, $1 - b_1 - b_2 \in \mathbf{S}_r(R)$. If $1 - b_1 - b_2$ is semicentral reduced in R, then $\{b_1, b_2, 1 - b_1 - b_2\}$ is a complete set of left triangulating idempotents.

We continue the process to obtain a descending chain in $\{cR \mid c \in \mathbf{S}_r(R)\}$, which is $(1 - b_1)R \supseteq (1 - b_1 - b_2)R \supseteq (1 - b_1 - b_2 - b_3)R \supseteq \dots$. By the DCC hypothesis of $\{cR \mid c \in \mathbf{S}_r(R)\}$, this chain becomes stationary after a finite steps, yielding a complete set of left triangulating idempotents.

The equivalence (vii)$\Leftrightarrow$(i) follows from Corollary 5.1.8, while the equivalence (i)$\Leftrightarrow$(viii) follows from Theorem 5.1.4. □

Corollary 5.2.6 *Let R be a ring with a complete set of left triangulating idempotents. Then for any $0 \neq e \in \mathbf{S}_\ell(R)$ (resp., $0 \neq e \in \mathbf{S}_r(R)$), the ring eRe also has a complete set of left (resp., right) triangulating idempotents.*

Proof Say $0 \neq e \in \mathbf{S}_\ell(R)$. Define

$$\lambda : \{bR \mid b \in \mathbf{S}_\ell(R)\} \to \{d(eRe) \mid d \in \mathbf{S}_\ell(eRe)\}$$

by $\lambda(bR) = (ebe)(eRe)$. From Lemma 5.1.5(ii), $ebe \in \mathbf{S}_\ell(eRe)$ for $b \in \mathbf{S}_\ell(R)$. If $bR = b_1R$ with $b, b_1 \in \mathbf{S}_\ell(R)$, then $bRe = b_1Re$, and so $ebeRe = eb_1eRe$ since $e \in \mathbf{S}_\ell(R)$. Thus λ is well-defined. As $\mathbf{S}_\ell(eRe) \subseteq \mathbf{S}_\ell(R)$ by Lemma 5.1.5(i), λ is onto. From Theorem 5.2.5, it follows that $\{bR \mid b \in \mathbf{S}_\ell(R)\}$ is finite. Furthermore, we get that $\{d(eRe) \mid d \in \mathbf{S}_\ell(eRe)\}$ is also finite. Again by Theorem 5.2.5, eRe has a complete set of left triangulating idempotents. Similarly, if $0 \neq e \in \mathbf{S}_r(R)$, then eRe has also a complete set of right triangulating idempotents. □

In Theorem 5.2.8, the uniqueness of a complete generalized triangular matrix representation will be established. For the proof of this theorem, we need the following result due to Azumaya [32, Theorem 3].

Lemma 5.2.7 *Let I be a quasi-regular ideal of a ring R. If $\{e_1, \dots, e_n\}$ and $\{f_1, \dots, f_n\}$ are two sets of orthogonal idempotents of R such that $\overline{e}_i = \overline{f}_i$ for each i with images $\overline{e}_i$ and $\overline{f}_i$ in R/I, then there is an invertible element $\alpha \in R$ with $f_i = \alpha^{-1}e_i\alpha$ for each i.*

Proof Let $e = \sum_{i=1}^n e_i$ and $f = \sum_{i=1}^n f_i$. Put $\beta = e + f - ef - \sum_{i=1}^n e_if_i$. Then $\alpha = 1 - \beta$ is invertible and $f_i = \alpha^{-1}e_i\alpha$ for each i. □

A nonzero central idempotent e of R is said to be *centrally primitive* if 0 and e are the only central idempotents in eRe. Let g be a nonzero central idempotent in R such that $g = g_1 + \cdots + g_t$, where $\{g_i \mid 1 \le i \le t\}$ is a set of centrally primitive orthogonal idempotents of R. Then t is uniquely determined (see Exercise 5.2.21.1). A ring R is said to have a *complete set of centrally primitive idempotents* if there exists a finite set of centrally primitive orthogonal idempotents whose sum is 1. It is routine to check that R has a complete set of centrally primitive idempotents if and only if R is a ring direct sum of indecomposable rings.

Theorem 5.2.8 (Uniqueness) *Let $\{b_1, \dots, b_n\}$ and $\{c_1, \dots, c_k\}$ each be a complete set of left triangulating idempotents of R. Then $n = k$ and there exist an invertible element $\alpha \in R$ and a permutation σ on $\{1, \dots, n\}$ such that $b_{\sigma(i)} = \alpha^{-1} c_i \alpha$ for each i. Thus for each i, $c_i R \cong b_{\sigma(i)} R$, as R-modules, and $c_i R c_i \cong b_{\sigma(i)} R b_{\sigma(i)}$, as rings.*

Proof Let $U = \sum_{i<j} b_i R b_j$. Then $U \trianglelefteq R$ and $U^n = 0$. Let $\overline{R} = R/U$ and denote by $\overline{x}$ the image of $x \in R$ in R/U. Since $b_i R b_i \cap U = 0$, for $i = 1, \dots, n$, $b_i R b_i \cong \overline{b_i}\, \overline{R}\, \overline{b_i}$ as rings. So $\overline{R}$ is a direct sum of the $\overline{b_i}\, \overline{R}\, \overline{b_i}$, and consequently $\{\overline{b}_1, \dots, \overline{b}_n\}$ is a complete set of centrally primitive idempotents of $\overline{R}$.

Clearly, $\overline{c}_1 \in \mathbf{S}_\ell(\overline{R})$. Further, $\overline{c}_1 \neq \overline{0}$. Indeed, if $\overline{c}_1 = \overline{0}$, then $c_1 \in U$, and so $c_1 = c_1^n \in U^n = 0$, a contradiction. Because $\overline{b}_i$ is semicentral reduced, $\overline{c}_1 \overline{b}_i \in \{\overline{0}, \overline{b}_i\}$. Therefore $\overline{c}_1 = \sum_{i=1}^n \overline{c}_1 \overline{b}_i = \sum \overline{b}_k$ for which $\overline{c}_1 \overline{b}_k \neq \overline{0}$. So $\overline{c}_1 \in \mathcal{B}(\overline{R})$. Now we note that $\overline{c}_2 \in \mathbf{S}_\ell((\overline{1} - \overline{c}_1)\overline{R}(\overline{1} - \overline{c}_1))$. As $\overline{1} - \overline{c}_1 \in \mathcal{B}(\overline{R})$, $\overline{c}_2 \in \mathbf{S}_\ell(\overline{R})$ by Lemma 5.1.5(i). Using the preceding argument, with $\overline{c}_2$ in place of $\overline{c}_1$, we obtain $\overline{c}_2 \in \mathcal{B}(\overline{R})$.

Continuing this procedure, we obtain that $\{\overline{c}_1, \dots, \overline{c}_k\}$ is a set of orthogonal nonzero central idempotents in $\overline{R}$. Hence $\overline{c}_i \overline{R}\, \overline{c}_j = \overline{0}$ for $i < j$. Thus $c_i R c_j \subseteq U$ for all $1 \le i < j \le k$.

Let $V = \sum_{i<j} c_i R c_j$. Then $V^k = 0$. By the preceding argument, $b_i R b_j \subseteq V$ for all $1 \le i < j \le n$. Hence, $U = V$ and so $\{\overline{b}_1, \dots, \overline{b}_n\}$ and $\{\overline{c}_1, \dots, \overline{c}_k\}$ are both complete sets of centrally primitive idempotents for $\overline{R}$. It is well known that for such sets of centrally primitive idempotents, $n = k$ and there is a permutation σ on $\{1, \dots, n\}$ such that $\overline{c}_i = \overline{b}_{\sigma(i)}$ (Exercises 5.2.21.1 and 5.2.21.2). As $U^n = 0$, U is a quasi-regular ideal of R.

From Lemma 5.2.7, there exists an invertible element $\alpha \in R$ such that $b_{\sigma(i)} = \alpha^{-1} c_i \alpha$ for every i. Thus, $c_i R \cong b_{\sigma(i)} R$ as R-modules. We observe that $\mathrm{End}(c_i R_R) \cong c_i R c_i$ and $\mathrm{End}(b_j R_R) \cong b_j R b_j$. So $c_i R c_i \cong b_{\sigma(i)} R b_{\sigma(i)}$. □

The following example shows that the isomorphism $c_i R \cong b_{\sigma(i)} R$, given in Theorem 5.2.8, cannot be sharpened to equality. This is in contrast to the result for a complete set of centrally primitive idempotents.

Example 5.2.9 Let $R = T_2(\mathbb{R})$. Consider

$$b_1 = \begin{bmatrix} 1 & 0 \\ 0 & 0 \end{bmatrix}, \; b_2 = \begin{bmatrix} 0 & 0 \\ 0 & 1 \end{bmatrix}$$

and let

$$c_1 = \begin{bmatrix} 1 & a \\ 0 & 0 \end{bmatrix}, \ c_2 = \begin{bmatrix} 0 & -a \\ 0 & 1 \end{bmatrix}, \ 0 \neq a \in \mathbb{R}.$$

Then $\{b_1, b_2\}$ and $\{c_1, c_2\}$ are complete sets of left triangulating idempotents for R. In this case, $b_1 R = c_1 R$ and $b_2 R \cong c_2 R$, but $b_2 R \neq c_2 R$.

Kaplansky raised the following question: Let A and B be two rings. If $\mathrm{Mat}_n(A) \cong \mathrm{Mat}_n(B)$ as rings, does it follow that $A \cong B$ as rings? (See [261, p. 35].) It is known that there are nonisomorphic semicentral reduced rings (e.g., simple Noetherian domains) which have isomorphic matrix rings (see [260] and [378]). The next result shows that this cannot happen for $n \times n$ $(n > 1)$ upper triangular matrix rings over semicentral reduced rings.

Corollary 5.2.10 *Let A and B be semicentral reduced rings. If $T_m(A) \cong T_n(B)$ as rings, then $m = n$ and $A \cong B$ as rings.*

Proof Let e_{ii} be the matrix in $T_m(A)$ with 1_A in the (i, i)-position and 0 elsewhere. As A is semicentral reduced, $\{e_{11}, \ldots, e_{mm}\}$ is a complete set of left triangulating idempotents for $T_m(A)$. A similar fact holds for $T_n(B)$. Because $T_m(A) \cong T_n(B)$, $m = n$ by Theorem 5.2.8.

Next, say $\lambda : T_n(A) \to T_n(B)$ is an isomorphism. Then $\{\lambda(e_{11}), \ldots, \lambda(e_{nn})\}$ is a complete set of left triangulating idempotents of $T_n(B)$. Let f_{ii} be the matrix in $T_n(B)$ with 1_B in the (i, i)-position and 0 elsewhere. Then because B is semicentral reduced, $\{f_{11}, \ldots, f_{nn}\}$ is also a complete set of left triangulating idempotents of $T_n(B)$.

By Theorem 5.2.8, $f_{11} T_n(B) f_{11} \cong \lambda(e_{jj}) T_n(B) \lambda(e_{jj})$ for some j. Therefore, $B \cong f_{11} T_n(B) f_{11} \cong \lambda(e_{jj}) T_n(B) \lambda(e_{jj}) \cong e_{jj} T_n(A) e_{jj} \cong A$. □

From Theorem 5.2.8, the number of elements in a complete set of left triangulating idempotents is unique for a given ring R (which has such a set). This is also the number of elements in any complete set of right triangulating idempotents of R by Corollary 5.1.8. So we are motivated to give the following definition.

Definition 5.2.11 A ring R is said to have *triangulating dimension* n, written $\mathrm{Tdim}(R) = n$, if R has a complete set of left triangulating idempotents with n elements. Note that R is semicentral reduced if and only if $\mathrm{Tdim}(R) = 1$. If R has no complete set of left triangulating idempotents, then we say that R has *infinite triangulating dimension*, denoted $\mathrm{Tdim}\,(R) = \infty$.

Lemma 5.2.12 *Let $\{e_1, \ldots, e_n\}$ be a complete set of primitive idempotents of R. If $0 \neq b \in \mathbf{S}_\ell(R) \cup \mathbf{S}_r(R)$, then there exists a nonempty subset P of $\{e_1, \ldots, e_n\}$ such that $\{b e_j b \mid e_j \in P\}$ forms a complete set of primitive idempotents of bRb.*

Proof Assume that $b \in \mathbf{S}_\ell(R)$. From $b = b(e_1 + \cdots + e_n)b = be_1b + \cdots + be_nb$, some $be_kb \neq 0$. Let P be the set of all e_j such that the elements be_jb are nonzero. Without loss of generality, let $P = \{e_1, \ldots, e_m\}$.

By Lemma 5.1.5(iii), the be_jb, $j = 1, \ldots, m$, are primitive idempotents in bRb. From $b = be_1b + \cdots + be_nb = be_1b + \cdots + be_mb$, $\{be_jb \mid 1 \leq j \leq m\}$ is a complete set of primitive idempotents for bRb. The proof for $b \in \mathbf{S}_r(R)$ is a right-sided version of the preceding proof. □

The next two results may be useful for studying many well known classes of rings via complete generalized triangular matrix representations and semicentral reduced rings from the same respective class.

Proposition 5.2.13 *Let a ring R satisfy any one of the following conditions.*

(i) *R has a complete set of primitive idempotents.*
(ii) *R is orthogonally finite.*
(iii) *R has DCC on idempotent generated (resp., principal, or finitely generated) ideals.*
(iv) *R has ACC on idempotent generated (resp., principal, or finitely generated) ideals.*
(v) *R has DCC on idempotent generated (resp., principal, or finitely generated) right ideals.*
(vi) *R has ACC on idempotent generated (resp., principal, or finitely generated) right ideals.*
(vii) *R is a semilocal ring.*
(viii) *R is a semiperfect ring.*
(ix) *R is a right perfect ring.*
(x) *R is a semiprimary ring.*

Then $\mathrm{Tdim}(R) < \infty$ *and*

$$R \cong \begin{bmatrix} R_1 & R_{12} & \cdots & R_{1n} \\ 0 & R_2 & \cdots & R_{2n} \\ \vdots & \vdots & \ddots & \vdots \\ 0 & 0 & \cdots & R_n \end{bmatrix},$$

where $n = \mathrm{Tdim}(R)$, each R_i is semicentral reduced, and satisfies the same condition as R. Further, each R_{ij} is an (R_i, R_j)-bimodule, and the rings $R_1, \ldots, R_n$ are uniquely determined by R up to isomorphism and permutation.

Proof (i) Let $\{f_1, \ldots, f_k\}$ be a complete set of primitive idempotents of R. Then for any $0 \neq b \in \mathbf{S}_\ell(R)$, $b = f_1b + \cdots + f_kb$. Each f_ib is an idempotent, as $b \in \mathbf{S}_\ell(R)$. Assume that $j = 1, \ldots, m$ is the set of all indices for which $f_jb \neq 0$.

Now we have that $bR \subseteq f_1bR + \cdots + f_mbR = bf_1bR + \cdots + bf_mbR \subseteq bR$, hence $bR = f_1bR + \cdots + f_mbR$. Primitivity of f_j implies that $f_jbR = f_jR$, whenever $f_jb \neq 0$. Hence, the total number of right ideals of the form bR, $b \in \mathbf{S}_\ell(R)$ cannot exceed 2^k. Thus, by Theorem 5.2.5, R has a complete set of left triangulating idempotents. So $\mathrm{Tdim}(R) < \infty$.

Let $\{e_1, \ldots, e_n\}$ be a complete set of left triangulating idempotents of R. Take $R_i = e_i R e_i$ and $R_{ij} = e_i R e_j$ for $i < j$. Then R_{ij} is an (R_i, R_j)-bimodule for $i < j$. Since $e_1 \in \mathbf{S}_\ell(R)$, $R_1 = e_1 R e_1$ has a complete set of primitive idempotents from Lemma 5.2.12. Also $1 - e_1 \in \mathbf{S}_r(R)$ by Proposition 1.2.2, $(1 - e_1)R(1 - e_1)$ has a complete set of primitive idempotents by Lemma 5.2.12. Next we see that $e_2 \in \mathbf{S}_\ell((1 - e_1)R(1 - e_1))$, again Lemma 5.2.12 yields that

$$R_2 = e_2 R e_2 = e_2((1 - e_1)R(1 - e_1))e_2$$

has a complete set of primitive idempotents, and so on. The uniqueness of the R_i follows from Theorem 5.2.8.

(ii) By part (i) and Proposition 1.2.15, we have a unique generalized triangular matrix representation. Further, each R_i is orthogonally finite.

(iii) Assume that R has DCC on idempotent generated (resp., principal, or finitely generated) ideals. Then $\{eR \mid e \in \mathbf{S}_\ell(R)\}$ has DCC since $eR = ReR$ for each e in $\mathbf{S}_\ell(R)$. Consider $\{Rf \mid f \in \mathbf{S}_r(R)\}$. Then $Rf = RfR$ for each $f \in \mathbf{S}_r(R)$. Thus $\{Rf \mid f \in \mathbf{S}_r(R)\}$ also has DCC. By Theorem 5.2.5, R has a complete set of left triangulating idempotents. So $\mathrm{Tdim}(R) < \infty$.

Now say $h^2 = h \in R$. Then hRh has DCC on idempotent generated (resp., principal, or finitely generated) ideals by using [259, Theorem 21.11].

(iv) By assumption, $\{eR \mid e \in \mathbf{S}_\ell(R)\}$ has ACC as $eR = ReR$. Also since $Rf = RfR$ for any $f \in \mathbf{S}_r(R)$, $\{Rf \mid f \in \mathbf{S}_r(R)\}$ has ACC. From Theorem 5.2.5, R has a complete set of left triangulating idempotents, so $\mathrm{Tdim} < \infty$. Say $h^2 = h \in R$. By using [259, Theorem 21.11], hRh has ACC on idempotent generated (resp., principal, or finitely generated) ideals.

(v) By Proposition 1.2.13, R is orthogonally finite. By part (ii), R has a complete set of left triangulating idempotents, so $\mathrm{Tdim}(R) < \infty$. Next, let $h^2 = h \in R$. Then hRh has DCC on idempotent generated (resp., principal or finitely generated) right ideals by using [259, Theorem 21.11].

(vi) The proof is similar to that of part (v) by Proposition 1.2.13 and using [259, Theorem 21.11].

(vii) and (viii) We note that, for each of these conditions, R is orthogonally finite. By part (ii), $\mathrm{Tdim}(R) < \infty$. Homomorphic images of a semilocal ring and a semiperfect ring are semilocal and semiperfect, respectively (see [259, Proposition 20.7] and [8, Corollary 27.9]). By Proposition 5.1.7(iii), if R is semilocal (resp., semiperfect), then each R_i is semilocal (resp., semiperfect).

(ix) If R is right perfect, then R is orthogonally finite. Thus part (ii) yields that $\mathrm{Tdim}(R) < \infty$. By 1.1.14, R has DCC on principal left ideals. Say $h^2 = h \in R$. Then by the left-sided version of the proof for part (v), hRh also has DCC on principal left ideals. So hRh is right perfect, and hence each R_i is right perfect.

(x) If R is semiprimary, then also R is orthogonally finite. Hence by part (ii), $\mathrm{Tdim}(R) < \infty$. Say $h^2 = h \in R$. It is well known that $J(hRh) = hJ(R)h$ (see [259, Theorem 21.10]). Hence if R is semiprimary, then so is hRh. Thus each R_i is semiprimary. □

Proposition 5.2.14 *Let P be a property of rings such that whenever a ring A satisfies P, then A/I ($I \trianglelefteq A$) or eAe ($e^2 = e \in A$) also satisfies P. Assume that R is a* *ring with* Tdim$(R) = n < \infty$ *and satisfies P. Then*

$$R \cong \begin{bmatrix} R_1 & R_{12} & \cdots & R_{1n} \\ 0 & R_2 & \cdots & R_{2n} \\ \vdots & \vdots & \ddots & \vdots \\ 0 & 0 & \cdots & R_n \end{bmatrix},$$

where each R_i is semicentral reduced and satisfies the property P. Further, each R_{ij} is an (R_i, R_j)-bimodule, and the rings $R_1, \dots, R_n$ are uniquely determined by R up to isomorphism and permutation.

Proof Since Tdim$(R) = n < \infty$, R has the indicated unique generalized triangular matrix representation by Theorems 5.1.4 and 5.2.8. Rings R_i have the form eRe, where $e^2 = e \in R$, also R_i are ring homomorphic images of R by Proposition 5.1.7(iii). By assumption each R_i has the property P. □

We remark that the following classes of rings determined by property P indicated in Proposition 5.2.14: Baer rings, right Rickart rings, quasi-Baer rings, right p.q.-Baer rings, right hereditary rings, right semihereditary rings, π-regular rings, PI-rings, and rings with bounded index (of nilpotency), etc.

By the next result, if Tdim$(R) < \infty$, central idempotents can be written as sums of elements in a complete set of left triangulating idempotents.

Proposition 5.2.15 *Assume that $\{b_1, \dots, b_n\}$ is a complete set of left triangulating idempotents for a ring R. If $c \in \mathcal{B}(R) \setminus \{0, 1\}$, then there exists $\emptyset \neq \Lambda \subsetneq \{1, \dots, n\}$ such that $c = \sum_{i \in \Lambda} b_i$.*

Proof Let $c \in \mathcal{B}(R) \setminus \{0, 1\}$. Then $c = c(b_1 + \cdots + b_n) = cb_1 + \cdots + cb_n$. We note that $cb_i \in \mathbf{S}_\ell(b_i R b_i)$ and $\mathbf{S}_\ell(b_i R b_i) = \{0, b_i\}$ for each i. Therefore, there exists $\emptyset \neq \Lambda \subsetneq \{1, \dots, n\}$ such that $c = \sum_{i \in \Lambda} b_i$. □

Theorem 5.2.16 *Let R be a ring. Consider the following conditions.*

(i) *R has a complete set of primitive idempotents.*
(ii) *R has a complete set of left triangulating idempotents.*
(iii) *R has a complete set of centrally primitive idempotents.*

Then (i)⇒(ii)⇒(iii).

Proof Proposition 5.2.13(i) yields the implication (i)⇒(ii). For (ii)⇒(iii), assume that R has a complete set of left triangulating idempotents for R. By Proposition 5.2.15, $\mathcal{B}(R)$ is a finite set. Now a standard argument yields that R has a complete set of centrally primitive idempotents. □

We remark that when R is commutative, conditions (i), (ii), and (iii) of Theorem 5.2.16 are equivalent. The next example shows that the converse of each of the implications in Theorem 5.2.16 does not hold.

Example 5.2.17 (i) There is a ring R with a complete set of left triangulating idempotents (i.e., $\text{Tdim}(R) < \infty$), but R does not have a complete set of primitive idempotents. Indeed, let V be an infinite dimensional right vector space over a field F and let $R = \text{End}_F(V)$. Then R is a prime ring, so $\text{Tdim}(R) = 1$. Since R is a regular ring which is not semisimple Artinian, R cannot have a complete set of primitive idempotents.

(ii) There is a ring R with a complete set of centrally primitive idempotents, but R does not have a complete set of left triangulating idempotents. For this, let R be the $\aleph_0 \times \aleph_0$ upper triangular row finite matrix ring over a field. Then $\{\mathbf{1}\}$ is a complete set of centrally primitive idempotents of R, where $\mathbf{1}$ is the identity of R. Let e_{ii} be the matrix in R with 1 in the (i,i)-position and 0 elsewhere. Then for any positive integer n, $e_{11} + \cdots + e_{nn} \in \mathbf{S}_\ell(R)$. As

$$(e_{11} + \cdots + e_{nn})R \subsetneq (e_{11} + \cdots + e_{nn} + e_{n+1n+1})R$$

for each n, Theorem 5.2.5 yields that R cannot have a complete set of left triangulating idempotents.

We need the next lemma for investigating $\text{Tdim}(R)$ of a ring R.

Lemma 5.2.18 *Let $\{b_1, \ldots, b_n\}$ be a set of left triangulating idempotents of a ring R and $\{b_{(i,1)}, \ldots, b_{(i,k_i)}\}$ a set of left triangulating idempotents of $b_i R b_i$. Then $\{b_{(1,1)}, \ldots, b_{(1,k_1)}, b_{(2,1)}, \ldots, b_{(2,k_2)}, \ldots, b_{(n,1)}, \ldots, b_{(n,k_n)}\}$ is a set of left triangulating idempotents of R.*

Proof Clearly $1 = \sum_{i=1}^{k_1} b_{(1,i)} + \cdots + \sum_{i=1}^{k_n} b_{(n,i)}$. Also $b_{(1,1)} \in \mathbf{S}_\ell(R)$ by Lemma 5.1.5(i). Let $c_{(i,j)} = 1 - \sum_{\alpha=1}^{i-1} b_\alpha - \sum_{\gamma=1}^{j} b_{(i,\gamma)}$, where $1 \le j < k_i$. Then $b_{(i,j+1)}(\sum_{\alpha=1}^{i-1} b_\alpha + \sum_{\gamma=1}^{j} b_{(i,\gamma)}) = 0$, and so $b_{(i,j+1)} c_{(i,j)} = b_{(i,j+1)}$. Similarly, $c_{(i,j)} b_{(i,j+1)} = b_{(i,j+1)}$. So $b_{(i,j+1)} \in c_{(i,j)} R c_{(i,j)}$. Note that $c_{(i,j)}^2 = c_{(i,j)}$.

We claim that $b_{(i,j+1)} \in \mathbf{S}_\ell(c_{(i,j)} R c_{(i,j)})$. Put $c_j = b_i - \sum_{\gamma=1}^{j} b_{(i,\gamma)}$. Then $b_{(i,j+1)} \in \mathbf{S}_\ell(c_j(b_i R b_i)c_j) = \mathbf{S}_\ell(c_j R c_j)$ and $c_{(i,j)} = 1 - \sum_{\alpha=1}^{i} b_\alpha + c_j$. Note that $b_{(i,j+1)} \in b_i R b_i$, $(\sum_{\alpha=1}^{i-1} b_\alpha) b_{(i,j+1)} = 0$, and $\{b_1, \ldots, b_n\}$ is a set of orthogonal idempotents. Hence,

$$\begin{aligned} b_{(i,j+1)} &= c_{(i,j)} b_{(i,j+1)} = (1 - \sum_{\alpha=1}^{i} b_\alpha + c_j) b_{(i,j+1)} \\ &= b_{(i,j+1)} - b_i b_{(i,j+1)} + c_j b_{(i,j+1)} = c_j b_{(i,j+1)} \end{aligned}$$

as $b_i b_{(i,j+1)} = b_{(i,j+1)}$. Similarly, $b_{(i,j+1)} = b_{(i,j+1)} c_{(i,j)} = b_{(i,j+1)} c_j$. For $r \in R$,

$$\begin{aligned} (c_{(i,j)} r c_{(i,j)}) b_{(i,j+1)} &= (1 - \sum_{\alpha=1}^{i} b_\alpha + c_j) r c_j b_{(i,j+1)} \\ &= (1 - \sum_{\alpha=1}^{i} b_\alpha) r c_j b_{(i,j+1)} + c_j r c_j b_{(i,j+1)}. \end{aligned}$$

From Proposition 5.1.7(i), $1 - \sum_{\alpha=1}^{i} b_\alpha \in \mathbf{S}_r(R)$. Therefore, we now obtain that $(1 - \sum_{\alpha=1}^{i} b_\alpha) r c_j b_{(i,j+1)} = (1 - \sum_{\alpha=1}^{i} b_\alpha) r (1 - \sum_{\alpha=1}^{i} b_\alpha) c_j b_{(i,j+1)} = 0$ since

$$\begin{aligned}(1 - \sum_{\alpha=1}^{i} b_\alpha) c_j b_{(i,j+1)} &= (1 - \sum_{\alpha=1}^{i} b_\alpha) b_{(i,j+1)} = (1 - \sum_{\alpha=1}^{i-1} b_\alpha - b_i) b_{(i,j+1)} \\ &= b_{(i,j+1)} - b_i b_{(i,j+1)} = 0.\end{aligned}$$

Thus,

$$\begin{aligned}(c_{(i,j)} r c_{(i,j)}) b_{(i,j+1)} &= (c_j r c_j) b_{(i,j+1)} = b_{(i,j+1)} (c_j r c_j) b_{(i,j+1)} \\ &= b_{(i,j+1)} (c_{(i,j)} r c_{(i,j)}) b_{(i,j+1)}.\end{aligned}$$

So $b_{(i,j+1)} \in \mathbf{S}_\ell(c_{(i,j)} R c_{(i,j)})$. Now routinely we obtain the desired result. □

Theorem 5.2.19 *Let $\{b_1, \ldots, b_n\}$ be a set of left triangulating idempotents of a ring R. Then* $\mathrm{Tdim}(R) = \sum_{i=1}^{n} \mathrm{Tdim}(b_i R b_i)$.

Proof If $\mathrm{Tdim}(R) = \infty$, then $\mathrm{Tdim}(b_j R b_j) = \infty$ for some $1 \leq j \leq n$, otherwise Lemma 5.2.18 yields a contradiction.

Let $\mathrm{Tdim}(R) < \infty$. By Corollary 5.2.6, $\mathrm{Tdim}(b_1 R b_1) < \infty$. From Proposition 1.2.2, $1 - b_1 \in \mathbf{S}_r(R)$. By Corollary 5.2.6, $\mathrm{Tdim}((1 - b_1) R (1 - b_1)) < \infty$. We see that $b_2 \in \mathbf{S}_\ell((1 - b_1) R (1 - b_1))$. Hence, Corollary 5.2.6 yields that $\mathrm{Tdim}(b_2 R b_2) < \infty$. This procedure, by using Corollary 5.2.6, can be continued to show that $\mathrm{Tdim}(b_i R b_i) < \infty$ for all $1 \leq i \leq n$. Lemma 5.2.18 yields that $\mathrm{Tdim}(R) = \sum_{i=1}^{n} \mathrm{Tdim}(b_i R b_i)$. □

Corollary 5.2.20 *Let R be a ring with a generalized triangular matrix representation*

$$\begin{bmatrix} R_1 & R_{12} & \cdots & R_{1n} \\ 0 & R_2 & \cdots & R_{2n} \\ \vdots & \vdots & \ddots & \vdots \\ 0 & 0 & \cdots & R_n \end{bmatrix}.$$

Then $\mathrm{Tdim}(R) = \sum_{i=1}^{n} \mathrm{Tdim}(R_i)$. *So,* $\mathrm{Tdim}(T_n(A)) = n\mathrm{Tdim}(A)$, *where A is a ring and n is a positive integer.*

Exercise 5.2.21

1. Assume that R is a ring and $0 \neq g \in \mathcal{B}(R)$ such that $g = g_1 + \cdots + g_t$, where $\{g_i \mid 1 \leq i \leq t\}$ is a set of orthogonal centrally primitive idempotents in R. Show that t is uniquely determined.
2. Let R be a ring, and let $\{e_1, \ldots, e_m\}$ and $\{f_1, \ldots, f_n\}$ be two complete sets of centrally primitive idempotents of R. Show that $m = n$ and there exists a permutation σ on $\{1, \ldots, n\}$ such that $e_i = f_{\sigma(i)}$.

3. Assume that M_R is a right R-module and $S = \mathrm{End}(M_R)$. Show that the following are equivalent.
   (i) S has a complete set of left triangulating idempotents.
   (ii) There exists a positive integer n such that:
      (1) $M = M_1 \oplus \cdots \oplus M_n$.
      (2) $\mathrm{Hom}(M_i, M_j) = 0$ for $i < j$.
      (3) Each M_i has no nontrivial fully invariant direct summands.
4. ([93, Birkenmeier, Park and Rizvi]) Assume that S is an overring of a ring R such that $R_R \leq^{\text{ess}} S_R$. (The ring S is called a right essential overing of R. See Chap. 7 for right essential overrings for more details.) Show that if R is right FI-extending, then $\mathrm{Tdim}(S) \leq \mathrm{Tdim}(R)$.
5. ([93, Birkenmeier, Park and Rizvi]) Let S be an overring of a ring R such that $R_R \leq^{\text{ess}} S_R$. Prove that if R is right extending and $\{e_1, \ldots, e_n\}$ is a complete set of primitive idempotents for R, then $\{e_1, \ldots, e_n\}$ is a complete set of primitive idempotents for S.
6. ([79, Birkenmeier, Kim, and Park]) Show that a ring R is left perfect if and only if R has a complete generalized triangular matrix representation, where each diagonal ring R_i is simple Artinian or left perfect with $(\mathrm{Soc}(R_{i\,R_i}))^2 = 0$.

5.3 Canonical Representations

We show that if a ring R has a set of left triangulating idempotents, then it has a canonical generalized triangular matrix representation, where the diagonal subrings are organized into blocks of square diagonal matrix rings. This canonical representation is then used to obtain a result on the right global dimension of rings with a set of left triangulating idempotents.

Let $\{b_1, \ldots, b_n\}$ be a set of left triangulating idempotents of R. If J is a subset of $\{1, \ldots, n\}$, we denote $\sigma_J = \sum_{i \in J} b_i$. Our first result shows that under certain conditions the ordering in a set of left triangulating idempotents can be changed to obtain a new set of left triangulating idempotents.

Proposition 5.3.1 *Let j and m be in $\{1, \ldots, n\}$ with $j < m \leq n$. If $\{b_1, \ldots, b_n\}$ is a set of left triangulating idempotents of a ring R such that $b_i R b_m = 0$ for each i with $j \leq i < m$, then*

$$\{b_1, \ldots, b_{j-1}, b_m, b_j, b_{j+1}, \ldots, b_{m-1}, b_{m+1}, \ldots, b_n\}$$

is a set of left triangulating idempotents of R.

Proof The proof follows routinely from Theorem 5.1.3. □

Proposition 5.3.1 is applied to obtain a canonical form for a generalized triangular matrix representation of R. Let $\{b_1, \ldots, b_n\}$ be a set of left triangulating idempotents. Recursively define the sets I_k and $J(k)$ as follows:

$$I_1 = \{i \mid b_i \in \mathbf{S}_\ell(R)\} \text{ and } J(1) = I_1;$$

and let

$$I_{k+1} = \{i \mid b_i \in \mathbf{S}_\ell((1-\sigma_{J(k)})R(1-\sigma_{J(k)}))\} \text{ and } J(k+1) = J(k) \cup I_{k+1},$$

whenever I_k and $J(k)$ are defined. This process terminates within n steps.

Let $S_j = \{b_i \mid i \in I_j\}$. Then $S_1, \ldots, S_q$ is a partition for $\{b_1, \ldots, b_n\}$ (we will show in the proof of Theorem 5.3.2 that this always occurs). Then reorder $\{1, \ldots, n\}$ so that each I_j has any (fixed) ordering and so that elements of I_j always precede elements in I_{j+1}. This can be thought of in terms of a permutation ψ on $\{1, \ldots, n\}$. Then the ordered set $\{b_{\psi(1)}, \ldots, b_{\psi(n)}\}$ is called a *canonical form* for $\{b_1, \ldots, b_n\}$.

Theorem 5.3.2 *Let $\{b_1, \ldots, b_n\}$ be a set of left triangulating idempotents. Then a canonical form for $\{b_1, \ldots, b_n\}$ exists, and any such canonical form is a set of left triangulating idempotents of R.*

Proof The proof involves repeated use of Propositions 5.3.1, as in the following discussion. We note that $b_1 \in S_1 = \mathbf{S}_\ell(R)$. If $b_m \in S_1$ and $m \neq 1$, then $b_i R b_m = b_i b_m R b_m = 0$ for all $i \neq m$. We use Proposition 5.3.1 to get that $\{b_m, b_1, \ldots, b_{m-1}, b_{m+1}, \ldots, b_n\}$ is a set of left triangulating idempotents of R. Continue this process using elements of S_1 until they are exhausted.

Following the procedure given in Proposition 5.3.1, there exists a permutation α on $\{1, \ldots, n\}$ such that $S_1 = \{b_{\alpha(1)}, \ldots, b_{\alpha(n_1)}\}$. Also, the ordered set $\{b_{\alpha(1)}, b_{\alpha(2)}, \ldots, b_{\alpha(n)}\}$ is a set of left triangulating idempotents of R.

If $n_1 = n$, then we are finished. So consider $n_1 < n$ and let $q = \alpha(n_1+1)$, where $\alpha(n_1+1)$ is the smallest positive integer i such that $b_i \notin S_1$. Observe that b_q is the first element in this new ordering which is not in S_1.

We show that $b_q \in S_2$. For this, let y be the sum of all elements in S_1. Thus, $y = b_{\alpha(1)} + \cdots + b_{\alpha(n_1)}$. Let g be the sum of all elements in $\{b_{\alpha(1)}, \ldots, b_{\alpha(n)}\}$ which are not in $\{b_q, b_{\alpha(1)}, \ldots, b_{\alpha(n_1)}\}$. Then $1 = y + b_q + g$. Thus $1 - y = b_q + g$, and therefore $b_q \in (1-y)R(1-y)$. Now for every $a \in R$, we can see that

$$\begin{aligned}(1-y)a(1-y)b_q &= (1-y)ab_q = (b_q+g)ab_q \\ &= b_q a b_q = b_q(1-y)a(1-y)b_q\end{aligned}$$

as $b_q(1-y) = b_q$, $(1-y)b_q = b_q$, and $gab_q = 0$. So $b_q \in \mathbf{S}_\ell((1-y)R(1-y))$.

Consequently, $q \in I_2$ and hence $b_q \in S_2$. Either this exhausts the elements in S_2 or (in the ordering given by α) there is an element $b_p \in S_2$ beyond b_q. Use Proposition 5.3.1 as before to obtain a set of left triangulating idempotents of R of the form $\{b_{\alpha(1)}, \ldots, b_{\alpha(n_1)}, b_p, b_q, b_{\alpha(n_1+2)}, \ldots, b_{\alpha(n)}\}$.

Repeat this process using elements of S_2 until they are exhausted. Then there exists a permutation γ on $\{1, \ldots, n\}$ such that

$$\{b_{\gamma(1)}, \ldots, b_{\gamma(n_1)}, b_{\gamma(n_1+1)}, \ldots, b_{\gamma(n_2)}, \ldots, b_{\gamma(n)}\}$$

forms a set of left triangulating idempotents, where $\gamma(i) = \alpha(i)$ for $1 \leq i \leq n_1$, $b_{\gamma(n_2)} = b_q$, and $\{b_{\gamma(n_1+1)}, \ldots, b_{\gamma(n_2)}\} = S_2$.

Now either $S_1 \cup S_2 = \{b_1, \ldots, b_n\}$ or we can continue the process on S_3, and so on. After k steps, $k \leq n$, the process terminates in a set of left triangulating idempotents of R in a canonical form. So we obtain a permutation ψ so that $S_1, \ldots, S_k$ is our desired partition of $\{b_1 \ldots, b_n\}$. □

Theorems 5.1.4 and 5.3.2 provide a tool for a generalized triangular matrix representation of R in a special canonical form, which we give next.

Corollary 5.3.3 (Canonical Representation) *Let $\{b_1, \ldots, b_n\}$, $S_1, \ldots, S_k$, and ψ be as before. Then using $0 = n_0 < n_1 < \cdots < n_k$, we have that $S_{j+1} = \{b_{\psi(n_j+1)}, \ldots, b_{\psi(n_{j+1})}\}$, $j = 0, 1, \ldots, k-1$, and R is isomorphic to the $n \times n$ matrix $[A(i, j)]$, where the $A(i, j)$ are $n_i \times n_j$ block matrices*

$$A(i+1, i+1) = \begin{bmatrix} b_{\psi(n_i+1)} R b_{\psi(n_i+1)} & 0 & \cdots & 0 \\ 0 & \ddots & & 0 \\ \vdots & & \ddots & \vdots \\ 0 & & 0 \cdots & b_{\psi(n_{i+1})} R b_{\psi(n_{i+1})} \end{bmatrix};$$

$$A(i+1, j+1) = \begin{bmatrix} b_{\psi(n_i+1)} R b_{\psi(n_j+1)} & \cdots & b_{\psi(n_i+1)} R b_{\psi(n_{j+1})} \\ \vdots & \ddots & \vdots \\ b_{\psi(n_{i+1})} R b_{\psi(n_j+1)} & \cdots & b_{\psi(n_{i+1})} R b_{\psi(n_{j+1})} \end{bmatrix},$$

for $i < j$; and $A(i, j) = 0$ for $i > j$, where $i, j = 0, 1, \ldots, k-1$.

For the proof of Theorem 5.3.5, we need the following lemma.

Lemma 5.3.4 *Let A and B be rings, and let M be an (A, B)-bimodule. Set $R = \begin{bmatrix} A & M \\ 0 & B \end{bmatrix}$, a generalized triangular matrix ring. Then*

$$\begin{aligned} \max\{\text{r.gl.dim}(A), \text{r.gl.dim}(B)\} &\leq \text{r.gl.dim}(R) \\ &\leq \max\{\text{r.gl.dim}(A) + \text{pd}(M_B) + 1, \text{r.gl.dim}(B)\}, \end{aligned}$$

where $\text{pd}(M_B)$ is the projective dimension of M_B.

Proof See [295, Proposition 7.5.1] for the proof. □

In Lemma 5.3.4, if $M = 0$, then $R = A \oplus B$ (ring direct sum). Also

$$\text{r.gl.dim}(R) \leq \max\{\text{r.gl.dim}(A) + \text{pd}(A_R), \text{r.gl.dim}(B) + \text{pd}(B_R)\}$$

from the proof of [295, Proposition 7.5.1]. As A_R and B_R are projective, it follows that $\text{pd}(A_R) = 0$ and $\text{pd}(B_R) = 0$, so

$$\text{r.gl.dim}(R) \leq \max\{\text{r.gl.dim}(A), \text{r.gl.dim}(B)\}.$$

Thus, $\text{r.gl.dim}(A \oplus B) = \max\{\text{r.gl.dim}(A), \text{r.gl.dim}(B)\}$ by Lemma 5.3.4.

As an application of canonical representation, we discuss the following result which exhibits a connection between the right global dimension of R and that of the sum of diagonal subrings.

Theorem 5.3.5 *Let* $\{b_1, \ldots, b_n\}$ *be a set of left triangulating idempotents of* R, *and* $S_1, \ldots, S_k$ *be as in Corollary* 5.3.3. *Then*

$$\text{r.gl.dim}(D) \leq \text{r.gl.dim}(R) \leq k\,(\text{r.gl.dim}(D)) + k - 1,$$

where $D = b_1Rb_1 + \cdots + b_nRb_n$. *Thereby,* $\text{r.gl.dim}(R) < \infty$ *if and only if* $\text{r.gl.dim}(D) < \infty$.

Proof The proof is given by induction on k. If $k = 1$, then $R = D$ by Theorem 5.3.2 and we are finished. Assume that $k \geq 2$. We take $A = \sum_{b_i \in S_1} b_i R b_i$, $M = \sum_{b_i \in S_1, b_j \in S_2 \cup \cdots \cup S_k} b_i R b_j$, and $B = (1 - \sum_{b_i \in S_1} b_i)\, R\, (1 - \sum_{b_i \in S_1} b_i)$. Then obviously $B = (\sum_{b_j \in S_2 \cup \cdots \cup S_k} b_j)\, R\, (\sum_{b_j \in S_2 \cup \cdots \cup S_k} b_j)$.

We note that $S_2 \cup \cdots \cup S_k$ is a set of left triangulating idempotents of B and $\{S_2, \ldots, S_k\}$ is a partition which establishes a canonical generalized triangular matrix representation for B. Let $D_1 = \sum_{b_j \in S_2 \cup \cdots \cup S_k} b_j R b_j$. Then by induction $\text{r.gl.dim}(D_1) \leq \text{r.gl.dim}(B) \leq (k-1)(\text{r.gl.dim}(D_1)) + k - 2$.

Because $D = A \oplus D_1$ from Theorem 5.3.2 or Corollary 5.3.3, it follows that $\text{r.gl.dim}(D) = \max\{\text{r.gl.dim}(A), \text{r.gl.dim}(D_1)\}$. Observe that $R = \begin{bmatrix} A & M \\ 0 & B \end{bmatrix}$ and M is an (A, B)-bimodule. Hence,

$$\begin{aligned} \max\{\text{r.gl.dim}(A), \text{r.gl.dim}(B)\} &\leq \text{r.gl.dim}(R) \\ &\leq \max\{\text{r.gl.dim}(A) + \text{pd}(M_B) + 1,\ \text{r.gl.dim}(B)\} \end{aligned}$$

from Lemma 5.3.4. Because $\text{r.gl.dim}(D_1) \leq \text{r.gl.dim}(B)$,

$$\begin{aligned} \text{r.gl.dim}(D) &= \max\{\text{r.gl.dim}(A), \text{r.gl.dim}(D_1)\} \\ &\leq \max\{\text{r.gl.dim}(A),\ \text{r.gl.dim}(B)\} \\ &\leq \text{r.gl.dim}(R). \end{aligned}$$

We observe that $\text{pd}(M_B) \leq \text{r.gl.dim}(B)$. Therefore,

$$\begin{aligned} \text{r.gl.dim}(R) &\leq \max\{\text{r.gl.dim}(A) + \text{pd}(M_B) + 1,\ \text{r.gl.dim}(B)\} \\ &\leq \max\{\text{r.gl.dim}(A) + \text{r.gl.dim}(B) + 1,\ \text{r.gl.dim}(B)\} \\ &= \text{r.gl.dim}(A) + \text{r.gl.dim}(B) + 1 \\ &\leq \text{r.gl.dim}(D) + [(k-1)\,(\text{r.gl.dim}(D_1)) + (k-2)] + 1 \\ &\leq \text{r.gl.dim}(D) + (k-1)\,(\text{r.gl.dim}(D)) + k - 1 \\ &= k\,(\text{r.gl.dim}(D)) + k - 1. \end{aligned}$$

Therefore, $\text{r.gl.dim}(D) \leq \text{r.gl.dim}(R) \leq k\,(\text{r.gl.dim}(D)) + k - 1$. Thereby, $\text{r.gl.dim}(R) < \infty$ if and only if $\text{r.gl.dim}(D) < \infty$. □

5.4 Piecewise Prime Rings and Piecewise Domains

In this section, a criterion for a ring with a complete set of triangulating idempotents to be quasi-Baer is provided. Also a structure theorem for a quasi-Baer ring with a complete set of triangulating idempotents is shown. Among the applications of this structure theorem, several well-known results are obtained as its consequences. These include Levy's decomposition theorem of semiprime right Goldie rings, Faith's characterization of semiprime right FPF rings with no infinite set of central orthogonal idempotents, Gordon and Small's characterization of piecewise domains, and Chatters' decomposition theorem of hereditary Noetherian rings. A result related to Michler's splitting theorem for right hereditary right Noetherian rings is also obtained as an application.

The next result provides a criterion for a ring with a complete set of left triangulating idempotents to be quasi-Baer.

Theorem 5.4.1 *Assume that a ring R has a complete set of left triangulating idempotents with* $\text{Tdim}(R) = n$. *Then the following are equivalent.*

(i) *R is quasi-Baer.*
(ii) *For any complete set of left triangulating idempotents $\{b_1, \ldots, b_n\}$ of R, if $b_i x b_j R b_j y b_k = 0$ for some $x, y \in R$ and some $1 \leq i, j, k \leq n$, then either $b_i x b_j = 0$ or $b_j y b_k = 0$.*
(iii) *There is a complete set of left triangulating idempotents $\{c_1, \ldots, c_n\}$ of R such that if $c_i x c_j R c_j y c_k = 0$ for some $x, y \in R$ and some $1 \leq i, j, k \leq n$, then either $c_i x c_j = 0$ or $c_j y c_k = 0$.*
(iv) *For any complete set of left triangulating idempotents $\{b_1, \ldots, b_n\}$, assume that $K b_j V = 0$ for some ideals K and V of R and some b_j, $1 \leq j \leq n$. Then either $K b_j = 0$ or $b_j V = 0$.*

Proof (i)⇒(ii) Assume that $b_i x b_j R b_j y b_k = 0$ for some $x, y \in R$ and some $1 \leq i, j, k \leq n$. Since R is quasi-Baer, $r_R(b_i x b_j R) = fR$ for some $f \in \mathbf{S}_\ell(R)$. By Lemma 5.1.5(ii), $b_j f b_j \in \mathbf{S}_\ell(b_j R b_j)$. As $\{b_1, \ldots, b_n\}$ is a complete set of left triangulating idempotents, $\mathbf{S}_\ell(b_j R b_j) = \{0, b_j\}$. So either $b_j f b_j = 0$ or $b_j f b_j = b_j$. If $b_j f b_j = 0$, then since $b_j y b_k \in r_R(b_i x b_j R) = fR$, we have that $b_j y b_k = f b_j y b_k$. So $b_j y b_k = b_j f b_j y b_k = 0$. On the other hand, if $b_j f b_j = b_j$, then $b_i x b_j = b_i x b_j f b_j = 0$ as $b_i x b_j f = 0$.

(ii)⇒(iii) It follows immediately because R has a complete set of left triangulating idempotents.

(iii)⇒(i) Say L is a left ideal of R. First, assume that $Rc_i \cap \ell_R(L) \neq 0$ for some i. Then we may assume that

$$Rc_1 \cap \ell_R(L) \neq 0, \ldots, Rc_m \cap \ell_R(L) \neq 0,$$

and

$$Rc_{m+1} \cap \ell_R(L) = 0, \ldots, Rc_n \cap \ell_R(L) = 0.$$

Thus $\ell_R(L)Rc_{m+1} = 0, \ldots$, and $\ell_R(L)Rc_n = 0$. Put $T = Rc_1 + \cdots + Rc_m$.

Say $v \in \ell_R(L)$. Then $v = v(c_1 + \cdots + c_n) = vc_1 + \cdots + vc_m \in T$. Therefore, $\ell_R(L) \subseteq T$. To show that $c_1 \in \ell_R(L)$, take $y \in L$. Since $Rc_1 \cap \ell_R(L) \neq 0$, there exists $x \in R$ such that $0 \neq xc_1 \in Rc_1 \cap \ell_R(L)$. So $xc_1Rc_1y = 0$. Now there is $c_kxc_1 \neq 0$ for some c_k because $1 = c_1 + \cdots + c_n$. Thus, $c_kxc_1Rc_1yc_j = 0$ for all j. Therefore $c_1yc_j = 0$ for all j, and so $c_1y = 0$. Hence, $c_1 \in \ell_R(L)$. Thus, $Rc_1 \subseteq \ell_R(L)$. Similarly, $Rc_2, \ldots, Rc_m \subseteq \ell_R(L)$. So $T \subseteq \ell_R(L)$. Therefore, $\ell_R(L) = T = Rc_1 + \cdots + Rc_m = R(c_1 + \cdots + c_m)$. Put $e = c_1 + \cdots + c_m$. Then $e^2 = e \in R$ and so $\ell_R(L) = Re$.

Next, assume that $Rc_i \cap \ell_R(L) = 0$ for all i. Then $\ell_R(L)Rc_i = 0$ for all i. So $\ell_R(L) = \ell_R(L)(Rc_1 + \cdots + Rc_n) = 0$. Therefore, R is quasi-Baer.

(ii)⇒(iv) Let $Kb_jV = 0$ and $b_jV \neq 0$ for some b_j. Say $y \in V$ with $b_jy \neq 0$. So $0 \neq b_jy = \sum_{t=1}^{n} b_jyb_t$, hence $b_jyb_k \neq 0$ for some b_k.

Let $x \in K$. Then $xb_jRb_jy = 0$. Hence $b_ixb_jRb_jyb_k = 0$ for each b_i. As $b_jyb_k \neq 0$, $b_ixb_j = 0$ for all b_i. Thus $xb_j = \sum_{i=1}^{n} b_ixb_j = 0$, so $Kb_j = 0$. If $Kb_j \neq 0$, similarly $b_jV = 0$.

(iv)⇒(ii) If $b_ixb_jRb_jyb_k = 0$, then $(Rb_ixb_jR)b_j(Rb_jyb_kR) = 0$. By assumption $Rb_ixb_jR = 0$ or $Rb_jyb_kR = 0$, so $b_ixb_j = 0$ or $b_jyb_k = 0$. □

Corollary 5.4.2 *If R has a complete set of primitive idempotents, then the following are equivalent.*

(i) *R is quasi-Baer.*
(ii) *For any given complete set of primitive idempotents $\{e_1, \ldots, e_n\}$, if $e_ixe_jRe_jye_k = 0$ for some $x, y \in R$ and some $1 \leq i, j, k \leq n$, then either $e_ixe_j = 0$ or $e_jye_k = 0$.*
(iii) *There is a complete set of primitive idempotents $\{f_1 \ldots, f_m\}$ of R such that if $f_ixf_jRf_jyf_k = 0$ for some $x, y \in R$ and some $1 \leq i, j, k \leq m$, then either $f_ixf_j = 0$ or $f_jyf_k = 0$.*
(iv) *For any complete set of primitive idempotents $\{g_1, \ldots, g_\ell\}$, assume that $Kg_jV = 0$ for some ideals K and V of R and for some g_j, $1 \leq j \leq \ell$. Then either $Kg_j = 0$ or $g_jV = 0$.*

Proof Let $f \in \mathbf{S}_\ell(R)$ and $0 \neq e^2 = e \in R$. Then $efe \in \mathbf{S}_\ell(eRe)$ by Lemma 5.1.5(ii). In particular, if e is primitive, then $\mathbf{S}_\ell(eRe) = \{0, e\}$. So either $efe = 0$ or $efe = e$. The proof can then be completed by using a similar argument as in the proof of Theorem 5.4.1. □

Definition 5.4.3 A ring R is called a *piecewise domain* (or simply, *PWD*) if there is a complete set of primitive idempotents $\{e_1, \ldots, e_n\}$ such that $xy = 0$ implies $x = 0$ or $y = 0$ whenever $x \in e_iRe_j$ and $y \in e_jRe_k$, for $1 \leq i, j, k \leq n$.

To avoid ambiguity, we sometimes say that R is a PWD with respect to a complete set $\{e_i\}_{i=1}^n$ of primitive idempotents. In light of Theorem 5.4.1 and Corollary 5.4.2, it is interesting to compare quasi-Baer rings having a complete set of left triangulating (or primitive) idempotents with PWDs. In fact, Definition 5.4.3 and the equivalence of (i) and (iii) in Theorem 5.4.1 and Corollary 5.4.2 suggest the following definition.

Definition 5.4.4 A quasi-Baer ring with a complete set of triangulating idempotents is called a *piecewise prime ring* (or simply, *PWP ring*).

The following result is somewhat of a right p.q.-Baer analogue of Theorem 3.1.25.

Proposition 5.4.5 *Let R be a right p.q.-Baer ring with* $\text{Tdim}(R) < \infty$. *Then R is a PWP ring.*

Proof Let I be a right ideal of R, and say $I = \sum_{i\in\Lambda} x_i R$ with $x_i \in R$. Then $r_R(I) = \bigcap_{i\in\Lambda} r_R(x_i R) = \bigcap_{i\in\Lambda} e_i R$ with $e_i \in \mathbf{S}_\ell(R)$ for each $i \in \Lambda$ because R is right p.q.-Baer. By Theorem 5.2.5 and Proposition 1.2.4(i), there exists $e \in \mathbf{S}_\ell(R)$ such that $\sum_{i\in\Lambda} e_i R = eR$. Therefore R is a PWP ring. □

The next question was posed by Gordon and Small (see [187, p. 554]): *Can a PWD R possess a complete set $\{f_i\}_{i=1}^m$ of primitive idempotents for which it is not true that $xy = 0$ implies $x = 0$ or $y = 0$ for some $x \in f_i R f_k$ and $y \in f_k R f_j$?* Theorem 5.4.1 and Corollary 5.4.2 show that if R is a PWP ring, then it is a PWP ring with respect to *any* complete set of left triangulating idempotents. Thereby for the case of PWP rings it provides an answer to the above question.

Proposition 5.4.6 *Any PWD is a PWP ring.*

Proof The result follows from Proposition 5.2.13 and Corollary 5.4.2. □

The following example illustrates that the converse of Proposition 5.4.6 does not hold true.

Example 5.4.7 (i) Let R be the ring in Example 3.2.7(ii). Then R is a PWP ring, but it is not a PWD.

(ii) Let R be the ring of Example 5.2.17(i). Then R is a prime ring, so it is a PWP ring. But R does not have a complete set of primitive idempotents. Thus, R is not a PWD.

Example 5.4.8 There is a PWD which is not Baer. Let R be a commutative domain which is not semihereditary (e.g., $\mathbb{Z}[x]$). Then $\text{Mat}_n(R)$ is a PWD for any positive integer $n > 1$, but it is not a Baer ring (see Theorem 6.1.4).

Proposition 5.4.9 *Let R be a ring and $\{e_1, \ldots, e_n\}$ be a complete set of primitive idempotents of R. Then the following are equivalent.*

(i) *R is a PWD with respect to $\{e_1, \ldots, e_n\}$.*

(ii) *Every nonzero element of* $\mathrm{Hom}(e_iR_R, e_jR_R)$ *is a monomorphism for all* i, j, $1 \le i, j \le n$.

(iii) *Every nonzero element of* $\mathrm{Hom}(e_iR_R, R_R)$ *is a monomorphism for all* i, $1 \le i \le n$.

Proof Exercise. □

Example 5.4.10 (i) It is routine to check that the ring of $n \times n$ matrices over a PWD is a PWD.

(ii) The polynomial ring over a PWD is a PWD. Indeed, say R is a PWD with respect to a complete set of primitive idempotent $\{e_1, \dots, e_n\}$. Then $\{e_1, \dots, e_n\}$ is a complete set of primitive idempotents of $R[x]$, and $R[x]$ is a PWD with respect to $\{e_1, \dots, e_n\}$.

(iii) A right Rickart ring with a complete set of primitive idempotents is a PWD. In fact, say R is a right Rickart ring with a complete set $\{e_1, \dots, e_n\}$ of primitive idempotents.

Suppose that $e_ixe_je_jye_k = 0$, where $x, y \in R$ and $1 \le i, j, k \le n$. Since R is right Rickart, $r_R(e_ixe_j) = fR$ for some $f^2 = f \in R$. So $1 - e_j = f(1 - e_j)$ since $1 - e_j \in r_R(e_ixe_j)$. Note that $1 - e_j = \sum_{k \ne j}^n e_k$, thus

$$\sum_{k \ne j} e_k = 1 - e_j = f(1 - e_j) = \sum_{k \ne j} fe_k.$$

Hence $e_k = fe_k$ for $k \ne j$ and $1 \le k \le n$. Therefore,

$$f = \sum_{k=1}^{n} fe_k = \sum_{k \ne j}^{n} fe_k + fe_j = \sum_{k \ne j}^{n} e_k + fe_j.$$

Thus $1 - f = 1 - \sum_{k \ne j}^n e_k - fe_j = e_j - fe_j = (1 - f)e_j$, so $R(1 - f) \subseteq Re_j$. Hence, it follows that $R(1 - f) = Re_j$ or $R(1 - f) = 0$ as e_j is a primitive idempotent.

If $R(1 - f) = Re_j$, then $e_jf = 0$. Because $e_ixe_je_jye_k = e_ixe_jye_k = 0$, we get that $ye_k \in r_R(e_ixe_j) = fR$, and $ye_k = fye_k$. Hence, $e_jye_k = e_jfye_k = 0$. Finally, assume that $R(1 - f) = 0$. Then $f = 1$, and thus $e_ixe_j = 0$. So R is a PWD.

If $R(1 - f) = Re_j$, then $e_jf = 0$. Because $e_ixe_je_jye_k = e_ixe_jye_k = 0$, we get $ye_k \in r_R(e_ixe_j) = fR$, and therefore $ye_k = fye_k$. Hence $e_jye_k = e_jfye_k = 0$.

Further, if $R(1 - f) = 0$, then $f = 1$, and thus $e_ixe_j = 0$. So R is a PWD.

(iv) There exists a PWD which is not right Rickart. Let $R = \mathrm{Mat}_2(\mathbb{Z}[x])$. Then R is a PWD by part (i), but R is not (right) Rickart (see Example 3.1.28).

(v) A right nonsingular ring which is a direct sum of uniform right ideals is a PWD. Indeed, let R be a right nonsingular ring such that $R = \oplus_{i=1}^n I_i$, where each I_i is a uniform right ideal of R. Then there is a complete set of primitive idempotents $\{e_1, \dots, e_n\}$ with $I_i = e_iR$ for each i. As $Z(R_R) = 0$, by Corollary 1.3.15 $E(R_R) = Q(R)$. Now $Q(R)$ is a regular ring from Theorem 2.1.31 and $Q(R) = e_1Q(R) \oplus \cdots \oplus e_nQ(R)$. Also each $e_iQ(R)_{Q(R)}$ is uniform, so $\{e_1, \dots, e_n\}$

is a complete set of primitive idempotents in $Q(R)$. Thus, $Q(R)$ is semisimple Artinian. Say $e_i x e_j e_j y e_k = 0$, where $x, y \in R$ and $1 \le i, j, k \le n$. Then since $Q(R)$ is a PWD with respect to $\{e_1, \dots, e_n\}$ by part (iii), either $e_i x e_j = 0$ or $e_j y e_k = 0$. So R is a PWD.

Proposition 5.4.11 *Let $\{b_1, \dots, b_n\}$ be a set of left triangulating idempotents of a ring R. Then the following are equivalent.*

(i) *P is a (minimal) prime ideal of R.*
(ii) *There exist m, $1 \le m \le n$, and a (minimal) prime ideal P_m of the ring $b_m R b_m$ such that $P = P_m + \sum_{k \neq m} b_k R b_k + \sum_{i \neq j} b_i R b_j$.*

Proof The proof is routine. □

Theorem 5.4.12 *Let R be a PWP ring with* $\text{Tdim}(R) = n$. *Then $R = A \bigoplus B$ (ring direct sum) such that:*

(i) *$A = \bigoplus_{i=1}^{k} A_i$ is a direct sum of prime rings A_i.*
(ii) *There exists a ring isomorphism*

$$B \cong \begin{bmatrix} B_1 & B_{12} & \cdots & B_{1m} \\ 0 & B_2 & \cdots & B_{2m} \\ \vdots & \vdots & \ddots & \vdots \\ 0 & 0 & \cdots & B_m \end{bmatrix},$$

where each B_i is a prime ring, and B_{ij} is a (B_i, B_j)-bimodule.
(iii) *$n = k + m$.*
(iv) *For each $i \in \{1, \dots, m\}$ there is $j \in \{1, \dots, m\}$ such that $B_{ij} \neq 0$ or $B_{ji} \neq 0$.*
(v) *The rings $B_1, \dots, B_m$ are uniquely determined by B up to isomorphism and permutation.*
(vi) *B has exactly m minimal prime ideals $P_1, \dots, P_m$, R has exactly n minimal prime ideals of the form $A \oplus P_i$ or $C_i \oplus B$ where $C_i = \bigoplus_{j \neq i} A_j$. Further, $P_1, \dots, P_m$ are comaximal, $P(R) = P(B)$, and $P(R)^m = 0$.*

Proof Say $E = \{b_1, b_2, \dots, b_n\}$ is a complete set of left triangulating idempotents of R.

(i) Let $\{e_1, \dots, e_k\} = E \cap \mathcal{B}(R)$. Take $A_i = e_i R$. By Proposition 3.2.5 and Theorem 3.2.10, each A_i is a prime ring.

(ii) Let $\{f_1, \dots, f_m\} = E \setminus \{e_1, \dots, e_k\}$, where the f_i are maintained in the same relative order as they were in E. Let $B_i = f_i B f_i$ and $B_{ij} = f_i B f_j$. Then each B_i is a prime ring by Proposition 3.2.5 and Theorem 3.2.10. Define ϕ by $\phi(b) = [f_i b f_j]$ for $b \in B$, as in the proof of Theorem 5.1.4. Then ϕ is a ring isomorphism.

(iii) The proof follows immediately from the proof of part (ii).

(iv) It is evident since $\{f_1, \dots, f_m\} = E \setminus \{e_1, \dots, e_k\}$.

(v) This is a consequence of Theorem 5.2.8.

(vi) The proof follows from a routine argument using Lemma 5.4.11. □

Corollary 5.4.13 (i) *Any semiprime PWP ring is a finite direct sum of prime rings.*
(ii) *Any biregular ring R with* $\mathrm{Tdim}(R) < \infty$ *is a finite direct sum of simple rings.*

Proof The proof follows from Theorems 5.4.12 and 3.2.22(ii). □

The next corollary is related to Michler's splitting theorem [299, Theorem 2.2] for right hereditary right Noetherian rings.

Corollary 5.4.14 *Let R be a right hereditary right Noetherian ring. Then*

$$R \cong \begin{bmatrix} R_1 & R_{12} & \cdots & R_{1n} \\ 0 & R_2 & \cdots & R_{2n} \\ \vdots & \vdots & \ddots & \vdots \\ 0 & 0 & \cdots & R_n \end{bmatrix},$$

where each R_i is a prime right hereditary, right Noetherian ring, and each R_{ij} is an (R_i, R_j)-bimodule.

Proof As R is right hereditary right Noetherian, R is Baer by Theorem 3.1.25. Thus the proof follows from Theorem 5.4.12 and Proposition 5.2.14. □

We will now see that Levy's decomposition theorem [279] for semiprime right Goldie right hereditary rings, follows as a consequence of Theorem 5.4.12.

Corollary 5.4.15 *Any semiprime right Goldie, right hereditary ring is a finite direct sum of prime right Goldie, right hereditary rings.*

Proof Let R be a semiprime right Goldie, right hereditary ring. Then R is orthogonally finite, so R is Baer by Theorem 3.1.25 and $\mathrm{Tdim}(R) < \infty$ from Proposition 5.2.13(ii). Corollary 5.4.13(i) and a routine verification yield that R is a finite direct sum prime right Goldie, right hereditary rings. □

A ring R is called *right FPF* if every faithful finitely generated right R-module generates the category Mod-R of right R-modules (see [156]). We may note that a semiprime right FPF ring is quasi-Baer (see [78, Corollary 1.19]). By Theorem 5.4.12, Faith's characterization of semiprime right FPF rings with no infinite set of central orthogonal idempotents (see [156, Theorem I.4]) is provided as follows.

Corollary 5.4.16 *Let R be a ring with no infinite set of central orthogonal idempotents. Then R is semiprime right FPF if and only if R is a finite direct sum of prime right FPF rings.*

Proof Let R be a semiprime right FPF ring with no infinite set of central orthogonal idempotents. Because R is semiprime, $\mathcal{B}(R) = \mathbf{S}_\ell(R)$ by Proposition 1.2.6(ii). Since

R has no infinite set of central orthogonal idempotents, we see that

$$\{eR \mid e \in \mathbf{S}_\ell(R)\} = \{eR \mid e \in \mathcal{B}(R)\}$$

has ACC and DCC. By Theorem 5.2.5, $\text{Tdim}(R) < \infty$, so R is a PWP ring. By Corollary 5.4.13(i), R is a finite direct sum of prime rings. Since ring direct summands of right FPF rings are right FPF, these prime rings are right FPF. The converse is immediate. □

A ring R for which the diagonal rings R_i in a complete generalized triangular matrix representation are simple Artinian, is called a *TSA ring*. Recall from 1.1.14 that if R is a right (or left) perfect ring, then $J(R) = P(R)$. Thus any prime right (or left) perfect ring is simple Artinian.

By Theorem 5.4.12, every quasi-Baer right (or left) perfect ring is a TSA ring. So Teply's result [391] given next follows from Theorem 5.4.12 since an orthogonally finite right Rickart ring is Baer by Theorem 3.1.25.

Corollary 5.4.17 *A right (or left) perfect right Rickart ring is a semiprimary TSA ring.*

For a π-regular Baer ring with only countably many idempotents, we obtain the following.

Corollary 5.4.18 *A π-regular Baer ring with only countably many idempotents is a semiprimary TSA ring.*

Proof Theorems 3.1.11, 3.1.26, and 5.4.12 yield the result. □

Corollary 5.4.19 *Assume that R is a PWP ring with* $\text{Tdim}(R) = n$. *Then the following are equivalent.*

(i) $\text{r.gl.dim}(R) < \infty$.
(ii) $\text{r.gl.dim}(R/P(R)) < \infty$.
(iii) $\text{r.gl.dim}(R_1 + \cdots + R_n) < \infty$, *where the R_i are the diagonal rings in the complete generalized triangular matrix representation of R.*

Proof (i)⇔(iii) is a direct consequence of Theorem 5.3.5. From Theorem 5.4.12, $R/P(R) \cong R_1 \oplus \cdots \oplus R_n$. Hence, (ii)⇔(iii) follows immediately. □

Theorem 5.4.20 *Let R be a right p.q.-Baer ring. Then* $\text{Tdim}(R) = n$ *if and only if R has exactly n minimal prime ideals.*

Proof Assume that $\text{Tdim}(R) = n$. By Proposition 5.4.5, R is a PWP ring. Thus from Theorem 5.4.12, R has exactly n minimal prime ideals.

Conversely, let R have exactly n minimal prime ideals. We proceed by induction on n. First, say $n = 1$. If $\text{Tdim}(R) \neq 1$, then R is not semicentral reduced. So there is $0 \neq b \in \mathbf{S}_\ell(R)$ with $b \neq 1$. Then bRb and $(1-b)R(1-b)$ each have at least one minimal prime ideal. Note that $\{b, 1-b\}$ is a set of left triangulating idempotents of R. Thus, by Proposition 5.4.11, R has at least two minimal prime ideals, a contradiction. Hence, $\text{Tdim}(R) = 1$.

Suppose that $n > 1$. If R is semicentral reduced, then R is prime by Proposition 3.2.25. So $n = 1$, a contradiction. Thus R is not semicentral reduced, hence there is $0 \neq d \in \mathbf{S}_\ell(R)$ and $d \neq 1$. By Theorem 3.2.34(i), both dRd and $(1-d)R(1-d)$ are right p.q.-Baer rings. We note that $\{d, 1-d\}$ is a set of left triangulating idempotents. From Proposition 5.4.11, there are some positive integers k_1 and k_2 such that dRd and $(1-d)R(1-d)$ have exactly k_1 and k_2 number of minimal prime ideals, respectively, where $k_1 + k_2 = n$.

By induction, $\text{Tdim}(dRd) + \text{Tdim}((1-d)R(1-d)) = k_1 + k_2 = n$. From Theorem 5.2.19, $\text{Tdim}(R) = n$. □

Corollary 5.4.21 *The PWP property is Morita invariant.*

Proof Assume that R and S are Morita equivalent rings. Suppose that R is a PWP ring and let $\text{Tdim}(R) = n$. By Theorem 5.4.20, R has exactly n minimal prime ideals. Since R is quasi-Baer, S is also quasi-Baer from Theorem 3.2.11. Now S has also exactly n minimal prime ideals because R and S are Morita equivalent (see [262, Proposition 18.44 and Corollary 18.45]). Thus $\text{Tdim}(S) = n$ by Theorem 5.4.20, so S is also a PWP ring. □

The next example illustrates that the right p.q.-Baer condition is not superfluous in Theorem 5.4.20.

Example 5.4.22 There exists a ring R such that:

(i) R has only two minimal prime ideals.
(ii) $\text{Tdim}(R) = 1$.

Indeed, we let $F\{X, Y\}$ be the free algebra over a field F, and we put $R = F\{X, Y\}/I$, where I is the ideal of $F\{X, Y\}$ generated by YX. Say $x = X + I$ and $y = Y + I$ in R. Then $R/RxR \cong F[y]$ and $R/RyR \cong F[x]$, so RxR and RyR are prime ideals of R. As $yx = 0$, we see that $(RyR)(RxR) = 0$. So, if P is a prime ideal, then either $RyR \subseteq P$ or $RxR \subseteq P$. Thus RxR and RyR are the only two minimal prime ideals of R. We can verify that all idempotents of R are only 0 and 1. In particular, R is semicentral reduced, so $\text{Tdim}(R) = 1$.

Let R be a quasi-Baer (resp., Baer) ring with $\text{Tdim}(R) < \infty$. Then $P(R)$ is nilpotent and $R/P(R)$ is a finite direct sum of prime (resp., Baer) rings from Theorem 5.4.12, so $R/P(R)$ is a quasi-Baer (resp., Baer) ring (cf. Example 3.2.42). There is a quasi-Baer ring R with $P(R)$ nilpotent, but $\text{Tdim}(R)$ is infinite. Let $R = T_2(\prod_{n=1}^{\infty} F_n)$, where F is a field, and $F_n = F, n = 1, 2, \ldots$. In this case, $P(R)^2 = 0$, but $\text{Tdim}(R) = \infty$.

An R-module M is said to satisfy the *restricted minimum condition* if, for every essential submodule N of M, the module M/N is Artinian.

Lemma 5.4.23 *Let R be a hereditary Noetherian ring. Then both R_R and ${}_RR$ satisfy the restricted minimum condition.*

Proof Assume that $J_R \leq^{\text{ess}} R_R$. Then J_R is finitely generated projective because R is right hereditary and right Noetherian. From Dual Basis lemma (see [262, Lemma 2.9]), there are $a_1, \dots, a_n \in J$ and $f_1, \dots, f_n \in \text{Hom}(J_R, R_R)$ such that $x = a_1 f_1(x) + \cdots + a_n f_n(x)$ for each $x \in J$. Because $Z(R_R) = 0$ from Proposition 3.1.18, $J_R \leq^{\text{den}} R_R$ by Proposition 1.3.14. Thus, it follows that $f_i \in Q(R)$ for $i = 1, \dots, n$, so $a_1 f_1 + \cdots + a_n f_n \in Q(R)$. We note that $a_1 f_1 + \cdots + a_n f_n = 1$ in $Q(R)$ as $a_1 f_1 + \cdots + a_n f_n$ is the identity map of J.

Put $D(J) = \text{Hom}\,(J_R, R_R)$. Then $Rf_1 + \cdots + Rf_n \subseteq D(J)$ because $D(J)$ is a left R-module. Let $q \in D(J)$. Then $qJ \subseteq R$ and so

$$q = q(a_1 f_1 + \cdots + a_n f_n) = qa_1 f_1 + \cdots + qa_n f_n \in Rf_1 + \cdots + Rf_n$$

since each $a_i \in J$. So $D(J) = Rf_1 + \cdots + Rf_n$.

Furthermore, $J = \{r \in R \mid D(J)r \subseteq R\}$. Indeed, first obviously we have that $J \subseteq \{r \in R \mid D(J)r \subseteq R\}$. Next, we take $r \in R$ such that $D(J)r \subseteq R$. Then

$$r = a_1 f_1 r + \cdots + a_n f_n r \in a_1 D(J)r + \cdots + a_n D(J)r \subseteq JR \subseteq J$$

since $1 = a_1 f_1 + \cdots + a_n f_n$ in $Q(R)$. So $J = \{r \in R \mid D(J)r \subseteq R\}$.

We show that R_R satisfies the restricted minimum condition. For this, we now let $I_1 \supseteq I_2 \supseteq \dots$ be a descending chain of right ideals of R all containing a fixed essential right ideal I of R. Then $D(I_1) \subseteq D(I_2) \subseteq \dots$ and all $D(I_i)$ are contained in the left R-module $D(I)$. By the preceding argument, $D(I)$ is finitely generated as a left R-module.

Since R is left Noetherian, $D(I)$ is Noetherian as a left R-module. So there exists a positive integer n such that $D(I_n) = D(I_{n+1}) = \dots$. Therefore, we have that $\{r \in R \mid D(I_n)r \subseteq R\} = \{r \in R \mid D(I_{n+1})r \subseteq R\} = \dots$. Hence $I_n = I_{n+1} = \dots$, so R_R satisfies the restricted minimum condition. Similarly, ${}_RR$ has the restricted minimum condition. □

As another application of Theorem 5.4.12, Chatters' decomposition theorem [117] for hereditary Noetherian rings is shown as follows.

Theorem 5.4.24 *If R is a hereditary Noetherian ring, then $R = A \oplus B$ (ring direct sum), where A is a finite direct sum of prime rings and B is an Artinian TSA ring.*

Proof Note that a hereditary Noetherian ring is Baer by Theorem 3.1.25. Thus R is a PWP ring. Therefore, $R = A \oplus B$ as in Theorem 5.4.12.

We claim that B is an Artinian TSA ring. For this, say $\{f_1, \dots, f_m\}$ is a complete set of left triangulating idempotents of B as in the proof of Theorem 5.4.12. We need to show that each B_i is simple Artinian. By Theorem 5.4.12, for given i, $1 \leq i \leq m$ there exists j, $1 \leq j \leq m$ such that either $B_{ij} \neq 0$ or $B_{ji} \neq 0$. We may assume that $B_{ij} \neq 0$ and $i < j$. Now $B_i = f_i B f_i$, $B_{ij} = f_i B f_j$, and $B_j = f_j B f_j$. Consider

$$S = (f_i + f_j)B(f_i + f_j) \cong \begin{bmatrix} B_i & B_{ij} \\ 0 & B_j \end{bmatrix}.$$

Then S is a hereditary Noetherian ring. Also $\{f_i, f_j\}$ is a complete set of left triangulating idempotents of S. Since B is Baer, so is S by Theorem 3.1.8. Therefore, S is a PWP ring.

We show that B_{ij} is a faithful left B_i-module. For this, let $f_ibf_i \in B_i$ with $b \in B$ such that $f_ibf_iB_{ij} = 0$. Since $f_iBf_j = B_{ij} \neq 0$, there exists $y \in B$ such that $f_iyf_j \neq 0$. Now $(f_ibf_i)(f_iBf_iyf_j) \subseteq (f_ibf_i)(f_iBf_j) = 0$, and so we have that $f_ibf_iBf_iyf_j = (f_ibf_i)(f_iBf_iyf_j) = 0$. Since $f_iyf_j \neq 0$, $f_ibf_i = 0$ from Theorem 5.4.1. Therefore, B_{ij} is a faithful left B_i-module. Similarly, B_{ij} is a faithful right B_j-module. Let

$$V_1 = \begin{bmatrix} 0 & B_{ij} \\ 0 & B_j \end{bmatrix} \text{ and } V_2 = \begin{bmatrix} B_i & B_{ij} \\ 0 & 0 \end{bmatrix}.$$

The ideal V_1 of S is right essential in S since B_{ij} is a faithful left B_i-module. Also the ideal V_2 of S is left essential in S. Since both S_S and ${}_SS$ satisfy the restricted minimum condition by Lemma 5.4.23, S/V_1 is a right Artinian S-module, while S/V_2 is a left Artinian S-module. Now to show that B_i is a right Artinian ring, we let $I_1 \supseteq I_2 \supseteq \ldots$ be a descending chain of right ideals of B_i. Put

$$K_\ell = \left\{ \begin{bmatrix} \alpha & 0 \\ 0 & 0 \end{bmatrix} + V_1 \in S/V_1 \mid \alpha \in I_\ell \right\}$$

for $\ell = 1, 2, \ldots$. Then we see that each K_ℓ is a right S-submodule of $(S/V_1)_S$ and $K_1 \supseteq K_2 \supseteq \ldots$. Since $(S/V_1)_S$ is Artinian, $K_t = K_{t+1} = \ldots$ for some positive integer t. So $I_t = I_{t+1} = \ldots$. Therefore, B_i is a right Artinian ring. Similarly, B_j is a left Artinian ring. Since B_i and B_j are prime rings by Theorem 5.4.12, B_i and B_j are simple Artinian rings.

The preceding argument is applied to show that all B_i are simple Artinian rings. Now $J(B) = \sum_{i \neq j} B_{ij}$ is nilpotent and $B/J(B) = B_1 \oplus \cdots \oplus B_m$. Hence, B is semiprimary Noetherian. So B is an Artinian TSA ring. □

To obtain a structure theorem for PWDs, we need the next lemma.

Lemma 5.4.25 *If R is a PWD and $0 \neq e \in \mathbf{S}_\ell(R) \cup \mathbf{S}_r(R)$, then the ring eRe is also a PWD.*

Proof Say $e \in \mathbf{S}_\ell(R)$. Let R be a PWD with respect to a complete set of primitive idempotents $\{e_1, \ldots, e_n\}$. Since $e \in \mathbf{S}_\ell(R)$, $e_ie = ee_ie$ is an idempotent for each i. As e_i is primitive and $e_ieR \subseteq e_iR$, either $e_ie = 0$ or $e_ieR = e_iR$. If necessary, rearrange $\{e_1, \ldots, e_n\}$ so that $J = \{1, \ldots, r\}$ is the set of all indices such that $e_ie \neq 0$ for all $i \in J$. Then $e = (e_1 + \cdots + e_n)e = e_1e + \cdots + e_re$ and

$$eR = e_1eR + \cdots + e_reR = e_1R + \cdots + e_rR.$$

Further, by Lemma 5.2.12, $\{ee_1e, \ldots, ee_re\}$ is a complete set of primitive idempotents in eRe.

Assume that $x \in (ee_ie)(eRe)(ee_je)$ and $y \in (ee_je)(eRe)(ee_ke)$ with $xy = 0$ for $1 \le i, j, k \le r$. Put $x = (ee_ie)(eae)(ee_je)$ and $y = (ee_je)(ebe)(ee_ke)$ with $a, b \in R$. Then $x = e_iae_je$ since $e \in \mathbf{S}_\ell(R)$. Similarly, $y = e_jbe_ke$. Thus $xy = e_iae_jee_jbe_ke = e_iae_jbe_ke = 0$. So $e_iae_je_jbe_keR = e_iae_je_jbe_kR = 0$ since $e_keR = e_kR$. Hence $(e_iae_j)(e_jbe_k) = 0$, so $e_iae_j = 0$ or $e_jbe_k = 0$ as R is a PWD. Thus $x = 0$ or $y = 0$. Therefore, eRe is a PWD with respect to the complete set of primitive idempotents $\{ee_1e, \ldots, ee_re\}$. Similarly, when $e \in \mathbf{S}_r(R)$, we see that eRe is a PWD. □

As yet another application of Theorem 5.4.12, we obtain the next theorem, due to Gordon and Small [187], which describes the structure of a PWD.

Theorem 5.4.26 *Assume that R is a PWD. Then*

$$R \cong \begin{bmatrix} R_1 & R_{12} & \cdots & R_{1n} \\ 0 & R_2 & \cdots & R_{2n} \\ \vdots & \vdots & \ddots & \vdots \\ 0 & 0 & \cdots & R_n \end{bmatrix},$$

where each R_i is a prime PWD and each R_{ij} is an (R_i, R_j)-bimodule. The integer n is unique and the ring R_i is unique up to isomorphism. Furthermore,

$$R_i \cong \begin{bmatrix} D_1 & \cdots & D_{1n_i} \\ \vdots & \ddots & \vdots \\ D_{n_i1} & \cdots & D_{n_i} \end{bmatrix},$$

where each D_i is a domain and each D_{jk} is isomorphic as a right D_k-module to a nonzero right ideal in D_k, and as a left D_j-module to a nonzero left ideal in D_j.

Proof Let R be a PWD. By Proposition 5.4.6, R is a PWP ring. The uniqueness of n and that of the ring R_i up to isomorphism follow from Theorem 5.2.8 or Theorem 5.4.12.

Say $\{b_1, \ldots, b_n\}$ is a complete set of left triangulating idempotents of R. By Theorem 5.4.12, each $R_i = b_iRb_i$ is a prime ring. From Lemma 5.4.25, $R_1 = b_1Rb_1$ and $(1 - b_1)R(1 - b_1)$ are PWDs.

We observe that $0 \ne b_2 \in \mathbf{S}_\ell((1 - b_1)R(1 - b_1))$. Thus, Lemma 5.4.25 yields that $R_2 = b_2Rb_2 = b_2(1 - b_1)R(1 - b_1)b_2$ is a PWD. By the same method, we see that each $R_i = b_iRb_i$ is a PWD. Hence, there exists a complete set of primitive idempotents $\{c_1, \ldots, c_{n_i}\}$ for R_i such that $c_jxc_kyc_q = 0$ implies that $c_jxc_k = 0$ or $c_kyc_q = 0$, for $x, y \in R_i$. Put $D_{jk} = c_jR_ic_k$ and $D_i = D_{ii}$. Then each D_i is a domain.

As R_i is a prime ring and $0 \ne c_k, 0 \ne c_j \in R_i$, it follows that $c_kR_ic_j \ne 0$. We let $0 \ne x \in c_kR_ic_j$. Then $c_jR_ic_k$ is isomorphic to a nonzero right ideal $xc_jR_ic_k$ of $c_kR_ic_k$ as a right $c_kR_ic_k$-module since R_i is a PWD with respect to the complete set of primitive idempotents $\{c_1, \ldots, c_{n_i}\}$. Similarly $c_jR_ic_k$ is isomorphic to a nonzero left ideal of $c_jR_ic_j$ as a left $c_jR_ic_j$-module. □

Exercise 5.4.27

1. Prove Propositions 5.4.9 and 5.4.11.
2. Show that if R is a PWD, then $\mathrm{Mat}_n(R)$ is a PWD for every positive integer n (see Example 5.4.10(i)).
3. ([66, Birkenmeier and Park]) Assume that R is a ring and X is a nonempty set of not necessarily commuting indeterminates. Show that R is quasi-Baer with $\mathrm{Tdim}(R) = n$ if and only if Γ is quasi-Baer with $\mathrm{Tdim}(\Gamma) = n$, where Γ is any of the following ring extensions of R.
(i) $R[X]$. (ii) $R[x, x^{-1}]$. (iii) $R[[x, x^{-1}]]$. (iv) $\mathrm{Mat}_k(R)$ for every positive integer k.
4. ([82, Birkenmeier, Kim, and Park]) Prove that the following conditions are equivalent for a ring R.
   (i) R is a TSA ring.
   (ii) R is a left perfect ring such that there exists a numbering of all the distinct prime ideals $P_1, P_2, \ldots, P_n$ of R such that $P_1 P_2 \cdots P_n = 0$.
   (iii) R is a left perfect ring such that some product of distinct prime ideals, without repetition, is zero.
5. Let R be a quasi-Baer ring such that $\mathbf{S}_\ell(R)$ is a countable set. Show that R is a PWP ring. Additionally, if R is also biregular, then R is a direct sum of simple rings (cf. Corollary 5.4.13(ii)).

5.5 A Sheaf Representation of Piecewise Prime Rings

After a brief discussion on certain ideals in a quasi-Baer ring, PWP rings with a sheaf representation will be studied in this section. Quasi-Baer rings with a nontrivial subdirect product representation will also be discussed.

The set of all prime ideals and the set of all minimal prime ideals of a ring R is denoted by *Spec* (R) and *MinSpec*(R), respectively. For a subset X of R, let $\mathrm{supp}(X) = \{P \in \mathrm{Spec}(R) \mid X \not\subseteq P\}$, which is called the support of X. In case, $X = \{s\}$, we write $\mathrm{supp}(s)$.

For any $P \in \mathrm{Spec}(R)$, there is $s \in R \setminus P$ and so $P \in \mathrm{supp}(s)$. Thus the family $\{\mathrm{supp}(s) \mid s \in R\}$ covers $\mathrm{Spec}(R)$. Also for $P \in \mathrm{supp}(x) \cap \mathrm{supp}(y)$, $d = xcy \notin P$ for some $c \in R$. So $P \in \mathrm{supp}(d) \subseteq \mathrm{supp}(x) \cap \mathrm{supp}(y)$. Therefore, $\{\mathrm{supp}(s) \mid s \in R\}$ forms a base (for open sets) on $\mathrm{Spec}(R)$. This induced topology on $\mathrm{Spec}(R)$ is called the hull-kernel topology on $\mathrm{Spec}(R)$.

For $P \in \mathrm{Spec}(R)$, let $O(P) = \{a \in R \mid aRs = 0 \text{ for some } s \in R \setminus P\}$. Then $O(P)$ is an ideal of R, $O(P) = \sum_{s \in R \setminus P} \ell_R(Rs)$, and $O(P) \subseteq P$. We let

$$\mathfrak{K}(R) = \bigcup_{P \in \mathrm{Spec}(R)} R/O(P)$$

be the *disjoint* union of the rings $R/O(P)$, where P ranges through $\mathrm{Spec}(R)$.

For $a \in R$, define $\widehat{a} : \mathrm{Spec}(R) \to \mathfrak{K}(R)$ by $\widehat{a}(P) = a + O(P)$. Then it can be verified that $\mathfrak{K}(R)$ is a sheaf of rings over $\mathrm{Spec}(R)$ with the topology on $\mathfrak{K}(R)$ generated by $\{\widehat{a}(\mathrm{supp}(s)) \mid a, s \in R\}$. By a *sheaf representation* of a ring R, we mean a sheaf representation whose base space is Spec(R) and whose stalks are the $R/O(P)$, where $P \in \text{Spec}(R)$. Let $\Gamma(\text{Spec}(R), \mathfrak{K}(R))$ be the set of all global sections. We remark that $\Gamma(\text{Spec}(R), \mathfrak{K}(R))$ becomes a ring (see [345, 3.1], [209], and [369] for more details).

It is well-known that $\widehat{a}$ is a global section for $a \in R$. Next, for $a, b \in R$ and $P \in \text{Spec}(R)$, $(\widehat{a}+\widehat{b})(P) = a + b + O(P)$ and $(\widehat{a}\widehat{b})(P) = ab + O(P)$. Therefore we see that the map

$$\theta : R \to \Gamma(\text{Spec}(R), \mathfrak{K}(R))$$

defined by $\theta(a) = \widehat{a}$ is a ring homomorphism, which is called the Gelfand homomorphism. Furthermore, $\text{Ker}(\theta) = \bigcap_{P\in\text{Spec}(R)} O(P)$, which is 0 (see Proposition 5.5.7). Thus θ is a monomorphism.

We discuss some relevant properties of $O(P)$ and $R/O(P)$ for the previously mentioned sheaf representation of PWP rings.

Proposition 5.5.1 *Let R be a quasi-Baer ring and P a prime ideal of R. Then $O(P) = \sum Rf$, where the sum is taken for all $f \in \mathbf{S}_r(R) \cap P$.*

Proof Note that $O(P) = \sum_{s\in R\setminus P} \ell_R(Rs)$. As R is quasi-Baer, $\ell_R(Rs) = Rf$ with $f \in \mathbf{S}_r(R)$. Then $f \in P$ because $fRs = 0$ and $s \notin P$. Next let $f \in \mathbf{S}_r(R) \cap P$. Then $f \in O(P)$ since $fR(1-f) = 0$ (Proposition 1.2.2) and $1 - f \in R \setminus P$. Thus, we get the desired result. □

Corollary 5.5.2 *Let R be a quasi-Baer ring. If P and Q are prime ideals such that $P \subseteq Q$, then $O(P) = O(Q)$.*

Proof From the definition, we see that $O(Q) \subseteq O(P)$. Proposition 5.5.1 yields that $O(P) \subseteq O(Q)$, so $O(P) = O(Q)$. □

We remark that Proposition 5.5.1 and Corollary 5.5.2 hold true when R is a left p.q.-Baer ring.

Proposition 5.5.3 *Assume that R is a PWP ring and P is a prime ideal. Then $O(P) = Re$ for some $e \in \mathbf{S}_r(R)$.*

Proof As R has a complete set of triangulating idempotents, $\{Rb \mid b \in \mathbf{S}_r(R)\}$ is a finite set by the left-sided version of Theorem 5.2.5. From Proposition 5.5.1, $O(P) = \sum Rf$, where the sum is taken for all $f \in \mathbf{S}_r(R) \cap P$. Therefore, $O(P) = Rf_1 + \cdots + Rf_k$ with $f_i \in \mathbf{S}_r(R)$. By Proposition 1.2.4(ii), $O(P) = Re$ for some $e \in \mathbf{S}_r(R)$. □

Let R be a ring and S be a multiplicatively closed subset of R (i.e., $1 \in S$ and $s, t \in S$ implies $st \in S$). A ring RS^{-1} is called a *right ring of fractions* of R with respect to S together with a ring homomorphism $\phi : R \to RS^{-1}$ if the following are satisfied:

(i) $\phi(s)$ is invertible for every $s \in S$.

(ii) Each element in RS^{-1} has the form $\phi(a)\phi(s)^{-1}$ with $a \in R$ and $s \in S$.
(iii) $\phi(a) = 0$ with $a \in R$ if and only if $as = 0$ for some $s \in S$.

Proposition 5.5.4 *Let R be a ring and S a multiplicatively closed subset of R. Then RS^{-1} exists if and only if S satisfies:*
S1. *If $s \in S$ and $a \in R$, then there exist $t \in S$ and $b \in R$ with $sb = at$.*
S2. *If $sa = 0$ with $a \in R$ and $s \in S$, then $at = 0$ for some $t \in S$.*

Proof See [382, Proposition 1.4, p. 51] for the proof. □

When RS^{-1} exists, it has the form $RS^{-1} = (R \times S)/\sim$, where $\sim$ is the equivalence relation defined as $(a, s) \sim (b, t)$ if there exist $c, d \in R$ such that $sc = td \in S$ and $ac = bd$. A multiplicatively closed subset with S1 and S2 is called a *right denominator set*. In particular, if R is a right Ore ring and S is the set of all nonzero-divisors in R, then S is a right denominator set. Thus RS^{-1} exists by Proposition 5.5.4 and $Q^r_{c\ell}(R) = RS^{-1}$ (see 1.1.17).

Proposition 5.5.5 *Assume that P is a prime ideal of a ring R and let $S_P = \{e \in \mathbf{S}_\ell(R) \mid e \notin P\}$. Then RS_P^{-1} exists.*

Proof Obviously $1 \in S_P$. To see that S_P is a multiplicatively closed subset, let $e, f \in S_P$. Then $ef \in \mathbf{S}_\ell(R)$ by Proposition 1.2.4(i). If $ef \in P$, then $efRf \subseteq P$. Therefore $eRf = efRf \subseteq P$, a contradiction. Thus, $ef \notin P$. So $ef \in S_P$ and hence S_P is a multiplicatively closed subset of R.

For $e \in S_P$ and $a \in R$, we have that $e(ae) = ae$. So the condition S1 is satisfied. Next for S2, take $e \in S_P$ and $a \in R$ such that $ea = 0$. Then

$$ae = (1 - e)ae = (1 - e)eae = 0,$$

so the condition S2 is satisfied. Hence S_P is a denominator set. Thus, RS_P^{-1} exists from Proposition 5.5.4. □

When R is a quasi-Baer ring, we obtain the next result for stalks $R/O(P)$.

Theorem 5.5.6 *Assume that R is a quasi-Baer ring and P is a prime ideal of R. Then $RS_P^{-1} \cong R/O(P)$.*

Proof First we show that $O(P) = \{a \in R \mid ae = 0$ for some $e \in S_P\}$. Indeed, if $a \in R$ such that $ae = 0$ with $e \in S_P$, then $aRe = aeRe = 0$ and so $a \in O(P)$. Thus $I := \{a \in R \mid ae = 0$ for some $e \in S_P\} \subseteq O(P)$. To see that $O(P) \subseteq I$, first we prove that $I \trianglelefteq R$. For this, say $a_1, a_2 \in I$ with $a_1e_1 = 0$ and $a_2e_2 = 0$ for some $e_1, e_2 \in S_P$. Then $(a_1 + a_2)e_1e_2 = a_2e_1e_2 = a_2e_2e_1e_2 = 0$. By Proposition 5.5.5, S_P is a multiplicatively closed set, hence $e_1e_2 \in S_P$. So $a_1 + a_2 \in I$. Let $a \in I$ and $r \in R$. Clearly $ra \in I$. Say $e \in S_P$ such that $ae = 0$. Then $are = aere = 0$, so $ar \in I$. Therefore $I \trianglelefteq R$.

Now say $f \in \mathbf{S}_r(R) \cap P$. Then $1 - f \notin P$ and $1 - f \in \mathbf{S}_\ell(R)$. Hence $1 - f \in S_P$, so $f \in I$. By Proposition 5.5.1, $O(P) \subseteq I$. Thus $O(P) = I$.

From Proposition 5.5.5, RS_P^{-1} exists and there is a ring homomorphism ϕ from R to RS_P^{-1}, where $RS_P^{-1} = \{\phi(a)\phi(e)^{-1} \mid a \in R \text{ and } e \in S_P\}$. Now we observe that $O(P) = I$, so $\text{Ker}(\phi) = O(P)$.

Further, for each $e \in S_P$, note that $\phi(e)^2 = \phi(e) \in RS_P^{-1}$, which is invertible. Thus $\phi(e) = 1$ for every $e \in S_P$. So $RS_P^{-1} = \phi(R)$ and $\text{Ker}(\phi) = O(P)$. Hence we get that $RS_P^{-1} \cong R/O(P)$. □

Recall that a ring R is a subdirect product of rings S_i, $i \in \Lambda$, if $S_i \cong R/K_i$, where $K_i \trianglelefteq R$ and $\bigcap_{i \in \Lambda} K_i = 0$. A subdirect product is *nontrivial* if $K_i \neq 0$ for all $i \in \Lambda$. Otherwise, it is *trivial*.

Proposition 5.5.7 *Let R be a ring. Then $\bigcap_{P \in \text{Spec}(R)} O(P) = 0$. Thus R has a subdirect product representation of $\{R/O(P) \mid P \in \text{Spec}(R)\}$.*

Proof Assume that $\cap_{P \in \text{Spec}(R)} O(P) \neq 0$. Let $0 \neq a \in \cap_{P \in \text{Spec}(R)} O(P)$. Then $r_R(aR)$ is a proper ideal of R. Let P_0 be a prime ideal such that $r_R(aR) \subseteq P_0$. Because $a \in \cap_{P \in \text{Spec}(R)} O(P) \subseteq O(P_0)$, $aRs = 0$ with $s \in R \setminus P_0$. Therefore $s \in r_R(aR) \subseteq P_0$, a contradiction. So $\cap_{P \in \text{Spec}(R)} O(P) = 0$. □

The following example shows that the subdirect product representation in Proposition 5.5.7 may be trivial.

Example 5.5.8 For a field F, let $R = T_2(F)$. Then R is quasi-Baer. Let $e_{ij} \in T_2(F)$ be the matrix with 1 in the (i, j)-position and 0 elsewhere. Put $P = Fe_{11} + Fe_{12}$ and $Q = Fe_{12} + Fe_{22}$. Then we see that R has only two prime ideals which are P and Q (see Proposition 5.4.11). Hence, $O(P) = 0$ and $O(Q) = Q$ by using Proposition 5.5.1.

Next, we consider the subdirect product representation of Proposition 5.5.7 for quasi-Baer rings. Corollary 5.5.2 suggests that we may be able to improve the subdirect product representation by reducing the number of components through using only the minimal prime ideals. So it is natural to consider suitable conditions under which $\cap_{P \in \text{MinSpec}(R)} O(P) = 0$. The next example illustrates that there is a ring R such that $\cap_{P \in \text{MinSpec}(R)} O(P) \neq 0$.

Example 5.5.9 Assume that R is the Dorroh extension of $S = \begin{bmatrix} \mathbb{Z}_2 & \mathbb{Z}_2 \\ 0 & 0 \end{bmatrix}$ by $\mathbb{Z}$ (i.e., the ring formed from $S \times \mathbb{Z}$ with componentwise addition and multiplication given by $(x, k)(y, m) = (xy + mx + ky, km)$). Let e_{ij} be the matrix in S with 1 in the (i, j)-position and 0 elsewhere.

Put $e = (e_{11}, 0) \in R$. Then $e \in \mathbf{S}_\ell(R)$, so $(1 - e)Re = 0$ by Proposition 1.2.2. Also $eRe = (\mathbb{Z}_2 e_{11}, 0)$, $(1_R - e)R(1_R - e) = \{(me_{11}, m) \mid m \in \mathbb{Z}\}$, and $P(R) = eR(1 - e) = (\mathbb{Z}_2 e_{12}, 0)$ (note that $1 := 1_R = (0, 1) \in R$). Since

$$R \cong \begin{bmatrix} eRe & eR(1-e) \\ 0 & (1-e)R(1-e) \end{bmatrix},$$

all the minimal prime ideals of R are $P_1 := Q_1 + eR(1-e) + (1-e)R(1-e)$ and $P_2 := eRe + eR(1-e) + Q_2$, where Q_1 and Q_2 are minimal prime ideals of eRe and $(1-e)R(1-e)$, respectively by Proposition 5.4.11.

As $eRe \cong \mathbb{Z}_2$ and $(1-e)R(1-e) \cong \mathbb{Z}$, $Q_1 = 0$ and $Q_2 = 0$. So

$$P_1 = \{(me_{11} + ne_{12}, m) \mid m, n \in \mathbb{Z}\} \text{ and } P_2 = (\mathbb{Z}_2 e_{11} + \mathbb{Z}_2 e_{12}, 0).$$

Take $\alpha = (e_{12}, 0) \in R$. Then $\alpha R = (\mathbb{Z}_2 e_{12}, 0)$. Now say $s_1 = e = (e_{11}, 0)$ and $s_2 = (0, 2)$. Then $\alpha R s_1 = 0$ with $s_1 \in R \setminus P_1$, and $\alpha R s_2 = 0$ with $s_2 \in R \setminus P_2$. Hence, $0 \neq \alpha \in O(P_1) \cap O(P_2) = \cap_{P \in \text{MinSpec}(R)} O(P)$.

In spite of Example 5.5.9, we have the following.

Lemma 5.5.10 *If R is a quasi-Baer ring, then $\bigcap_{P \in \text{MinSpec}(R)} O(P) = 0$.*

Proof For a minimal prime ideal P of R, $O(P) = O(Q)$ for every prime ideal Q of R containing P by Corollary 5.5.2. Thus, $\cap_{P \in \text{MinSpec}(R)} O(P) = 0$ by Proposition 5.5.7. □

Theorem 5.5.11 *Let R be a semiprime ring, which is not prime. If R is quasi-Baer, then R has a nontrivial representation as a subdirect product of $R/O(P)$, where P ranges through all minimal prime ideals.*

Proof As R is a nonprime quasi-Baer ring, R is not semicentral reduced by Proposition 3.2.5. So there is $e \in \mathbf{S}_\ell(R)$ with $e \neq 0$ and $e \neq 1$. By Proposition 1.2.6(ii), $e \in \mathcal{B}(R)$ since R is semiprime. Suppose that there exists a minimal prime ideal P with $O(P) = 0$. Since R is not prime, $P \neq 0$. As $(1-e)Re = 0$, $e \in P$ or $1 - e \in P$. If $e \in P$, then $1 - e \notin P$ and $eR(1-e) = 0$, so $e \in O(P)$, a contradiction. Similarly, if $1 - e \in P$, then we get a contradiction. Thus $O(P) \neq 0$ for every minimal prime ideal P of R. Lemma 5.5.10 yields the desired result. □

Corollary 5.5.12 *Let R be a semiprime ring, which is not prime. If R is quasi-Baer, then R has a nontrivial representation as a subdirect product of RS_P^{-1}, where P ranges through all minimal prime ideals.*

Proof It is a direct consequence of Theorems 5.5.6 and 5.5.11. □

Definition 5.5.13 For a ring R, a left (resp., right) semicentral idempotent $e\,(\neq 1)$ is called *maximal* if $eR \subseteq fR$ (resp., $Re \subseteq Rf$) with $f \in \mathbf{S}_\ell(R)$ (resp., $f \in \mathbf{S}_r(R)$), then $fR = eR$ or $fR = R$ (resp., $Rf = Re$ or $Rf = R$).

Hofmann showed in [209, Theorem 1.17] that $\theta : R \cong \Gamma(\text{Spec}(R), \mathfrak{K}(R))$ when R is a semiprime ring. This result motivates the following question: *If a quasi-Baer ring R has such the sheaf representation, then is R semiprime?* Theorem 5.5.14 provides an affirmative partial answer to the question by giving a characterization of a certain class of quasi-Baer rings having such the sheaf representation.

Theorem 5.5.14 *The following are equivalent for a ring R.*

(i) *R is a PWP ring and $\theta : R \cong \Gamma(\mathrm{Spec}(R), \mathfrak{K}(R))$.*
(ii) *R is a finite direct sum of prime rings.*
(iii) *R is a semiprime PWP ring.*

Proof (i)⇒(ii) Let $\mathrm{Tdim}(R) = n$. If $n = 1$, then R is semicentral reduced, so R is prime by Proposition 3.2.5, and hence we are done. So suppose that $n \geq 2$. By Theorem 5.4.20, there are exactly n minimal prime ideals of R, say $P_1, P_2, \ldots, P_n$ and from Theorem 5.4.12 these are comaximal (i.e., $P_i + P_j = R$ for $i \neq j$).

For each $i = 1, 2, \ldots, n$, we let $\mathfrak{A}_i = \{P \in \mathrm{Spec}(R) \mid P_i \subseteq P\}$. Then it follows that $\mathrm{Spec}(R) = \mathfrak{A}_1 \cup \mathfrak{A}_2 \cup \cdots \cup \mathfrak{A}_n$ since $\{P_1, P_2, \ldots, P_n\}$ is the set of all minimal prime ideals. Also because $P_i + P_j = R$ for $i \neq j$, $\mathfrak{A}_i \cap \mathfrak{A}_j = \emptyset$ for $i \neq j$. By the hull-kernel topology on $\mathrm{Spec}(R)$, each $\mathfrak{A}_i$ is a closed subset of $\mathrm{Spec}(R)$. Hence for $i = 1, 2, \ldots, n$, $\mathfrak{A}_1 \cup \cdots \cup \mathfrak{A}_{i-1} \cup \mathfrak{A}_{i+1} \cup \cdots \cup \mathfrak{A}_n$ is closed, and so each $\mathfrak{A}_i$ is open.

Define $f : \mathrm{Spec}(R) \to \mathfrak{K}(R)$ such that $f(P) = 1 + O(P)$ for $P \in \mathfrak{A}_1$, and $f(P) = 0 + O(P)$ for $P \in \mathfrak{A}_k$ with $k \neq 1$. We claim that f is a continuous function. For this, first take $P \in \mathfrak{A}_1$. Then $f(P) = 1 + O(P) \in \mathfrak{K}(R)$. Consider a basic neighborhood $\widehat{r}(\mathrm{supp}(s))$ (with $r, s \in R$) containing $f(P) = 1 + O(P)$ in $\mathfrak{K}(R)$. Then $\mathrm{supp}(s) \cap \mathfrak{A}_1$ is an open subset of $\mathrm{Spec}(R)$ with $P \in \mathrm{supp}(s) \cap \mathfrak{A}_1$.

For $M \in \mathrm{supp}(s) \cap \mathfrak{A}_1$, $f(M) = 1 + O(M) \in R/O(M)$. Hence we obtain that $1 + O(P) = r + O(P)$ and so $r - 1 \in O(P)$ as $1 + O(P) \in \widehat{r}(\mathrm{supp}(s))$. Now we note that $O(P_1) = O(P) = O(M)$ from Corollary 5.5.2, hence $r - 1 \in O(M)$. Thus,

$$f(M) = 1 + O(M) = r + O(M) \in \widehat{r}(\mathrm{supp}(s)).$$

So $f(\mathrm{supp}(s) \cap \mathfrak{A}_1) \subseteq \widehat{r}(\mathrm{supp}(s))$.

For $P \in \mathfrak{A}_k$ with $k \neq 1$, assume that $f(P) = 0 + O(P) \in \widehat{r}(\mathrm{supp}(s))$ for some $r, s \in R$. Then we also see that $f(\mathrm{supp}(s) \cap \mathfrak{A}_k) \subseteq \widehat{r}(\mathrm{supp}(s))$. Therefore, f is a continuous function.

Next, consider $\pi : \mathfrak{K}(R) \to \mathrm{Spec}(R)$ defined by $\pi(r + O(P)) = P$ for $r \in R$ and $P \in \mathrm{Spec}(R)$. Then we see that $\pi(f(P)) = P$ for all $P \in \mathrm{Spec}(R)$. Thus, it follows that $f \in \Gamma(\mathrm{Spec}(R), \mathfrak{K}(R))$ as f is a continuous function.

Since $R \cong \Gamma(\mathrm{Spec}(R), \mathfrak{K}(R))$, there exists $a \in R$ with $f = \widehat{a}$. Therefore

$$a + O(P_1) = 1 + O(P_1) \text{ and } a + O(P_k) = 0 + O(P_k) \text{ for each } k \neq 1.$$

So $1 - a \in O(P_1)$ and $a \in O(P_k)$ for each $k \neq 1$. Thus $O(P_1) + O(P_k) = R$ for each $k \neq 1$. Similarly, $O(P_i) + O(P_j) = R$ for $i \neq j$, $1 \leq i, j \leq n$. By Lemma 5.5.10, we obtain that $O(P_1) \cap \cdots \cap O(P_n) = 0$, hence

$$R \cong R/O(P_1) \oplus \cdots \oplus R/O(P_n)$$

by Chinese Remainder Theorem. From Proposition 5.5.3, $O(P_1) = Re$ with $e \in \mathbf{S}_r(R)$, so $eR(1 - e) = 0$. Hence $R/O(P_1) \cong (1 - e)R(1 - e)$.

Our claim is that $(1-e)R(1-e)$ is semicentral reduced. For this, assume on the contrary that $(1-e)R(1-e)$ is not semicentral reduced. By Theorem 3.2.10, $(1-e)R(1-e)$ is a quasi-Baer ring. Hence, $(1-e)R(1-e)$ is a PWP ring by Theorem 5.2.19.

From Theorem 5.2.5, there is a maximal right semicentral idempotent in the ring $(1-e)R(1-e)$, say $(1-e)b(1-e)$. Because $(1-e)R(1-e)$ is not semicentral reduced,

$$[(1-e)R(1-e)](1-e)b(1-e)$$

is a nonzero proper ideal of $(1-e)R(1-e)$. Since $e \in \mathbf{S}_r(R)$,

$$e+(1-e)b(1-e) \in \mathbf{S}_r(R).$$

Put $g = e+(1-e)b(1-e)$. We show that g is a maximal right semicentral idempotent of R. Take $\alpha \in \mathbf{S}_r(R)$ such that $Rg \subseteq R\alpha$ and $\alpha \neq 1$. Because

$$R = \begin{bmatrix} eRe & 0 \\ (1-e)Re & (1-e)R(1-e) \end{bmatrix}$$

and $g = e+(1-e)b(1-e)$, we have that $\alpha = e+k+h$ with $k \in (1-e)Re$ and $h \in \mathbf{S}_r((1-e)R(1-e))$.

Since $Rg \subseteq R\alpha$, $(1-e)R(1-e)(1-e)b(1-e) \subseteq (1-e)R(1-e)h$. From the maximality of $(1-e)b(1-e)$ and $h \neq 1-e$ (because $\alpha \neq 1$), we have that $(1-e)R(1-e)(1-e)b(1-e) = (1-e)R(1-e)h$, and thus $h(1-e)b(1-e) = h$. Further, $ke = k$ since $k \in (1-e)Re$. Hence,

$$\alpha g = \begin{bmatrix} e & 0 \\ k & h \end{bmatrix} \begin{bmatrix} e & 0 \\ 0 & (1-e)b(1-e) \end{bmatrix} = \begin{bmatrix} e & 0 \\ k & h \end{bmatrix} = \alpha.$$

Thus, $R\alpha \subseteq Rg$. Therefore, g is a maximal right semicentral idempotent of R.

Next, note that $\{1, 1-g\}$ forms a multiplicatively closed subset of R. By Zorn's lemma, there is an ideal Q of R maximal with respect to being disjoint with $\{1, 1-g\}$. Then Q is a prime ideal of R. Since $gR(1-g) = 0$ and $1-g \notin Q$, it follows $g \in O(Q)$. Also, since g is a maximal right semicentral idempotent of R and $g \in O(Q)$, $O(Q) = Rg$ from Proposition 5.5.3. We observe that $O(P_1) = Re \neq Rg$ as $(1-e)b(1-e) \neq 0$. Hence, $Q \notin \mathfrak{A}_1$ by Corollary 5.5.2. So $Q \in \mathfrak{A}_k$ for some $k \neq 1$. So $O(Q) = O(P_k)$ from Corollary 5.5.2. Now $R = O(P_1)+O(P_k) = Re+Rg$, a contradiction since $(1-e)b(1-e) \neq 1-e$. Thus, the ring $(1-e)R(1-e)$ is a semicentral reduced quasi-Baer ring. So $(1-e)R(1-e)$ is a prime ring by Proposition 3.2.5, thus $R/O(P_1)$ is a prime ring because $R/O(P_1) \cong (1-e)R(1-e)$. Similarly, $R/O(P_i)$ is a prime ring for each $i = 2, \dots, n$. Therefore $R \cong R/O(P_1) \oplus \cdots \oplus R/O(P_n)$, which is a finite direct sum of prime rings. Further, note that $O(P_i) = P_i$ for each $i = 1, \dots, n$, so $R \cong R/P_1 \oplus \cdots \oplus R/P_n$.

(ii)⇒(iii) It is evident.

(iii)⇒(i) The proof follows from [209, Theorem 1.17]. □

We obtain the next corollary from Proposition 5.4.6, Lemma 5.4.25, and Theorem 5.5.14.

Corollary 5.5.15 *The following are equivalent.*

(i) *R is a PWD with $\theta : R \cong \Gamma(\mathrm{Spec}(R), \mathfrak{K}(R))$.*
(ii) *R is a finite direct sum of prime PWDs.*
(iii) *R is a semiprime PWD.*

Exercise 5.5.16

1. ([74, Birkenmeier, Kim, and Park]) Assume that R is a (quasi-)Baer ring with $\mathrm{Tdim}(R) < \infty$ and P is a prime ideal of R. Prove that $R/O(P)$ is a (quasi-)Baer ring.
2. ([74, Birkenmeier, Kim, and Park]) Let R be a Baer ring and P be a prime ideal of R. Show that $R/O(P)$ is a right Rickart ring.
3. ([74, Birkenmeier, Kim, and Park]) Assume that R is a quasi-Baer ring and P is a prime ideal of R. Prove that $\mathrm{r.gl.dim}(R/O(P)) \le \mathrm{r.gl.dim}(R)$.

5.6 Triangular Matrix Ring Extensions

Our focus in this section is the study of the Baer, the quasi-Baer, and the (strongly) FI-extending properties of upper triangular and generalized triangular matrix ring extensions. The study of full matrix ring extensions will be considered in Chap. 6.

Theorem 5.6.1 *Let R be a ring. Then the following are equivalent.*

(i) *R is regular and right self-injective.*
(ii) *$T_n(R)$ is right nonsingular right extending for every positive integer n.*
(iii) *$T_k(R)$ is right nonsingular right extending for some integer $k > 1$.*
(iv) *$T_2(R)$ is right nonsingular right extending.*

Proof (i)⇒(ii) The proof follows from [3, Corollary 2.8(3)] and [1, Proposition 1.8(ii)].

(ii)⇒(iii) It is evident.

(iii)⇒(i) [3, Corollary 2.8(2) and Proposition 1.6(2)] yield this implication.

(i)⇔(iv) This equivalence follows from [393, Theorem 3.4] (see also Theorem 5.6.9). □

Theorem 5.6.2 *Let R be an orthogonally finite Abelian ring. Then the following are equivalent.*

(i) *R is a direct sum of division rings.*
(ii) *$T_n(R)$ is a Baer (resp., right Rickart) ring for every positive integer n.*
(iii) *$T_k(R)$ is a Baer (resp., right Rickart) ring for some integer $k > 1$.*

(iv) *$T_2(R)$ is a Baer (resp., right Rickart) ring.*

Proof (i)⇒(ii) The proof follows from Theorems 5.6.1, 3.3.1, and 3.1.25.

(ii)⇒(iii) It is evident.

(iii)⇒(iv) The proof follows from Theorems 3.1.8 and 3.1.22(i).

(iv)⇒(i) Let $T_2(R)$ be Baer (resp., right Rickart). By Proposition 1.2.15, R has a complete set of primitive idempotents. As R is Abelian, $R = \oplus_{i=1}^{m} R_i$ (ring direct sum), for some positive integer m, where each R_i is indecomposable as a ring. Then each $T_2(R_i)$ is a Baer (resp., right Rickart) ring by Proposition 3.1.5(i) (resp., Proposition 3.1.21). From Theorem 3.1.8 (resp., Theorem 3.1.22(i)), each R_i is a Baer (resp., right Rickart) ring. If R_i is Baer or right Rickart, R_i is a domain (see Example 3.1.4(ii)). From [246, Exercise 2, p. 16] or [262, Exercise 25, p. 271], each R_i is a division ring. □

Notation 5.6.3 Let S and R be rings, and let ${}_SM_R$ be an (S, R)-bimodule. For the remainder of this section, we let

$$T = \begin{bmatrix} S & M \\ 0 & R \end{bmatrix}$$

denote a generalized triangular matrix ring.

Lemma 5.6.4 *Let T be the ring as in Notation 5.6.3. Say*

$$e = \begin{bmatrix} e_1 & k \\ 0 & e_2 \end{bmatrix} \in \mathbf{S}_\ell(T) \text{ and } f = \begin{bmatrix} e_1 & 0 \\ 0 & e_2 \end{bmatrix}.$$

Then we have the following.

(i) *$e_1 \in \mathbf{S}_\ell(S)$, $e_2 \in \mathbf{S}_\ell(R)$, and $f \in \mathbf{S}_\ell(T)$*

(ii) *$eT = fT$.*

Proof (i) It can be easily checked that $e_1 \in \mathbf{S}_\ell(S)$ and $e_2 \in \mathbf{S}_\ell(R)$. Also we see that $e_1me_2 = me_2$ for all $m \in M$. Thus, $f \in \mathbf{S}_\ell(T)$.

(ii) Since $e_1me_2 = me_2$ for all $m \in M$, in particular $e_1ke_2 = ke_2$ and so $f = e\begin{bmatrix} e_1 & -ke_2 \\ 0 & e_2 \end{bmatrix}$. Hence $fT \subseteq eT$. As $e \in \mathbf{S}_\ell(T)$, $\begin{bmatrix} 1 & 0 \\ 0 & 0 \end{bmatrix} e = e \begin{bmatrix} 1 & 0 \\ 0 & 0 \end{bmatrix} e$, so $k = e_1k$. Thus, $e = f\begin{bmatrix} 1 & k \\ 0 & 1 \end{bmatrix} \in fT$. Therefore $eT \subseteq fT$, and so $eT = fT$. □

Next, we characterize the quasi-Baer property for the ring T.

Theorem 5.6.5 *Let T be the ring as in Notation 5.6.3. Then the following are equivalent.*

(i) *T is a quasi-Baer ring.*

(ii) (1) *R and S are quasi-Baer rings.*

(2) $r_M(I) = r_S(I)M$ *for all* $I \trianglelefteq S$.
(3) *For any* ${}_SN_R \leq {}_SM_R$, $r_R(N) = gR$ *for some* $g^2 = g \in R$.

Proof (i)⇒(ii) By Theorem 3.2.10, R and S are quasi-Baer. Let $I \trianglelefteq S$. Then $A := \begin{bmatrix} I & M \\ 0 & 0 \end{bmatrix} \trianglelefteq T$. Hence, $r_T(A) = eT$ for some $e^2 = e \in T$. Because $A \trianglelefteq T$, $e \in \mathbf{S}_\ell(T)$ by Proposition 1.2.2. Put $e = \begin{bmatrix} e_1 & k \\ 0 & e_2 \end{bmatrix}$ and $f = \begin{bmatrix} e_1 & 0 \\ 0 & e_2 \end{bmatrix}$. From Lemma 5.6.4, $e_1 \in \mathbf{S}_\ell(S)$, $e_2 \in \mathbf{S}_\ell(R)$, $f \in \mathbf{S}_\ell(T)$, and $eT = fT$. Thus it is routine to check that $r_S(I) = e_1 S$ and $r_M(I) = e_1 M = e_1 SM = r_S(I)M$.

Next, let ${}_SN_R \leq {}_SM_R$. Then $K := \begin{bmatrix} 0 & N \\ 0 & 0 \end{bmatrix} \trianglelefteq T$. So $r_T(K) = hT$ for some $h \in \mathbf{S}_\ell(T)$. Say $h = \begin{bmatrix} g_1 & m \\ 0 & g_2 \end{bmatrix}$. Then $r_R(N) = g_2 R$, where $g_2 \in \mathbf{S}_\ell(R)$. Take $g = g_2$. Then $r_R(N) = gR$ and $g^2 = g \in R$.

(ii)⇒(i) Let $K \trianglelefteq T$. Then we see that $K = \begin{bmatrix} I & N \\ 0 & J \end{bmatrix}$, where $I \trianglelefteq S$, $J \trianglelefteq R$, ${}_SN_R \leq {}_SM_R$, and $IM + MJ \subseteq N$. Because S and R are quasi-Baer, there are $e_1 \in \mathbf{S}_\ell(S)$, $f \in \mathbf{S}_\ell(R)$ satisfying $r_S(I) = e_1 S$ and $r_R(J) = fR$. By assumption, $r_M(I) = r_S(I)M = e_1 M$ and $r_R(N) = gR$ for some $g^2 = g \in R$. As $r_R(N) = gR \trianglelefteq R$, $g \in \mathbf{S}_\ell(R)$ by Proposition 1.2.2. From Proposition 1.2.4(i), $gf \in \mathbf{S}_\ell(R)$. Put $e = \begin{bmatrix} e_1 & 0 \\ 0 & gf \end{bmatrix} \in T$. Then $e^2 = e$ and $r_T(K) = eT$. Thus, T is quasi-Baer. □

Corollary 5.6.6 *Let* $S = \text{End}(M_R)$ *and let* T *be the ring as in Notation* 5.6.3. *Then the following are equivalent.*

(i) T *is a quasi-Baer ring.*
(ii) (1) R *is a quasi-Baer ring.*
(2) M_R *is a quasi-Baer module.*
(3) *If* $N_R \trianglelefteq M_R$, *then* $r_R(N) = gR$ *for some* $g^2 = g \in R$.

Proof (i)⇒(ii) Assume that T is a quasi-Baer ring. Then M_R is a quasi-Baer module by Proposition 4.6.3 and Theorem 5.6.5. So we get (ii).

(ii)⇒(i) As M_R is a quasi-Baer module, S is a quasi-Baer ring by Theorem 4.6.16. Let $I \trianglelefteq S$. Then $r_S(I) = fS$ for some $f^2 = f \in S$. Also $r_M(I) = hM$ for some $h^2 = h \in S$ by Proposition 4.6.3. Since $If = 0$, $IfM = 0$, and so $fM \subseteq r_M(I) = hM$. As $IhM = 0$, $Ih = 0$, and hence $h \in r_S(I) = fS$. Thus, $hM \subseteq fSM = fM$. Therefore $hM = fM = fSM = r_S(I)M$. So T is a quasi-Baer ring by Theorem 5.6.5. □

We observe that in contrast to Theorem 5.6.2, the next two results hold true without any additional assumption on R.

Theorem 5.6.7 *The following are equivalent for a ring* R.

(i) *R is a quasi-Baer ring.*
(ii) *$T_n(R)$ is a quasi-Baer ring for every positive integer n.*
(iii) *$T_k(R)$ is a quasi-Baer ring for some integer $k > 1$.*
(iv) *$T_2(R)$ is a quasi-Baer ring.*

Proof (i)⇒(ii) We use induction on n. As R is quasi-Baer, $T_2(R)$ is quasi-Baer by applying Corollary 5.6.6.

Let $T_n(R)$ be quasi-Baer. We show that $T_{n+1}(R)$ is quasi-Baer. Write

$$T_{n+1}(R) = \begin{bmatrix} R & M \\ 0 & T_n(R) \end{bmatrix},$$

where $M = [R, \dots, R]$ (n-tuple). To apply Theorem 5.6.5, let $I \trianglelefteq R$. Then $r_R(I) = eR$ for some $e^2 = e \in R$. Also $r_M(I) = eM = r_R(I)M$.

Next, say ${}_RN_{T_n(R)} \leq {}_RM_{T_n(R)}$. Note that $\begin{bmatrix} 0 & N \\ 0 & 0 \end{bmatrix} \trianglelefteq T_{n+1}(R)$. Therefore, we have that $N = [N_1, \dots, N_n]$, where $N_i \trianglelefteq R$ for each i and $N_1 \subseteq \cdots \subseteq N_n$. As R is quasi-Baer, $r_R(N_i) = f_iR$ with $f_i^2 = f_i \in R$ for each i.

Let $e_{ij} \in T_n(R)$ be the matrix with 1 in the (i, j)-position and 0 elsewhere. Put $g = f_1e_{11} + \cdots + f_ne_{nn} \in T_n(R)$. Then $g^2 = g$ and $r_{T_n(R)}(N) = gT_n(R)$. By Theorem 5.6.5, $T_{n+1}(R)$ is a quasi-Baer.

(ii)⇒(iii) is obvious. For (iii)⇒(iv), let $e_{ij} \in T_k(R)$ be the matrix with 1 in the (i, j)-position and 0 elsewhere. Set $f = e_{11} + e_{22}$. Then $f^2 = f \in T_k(R)$ and $T_2(R) \cong fT_k(R)f$. By Theorem 3.2.10, $T_2(R)$ is quasi-Baer. Similarly, (iv)⇒(i) follows from Theorem 3.2.10. □

Proposition 5.6.8 *The following are equivalent for a ring R.*

(i) *R is a right p.q.-Baer ring.*
(ii) *$T_n(R)$ is a right p.q.-Baer ring for every positive integer n.*
(iii) *$T_k(R)$ is a right p.q.-Baer ring for some integer $k > 1$.*
(iv) *$T_2(R)$ is a right p.q.-Baer ring.*

Proof (i)⇒(ii) Put $T = T_n(R)$. Let e_{ij} be the matrix in T with 1 in the (i, j)-position and 0 elsewhere. Say $[a_{ij}] \in T$ and consider the right ideal $[a_{ij}]T$. Take $\alpha = [\alpha_{ij}] \in r_T([a_{ij}]T)$. Since R is right p.q.-Baer, for $i \leq j$, $r_R(a_{ij}R) = f_{ij}R$ with $f_{ij}^2 = f_{ij} \in R$. Then $f_{ij} \in \mathbf{S}_\ell(R)$ from Proposition 1.2.2 because $f_{ij}R \trianglelefteq R$.

Now observe that $\alpha_{1\ell} \in r_R(a_{11}R) = f_{11}R$ for $\ell = 1, \dots, n$. Also we see that $\alpha_{2\ell} \in r_R(a_{11}R) \cap r_R(a_{12}R) \cap r_R(a_{22}R) = f_{11}R \cap f_{12}R \cap f_{22}R = f_{11}f_{12}f_{22}R$ for $\ell = 2, \dots, n$, and $f_{11}f_{12}f_{22} \in \mathbf{S}_\ell(R)$ (see Proposition 1.2.4(i)). In general, $\alpha_{k\ell} \in (f_{11} \cdots f_{1k})(f_{22} \cdots f_{2k}) \cdots (f_{k-1k-1}f_{k-1k})f_{kk}R$ for $\ell = k, \dots, n$.

Put $g_k = (f_{11} \cdots f_{1k})(f_{22} \cdots f_{2k}) \cdots (f_{k-1k-1}f_{k-1k})f_{kk}$ for $k = 1, \dots, n$. Then $g_k \in \mathbf{S}_\ell(R)$ by Proposition 1.2.4(i). Note that $g_k\alpha_{k\ell} = \alpha_{k\ell}$ for $\ell = k, \dots, n$.

Let $e = g_1e_{11} + \cdots + g_ne_{nn} \in T$. Then $e^2 = e$ and $r_R([a_{ij}]T) = eT$. Therefore, $T = T_n(R)$ is right p.q.-Baer.

(ii)⇒(iii) It is evident.

(iii)⇒(iv) Let $f = e_{11} + e_{22} \in T_k(R)$. Then we see that $f^2 = f \in T_k(R)$ and $T_2(R) \cong fT_k(R)f$, so $T_2(R)$ is right p.q.-Baer by Theorem 3.2.34(i).

(iv)⇒(i) It follows also from Theorem 3.2.34(i). □

The following result, due to Tercan in [393], characterizes the generalized triangular matrix ring T (see Notation 5.6.3) to be a right nonsingular right extending ring (hence T is Baer and right cononsingular by Theorem 3.3.1) when ${}_SM$ is faithful.

Theorem 5.6.9 *Let T be the ring as in Notation* 5.6.3 *and ${}_SM$ be faithful. Then the following are equivalent.*

(i) *T is right nonsingular and right extending.*
(ii) (1) *For each complement K_R in M_R there is $e^2 = e \in S$ with $K = eM$.*
   (2) *R is right nonsingular and right extending.*
   (3) *M_R is nonsingular and injective.*

In the next result, a characterization for T to be right FI-extending is presented. This will be used to consider the FI-extending triangular matrix ring extensions.

Theorem 5.6.10 *Let T be the ring as in Notation* 5.6.3. *Then the following are equivalent.*

(i) *T_T is FI-extending.*
(ii) (1) *For ${}_SN_R \le {}_SM_R$ and $I \trianglelefteq S$ with $IM \subseteq N$, there is $f^2 = f \in S$ such that $I \subseteq fS$, $N_R \le^{\text{ess}} fM_R$, and $(I \cap \ell_S(M))_S \le^{\text{ess}} (fS \cap \ell_S(M))_S$.*
   (2) *R_R is FI-extending.*

Proof Throughout the proof, we let $e_{11} = \begin{bmatrix} 1 & 0 \\ 0 & 0 \end{bmatrix} \in T$.

(i)⇒(ii) First, we claim that $\ell_S(M) = eS$ for some $e^2 = e \in S$. Observe that $T_T = e_{11}T_T \oplus (1 - e_{11})T_T$ and $e_{11} \in \mathbf{S}_\ell(T)$. From Proposition 2.3.11(i), $e_{11}T_T = \begin{bmatrix} S & M \\ 0 & 0 \end{bmatrix}_T$ is FI-extending. First, to see that $\ell_S(M) = eS$ for some $e^2 = e \in S$, put $U = \begin{bmatrix} \ell_S(M) & 0 \\ 0 & 0 \end{bmatrix}$. Then $U_T \trianglelefteq e_{11}T_T$ because $\ell_S(M) \trianglelefteq S$ and $\text{End}(e_{11}T_T) \cong e_{11}Te_{11} = \begin{bmatrix} S & 0 \\ 0 & 0 \end{bmatrix}$. Because $e_{11}T_T$ is FI-extending, we have that $U_T \le^{\text{ess}} \begin{bmatrix} e & 0 \\ 0 & 0 \end{bmatrix} e_{11}T_T$ for some $e^2 = e \in S$. So $\begin{bmatrix} \ell_S(M) & 0 \\ 0 & 0 \end{bmatrix}_T \le^{\text{ess}} \begin{bmatrix} eS & eM \\ 0 & 0 \end{bmatrix}_T$. Thus, $\ell_S(M) \subseteq eS$. For any $m \in M$, $em = 0$ because $U \cap \begin{bmatrix} 0 & em \\ 0 & 0 \end{bmatrix} T = 0$. Hence $eM = 0$, so $e \in \ell_S(M)$. Thus $eS \subseteq \ell_S(M)$, and hence $\ell_S(M) = eS$.

For condition (1), let ${}_SN_R \le {}_SM_R$ and $I \trianglelefteq S$ such that $IM \subseteq N$. Then $V := \begin{bmatrix} I & N \\ 0 & 0 \end{bmatrix}_T \trianglelefteq e_{11}T_T = \begin{bmatrix} S & M \\ 0 & 0 \end{bmatrix}_T$. Since $e_{11}T_T$ is FI-extending, we have that

$\begin{bmatrix} I & N \\ 0 & 0 \end{bmatrix}_T \leq^{\text{ess}} \begin{bmatrix} fS & fM \\ 0 & 0 \end{bmatrix}_T$ for some $f^2 = f \in S$, therefore $I \subseteq fS$ and $N_R \leq^{\text{ess}} fM_R$. Next, for $0 \neq fs \in fS \cap eS = fS \cap \ell_S(M)$ with $s \in S$, we see that $V \cap \begin{bmatrix} fs & 0 \\ 0 & 0 \end{bmatrix} T = V \cap \begin{bmatrix} fsS & 0 \\ 0 & 0 \end{bmatrix} \neq 0$. Hence, $fsS \cap (I \cap eS) = fsS \cap I \neq 0$ because $fsS \subseteq eS$. Therefore, we have that $(I \cap eS)_S \leq^{\text{ess}} (fS \cap eS)_S$.

Since $e_{11} \in \mathbf{S}_\ell(T)$, Proposition 2.3.11(ii) yields condition (2) immediately.

(ii)⇒(i) By condition (2), $(1 - e_{11})T_T$ is FI-extending. To show that $e_{11}T_T$ is FI-extending, let $V_T \trianglelefteq e_{11}T_T$. Since $e_{11} \in \mathbf{S}_\ell(T)$, $e_{11}T_T \trianglelefteq T_T$ from Proposition 1.2.2, and so $V_T \trianglelefteq T_T$ by Proposition 2.3.3(ii). Thus $V = \begin{bmatrix} I & N \\ 0 & 0 \end{bmatrix}$ with $I \trianglelefteq S$, ${}_SN_R \leq {}_SM_R$, and $IM \subseteq N$. By condition (1), there is $f^2 = f \in S$ such that $I \subseteq fS$, $N_R \leq^{\text{ess}} fM_R$, and $(I \cap \ell_S(M))_S \leq^{\text{ess}} (fS \cap \ell_S(M))_S$. Thus, it follows that $V \subseteq \begin{bmatrix} f & 0 \\ 0 & 0 \end{bmatrix}\begin{bmatrix} S & M \\ 0 & 0 \end{bmatrix} = \begin{bmatrix} fS & fM \\ 0 & 0 \end{bmatrix}$. Let $W = \begin{bmatrix} fS & fM \\ 0 & 0 \end{bmatrix}$. Then W_T is a direct summand of $e_{11}T_T$ because $f^2 = f \in S \cong \text{End}(e_{11}T_T)$.

We prove that $V_T \leq^{\text{ess}} W_T$. For this, take $0 \neq w = \begin{bmatrix} fs & fm \\ 0 & 0 \end{bmatrix} \in W$, where $s \in S$ and $m \in M$. If $fm \neq 0$, then $V \cap wT \neq 0$ since $N_R \leq^{\text{ess}} fM_R$. Next, assume that $fm = 0$. Then $fs \neq 0$. Hence $wT = \begin{bmatrix} fsS & fsM \\ 0 & 0 \end{bmatrix}$.

If $fsM \neq 0$, clearly $V \cap wT \neq 0$ since $N_R \leq^{\text{ess}} fM_R$. If $fsM = 0$, then

$$fs \in \ell_S(M), \text{ so } 0 \neq fs \in fS \cap \ell_S(M).$$

Since $(I \cap \ell_S(M))_S \leq^{\text{ess}} (fS \cap \ell_S(M))_S$, $fsS \cap (I \cap \ell_S(M)) \neq 0$, so $V \cap wT \neq 0$. Therefore $V_T \leq^{\text{ess}} W_T$, thus $e_{11}T_T$ is FI-extending. Hence T_T is FI-extending by Theorem 2.3.5. □

Corollary 5.6.11 *Let T be the ring as in Notation 5.6.3. Assume that ${}_SM$ is faithful. Then the following are equivalent.*

(i) *T_T is FI-extending.*
(ii) (1) *For ${}_SN_R \leq {}_SM_R$, there is $f^2 = f \in S$ with $N_R \leq^{\text{ess}} fM_R$.*
  (2) *R_R is FI-extending.*

Proof (i)⇒(ii) Assume that T_T is FI-extending. As ${}_SM$ is faithful, $\ell_S(M) = 0$. By taking $I = 0$ in Theorem 5.6.10, we obtain part (ii).

(ii)⇒(i) Let ${}_SN_R \leq {}_SM_R$ and $I \trianglelefteq S$ such that $IM \subseteq N$. By (1), there exists $f^2 = f \in S$ such that $N_R \leq^{\text{ess}} fM_R$. Since $IM \subseteq N \subseteq fM$, $fn = n$ for all $n \in N$, in particular $fsm = sm$ for any $s \in I$ and $m \in M$. Therefore, $(s - fs)M = 0$, so $s - fs = 0$ for any $s \in I$ because ${}_SM$ is faithful. Hence, $I = fI \subseteq fS$. Thus, T_T is FI-extending by Theorem 5.6.10. □

Corollary 5.6.12 *Let M_R be a right R-module. Then the ring*

$$T = \begin{bmatrix} \mathrm{End}_R(M) & M \\ 0 & R \end{bmatrix}$$

is right FI-extending if and only if M_R and R_R are FI-extending.

Proof It follows immediately from Corollary 5.6.11. □

We remark that if R is a right FI-extending ring, then $T_2(R)$ is right FI-extending by taking $M = R_R$ in Corollary 5.6.12. When $n \geq 2$, we obtain the FI-extending property of $T_n(R)$ in Theorem 5.6.19 precisely when R is right FI-extending. By our previous results, we establish a class of rings which are right FI-extending, but not left FI-extending as the next example illustrates.

Example 5.6.13 Let R be a right self-injective ring with $J(R) \neq 0$. Put

$$T = \begin{bmatrix} R/J(R) & R/J(R) \\ 0 & R \end{bmatrix}.$$

Then the ring $R/J(R)$ is right self-injective by Corollary 2.1.30. Further, $\mathrm{End}_R(R/J(R)) \cong R/J(R)$. Also $R/J(R)$ is an FI-extending right R-module. Thus the ring T is right FI-extending by Corollary 5.6.12. If T is left FI-extending, then $r_R((R/J(R))_R) = J(R) = Rf$ for some $f \in \mathbf{S}_r(R)$ from the left-sided version of the proof for (i)⇒(ii) of Theorem 5.6.10. Thus $f = 0$ and hence $J(R) = 0$, a contradiction. Thus, T cannot be left FI-extending.

Definition 5.6.14 Let $N_R \leq M_R$. We say that N_R has a *direct summand cover* $\mathcal{D}(N_R)$ if there is $e^2 = e \in \mathrm{End}_R(M)$ with $N_R \leq^{\mathrm{ess}} eM_R = \mathcal{D}(N_R)$.

If M_R is a strongly FI-extending module, then every fully invariant submodule has a unique direct summand cover from Lemma 2.3.22. For $N_R \leq M_R$, let $(N_R : M_R) = \{a \in R \mid Ma \subseteq N\}$. Then $(N_R : M_R) \trianglelefteq R$.

We use $\mathcal{D}[(N_R : M_R)_R]$ to denote a direct summand cover of the right ideal $(N_R : M_R)$ in R_R. Let M be an (S, R)-bimodule and ${}_SN_R \leq {}_SM_R$. If there exists $e^2 = e \in \mathbf{S}_\ell(S)$ such that $N_R \leq^{\mathrm{ess}} eM_R$, then we write $\mathcal{D}_S(N_R) = eM$.

In the next result, we obtain a necessary and sufficient condition for a 2×2 generalized triangular matrix ring to be right strongly FI-extending. Some applications of this characterization will also be presented.

Theorem 5.6.15 *Let T be as in Notation* 5.6.3. *Then the following are equivalent.*

(i) *T_T is strongly FI-extending.*
(ii) (1) *For ${}_SN_R \leq {}_SM_R$ and $I \trianglelefteq S$ with $IM \subseteq N$, there is $e \in \mathbf{S}_\ell(S)$ such that $I \subseteq eS$, $N_R \leq^{\mathrm{ess}} eM_R$ and $(I \cap \ell_S(M))_S \leq^{\mathrm{ess}} (eS \cap \ell_S(M))_S$.*
(2) *R_R is strongly FI-extending.*

(3) $\mathcal{D}_S(N_R)\mathcal{D}[(N_R : M_R)_R] = M\mathcal{D}[(N_R : M_R)_R]$ *for* ${}_SN_R \leq {}_SM_R$.

Proof (i)⇒(ii) We let $e_{11} = \begin{bmatrix} 1 & 0 \\ 0 & 0 \end{bmatrix} \in T$. Assume that T_T is strongly FI-extending. By Theorem 2.3.19, $(1 - e_{11})T_T$ is strongly FI-extending, so R_R is strongly FI-extending, which is condition (2).

For condition (1), let ${}_SN_R \leq {}_SM_R$ and $I \trianglelefteq S$ with $IM \subseteq N$. Then $V := \begin{bmatrix} I & N \\ 0 & 0 \end{bmatrix}_T \trianglelefteq e_{11}T_T = \begin{bmatrix} S & M \\ 0 & 0 \end{bmatrix}_T$. Since $e_{11}T_T$ is strongly FI-extending, there exists $e^2 = e \in \mathbf{S}_\ell(S)$ such that $V_T \leq^{\text{ess}} \begin{bmatrix} eS & eM \\ 0 & 0 \end{bmatrix}_T$. So $I \subseteq eS$ and $N_R \leq^{\text{ess}} eM_R$.

Next, say $0 \neq es \in eS \cap \ell_S(M)$ with $s \in S$. There is $\begin{bmatrix} s_1 & m_1 \\ 0 & r_1 \end{bmatrix} \in T$ such that

$$0 \neq \begin{bmatrix} es & 0 \\ 0 & 0 \end{bmatrix} \begin{bmatrix} s_1 & m_1 \\ 0 & r_1 \end{bmatrix} = \begin{bmatrix} ess_1 & 0 \\ 0 & 0 \end{bmatrix} \in V.$$

Thus $0 \neq ess_1 \in I \cap \ell_S(M)$. Therefore $(I \cap \ell_S(M))_S \leq^{\text{ess}} (eS \cap \ell_S(M))_S$.

For condition (3), let ${}_SN_R \leq {}_SM_R$ and put $A = (N_R : M_R)$. Take $I = 0$ in condition (1). There exists $e \in \mathbf{S}_\ell(S)$ with $\mathcal{D}_S(N_R) = eM$. By condition (2), $\mathcal{D}(A_R) = fR$ for some $f \in \mathbf{S}_\ell(R)$. Since $MA \subseteq N$, $W := \begin{bmatrix} 0 & N \\ 0 & A \end{bmatrix} \trianglelefteq T$, and $W_T \leq^{\text{ess}} wT_T$ for some $w \in \mathbf{S}_\ell(T)$. By Lemma 5.6.4, there exist $e_0 \in \mathbf{S}_\ell(S)$ and $f_0 \in \mathbf{S}_\ell(R)$ such that $wT = \begin{bmatrix} e_0 & 0 \\ 0 & f_0 \end{bmatrix} T$. We put $w_0 = \begin{bmatrix} e_0 & 0 \\ 0 & f_0 \end{bmatrix} \in \mathbf{S}_\ell(T)$. Hence $N_R \leq^{\text{ess}} e_0M_R$ and $A_R \leq^{\text{ess}} f_0R_R$. So $\mathcal{D}_S(N_R) = eM = e_0M$ by Lemma 2.3.22 as $e_0 \in \mathbf{S}_\ell(S)$. Also $\mathcal{D}(A_R) = fR = f_0R$.

Note that $Mf_0 = e_0Mf_0$ since $w_0 \in \mathbf{S}_\ell(T)$. Thus, $e_0Mf_0R = Mf_0R$. Therefore, $\mathcal{D}_S(N_R)\mathcal{D}[(N_R : M_R)_R] = M\mathcal{D}[(N_R : M_R)_R]$.

(ii)⇒(i) Assume that $K \trianglelefteq T$. Then

$$K = \begin{bmatrix} I & N \\ 0 & B \end{bmatrix} \trianglelefteq T,$$

where ${}_SN_R \leq {}_SM_R$, $I \trianglelefteq S$, $IM + MB \subseteq N$, and $B \trianglelefteq R$.

From condition (1), there exists $e \in \mathbf{S}_\ell(S)$ with

$$I \subseteq eS, \ \mathcal{D}_S(N_R) = eM, \ \text{ and } \ (I \cap \ell_S(M))_S \leq^{\text{ess}} (eS \cap \ell_S(M))_S.$$

Since $B \trianglelefteq R$, by condition (2), there exists $f \in \mathbf{S}_\ell(R)$ with $\mathcal{D}(B_R) = fR$. Also, from condition (2), $\mathcal{D}[(N_R : M_R)_R] = f_0R$ for some $f_0 \in \mathbf{S}_\ell(R)$.

As $MB \subseteq N$, $B \subseteq (N_R : M_R)$. Thus,

$$B_R \leq^{\text{ess}} (fR \cap f_0R)_R = f_0fR$$

with $f_0f \in \mathbf{S}_\ell(R)$ (see Proposition 1.2.4(i)). So $\mathcal{D}(B_R) = f_0fR$. By Lemma 2.3.22, we get that $fR = f_0fR$.

By condition (3), $eMf_0R = Mf_0R$. Because $f \in \mathbf{S}_\ell(R)$ and $f_0fR = fR$, $eMf_0Rf = eMf_0fRf = eMfRf = eMRf = eMf$. Similarly, we have that $Mf_0Rf = Mf$. As $eMf_0R = Mf_0R$, $eMf_0Rf = Mf_0Rf$ and so $eMf = Mf$.

Since $(I \cap \ell_S(M))_S \leq^{\text{ess}} (eS \cap \ell_S(M))_S$ and $N_R \leq^{\text{ess}} eM_R$, we see that $\begin{bmatrix} I & N \\ 0 & 0 \end{bmatrix}_T \leq^{\text{ess}} \begin{bmatrix} e & 0 \\ 0 & 0 \end{bmatrix} T_T$. So $\begin{bmatrix} 0 & 0 \\ 0 & B \end{bmatrix}_T \leq^{\text{ess}} \begin{bmatrix} 0 & 0 \\ 0 & f \end{bmatrix} T_T$ because $B_R \leq^{\text{ess}} fR_R$.

Thus $K_T \leq^{\text{ess}} \begin{bmatrix} e & 0 \\ 0 & f \end{bmatrix} T_T$. As $Mf = eMf$, $mf = emf$ for each $m \in M$. Hence $\begin{bmatrix} e & 0 \\ 0 & f \end{bmatrix} \in \mathbf{S}_\ell(T)$. Therefore, T_T is strongly FI-extending. □

Corollary 5.6.16 *Let T be the ring as in Notation* 5.6.3 *with ${}_SM$ faithful. Then the following are equivalent.*

(i) *T_T is strongly FI-extending.*
(ii) (1) *For ${}_SN_R \leq {}_SM_R$, there is $e \in \mathbf{S}_\ell(S)$ with $N_R \leq^{\text{ess}} eM_R$.*
  (2) *R_R is strongly FI-extending.*
  (3) *$\mathcal{D}_S(N_R)\mathcal{D}[(N_R : M_R)_R] = M\mathcal{D}[(N_R : M_R)_R]$ for ${}_SN_R \leq {}_SM_R$.*

Proof (i)⇒(ii) The proof follows from Theorem 5.6.15 by taking $I = 0$. For (ii)⇒(i), let ${}_SN_R \leq {}_SM_R$ and $I \trianglelefteq S$ such that $IM \subseteq N$. By condition (1), there is $e \in \mathbf{S}_\ell(S)$ with $N_R \leq^{\text{ess}} eM_R$. As $IM \subseteq N \subseteq eM$, $n = en$ for all $n \in N$, in particular $sm = esm$ for any $s \in I$ and $m \in M$. Thus $(s - es)M = 0$, so $s - es = 0$ for any $s \in I$, as ${}_SM$ is faithful. So $I = eI \subseteq eS$. Thus T_T is strongly FI-extending by Theorem 5.6.15. □

Corollary 5.6.17 *Let M_R be a right R-module and $T = \begin{bmatrix} \text{End}_R(M) & M \\ 0 & R \end{bmatrix}$. Then the following are equivalent.*

(i) *T_T is strongly FI-extending.*
(ii) (1) *M_R is strongly FI-extending.*
  (2) *R_R is strongly FI-extending.*
  (3) *For any $N_R \trianglelefteq M_R$, $\mathcal{D}(N_R)\mathcal{D}[(N_R : M_R)_R] = M\mathcal{D}[(N_R : M_R)_R]$.*

Proof It follows immediately from Corollary 5.6.16. □

Theorem 5.6.18 *Let R be a ring. Then the following are equivalent.*

(i) *R is right strongly FI-extending.*
(ii) *$T_n(R)$ is right strongly FI-extending for every positive integer n.*
(iii) *$T_k(R)$ is right strongly FI-extending for some integer $k > 1$.*
(iv) *$T_2(R)$ is right strongly FI-extending.*

Proof (i)⇒(ii) Assume that R is right strongly FI-extending. We proceed by induction on n. Let $n = 2$. Take $M = R$ in Corollary 5.6.17. Let $N_R \trianglelefteq M_R$. Since R_R

is strongly FI-extending, there exists $e^2 = e \in \mathbf{S}_\ell(R)$ such that $N_R \le^{\text{ess}} eM_R$. We observe that $(N_R : M_R) = N_R \le^{\text{ess}} eR_R$. Therefore we have that

$$\mathcal{D}_R(N_R)\mathcal{D}[(N_R : M_R)_R] = eReR = ReR = M\mathcal{D}[(N_R : M_R)_R].$$

Hence, $T_2(R)$ is a right strongly FI-extending ring by Corollary 5.6.17.

Assume that $T_n(R)$ is right strongly FI-extending. Then we show that $T_{n+1}(R)$ is right strongly FI-extending. Now

$$T_{n+1}(R) = \begin{bmatrix} R & M \\ 0 & T_n(R) \end{bmatrix},$$

where $M = [R, \ldots, R]$ (n-tuple). Let ${}_RN_{T_n(R)} \le {}_RM_{T_n(R)}$. As in the proof of Theorem 5.6.7, $N = [N_1, \ldots, N_n]$, where $N_i \trianglelefteq R$ for each i and $N_1 \subseteq \cdots \subseteq N_n$. As R_R is strongly FI-extending, there is $e \in \mathbf{S}_\ell(R)$ with $N_{nR} \le^{\text{ess}} eR_R$, so $N = [N_1, \ldots, N_n]_{T_n(R)} \le^{\text{ess}} e[R, \ldots, R]_{T_n(R)} = eM$. Thus,

$$(N_{T_n(R)} : M_{T_n(R)}) = \begin{bmatrix} N_1 & N_2 & \cdots & N_n \\ 0 & N_2 & \cdots & N_n \\ \vdots & \vdots & \ddots & \vdots \\ 0 & 0 & \cdots & N_n \end{bmatrix}_{T_n(R)} \le^{\text{ess}} (e\mathbf{1})T_n(R)_{T_n(R)},$$

where $\mathbf{1}$ is the identity matrix in $T_n(R)$. Hence, we have that

$$\mathcal{D}_R(N_{T_n(R)})\mathcal{D}[(N_{T_n(R)} : M_{T_n(R)})_{T_n(R)}] = eM(e\mathbf{1})T_n(R) = M(e\mathbf{1})T_n(R),$$

since $e \in \mathbf{S}_\ell(R)$. Note that $M\mathcal{D}[(N_{T_n(R)} : M_{T_n(R)})_{T_n(R)}] = M(e\mathbf{1})T_n(R)$. So $M\mathcal{D}[(N_{T_n(R)} : M_{T_n(R)})_{T_n(R)}] = \mathcal{D}_R(N_{T_n(R)})\mathcal{D}[(N_{T_n(R)} : M_{T_n(R)})_{T_n(R)}]$. Thus by Corollary 5.6.16, $T_{n+1}(R)$ is a right strongly FI-extending ring.

(ii)⇒(iii) is obvious, and (iii)⇒(i) is a consequence of Theorem 5.6.15.

(i)⇒(iv) follows from the proof of (i)⇒(ii) for the case when $n = 2$, and (iv)⇒(i) follows from Theorem 5.6.15. □

Theorem 5.6.19 *Let R be a ring. Then the following are equivalent.*

(i) *R is right FI-extending.*
(ii) *$T_n(R)$ is right FI-extending for every positive integer n.*
(iii) *$T_k(R)$ is right FI-extending for some integer $k > 1$.*
(iv) *$T_2(R)$ is right FI-extending.*

Proof The proof follows by using Corollary 5.6.11 and an argument similar to that used in the proof of Theorem 5.6.18. □

Theorem 5.6.19 provides a full characterization of $T_n(R)$ to be right FI-extending for any positive integer n. Let R be a commutative domain which is not a field. Say n is an integer such that $n > 1$. Then $T_n(R)$ is right strongly FI-extending (hence right FI-extending) by Theorem 5.6.18. Observe that $T_n(R)$ is not Baer from Theorem 5.6.2. Thus by Corollary 3.3.3, $T_n(R)$ is neither right nor left extending. Corollary 5.6.16 and Theorem 5.6.18 are now applied to show that the strongly FI-extending property for rings is not left-right symmetric.

Example 5.6.20 Let R be a commutative domain and let $M = \begin{bmatrix} 0 & R \\ 0 & 0 \end{bmatrix}$. Then naturally M can be considered as an $(R, T_2(R))$-bimodule. We show that the generalized triangular matrix ring $T = \begin{bmatrix} R & M \\ 0 & T_2(R) \end{bmatrix}$ is right strongly FI-extending, but it is not left strongly FI-extending. For this, note that ${}_RM$ is faithful. Because R is right strongly FI-extending, $T_2(R)$ is right strongly FI-extending from Theorem 5.6.18. Say ${}_RN_{T_2(R)} \leq {}_RM_{T_2(R)}$. If $N = 0$, then $\mathcal{D}_R(N_{T_2(R)})\mathcal{D}[(N_{T_2(R)} : M_{T_2(R)})_{T_2(R)}] = 0 = M\mathcal{D}[(N_{T_2(R)} : M_{T_2(R)})_{T_2(R)}]$. So assume that $N \neq 0$. Then there is $0 \neq I \trianglelefteq R$ with $N = \begin{bmatrix} 0 & I \\ 0 & 0 \end{bmatrix}$. Then $I_R \leq^{\text{ess}} R_R$, hence $\mathcal{D}_R(N_{T_2(R)}) = \begin{bmatrix} 0 & R \\ 0 & 0 \end{bmatrix} = M$. Therefore,

$$\mathcal{D}_R(N_{T_2(R)})\mathcal{D}[(N_{T_2(R)} : M_{T_2(R)})_{T_2(R)}] = M\mathcal{D}[(N_{T_2(R)} : M_{T_2(R)})_{T_2(R)}].$$

Thus, T_T is strongly FI-extending by Corollary 5.6.16.

We may note that $r_{T_2(R)}(M)$ is not generated, as a left ideal, by an idempotent in $T_2(R)$. Thus, ${}_TT$ is not FI-extending by the left-sided version of the proof for (i)⇒(ii) of Theorem 5.6.10. So ${}_TT$ is not strongly FI-extending.

Exercise 5.6.21

1. Assume that R is a PWP ring. Show that $T_n(R)$ is a PWP ring for each positive integer n.
2. ([85, Birkenmeier, Park, and Rizvi]) Let R be a prime ring with P a nonzero prime ideal. Prove that the ring $\begin{bmatrix} R/P & R/P \\ 0 & R \end{bmatrix}$ is right FI-extending, but not left FI-extending.
3. ([85, Birkenmeier, Park, and Rizvi]) Let R be a commutative PID and let I be a nonzero proper ideal of R. Show that the ring $\begin{bmatrix} R/I & R/I \\ 0 & R \end{bmatrix}$ is right FI-extending, but not left FI-extending.
4. ([64, Birkenmeier and Lennon]) Let T be the ring as in Notation 5.6.3. Prove that T_T is FI-extending if and only if the following conditions hold.
   (1) $\ell_S(M) = eS$, where $e \in \mathbf{S}_\ell(S)$, and eS_S is FI-extending.
   (2) For ${}_SN_R \leq {}_SM_R$, there is $f^2 = f \in S$ with $N_R \leq^{\text{ess}} fM_R$.
   (3) R_R is FI-extending.
5. Let T be the ring as in Notation 5.6.3. Characterize T being right p.q.-Baer in terms of conditions on S, M, and R. (Hint: see [78, Birkenmeier, Kim, and Park].)

Historical Notes Some of the diverse applications associated with generalized triangular matrix representations appear in the study of operator theory [212], quasitriangular Hopf algebras [113], and various Lie algebras [303]. Also many authors have studied a variety of conditions on generalized triangular matrix rings (e.g., [37, 189–191, 196, 228, 280], and [416]). Most results from Sects. 5.1, 5.2, and 5.3 are due to Birkenmeier, Heatherly, Kim, and Park [70]. Results 5.2.18–5.2.20 appear in [66]. Some of the motivating ideas for defining triangulating idempotents originated with [55]. Lemma 5.3.4 is due to Fields [164].

Theorem 5.4.1, Corollary 5.4.2, and Definition 5.4.4 appear in [70]. Piecewise domains (PWDs) were defined and investigated by Gordon and Small [187]. Proposition 5.4.6 is in [70]. Proposition 5.4.9 and Example 5.4.10(i)–(iii) and (v) are taken from [187]. Theorem 5.4.12 from [70] is a structure theorem for a PWP ring. Results 5.4.13–5.4.16 and Corollary 5.4.19 appear in [70]. Theorem 5.4.20 and Corollary 5.4.21 are taken from [66]. Examples 5.4.22 appears in [103] and [68]. In [118], Theorem 5.4.24 has been improved to the case when R is a Noetherian Rickart ring. Lemma 5.4.25 is in [70].

Results 5.5.1–5.5.3, Proposition 5.5.5, and Theorem 5.5.6 appear in [74]. Proposition 5.5.7 is in [369]. Examples 5.5.8, 5.5.9, Results 5.5.10–5.5.12 are taken from [74]. Theorem 5.5.14 is due to Birkenmeier, Kim, and Park [74]. Koh ([255] and [256]), Lambek [265], Shin [369], and Sun [388] showed that the Gelfand homomorphism θ is an isomorphism for various classes of rings.

Theorem 5.6.1 is due to Akalan, Birkenmeier, and Tercan (see [1, 3], and [393]). Theorem 5.6.2 appears to be a new result which is due to the authors. Results 5.6.4–5.6.6 appear in [85]. Theorem 5.6.7 was obtained by Pollingher and Zaks in [347], but we give the proof in a different way by applying Theorem 5.6.5. Proposition 5.6.8 is from [78]. Theorem 5.6.9 is completely generalized in [3]. Results 5.6.10–5.6.13 and Definition 5.6.14 appear in [85]. A characterization of generalized triangular right FI-extending rings are also considered in [64] (see Exercise 5.6.21.4). Results 5.6.15–5.6.18 appear in [85]. Theorem 5.6.19 was shown in [83], while Example 5.6.20 was given in [85]. Further related references include [51, 81, 91, 116, 122, 125, 135, 160], and [387].

Chapter 6
Matrix, Polynomial, and Group Ring Extensions

It was shown in Chap. 3 that the quasi-Baer property transfers to matrix ring extensions readily and without any additional requirements (Theorem 3.2.12). We continue in this chapter our discussions on the transference of various properties not only to matrix ring extensions, but also to the other extensions listed in the title of this chapter. To do so, we shall use the results which have been developed in the previous chapters. For earlier results on matrix extensions, see Sects. 3.1, 3.2, 4.3, and 4.6.

The transference of the Baer property to the ring extensions of the title is somewhat restricted (see Example 3.1.28). In this chapter, we show that this happens only under special conditions (e.g., see Theorems 6.1.3, 6.1.4, 6.1.12, and Corollary 6.2.6). However, for the case of the quasi-Baer property, we shall see that the property transfers to various matrix and polynomial ring extensions without any additional assumptions (e.g., see Theorems 3.2.12, 6.1.16, 6.2.3, and 6.2.4). Furthermore, we explore in detail the transference of the aforementioned properties, as well as, the Rickart, extending, p.q.-Baer, and FI-extending properties to various matrix (both finite and infinite), polynomial, Ore, and group ring extensions. In particular, a characterization of semiprime quasi-Baer group algebras is included (see Theorem 6.3.2).

6.1 Matrix Ring Extensions

Results shown in previous chapters are applied in this section to study the Baer, the quasi-Baer, and the FI-extending properties of a ring R to various matrix extensions of R.

For a nonempty ordered set Γ and a ring R, recall that $\mathrm{CFM}_\Gamma(R)$ and $\mathrm{RFM}_\Gamma(R)$ denote the $\Gamma \times \Gamma$ column finite matrix ring and the $\Gamma \times \Gamma$ row finite matrix ring over R, respectively. Note that $\mathrm{CFM}_\Gamma(R) \cong \mathrm{End}(R_R^{(\Gamma)})$ and $\mathrm{RFM}_\Gamma(R) \cong \mathrm{End}({}_R R^{(\Gamma)})$.

The next result is obtained as an application of Baer module theory presented in Chap. 4.

G.F. Birkenmeier et al., *Extensions of Rings and Modules*,

DOI 10.1007/978-0-387-92716-9_6,

Theorem 6.1.1 *The following are equivalent for a ring R.*

(i) *R is a semiprimary hereditary (hence Baer) ring.*
(ii) $\mathrm{CFM}_\Gamma(R)$ *is a Baer ring for any nonempty ordered set Γ.*
(iii) $\mathrm{RFM}_\Gamma(R)$ *is a Baer ring for any nonempty ordered set Γ.*

Proof (i)⇔(ii) The equivalence follows from Theorems 4.2.8 and 4.3.3 because any free right R-module is retractable.

(i)⇔(iii) By Proposition 4.3.2 and Theorem 4.3.3, a ring R is semiprimary hereditary if and only if every free left R-module is Baer. So the proof follows from the left-sided version of (i)⇔(ii). □

For a characterization of a ring R for which $\mathrm{Mat}_n(R)$ is a Baer ring for every positive integer n, we need the following result.

Theorem 6.1.2 (i) *Any finitely presented flat module is projective.*

(ii) *A ring R is left semihereditary if and only if all torsionless right R-modules are flat.*

Proof See [262, Theorem 4.30] and [262, Theorem 4.67]. □

Recall that a ring R is called *right Π-coherent* if every finitely generated torsionless right R-module is finitely presented. A left Π-coherent ring is defined similarly.

Theorem 6.1.3 *The following are equivalent for a ring R.*

(i) $\mathrm{Mat}_n(R)$ *is a Baer ring for every positive integer n.*
(ii) *R is left semihereditary and right Π-coherent.*
(iii) *R is right semihereditary and left Π-coherent.*

Proof (i)⇒(ii) Assume that (i) holds. Let n be a positive integer. From Theorem 4.2.8, $R_R^{(n)}$ is a Baer module since $R_R^{(n)}$ is retractable. Let I be a finitely generated left ideal of R. Then ${}_RI$ is torsionless. Thus ${}_RI$ is projective by Theorem 4.3.5. Therefore R is left semihereditary. Moreover, since all finitely generated torsionless right R-modules are projective from Theorem 4.3.5, they are finitely presented. Thus, R is right Π-coherent.

(ii)⇒(i) Assume that R is a left semihereditary and right Π-coherent ring. Let V be a finitely generated torsionless right R-module. From Theorem 6.1.2, V is flat. Note that V is finitely presented because R is right Π-coherent. Now Theorem 6.1.2 yields that V is projective. Therefore, $\mathrm{Mat}_n(R)$ is a Baer ring for every positive integer n by Theorem 4.3.5.

(i)⇔(iii) From Theorem 4.3.5, $\mathrm{Mat}_n(R)$ is a Baer ring for every positive integer n if and only if every finitely generated torsionless left R-module is projective. Thus, (i)⇔(iii) follows from the left-sided version of (i)⇔(ii). □

Assume that R is a commutative domain with the field K of fractions. For $I_R \leq K_R$, let $I^{-1} = \{k \in K \mid kI \subseteq R\}$, which is called the inverse of I (see [248,

Definitions, p. 37]). The R-submodule I is called *invertible* if $II^{-1} = R$. It is well-known that I_R is an invertible submodule of K_R if and only if I_R is projective (see [382, Proposition 4.3, p. 59]). For an ideal J of a commutative local domain R, if J is invertible, then J is principal ([248, Theorem 59]).

A *Prüfer domain* is a commutative domain in which every nonzero finitely generated ideal is invertible. Equivalently, a Prüfer domain R is a commutative domain which is semihereditary. A commutative domain is called *Bezout* if every finitely generated ideal is principal. It is well known that a commutative domain R is a Prüfer domain if and only if the localization R_P (of R at P) is a Bezout domain for every prime ideal P of R ([248, Theorems 63 and 64]).

Assume that R is a Prüfer domain and M is a finitely generated torsionless R-module. Then M is torsion-free. By [363, Theorem 4.32], M is projective, and hence M is finitely presented. So R is Π-coherent. Thereby, a commutative domain R is Prüfer if and only if R is semihereditary if and only if R is semihereditary and Π-coherent. By Theorem 6.1.3, a commutative domain R is Prüfer if and only if $\text{Mat}_n(R)$ is a Baer ring for every positive integer n.

In the next theorem, we can say more for a commutative domain. Indeed, it is shown that R is Prüfer if and only if $\text{Mat}_n(R)$ is a Baer ring for every positive integer n if and only if $\text{Mat}_2(R)$ is a Baer ring (cf. Example 6.1.6).

Theorem 6.1.4 *Let R be a commutative domain. Then the following are equivalent.*

(i) *R is a Prüfer domain.*
(ii) $\text{Mat}_n(R)$ *is a Baer (Rickart) ring for every positive integer n.*
(iii) $\text{Mat}_k(R)$ *is a Baer (Rickart) ring for some integer $k > 1$.*
(iv) $\text{Mat}_2(R)$ *is a Baer (Rickart) ring.*

Further, in (ii), (iii), *and* (iv), *"Baer (Rickart) ring" can be replaced by "(right) extending ring".*

Proof As $\text{Mat}_n(R)$ is orthogonally finite, Theorem 3.1.25 yields that $\text{Mat}_n(R)$ is Baer if and only if $\text{Mat}_n(R)$ is Rickart for each positive integer n.

(i)$\Rightarrow$(ii) As R is semihereditary Π-coherent, $\text{Mat}_n(R)$ is Baer for every positive integer n by Theorem 3.1.29 or Theorem 6.1.3.

(ii)$\Rightarrow$(iii) is evident. For (iii)$\Rightarrow$(iv), let e_{ii} be the matrix in $\text{Mat}_k(R)$ with 1 in the (i,i)-position and 0 elsewhere. Take $f = e_{11} + e_{22}$. Then by Theorem 3.1.8, $f\text{Mat}_k(R)f \cong \text{Mat}_2(R)$ is a Baer ring.

(iv)$\Rightarrow$(i) Let P be a prime ideal of R and R_P be the localization of R at P. Take $a_1s_1^{-1}, a_2s_2^{-1} \in R_P$ (note that $a_i, s_i \in R$ and $s_i \notin P$). We show that the ideal $a_1s_1^{-1}R_P + a_2s_2^{-1}R_P = a_1R_P + a_2R_P$ is principal.

As $\text{Mat}_2(R)$ is right Rickart, $a_1R + a_2R$ is a projective right R-module by the proof of Theorem 3.1.29. So $a_1R + a_2R$ is R-isomorphic to a direct summand of $R^{(2)}$. Hence, $a_1R_P + a_2R_P$ is R_P-isomorphic to a direct summand of $R_P^{(2)}$. Thus, $a_1R_P + a_2R_P$ is projective as a right R_P-module, so it is an invertible ideal of the local domain R_P. Hence, $a_1R_P + a_2R_P$ is a principal ideal of R_P because R_P is local (see [248, Theorem 59]). Inductively, every finitely generated ideal of R_P is principal. So R_P is a Bezout domain for each prime ideal P of R. Therefore R is a Prüfer domain.

From Corollary 3.3.3, in (ii), (iii), and (iv), "Baer (Rickart) ring" can be replaced with "(right) extending ring". □

We obtain the next corollary which is an extension of Theorem 6.1.4 to an orthogonally finite commutative ring case.

Corollary 6.1.5 *Let R be a commutative orthogonally finite ring. Then the following are equivalent.*

(i) *R is a finite direct sum of Prüfer domains.*
(ii) $\mathrm{Mat}_n(R)$ *is a Baer (Rickart) ring for every positive integer n.*
(iii) $\mathrm{Mat}_k(R)$ *is a Baer (Rickart) ring for some integer $k > 1$.*
(iv) $\mathrm{Mat}_2(R)$ *is a Baer (Rickart) ring.*

Further, in (ii), (iii), *and* (iv), *"Baer (Rickart) ring" can be replaced by "(right) extending ring".*

Proof (i)⇒(ii) Let $R = R_1 \oplus \cdots \oplus R_m$, where each R_i is a Prüfer domain. By Theorem 6.1.4, $\mathrm{Mat}_n(R_i)$ is Baer for each i. So $\mathrm{Mat}_n(R) = \oplus_{i=1}^m \mathrm{Mat}_n(R_i)$ is Baer by Proposition 3.1.5(i).

(ii)⇒(iii) is evident, and (iii)⇒(iv) is a consequence of Theorem 3.1.8.

(iv)⇒(i) Let $\mathrm{Mat}_2(R)$ be a Baer ring. Say $f \in \mathrm{Mat}_2(R)$ with 1 in the (1, 1)-position and 0 elsewhere. Then $R \cong f\mathrm{Mat}_2(R)f$ is a Baer ring by Theorem 3.1.8. Thus, R is nonsingular by Proposition 3.1.18, so R is semiprime because R is commutative. Next, as R is orthogonally finite, R is a finite direct sum of commutative domains by Corollary 5.4.13(i). Say $R = R_1 \oplus \cdots \oplus R_m$, with each R_i a commutative domain. As $\mathrm{Mat}_2(R_1) \oplus \cdots \oplus \mathrm{Mat}_2(R_m)$ is Baer, each $\mathrm{Mat}_2(R_i)$ is Baer from Proposition 3.1.5(i). So each R_i is a Prüfer domain by Theorem 6.1.4. Thus, R is a finite direct sum of Prüfer domains.

By Corollary 3.3.3, in (ii), (iii), and (iv), "Baer (Rickart) ring" can be replaced by "(right) extending ring". □

We remark that when R is a commutative domain, any one of the conditions of Theorem 6.1.4 is equivalent to the condition that every finitely generated free (projective) R-module is a Baer (Rickart) module if and only if $(R \oplus R)_R$ is a Baer (Rickart) module. The following example illustrates that there is a domain R such that $\mathrm{Mat}_2(R)$ is Baer, but not so $\mathrm{Mat}_3(R)$. Hence, R being commutative is not superfluous in Theorem 6.1.4.

Example 6.1.6 Let R be the K-algebra (K is a field) on the generators x_i, y_i, $i = 1, 2, 3$, with the defining relation $x_1y_1 + x_2y_2 + x_3y_3 = 0$ as in Example 4.3.13. Then $R_R^{(2)}$ is a Baer module from Theorem 4.3.12, so $\mathrm{Mat}_2(R)$ is a Baer ring by Theorem 4.2.8.

But from Example 4.3.13, $R_R^{(3)}$ is not Baer. So by Theorem 4.2.8, $\mathrm{Mat}_3(R)$ is not a Baer ring. Note that R is a domain as R is a 2-fir. By Theorem 3.1.8, whenever $n \geq 3$, $\mathrm{Mat}_n(R)$ is not a Baer ring. Finally, we recall from Theorem 6.1.4 that, when R is a commutative domain, $\mathrm{Mat}_2(R)$ is a Baer ring if and only if $\mathrm{Mat}_n(R)$ is a Baer ring for every positive integer n.

For a ring R and a nonempty ordered set Γ, we use $\mathrm{CRFM}_\Gamma(R)$ to denote the ring of $\Gamma \times \Gamma$ column and row finite matrices over R. We consider another application of generalized triangular matrix representations developed in Chap. 5 to show that if $\mathrm{CRFM}_{\mathbb{N}}(R)$ is Baer, then R is semisimple Artinian (Theorem 6.1.10), where $\mathbb{N}$ is the set of positive integers.

For convenience, we put $M(R) = \mathrm{CRFM}_{\mathbb{N}}(R)$ and $N(R) = \mathrm{RFM}_{\mathbb{N}}(R)$. First, we need the next lemma (see [371, Proposition 2.2] for its proof).

Lemma 6.1.7 *Let R be a ring and $g^2 = g \in M(R)$. Then ${}_{N(R)}N(R)g \cong {}_{N(R)}N(R)$ if and only if ${}_{M(R)}M(R)g \cong {}_{M(R)}M(R)$.*

We use a left module argument in the next lemma for convenience.

Lemma 6.1.8 *Assume that R is a domain such that $\mathrm{CRFM}_{\mathbb{N}}(R)$ is a Baer ring. Take $0 \neq d \in R$ and put*

$$a = \begin{bmatrix} d & 0 & \cdots & \cdots & \\ 1 & 1 & 0 & \cdots & \\ 0 & d & 0 & \cdots & \\ 0 & 1 & 1 & 0 & \cdots \\ & & \ddots & \ddots & \end{bmatrix} \in \mathrm{CRFM}_{\mathbb{N}}(R).$$

Let $\{a_i\}$ be the set of rows of a viewed as elements of ${}_R R^{(\mathbb{N})}$. Then there exist $j_0 \in \mathbb{N}$ and a direct summand K of ${}_R R^{(\mathbb{N})}$, such that $R^{(\mathbb{N})}a = Ra_1 \oplus K$ and $a_{j_0+k} \in K$ for all positive integers k.

Proof Set $M(R) = \mathrm{CRFM}_{\mathbb{N}}(R)$. Let e_{ij} (or $e_{i,j}$) be the matrix in $M(R)$ with 1 in the (i, j)-position and 0 elsewhere. Since $M(R)$ is a Baer ring, $M(R)$ is a left Rickart ring. So $M(R)a$ is projective as a left $M(R)$-module by Proposition 3.1.17, hence ${}_{M(R)}M(R)a \cong {}_{M(R)}M(R)g$ for some $g^2 = g \in M(R)$.

Let $x = \sum_{n=1}^{\infty} e_{n,4n-2}$ and $y = \sum_{n=1}^{\infty} e_{2n-1,n}$. Then we see that $xay = 1$. Define $\mu : M(R)a \to M(R)$ by $\mu(va) = vay$ for $v \in M(R)$. Furthermore, we define $\nu : M(R) \to M(R)a$ by $\nu(w) = wxa$ for $w \in M(R)$. Then μ and ν are left $M(R)$-homomorphisms, and $\mu \circ \nu$ is the identity map of $M(R)$ as $xay = 1$.

Note that $\mathrm{Image}(\nu) \cap \mathrm{Ker}(\mu) = 0$. Next, for $wa \in M(R)a$ with $w \in M(R)$, $wayxa \in \mathrm{Image}(\nu)$ and $wa - wayxa \in \mathrm{Ker}(\mu)$. Thus,

$$M(R)a = \mathrm{Image}(\nu) \oplus \mathrm{Ker}(\mu).$$

Therefore $M(R)g \cong M(R)a = \text{Image}(\nu) \oplus \text{Ker}(\mu) \cong M(R) \oplus \text{Ker}(\mu)$, and thus $R^{(\mathbb{N})} \otimes_{M(R)} M(R)g \cong [R^{(\mathbb{N})} \otimes_{M(R)} M(R)] \oplus [R^{(\mathbb{N})} \otimes_{M(R)} \text{Ker}(\mu)]$. Consequently, we get $R^{(\mathbb{N})}g \cong R^{(\mathbb{N})} \oplus M_0$ as left R-modules, where $M_0 = R^{(\mathbb{N})} \otimes_{M(R)} \text{Ker}(\mu)$.

Further, ${}_RR^{(\mathbb{N})} = {}_RR^{(\mathbb{N})}g \oplus {}_RR^{(\mathbb{N})}(1-g) \cong {}_RR^{(\mathbb{N})} \oplus {}_RM_0 \oplus {}_RR^{(\mathbb{N})}(1-g)$. Put $U = R^{(\mathbb{N})} \oplus R^{(\mathbb{N})}(1-g)$. Then $R^{(\mathbb{N})} \cong M_0 \oplus U$. Now

$$\begin{aligned} R^{(\mathbb{N})} &\cong (R^{(\mathbb{N})})^{(\mathbb{N})} \cong (M_0 \oplus U) \oplus (M_0 \oplus U) \oplus (M_0 \oplus U) \oplus \cdots \\ &= M_0 \oplus (U \oplus M_0) \oplus (U \oplus M_0) \oplus \cdots \\ &= M_0 \oplus (R^{(\mathbb{N})})^{(\mathbb{N})} \cong M_0 \oplus R^{(\mathbb{N})} \cong R^{(\mathbb{N})}g. \end{aligned}$$

So $\text{Hom}({}_RR^{(\mathbb{N})}, {}_RR^{(\mathbb{N})}g) \cong \text{Hom}({}_RR^{(\mathbb{N})}, {}_RR^{(\mathbb{N})})$ because ${}_RR^{(\mathbb{N})} \cong {}_RR^{(\mathbb{N})}g$, and thus ${}_{N(R)}N(R)g \cong {}_{N(R)}N(R)$, where $N(R) = \text{RFM}_{\mathbb{N}}(R)$. By Lemma 6.1.7, ${}_{M(R)}M(R)g \cong {}_{M(R)}M(R)$. Hence, ${}_{M(R)}M(R)a \cong {}_{M(R)}M(R)$.

Let $\theta : M(R)a \to M(R)$ be an $M(R)$-isomorphism. Set $b = \theta^{-1}(1)$ in $M(R)a$, where 1 is the identity of $M(R)$. Say $b = sa$, where $s \in M(R)$. Thus, we obtain that $R^{(\mathbb{N})}b = R^{(\mathbb{N})}sa \subseteq R^{(\mathbb{N})}a$.

Next, $M(R)b = M(R)\theta^{-1}(1) = \theta^{-1}(M(R)) = M(R)a$, thus $a \in M(R)b$. Hence, there exists $c \in M(R)$ such that $a = cb$. So $R^{(\mathbb{N})}a = R^{(\mathbb{N})}cb \subseteq R^{(\mathbb{N})}b$. Therefore, $R^{(\mathbb{N})}b = R^{(\mathbb{N})}a$.

The rows of b, say $\{b_i\}$, are R-independent. For, if $r_1b_1 + \cdots + r_kb_k = 0$, where $r_1, \ldots, r_k \in R$. Then $0 = r_1e_{11}b + \cdots + r_ke_{kk}b = (r_1e_{11} + \cdots + r_ke_{kk})b$ because $e_{ii}b = b_i$ for i, $1 \leq i \leq k$. Thus

$$\begin{aligned} 0 &= \theta((r_1e_{11} + \cdots + r_ke_{kk})b) = (r_1e_{11} + \cdots + r_ke_{kk})\theta(b) \\ &= r_1e_{11} + \cdots + r_ke_{kk}. \end{aligned}$$

So $r_1 = 0, \ldots, r_k = 0$, and hence $\{b_i\}$ are R-independent. Therefore, $\{b_i\}$ is an R-basis for $R^{(\mathbb{N})}a$ because $R^{(\mathbb{N})}a = R^{(\mathbb{N})}b$.

We note that $a_1 \in R^{(\mathbb{N})}a = R^{(\mathbb{N})}b$. Say $a_1 \in \sum_{i=1}^n Rb_i$. As $c \in M(R)$ (recall that $a = cb$, where $c \in M(R)$), there is a positive integer j_0 such that $e_{j_0+k,j_0+k}c(\sum_{i=1}^n e_{ii}) = 0$ for all positive integers k. Note that

$$a_{j_0+k} = e_{j_0+k,j_0+k}a = e_{j_0+k,j_0+k}cb.$$

So $a_{j_0+k} \in \sum_{i>n} Rb_i$ for all positive integers k.

Let $V = \sum_{i=1}^n Rb_i$ and $W = \sum_{i>n} Rb_i$. Then $R^{(\mathbb{N})}a = V \oplus W$ with $a_1 \in V$ and $a_{j_0+k} \in W$ for all positive integers k. Now $\{a_1, a_2, a_4, a_6, \ldots\}$ is an R-basis for $R^{(\mathbb{N})}a$, so Ra_1 is a direct summand of $R^{(\mathbb{N})}a$. By the modular law, $V = Ra_1 \oplus V'$, for some $V' \leq V$, as $a_1 \in V$ and Ra_1 is a direct summand of $R^{(\mathbb{N})}a$. Therefore, $R^{(\mathbb{N})}a = V \oplus W = Ra_1 \oplus V' \oplus W$. Put $K = V' \oplus W$. Then $R^{(\mathbb{N})}a = Ra_1 \oplus K$ and $a_{j_0+k} \in K$ for all positive integers k. □

Lemma 6.1.9 *Let R be a domain such that* $\text{CRFM}_{\mathbb{N}}(R)$ *is a Baer ring. Then R is a division ring.*

Proof We use the notation of the preceding lemma. Recall that the first row a_1 of the matrix in Lemma 6.1.8 is $a_1 = (d, 0, 0, \dots)$.

By Lemma 6.1.8, we see that $R^{(\mathbb{N})}a = Ra_1 \oplus K$ such that $a_{j_0+k} \in K$ for all positive integers k. Let $\varphi : R^{(\mathbb{N})}a \to Ra_1$ be the canonical projection. For each positive integer i, let $x_i \in R$ such that $\varphi(a_i) = x_i a_1$.

We observe that $a_{j_0+k} \in K$ for all positive integers k, so it follows that $0 = \varphi(a_{j_0+k}) = x_{j_0+k}a_1 = (x_{j_0+k}d, 0, 0, \dots)$. Thus, $x_{j_0+k} = 0$ for all positive integers k. Note that $da_{2n} = a_{2n-1} + a_{2n+1}$, so $\varphi(da_{2n}) = \varphi(a_{2n-1}) + \varphi(a_{2n+1})$. Thus, we get that $dx_{2n}d = x_{2n-1}d + x_{2n+1}d$. Since R is a domain and $d \neq 0$,

$$dx_{2n} = x_{2n-1} + x_{2n+1}.$$

We see that $x_1 = 1$ because $\varphi(a_1) = a_1$. If $x_3 = 0$, then $dx_2 = x_1 = 1$ from $dx_2 = x_1 + x_3$. Now $(x_2 d)(x_2 d) = x_2 d$ and $x_2 d \neq 0$, so $x_2 d = 1$ as R is a domain. Thus, d is invertible, so we are done. If $x_3 \neq 0$, then we have that $x_3 = -1 + dx_2$. So $dx_4 = x_3 + x_5 = -1 + dx_2 + x_5$, and $x_5 = 1 + d(x_4 - x_2)$. In this case, if $x_5 = 0$, then d is invertible.

Otherwise, we continue the procedure. In general, suppose that $x_{2r+1} \neq 0$ with $x_{2r+1} = 1 + dm_r$ or $x_{2r+1} = -1 + dm_r$ for some $m_r \in R$. From the equality $dx_{2(r+1)} = x_{2r+1} + x_{2(r+1)+1}$, we see that $x_{2(r+1)+1} = 0$ implies that d is invertible, and $x_{2(r+1)+1} \neq 0$ implies that $x_{2(r+1)+1} = 1 + dm_{r+1}$ or $x_{2(r+1)+1} = -1 + dm_{r+1}$. Since $x_{j_0+k} = 0$ for all positive integers k, there exists some r with $x_{2(r+1)+1} = 0$, and so d is invertible. Therefore, R is a division ring. □

Given a matrix α, we use $\alpha(i, j)$ to denote its (i, j)-position element.

Theorem 6.1.10 *Let R be a ring such that* $\mathrm{CRFM}_{\mathbb{N}}(R)$ *is a Baer ring. Then R is semisimple Artinian.*

Proof Put $M(R) = \mathrm{CRFM}_{\mathbb{N}}(R)$, $N(R) = \mathrm{RFM}_{\mathbb{N}}(R)$, and $W(R) = \mathrm{CFM}_{\mathbb{N}}(R)$. Let e_{ij} be the matrix in $M(R)$ with 1 in the (i, j)-position and 0 elsewhere. Set $e_i = e_{ii}$.

Step 1. R is orthogonally finite. Assume on the contrary that there exists an infinite set of nonzero orthogonal idempotents, say $\{\alpha_i\}_{i=1}^{\infty} \subseteq R$. Consider an idempotent matrix in $N(R) = \mathrm{RFM}_{\mathbb{N}}(R)$, which is

$$a = \begin{bmatrix} \alpha_1 & 0 & \cdots & & \\ \alpha_2 & \alpha_2 & 0 & \cdots & \\ \alpha_3 & \alpha_3 & \alpha_3 & 0 & \cdots \\ & & \vdots & & \end{bmatrix} \in N(R).$$

Let

$$b = \begin{bmatrix} 1-\alpha_1 & 0 & \cdots & & & \\ -1 & 1-\alpha_2 & 0 & \cdots & & \\ 0 & -1 & 1-\alpha_3 & 0 & \cdots & \\ \vdots & 0 & \ddots & & & \\ \vdots & \vdots & & & & \end{bmatrix} \in M(R).$$

Then $ab = 0$, so $r_{M(R)}(a) \neq 0$. We can check that

$$r_{M(R)}(a) = \cap_{i=1}^{\infty} r_{M(R)}(e_i a) = r_{M(R)}(\{e_i a\}_{i=1}^{\infty})$$

and $\{e_i a\}_{i=1}^{\infty} \subseteq M(R)$. Now $r_{M(R)}(a) = fM(R)$ for some $0 \neq f^2 = f \in M(R)$ as $M(R)$ is Baer and $r_{M(R)}(a) \neq 0$. Thus, $fb = b$ since $b \in r_{M(R)}(a) = fM(R)$.

We consider the matrix

$$c = \begin{bmatrix} 0 & 1 & 1-\alpha_2 & (1-\alpha_2)(1-\alpha_3) & \cdots \\ 0 & 0 & 1 & 1-\alpha_3 & \cdots \\ \vdots & \vdots & 0 & 1 & \ddots \\ & & \vdots & 0 & \ddots \\ & & & \vdots & \end{bmatrix} \in W(R) = \mathrm{CFM}_{\mathbb{N}}(R).$$

As $f, b \in M(R)$ and $c \in W(R)$, $f(bc) = (fb)c$ by direct computation, and so $f(bc) = bc$. We see that

$$bc = \begin{bmatrix} 0 & 1-\alpha_1 & (1-\alpha_1)(1-\alpha_2) & \cdots & \\ 0 & -1 & 0 & \cdots & \\ 0 & 0 & -1 & 0 & \cdots \\ \vdots & \vdots & \vdots & \ddots & \end{bmatrix}.$$

Since $f \in M(R)$, there is a positive integer n such that $e_{n+k} f e_1 = 0$ for all positive integers k. Hence, $e_{n+k} f = e_{n+k} f(1 - e_1)$. Now we observe that $(1 - e_1)bc = e_1 - 1$. Thus, $e_{n+k} fbc = e_{n+k} f(1 - e_1)bc = e_{n+k} f(e_1 - 1) = -e_{n+k} f$. Also $e_{n+k} fbc = e_{n+k} bc = -e_{n+k}$ for all positive integers k, because $fbc = bc$. Thus, $e_{n+k} f = e_{n+k}$ for all positive integers k.

Because $f \in M(R)$, there exists a positive integer m such that $m > n$ and

$$(e_1 + \cdots + e_n) f e_{m+k} = 0$$

for all positive integers k. Thus, $e_1 f e_{m+k} = 0, \ldots, e_n f e_{m+k} = 0$. So

$$f(1, m+k) = 0, \ldots, f(n, m+k) = 0$$

(recall that $f(i, j)$ denotes the (i, j)-position of the matrix f).

On the other hand, for $i > n$, $e_i f e_{m+k} = e_i e_{m+k}$ because $e_{n+k} f = e_{n+k}$ for all positive integers k. Therefore,

$$f(i, m+k) = 0 \text{ if } i \neq m+k \text{ and } i > n, \text{ and } f(m+k, m+k) = 1.$$

Thus, $f(i, m+k) = 0$ if $i \neq m+k$ since $f(1, m+k) = 0, \ldots, f(n, m+k) = 0$. So the $(m+k)$-column of f is equal to that of e_{m+k}. Hence, $fe_{m+k} = e_{m+k}$.

Now $af \neq 0$ because $af e_{m+k} = ae_{m+k} \neq 0$ from definition of a. This contradicts the fact that $af = 0$. Therefore, R is orthogonally finite.

Step 2. $\mathbf{S}_r(R) = \mathbf{S}_\ell(R) = \mathcal{B}(R)$. Let $e \in \mathbf{S}_r(R)$. By Proposition 1.2.2,

$$R \cong \begin{bmatrix} eRe & 0 \\ (1-e)Re & (1-e)R(1-e) \end{bmatrix}.$$

Thus

$$M(R) \cong \begin{bmatrix} M(eRe) & 0 \\ M((1-e)Re) & M((1-e)R(1-e)) \end{bmatrix}.$$

To prove that $e \in \mathcal{B}(R)$, we show that $(1-e)Re = 0$. For this, assume on the contrary that $(1-e)Re \neq 0$. Take $0 \neq s \in (1-e)Re$.

- Let σ be the matrix such that $\sigma(i, j) = s$ if $i \geq j$ and $\sigma(i, j) = 0$ otherwise (i.e., σ is the lower triangular matrix with constant term s). Then it follows that $\sigma \in N((1-e)Re) \setminus M((1-e)Re)$.
- Let τ be the constant diagonal matrix with s on the diagonal. Then we note that $\tau \in M((1-e)Re)$.
- Let δ be the constant diagonal matrix with $1-e$ on the diagonal. Then we get that $\delta \in M((1-e)R(1-e))$.
- Let γ be the matrix such that $\gamma(i, j) = e$ if $i = j$, $\gamma(i, j) = -e$ if $i = j+1$, and $\gamma(i, j) = 0$ otherwise. Then $\gamma \in M(eRe)$.

We note that, for any $w \in M(eRe)$, $w\gamma = 0$ implies that $w = 0$. Put $\beta = \begin{bmatrix} 0 & 0 \\ \sigma & \delta \end{bmatrix}$ and $x = \begin{bmatrix} \gamma & 0 \\ -\tau & 0 \end{bmatrix}$. Then $\beta x = 0$ as $\sigma\gamma = \tau$ and $\delta\tau = \tau$. Since $M(R)$ is a Baer ring, $\ell_{M(R)}(x) = M(R)g$ for some $g^2 = g \in M(R)$. Note that $\beta \notin M(R)$. But, for any e_i, $e_i\beta \in M(R)$ and so $e_i\beta \in \ell_{M(R)}(x) = M(R)g$. Hence, $e_i\beta = e_i\beta g$. Therefore, $\beta = \beta g$. Now we let

$$g = \begin{bmatrix} v & 0 \\ t & y \end{bmatrix}$$

with $v \in M(eRe)$, $t \in M((1-e)Re)$, and $y \in M((1-e)R(1-e))$. Then $v\gamma = 0$ since $gx = 0$. We see that $v = 0$ because $v \in M(eRe)$ and $v\gamma = 0$.

From $\beta g = \beta$, it follows that $\delta t = \sigma$. Because $t \in M((1-e)Re)$, $\delta t = t$ by definition of δ. Hence,

$$\sigma = t \in M((1-e)Re),$$

a contradiction because $\sigma \notin M((1-e)Re)$. Therefore $(1-e)Re = 0$. So we obtain that $e \in \mathbf{S}_\ell(R)$ from Proposition 1.2.2. Consequently, $e \in \mathcal{B}(R)$ by Proposition 1.2.6(i), and hence $\mathbf{S}_r(R) = \mathcal{B}(R)$.

If $h \in \mathbf{S}_\ell(R)$, then $1-h \in \mathbf{S}_r(R)$ by Proposition 1.2.2. Hence, $1-h \in \mathcal{B}(R)$ by the preceding argument, thus $h \in \mathcal{B}(R)$. So $\mathbf{S}_r(R) = \mathbf{S}_\ell(R) = \mathcal{B}(R)$.

Step 3. R is a finite direct sum of prime Baer rings. Since $M(R)$ is a Baer ring and $R \cong e_1M(R)e_1$, R is a Baer ring from Theorem 3.1.8. As R is orthogonally finite by Step 1, $\text{Tdim}(R) = n < \infty$ by Proposition 5.2.13(ii). Say $\{b_1, \ldots, b_n\}$ is a complete set of left triangulating idempotents of R. Then $b_1 \in \mathbf{S}_\ell(R)$, so

$b_1 \in \mathcal{B}(R)$ from Step 2. By Proposition 5.1.7(ii), $b_1 + b_2 \in \mathbf{S}_\ell(R)$. So by Step 2, $b_1 + b_2 \in \mathcal{B}(R)$, and hence $b_2 \in \mathcal{B}(R)$. Also by Proposition 5.1.7(ii), $b_1 + b_2 + b_3 \in \mathbf{S}_\ell(R)$, so $b_1 + b_2 + b_3 \in \mathcal{B}(R)$ from Step 2. Thus $b_3 \in \mathcal{B}(R)$ since $b_1 + b_2 \in \mathcal{B}(R)$. By continuing this procedure, $b_i \in \mathcal{B}(R)$ for all i. From Theorems 3.1.8 and 5.4.12, R is a finite direct sum of prime Baer rings.

Step 4. R is semisimple Artinian. By Step 3, we may assume that R is a prime Baer ring. As R is orthogonally finite, R has a complete set of primitive idempotents (see Proposition 1.2.15), say $\{h_1, h_2 \dots, h_n\}$. Then it follows that $R = h_1 R \oplus \cdots \oplus h_n R$ as right R-modules. By Theorem 3.1.8, each $h_i R h_i$ is a Baer ring. Thus from Example 3.1.4(ii), each $h_i R h_i$ is a domain. In matrix notation, $R \cong [h_i R h_j]$, so $M(R) \cong [M(h_i R h_j)]$. Since $M(R)$ is Baer, so is $M(h_i R h_i)$ for each i from Theorem 3.1.8. Hence by Lemma 6.1.9, each $h_i R h_i$ is a division ring.

We claim that $h_i R_R$ is simple. Indeed, let $0 \neq w \in h_i R$. Then $w = h_i w$. Since R is a prime ring, $h_i w R h_i \neq 0$. Take $h_i w u h_i \neq 0$, where $u \in R$. There exists $v \in R$ with $h_i = (h_i w u h_i)(h_i v h_i) = w u h_i v h_i$ because $h_i R h_i$ is a division ring with identity h_i and $w = h_i w$. Hence, $h_i R \subseteq wR$, so $h_i R = wR$. Therefore $h_i R_R$ is simple, and hence R is semisimple Artinian. Since R is prime, R is simple Artinian. □

For a division ring R, $\mathrm{CRFM}_\Gamma(R)$ has been investigated by Mackey [286] and Ornstein [321] under the topological notion of continuous endomorphisms of $\mathrm{End}(R_R^{(\Gamma)})$. As is mentioned in [246, p. 5], the following lemma has been shown by Mackey [286] and Ornstein [321].

Lemma 6.1.11 *If R is a division ring, then $\mathrm{CRFM}_\Gamma(R)$ is Baer for any infinite ordered set Γ.*

Theorem 6.1.12 *The following are equivalent for a ring R.*

(i) $\mathrm{CRFM}_\Gamma(R)$ *is Baer for any nonempty ordered set* Γ.
(ii) $\mathrm{CRFM}_{\mathbb{N}}(R)$ *is Baer.*
(iii) R *is semisimple Artinian.*

Proof (i)⇒(ii) It is clear. Theorem 6.1.10 yields that (ii)⇒(iii). The implication (iii)⇒(i) follows from Lemma 6.1.11. □

When R is a commutative ring, Theorems 6.1.1 and 6.1.12 yield the next corollary. It is interesting to compare Corollary 6.1.13 with Corollary 6.1.5.

Corollary 6.1.13 *Let R be a commutative ring. Then the following are equivalent.*

(i) $\mathrm{CFM}_\Gamma(R)$ *is a Baer ring for any nonempty ordered set* Γ.
(ii) $\mathrm{RFM}_\Gamma(R)$ *is a Baer ring for any nonempty ordered set* Γ.
(iii) $\mathrm{CRFM}_\Gamma(R)$ *is a Baer ring for any nonempty ordered set* Γ.
(iv) $\mathrm{CRFM}_{\mathbb{N}}(R)$ *is a Baer ring.*
(v) R *is a finite direct sum of fields.*

Proof (i)⇔(ii) follows from Theorem 6.1.1. To see that (i)⇒(v), assume that (i) holds. By Theorem 6.1.1, R is semiprimary hereditary. Thus, R is orthogonally finite. Also R is Baer. By Corollary 5.4.13(i), R is a finite direct sum of fields since R is semiprime and semiprimary. (v)⇒(i) follows from Theorem 6.1.1. Theorem 6.1.12 yields that (iii)⇔(iv)⇔(v). □

For the Rickart ring property, we obtain the following.

Theorem 6.1.14 *Let R be a ring. Then the following are equivalent.*

(i) *R is right hereditary.*
(ii) $\mathrm{CFM}_\Gamma(R)$ *is right Rickart for any nonempty ordered set Γ.*

Proof Since any free right R-module is k-local retractable (see Exercise 4.5.16.2), Theorems 4.5.6 and 4.5.9 yield the result. □

In the next theorem, we obtain a characterization of rings for which every finitely generated free module is Rickart.

Proposition 6.1.15 *Let R be a ring. Then the following are equivalent.*

(i) *Every finitely generated free (projective) right R-module is a Rickart module.*
(ii) *Every finite direct sum of copies of $R_R^{(k)}$ is a Rickart R-module for some positive integer k.*
(iii) $\mathrm{Mat}_n(R)$ *a right Rickart ring for any positive integer n.*
(iv) *R is a right semihereditary ring.*

Proof (i)⇒(ii) is clear. For (ii)⇒(i), say $R_R^{(m)}$ is a finitely generated free R-module. Then $R_R^{(m)} \leq^{\oplus} R_R^{(km)} = \oplus^m R_R^{(k)}$. By hypothesis, $\oplus^m R_R^{(k)}$ is a Rickart module. By Proposition 4.5.4(i), $R_R^{(m)}$ is a Rickart module.

(i)⇔(iii) follows from Theorem 4.5.6 because every free module is k-local retractable (see Exercise 4.5.16.2). (iii)⇔(iv) is Theorem 3.1.29. □

Some special conditions are required on a Baer ring R for the transference of the Baer property to $\mathrm{Mat}_n(R)$, $\mathrm{CFM}_\Gamma(R)$, $\mathrm{RFM}_\Gamma(R)$, and $\mathrm{CRFM}_\Gamma(R)$ in Theorems 6.1.1, 6.1.3, 6.1.4, and 6.1.12. However, the next result once again shows that the quasi-Baer property always transfers to these types of infinite matrix ring extensions without any additional assumptions.

Theorem 6.1.16 *The following are equivalent for a ring R.*

(i) *R is quasi-Baer.*
(ii) $\mathrm{Mat}_n(R)$ *is quasi-Baer for every positive integer n.*
(iii) $\mathrm{CFM}_\Gamma(R)$ *is quasi-Baer for any nonempty ordered set Γ.*
(iv) $\mathrm{RFM}_\Gamma(R)$ *is quasi-Baer for any nonempty ordered set Γ.*
(v) $\mathrm{CRFM}_\Gamma(R)$ *is quasi-Baer for any nonempty ordered set Γ.*

Proof The equivalence (i)$\Leftrightarrow$(ii) follows from Theorem 3.2.12. Also the equivalence (i)$\Leftrightarrow$(iii)$\Leftrightarrow$(iv) is Corollary 4.6.20.

(i)$\Leftrightarrow$(v) Assume that R is quasi-Baer. Put $M(R) = \mathrm{CRFM}_\Gamma(R)$. We let e_{ij} denote the matrix in $M(R)$ with 1 in the (i, j)-position and 0 elsewhere. For $\omega \in M(R)$, ω_{ij} denotes the (i, j)-position element of ω.

Say $I \trianglelefteq M(R)$. Let A be the set of entries of matrices in I. Then $I \trianglelefteq R$. In fact, let $x, y \in A$. Then there exist $\alpha, \beta \in I$ such that $x = \alpha_{ki}$ and $y = \beta_{\ell j}$ for some $k, i, \ell, j \in \Gamma$. Then $\xi := \alpha + e_{k\ell}\beta e_{ji} \in I$ and $\xi_{ki} = x + y$, so $x + y \in A$. Further, for $r \in R$, as $e_{kk}\alpha(re_{ii}) \in I$ and $(re_{kk})\alpha e_{ii} \in I$, $xr \in A$ and $rx \in A$. Therefore, $A \trianglelefteq R$.

Because R is quasi-Baer, there exists $e^2 = e \in R$ with $r_R(A) = eR$. So $(e\mathbf{1}_{M(R)})M(R) \subseteq r_{M(R)}(I)$, where $\mathbf{1}_{M(R)}$ is the identity matrix in $M(R)$. Next to see that $r_{M(R)}(I) \subseteq (e\mathbf{1}_{M(R)})M(R)$, let $[b_{mn}] \in r_{M(R)}(I)$. We claim that $Ab_{ij} = 0$ for all $i, j \in \Gamma$. Let $a \in A$ and choose $i, j \in \Gamma$. There exists $\alpha \in I$ such that $a = \alpha_{k\ell}$, where $k, \ell \in \Gamma$. Then $e_{kk}\alpha e_{\ell\ell}e_{\ell i} \in I$, therefore we have that $(e_{kk}\alpha e_{\ell\ell}e_{\ell i})[b_{mn}]e_{jj} = 0$ since $[b_{mn}] \in r_{M(R)}(I)$. We observe that ab_{ij} is the (k, j)-position element of $(e_{kk}\alpha e_{\ell\ell}e_{\ell i})[b_{mn}]e_{jj}$, and hence $ab_{ij} = 0$. Since i and j are chosen arbitrarily, $Ab_{ij} = 0$ for all $i, j \in \Gamma$.

Hence all $b_{ij} \in r_R(A) = eR$, and thus $b_{ij} = eb_{ij}$ for all $i, j \in \Gamma$. Therefore, we have that $[b_{mn}] = [eb_{mn}] = (e\mathbf{1}_{M(R)})[b_{mn}] \in (e\mathbf{1}_{M(R)})M(R)$. Consequently, it follows that $r_{M(R)}(I) = (e\mathbf{1}_{M(R)})M(R)$, so $M(R)$ is quasi-Baer.

Conversely, if $\mathrm{CRFM}_\Gamma(R)$ is quasi-Baer, then R is a quasi-Baer ring by Theorem 3.2.10. □

The transference of the FI-extending property of a ring R to infinite matrix rings over R, is shown in the next result.

Theorem 6.1.17 *Let R be a right FI-extending ring. Then* $\mathrm{CFM}_\Gamma(R)$ *and* $\mathrm{CRFM}_\Gamma(R)$ *are right FI-extending rings for any nonempty ordered set* Γ.

Proof Let S denote either $\mathrm{CFM}_\Gamma(R)$ or $\mathrm{CRFM}_\Gamma(R)$. We let e_{ij} denote the matrix in S with 1 in the (i, j)-position and 0 elsewhere. For $\omega \in S$, ω_{ij} denotes the (i, j)-position element of ω.

Say $I \trianglelefteq S$. Let A be the set of all entries of matrices in I. Then $A \trianglelefteq R$ as in the proof of Theorem 6.1.16, so $A_R \leq^{\text{ess}} eR_R$ for some $e^2 = e \in R$ since R is right FI-extending.

Let $f = e\mathbf{1}_S \in S$, where $\mathbf{1}_S$ is the identity matrix of S. To show that $I_S \leq^{\text{ess}} fS_S$, take $0 \neq \beta \in fS$. Then there is j_0 such that the j_0-th column of β is nonzero. Let $\{er_1, \dots, er_n\}$ be the set of all nonzero entries in the j_0-th column, where $r_i \in R$ for $i = 1, \dots, n$. Then there exists $s \in R$ such that $\{er_1 s, \dots, er_n s\} \subseteq A$ and $er_k s \neq 0$ for some $k \in \{1, \dots, n\}$ because $A_R \leq^{\text{ess}} eR_R$. Let $\xi = se_{j_0 j_0} \in S$, which is the matrix with s in the (j_0, j_0)-position and 0 elsewhere. Then the j_0-th column of $\beta\xi$ is nonzero with the entries $\{er_1 s, \dots, er_n s\}$. All other columns of $\beta\xi$ are zero.

Say $er_m s$ is the (k_m, j_0)-position element of the matrix $\beta\xi$ for $1 \leq m \leq n$. As $er_1 s \in A$, there is $[a_{ij}] \in I$ with $a_{i_1 j_1} = er_1 s$. Let $[b_{ij}] = e_{i_1 i_1}[a_{ij}]e_{j_1 j_1}$. Then we see that $[b_{ij}] = (er_1s)e_{i_1 j_1} \in I$. Now put

$$\alpha_1 = e_{k_1 i_1}(er_1s)e_{i_1 j_1}e_{j_1 j_0} \in I.$$

Then $\alpha_1 = (er_1s)e_{k_1 j_0}$. Similarly, $\alpha_2 := (er_2s)e_{k_2 j_0} \in I$, and so on. Therefore, we obtain that $0 \neq \beta\xi = \alpha_1 + \alpha_2 + \cdots + \alpha_n \in I$, hence $I_S \leq^{\text{ess}} fS_S$. Thus S is right FI-extending. □

We remark that if R is right strongly FI-extending, then $\text{CFM}_\Gamma(R)$ is right strongly FI-extending for any nonempty ordered set Γ (see Exercise 6.1.18.3).

Exercise 6.1.18

1. ([392, Tercan] and [121, Chatters and Khuri]) Let R be a ring and n a positive integer. Show that $\text{Mat}_n(R)$ is a right extending ring if and only if $R_R^{(n)}$ is an extending module.
2. ([271, Lee, Rizvi, and Roman]) Let R be a commutative domain. Show that the following are equivalent.
   (i) Every finitely generated free (projective) R-module has the SIP.
   (ii) The free R-module $R_R^{(k)}$ has the SIP for some integer $k \geq 3$.
   (iii) The free R-module $R_R^{(3)}$ has the SIP.
   (iv) R is a Prüfer domain.
3. ([84, Birkenmeier, Park, and Rizvi]) Show that if R right strongly FI-extending, then so is $\text{CFM}_\Gamma(R)$ for any nonempty ordered set Γ.

6.2 Polynomial Ring Extensions and Ore Extensions

We saw in Example 3.1.28 that in general the Baer and Rickart properties do not transfer from R to $R[x]$. Motivated by that example, we now prove that this is not so for the quasi-Baer property. In particular, Theorem 6.2.4 shows that a polynomial ring over a quasi-Baer ring inherits the property. In Theorem 6.2.7, we show that an analogous result holds for right p.q.-Baer rings. It will be shown that Ore extensions of a quasi-Baer ring are also quasi-Baer under certain conditions. As an application, we will see that the quantum n-space over a quasi-Baer ring is always quasi-Baer.

Definition 6.2.1 A monoid G is called a *u.p.-monoid* (unique product monoid) if for any two nonempty finite subsets $A, B \subseteq G$ there exists an element $x \in G$ uniquely presented in the form ab where $a \in A$ and $b \in B$.

Every u.p.-monoid has no non-identity element of finite order. In [341], group algebras of a u.p.-group are studied in detail (see also [320]).

Lemma 6.2.2 *Assume that G is a u.p.-monoid. Then G is cancellative (i.e., for $g, h, x \in G$, if $gx = hx$ or $xg = xh$, then $g = h$).*

Proof Let $gx = hx$ with $g, h, x \in G$. Consider $A = \{g, h\}$ and $B = \{x\}$. If gx is uniquely presented, then $g = h$. If hx is uniquely presented, then $g = h$. Similarly, if $xg = xh$, then $g = h$. □

The following result shows that the quasi-Baer and the right p.q.-Baer properties transfer between a ring R and the monoid ring $R[G]$ of a u.p.-monoid G over R.

Theorem 6.2.3 *Let $R[G]$ be the monoid ring of a u.p.-monoid G over a ring R. Then*:

(i) *R is right p.q.-Baer if and only if $R[G]$ is right p.q.-Baer.*

(ii) *R is quasi-Baer if and only if $R[G]$ is quasi-Baer.*

Proof (i) Let R be right p.q.-Baer. Say $\alpha = a_1g_1 + \cdots + a_ng_n \in R[G]$ with $a_s \in R$ and $g_s \in G$, $s = 1, \ldots, n$. There is $e_s \in \mathbf{S}_\ell(R)$ with $r_R(a_sR) = e_sR$ for each s. Then $e = e_1 \cdots e_n \in \mathbf{S}_\ell(R)$ from Proposition 1.2.4(i). We see that $eR = \cap_{s=1}^n r_R(a_sR)$. Hence $eR[G] \subseteq r_{R[G]}(\alpha R[G])$.

Note that $r_{R[G]}(\alpha R[G]) \subseteq r_{R[G]}(\alpha R)$. We claim that

$$r_{R[G]}(\alpha R) \subseteq eR[G]$$

so that $r_{R[G]}(\alpha R[G]) = eR[G]$. For this, let $\gamma = c_1h_1 + \cdots + c_mh_m \in r_{R[G]}(\alpha R)$ with $c_s \in R$ and $h_s \in G$, $s = 1, 2, \ldots, m$. Then $\alpha R\gamma = 0$, and hence we have that $(a_1g_1 + \cdots + a_ng_n)R(c_1h_1 + \cdots + c_mh_m) = 0$.

We proceed by induction on n to show that $\gamma \in eR[G]$. If $n = 1$, then $\alpha = a_1g_1$, so $(a_1g_1)b(c_1h_1 + \cdots + c_mh_m) = a_1bc_1g_1h_1 + \cdots + a_1bc_mg_1h_m = 0$ for every $b \in R$. By Lemma 6.2.2, $g_1h_s \neq g_1h_t$ for $s \neq t$. Thus $a_1bc_s = 0$ for all s. Hence, $a_1Rc_s = 0$ for all s and so $c_s \in r_R(a_1R) = e_1R$ for all s.

As G is a u.p.-monoid, there *exist fixed* i, j with $1 \leq i \leq n$ and $1 \leq j \leq m$ such that g_ih_j is *uniquely presented* by considering two subsets

$$A = \{g_1, g_2, \ldots, g_n\} \text{ and } B = \{h_1, h_2, \ldots, h_m\} \text{ of } G.$$

From $(a_1g_1 + \cdots + a_ng_n)R(c_1h_1 + \cdots + c_mh_m) = 0$, $a_iRc_jg_ih_j = 0$ and so $a_iRc_j = 0$. Therefore $c_j \in r_R(a_iR) = e_iR$. Now, for every $b \in R$,

$$\begin{aligned}
0 &= (a_1g_1 + \cdots + a_ng_n)be_i(c_1h_1 + \cdots + c_mh_m) \\
&= (a_1g_1 + \cdots + a_{i-1}g_{i-1} + a_{i+1}g_{i+1} + \cdots + a_ng_n)be_i(c_1h_1 + \cdots + c_mh_m) \\
&= (a_1g_1 + \cdots + a_{i-1}g_{i-1} + a_{i+1}g_{i+1} + \cdots + a_ng_n)b(e_ic_1h_1 + \cdots + e_ic_mh_m).
\end{aligned}$$

By induction, $e_ic_k \in \cap_{s\neq i}^n r_R(a_sR) = \cap_{s\neq i}^n e_sR$ for k, $1 \leq k \leq m$. In particular, $c_j = e_ic_j \in \cap_{s\neq i}^n e_sR$. So $c_j \in \cap_{s=1}^n e_sR = \cap_{s=1}^n r_R(a_sR) = eR$. Thus, we have that $(\sum_{s=1}^n a_sg_s)Rc_jh_j = \sum_{s=1}^n a_sRc_jg_sh_j = 0$.

From $(a_1g_1 + \cdots + a_ng_n)R(c_1h_1 + \cdots + c_mh_m) = 0$, it follows that

$$(\sum_{s=1}^n a_sg_s)\, R\, (\sum_{k=1, k\neq j}^m c_kh_k) = 0.$$

By iteration again, there is $\ell \in \{1, 2, \ldots, j-1, j+1, \ldots, m\}$ such that $c_\ell \in eR$ to get $(\sum_{s=1}^{n} a_s g_s)\, R\, (\sum_{k=1, k\neq \ell, k\neq j}^{m} c_k h_k) = 0$.

Continuing this procedure yields that all $c_1, \ldots, c_m$ are in eR, so we have that $\gamma = c_1 h_1 + \cdots + c_m h_m \in eR[G]$. Thus $r_{R[G]}(\alpha R) \subseteq eR[G]$, hence

$$eR[G] \subseteq r_{R[G]}(\alpha R[G]) \subseteq r_{R[G]}(\alpha R) \subseteq eR[G].$$

Therefore, $r_{R[G]}(\alpha R[G]) = eR[G]$. So $R[G]$ is a right p.q.-Baer ring.

Conversely, let $R[G]$ be right p.q.-Baer. Take $a \in R$. Then

$$r_{R[G]}(aR[G]) = eR[G]$$

for some $e^2 = e \in R[G]$. Write $e = e_0\mu + e_2 g_2 + \cdots + e_n g_n$, where μ is the identity of G. Let $b \in r_R(aR)$. Since $r_R(aR) \subseteq r_{R[G]}(aR[G]) = eR[G]$, $eb = b$, so $e_0 b = b$. Thus $b \in e_0 R$, and hence $r_R(aR) \subseteq e_0 R$. As $r_{R[G]}(aR[G]) = eR[G]$, $aRe = 0$ and thus $aRe_0 = 0$. Hence, $e_0 \in r_R(aR)$ and $e_0 R \subseteq r_R(aR)$. So $r_R(aR) = e_0 R$. From $e_0 R = r_R(aR) \subseteq r_{R[G]}(aR[G]) = eR[G]$, $ee_0 = e_0$ and hence $e_0^2 = e_0$. Therefore, R is a right p.q.-Baer ring.

(ii) Assume that R is a quasi-Baer ring. Let $I \trianglelefteq R[G]$ and let I_0 be the set of all coefficients in R of elements of I. Then $I_0 \trianglelefteq R$. In fact, say $a, b \in I_0$. Then there exist $\alpha = a_1 g_1 + \cdots + a_m g_m$, $\beta = b_1 h_1 + \cdots + b_n h_n \in I$ such that $a = a_1$ and $b = b_1$, where $a_i, b_j \in R$ and $g_i, h_j \in G$ for $i = 1, \ldots, m$ and $j = 1, \ldots, n$. The $h_1\alpha + \beta g_1 \in I$ and

$$h_1\alpha + \beta g_1 = (a+b)h_1 g_1 + a_2 h_1 g_2 + \cdots + a_m h_1 g_m + b_2 h_2 g_1 + \cdots + b_n h_n g_1.$$

Note that $h_1 g_1 \neq h_1 g_k$ for $k = 2, \ldots, m$ and $h_1 g_1 \neq h_\ell g_1$ for $\ell = 2, \ldots, m$ by Lemma 6.2.2. Thus, $a + b$ is the coefficient of $h_1\alpha + \beta g_1$. So $a + b \in I_0$. Obviously, $ar, ra \in I_0$ for all $r \in R$. Thus, I_0 is an ideal of R. Because R is quasi-Baer, $r_R(I_0) = eR$ for some $e^2 = e \in R$.

We claim that $r_{R[G]}(I) = eR[G]$. Indeed, since $I_0 e = 0$, $Ie = 0$ and so $e \in r_{R[G]}(I)$. Thus, $eR[G] \subseteq r_{R[G]}(I)$. Take $c_1 h_1 + \cdots + c_m h_m \in r_{R[G]}(I)$. To show that each $c_s \in r_R(I_0) = eR$, let $a \in I_0$. Then there is $a_1 g_1 + \cdots + a_n g_n \in I$ with $a = a_1$. Since $c_1 h_1 + \cdots + c_m h_m \in r_{R[G]}(I)$, it follows that

$$(a_1 g_1 + \cdots + a_n g_n) R (c_1 h_1 + \cdots + c_m h_m) = 0.$$

As in the proof of the necessity in part (i), for each s, $1 \leq s \leq m$, we can see that all $c_s \in r_R(a_1 R) \cap \cdots \cap r_R(a_n R)$. Thus, $ac_s = a_1 c_s = 0$, and hence $c_s \in r_R(I_0) = eR$ for s, $1 \leq s \leq m$. Therefore, $c_1 h_1 + \cdots + c_m h_m \in eR[G]$. So $r_{R[G]}(I) \subseteq eR[G]$. Since $eR[G] \subseteq r_{R[G]}(I)$, $r_{R[G]}(I) = eR[G]$. Thus $R[G]$ is a quasi-Baer ring.

Conversely, assume that $R[G]$ is a quasi-Baer ring. Let $I \trianglelefteq R$. Then as in the proof of the sufficiency of part (i), $r_R(I)$ is generated by an idempotent of R. So R is a quasi-Baer ring. □

We remark that the converse of Theorem 6.2.3(i) and (ii) are true for an arbitrary (not necessarily u.p.-) monoid G. By Theorem 6.2.3, we obtain the following results for transference of the quasi-Baer and p.q.-Baer properties of rings between R and its polynomial ring extensions.

Theorem 6.2.4 *Let R be a ring and X a nonempty set of not necessarily commuting indeterminates. Then the following are equivalent.*

(i) *R is a quasi-Baer ring.*
(ii) *$R[X]$ is a quasi-Baer ring.*
(iii) *$R[[X]]$ is a quasi-Baer ring.*
(iv) *$R[x, x^{-1}]$ is a quasi-Baer ring.*
(v) *$R[[x, x^{-1}]]$ is a quasi-Baer ring.*

Proof We note that $R[X]$ is a monoid ring over the u.p.-monoid generated by X. Also the Laurent polynomial ring $R[x, x^{-1}]$ is a u.p.-monoid ring. Indeed, $R[x, x^{-1}] \cong R[C_\infty]$, where $R[C_\infty]$ is the group ring of the infinite cyclic group C_∞ over R. Thus the equivalence of (i), (ii), and (iv) follows immediately from Theorem 6.2.3(ii).

(i)⇒(v) Let $T = R[[x, x^{-1}]]$ and $I \trianglelefteq T$. We claim that $\ell_T(I) = Te$ for some $e^2 = e \in T$. If $I = 0$, then we are done. So suppose that $I \neq 0$. Let I_0 be the set of nonzero coefficients of the lowest degree terms of nonzero elements in I together with 0. Then $I_0 \trianglelefteq R$, so $\ell_R(I_0) = Re$ for some $e^2 = e \in R$ as R is quasi-Baer.

First, to see that $Te \subseteq \ell_T(I)$, take $h(x) \in I$. If $h(x) = 0$, then $eh(x) = 0$. So assume that $h(x) \neq 0$. Then $h(x)x^m = a_0 + a_1x + \cdots + a_nx^n + \cdots \in I$, for some integer $m \geq 0$, with each $a_i \in R$. If $a_0 \neq 0$, then $a_0 \in I_0$ and $ea_0 = 0$, so $eh(x)x^m = ea_1x + \cdots + ea_nx^n + \cdots \in I$. If $ea_1 \neq 0$, then $ea_1 \in I_0$. But $ea_1 = e(ea_1) = 0$, a contradiction. Similarly, $ea_k = 0$ for all $k \geq 2$. Hence $eh(x)x^m = 0$, so $eh(x) = 0$. Therefore $e \in \ell_T(I)$, thus $Te \subseteq \ell_T(I)$.

Next, we show that $\ell_T(I) \subseteq Te$. Let $g(x) \in \ell_T(I)$. Then we see that $g(x)x^k = b_0 + b_1x + \cdots + b_sx^s + \cdots$, for some integer $k \geq 0$, with each $b_i \in R$. We prove that $g(x)x^k = g(x)x^ke$. Let $0 \neq c \in I_0$. Then there exists $0 \neq f(x) \in I$ and an integer $t \geq 0$ so that $f(x)x^t = c_0x^\ell + c_1x^{\ell+1} + \cdots \in I$ for some integer $\ell \geq 0$, where $c_0 = c$. So $g(x)x^kf(x)x^t = g(x)f(x)x^{k+t} = 0$, hence $b_0c_0 = 0$. Therefore, we obtain that $b_0 \in \ell_R(I_0) = Re \subseteq \ell_T(I)$, thus $b_0 = b_0e$.

As $b_0 \in \ell_T(I)$ and $g(x)x^k \in \ell_T(I)$, $g(x)x^k - b_0 = b_1x + b_2x^2 + \cdots \in \ell_T(I)$. By the preceding argument, $b_1 \in \ell_R(I_0) \subseteq \ell_T(I)$, so $b_1 = b_1e$. As $b_1 \in \ell_T(I)$, $b_1x \in \ell_T(I)$ and $g(x)x^k - b_0 - b_1x = b_2x^2 + b_3x^3 + \cdots \in \ell_T(I)$. Similarly, we see that $b_2 \in \ell(I_0) = Re \subseteq \ell_T(I)$, hence $b_2 = b_2e$, and so on. Whence we have that $g(x)x^k = g(x)x^ke$, so $g(x) = g(x)e \in Te$. Thus $\ell_T(I) \subseteq Te$. Therefore $\ell_T(I) = Te$, hence T is quasi-Baer.

(v)⇒(i) Let $T = R[[x, x^{-1}]]$ and I a right ideal of R. Then $r_T(IT) = e(x)T$ for some $e(x) \in \mathbf{S}_\ell(T)$. Let e_0 be the constant term of $e(x)$. Since $Ie(x) = 0$, it follows that $Ie_0 = 0$ and hence $e_0R \subseteq r_R(I)$. Next, let $b \in r_R(I)$. Then $Ib = 0$ and so $ITb = 0$. Thus $b \in r_T(IT)$. Then $b \in e(x)T$ and so $b = e(x)b$. Hence $b = e_0b \in e_0R$, and thus $r_R(I) = e_0R$. As $e_0R = r_R(I) \subseteq r_T(IT) = e(x)T$, $e(x)e_0 = e_0$ and so $e_0^2 = e_0$. Hence, $r_R(I) = e_0R$ and $e_0^2 = e_0$. So R is quasi-Baer. Similarly, (i)⇔(iii). □

Corollary 6.2.5 *Let R be a semiprime ring and X a nonempty set of not necessarily commuting indeterminates. Then the following are equivalent.*

(i) *R is (strongly) FI-extending.*
(ii) *$R[X]$ is (strongly) FI-extending.*
(iii) *$R[[X]]$ is (strongly) FI-extending.*
(iv) *$R[x, x^{-1}]$ is (strongly) FI-extending.*
(v) *$R[[x, x^{-1}]]$ is (strongly) FI-extending.*

Proof It is a consequence of Theorems 6.2.4 and 3.2.37. □

Corollary 6.2.6 *Let R be a reduced ring and X a nonempty set of not necessarily commuting indeterminates. Then the following are equivalent.*

(i) *R is a Baer ring.*
(ii) *$R[X]$ is a Baer ring.*
(iii) *$R[[X]]$ is a Baer ring.*
(iv) *$R[x, x^{-1}]$ is a Baer ring.*
(v) *$R[[x, x^{-1}]]$ is a Baer ring.*

Proof The proof follows from Theorem 6.2.4 because a reduced quasi-Baer ring is a Baer ring. □

Theorem 6.2.7 *Let R be a ring and X a nonempty set of not necessarily commuting indeterminates. Then the following are equivalent.*

(i) *R is a right p.q.-Baer ring.*
(ii) *$R[X]$ is a right p.q.-Baer ring.*
(iii) *$R[x, x^{-1}]$ is a right p.q.-Baer ring.*

Proof Theorem 6.2.3(i) yields the result as $R[x, x^{-1}] \cong R[C_\infty]$. □

Corollary 6.2.8 *Let R be a reduced ring and X a nonempty set of not necessarily commuting indeterminates. Then the following are equivalent.*

(i) *R is a (right) Rickart ring.*
(ii) *$R[X]$ is a (right) Rickart ring.*
(iii) *$R[x, x^{-1}]$ is a (right) Rickart ring.*

Proof The result follows from Theorem 6.2.7 since a reduced right p.q.-Baer ring is a Rickart ring. □

In Theorem 6.2.7 or in Corollary 6.2.8, $R[X]$ cannot be replaced by $R[[X]]$. Indeed, Theorem 6.2.7(ii) and Corollary 6.2.8(ii) do not hold true for $R[[X]]$, when R is a right p.q.-Baer ring or R is a (right) Rickart ring. In the following, there is a commutative regular ring R (hence reduced Rickart and p.q.-Baer) for which the ring $R[[x]]$ is not Rickart and not p.q.-Baer.

Example 6.2.9 Let R be the ring in Example 3.1.14(ii). Then R is a commutative regular ring. Let $f(x) = \alpha_0 + \alpha_1 x + \alpha_2 x^2 + \cdots \in R[[x]]$, where

$$\alpha_0 = (0,1,0,0,\dots),\ \alpha_1 = (0,1,0,1,0,0,\dots),\ \alpha_2 = (0,1,0,1,0,1,0,0,\dots),$$

and so on. If $R[[x]]$ is Rickart, then there is $e(x)^2 = e(x) \in R[[x]]$ such that $r_{R[[x]]}(f(x)) = e(x)R[[x]]$. Let e_0 be the constant term of $e(x)$. Then by computation, $e(x) = e_0$. So $f(x)e_0 = 0$, hence $\alpha_i e_0 = 0$ for $i = 0, 1, 2, \dots$. Therefore $e_0 = (b_1, 0, b_3, 0, \dots, b_{2n+1}, 0, 0, 0, 0 \dots)$ for some positive integer n and some b_i, $i = 1, 3, \dots, 2n+1$. Let $g(x) = \beta_0 + \beta_1 x + \beta_2 x^2 + \cdots \in R[[x]]$ with

$$\beta_0 = (1,0,0,\dots),\ \beta_1 = (1,0,1,0,0,0,\dots),\ \beta_2 = (1,0,1,0,1,0,0,0,0,\dots),$$

and so on. Then $f(x)g(x) = 0$, but $g(x) \notin e_0 R[[x]] = e(x)R[[x]]$, a contradiction. Hence $R[[x]]$ is not Rickart, and also $R[[x]]$ is not p.q.-Baer.

Let $E = \{e_1, e_2, \dots\}$ be a countable subset of $\mathbf{S}_r(R)$. Then $e \in \mathbf{S}_r(R)$ is called a *generalized countable join* of E if the following conditions hold.

(1) $e_i e = e_i$ for all $i = 1, 2, \dots$.
(2) If $a \in R$ and $e_i a = e_i$ for all $i = 1, 2, \dots$, then $ea = e$.

Theorem 6.2.10 *The ring $R[[x]]$ is right p.q.-Baer if and only if R is right p.q.-Baer and each countable subset of $\mathbf{S}_r(R)$ has a generalized countable join.*

Proof See [124, Theorem 5]. □

For idempotents e and f in a reduced ring R, $e \le f$ means that $ef = e$. When R is a reduced Rickart ring and $E = \{e_1, e_2, \dots\}$ is a countable set of idempotents of R, it is proved in [124] that an idempotent $e \in R$ is a generalized countable join of E if and only if $e = \sup(E)$. The next result follows from this fact and Theorem 6.2.10.

Corollary 6.2.11 *The ring $R[[x]]$ is reduced Rickart if and only if R is a reduced Rickart ring and for any countable set E of idempotents in R, there exists $e^2 = e \in R$ such that $e = \sup(E)$.*

Let R be a ring with α a ring endomorphism of R. Then a map $\delta : R \to R$ is called an *α-derivation* of R if

$$\delta(a+b) = \delta(a) + \delta(b) \text{ and } \delta(ab) = \delta(a)b + \alpha(a)\delta(b)$$

for all $a, b \in R$. We denote by $R[x; \alpha, \delta]$ the *Ore extension* of R whose elements are polynomials over R, the addition is defined as usual and the multiplication is subject to the relation

$$xa = \alpha(a)x + \delta(a)$$

for each $a \in R$. When $\delta = 0$, we use $R[x;\alpha]$ (the skew polynomial ring) for $R[x;\alpha,\delta]$. If α is the identity map, we write $R[x;\delta]$ for $R[x;\alpha,\delta]$. The next result is on the transference of the quasi-Baer property from R to its Ore extensions.

Theorem 6.2.12 *Let R be a ring, α a ring automorphism of R, and let δ be an α-derivation of R. If R is quasi-Baer, then $R[x;\alpha,\delta]$ is quasi-Baer.*

Proof Let $I \trianglelefteq S := R[x;\alpha,\delta]$, and let I_0 be the set of all leading coefficients of elements of I together with 0. Then $I_0 \trianglelefteq R$ because α is a ring automorphism of R. Thus, $\ell_R(I_0) = Re$ with $e^2 = e \in R$.

We claim that $\ell_S(I) = Se$. For $Se \subseteq \ell_S(I)$, let $f(x) = a_0 + \cdots + a_n x^n \in I$. Then $ea_n = 0$ since $a_n \in I_0$. But $ef(x) = ea_0 + \cdots + ea_{n-1}x^{n-1} \in I$.

Hence $ea_{n-1} = eea_{n-1} = 0$ because $ea_{n-1} \in I_0$. Continuing this procedure, we obtain $ea_i = 0$ for all i. Thus $ef(x) = 0$, so $Se \subseteq \ell_S(I)$.

Next, we prove that each $g(x) \in \ell_S(I)$ satisfies $g(x) = g(x)e$ so that $\ell_S(I) \subseteq Se$. We may assume that $g(x) \neq 0$. The proof proceeds by induction on $n = \deg(g(x))$, the degree of $g(x)$. Suppose that $\deg(g(x)) = 0$. Take $d \in I_0$. Then there exists $p(x) = d_0 + \cdots + d_k x^k \in I$ with $d = d_k$. Since $g(x)p(x) = 0$, $g(x)d = 0$ and so $g(x) \in \ell_R(I_0) = Re$. Hence, $g(x) = g(x)e$. Assume inductively that the assertion is true for all $k(x) \in \ell_S(I)$ with $\deg(k(x)) < n$.

Say $g(x) = b_0 + \cdots + b_n x^n \in \ell_S(I)$. Since α is an automorphism, $b_n = \alpha^n(c)$ for some $c \in R$. Take $v \in I_0$. There is $h(x) = v_0 + \cdots + v_{m-1}x^{m-1} + vx^m \in I$ and $g(x)h(x) = 0$. Thus, $b_n\alpha^n(v) = \alpha^n(c)\alpha^n(v) = 0$ and $cv = 0$. So it follows that $c \in \ell_R(I_0) = Re$, so $c = ce$.

We observe that $b_n = \alpha^n(c) = \alpha^n(ce) = \alpha^n(c)\alpha^n(e) = b_n\alpha^n(e)$. Therefore, $g(x) = b_0 + \cdots + b_{n-1}x^{n-1} + b_n\alpha^n(e)x^n = b_0 + \cdots + b_{n-1}x^{n-1} + b_n x^n e + t(x)$ for some $t(x) \in S$ such that $\deg(t(x)) \leq n-1$ or $t(x) = 0$. Thus,

$$\begin{aligned} 0 = g(x)I &= (b_0 + \cdots + b_{n-1}x^{n-1} + b_n x^n e + t(x))I \\ &= (b_0 + \cdots + b_{n-1}x^{n-1} + t(x))I \end{aligned}$$

because $eI = 0$. Put $k(x) = b_0 + \cdots + b_{n-1}x^{n-1} + t(x)$.

Note that $g(x) = k(x) + b_n x^n e$. If $k(x) = 0$, then $g(x) = b_n x^n e$ and thus $g(x) = g(x)e$. Next, assume that $k(x) \neq 0$. By induction hypothesis, $k(x) = k(x)e$ as $k(x) \in \ell_S(I)$. So $g(x) = k(x) + b_n x^n e = k(x)e + b_n x^n e = g(x)e$, hence $\ell_S(I) \subseteq Se$. Therefore, $\ell_S(I) = Se$. □

Theorem 6.2.12 yields the following corollary immediately.

Corollary 6.2.13 *Let R be a ring and α a ring automorphism of R. If R is a quasi-Baer ring, then $R[x;\alpha]$ is a quasi-Baer ring.*

The next example provides a domain R which shows that the condition "α is a ring automorphism" cannot be relaxed to "α is a ring endomorphism" in Corollary 6.2.13.

Example 6.2.14 Let F be a field and $R = F[t]$ the polynomial ring with the endomorphism σ given by $\sigma(f(t)) = f(0)$ for $f(t) \in R$.

We see that $R[x;\sigma]$ is not a domain because $xt = \sigma(t)x = 0$. Consider a right ideal $xR[x;\sigma]$. We note that $R[x;\sigma]$ has 0 and 1 as the only idempotents by simple computation. Also note that $r_{R[x;\sigma]}(xR[x;\sigma]) \neq R[x;\sigma]$ since $1 \notin r_{R[x;\sigma]}(xR[x;\sigma])$. Further, $g(x)t = 0$ for all $g(x) \in xR[x;\sigma]$. Hence $r_{R[x;\sigma]}(xR[x;\sigma]) \neq 0$. Thus, $r_{R[x;\sigma]}(xR[x;\sigma])$ is not generated by an idempotent and so $R[x;\sigma]$ is not a quasi-Baer ring.

Theorem 6.2.12 can be applied to an iterated Ore extension, that is, a ring of the form $R[x_1;\alpha_1,\delta_1][x_2;\alpha_2,\delta_2]\cdots[x_n;\alpha_n,\delta_n]$, where α_1 is an automorphism of R and δ_1 is an α_1-derivation of R; while α_2 is an automorphism of $R[x_1;\alpha_1,\delta_1]$, δ_2 is an α_2-derivation of $R[x_1;\alpha_1,\delta_1]$, and so on.

Let $\delta_1,\dots,\delta_n$ be a commuting derivations on a ring R. Define a map D_2 on $R[x_1;\delta_1]$ by $D_2(\sum a_i x_1^i) = \sum \delta_2(a_i)x_1^i$. Since δ_2 commutes with δ_1, D_2 is a derivation on $R[x_1;\delta_1]$, and we can form the ring $R[x_1;\delta_1][x_2;D_2]$. Similarly, δ_3 can be extended to a derivation D_3 on $R[x_1;\delta_1][x_2;D_2]$, then we can form $R[x_1;\delta_1][x_2;D_2][x_3;D_3]$, and so on. Finally, we obtain the differential operator ring $R[x_1;\delta_1][x_2;D_2]\cdots[x_n;D_n]$, which is denoted by $R[x_1,\dots,x_n;\delta_1,\dots,\delta_n]$. See [185, pp. 12–20] for more details on iterated Ore extensions. The next corollary follows immediately from Theorem 6.2.12.

Corollary 6.2.15 *Let R be a quasi-Baer ring. If $\delta_1,\dots,\delta_n$ are commuting derivations of R, then $R[x_1,\dots,x_n;\delta_1,\dots,\delta_n]$ is a quasi-Baer ring.*

There is a ring R with a derivation δ such that $R[x;\delta]$ is a quasi-Baer ring, but R is not a quasi-Baer ring as the next example shows.

Example 6.2.16 Let $R = \mathbb{Z}_2[x]/A$ with a derivation δ such that $\delta(\overline{x}) = 1$, where $A = x^2\mathbb{Z}_2[x]$ and $\overline{x} = x + A$. Consider the Ore extension $R[y;\delta]$.

If we set $e_{11} = \overline{x}\,y$, $e_{12} = \overline{x}$, $e_{21} = \overline{x}\,y^2 + y$, and $e_{22} = 1 + \overline{x}\,y$, then they form a set of matrix units in $R[y;\delta]$ (see 1.1.16). By direct computation, we see that $\mathbb{Z}_2[y^2] = \{v \in R[y;\delta] \mid ve_{ij} = e_{ij}v \text{ for } i,j = 1,2\}$. By 1.1.16, we have that $R[y;\delta] \cong \mathrm{Mat}_2(\mathbb{Z}_2[y^2])$. Hence, $R[y;\delta]$ is Baer from Theorem 6.1.4 as $\mathbb{Z}_2[y^2]$ is a Prüfer domain. But R is not quasi-Baer.

Let k be a ring and let $q \in k$ be a central invertible element. Then the *quantum n-space* is the ring $\mathcal{O}_q(k^n)$, generated by k together with n additional elements $x_1,\dots,x_n$ which commute with all elements of k, and $x_jx_i = qx_ix_j$ for all $i < j$. Clearly $\mathcal{O}_q(k^2) = k[x_1][x_2;\alpha_2]$, where $k[x_1]$ is an ordinary polynomial ring and α_2 is the ring automorphism of $k[x_1]$ such that $\alpha_2(a) = a$ for $a \in k$ and $\alpha_2(x_1) = qx_1$. Also $\mathcal{O}_q(k^3) = k[x_1][x_2;\alpha_2][x_3;\alpha_3]$, where α_3 is the ring automorphism of $k[x_1][x_2;\alpha_2]$ such that $\alpha_3(a) = a$ for $a \in k$, $\alpha_3(x_1) = qx_1$ and $\alpha_3(x_2) = qx_2$, and so on.

Corollary 6.2.17 *Let k be a quasi-Baer ring. Then the quantum n-space $\mathcal{O}_q(k^n)$ is a quasi-Baer ring.*

Exercise 6.2.18

1. ([66, Birkenmeier and Park]) Let R be a ring and G be a u.p.-monoid. Show that R is quasi-Baer with $\text{Tdim}(R) = n$ if and only if $R[G]$ is quasi-Baer with $\text{Tdim}(R[G]) = n$.
2. ([193, Han, Hirano, and Kim]) Let $S = \prod_{i\in\mathbb{Z}} T_i$ with $T_i = \mathbb{Q}$, for $i \in \mathbb{Z}$. We consider the ring automorphism σ of S defined by $\sigma((a_i)_{i\in\mathbb{Z}}) = (a_{i+1})_{i\in\mathbb{Z}}$. Let $R = \oplus_{i\in\mathbb{Z}} T_i + \mathbb{Q}1_S$, where 1_S is the identity of S. Then R is a reduced Rickart ring and so is a p.q.-Baer ring. Clearly, the restriction α of σ to R is a ring automorphism of R. Show that the ring $R[x;\alpha]$ is not right p.q.-Baer.
3. ([193, Han, Hirano, and Kim]) Let R be a reduced Rickart ring with a ring automorphism σ of finite order. Prove that the ring $R[x;\sigma]$ is p.q.-Baer.
4. ([312, Nasr-Isfahani and Moussavi]) Let R be a ring with a derivation δ. Show that if R is right p.q.-Baer, then $R[x;\delta]$ is right p.q.-Baer.

6.3 Group Ring Extensions

A semiprime quasi-Baer group algebra over a field is the focus of our discussion in this section. As a byproduct, it is shown that every semiprime right Noetherian group algebra over a field is quasi-Baer (hence FI-extending). In particular, any group algebra of a polycyclic-by-finite group over a field with characteristic zero is quasi-Baer. Examples of several classes of Baer or quasi-Baer group rings are also shown.

In a semiprime ring R, an ideal A of R is called an annihilator ideal if $A = r_R(V)$ for some $V \trianglelefteq R$ (also $A = \ell_R(V)$ in this case). We start with the following result (which is due to M. Smith [373]).

Theorem 6.3.1 *Let $F[G]$ be a semiprime group algebra of a group G over a field F and A an annihilator ideal of $F[G]$. If $\alpha \in A$, then there exists a central idempotent e of $F[G]$ such that $e \in A$ and $\alpha = e\alpha$.*

Proof See [341, Theorem 3.18, pp. 143–144]. □

The next result describes semiprime quasi-Baer group algebras over a field.

Theorem 6.3.2 *Let $R = F[G]$ be a semiprime group algebra of a group G over a field F. Then R is quasi-Baer if and only if each annihilator ideal of R is finitely generated.*

Proof Let R be quasi-Baer and let A be an annihilator ideal of R. Then there exists $V \trianglelefteq R$ such that $A = r_R(V)$. As R is quasi-Baer, $r_R(V) = eR$ for some $e \in \mathbf{S}_\ell(R)$.

As R is semiprime, $e \in \mathcal{B}(R)$ by Proposition 1.2.6(ii). Thus $A = eR = ReR$, so A is finitely generated.

Conversely, assume that each annihilator ideal of R is finitely generated. Let A be an annihilator ideal of R. Then $A = \sum_{i=1}^{n} R\alpha_i R$ for some $\alpha_1, \dots, \alpha_n$ in R. By Theorem 6.3.1, there is a central idempotent $e_i \in A$ with $\alpha_i \in e_i R$ for each i. So $A = \sum_{i=1}^{n} R\alpha_i R \subseteq \sum_{i=1}^{n} e_i R \subseteq A$. Hence, $A = \sum_{i=1}^{n} e_i R$. Thus, there is a central idempotent $e \in R$ with $A = eR$. So R is quasi-Baer. □

Corollary 6.3.3 *Any semiprime right Noetherian group algebra $F[G]$ of a group G over a field F is quasi-Baer.*

Proof Say $R = F[G]$. Let A be an annihilator ideal of R. Then as R is right Noetherian, $A = a_1 R + \cdots + a_n R$ for some $a_1, \dots, a_n \in R$. Since $A \trianglelefteq R$, $A = Ra_1 R + \cdots + Ra_n R$. Thus R is quasi-Baer from Theorem 6.3.2. □

The group algebra of a polycyclic-by finite group over a field is a Noetherian ring (see [341, Corollary 2.8, p. 425]).

Corollary 6.3.4 *The group algebra $F[G]$ of a polycyclic-by-finite group G over a field F with characteristic zero is quasi-Baer.*

In the following example, there exist a polycyclic-by-finite group G and a field F with characteristic zero such that the group algebra $F[G]$ is not Baer.

Example 6.3.5 Let F be a field with characteristic zero and G be the group $D_\infty \times C_\infty$, where D_∞ is the infinite dihedral group and C_∞ is the infinite cyclic group. Then G is polycyclic-by-finite, so $F[G]$ is quasi-Baer by Corollary 6.3.4. But $F[G]$ is not Baer (see [39, Theorem 3.3.10]).

We get the next two corollaries from Theorems 3.2.37 and 6.3.2.

Corollary 6.3.6 *Let $R = F[G]$ be a semiprime group algebra of a group G over a field F. Then R is (strongly) FI-extending if and only if each annihilator ideal of R is finitely generated.*

Corollary 6.3.7 *Any semiprime right Noetherian group algebra is (strongly) FI-extending. In particular, the group algebra $F[G]$ of a polycyclic-by-finite group G over a field F with characteristic zero is (strongly) FI-extending.*

We note that if $F[G]$ is (right) FI-extending, where G is Abelian, then $F[G]$ is (right) extending. Thus, the group algebra of a polycyclic-by-finite Abelian group over a field with characteristic zero is extending (see [227] for other examples of extending group rings).

The *support* of an element $\sum_{g \in G} a_g g \in R[G]$ is the set $\{h \in G \mid a_h \neq 0\}$.

Theorem 6.3.8 *Let $R[G]$ be the group ring of a group G over a ring R. If $R[G]$ is Baer, then $R[H]$ is Baer for any subgroup H of G. In particular, R is Baer.*

Proof Let H be a subgroup of G and let $\emptyset \neq X \subseteq R[H]$. Because $R[G]$ is Baer, $\ell_{R[G]}(X) = R[G]e$ for some $e^2 = e \in R[G]$. Write $e = \sum_{h\in H} a_h h + \sum_{g\notin H} b_g g$ with $a_h, b_g \in R$. Then for $\beta \in X$, $e\beta = (\sum_{h\in H} a_h h)\beta + (\sum_{g\notin H} b_g g)\beta = 0$.

We observe that if $g \notin H$ and $h \in H$, then $gh \notin H$. So the support of $(\sum_{g\notin H} b_g g)\beta$ is contained in $G \setminus H$ because $\beta \in R[H]$. Thus, it follows that $(\sum_{h\in H} a_h h)\beta = 0$, so $\sum_{h\in H} a_h h \in \ell_{R[H]}(X) \subseteq \ell_{R[G]}(X) = R[G]e$. Hence, $\sum_{h\in H} a_h h = (\sum_{h\in H} a_h h)e = (\sum_{h\in H} a_h h)^2 + (\sum_{h\in H} a_h h)(\sum_{g\notin H} b_g g)$. Say $\alpha = \sum_{h\in H} a_h h$. Then $\alpha^2 = \alpha$ and $R[H]\alpha \subseteq \ell_{R[H]}(X)$.

Assume that $\gamma \in \ell_{R[H]}(X) \subseteq \ell_{R[G]}(X) = R[G]e$. Then we obtain

$$\gamma = \gamma e = \gamma\left(\sum_{h\in H} a_h h\right) + \gamma\left(\sum_{g\notin H} b_g g\right), \text{ so } \gamma = \gamma\left(\sum_{h\in H} a_h h\right) = \gamma\alpha \in R[H]\alpha.$$

Therefore $\ell_{R[H]}(X) = R[H]\alpha$. As a consequence, $R[H]$ is Baer. □

Proposition 6.3.9 *Let R be a subring of a ring S such that both share the same identity 1. Suppose that S is a free left R-module with a basis G such that $1 \in G$ and $ag = ga$ for all $a \in R$ and all $g \in G$. If S is quasi-Baer (resp., Baer), then R is quasi-Baer (resp., Baer).*

Proof Let $I \trianglelefteq R$. Since S is quasi-Baer, $\ell_S(SI) = Se$ for some $e^2 = e \in S$. Let $e = a_1 g_1 + \cdots + a_n g_n$, where $a_i \in R$, $g_i \in G$, $i = 1, \ldots, n$, and $g_1 = 1$. Then for all $a \in I$, $0 = ea = (a_1 g_1 + \cdots + a_n g_n)a = a_1 a g_1 + \cdots + a_n a g_n$. Thus, $a_i a = 0$ for $i = 1, \ldots, n$, so $a_i I = 0$ for $i = 1, \ldots, n$. Therefore, $a_i SI \subseteq \sum_{g\in G} a_i I g = 0$, and hence $a_i \in \ell_S(SI) = Se$, which implies that $a_i = a_i e$. Thus, $a_1^2 = a_1 \in R$. Since $a_1 I = 0$, $Ra_1 \subseteq \ell_R(I)$. Next, if $r \in \ell_R(I)$, then $rSI \subseteq \sum_{g\in G} rIg = 0$. So $r \in \ell_S(SI) = Se$, thus $r = re = r(a_1 g_1 + \cdots + a_n g_n) = ra_1 g_1 + \cdots + ra_n g_n$. Hence, $r = ra_1 \in Ra_1$. Therefore, $\ell_R(I) = Ra_1$. Thus, R is quasi-Baer. The case when R is Baer follows by similar arguments. □

Let $R[G]$ be the group ring of a group G over a ring R. Consider the ideal $\omega(R[G]) = \{\sum_{g\in G} a_g g \mid a_g \in R \text{ and } \sum_{g\in G} a_g = 0\}$. If G finite, then $\ell_{R[G]}(\omega(R[G])) = R[G]\widehat{G}$, where $\widehat{G} = \sum_{g\in G} g$ (see [264, Lemma 2, p. 154] or [341, Lemma 1.2, p. 68]).

In Theorem 6.2.3, we have seen that for the case of a u.p.-monoid G, the ring R is quasi-Baer if and only if the monoid ring $R[G]$ is quasi-Baer. On the contrary, if G is a group, then we only obtain a one-sided implication as shown in the next result. Example 6.3.11 indicates that the converse of the other implication does not hold true, in general (see also Example 6.3.16).

Theorem 6.3.10 *Let $R[G]$ be the group ring of a group G over a ring R.*

(i) *If* $R[G]$ *is quasi-Baer, then so is* R.
(ii) *If* $R[G]$ *is quasi-Baer and* G *is finite, then* $|G|^{-1} \in R$.

Proof (i) It follows immediately from Proposition 6.3.9.

(ii) Since $R[G]$ is quasi-Baer, $\ell_{R[G]}(\omega(R[G])) = R[G]\widehat{G} = R[G]e$ for some $e^2 = e \in R[G]$. So there is $\sum r_g g \in R[G]$ such that $e = (\sum r_g g)\widehat{G} = (\sum r_g)\widehat{G}$. Put $r = \sum r_g$ and $n = |G|$. Then $e = r\widehat{G}$ and $e = e^2 = r\widehat{G}r\widehat{G} = nr^2\widehat{G}$. Hence it follows that $r = nr^2$. As $\widehat{G} \in R[G]\widehat{G} = R[G]e$, there is $\sum s_g g \in R[G]$ such that $\widehat{G} = (\sum s_g g)e$. Thus, $\widehat{G} = (\sum s_g g)e = (\sum s_g g)r\widehat{G} = (\sum s_g r)\widehat{G}$. Thus, $1 = \sum s_g r$. Therefore, $1 = (\sum s_g)r = (\sum s_g)nr^2 = n(\sum s_g r)r = nr$ because $r = nr^2$. So $|G|^{-1} \in R$. □

The next example follows immediately from Theorem 6.3.10(ii).

Example 6.3.11 The group ring $\mathbb{Z}[G]$ is not quasi-Baer for any nontrivial finite group G.

Assume that R is a ring and G is a finite group. If the group ring $R[G]$ is (quasi-) Baer, then Theorems 6.3.8 and 6.3.10 yield that R is (quasi-)Baer and $|G|^{-1} \in R$. Thus, it is natural to ask whether the converse of this observation also holds true. This question for the quasi-Baer case, was raised by Hirano (see [205]). The answer to the question for both the Baer and the quasi-Baer ring cases, is in the negative. Counterexamples to this question are provided.

In the next example, there exist a commutative domain R and a finite group G such that $|G|^{-1} \in R$, but the group ring $R[G]$ is neither Baer nor right extending.

Example 6.3.12 Let $R = F[x, y]$, where F is an algebraically closed field with characteristic zero, and let G be a finite non-Abelian group. Then we have that $F[G] \cong \text{Mat}_{k_1}(D_1) \oplus \cdots \oplus \text{Mat}_{k_n}(D_n)$, for some division rings D_i and some positive integers k_i, $1 \le i \le n$ by Maschke's Theorem [203, Theorem 1.4.1]. Since each D_i is finite dimensional over F and F is algebraically closed, $D_i = F$ for each i. As G is not Abelian, there is k_i, say k_1, such that $k_1 \ge 2$.

We note that $R[G] \cong \text{Mat}_{k_1}(F[x, y]) \oplus \cdots \oplus \text{Mat}_{k_n}(F[x, y])$. If $R[G]$ is Baer, then $\text{Mat}_{k_1}(F[x, y])$ is Baer from Theorem 3.1.8, which is a contradiction by Theorem 6.1.4, as the commutative domain $F[x, y]$ is not Prüfer. Also from Theorem 6.1.4, $\text{Mat}_{k_1}(F[x, y])$ cannot be right extending. So $R[G]$ is not right extending.

Lemma 6.3.13 *Assume that* R *is a ring such that* $2^{-1} \in R$ *and* C_2 *is the group of order* 2. *Then* $R[C_2] \cong R \oplus R$ *as a ring direct sum.*

Proof Write $C_2 = \{1, g\}$. Since $2^{-1} \in R$, the map $f : R[C_2] \to R \oplus R$ defined by $f(a + bg) = (a + b, a - b)$ (ring direct sum) is a ring isomorphism. □

Proposition 6.3.14 *Let* R *be a ring. Then* $R[C_2]$ *is quasi-Baer* (*resp.*, *Baer*) *if and only if* R *is quasi-Baer* (*resp.*, *Baer*) *and* $2^{-1} \in R$.

Proof Theorem 6.3.8, Theorem 6.3.10, Lemma 6.3.13, Proposition 3.1.5, and Proposition 3.2.8 yield the result immediately. □

Proposition 6.3.15 *Let R be a ring with $2^{-1} \in R$ and let C_4 be the cyclic group of order 4. Then $R[C_4] \cong R \oplus R \oplus R[x]/(x^2+1)R[x]$ (ring direct sum).*

Proof Put $C_4 = \{1, g, g^2, g^3\}$, and let $e = (1+g^2)/2$. Then it follows that $R[C_4] = R[C_4]e \oplus R[C_4](1-e)$ (ring direct sum) since e is a central idempotent of $R[C_4]$. Next, it can be checked that $R[C_4]e = \{ae + bge \mid a, b \in R\}$ and $R[C_4](1-e) = \{a(1-e) + bg(1-e) \mid a, b \in R\}$. Now the map

$$\alpha : R[C_4]e \to R[x]/(x^2-1)R[x]$$

defined by $\alpha(ae + bge) = a + bx + (x^2-1)R[x]$ is a ring isomorphism. Also, the map

$$\beta : R[C_4](1-e) \to R[x]/(x^2+1)R[x]$$

given by $\beta(a(1-e) + bg(1-e)) = a + bx + (x^2+1)R[x]$ is a ring isomorphism. Further, $R[x]/(x^2-1)R[x] \cong R[C_2]$. By Lemma 6.3.13, $R[C_2] \cong R \oplus R$. Hence, $R[x]/(x^2-1)R[x] \cong R \oplus R$. So $R[C_4] \cong R \oplus R \oplus R[x]/(x^2+1)R[x]$. □

The next example shows the existence of a quasi-Baer ring R and a finite group G such that $|G|^{-1} \in R$, but the group ring $R[G]$ is not quasi-Baer. Thereby, Hirano's question is answered in the negative. The example shows that R is right (FI-)extending but $R[G]$ is not right FI-extending.

Example 6.3.16 Let $A = \{n/2^k \mid n \text{ and } k \text{ are integers}\}$, which is a subring of $\mathbb{Q}$. Set $R = \{a + 3b\mathbf{i} \mid a, b \in A\}$, where $\mathbf{i}$ is the imaginary unit. Then R is a subring of $\mathbb{C}$. Since R is a commutative domain, R is Baer (hence quasi-Baer). Also $|C_4|^{-1} \in R$. We claim that $R[C_4]$ is not quasi-Baer. Let F be the field of fractions of R. Define

$$\sigma : R[x]/(x^2+1)R[x] \to F \oplus F$$

by $\sigma(h(x) + (x^2+1)R[x]) = (h(\mathbf{i}),\ h(-\mathbf{i}))$. Then σ is a ring monomorphism.

Note that $(3\mathbf{i} + 3x + (x^2+1)R[x])(-3\mathbf{i} + 3x + (x^2+1)R[x]) = 0$. Therefore, $R[x]/(x^2+1)R[x]$ is not a domain. So if $R[x]/(x^2+1)R[x]$ is quasi-Baer (equivalently, Baer), then it has a nontrivial idempotent (see Proposition 3.2.5).

Since $\{(0,0), (1,0), (0,1), (1,1)\}$ is the set of all idempotents of $F \oplus F$, $\sigma(R[x]/(x^2+1)R[x])$ contains $(1,0)$ or $(0,1)$ (hence all of $(1,0)$ and $(0,1)$). Thus, there exist $r, s \in R$ such that $r + s\mathbf{i} = 1$ and $r - s\mathbf{i} = 0$. So $r = 1/2$ and $s = -\mathbf{i}/2$. But this is a contradiction because $s = -\mathbf{i}/2 \notin R$. Therefore $R[x]/(x^2+1)R[x]$ is not quasi-Baer. From Propositions 6.3.15 and 3.2.8, the ring $R[C_4]$ is not quasi-Baer.

Further, $R[x]/(x^2+1)R[x]$ is semiprime since σ is a ring monomorphism. By Proposition 6.3.15, $R[C_4]$ is a semiprime ring. Clearly, R is right FI-extending. But $R[C_4]$ is not right FI-extending by Theorem 3.2.37.

The next result provides an interesting class of quasi-Baer group rings.

Proposition 6.3.17 *The group ring $R[D_\infty]$ is quasi-Baer if and only if R is a quasi-Baer ring, where D_∞ is the infinite dihedral group.*

Proof If $R[D_\infty]$ is quasi-Baer, then R is quasi-Baer from Theorem 6.3.10(i). Conversely, let R be quasi-Baer. We note that the infinite dihedral group D_∞ is generated by t and y such that $|t| = 2$, $|y| = \infty$, and $t^{-1}yt = y^{-1}$.

Let $S = R[y, y^{-1}]$ and let σ be the ring automorphism of S satisfying that $\sigma(y) = t^{-1}yt = y^{-1}$ and $\sigma(r) = r$ for all $r \in R$. Then $x^2 - 1$ is a central polynomial of the skew polynomial ring $S[x; \sigma]$. Further, we see that

$$R[D_\infty] \cong S[x; \sigma]/(x^2 - 1)S[x; \sigma].$$

Put $T = S[x; \sigma]/(x^2 - 1)S[x; \sigma]$. We prove that T is a quasi-Baer ring. Let V be a nonzero ideal of T and put

$$I = \{a \in S \mid a + b\overline{x} \in V \text{ for some } b \in S\}$$

and

$$J = \{b \in S \mid a + b\overline{x} \in V \text{ for some } a \in S\},$$

where $\overline{x} = x + (x^2 - 1)S[x; \sigma]$. Then $I = J$ is an ideal of S. From Theorem 6.2.4, $S = R[y, y^{-1}]$ is quasi-Baer since R is quasi-Baer. Thus, $\ell_S(I) = Se$ for some $e^2 = e \in S$.

We show that $\ell_T(V) = Te$. As $eI = 0$ and $I = J$, we have that $eV = 0$ and hence $Te \subseteq \ell_T(V)$. Next, let $c + d\overline{x} \in \ell_T(V)$, where $c, d \in S$. To prove that $c + d\overline{x} \in Te$, first take $a_0 \in I$. Then there exists $b_0 \in I$ such that $a_0 + b_0\overline{x} \in V$ because $I = J$. Therefore, for all $a \in S$, we have that

$$\begin{aligned} 0 &= (c + d\,\overline{x})\,a\,(a_0 + b_0\overline{x}) = (c + d\,\overline{x})\,(aa_0 + ab_0\overline{x}) \\ &= (caa_0 + d\sigma(a)\sigma(b_0)) + (cab_0 + d\sigma(a)\sigma(a_0))\overline{x}. \end{aligned}$$

Thus, for all $a \in S$, $caa_0 + d\sigma(a)\sigma(b_0) = 0$ and $cab_0 + d\sigma(a)\sigma(a_0) = 0$. We take $a = y^n$ ($n \in \mathbb{Z}$). Then

$$cy^n a_0 + dy^{-n}\sigma(b_0) = 0 \text{ and } cy^n b_0 + dy^{-n}\sigma(a_0) = 0.$$

Since y^n is in the center of S, $y^{2n}ca_0 = -d\sigma(b_0)$ and $y^{2n}cb_0 = -d\sigma(a_0)$. This holds for all $n \in \mathbb{Z}$ and c, d, a_0, b_0 are fixed elements of S. Thus, $ca_0 = 0$ and $d\sigma(a_0) = 0$. So $0 = \sigma^{-1}(d\sigma(a_0)) = \sigma^{-1}(d)a_0$. Since a_0 is an arbitrary element of I, it follows that $c, \sigma^{-1}(d) \in \ell_S(I) = Se$. Write $c = s_1e$ and $\sigma^{-1}(d) = s_2e$, where $s_1, s_2 \in S$. Then $d = \sigma(s_2)\sigma(e)$.

Now $c + d\overline{x} = s_1e + \sigma(s_2)\sigma(e)\overline{x} = (s_1 + \sigma(s_2)\overline{x})e \in Te$. So $\ell_T(V) \subseteq Te$. Hence, $\ell_T(V) = Te$. Therefore, T is a quasi-Baer ring. □

Remark 6.3.18 (i) The group ring $\mathbb{Z}[D_\infty]$ is not Baer. For, if $\mathbb{Z}[D_\infty]$ is Baer, then $\mathbb{Z}[C_2]$ is Baer from Theorem 6.3.8. However, this is impossible by Theorem 6.3.10(ii).

(ii) From Proposition 6.3.17, $\mathbb{Z}[D_\infty]$ is quasi-Baer. But $\mathbb{Z}[C_2]$ is not quasi-Baer by Theorem 6.3.10(ii). Thus, the quasi-Baer analogue of Theorem 6.3.8 does not hold.

Exercise 6.3.19

1. ([418, Yi and Zhou]) Assume that A is the ring of numbers $n/3^k$, where n is an integer and k is a nonnegative integer. Let $R = \{a + b\sqrt{3}\mathbf{i} \mid a, b \in A\}$. Then R is Baer and $|C_3|^{-1} \in R$. Prove that the group ring $R[C_3]$ is not quasi-Baer.
2. ([123, Chen, Li, and Zhou]) Assume that R is a ring for which $6^{-1} \in R$. Show that $R[S_3] \cong R \oplus R \oplus \mathrm{Mat}_2(R)$. Thus, $R[S_3]$ is quasi-Baer if and only if R is quasi-Baer. Moreover, if R is a Prüfer domain, then $R[S_3]$ is Baer.
3. ([418, Yi and Zhou]) Let G be a finite group and n be an integer such that $n > 1$. Show that the following are equivalent.
   (i) $\mathbb{Z}_n[G]$ is Baer.
   (ii) $\mathbb{Z}_n[G]$ is quasi-Baer.
   (iii) $|G|$ and n are relatively prime, and n is square free.

Historical Notes Theorems 6.1.1 and 6.1.3 are due to Rizvi and Roman [360]. Theorem 6.1.2(ii) is due to Chase [114]. Theorem 6.1.4 appears in [414, 419], and [271]. See also [194] for Theorem 6.1.4. Lemmas 6.1.8, 6.1.9, and Theorem 6.1.10 are due to Camillo, Costa-Cano, and Simón in [110]. Herein their proof of Theorem 6.1.10 has been reorganized by using Proposition 5.2.13, Theorem 5.4.12, and semicentral idempotents. Theorem 6.1.14 was obtained in [277] and [271]. Proposition 6.1.15 appears in [271]. Theorem 6.1.16 and Theorem 6.1.17 are taken from [100].

Theorem 6.2.3 is due to Birkenmeier and Park in [66]. Ordered monoid rings over a quasi-Baer ring were studied by Hirano [206]. The Baer and Rickart properties of the monoid ring $R[G]$ of a u.p.-monoid G over a reduced ring R were investigated in [188]. Also the Rickart property for polynomial rings has been studied in [239]. Theorem 6.2.4 appears in [77] and [422]. Corollary 6.2.6 generalizes a well-known result of Armendariz [20]. Theorem 6.2.7 is taken from [66]. Example 6.2.9 is in [75], while Theorem 6.2.10 appears in [124]. Corollary 6.2.11 is in [171]. For the right p.q.-Baer property of $R[[x]]$, see also [281]. Theorem 6.2.12, Corollary 6.2.13, Corollary 6.2.15, and Corollary 6.2.17 appear in [312]. Example 6.2.14 is in [193] (see [295, Example (i), p. 18]) and Example 6.2.16 appears in [28]. A. Moussavi and his colleagues have studied various generalizations of the Baer and related properties and have investigated these properties with respect to several types of ring extensions (see [304] and [312]).

Theorem 6.3.2 is in [66], while Corollaries 6.3.6 and 6.3.7 are from [84]. Results 6.3.8–6.3.11 appear in [418]. Example 6.3.12 is in [97]. Also Results 6.3.13–6.3.17 and Remark 6.3.18 are due to Yi and Zhou [418]. See [39, 72], and [227] for Baer or extending group algebras. Some related results can be found also in [143, 283, 313, 319], and [421].

Chapter 7
Essential Overring Extensions-Beyond the Maximal Ring of Quotients

This chapter is mainly concerned with the study of right essential overrings of a ring and their properties. We provide various stimulating results and examples of right essential overrings which are not right rings of quotients of a ring R. We describe all possible right essential overrings of the ring

$$R = \begin{bmatrix} \mathbb{Z}_4 & 2\mathbb{Z}_4 \\ 0 & \mathbb{Z}_4 \end{bmatrix}.$$

The importance of this ring R is that it was used by Osofsky in [327] to show that there are rings whose injective hulls have no compatible ring structures.

The right essential overrings of R possess many interesting properties as we shall see. Furthermore, Osofsky compatibility, more specifically a class of rings R for which $E(R_R)$ has distinct compatible ring structures will be discussed. Results and examples in this chapter also provide a motivation for the definition of ring hulls (see Definition 8.2.1).

7.1 Compatibility of Ring Structures

In this section, we discuss the right essential overrings of a ring. Also we consider the following class of rings R. Let A be a commutative local QF-ring with nonzero Jacobson radical and let

$$R = \begin{bmatrix} A & \mathrm{Soc}(A) \\ 0 & A \end{bmatrix},$$

where $\mathrm{Soc}(A)$ is the socle of A. We note that $R = Q(R)$ as R is right Kasch by Proposition 1.3.18. We explicitly describe an injective hull of R_R and discuss ring structures on this injective hull. Then we consider an intermediate R-module S_R between R_R and $E(R_R)$ with two *nonisomorphic compatible structures* $(S, +, \circ)$ and $(S, +, \star)$, where $\circ$ and $\star$ are ring multiplications, each of which is compatible with the R-module scalar multiplication of S_R. Further, $(S, +, \circ)$ is right self-injective

G.F. Birkenmeier et al., *Extensions of Rings and Modules*,
DOI 10.1007/978-0-387-92716-9_7,

(in fact, QF), while $(S,+,\star)$ is not right self-injective. Actually, $(S,+,\star)$ is not even right FI-extending.

Definition 7.1.1 An overring S of a ring R is called a *right essential overring* of R if $R_R \leq^{\text{ess}} S_R$.

We notice from Definition 1.3.10 that an overring T of ring R is a right ring of quotients of R if and only if $R_R \leq^{\text{den}} T_R$. Since every dense overmodule is an essential overmodule, any right ring of quotients of R is a right essential overring of R. The next example presents a right essential overring of a ring R which is not a right ring of quotients of R. Indeed, R has a right essential overring which is incomparable with its maximal right ring of quotients.

Example 7.1.2 Assume that R_1 is a right Kasch ring with a right essential overring S_1 such that R_1 is a proper subring of S_1. Also assume that R_2 is a ring with a right ring of quotients S_2 that is properly intermediate between R_2 and $Q(R_2)$. Let $R = R_1 \oplus R_2$. Then $S = S_1 \oplus S_2$ is a proper right essential overring of R such that $Q(R)$ is not contained in S and S is not contained in $Q(R)$. For a concrete example, take $R_1 = \begin{bmatrix} \mathbb{Z}_4 & 2\mathbb{Z}_4 \\ 0 & \mathbb{Z}_4 \end{bmatrix}$ (see Example 7.1.8) and $R_2 = \mathbb{Z}$.

The following result provides information on a right ring of quotients which is right self-injective.

Theorem 7.1.3 *Let T be a right ring of quotients of a ring R. Then T is right self-injective if and only if $T = E(R_R)$.*

Proof Assume that T is right self-injective. Note that $R_R \leq^{\text{den}} T_R$. We claim that T_R is injective. For this, let $I_R \leq R_R$ and $f \in \text{Hom}(I_R, T_R)$. Define $g : IT \to T$ by $g(\sum_{i=1}^n a_i t_i) = \sum_{i=1}^n f(a_i)t_i$, where $a_i \in I$ and $t_i \in T$ for $i = 1, \ldots, n$.

To see that g is well-defined, say $\sum_{i=1}^n a_i t_i = 0$. If $\sum_{i=1}^n f(a_i)t_i \neq 0$, then there is $r_1 \in R$ such that $t_1 r_1 \in R$ and $\sum_{i=1}^n f(a_i)t_i r_1 \neq 0$ since $R_R \leq^{\text{den}} T_R$. Again there is $r_2 \in R$ with $t_2 r_1 r_2 \in R$ and $\sum_{i=1}^n f(a_i)t_i r_1 r_2 \neq 0$. Note that $t_1 r_1 r_2 \in R$ and $t_2 r_1 r_2 \in R$. Continuing this procedure, there is $r \in R$ with $t_i r \in R$ for $i = 1, 2, \ldots, n$ and $\sum_{i=1}^n f(a_i)t_i r \neq 0$.

But $\sum_{i=1}^n f(a_i)t_i r = \sum_{i=1}^n f(a_i t_i r) = f(\sum_{i=1}^n a_i t_i r) = f(0) = 0$, which is a contradiction. Hence $\sum_{i=1}^n f(a_i)t_i = 0$, so g is well-defined. Obviously, $g \in \text{Hom}(IT_T, T_T)$. As T_T is injective, there is $t_0 \in T$ such that $g(x) = t_0 x$ for every $x \in IT$ by Baer's Criterion. Thus $f(a) = g(a) = t_0 a$ for each $a \in I$. Hence T_R is injective, so $T = E(R_R)$.

The converse follows from Proposition 2.1.32 and Baer's Criterion. □

Remark 7.1.4 Theorem 7.1.3 does not hold true in general for the case when T is a right essential overring of R. Theorems 7.1.21 and 7.1.22 will show that there is a ring R such that R has a right essential overring S which is right self-injective, but $S \neq E(R_R)$.

Definition 7.1.5 Let R be a ring and S_R be a right R-module such that $R_R \leq S_R$. Then we say that a ring structure $(S, +, \circ)$ on S is *compatible* if the ring multiplication $\circ$ extends the R-module scalar multiplication of S over R.

Proposition 7.1.6 *Let R be a ring and $R_R \leq^{\text{den}} T_R$. If T has a compatible ring structure, then all of the compatible ring structures on T coincide with each other.*

Proof Let $(T, +, \circ_1)$ and $(T, +, \circ_2)$ be two compatible ring structures on T. Assume on the contrary that there are $x, y \in T$ with $x \circ_1 y - x \circ_2 y \neq 0$. Then there exists $r \in R$ such that $yr \in R$ and $(x \circ_1 y - x \circ_2 y)r \neq 0$ because $R_R \leq^{\text{den}} T_R$. Thus $(x \circ_1 y - x \circ_2 y)r = x \circ_1 (yr) - x \circ_2 (yr) = x(yr) - x(yr) = 0$, a contradiction. Thus, $\circ_1 = \circ_2$. □

In the remainder of this section, we are concerned with right essential overrings. For this, we start with a lemma as follows.

Lemma 7.1.7 *Let R be a ring and S a right essential overring of R. Then $1_R = 1_S$, where 1_R and 1_S are identities of R and S, respectively.*

Proof Assume on the contrary that $1_S - 1_R \neq 0$. Then there is $r \in R$ such that $0 \neq (1_S - 1_R)r \in R$. Thus $0 \neq 1_S r - 1_R r = r - r$, a contradiction. □

The next example illustrates that Proposition 7.1.6 does not hold true if T is a right essential overring of R that is not a right ring of quotients of R.

Example 7.1.8 Let $R = \begin{bmatrix} \mathbb{Z}_4 & 2\mathbb{Z}_4 \\ 0 & \mathbb{Z}_4 \end{bmatrix}$ and $T = \begin{bmatrix} \mathbb{Z}_4 & \mathbb{Z}_4 \\ 0 & \mathbb{Z}_4 \end{bmatrix}$. Then $R_R \leq^{\text{ess}} T_R$. The addition on T is the usual addition. For $t_1 = \begin{bmatrix} a_1 & b_1 \\ 0 & c_1 \end{bmatrix}$, $t_2 = \begin{bmatrix} a_2 & b_2 \\ 0 & c_2 \end{bmatrix} \in T$, let

$$t_1 \diamond_1 t_2 = \begin{bmatrix} a_1a_2 & a_1b_2 + b_1c_2 \\ 0 & c_1c_2 \end{bmatrix},$$

the usual multiplication. Next, define another multiplication $\diamond_2$ on T by

$$t_1 \diamond_2 t_2 = \begin{bmatrix} a_1a_2 & a_1b_2 + 2b_1b_2 + b_1c_2 \\ 0 & c_1c_2 + 2a_1b_2 + 2c_1b_2 \end{bmatrix}.$$

We show that $(T, +, \diamond_1)$ and $(T, +, \diamond_2)$ are all possible compatible ring structures on T. Furthermore, $\diamond_1 \neq \diamond_2$, but $(T, +, \diamond_1) \cong (T, +, \diamond_2)$.

Assume that T has a compatible ring structure. Let $e_1 = \begin{bmatrix} 1 & 0 \\ 0 & 0 \end{bmatrix}$ and $e_2 = \begin{bmatrix} 0 & 0 \\ 0 & 1 \end{bmatrix}$. By Lemma 7.1.7, $1_T = 1_R = e_1 + e_2$. Then $e_1^2 = e_1$, $e_2^2 = e_2$, and $e_1e_2 = e_2e_1 = 0$. Also, $T = e_1Te_1 + e_1Te_2 + e_2Te_1 + e_2Te_2$. Put $A = \mathbb{Z}_4$. By direct computation,

$$(1)\ e_1Te_1 = \begin{bmatrix} A & 0 \\ 0 & 0 \end{bmatrix} \text{ and } (2)\ e_2Te_1 = 0.$$

Claim 1 $e_2Te_2 = \begin{bmatrix} 0 & 0 \\ 0 & A \end{bmatrix}$.

Proof of Claim 1 Say $w := e_2\begin{bmatrix} 0 & 1 \\ 0 & 0 \end{bmatrix} = \begin{bmatrix} x & y \\ 0 & z \end{bmatrix} \in e_2Te_2$. As $w = we_2$, we have that $w = \begin{bmatrix} 0 & y \\ 0 & z \end{bmatrix}$. Note that $2w = e_2\begin{bmatrix} 0 & 2 \\ 0 & 0 \end{bmatrix} = 0$, $2\begin{bmatrix} 0 & y \\ 0 & z \end{bmatrix} = 0$ and therefore $2y = 2z = 0$. Thus $y = 2y_0$ and $z = 2z_0$ for some $y_0, z_0 \in A$. We observe that $e_1w = e_1\begin{bmatrix} 0 & 2y_0 \\ 0 & 2z_0 \end{bmatrix} = \begin{bmatrix} 0 & 2y_0 \\ 0 & 0 \end{bmatrix}$, so $y = 2y_0 = 0$ and hence $w = \begin{bmatrix} 0 & 0 \\ 0 & z \end{bmatrix} \in \begin{bmatrix} 0 & 0 \\ 0 & A \end{bmatrix}$. Since $e_2e_1e_2 = 0$ and $e_2e_2e_2 = e_2$, $e_2Te_2 = \begin{bmatrix} 0 & 0 \\ 0 & A \end{bmatrix}$.

Claim 2 $e_1Te_2 = \begin{bmatrix} 0 & A \\ 0 & 0 \end{bmatrix}$ *or* $e_1Te_2 = \left\{0, \begin{bmatrix} 0 & 1 \\ 0 & 2 \end{bmatrix}, \begin{bmatrix} 0 & 2 \\ 0 & 0 \end{bmatrix}, \begin{bmatrix} 0 & 3 \\ 0 & 2 \end{bmatrix}\right\}$.

Proof of Claim 2 If $\begin{bmatrix} u & v \\ 0 & w \end{bmatrix} \in e_1Te_2$, then $\begin{bmatrix} u & v \\ 0 & w \end{bmatrix} = \begin{bmatrix} u & v \\ 0 & w \end{bmatrix} e_2 = \begin{bmatrix} 0 & v \\ 0 & w \end{bmatrix}$. Because $T = e_1Te_1 + e_1Te_2 + e_2Te_1 + e_2Te_2$ and $e_2Te_1 = 0$, by (1), (2), and Claim 1, we see that

$$\begin{bmatrix} 0 & 1 \\ 0 & 0 \end{bmatrix} = \begin{bmatrix} a & 0 \\ 0 & 0 \end{bmatrix} + \begin{bmatrix} 0 & b \\ 0 & c \end{bmatrix} + \begin{bmatrix} 0 & 0 \\ 0 & d \end{bmatrix},$$

where $\begin{bmatrix} 0 & b \\ 0 & c \end{bmatrix} \in e_1Te_2$. Thus $a = 0$, $b = 1$ and $d = -c$. Hence, it follows that $\begin{bmatrix} 0 & 1 \\ 0 & 0 \end{bmatrix} = \begin{bmatrix} 0 & 1 \\ 0 & c \end{bmatrix} + \begin{bmatrix} 0 & 0 \\ 0 & -c \end{bmatrix}$ with $\begin{bmatrix} 0 & 1 \\ 0 & c \end{bmatrix} \in e_1Te_2$. Therefore, we have the following four cases.

(α) $c = 0$. In this case, $\begin{bmatrix} 0 & 1 \\ 0 & 0 \end{bmatrix} \in e_1Te_2$. Thus $\begin{bmatrix} 0 & A \\ 0 & 0 \end{bmatrix} \subseteq e_1Te_2$, and hence we get that $e_1Te_2 = \begin{bmatrix} 0 & A \\ 0 & 0 \end{bmatrix}$ because $|e_1Te_2| = 4$.

(β) $c = 1$. There is $\begin{bmatrix} x & 2y \\ 0 & z \end{bmatrix} \in R$ with $0 \neq \begin{bmatrix} 0 & 1 \\ 0 & 1 \end{bmatrix}\begin{bmatrix} x & 2y \\ 0 & z \end{bmatrix} = \begin{bmatrix} 0 & z \\ 0 & z \end{bmatrix} \in R$ since $R_R \leq^{\text{ess}} T_R$. As $\begin{bmatrix} 0 & 1 \\ 0 & 1 \end{bmatrix} \in e_1Te_2 \subseteq e_1T$, $0 \neq \begin{bmatrix} 0 & z \\ 0 & z \end{bmatrix} \in e_1R$, a contradiction. So this case cannot happen.

(γ) $c = 2$. We note that $|e_1Te_2| = 4$. Since $\begin{bmatrix} 0 & 1 \\ 0 & 2 \end{bmatrix} \in e_1Te_2$, it follows that $e_1Te_2 = \left\{0, \begin{bmatrix} 0 & 1 \\ 0 & 2 \end{bmatrix}, \begin{bmatrix} 0 & 2 \\ 0 & 0 \end{bmatrix}, \begin{bmatrix} 0 & 3 \\ 0 & 2 \end{bmatrix}\right\}$.

(δ) $c = 3$. As in Case (β), this case cannot happen.

By (1), (2), Claim 1, and Claim 2, we get the following Cases 1 and 2.

Case 1. $e_1Te_1 = \begin{bmatrix} A & 0 \\ 0 & 0 \end{bmatrix}$, $e_1Te_2 = \begin{bmatrix} 0 & A \\ 0 & 0 \end{bmatrix}$, $e_2Te_1 = 0$, and $e_2Te_2 = \begin{bmatrix} 0 & 0 \\ 0 & A \end{bmatrix}$.

In this case, $\begin{bmatrix}1&0\\0&0\end{bmatrix}\begin{bmatrix}0&1\\0&0\end{bmatrix}=\begin{bmatrix}0&1\\0&0\end{bmatrix}$ since $e_1 \in e_1Te_1$ and $\begin{bmatrix}0&1\\0&0\end{bmatrix} \in e_1Te_2$. Also $\begin{bmatrix}0&1\\0&0\end{bmatrix}\begin{bmatrix}0&1\\0&0\end{bmatrix} \in e_1Te_2e_1Te_2 = 0$ and $\begin{bmatrix}0&0\\0&1\end{bmatrix}\begin{bmatrix}0&1\\0&0\end{bmatrix} \in e_2e_1Te_2 = 0$. So there exists a multiplication on T such that T has a compatible ring structure under this multiplication $\diamond_1$ given by

$$\begin{bmatrix}a_1&b_1\\0&c_1\end{bmatrix}\diamond_1\begin{bmatrix}a_2&b_2\\0&c_2\end{bmatrix}=\begin{bmatrix}a_1a_2&a_1b_2+b_1c_2\\0&c_1c_2\end{bmatrix}.$$

Case 2. $e_1Te_1=\begin{bmatrix}A&0\\0&0\end{bmatrix}$, $e_1Te_2=\left\{0,\begin{bmatrix}0&1\\0&2\end{bmatrix},\begin{bmatrix}0&2\\0&0\end{bmatrix},\begin{bmatrix}0&3\\0&2\end{bmatrix}\right\}$, $e_2Te_1=0$,

and $e_2Te_2=\begin{bmatrix}0&0\\0&A\end{bmatrix}$.

In this case, there is another compatible ring structure on T as shown in the following steps.

Step 1. $\begin{bmatrix}1&0\\0&0\end{bmatrix}\begin{bmatrix}0&1\\0&0\end{bmatrix}=\begin{bmatrix}0&1\\0&2\end{bmatrix}$.

We note that $\begin{bmatrix}1&0\\0&0\end{bmatrix}\begin{bmatrix}0&1\\0&2\end{bmatrix}=\begin{bmatrix}0&1\\0&2\end{bmatrix}$ since $\begin{bmatrix}0&1\\0&2\end{bmatrix} \in e_1Te_2$ and $e_1=\begin{bmatrix}1&0\\0&0\end{bmatrix}$. Therefore we obtain $\begin{bmatrix}0&1\\0&2\end{bmatrix}=\begin{bmatrix}1&0\\0&0\end{bmatrix}\begin{bmatrix}0&1\\0&2\end{bmatrix}=\begin{bmatrix}1&0\\0&0\end{bmatrix}\begin{bmatrix}0&1\\0&0\end{bmatrix}+\begin{bmatrix}1&0\\0&0\end{bmatrix}\begin{bmatrix}0&0\\0&2\end{bmatrix}=\begin{bmatrix}1&0\\0&0\end{bmatrix}\begin{bmatrix}0&1\\0&0\end{bmatrix}$.

Step 2. $\begin{bmatrix}0&0\\0&1\end{bmatrix}\begin{bmatrix}0&1\\0&0\end{bmatrix}=\begin{bmatrix}0&0\\0&2\end{bmatrix}$.

Step 2 can be checked similarly from $\begin{bmatrix}0&0\\0&1\end{bmatrix}\begin{bmatrix}0&1\\0&2\end{bmatrix} \in e_2e_1Te_2 = 0$. Indeed,

$$0=\begin{bmatrix}0&0\\0&1\end{bmatrix}\begin{bmatrix}0&1\\0&2\end{bmatrix}=\begin{bmatrix}0&0\\0&1\end{bmatrix}\begin{bmatrix}0&1\\0&0\end{bmatrix}+\begin{bmatrix}0&0\\0&1\end{bmatrix}\begin{bmatrix}0&0\\0&2\end{bmatrix}=\begin{bmatrix}0&0\\0&1\end{bmatrix}\begin{bmatrix}0&1\\0&0\end{bmatrix}+\begin{bmatrix}0&0\\0&2\end{bmatrix}.$$

Thus $\begin{bmatrix}0&0\\0&1\end{bmatrix}\begin{bmatrix}0&1\\0&0\end{bmatrix}=\begin{bmatrix}0&0\\0&2\end{bmatrix}$.

Step 3. $\begin{bmatrix}0&1\\0&0\end{bmatrix}\begin{bmatrix}0&1\\0&0\end{bmatrix}=\begin{bmatrix}0&2\\0&0\end{bmatrix}$.

Note that $\begin{bmatrix}0&1\\0&2\end{bmatrix}\begin{bmatrix}0&1\\0&2\end{bmatrix} \in e_1Te_2e_1Te_2 = 0$ and $\begin{bmatrix}0&1\\0&2\end{bmatrix}\begin{bmatrix}0&0\\0&2\end{bmatrix}=\begin{bmatrix}0&2\\0&0\end{bmatrix}$. Thus

$$0=\begin{bmatrix}0&1\\0&2\end{bmatrix}\begin{bmatrix}0&1\\0&2\end{bmatrix}=\begin{bmatrix}0&1\\0&2\end{bmatrix}\begin{bmatrix}0&1\\0&0\end{bmatrix}+\begin{bmatrix}0&1\\0&2\end{bmatrix}\begin{bmatrix}0&0\\0&2\end{bmatrix}=\begin{bmatrix}0&1\\0&2\end{bmatrix}\begin{bmatrix}0&1\\0&0\end{bmatrix}+\begin{bmatrix}0&2\\0&0\end{bmatrix}$$

$$=\begin{bmatrix}0&1\\0&0\end{bmatrix}\begin{bmatrix}0&1\\0&0\end{bmatrix}+\begin{bmatrix}0&0\\0&2\end{bmatrix}\begin{bmatrix}0&1\\0&0\end{bmatrix}+\begin{bmatrix}0&2\\0&0\end{bmatrix}=\begin{bmatrix}0&1\\0&0\end{bmatrix}\begin{bmatrix}0&1\\0&0\end{bmatrix}+\begin{bmatrix}0&2\\0&0\end{bmatrix}$$

because $\begin{bmatrix} 0 & 0 \\ 0 & 2 \end{bmatrix}\begin{bmatrix} 0 & 1 \\ 0 & 0 \end{bmatrix} = 2\begin{bmatrix} 0 & 0 \\ 0 & 1 \end{bmatrix}\begin{bmatrix} 0 & 1 \\ 0 & 0 \end{bmatrix} = 2\begin{bmatrix} 0 & 0 \\ 0 & 2 \end{bmatrix} = 0$ by Step 2. Therefore, we have that $\begin{bmatrix} 0 & 1 \\ 0 & 0 \end{bmatrix}\begin{bmatrix} 0 & 1 \\ 0 & 0 \end{bmatrix} = \begin{bmatrix} 0 & 2 \\ 0 & 0 \end{bmatrix}$.

By Steps 1, 2, and 3 of Case 2, there is also a multiplication $\diamond_2$ on T such that T has a compatible ring structure under this multiplication:

$$\begin{bmatrix} a_1 & b_1 \\ 0 & c_1 \end{bmatrix} \diamond_2 \begin{bmatrix} a_2 & b_2 \\ 0 & c_2 \end{bmatrix} = \begin{bmatrix} a_1a_2 & a_1b_2 + 2b_1b_2 + b_1c_2 \\ 0 & c_1c_2 + 2a_1b_2 + 2c_1b_2 \end{bmatrix}.$$

Finally, define $f : (T, +, \diamond_1) \to (T, +, \diamond_2)$ by $f\begin{bmatrix} a & b \\ 0 & c \end{bmatrix} = \begin{bmatrix} a & b \\ 0 & 2b + c \end{bmatrix}$. Then f is a ring isomorphism.

In honor of Osofsky's contributions and pioneering work on the study of injective hulls of rings and their ring structures, we give the next definition.

Definition 7.1.9 We say that a ring R is *right Osofsky compatible* if some injective hull $E(R_R)$ of R_R has a ring structure, where the ring multiplication extends the R-module scalar multiplication of $E(R_R)$ over R. A left Osofsky compatible ring is defined similarly.

Every ring R satisfying $Q(R) = E(R_R)$ is right Osofsky compatible (e.g., when R is right nonsingular). The next result shows that if one injective hull of R_R has a compatible ring structure, then every injective hull of R_R has a compatible ring structure.

Proposition 7.1.10 *The following are equivalent for a ring R.*

(i) *R is right Osofsky compatible.*
(ii) *Every injective hull of R_R has a compatible ring structure.*

Proof For (i)⇒(ii), assume that there is an injective hull $E(R_R)$ of R_R such that $(E(R_R), +, \star)$ is a compatible ring structure. Let E_R be an arbitrary injective hull of R_R. Then there exists an isomorphism $\phi : E_R \to E(R_R)$ such that $\phi(r) = r$ for each $r \in R$.

Define $x \circ y = \phi^{-1}(\phi(x) \star \phi(y))$ for $x, y \in E_R$. Then $(E_R, +, \circ)$ is a ring. Further,

$$\begin{aligned} x \circ r &= \phi^{-1}(\phi(x) \star \phi(r)) = \phi^{-1}(\phi(x) \star r) \\ &= \phi^{-1}(\phi(x)r) = \phi^{-1}(\phi(xr)) \\ &= xr, \end{aligned}$$

for $x \in E_R$ and $r \in R$. Thus $(E_R, +, \circ)$ is a compatible ring structure (also, $(E_R, +, \circ) \cong (E(R_R), +, \star)$ via ϕ). (ii)⇒(i) is evident. □

The next result exhibits a relationship between $Q(R)$ and an injective hull of R_R when R is right Osofsky compatible. Recall from Sect. 1.3, that the injective hull $E = E(R_R)$ is a $(H, Q(R))$-bimodule, where $H = \mathrm{End}(E_R)$. Then $\mathrm{End}(E_R) = \mathrm{End}(E_{Q(R)})$ (see the proof of Theorem 2.1.31).

Proposition 7.1.11 *Let S be a right essential overring of R. Then $Q(R) \cap S$ is a subring of both $Q(R)$ and S. Thus, if R is right Osofsky compatible, then $Q(R)$ is a subring of $E(R_R)$.*

Proof Let $\cdot$ and $\circ$ denote the ring multiplications of $Q(R)$ and S, respectively. Put $E = E(R_R)$. Take $s \in S$ and define $f_s : S \to E$ by $f_s(x) = s \circ x$ for $x \in S$. We see that f_s is an R-homomorphism and extends to $\overline{f}_s \in \mathrm{End}(E_R)$. Because $\mathrm{End}(E_R) = \mathrm{End}(E_{Q(R)})$, $\overline{f}_s$ is a $Q(R)$-homomorphism. Let juxtaposition denote scalar multiplication (by R or $Q(R)$). For $q_1, q_2 \in Q(R) \cap S$, $q_1 \circ q_2 = f_{q_1}(q_2) = \overline{f}_{q_1}(1)q_2 = (q_1 \circ 1)q_2 = q_1 q_2 = q_1 \cdot q_2$. □

Let A be a commutative ring. For $f \in \mathrm{Hom}(\mathrm{Soc}(A)_A, A_A)$ and $x \in A$, let $f \cdot x \in \mathrm{Hom}(\mathrm{Soc}(A)_A, A_A)$ be defined by $(f \cdot x)(v) = f(xv)$ for every $v \in \mathrm{Soc}(A)$. Similarly, $x \cdot f \in \mathrm{Hom}(\mathrm{Soc}(A)_A, A_A)$ where $(x \cdot f)(v) = xf(v)$ for $v \in \mathrm{Soc}(A)$. Note that $f \cdot x = x \cdot f$.

Proposition 7.1.12 *Assume that A is a commutative self-injective ring, and let $f_0 \in \mathrm{Hom}(\mathrm{Soc}(A)_A, A_A)$ such that $f_0(a) = a$ for every $a \in \mathrm{Soc}(A)$. Then $\mathrm{Hom}(\mathrm{Soc}(A)_A, A_A) = f_0 \cdot A$.*

Proof Let $f \in \mathrm{Hom}(\mathrm{Soc}(A)_A, A_A)$. As A_A is injective, there is $\overline{f} \in \mathrm{End}(A_A)$ with $\overline{f}|_{\mathrm{Soc}(A)} = f$. For $a \in \mathrm{Soc}(A)$, $f(a) = \overline{f}(a) = ra$, where $r = \overline{f}(1) \in A$. Then $f = f_0 \cdot r$. Hence, $\mathrm{Hom}(\mathrm{Soc}(A)_A, A_A) = f_0 \cdot A$. □

Osofsky [327] showed that the ring R in the next example is not right Osofsky compatible without explicitly constructing an injective hull. We will discuss Example 7.1.13 in detail in Sect. 7.2.

Example 7.1.13 The ring $R = \begin{bmatrix} \mathbb{Z}_4 & 2\mathbb{Z}_4 \\ 0 & \mathbb{Z}_4 \end{bmatrix}$ is not right Osofsky compatible.

Motivated by Example 7.1.13, we now look at a more general class of rings. For this, let A be a commutative local QF-ring with $J(A) \neq 0$ and let

$$R = \begin{bmatrix} A & \mathrm{Soc}(A) \\ 0 & A \end{bmatrix}.$$

We see that the ring in Example 7.1.13 is a particular case of the ring R. Next, we construct an injective hull of R_R explicitly as follows.

Theorem 7.1.14 *Let* $E = \begin{bmatrix} A \oplus \mathrm{Hom}(\mathrm{Soc}(A)_A, A_A) & A \\ \mathrm{Hom}(\mathrm{Soc}(A)_A, A_A) & A \end{bmatrix}$, *where the addition is componentwise and the R-module scalar multiplication is given by*

$$\begin{bmatrix} a+f & b \\ g & c \end{bmatrix} \begin{bmatrix} x & y \\ 0 & z \end{bmatrix} = \begin{bmatrix} ax + f \cdot x & ay + f(y) + bz \\ g \cdot x & g(y) + cz \end{bmatrix}$$

for $\begin{bmatrix} a+f & b \\ g & c \end{bmatrix} \in E$ *and* $\begin{bmatrix} x & y \\ 0 & z \end{bmatrix} \in R$. *Then* E_R *is an injective hull of* R_R.

Proof We see that E is a right R-module by computation. Clearly, $R_R \leq E_R$. As A_A is uniform and $\mathrm{Soc}(A)^2 \subseteq \mathrm{Soc}(A)J(A) = 0$, for each $0 \neq v \in E$ there is $r \in R$ with $0 \neq vr \in R$ by a routine argument. So $R_R \leq^{\mathrm{ess}} E_R$. Thus, to show that E_R is an injective hull of R_R, we need to prove that E_R is an injective R-module.

As R is Artinian, $\mathrm{Soc}(R_R) = \ell_R(J(R))$, thus $\mathrm{Soc}(R_R) = \begin{bmatrix} \mathrm{Soc}(A) & \mathrm{Soc}(A) \\ 0 & \mathrm{Soc}(A) \end{bmatrix}$. Let $I_R \leq^{\mathrm{ess}} R_R$. Then $\mathrm{Soc}(R) \subseteq I$. Because $\mathrm{Soc}(A)$ is the smallest nonzero ideal of A, $I = \begin{bmatrix} B & \mathrm{Soc}(A) \\ 0 & D \end{bmatrix}$ with B and D nonzero ideals of A (Exercise 7.1.27.2). Put

$$V = \begin{bmatrix} A \oplus \mathrm{Hom}(\mathrm{Soc}(A)_A, A_A) & A \\ 0 & 0 \end{bmatrix} \text{ and } W = \begin{bmatrix} 0 & 0 \\ \mathrm{Hom}(\mathrm{Soc}(A)_A, A_A) & A \end{bmatrix}.$$

Then V and W are right R-modules such that $E = V \oplus W$.

We claim that V is an injective R-module. For this, let $\varphi \in \mathrm{Hom}(I_R, V_R)$. We show that φ can be extended to a homomorphism from R_R to V_R. First, recall that $\mathrm{Hom}(\mathrm{Soc}(A)_A, A_A) = f_0 \cdot A$ from Proposition 7.1.12. Consider the following two possibilities for I.

Case 1. $B = A$. Let $\varphi \begin{bmatrix} 1 & 0 \\ 0 & 0 \end{bmatrix} = \begin{bmatrix} a_0 + f_0 \cdot r_0 & b_0 \\ 0 & 0 \end{bmatrix}$, where a_0, r_0, $b_0 \in A$. Then $\varphi \begin{bmatrix} 1 & 0 \\ 0 & 0 \end{bmatrix} = \begin{bmatrix} a_0 + f_0 \cdot r_0 & 0 \\ 0 & 0 \end{bmatrix}$. *We fix* $0 \neq s \in \mathrm{Soc}(A)$. So $\mathrm{Soc}(A) = sA$ because $\mathrm{Soc}(A)$ is the smallest nonzero ideal of A.

Now $\varphi \begin{bmatrix} 0 & s \\ 0 & 0 \end{bmatrix} = \left(\varphi \begin{bmatrix} 1 & 0 \\ 0 & 0 \end{bmatrix} \right) \begin{bmatrix} 0 & s \\ 0 & 0 \end{bmatrix} = \begin{bmatrix} 0 & a_0 s + r_0 s \\ 0 & 0 \end{bmatrix}$. For $d \in D$, let

$$\varphi \begin{bmatrix} 0 & 0 \\ 0 & d \end{bmatrix} = \begin{bmatrix} x_d + f_0 \cdot y_d & z_d \\ 0 & 0 \end{bmatrix}$$

for some x_d, y_d, $z_d \in A$. Then we obtain

$$0 = \varphi \left(\begin{bmatrix} 0 & 0 \\ 0 & d \end{bmatrix} \begin{bmatrix} 1 & 0 \\ 0 & 0 \end{bmatrix} \right) = \left(\varphi \begin{bmatrix} 0 & 0 \\ 0 & d \end{bmatrix} \right) \begin{bmatrix} 1 & 0 \\ 0 & 0 \end{bmatrix},$$

so $x_d + f_0 \cdot y_d = 0$. Therefore $\varphi \begin{bmatrix} 0 & 0 \\ 0 & d \end{bmatrix} = \begin{bmatrix} 0 & z_d \\ 0 & 0 \end{bmatrix}$. Now, $\lambda : D \to A$ defined by $\lambda(d) = z_d$ is an A-homomorphism. Since A is self-injective, there is $z_0 \in A$ such that $\lambda(d) = z_d = z_0 d$ for all $d \in D$. So $\varphi \begin{bmatrix} 0 & 0 \\ 0 & d \end{bmatrix} = \begin{bmatrix} 0 & z_d \\ 0 & 0 \end{bmatrix} = \begin{bmatrix} 0 & z_0 d \\ 0 & 0 \end{bmatrix}$. Take

$$v_1 = \begin{bmatrix} a_0 + f_0 \cdot r_0 & z_0 \\ 0 & 0 \end{bmatrix} \in V.$$

Then φ has an extension $\overline{\varphi} \in \mathrm{Hom}(R_R, V_R)$ such that $\overline{\varphi}(1) = v_1$.

Case 2. $B \neq A$. Then B is a nonzero proper ideal of A. For $b \in B$, let

$$\varphi \begin{bmatrix} b & 0 \\ 0 & 0 \end{bmatrix} = \begin{bmatrix} x_b + f_0 \cdot r_b & y_b \\ 0 & 0 \end{bmatrix},$$

where $x_b, r_b, y_b \in A$. As in Case 1, $\varphi \begin{bmatrix} b & 0 \\ 0 & 0 \end{bmatrix} = \begin{bmatrix} x_b + f_0 \cdot r_b & 0 \\ 0 & 0 \end{bmatrix}$. Recall that $\mathrm{Soc}(A) = sA$ and $B \subseteq J(A)$. Thus $bs = 0$, so

$$0 = \varphi\left(\begin{bmatrix} b & 0 \\ 0 & 0 \end{bmatrix}\begin{bmatrix} s & 0 \\ 0 & 0 \end{bmatrix}\right) = \left(\varphi\begin{bmatrix} b & 0 \\ 0 & 0 \end{bmatrix}\right)\begin{bmatrix} s & 0 \\ 0 & 0 \end{bmatrix} = \begin{bmatrix} x_b s & 0 \\ 0 & 0 \end{bmatrix}.$$

Hence $x_b s = 0$.

On the other hand, since $bs = 0$,

$$0 = \varphi\left(\begin{bmatrix} b & 0 \\ 0 & 0 \end{bmatrix}\begin{bmatrix} 0 & s \\ 0 & 0 \end{bmatrix}\right) = \left(\varphi\begin{bmatrix} b & 0 \\ 0 & 0 \end{bmatrix}\right)\begin{bmatrix} 0 & s \\ 0 & 0 \end{bmatrix} = \begin{bmatrix} 0 & x_b s + r_b s \\ 0 & 0 \end{bmatrix}.$$

So $x_b s + r_b s = 0$, and hence $r_b s = 0$ because $x_b s = 0$. Therefore $f_0 \cdot r_b = 0$ since $\mathrm{Soc}(A) = sA$. Thus, $\varphi \begin{bmatrix} b & 0 \\ 0 & 0 \end{bmatrix} = \begin{bmatrix} x_b & 0 \\ 0 & 0 \end{bmatrix}$. As in Case 1, define $\mu : B \to A$ by $\mu(b) = x_b$. Then μ is an A-homomorphism. As A is self-injective, there is $x_0 \in A$ with $x_b = x_0 b$ for all $b \in B$. So $\varphi \begin{bmatrix} b & 0 \\ 0 & 0 \end{bmatrix} = \begin{bmatrix} x_0 b & 0 \\ 0 & 0 \end{bmatrix}$ for every $b \in B$. Next, say

$$\varphi \begin{bmatrix} 0 & s \\ 0 & 0 \end{bmatrix} = \begin{bmatrix} z + f_0 \cdot t & y \\ 0 & 0 \end{bmatrix}, \text{ where } z, t, y \in A.$$

Then $0 = \varphi\left(\begin{bmatrix} 0 & s \\ 0 & 0 \end{bmatrix}\begin{bmatrix} 1 & 0 \\ 0 & 0 \end{bmatrix}\right) = \left(\varphi\begin{bmatrix} 0 & s \\ 0 & 0 \end{bmatrix}\right)\begin{bmatrix} 1 & 0 \\ 0 & 0 \end{bmatrix}$, so $z + f_0 \cdot t = 0$.

For $a \in J(A)$, $0 = \varphi\left(\begin{bmatrix} 0 & s \\ 0 & 0 \end{bmatrix}\begin{bmatrix} 0 & 0 \\ 0 & a \end{bmatrix}\right) = \left(\varphi\begin{bmatrix} 0 & s \\ 0 & 0 \end{bmatrix}\right)\begin{bmatrix} 0 & 0 \\ 0 & a \end{bmatrix}$, thus we obtain that $ya = 0$ and $y \in \mathrm{Soc}(A) = sA$. Hence $y = su$ with $u \in A$.

So $\varphi \begin{bmatrix} 0 & s \\ 0 & 0 \end{bmatrix} = \begin{bmatrix} 0 & su \\ 0 & 0 \end{bmatrix}$. Finally, as in Case 1, there exists $\alpha_0 \in A$ with $\varphi \begin{bmatrix} 0 & 0 \\ 0 & d \end{bmatrix} = \begin{bmatrix} 0 & \alpha_0 d \\ 0 & 0 \end{bmatrix}$ for every $d \in D$. Let $v_2 = \begin{bmatrix} x_0 + f_0 \cdot (u - x_0) & \alpha_0 \\ 0 & 0 \end{bmatrix} \in V$. Then φ can be extended to $\overline{\varphi} \in \mathrm{Hom}(R_R, V_R)$ with $\overline{\varphi}(1) = v_2$.

By Cases 1 and 2, V_R is an injective R-module by Baer's Criterion. Similarly, W_R is an injective R-module, and thus $E_R = V_R \oplus W_R$ is an injective R-module. So $E_R = V_R \oplus W_R$ is an injective hull of R_R. □

Henceforth, in the remainder of this section, let A be a commutative local QF-ring with $J(A) \neq 0$ and

$$R = \begin{bmatrix} A & \mathrm{Soc}(A) \\ 0 & A \end{bmatrix}$$

as in Theorem 7.1.14. *Set*

$$S = \begin{bmatrix} A & \mathrm{Soc}(A) \\ \mathrm{Hom}(\mathrm{Soc}(A)_A, A_A) & A \end{bmatrix},$$

where the addition is componentwise and the R-module scalar multiplication of S over R is given by

$$\begin{bmatrix} a & d \\ f & c \end{bmatrix} \begin{bmatrix} x & t \\ 0 & z \end{bmatrix} = \begin{bmatrix} ax & at + dz \\ f \cdot x & f(t) + cz \end{bmatrix}$$

for $\begin{bmatrix} a & d \\ f & c \end{bmatrix} \in S$ *and* $\begin{bmatrix} x & t \\ 0 & z \end{bmatrix} \in R$. Then obviously $R_R \leq^{\text{ess}} S_R \leq^{\text{ess}} E_R$.

For inducing a compatible ring structure on S, two (A, A)-bimodule homomorphisms called pairings are considered:

$$(-,-) : \mathrm{Soc}(A) \otimes_A \mathrm{Hom}(\mathrm{Soc}(A)_A, A_A) \to A$$

and

$$[-,-] : \mathrm{Hom}(\mathrm{Soc}(A)_A, A_A) \otimes_A \mathrm{Soc}(A) \to A,$$

with $(x, f) = f(x)$, $[f, x] = f(x)$ for $x \in \mathrm{Soc}(A)$, and $f \in \mathrm{Hom}(\mathrm{Soc}(A)_A, A_A)$. Then $(-,-)$ and $[-,-]$ are (A, A)-bimodule homomorphisms. Moreover, for $x, y \in \mathrm{Soc}(A)$ and $f, g \in \mathrm{Hom}(\mathrm{Soc}(A)_A, A_A)$, we see that $(x, f)y = x[f, y]$ and $[f, x] \cdot g = f \cdot (x, g)$. Hence, there is a Morita context

$$(A,\ \mathrm{Soc}(A),\ \mathrm{Hom}(\mathrm{Soc}(A)_A, A_A),\ A)$$

with the pairings $(-,-)$ and $[-,-]$.

Consider $(S, +, \circ)$, where the addition $+$ is componentwise and the multiplication $\circ$ is defined by

$$\begin{bmatrix} a_1 & s_1 \\ f_1 & c_1 \end{bmatrix} \circ \begin{bmatrix} a_2 & s_2 \\ f_2 & c_2 \end{bmatrix} = \begin{bmatrix} a_1a_2 + (s_1, f_2) & a_1s_2 + s_1c_2 \\ f_1 \cdot a_2 + c_1 \cdot f_2 & [f_1, s_2] + c_1c_2 \end{bmatrix}.$$

We see that $\circ$ extends the R-module scalar multiplication of S over R. Therefore $(S, +, \circ)$ is a compatible ring structure.

We fix an element $0 \neq s \in \mathrm{Soc}(A)$. Then $\mathrm{Soc}(A) = sA$ since $\mathrm{Soc}(A)$ is the smallest nonzero ideal of A. So any element of S can be written as

$$\begin{bmatrix} a & sb \\ f_0 \cdot r & c \end{bmatrix}$$

with $a, b, r, c \in A$, because $\mathrm{Hom}(\mathrm{Soc}(A)_A, A_A) = f_0 \cdot A$ by Proposition 7.1.12. As $(s_1, f_2) = f_2(s_1)$ and $[f_1, s_2] = f_1(s_2)$ for $f_1, f_2 \in \mathrm{Hom}(\mathrm{Soc}(A)_A, A_A)$ and $s_1, s_2 \in \mathrm{Soc}(A)$, it follows that

$$\begin{bmatrix} a_1 & sb_1 \\ f_0 \cdot r_1 & c_1 \end{bmatrix} \circ \begin{bmatrix} a_2 & sb_2 \\ f_0 \cdot r_2 & c_2 \end{bmatrix} = \begin{bmatrix} a_1a_2 + sb_1r_2 & sa_1b_2 + sb_1c_2 \\ f_0 \cdot r_1a_2 + f_0 \cdot c_1r_2 & sr_1b_2 + c_1c_2 \end{bmatrix}.$$

For our results, other ring multiplications on S are provided. Let $k \in A$ and $t \in \mathrm{Soc}(A)$. Say $\begin{bmatrix} a_1 & sb_1 \\ f_0 \cdot r_1 & c_1 \end{bmatrix}, \begin{bmatrix} a_2 & sb_2 \\ f_0 \cdot r_2 & c_2 \end{bmatrix} \in S$. Define

$$\begin{bmatrix} a_1 & sb_1 \\ f_0 \cdot r_1 & c_1 \end{bmatrix} \circ_{(k,t)} \begin{bmatrix} a_2 & sb_2 \\ f_0 \cdot r_2 & c_2 \end{bmatrix}$$

$$= \begin{bmatrix} a_1a_2 - ta_1r_2 + ksb_1r_2 + tc_1r_2 & sa_1b_2 + sb_1c_2 \\ f_0 \cdot r_1a_2 + f_0 \cdot c_1r_2 & sr_1b_2 + c_1c_2 \end{bmatrix}.$$

Then $\circ_{(k,t)}$ is associative because $\mathrm{Soc}(A)^2 = 0$ and $f_0 \cdot y = 0$ for any $y \in J(A)$. Also it is left and right distributive over the addition. Other ring axioms are checked routinely. Further, the ring structure $(S, +, \circ_{(k,t)})$ is a compatible ring structure, and $\circ_{(1,0)} = \circ$. Let $\star = \circ_{(0,0)}$. Then for any $k \in J(A)$, $\star = \circ_{(k,0)}$, and

$$\begin{bmatrix} a_1 & sb_1 \\ f_0 \cdot r_1 & c_1 \end{bmatrix} \star \begin{bmatrix} a_2 & sb_2 \\ f_0 \cdot r_2 & c_2 \end{bmatrix} = \begin{bmatrix} a_1a_2 & sa_1b_2 + sb_1c_2 \\ f_0 \cdot r_1a_2 + f_0 \cdot c_1r_2 & sr_1b_2 + c_1c_2 \end{bmatrix}.$$

Remark 7.1.15 The motivation for defining multiplications $\circ_{(k,t)}$ on S, where $k \in A$ and $t \in \mathrm{Soc}(A)$, comes naturally from the case when $A = \mathbb{Z}_{p^n}$ with p a prime integer and $n \geq 2$ in Corollary 7.1.23.

Let u be a fixed invertible element of A, and we put

$$S = (S, +, \circ_{(u,0)}).$$

We prove that S is a QF-ring (Theorem 7.1.21). For this, we need the following series of lemmas. Let

$$U_S = \begin{bmatrix} A & \mathrm{Soc}(A) \\ 0 & 0 \end{bmatrix} \text{ and } W_S = \begin{bmatrix} 0 & 0 \\ \mathrm{Hom}(\mathrm{Soc}(A)_A, A_A) & A \end{bmatrix}.$$

We observe that $S_S = U_S \oplus W_S$.

Lemma 7.1.16 *Let Δ be an intermediate ring between R and S, and assume that $J_\Delta \leq S_\Delta$, $\gamma \in \mathrm{Hom}(J_\Delta, U_\Delta)$, and $\delta \in \mathrm{Hom}(J_\Delta, W_\Delta)$.*

(i) *For $x, y \in A$, if $\begin{bmatrix} x & 0 \\ f_0 \cdot y & 0 \end{bmatrix} \in J$, then there exist $a, b \in A$ such that*

$$\gamma \begin{bmatrix} x & 0 \\ f_0 \cdot y & 0 \end{bmatrix} = \begin{bmatrix} a & 0 \\ 0 & 0 \end{bmatrix} \text{ and } \delta \begin{bmatrix} x & 0 \\ f_0 \cdot y & 0 \end{bmatrix} = \begin{bmatrix} 0 & 0 \\ f_0 \cdot b & 0 \end{bmatrix}.$$

(ii) *For* $x, y \in A$, *if* $\begin{bmatrix} 0 & sx \\ 0 & y \end{bmatrix} \in J$, *then there exist* $c, d \in A$ *such that*

$$\gamma \begin{bmatrix} 0 & sx \\ 0 & y \end{bmatrix} = \begin{bmatrix} 0 & sc \\ 0 & 0 \end{bmatrix} \text{ and } \delta \begin{bmatrix} 0 & sx \\ 0 & y \end{bmatrix} = \begin{bmatrix} 0 & 0 \\ 0 & d \end{bmatrix}.$$

Proof The proof is routine. □

Lemma 7.1.17 *Let* Ω *be an overring of a ring* Φ *and* M_Ω *an* Ω*-module. If* M_Φ *is injective and* $\text{Hom}(\Omega_\Phi, M_\Phi) = \text{Hom}(\Omega_\Omega, M_\Omega)$, *then* M_Ω *is injective.*

Proof Exercise. □

Lemma 7.1.18 $W_S = \begin{bmatrix} 0 & 0 \\ \text{Hom}(\text{Soc}(A)_A, A_A) & A \end{bmatrix}_S$ *is injective.*

Proof Since $E_R = \begin{bmatrix} A \oplus \text{Hom}(\text{Soc}(A)_A, A_A) & A \\ 0 & 0 \end{bmatrix}_R \oplus W_R$ is injective by Theorem 7.1.14, W_R is injective. By direct computation and using Lemma 7.1.16, $\text{Hom}(S_R, W_R) \subseteq \text{Hom}(S_S, W_S)$. So $\text{Hom}(S_R, W_R) = \text{Hom}(S_S, W_S)$. Since W_R is injective, W_S is injective by Lemma 7.1.17. □

Let

$$\Gamma = \begin{bmatrix} A & 0 \\ \text{Hom}(\text{Soc}(A)_A, A_A) & A \end{bmatrix}.$$

Then Γ is a subring of S. To prove that $U_S = \begin{bmatrix} A & \text{Soc}(A) \\ 0 & 0 \end{bmatrix}_S$ is injective in Lemma 7.1.20, first we show that U_Γ is injective. For this, we need the following lemma.

Lemma 7.1.19 *All essential right ideals of* Γ *are precisely given by*

$$\begin{bmatrix} B & 0 \\ \text{Hom}(\text{Soc}(A)_A, A_A) & C \end{bmatrix},$$

where $\text{Soc}(A) \subseteq B \trianglelefteq A$ *and* $\text{Soc}(A) \subseteq C \trianglelefteq A$.

Proof We see that $J(\Gamma) = \begin{bmatrix} J(A) & 0 \\ \text{Hom}(\text{Soc}(A)_A, A_A) & J(A) \end{bmatrix}$. Since Γ is right Artinian, $\text{Soc}(\Gamma_\Gamma) = \ell_\Gamma(J(\Gamma))$, so $\text{Soc}(\Gamma_\Gamma) = \begin{bmatrix} \text{Soc}(A) & 0 \\ \text{Hom}(\text{Soc}(A)_A, A_A) & \text{Soc}(A) \end{bmatrix}$, which is the smallest essential right ideal of Γ. Let K be an essential right ideal of Γ. Then $\text{Soc}(\Gamma_\Gamma) \subseteq K$. Say B is the set of elements $b \in A$ so that there exists $\begin{bmatrix} b & 0 \\ f & y \end{bmatrix} \in K$

with $f \in \mathrm{Hom}(\mathrm{Soc}(A)_A, A_A)$ and $y \in A$. Then $B \trianglelefteq A$. We note that

$$\begin{bmatrix} b & 0 \\ 0 & y \end{bmatrix} = \begin{bmatrix} b & 0 \\ f & y \end{bmatrix} - \begin{bmatrix} 0 & 0 \\ f & 0 \end{bmatrix} \in K \text{ as } \begin{bmatrix} 0 & 0 \\ f & 0 \end{bmatrix} \in \mathrm{Soc}(\Gamma_\Gamma) \subseteq K.$$

So $\begin{bmatrix} b & 0 \\ 0 & 0 \end{bmatrix} = \begin{bmatrix} b & 0 \\ 0 & y \end{bmatrix}\begin{bmatrix} 1 & 0 \\ 0 & 0 \end{bmatrix} \in K$. Therefore, $\begin{bmatrix} B & 0 \\ 0 & 0 \end{bmatrix} \subseteq K$.

Let C be the set of elements $c \in A$ such that $\begin{bmatrix} a & 0 \\ g & c \end{bmatrix} \in K$ for some $a \in A$ and $g \in \mathrm{Hom}(\mathrm{Soc}(A)_A, A_A)$. Then as in the preceding argument, we see that $C \trianglelefteq A$ and $\begin{bmatrix} 0 & 0 \\ 0 & C \end{bmatrix} \subseteq K$. Thus $\begin{bmatrix} B & 0 \\ \mathrm{Hom}(\mathrm{Soc}(A)_A, A_A) & C \end{bmatrix} \subseteq K$. Obviously

$$K \subseteq \begin{bmatrix} B & 0 \\ \mathrm{Hom}(\mathrm{Soc}(A)_A, A_A) & C \end{bmatrix}. \text{ Hence } K = \begin{bmatrix} B & 0 \\ \mathrm{Hom}(\mathrm{Soc}(A)_A, A_A) & C \end{bmatrix}. \quad \square$$

With these preparations, we prove the following.

Lemma 7.1.20 *U_S is injective.*

Proof First, we show that U_Γ is injective. For this, let K be an essential right ideal of Γ. By Lemma 7.1.19, $K = \begin{bmatrix} B & 0 \\ \mathrm{Hom}(\mathrm{Soc}(A)_A, A_A) & C \end{bmatrix}$ with $\mathrm{Soc}(A) \subseteq B \trianglelefteq A$ and $\mathrm{Soc}(A) \subseteq C \trianglelefteq A$.

Take $\varphi \in \mathrm{Hom}(K_\Gamma, U_\Gamma)$. For $b \in B$, then $\varphi \begin{bmatrix} b & 0 \\ 0 & 0 \end{bmatrix} = \begin{bmatrix} x_b & 0 \\ 0 & 0 \end{bmatrix}$ for some $x_b \in A$ (see Lemma 7.1.16(i)). Consider $g \in \mathrm{Hom}(B_A, A_A)$ defined by $g(b) = x_b$. As A is self-injective, there exists $x_0 \in A$ with $g(b) = x_b = x_0 b$ for $b \in B$. Therefore, $\varphi \begin{bmatrix} b & 0 \\ 0 & 0 \end{bmatrix} = \begin{bmatrix} x_0 b & 0 \\ 0 & 0 \end{bmatrix}$ for all $b \in B$.

From Lemma 7.1.16(i), $\varphi \begin{bmatrix} 0 & 0 \\ f_0 & 0 \end{bmatrix} = \begin{bmatrix} a_0 & 0 \\ 0 & 0 \end{bmatrix}$ for some a_0. Take $y \in J(A)$. Then $0 = \varphi(0) = \varphi\left(\begin{bmatrix} 0 & 0 \\ f_0 & 0 \end{bmatrix}\begin{bmatrix} y & 0 \\ 0 & 0 \end{bmatrix}\right) = \begin{bmatrix} a_0 & 0 \\ 0 & 0 \end{bmatrix}\begin{bmatrix} y & 0 \\ 0 & 0 \end{bmatrix} = \begin{bmatrix} a_0 y & 0 \\ 0 & 0 \end{bmatrix}$. Therefore $a_0 y = 0$ for each $y \in J(A)$, hence $a_0 \in \mathrm{Soc}(A) = suA$, where s is a nonzero fixed element of $\mathrm{Soc}(A)$ (recall that $S = (S, +, \circ_{(u,0)})$ and u is the *fixed invertible element* in A). So $a_0 = sua_1$ for some $a_1 \in A$. Thus,

$$\varphi \begin{bmatrix} 0 & 0 \\ f_0 & 0 \end{bmatrix} = \begin{bmatrix} sua_1 & 0 \\ 0 & 0 \end{bmatrix}.$$

For $c \in C$, there is $y_c \in A$ with $\varphi \begin{bmatrix} 0 & 0 \\ 0 & c \end{bmatrix} = \begin{bmatrix} 0 & sy_c \\ 0 & 0 \end{bmatrix}$ by Lemma 7.1.16(ii). Now $\varphi\left(\begin{bmatrix} 0 & 0 \\ 0 & c \end{bmatrix}\begin{bmatrix} 0 & 0 \\ f_0 & 0 \end{bmatrix}\right) = \left(\varphi\begin{bmatrix} 0 & 0 \\ 0 & c \end{bmatrix}\right)\begin{bmatrix} 0 & 0 \\ f_0 & 0 \end{bmatrix} = \begin{bmatrix} 0 & sy_c \\ 0 & 0 \end{bmatrix}\begin{bmatrix} 0 & 0 \\ f_0 & 0 \end{bmatrix} = \begin{bmatrix} usy_c & 0 \\ 0 & 0 \end{bmatrix}$. On the other hand, we see that

$$\varphi\left(\begin{bmatrix} 0 & 0 \\ 0 & c \end{bmatrix}\begin{bmatrix} 0 & 0 \\ f_0 & 0 \end{bmatrix}\right) = \varphi\begin{bmatrix} 0 & 0 \\ f_0 \cdot c & 0 \end{bmatrix} = \varphi\left(\begin{bmatrix} 0 & 0 \\ f_0 & 0 \end{bmatrix}\begin{bmatrix} c & 0 \\ 0 & 0 \end{bmatrix}\right)$$

$$= \left(\varphi\begin{bmatrix} 0 & 0 \\ f_0 & 0 \end{bmatrix}\right)\begin{bmatrix} c & 0 \\ 0 & 0 \end{bmatrix} = \begin{bmatrix} sua_1 & 0 \\ 0 & 0 \end{bmatrix}\begin{bmatrix} c & 0 \\ 0 & 0 \end{bmatrix} = \begin{bmatrix} sua_1c & 0 \\ 0 & 0 \end{bmatrix}.$$

Hence, $usy_c = sua_1c$, so $sy_c = sa_1c$ for each $c \in C$. Thus, $\varphi \begin{bmatrix} 0 & 0 \\ 0 & c \end{bmatrix} = \begin{bmatrix} 0 & sa_1c \\ 0 & 0 \end{bmatrix}$ for all $c \in C$. Take $v = \begin{bmatrix} x_0 & sa_1 \\ 0 & 0 \end{bmatrix} \in U$. Then we see that there exists an extension $\overline{\varphi} \in \mathrm{Hom}(\Gamma_\Gamma, U_\Gamma)$ of φ such that $\overline{\varphi}(1) = v$. Therefore, U_Γ is injective by Baer's Criterion. Next, we see that $\mathrm{Hom}(S_\Gamma, U_\Gamma) = \mathrm{Hom}(S_S, U_S)$ by direct computation. By Lemma 7.1.17, U_S is injective. □

In contrast to Theorem 7.1.3, the following result provides a right self-injective right essential overring S of the ring R, but $S_R \neq E(R_R)$. Recall that $\circ = \circ_{(1,0)}$.

Theorem 7.1.21 *Let $u \in A$ be invertible. Then the ring $(S, +, \circ_{(u,0)})$ is QF. In particular, $(S, +, \circ)$ is QF.*

Proof Let $S = (S, +, \circ_{(u,0)})$. Then $S_S = U_S \oplus W_S$ is injective by Lemmas 7.1.18 and 7.1.20. Since R is right Artinian, S_R is Artinian and so S_S is Artinian. Therefore, S is QF. As $\circ = \circ_{(1,0)}$, $(S, +, \circ)$ is QF. □

The next result shows that there exist a number of distinct right essential overrings of R beyond $Q(R)$.

Theorem 7.1.22 *Let R be the ring as specified immediately preceding Theorem* 7.1.14. *Then:*

(i) $R = Q(R) \neq E(R_R)$.
(ii) *There exist $|\mathrm{Soc}(A)|^2$ distinct compatible right essential overrings $(S, +, \circ_{(k,t)})$, where $k \in A$ and $t \in \mathrm{Soc}(A)$, on S_R satisfying the following properties.*
  (1) *$(S, +, \circ_{(k,t)}) \cong (S, +, \circ_{(k,0)})$ for all $t \in \mathrm{Soc}(A)$.*
  (2) *$(S, +, \circ_{(k,t)})$ is QF if and only if k is invertible.*
  (3) *If k is not invertible, then $(S, +, \circ_{(k,t)}) = (S, +, \circ_{(0,t)})$ are all isomorphic to $(S, +, \star)$, hence they are not even right FI-extending.*
  (4) *If $|\mathrm{Soc}(A)|$ is finite, then there are $|\mathrm{Soc}(A)|^2 - |\mathrm{Soc}(A)|$ distinct right essential overring structures on S_R which are QF, and the other $|\mathrm{Soc}(A)|$ distinct right essential overring structures are not even right FI-extending.*

Proof (i) We easily see that the left annihilator of each maximal right ideal of R is nonzero. Thus from Proposition 1.3.18, the ring R is right Kasch, so $R = Q(R)$. Further, $R \neq E(R_R)$ by Theorem 7.1.14.

(ii) Note that $|\{\circ_{(k,t)} \mid k \in A,\ t \in \mathrm{Soc}(A)\}| = |A/J(A)|\,|\mathrm{Soc}(A)|$. Recall that $\mathrm{Soc}(A) = sA$, where $0 \neq s \in \mathrm{Soc}(A)$. Define $f : A \to \mathrm{Soc}(A)$ by $f(a) = sa$ for $a \in A$. Then f is an R-epimorphism with $\mathrm{Ker}(f) = J(A)$. Thus we have that $A/J(A) \cong \mathrm{Soc}(A)$ as A-modules. So $|A/J(A)| = |\mathrm{Soc}(A)|$. Hence it follows that $|\{\circ_{(k,t)} \mid k \in A, t \in \mathrm{Soc}(A)\}| = |\mathrm{Soc}(A)|^2$. So there exist $|\mathrm{Soc}(A)|^2$ distinct right essential overrings $(S, +, \circ_{(k,t)})$ on S_R.

(1) Take $k \in A$ and $t \in \mathrm{Soc}(A)$. Let $\mu_{(k,t)} : (S, +, \circ_{(k,t)}) \to (S, +, \circ_{(k,0)})$ defined by

$$\mu_{(k,t)} \begin{bmatrix} a & sb \\ f_0 \cdot r & c \end{bmatrix} = \begin{bmatrix} a - tr & sb \\ f_0 \cdot r & c \end{bmatrix}.$$

Then $\mu_{(k,t)}$ is a ring isomorphism by routine computation as $\mathrm{Soc}(A)^2 = 0$.

(2) and (3) If $k \in A$ is invertible, then $(S, +, \circ_{(k,0)})$ is QF by Theorem 7.1.21. Hence, if k is invertible, then $(S, +, \circ_{(k,t)})$ is QF for all $t \in \mathrm{Soc}(A)$ by part (1). If k is not invertible, then $k \in J(A)$, and so $k\,\mathrm{Soc}(A) = 0$. Therefore,

$$(S, +, \circ_{(k,t)}) = (S, +, \circ_{(0,t)}) \cong (S, +, \star)$$

by part (1) (recall that $\star = \circ_{(0,0)}$) and the definition of $\circ_{(k,t)}$.

We show that $(S, +, \star)$ is not right FI-extending. For this, let $T = (S, +, \star)$. Take $J = \begin{bmatrix} \mathrm{Soc}(A) & 0 \\ 0 & 0 \end{bmatrix}$. Then $J \trianglelefteq T$. But there is no $K_T \leq^{\oplus} T_T$ such that $J_T \leq^{\text{ess}} K_T$. Thus T is not even right FI-extending.

(4) Assume that $|\mathrm{Soc}(A)| < \infty$. Let $\mathcal{S}$ be the set of all $(S, +, \circ_{(k,t)})$, where $k \in A$ is not invertible, and $t \in \mathrm{Soc}(A)$. Since $(S, +, \circ_{(k,t)}) = (S, +, \circ_{(0,t)})$ by part (3), $|\mathcal{S}| = |\{(S, +, \circ_{(0,t)}) \mid t \in \mathrm{Soc}(A)\}| = |\mathrm{Soc}(A)|$. From part (3) there are $|\mathrm{Soc}(A)|$ distinct right essential overring structures on S_R which are not right FI-extending. The other $|\mathrm{Soc}(A)|^2 - |\mathrm{Soc}(A)|$ distinct right essential overring structures $(S, +, \circ_{(k,t)})$, with k invertible, are QF. □

In the next corollary, for a certain ring A, all possible compatible ring structures on S_R in Theorem 7.1.22 are described. From the construction of those compatible ring structures, we obtain a motivation for defining $\circ_{(k,t)}$ on S_R with $k \in A$ and $t \in \mathrm{Soc}(A)$ (see Remark 7.1.15).

Corollary 7.1.23 *Let $A = \mathbb{Z}_{p^n}$, where p is a prime integer and n is an integer such that $n \geq 2$. Then S_R has exactly $p^2 = |\mathrm{Soc}(A)|^2$ distinct compatible ring structures such that:*

(1) *$p^2 - p$ compatible ring structures are QF.*
(2) *the other p compatible ring structures are not even right FI-extending.*

Proof Let $s = p^{n-1} \in A$. Then $\mathrm{Soc}(A) = sA$. Suppose that S has a compatible ring structure. From Lemma 7.1.7, $1_R = 1_S$. Let $e_1 = \begin{bmatrix} 1 & 0 \\ 0 & 0 \end{bmatrix}$ and $e_2 = \begin{bmatrix} 0 & 0 \\ 0 & 1 \end{bmatrix}$. Then

$S = e_1Se_1 + e_1Se_2 + e_2Se_1 + e_2Se_2$, $e_1^2 = e_1$, $e_2^2 = e_2$, and $e_1e_2 = e_2e_1 = 0$. We see routinely that

$$e_1Se_1 = \begin{bmatrix} A & 0 \\ 0 & 0 \end{bmatrix},\ e_1Se_2 = \begin{bmatrix} 0 & \mathrm{Soc}(A) \\ 0 & 0 \end{bmatrix},\ \text{and } e_2Se_2 = \begin{bmatrix} 0 & 0 \\ 0 & A \end{bmatrix}.$$

Moreover, we show that $e_2Se_1 = \left\{ \begin{bmatrix} t_0 r & 0 \\ f_0 \cdot r & 0 \end{bmatrix} \mid r \in A \right\}$ with $t_0 \in \mathrm{Soc}(A)$. Indeed, let $\begin{bmatrix} a & sb \\ f_0 \cdot r & c \end{bmatrix} \in e_2Se_1$. Then we see that

$$\begin{bmatrix} a & sb \\ f_0 \cdot r & c \end{bmatrix} = e_2 \begin{bmatrix} a & sb \\ f_0 \cdot r & c \end{bmatrix} e_1 = e_2 \begin{bmatrix} a & 0 \\ f_0 \cdot r & 0 \end{bmatrix}$$

$$= e_2 \begin{bmatrix} a & 0 \\ 0 & 0 \end{bmatrix} + e_2 \begin{bmatrix} 0 & 0 \\ f_0 \cdot r & 0 \end{bmatrix} = e_2 \begin{bmatrix} 0 & 0 \\ f_0 \cdot r & 0 \end{bmatrix}.$$

Therefore, $e_2Se_1 = \left\{ e_2 \begin{bmatrix} 0 & 0 \\ f_0 \cdot r & 0 \end{bmatrix} \mid r \in A \right\} = \left\{ e_2 \begin{bmatrix} 0 & 0 \\ f_0 & 0 \end{bmatrix} \begin{bmatrix} r & 0 \\ 0 & 0 \end{bmatrix} \mid r \in A \right\}$.

First, consider $w := e_2 \begin{bmatrix} 0 & 0 \\ f_0 & 0 \end{bmatrix} \in e_2Se_1$. As $w = we_1$, $w = \begin{bmatrix} a_0 & 0 \\ f_0 \cdot y_0 & 0 \end{bmatrix}$ for some $a_0, y_0 \in A$. For $y \in J(A)$, $w \begin{bmatrix} y & 0 \\ 0 & 0 \end{bmatrix} = e_2 \begin{bmatrix} 0 & 0 \\ f_0 & 0 \end{bmatrix} \begin{bmatrix} y & 0 \\ 0 & 0 \end{bmatrix} = 0$. On the other hand, $w \begin{bmatrix} y & 0 \\ 0 & 0 \end{bmatrix} = \begin{bmatrix} a_0 & 0 \\ f_0 \cdot y_0 & 0 \end{bmatrix} \begin{bmatrix} y & 0 \\ 0 & 0 \end{bmatrix} = \begin{bmatrix} a_0 y & 0 \\ 0 & 0 \end{bmatrix}$. Thus $a_0 y = 0$ for $y \in J(A)$, hence $a_0 \in \mathrm{Soc}(A)$. So $w = e_2 \begin{bmatrix} 0 & 0 \\ f_0 & 0 \end{bmatrix} = \begin{bmatrix} a_0 & 0 \\ f_0 \cdot y_0 & 0 \end{bmatrix}$ with $a_0 \in \mathrm{Soc}(A)$. If $f_0 \cdot y_0 = 0$, then $w = \begin{bmatrix} a_0 & 0 \\ 0 & 0 \end{bmatrix} \in e_1Se_1 \cap e_2Se_1 = 0$. Hence, $e_2Se_1 = 0$. Therefore

$$|S| = |e_1Se_1|\,|e_1Se_2|\,|e_2Se_1|\,|e_2Se_2| = p^{2n+1} < p^{2n+2} = |S|,$$

a contradiction. Thus, $f_0 \cdot y_0 \neq 0$, and so y_0 is invertible.

Note that $w \begin{bmatrix} r & 0 \\ 0 & 0 \end{bmatrix} = \begin{bmatrix} a_0 & 0 \\ f_0 \cdot y_0 & 0 \end{bmatrix} \begin{bmatrix} r & 0 \\ 0 & 0 \end{bmatrix} = \begin{bmatrix} a_0 r & 0 \\ f_0 \cdot y_0 r & 0 \end{bmatrix}$ and $a_0 \in \mathrm{Soc}(A)$, so

$$e_2Se_1 = \left\{ w \begin{bmatrix} r & 0 \\ 0 & 0 \end{bmatrix} \mid r \in A \right\} = \left\{ \begin{bmatrix} a_0 r & 0 \\ f_0 \cdot y_0 r & 0 \end{bmatrix} \mid r \in A \right\}.$$

Hence, we have that $e_2Se_1 = \left\{ \begin{bmatrix} a_0 y_0^{-1} r & 0 \\ f_0 \cdot r & 0 \end{bmatrix} \mid r \in A \right\} = \left\{ \begin{bmatrix} t_0 r & 0 \\ f_0 \cdot r & 0 \end{bmatrix} \mid r \in A \right\}$, where $t_0 = a_0 y_0^{-1} \in \mathrm{Soc}(A)$. By routine computation,

$$\begin{bmatrix} 1 & 0 \\ 0 & 0 \end{bmatrix} \begin{bmatrix} 0 & 0 \\ f_0 & 0 \end{bmatrix} = \begin{bmatrix} -t_0 & 0 \\ 0 & 0 \end{bmatrix},\ \begin{bmatrix} 0 & 0 \\ f_0 & 0 \end{bmatrix} \begin{bmatrix} 0 & 0 \\ f_0 & 0 \end{bmatrix} = 0,\ \text{and } \begin{bmatrix} 0 & 0 \\ 0 & 1 \end{bmatrix} \begin{bmatrix} 0 & 0 \\ f_0 & 0 \end{bmatrix} = \begin{bmatrix} t_0 & 0 \\ f_0 & 0 \end{bmatrix}.$$

For example, $\begin{bmatrix}1&0\\0&0\end{bmatrix}\begin{bmatrix}t_0&0\\f_0&0\end{bmatrix}\in e_1e_2Se_1=0$, so

$$0=\begin{bmatrix}1&0\\0&0\end{bmatrix}\begin{bmatrix}t_0&0\\f_0&0\end{bmatrix}=\begin{bmatrix}1&0\\0&0\end{bmatrix}\begin{bmatrix}0&0\\f_0&0\end{bmatrix}+\begin{bmatrix}1&0\\0&0\end{bmatrix}\begin{bmatrix}t_0&0\\0&0\end{bmatrix}=\begin{bmatrix}1&0\\0&0\end{bmatrix}\begin{bmatrix}0&0\\f_0&0\end{bmatrix}+\begin{bmatrix}t_0&0\\0&0\end{bmatrix}.$$

Therefore, $\begin{bmatrix}1&0\\0&0\end{bmatrix}\begin{bmatrix}0&0\\f_0&0\end{bmatrix}=\begin{bmatrix}-t_0&0\\0&0\end{bmatrix}$.

We claim that $\begin{bmatrix}0&s\\0&0\end{bmatrix}\begin{bmatrix}0&0\\f_0&0\end{bmatrix}=\begin{bmatrix}sk_0&0\\0&0\end{bmatrix}$ for some $k_0\in A$. For this, note that $\begin{bmatrix}0&s\\0&0\end{bmatrix}\begin{bmatrix}t_0&0\\f_0&0\end{bmatrix}\in e_1Se_2e_2Se_1\subseteq e_1Se_1$. Hence, there is $q\in A$ such that

$$\begin{bmatrix}0&s\\0&0\end{bmatrix}\begin{bmatrix}t_0&0\\f_0&0\end{bmatrix}=\begin{bmatrix}q&0\\0&0\end{bmatrix}.$$

Say $y\in J(A)$. Then

$$\begin{bmatrix}qy&0\\0&0\end{bmatrix}=\begin{bmatrix}q&0\\0&0\end{bmatrix}\begin{bmatrix}y&0\\0&0\end{bmatrix}=\left(\begin{bmatrix}0&s\\0&0\end{bmatrix}\begin{bmatrix}t_0&0\\f_0&0\end{bmatrix}\right)\begin{bmatrix}y&0\\0&0\end{bmatrix}=\begin{bmatrix}0&s\\0&0\end{bmatrix}\left(\begin{bmatrix}t_0&0\\f_0&0\end{bmatrix}\begin{bmatrix}y&0\\0&0\end{bmatrix}\right)=0$$

because $y\in J(A)$, $t_0\in \mathrm{Soc}(A)$, and $f_0\cdot y=0$. Hence, $qy=0$ for all $y\in J(A)$. Thus $q\in\mathrm{Soc}(A)=sA$, so $q=sk_0$ for some $k_0\in A$. Therefore,

$$\begin{bmatrix}sk_0&0\\0&0\end{bmatrix}=\begin{bmatrix}0&s\\0&0\end{bmatrix}\begin{bmatrix}t_0&0\\f_0&0\end{bmatrix}=\begin{bmatrix}0&s\\0&0\end{bmatrix}\begin{bmatrix}t_0&0\\0&0\end{bmatrix}+\begin{bmatrix}0&s\\0&0\end{bmatrix}\begin{bmatrix}0&0\\f_0&0\end{bmatrix}=\begin{bmatrix}0&s\\0&0\end{bmatrix}\begin{bmatrix}0&0\\f_0&0\end{bmatrix},$$

since $\begin{bmatrix}0&s\\0&0\end{bmatrix}\begin{bmatrix}t_0&0\\0&0\end{bmatrix}\in e_1Se_2e_1Se_1=0$. Thus, if there exists a ring multiplication on S which extends the R-module scalar multiplication, then

$$\begin{bmatrix}a_1&sb_1\\f_0\cdot r_1&c_1\end{bmatrix}\begin{bmatrix}a_2&sb_2\\f_0\cdot r_2&c_2\end{bmatrix}=$$

$$\begin{bmatrix}a_1a_2-t_0a_1r_2+k_0sb_1r_2+t_0c_1r_2 & sa_1b_2+sb_1c_2\\ f_0\cdot r_1a_2+f_0\cdot c_1r_2 & sr_1b_2+c_1c_2\end{bmatrix}.$$

So there exist *exactly* $|\mathrm{Soc}(A)|^2$ compatible ring structures $(S,+,\circ_{(k,t)})$ with $k\in A$ and $t\in\mathrm{Soc}(A)$. Theorem 7.1.22 yields the remaining statements. □

In the following example, we show that there exist two compatible QF-ring structures on S_R which are not isomorphic.

Example 7.1.24 Let $A=\mathbb{Z}_{p^n}$, where p is a prime integer and n is an integer such that $n\geq 2$. Then $\mathrm{Soc}(A)=sA$, where $s=p^{n-1}\in A$. Assume that $u\in A$ is an invertible element such that $u-1$ is also invertible. Then both $(S,+,\circ_{(u,0)})$ and $(S,+,\circ_{(1,0)})$ are QF compatible ring structures by Theorem 7.1.21. But $(S,+,\circ_{(u,0)})\ncong(S,+,\circ_{(1,0)})=(S,+,\circ)$.

Assume on the contrary that there exists a ring isomorphism

$$\theta : (S, +, \circ_{(u,0)}) \to (S, +, \circ).$$

Now we let $\theta \begin{bmatrix} 1 & 0 \\ 0 & 0 \end{bmatrix} = \begin{bmatrix} a_1 & sb_1 \\ f_0 \cdot r_1 & c_1 \end{bmatrix}$ and $\theta \begin{bmatrix} 0 & 0 \\ 0 & 1 \end{bmatrix} = \begin{bmatrix} a & sb \\ f_0 \cdot r & c \end{bmatrix}$, for some $a_1, b_1, r_1, c_1, a, b, r, c \in A$. Then $\theta \begin{bmatrix} s & 0 \\ 0 & 0 \end{bmatrix} = s\left(\theta \begin{bmatrix} 1 & 0 \\ 0 & 0 \end{bmatrix}\right) = \begin{bmatrix} sa_1 & 0 \\ 0 & sc_1 \end{bmatrix}$. Similarly $\theta \begin{bmatrix} 0 & 0 \\ 0 & s \end{bmatrix} = \begin{bmatrix} sa & 0 \\ 0 & sc \end{bmatrix}$. Also $p\left(\theta \begin{bmatrix} 0 & s \\ 0 & 0 \end{bmatrix}\right) = 0$ and $p\left(\theta \begin{bmatrix} 0 & 0 \\ f_0 & 0 \end{bmatrix}\right) = 0$. Thus,

$$\theta \begin{bmatrix} 0 & s \\ 0 & 0 \end{bmatrix} = \begin{bmatrix} sa_2 & sb_2 \\ f_0 \cdot r_2 & sc_2 \end{bmatrix} \text{ and } \theta \begin{bmatrix} 0 & 0 \\ f_0 & 0 \end{bmatrix} = \begin{bmatrix} sa_3 & sb_3 \\ f_0 \cdot r_3 & sc_3 \end{bmatrix}$$

for some a_2, b_2, r_2, c_2, a_3, b_3, r_3, $c_3 \in A$. Now

$$\theta \begin{bmatrix} us & 0 \\ 0 & 0 \end{bmatrix} = \theta \left(\begin{bmatrix} 0 & s \\ 0 & 0 \end{bmatrix} \circ_{(u,0)} \begin{bmatrix} 0 & 0 \\ f_0 & 0 \end{bmatrix} \right) = \theta \begin{bmatrix} 0 & s \\ 0 & 0 \end{bmatrix} \circ \theta \begin{bmatrix} 0 & 0 \\ f_0 & 0 \end{bmatrix}$$
$$= \begin{bmatrix} sa_2 & sb_2 \\ f_0 \cdot r_2 & sc_2 \end{bmatrix} \circ \begin{bmatrix} sa_3 & sb_3 \\ f_0 \cdot r_3 & sc_3 \end{bmatrix} = \begin{bmatrix} sb_2r_3 & 0 \\ 0 & sr_2b_3 \end{bmatrix}.$$

On the other hand, $\theta \begin{bmatrix} 0 & 0 \\ 0 & us \end{bmatrix} = u\left(\theta \begin{bmatrix} 0 & 0 \\ 0 & s \end{bmatrix}\right) = \begin{bmatrix} usa & 0 \\ 0 & usc \end{bmatrix}$. Thus,

$$\begin{bmatrix} us & 0 \\ 0 & us \end{bmatrix} = us\left(\theta \begin{bmatrix} 1 & 0 \\ 0 & 1 \end{bmatrix}\right) = \theta \begin{bmatrix} us & 0 \\ 0 & us \end{bmatrix} = \begin{bmatrix} sb_2r_3 + usa & 0 \\ 0 & sr_2b_3 + usc \end{bmatrix}.$$

So $us = sb_2r_3 + usa$ and $us = sr_2b_3 + usc$. Note that

$$\begin{bmatrix} sa & 0 \\ 0 & sc \end{bmatrix} = \theta \begin{bmatrix} 0 & 0 \\ 0 & s \end{bmatrix} = \theta \left(\begin{bmatrix} 0 & 0 \\ f_0 & 0 \end{bmatrix} \circ_{(u,0)} \begin{bmatrix} 0 & s \\ 0 & 0 \end{bmatrix} \right) = \theta \begin{bmatrix} 0 & 0 \\ f_0 & 0 \end{bmatrix} \circ \theta \begin{bmatrix} 0 & s \\ 0 & 0 \end{bmatrix}$$
$$= \begin{bmatrix} sa_3 & sb_3 \\ f_0 \cdot r_3 & sc_3 \end{bmatrix} \circ \begin{bmatrix} sa_2 & sb_2 \\ f_0 \cdot r_2 & sc_2 \end{bmatrix} = \begin{bmatrix} sb_3r_2 & 0 \\ 0 & sr_3b_2 \end{bmatrix}.$$

Hence we have that $sa = sb_3r_2$ and $sc = sr_3b_2$. Since $us = sb_2r_3 + usa$ and $us = sr_2b_3 + usc$, $us = sc + usa$ and $us = sa + usc$. So $sc + usa = sa + usc$, and thus $(u - 1)sa = (u - 1)sc$. Now $u - 1$ is invertible by assumption, thus $sa = sc$, and so $(c - a)s = 0$. Hence, $c - a \in J(A) = pA$. Thus, $c = a + py_0$ for some $y_0 \in A$. Therefore, $\theta \begin{bmatrix} 0 & 0 \\ 0 & 1 \end{bmatrix} = \begin{bmatrix} a & sb \\ f_0 \cdot r & c \end{bmatrix} = \begin{bmatrix} a & sb \\ f_0 \cdot r & a + py_0 \end{bmatrix}$. Let $e = \theta \begin{bmatrix} 0 & 0 \\ 0 & 1 \end{bmatrix}$. Since $e = e \circ e$, we have the following relations:

$$a = a^2 + sbr,\ sb = 2sab,\ f_0 \cdot r = 2f_0 \cdot ar, \text{ and } a + py_0 = srb + a^2 + 2apy_0 + p^2y_0^2.$$

Case 1. $sb = 0$. Then $a = a^2$. Hence, either $a = 0$ or $a = 1$. If $a = 0$, then $py_0 = (py_0)^2 \in J(A)$. So $py_0 = 0$ and hence e is the zero matrix, a contradiction. If

$a = 1$, then $f_0 \cdot r = 0$, so $1 + py_0 = (1 + py_0)^2$. Thus, $1 + py_0 = 0$ or $1 + py_0 = 1$. If $1 + py_0 = 0$, then $1 = -py_0 \in J(A)$, a contradiction. Hence $1 + py_0 = 1$, so e is the identity matrix. Thus $\theta \begin{bmatrix} 0 & 0 \\ 0 & 1 \end{bmatrix} = \begin{bmatrix} 1 & 0 \\ 0 & 1 \end{bmatrix}$, also a contradiction.

Case 2. $sb \neq 0$. Then $b \notin J(A)$, hence b is invertible. From $sb = 2sab$, we see that $s(1 - 2a) = 0$. Hence, $1 - 2a \in J(A) = pA$. Therefore, there exists $y \in A$ such that $1 = 2a + py$.

Now we have the following two subcases.

Subcase 2.1. $f_0 \cdot r \neq 0$. From $a = a^2 + sbr$, $a(1 - a) = sbr \in J(A)$. Since A is local, either $a \in J(A)$ or $1 - a \in J(A)$. Suppose that $a \in J(A)$. Because $f_0 \cdot r = 2f_0 \cdot ar$, $f_0 \cdot r = 0$, a contradiction. If $1 - a \in J(A)$, then $1 - a = py_1$ for some $y_1 \in A$. Thus, $1 = a + py_1$. As $1 = 2a + py$ in above, $a = p(y_1 - y) \in J(A)$, and so $1 = a + (1 - a) \in J(A)$, also a contradiction.

Subcase 2.2. $f_0 \cdot r = 0$. Then $r \in J(A)$, so $sbr = 0$. Hence $a = a^2$. Thus, $a = 0$ or $a = 1$. If $a = 0$, then $sb = 2sab = 0$, a contradiction. Suppose that $a = 1$. Since $1 = 2a + py$ in above, $py = -1$, a contradiction.

Consequently, by Cases 1 and 2, $(S, +, \circ_{(u,0)}) \ncong (S, +, \circ)$. For example, when $p \geq 3$ and $n \geq 2$, it follows that for any integer k with $1 < k < p$, $(S, +, \circ_{(k,0)})$ and $(S, +, \circ)$ are QF, but $(S, +, \circ_{(k,0)}) \ncong (S, +, \circ)$.

When $A = \mathbb{Z}_{3^n}$ with $n \geq 2$, we can explicitly describe and classify all compatible ring structures on S_R as follows.

Example 7.1.25 In particular, assume that $A = \mathbb{Z}_{3^n}$ with $n \geq 2$. By Theorem 7.1.22 and Corollary 7.1.23, S_R has exactly 3^2 compatible ring structures $(S, +, \circ_{(k,t)})$, with $k \in A$ and $t \in \text{Soc}(A)$, satisfying the following property, for $s = 3^{n-1} \in \text{Soc}(A)$.

(1) $(S, +, \circ_{(1,t)}) \cong (S, +, \circ_{(1,0)}) = (S, +, \circ)$ which are three QF-ring structures, for $t = 0, s, 2s$.
(2) $(S, +, \circ_{(2,t)}) \cong (S, +, \circ_{(2,0)})$ which are three QF-ring structures, for $t = 0, s, 2s$.
(3) $(S, +, \circ_{(0,t)}) \cong (S, +, \circ_{(0,0)}) = (S, +, \star)$ are three ring structures which are not right FI-extending, for $t = 0, s, 2s$.
(4) $(S, +, \circ_{(1,0)})$, $(S, +, \circ_{(2,0)})$, and $(S, +, \circ_{(0,0)})$ are mutually nonisomorphic.

As an application of Theorems 7.1.14 and 7.1.22, the next result characterizes a certain class of rings R such that R_R has an essential extension with nonisomorphic compatible ring structures.

Theorem 7.1.26 *Let Λ be a commutative QF-ring and*

$$R = \begin{bmatrix} \Lambda & \text{Soc}(\Lambda) \\ 0 & \Lambda \end{bmatrix}.$$

Then the following are equivalent.

(i) $J(\Lambda) \neq 0$.
(ii) *There exists an essential extension S_R of R_R which has two nonisomorphic compatible ring structures $(S, +, \circ)$ and $(S, +, \star)$.*

Furthermore, in this case, $(S, +, \circ)$ is right self-injective (in fact, QF), while $(S, +, \star)$ is not even right FI-extending.

Proof (i)⇒(ii) Assume that $J(\Lambda) \neq 0$. Then $\Lambda = A \oplus B$, where $A = \oplus_{i=1}^m A_i$ with A_i local, $J(A_i) \neq 0$ for each i, and $J(B) = 0$. Let $R_i = \begin{bmatrix} A_i & \mathrm{Soc}(A_i) \\ 0 & A_i \end{bmatrix}$ for $i = 1, \ldots, m$. Then $R = (\oplus_{i=1}^m R_i) \oplus T_2(B)$. Also let

$$S_i = \begin{bmatrix} A_i & \mathrm{Soc}(A_i) \\ \mathrm{Hom}(\mathrm{Soc}(A_i)_{A_i}, A_{iA_i}) & A_i \end{bmatrix}$$

and

$$E_i = \begin{bmatrix} A_i \oplus \mathrm{Hom}(\mathrm{Soc}(A_i)_{A_i}, A_{iA_i}) & \mathrm{Soc}(A_i) \\ \mathrm{Hom}(\mathrm{Soc}(A_i)_{A_i}, A_{iA_i}) & A_i \end{bmatrix}$$

for each i, $1 \leq i \leq m$.

Let $S = (\oplus_{i=1}^m S_i) \oplus \mathrm{Mat}_2(B)$. Then $R_R \leq S_R \leq [(\oplus_{i=1}^m E_i) \oplus \mathrm{Mat}_2(B)]_R$. Further, $\oplus_{i=1}^m E_i$ is an injective right $\oplus_{i=1}^m R_i$-module by Theorem 7.1.14, while $\mathrm{Mat}_2(B)$ is an injective right $T_2(B)$-module. Hence, $[(\oplus_{i=1}^m E_i) \oplus \mathrm{Mat}_2(B)]_R$ is an injective hull of R_R, so $R_R \leq^{\mathrm{ess}} S_R$.

Each S_i has two compatible ring multiplications $\circ_i$ and $\star_i$ such that $(S_i, +, \circ_i)$ is QF, while $(S_i, +, \star_i)$ is not right FI-extending by Theorem 7.1.22. Let $(x_1, \ldots, x_m, b), (y_1, \ldots, y_m, c) \in S$ with $x_i, y_i \in S_i$ and $b, c \in \mathrm{Mat}_2(B)$.

Define

$$(x_1, \ldots, x_m, b) \circ (y_1, \ldots, y_m, c) = (x_1 \circ_1 y_1, \ldots, x_m \circ_m y_m, bc)$$

and

$$(x_1, \ldots, x_m, b) \star (y_1, \ldots, y_m, c) = (x_1 \star_1 y_1, \ldots, x_m \star_m y_m, bc).$$

Then $(S, +, \circ)$ and $(S, +, \star)$ are compatible ring structures on S from Theorem 7.1.22. Moreover, $(S, +, \circ)$ is right self-injective since each $(S_i, +, \circ_i)$ is right self-injective by Theorem 7.1.22 and $\mathrm{Mat}_2(B)$ is also right self-injective. But $(S, +, \star)$ is not right FI-extending because each $(S_i, +, \star_i)$ is not right FI-extending by Theorem 7.1.22. Thus, $(S, +, \circ)$ is not isomorphic to $(S, +, \star)$.

(ii)⇒(i) Assume on the contrary that $J(\Lambda) = 0$. Then Λ is semisimple Artinian, so R is right nonsingular. Thus, every essential extension of R_R is a rational extension by Proposition 1.3.14. So every essential extension of R_R has a unique compatible ring structure if it exists by Proposition 7.1.6. □

Exercise 7.1.27

1. In Example 7.1.8, show explicitly that $e_1 T e_1 = \begin{bmatrix} A & 0 \\ 0 & 0 \end{bmatrix}$ and $e_2 T e_1 = 0$.

2. Assume that R is the ring in Theorem 7.1.14 and $I_R \leq^{\text{ess}} R_R$. Show that $I = \begin{bmatrix} B & \text{Soc}(A) \\ 0 & D \end{bmatrix}$, for some nonzero ideals B and D of A.
3. Assume that $W = \begin{bmatrix} 0 & 0 \\ \text{Hom}(\text{Soc}(A)_A, A_A) & A \end{bmatrix}$ is the R-module in the proof of Theorem 7.1.14. Show that W_R is injective.
4. Prove Lemma 7.1.17.
5. Let S be as in Corollary 7.1.23. Also let e_1 and e_2 be as in the proof of Corollary 7.1.23. Show that $e_1Se_1 = \begin{bmatrix} A & 0 \\ 0 & 0 \end{bmatrix}$, $e_1Se_2 = \begin{bmatrix} 0 & \text{Soc}(A) \\ 0 & 0 \end{bmatrix}$, and $e_2Se_2 = \begin{bmatrix} 0 & 0 \\ 0 & A \end{bmatrix}$ (observe that Example 7.1.13 is of this form).

7.2 An Example of Osofsky and Essential Overrings

As we mentioned in Example 7.1.13, Osofsky constructed a ring R which is not right Osofsky compatible. Let us now revisit that example. Say $A = \mathbb{Z}_4$ and let

$$R = \begin{bmatrix} A & 2A \\ 0 & A \end{bmatrix},$$

which is a subring of $T_2(A)$. We note that $Q(R) = R$ since R is right Kasch by Proposition 1.3.18. In this section, we determine *all possible* right essential overrings of R and study their interrelationships and various properties. Results from Sect. 7.1 will be used to determine these essential overrings.

For $f \in \text{Hom}(2A_A, A_A)$ and $x \in A$, let $(f \cdot x)(s) = f(xs)$ for all $s \in 2A$. We put

$$E = \begin{bmatrix} A \oplus \text{Hom}(2A_A, A_A) & A \\ \text{Hom}(2A_A, A_A) & A \end{bmatrix},$$

where the addition on E is componentwise and the R-module scalar multiplication of E over R is given by

$$\begin{bmatrix} a+f & b \\ g & c \end{bmatrix}\begin{bmatrix} x & y \\ 0 & z \end{bmatrix} = \begin{bmatrix} ax + f \cdot x & ay + f(y) + bz \\ g \cdot x & g(y) + cz \end{bmatrix}$$

for $\begin{bmatrix} a+f & b \\ g & c \end{bmatrix} \in E$ and $\begin{bmatrix} x & y \\ 0 & z \end{bmatrix} \in R$, where a, b, c, x, y, z are in A and f, g are in Hom $(2A_A, A_A)$. Then E is an injective hull of R_R by Theorem 7.1.14. We now put $f_0 \in \text{Hom}(2A_A, A_A)$ such that $f_0(2a) = 2a$ for $a \in A$. Then

$$\text{Hom}(2A_A, A_A) = f_0 \cdot A$$

from Proposition 7.1.12. Thus if $f \in \text{Hom}(2A_A, A_A)$, then $f = f_0 \cdot r$ for some $r \in A$. Note that $\text{Hom}(2A_A, A_A) = \{0, f_0\}$. Therefore, we see that *all possible* intermediate R-modules between R_R and E_R are:

$$E,\ V = \begin{bmatrix} A \oplus \mathrm{Hom}\,(2A_A, A_A) & 2A \\ \mathrm{Hom}\,(2A_A, A_A) & A \end{bmatrix},$$

$$Y = \begin{bmatrix} A \oplus \mathrm{Hom}\,(2A_A, A_A) & A \\ 0 & A \end{bmatrix},\ W = \begin{bmatrix} A & A \\ \mathrm{Hom}\,(2A_A, A_A) & A \end{bmatrix},$$

$$S = \begin{bmatrix} A & 2A \\ \mathrm{Hom}\,(2A_A, A_A) & A \end{bmatrix},\ U = \begin{bmatrix} A \oplus \mathrm{Hom}\,(2A_A, A_A) & 2A \\ 0 & A \end{bmatrix},$$

$$T = \begin{bmatrix} A & A \\ 0 & A \end{bmatrix},\ \text{and } R.$$

For the convenience of the reader, we list the following multiplications on the aforementioned overmodules of R_R which will be used in Theorems 7.2.1 and 7.2.2 to describe all of the right essential overrings of R.

Multiplications on V. For

$$v_1 = \begin{bmatrix} a_1 + f_0 \cdot r_1 & 2b_1 \\ f_0 \cdot s_1 & c_1 \end{bmatrix},\ v_2 = \begin{bmatrix} a_2 + f_0 \cdot r_2 & 2b_2 \\ f_0 \cdot s_2 & c_2 \end{bmatrix} \text{ in } V,$$

define multiplications $\bullet_1$, $\bullet_2$, $\bullet_3$, and $\bullet_4$:

$$v_1 \bullet_1 v_2 = \begin{bmatrix} x & y \\ z & w \end{bmatrix},$$

where

$$x = a_1a_2 + f_0 \cdot r_1a_2 + f_0 \cdot a_1r_2 + f_0 \cdot r_1r_2,$$
$$y = 2a_1b_2 + 2r_1b_2 + 2b_1c_2,$$
$$z = f_0 \cdot s_1a_2 + f_0 \cdot s_1r_2 + f_0 \cdot c_1s_2,\ \text{and } w = 2s_1b_2 + c_1c_2.$$

$$v_1 \bullet_2 v_2 = \begin{bmatrix} x & y \\ z & w \end{bmatrix},$$

where

$$x = a_1a_2 + 2r_1r_2 + f_0 \cdot r_1a_2 + f_0 \cdot a_1r_2 + f_0 \cdot r_1r_2,$$
$$y = 2a_1b_2 + 2r_1b_2 + 2b_1c_2,$$
$$z = f_0 \cdot s_1a_2 + f_0 \cdot s_1r_2 + f_0 \cdot c_1s_2,\ \text{and } w = 2s_1b_2 + c_1c_2.$$

$$v_1 \bullet_3 v_2 = \begin{bmatrix} x & y \\ z & w \end{bmatrix},$$

where

$$x = a_1a_2 + 2s_1r_2 + 2a_1s_2 + 2c_1s_2 + f_0 \cdot r_1a_2 + f_0 \cdot a_1r_2 + f_0 \cdot r_1r_2,$$
$$y = 2a_1b_2 + 2r_1b_2 + 2b_1c_2,$$
$$z = f_0 \cdot s_1a_2 + f_0 \cdot s_1r_2 + f_0 \cdot c_1s_2, \text{ and } w = 2s_1b_2 + c_1c_2.$$

$$v_1 \bullet_4 v_2 = \begin{bmatrix} x & y \\ z & w \end{bmatrix},$$

where

$$x = a_1a_2 + 2r_1r_2 + 2s_1r_2 + 2a_1s_2 + 2c_1s_2 + f_0 \cdot r_1a_2 + f_0 \cdot a_1r_2 + f_0 \cdot r_1r_2,$$
$$y = 2a_1b_2 + 2r_1b_2 + 2b_1c_2,$$
$$z = f_0 \cdot s_1a_2 + f_0 \cdot s_1r_2 + f_0 \cdot c_1s_2, \text{ and } w = 2s_1b_2 + c_1c_2.$$

Multiplications on S. For $s_1 = \begin{bmatrix} a_1 & 2b_1 \\ f_0 \cdot r_1 & c_1 \end{bmatrix}$, $s_2 = \begin{bmatrix} a_2 & 2b_2 \\ f_0 \cdot r_2 & c_2 \end{bmatrix}$ in S, define multiplications $\circ_{(k,t)}$, where $k \in A$ and $t \in 2A$:

$$s_1 \circ_{(k,t)} s_2 = \begin{bmatrix} a_1a_2 - ta_1r_2 + 2kb_1r_2 + tc_1r_2 & 2a_1b_2 + 2b_1c_2 \\ f_0 \cdot r_1a_2 + f_0 \cdot c_1r_2 & 2r_1b_2 + c_1c_2 \end{bmatrix}.$$

Multiplications on U. For

$$u_1 = \begin{bmatrix} a_1 + f_0 \cdot r_1 & 2b_1 \\ 0 & c_1 \end{bmatrix}, \quad u_2 = \begin{bmatrix} a_2 + f_0 \cdot r_2 & 2b_2 \\ 0 & c_2 \end{bmatrix} \text{ in } U,$$

define multiplications $\circledcirc_1$ and $\circledcirc_2$:

$$u_1 \circledcirc_1 u_2 = \begin{bmatrix} x & y \\ 0 & w \end{bmatrix},$$

where

$$x = a_1a_2 + f_0 \cdot a_1r_2 + f_0 \cdot r_1a_2 + f_0 \cdot r_1r_2,$$
$$y = 2a_1b_2 + 2r_1b_2 + 2b_1c_2, \text{ and } w = c_1c_2.$$

$$u_1 \circledcirc_2 u_2 = \begin{bmatrix} x & y \\ 0 & w \end{bmatrix},$$

where

$$x = a_1a_2 + 2r_1r_2 + f_0 \cdot a_1r_2 + f_0 \cdot r_1a_2 + f_0 \cdot r_1r_2,$$
$$y = 2a_1b_2 + 2r_1b_2 + 2b_1c_2, \text{ and } w = c_1c_2.$$

Multiplications on T. For $t_1 = \begin{bmatrix} a_1 & b_1 \\ 0 & c_1 \end{bmatrix}$, $t_2 = \begin{bmatrix} a_2 & b_2 \\ 0 & c_2 \end{bmatrix}$ in T, define multiplications $\diamond_1$ and $\diamond_2$:

$$t_1 \diamond_1 t_2 = \begin{bmatrix} a_1a_2 & a_1b_2 + b_1c_2 \\ 0 & c_1c_2 \end{bmatrix}$$

and

$$t_1 \diamond_2 t_2 = \begin{bmatrix} a_1a_2 & a_1b_2 + 2b_1b_2 + b_1c_2 \\ 0 & c_1c_2 + 2a_1b_2 + 2c_1b_2 \end{bmatrix}.$$

Theorem 7.2.1 (i) *There are exactly thirteen right essential overrings of* R*, namely*:

$$(V, +, \bullet_1), (V, +, \bullet_2), (V, +, \bullet_3), (V, +, \bullet_4),$$

$$(S, +, \circ_{(0,0)}), (S, +, \circ_{(0,2)}), (S, +, \circ_{(1,0)}), (S, +, \circ_{(1,2)}),$$

$$(U, +, \circledcirc_1), (U, +, \circledcirc_2), (T, +, \diamond_1), (T, +, \diamond_2), \text{ and } R \text{ itself}$$

such that

$$(V, +, \bullet_1) \cong (V, +, \bullet_2) \cong (V, +, \bullet_3) \cong (V, +, \bullet_4),$$

$$(S, +, \circ_{(0,0)}) \cong (S, +, \circ_{(0,2)}), \ (S, +, \circ_{(1,0)}) \cong (S, +, \circ_{(1,2)}),$$

$$(S, +, \circ_{(0,0)}) \ncong (S, +, \circ_{(1,0)}),$$

$$(U, +, \circledcirc_1) \cong (U, +, \circledcirc_2), \text{ and } (T, +, \diamond_1) \cong (T, +, \diamond_2).$$

(ii) $(S, +, \circ_{(0,0)})$ *is a subring of both* $(V, +, \bullet_1)$ *and* $(V, +, \bullet_2)$*, while* $(S, +, \circ_{(0,2)})$ *is a subring of both* $(V, +, \bullet_3)$ *and* $(V, +, \bullet_4)$.

(iii) $(U, +, \circledcirc_1)$ *is a subring of both* $(V, +, \bullet_1)$ *and* $(V, +, \bullet_3)$*, while* $(U, +, \circledcirc_2)$ *is a subring of both* $(V, +, \bullet_2)$ *and* $(V, +, \bullet_4)$.

Proof (i) The proof proceeds by the following steps.

1. *R is not right Osofsky compatible.*

Proof. Assume that E has a compatible ring structure. Then

$$\begin{bmatrix} 0 & 2 \\ 0 & 0 \end{bmatrix} = \begin{bmatrix} f_0 & 0 \\ 0 & 0 \end{bmatrix}\begin{bmatrix} 0 & 2 \\ 0 & 0 \end{bmatrix} = 2\begin{bmatrix} f_0 & 0 \\ 0 & 0 \end{bmatrix}\begin{bmatrix} 0 & 1 \\ 0 & 0 \end{bmatrix} = \begin{bmatrix} f_0 \cdot 2 & 0 \\ 0 & 0 \end{bmatrix}\begin{bmatrix} 0 & 1 \\ 0 & 0 \end{bmatrix} = 0,$$

a contradiction. Thus E cannot have a compatible ring structure.

2. *Neither Y nor W has a compatible ring structure.*

Proof. It follows from a similar argument as in the proof of 1.

3. *Compatible ring structures on* V: There are exactly four compatible ring structures and they are isomorphic.

Proof. Assume that V has a compatible ring structure. First, we observe that $R_R \leq^{\text{ess}} V_R$. By Lemma 7.1.7, $1_V = 1_R$. Let $e_1 = \begin{bmatrix} 1 & 0 \\ 0 & 0 \end{bmatrix}$ and $e_2 = \begin{bmatrix} 0 & 0 \\ 0 & 1 \end{bmatrix}$. Then it follows that $V = e_1Ve_1 + e_1Ve_2 + e_2Ve_1 + e_2Ve_2$.

By direct (but technical) argument, we have the following two cases:

Case 1. $e_1Ve_1 = \begin{bmatrix} A \oplus \mathrm{Hom}(2A_A, A_A) & 0 \\ 0 & 0 \end{bmatrix}$, $e_1Ve_2 = \begin{bmatrix} 0 & 2A \\ 0 & 0 \end{bmatrix}$,

$$e_2Ve_1 = \begin{bmatrix} 0 & 0 \\ \mathrm{Hom}(2A_A, A_A) & 0 \end{bmatrix},\ e_2Ve_2 = \begin{bmatrix} 0 & 0 \\ 0 & A \end{bmatrix}.$$

Case 2. $e_1Ve_1 = \begin{bmatrix} A \oplus \mathrm{Hom}(2A_A, A_A) & 0 \\ 0 & 0 \end{bmatrix}$, $e_1Ve_2 = \begin{bmatrix} 0 & 2A \\ 0 & 0 \end{bmatrix}$,

$$e_2Ve_1 = \left\{0, \begin{bmatrix} 2 & 0 \\ f_0 & 0 \end{bmatrix}\right\},\ e_2Ve_2 = \begin{bmatrix} 0 & 0 \\ 0 & A \end{bmatrix}.$$

We can check that Case 1 is subdivided into the following:

Subcase 1.1. $\begin{bmatrix} f_0 & 0 \\ 0 & 0 \end{bmatrix}\begin{bmatrix} f_0 & 0 \\ 0 & 0 \end{bmatrix} = \begin{bmatrix} f_0 & 0 \\ 0 & 0 \end{bmatrix}$.

In this case, we have the following by direct computation.

1. $\begin{bmatrix} 1 & 0 \\ 0 & 0 \end{bmatrix}\begin{bmatrix} f_0 & 0 \\ 0 & 0 \end{bmatrix} = \begin{bmatrix} f_0 & 0 \\ 0 & 0 \end{bmatrix}$, 2. $\begin{bmatrix} 0 & 2 \\ 0 & 0 \end{bmatrix}\begin{bmatrix} f_0 & 0 \\ 0 & 0 \end{bmatrix} = 0$, 3. $\begin{bmatrix} 0 & 0 \\ f_0 & 0 \end{bmatrix}\begin{bmatrix} f_0 & 0 \\ 0 & 0 \end{bmatrix} = \begin{bmatrix} 0 & 0 \\ f_0 & 0 \end{bmatrix}$,

4. $\begin{bmatrix} 0 & 0 \\ 0 & 1 \end{bmatrix}\begin{bmatrix} f_0 & 0 \\ 0 & 0 \end{bmatrix} = 0$, 5. $\begin{bmatrix} 1 & 0 \\ 0 & 0 \end{bmatrix}\begin{bmatrix} 0 & 0 \\ f_0 & 0 \end{bmatrix} = 0$, 6. $\begin{bmatrix} f_0 & 0 \\ 0 & 0 \end{bmatrix}\begin{bmatrix} 0 & 0 \\ f_0 & 0 \end{bmatrix} = 0$,

7. $\begin{bmatrix} 0 & 2 \\ 0 & 0 \end{bmatrix}\begin{bmatrix} 0 & 0 \\ f_0 & 0 \end{bmatrix} = 0$, 8. $\begin{bmatrix} 0 & 0 \\ f_0 & 0 \end{bmatrix}\begin{bmatrix} 0 & 0 \\ f_0 & 0 \end{bmatrix} = 0$, 9. $\begin{bmatrix} 0 & 0 \\ 0 & 1 \end{bmatrix}\begin{bmatrix} 0 & 0 \\ f_0 & 0 \end{bmatrix} = \begin{bmatrix} 0 & 0 \\ f_0 & 0 \end{bmatrix}$.

Let $v_1 = \begin{bmatrix} a_1 + f_0 \cdot r_1 & 2b_1 \\ f_0 \cdot s_1 & c_1 \end{bmatrix}$, $v_2 = \begin{bmatrix} a_2 + f_0 \cdot r_2 & 2b_2 \\ f_0 \cdot s_2 & c_2 \end{bmatrix} \in V$. Then by 1–9, there is a multiplication on V which extends the R-module scalar multiplication of V over R such that

$$v_1v_2 = \begin{bmatrix} a_1a_2 + f_0 \cdot r_1a_2 + f_0 \cdot a_1r_2 + f_0 \cdot r_1r_2 & 2a_1b_2 + 2r_1b_2 + 2b_1c_2 \\ f_0 \cdot s_1a_2 + f_0 \cdot s_1r_2 + f_0 \cdot c_1s_2 & 2s_1b_2 + c_1c_2 \end{bmatrix}.$$

Let $\bullet_1$ denote this multiplication on V.

Subcase 1.2. $\begin{bmatrix} f_0 & 0 \\ 0 & 0 \end{bmatrix}\begin{bmatrix} f_0 & 0 \\ 0 & 0 \end{bmatrix} = \begin{bmatrix} 2 + f_0 & 0 \\ 0 & 0 \end{bmatrix}$.

In this case, we have the following.

1. $\begin{bmatrix} 1 & 0 \\ 0 & 0 \end{bmatrix}\begin{bmatrix} f_0 & 0 \\ 0 & 0 \end{bmatrix} = \begin{bmatrix} f_0 & 0 \\ 0 & 0 \end{bmatrix}$, 2. $\begin{bmatrix} 0 & 2 \\ 0 & 0 \end{bmatrix}\begin{bmatrix} f_0 & 0 \\ 0 & 0 \end{bmatrix} = 0$, 3. $\begin{bmatrix} 0 & 0 \\ f_0 & 0 \end{bmatrix}\begin{bmatrix} f_0 & 0 \\ 0 & 0 \end{bmatrix} = \begin{bmatrix} 0 & 0 \\ f_0 & 0 \end{bmatrix}$,

4. $\begin{bmatrix} 0 & 0 \\ 0 & 1 \end{bmatrix}\begin{bmatrix} f_0 & 0 \\ 0 & 0 \end{bmatrix} = 0$, 5. $\begin{bmatrix} 1 & 0 \\ 0 & 0 \end{bmatrix}\begin{bmatrix} 0 & 0 \\ f_0 & 0 \end{bmatrix} = 0$, 6. $\begin{bmatrix} f_0 & 0 \\ 0 & 0 \end{bmatrix}\begin{bmatrix} 0 & 0 \\ f_0 & 0 \end{bmatrix} = 0$,

7. $\begin{bmatrix} 0 & 2 \\ 0 & 0 \end{bmatrix}\begin{bmatrix} 0 & 0 \\ f_0 & 0 \end{bmatrix} = 0$, 8. $\begin{bmatrix} 0 & 0 \\ f_0 & 0 \end{bmatrix}\begin{bmatrix} 0 & 0 \\ f_0 & 0 \end{bmatrix} = 0$, 9. $\begin{bmatrix} 0 & 0 \\ 0 & 1 \end{bmatrix}\begin{bmatrix} 0 & 0 \\ f_0 & 0 \end{bmatrix} = \begin{bmatrix} 0 & 0 \\ f_0 & 0 \end{bmatrix}$.

Therefore, in this case, there exists a multiplication $\bullet_2$ on V which extends the R-module scalar multiplication of V over R:

$$\begin{bmatrix} a_1 + f_0 \cdot r_1 & 2b_1 \\ f_0 \cdot s_1 & c_1 \end{bmatrix} \bullet_2 \begin{bmatrix} a_2 + f_0 \cdot r_2 & 2b_2 \\ f_0 \cdot s_2 & c_2 \end{bmatrix} =$$

$$\begin{bmatrix} a_1a_2 + 2r_1r_2 + f_0 \cdot r_1a_2 + f_0 \cdot a_1r_2 + f_0 \cdot r_1r_2 & 2a_1b_2 + 2r_1b_2 + 2b_1c_2 \\ f_0 \cdot s_1a_2 + f_0 \cdot s_1r_2 + f_0 \cdot c_1s_2 & 2s_1b_2 + c_1c_2 \end{bmatrix}.$$

Case 2. As in Case 1,

$$\begin{bmatrix} f_0 & 0 \\ 0 & 0 \end{bmatrix}\begin{bmatrix} f_0 & 0 \\ 0 & 0 \end{bmatrix} = \begin{bmatrix} f_0 & 0 \\ 0 & 0 \end{bmatrix} \text{ or } \begin{bmatrix} f_0 & 0 \\ 0 & 0 \end{bmatrix}\begin{bmatrix} f_0 & 0 \\ 0 & 0 \end{bmatrix} = \begin{bmatrix} 2+f_0 & 0 \\ 0 & 0 \end{bmatrix}.$$

Thus we have two subcases, Subcase 2.1 and Subcase 2.2, which are:

Subcase 2.1. $\begin{bmatrix} f_0 & 0 \\ 0 & 0 \end{bmatrix}\begin{bmatrix} f_0 & 0 \\ 0 & 0 \end{bmatrix} = \begin{bmatrix} f_0 & 0 \\ 0 & 0 \end{bmatrix}$.

As in Subcase 1.1, we have the following.

1. $\begin{bmatrix} 1 & 0 \\ 0 & 0 \end{bmatrix}\begin{bmatrix} f_0 & 0 \\ 0 & 0 \end{bmatrix} = \begin{bmatrix} f_0 & 0 \\ 0 & 0 \end{bmatrix}$, 2. $\begin{bmatrix} 0 & 2 \\ 0 & 0 \end{bmatrix}\begin{bmatrix} f_0 & 0 \\ 0 & 0 \end{bmatrix} = 0$, 3. $\begin{bmatrix} 0 & 0 \\ f_0 & 0 \end{bmatrix}\begin{bmatrix} f_0 & 0 \\ 0 & 0 \end{bmatrix} = \begin{bmatrix} 2 & 0 \\ f_0 & 0 \end{bmatrix}$,
4. $\begin{bmatrix} 0 & 0 \\ 0 & 1 \end{bmatrix}\begin{bmatrix} f_0 & 0 \\ 0 & 0 \end{bmatrix} = 0$, 5. $\begin{bmatrix} 1 & 0 \\ 0 & 0 \end{bmatrix}\begin{bmatrix} 0 & 0 \\ f_0 & 0 \end{bmatrix} = \begin{bmatrix} 2 & 0 \\ 0 & 0 \end{bmatrix}$, 6. $\begin{bmatrix} f_0 & 0 \\ 0 & 0 \end{bmatrix}\begin{bmatrix} 0 & 0 \\ f_0 & 0 \end{bmatrix} = 0$,
7. $\begin{bmatrix} 0 & 2 \\ 0 & 0 \end{bmatrix}\begin{bmatrix} 0 & 0 \\ f_0 & 0 \end{bmatrix} = 0$, 8. $\begin{bmatrix} 0 & 0 \\ f_0 & 0 \end{bmatrix}\begin{bmatrix} 0 & 0 \\ f_0 & 0 \end{bmatrix} = 0$, 9. $\begin{bmatrix} 0 & 0 \\ 0 & 1 \end{bmatrix}\begin{bmatrix} 0 & 0 \\ f_0 & 0 \end{bmatrix} = \begin{bmatrix} 2 & 0 \\ f_0 & 0 \end{bmatrix}$.

Hence there is a multiplication $\bullet_3$ on V which extends the R-module scalar multiplication of V over R:

$$\begin{bmatrix} a_1 + f_0 \cdot r_1 & 2b_1 \\ f_0 \cdot s_1 & c_1 \end{bmatrix} \bullet_3 \begin{bmatrix} a_2 + f_0 \cdot r_2 & 2b_2 \\ f_0 \cdot s_2 & c_2 \end{bmatrix} = \begin{bmatrix} x & y \\ z & w \end{bmatrix},$$

where

$$x = a_1a_2 + 2s_1r_2 + 2a_1s_2 + 2c_1s_2 + f_0 \cdot r_1a_2 + f_0 \cdot a_1r_2 + f_0 \cdot r_1r_2,$$

$$y = 2a_1b_2 + 2r_1b_2 + 2b_1c_2,$$

$$z = f_0 \cdot s_1a_2 + f_0 \cdot s_1r_2 + f_0 \cdot c_1s_2, \text{ and } w = 2s_1b_2 + c_1c_2.$$

Subcase 2.2. $\begin{bmatrix} f_0 & 0 \\ 0 & 0 \end{bmatrix}\begin{bmatrix} f_0 & 0 \\ 0 & 0 \end{bmatrix} = \begin{bmatrix} 2+f_0 & 0 \\ 0 & 0 \end{bmatrix}$.

As in 1–9 of Subcase 1.1, we have the following.

1. $\begin{bmatrix} 1 & 0 \\ 0 & 0 \end{bmatrix}\begin{bmatrix} f_0 & 0 \\ 0 & 0 \end{bmatrix} = \begin{bmatrix} f_0 & 0 \\ 0 & 0 \end{bmatrix}$, 2. $\begin{bmatrix} 0 & 2 \\ 0 & 0 \end{bmatrix}\begin{bmatrix} f_0 & 0 \\ 0 & 0 \end{bmatrix} = 0$, 3. $\begin{bmatrix} 0 & 0 \\ f_0 & 0 \end{bmatrix}\begin{bmatrix} f_0 & 0 \\ 0 & 0 \end{bmatrix} = \begin{bmatrix} 2 & 0 \\ f_0 & 0 \end{bmatrix}$,
4. $\begin{bmatrix} 0 & 0 \\ 0 & 1 \end{bmatrix}\begin{bmatrix} f_0 & 0 \\ 0 & 0 \end{bmatrix} = 0$, 5. $\begin{bmatrix} 1 & 0 \\ 0 & 0 \end{bmatrix}\begin{bmatrix} 0 & 0 \\ f_0 & 0 \end{bmatrix} = \begin{bmatrix} 2 & 0 \\ 0 & 0 \end{bmatrix}$, 6. $\begin{bmatrix} f_0 & 0 \\ 0 & 0 \end{bmatrix}\begin{bmatrix} 0 & 0 \\ f_0 & 0 \end{bmatrix} = 0$,

7. $\begin{bmatrix} 0 & 2 \\ 0 & 0 \end{bmatrix}\begin{bmatrix} 0 & 0 \\ f_0 & 0 \end{bmatrix} = 0$, 8. $\begin{bmatrix} 0 & 0 \\ f_0 & 0 \end{bmatrix}\begin{bmatrix} 0 & 0 \\ f_0 & 0 \end{bmatrix} = 0$, 9. $\begin{bmatrix} 0 & 0 \\ 0 & 1 \end{bmatrix}\begin{bmatrix} 0 & 0 \\ f_0 & 0 \end{bmatrix} = \begin{bmatrix} 2 & 0 \\ f_0 & 0 \end{bmatrix}$.

Thus there is a multiplication $\bullet_4$ on V which extends the R-module scalar multiplication of V over R:

$$\begin{bmatrix} a_1 + f_0 \cdot r_1 & 2b_1 \\ f_0 \cdot s_1 & c_1 \end{bmatrix} \bullet_4 \begin{bmatrix} a_2 + f_0 \cdot r_2 & 2b_2 \\ f_0 \cdot s_2 & c_2 \end{bmatrix} = \begin{bmatrix} x & y \\ z & w \end{bmatrix},$$

where

$$x = a_1a_2 + 2r_1r_2 + 2s_1r_2 + 2a_1s_2 + 2c_1s_2 + f_0 \cdot r_1a_2 + f_0 \cdot a_1r_2 + f_0 \cdot r_1r_2,$$

$$y = 2a_1b_2 + 2r_1b_2 + 2b_1c_2,$$

$$z = f_0 \cdot s_1a_2 + f_0 \cdot s_1r_2 + f_0 \cdot c_1s_2, \text{ and } w = 2s_1b_2 + c_1c_2.$$

The multiplications $\bullet_1$, $\bullet_2$, $\bullet_3$, and $\bullet_4$ are well-defined and each extends the R-module scalar multiplication of V over R. Therefore it follows that $(V,+,\bullet_1)$, $(V,+,\bullet_2)$, $(V,+,\bullet_3)$, and $(V,+,\bullet_4)$ are precisely all of the possible compatible ring structures on V. Define $\theta_2 : (V,+,\bullet_2) \to (V,+,\bullet_1)$ by

$$\theta_2 \begin{bmatrix} a + f_0 \cdot r & 2b \\ f_0 \cdot s & c \end{bmatrix} = \begin{bmatrix} a + 2r + f_0 \cdot r & 2b \\ f_0 \cdot s & c \end{bmatrix}.$$

Then θ_2 is a ring isomorphism.

Also define $\theta_3 : (V,+,\bullet_3) \to (V,+,\bullet_1)$ and $\theta_4 : (V,+,\bullet_4) \to (V,+,\bullet_1)$ by

$$\theta_3 \begin{bmatrix} a + f_0 \cdot r & 2b \\ f_0 \cdot s & c \end{bmatrix} = \begin{bmatrix} a + 2s + f_0 \cdot r & 2b \\ f_0 \cdot s & c \end{bmatrix},$$

and

$$\theta_4 \begin{bmatrix} a + f_0 \cdot r & 2b \\ f_0 \cdot s & c \end{bmatrix} = \begin{bmatrix} a + 2r + 2s + f_0 \cdot r & 2b \\ f_0 \cdot s & c \end{bmatrix}.$$

Then θ_3 and θ_4 are also ring isomorphisms. Therefore

$$(V,+,\bullet_1) \cong (V,+,\bullet_2) \cong (V,+,\bullet_3) \cong (V,+,\bullet_4).$$

4. *Compatible ring structures on S*: There are exactly four compatible ring structures on S such that

(1) two compatible ring structures are QF and they are isomorphic.

(2) the other two compatible ring structures are not even right FI-extending but they are isomorphic.

Proof. By Theorem 7.1.22 and Corollary 7.1.23, there are exactly four ring multiplications: For $s_1 = \begin{bmatrix} a_1 & 2b_1 \\ f_0 \cdot r_1 & c_1 \end{bmatrix}$, $s_2 = \begin{bmatrix} a_2 & 2b_2 \\ f_0 \cdot r_2 & c_2 \end{bmatrix}$ in S,

$$s_1 \circ_{(k,t)} s_2 = \begin{bmatrix} a_1a_2 - ta_1r_2 + 2kb_1r_2 + tc_1r_2 & 2a_1b_2 + 2b_1c_2 \\ f_0 \cdot r_1a_2 + f_0 \cdot c_1r_2 & 2r_1b_2 + c_1c_2 \end{bmatrix},$$

where $k \in A$ and $t \in \mathrm{Soc}(A)$. Thus $\circ_{(0,0)} = \circ_{(2,0)}$, $\circ_{(1,0)} = \circ_{(3,0)}$, $\circ_{(0,2)} = \circ_{(2,2)}$, and $\circ_{(1,2)} = \circ_{(3,2)}$. Therefore all possible ring multiplications are precisely $\circ_{(0,0)}$, $\circ_{(1,0)}$, $\circ_{(0,2)}$, and $\circ_{(1,2)}$. By Theorem 7.1.22,

$$(S, +, \circ_{(0,2)}) \cong (S, +, \circ_{(0,0)}) \text{ and } (S, +, \circ_{(1,0)}) \cong (S, +, \circ_{(1,2)}).$$

Again by Theorem 7.1.22, $(S, +, \circ_{(1,0)})$ (hence $(S, +, \circ_{(1,2)})$) is QF, while $(S, +, \circ_{(0,0)})$ (hence $(S, +, \circ_{(0,2)})$) is not even right FI-extending. So it follows that $(S, +, \circ_{(1,0)}) \not\cong (S, +, \circ_{(0,0)})$.

5. *Compatible ring structures on* U: There are exactly two compatible ring structures $(U, +, \circledcirc_1)$ and $(U, +, \circledcirc_2)$ on U. Further, they are isomorphic.

Proof. The proof follows from a similar argument as in the proof for the case of V. Now define $\psi : (U, +, \circledcirc_2) \to (U, +, \circledcirc_1)$ by

$$\psi \begin{bmatrix} a + f_0 \cdot r & 2b \\ 0 & c \end{bmatrix} = \begin{bmatrix} a + 2r + f_0 \cdot r & 2b \\ 0 & c \end{bmatrix}.$$

Then ψ is a ring isomorphism. Thus, $(U, +, \circledcirc_1) \cong (U, +, \circledcirc_2)$.

6. *Compatible ring structures on* T: By Example 7.1.8, there are exactly two compatible ring structures $(T, +, \diamond_1)$ and $(T, +, \diamond_2)$ on T and they are isomorphic.

Consequently, part (i) can be proved. Parts (ii) and (iii) can be routinely checked. □

Theorem 7.2.2 *Let* R, V, S, U, *and* T *be as in Theorem* 7.2.1. *Then*:

(i) *R is not right FI-extending.*
(ii) *$(V, +, \bullet_1)$ (hence $(V, +, \bullet_2)$, $(V, +, \bullet_3)$, and $(V, +, \bullet_4)$) are right extending, but not right quasi-continuous.*
(iii) *$(S, +, \circ_{(1,0)})$ (so $(S, +, \circ_{(1,2)})$) are right self-injective, but $(S, +, \circ_{(0,0)})$ (so $(S, +, \circ_{(0,2)})$) is not even right FI-extending.*
(iv) *$(U, +, \circledcirc_1)$ (hence $(U, +, \circledcirc_2)$) are right FI-extending, but not right extending.*
(v) *$(T, +, \diamond_1)$ (hence $(T, +, \diamond_2)$) are right FI-extending, but not right extending.*

Proof (i) Let $I = \begin{bmatrix} 0 & 2A \\ 0 & 0 \end{bmatrix}$. Then $I \trianglelefteq R$. There is no idempotent $e \in R$ such that $I_R \leq^{\text{ess}} eR_R$. Thus R is not right FI-extending.

(ii) Let $\Delta := (V, +, \bullet_1)$. Recall that $e_1 = \begin{bmatrix} 1 & 0 \\ 0 & 0 \end{bmatrix}$ and $e_2 = \begin{bmatrix} 0 & 0 \\ 0 & 1 \end{bmatrix} \in \Delta$. Let $g_1 = \begin{bmatrix} f_0 & 0 \\ 0 & 0 \end{bmatrix}$ and $g_2 = \begin{bmatrix} 1 + f_0 & 0 \\ 0 & 0 \end{bmatrix}$. Then $e_1\Delta_\Delta = g_1\Delta_\Delta \oplus g_2\Delta_\Delta$. We can see that each $g_i\Delta_\Delta$ is uniform, hence it is extending. Also $g_i\Delta_\Delta$ and $g_j\Delta_\Delta$ are relatively injective for $i \neq j$. Thus $e_1\Delta_\Delta$ is extending by Theorem 2.2.18. Since $e_2\Delta_\Delta$ is uniform, $e_2\Delta_\Delta$ is extending.

By computation, all the possible right ideals I of Δ such that $I \cap e_1\Delta = 0$ are:

$$0, \begin{bmatrix} 0 & 0 \\ 0 & 2 \end{bmatrix}\Delta, \begin{bmatrix} 0 & 0 \\ \mathrm{Hom}(2A_A, A_A) & 2A \end{bmatrix}, \begin{bmatrix} 0 & 0 \\ 0 & 1 \end{bmatrix}\Delta, \begin{bmatrix} 0 & 2 \\ 0 & 2 \end{bmatrix}\Delta,$$

$$\begin{bmatrix} f_0 & 0 \\ f_0 & 0 \end{bmatrix}\Delta, \begin{bmatrix} 0 & 2 \\ 0 & 1 \end{bmatrix}\Delta, \text{ and } \begin{bmatrix} 0 & 2 \\ f_0 & 1 \end{bmatrix}\Delta.$$

Among these, all of the closed right ideals are:

$$0, \begin{bmatrix} 0 & 0 \\ 0 & 1 \end{bmatrix}\Delta, \begin{bmatrix} f_0 & 0 \\ f_0 & 0 \end{bmatrix}\Delta, \begin{bmatrix} 0 & 2 \\ 0 & 1 \end{bmatrix}\Delta, \text{ and } \begin{bmatrix} 0 & 2 \\ f_0 & 1 \end{bmatrix}\Delta,$$

which are direct summands of Δ_Δ. Thus, every closed right ideal I of Δ such that $I \cap e_1\Delta = 0$ is a direct summand of Δ_Δ.

Moreover, all the right ideals K of Δ such that $K \cap e_2\Delta = 0$ are:

$$0, \begin{bmatrix} 0 & 2A \\ 0 & 0 \end{bmatrix}, \begin{bmatrix} 2A & 0 \\ 0 & 0 \end{bmatrix}, \begin{bmatrix} 1+f_0 & 0 \\ 0 & 0 \end{bmatrix}\Delta, \begin{bmatrix} f_0 & 0 \\ 0 & 0 \end{bmatrix}\Delta, \begin{bmatrix} 2A & 2A \\ 0 & 0 \end{bmatrix}, \begin{bmatrix} 1+f_0 & 2 \\ 0 & 0 \end{bmatrix}\Delta,$$

$$\begin{bmatrix} 2+f_0 & 0 \\ 0 & 0 \end{bmatrix}\Delta, \begin{bmatrix} 1 & 0 \\ 0 & 0 \end{bmatrix}\Delta, \begin{bmatrix} 0 & 2 \\ 0 & 2 \end{bmatrix}\Delta, \begin{bmatrix} 0 & 2A \\ 0 & 2A \end{bmatrix}, \begin{bmatrix} 2 & 2 \\ 0 & 2 \end{bmatrix}\Delta, \begin{bmatrix} 1+f_0 & 2 \\ 0 & 2 \end{bmatrix}\Delta,$$

$$\begin{bmatrix} f_0 & 0 \\ f_0 & 0 \end{bmatrix}\Delta, \begin{bmatrix} 2+f_0 & 0 \\ f_0 & 0 \end{bmatrix}\Delta, \text{ and } \begin{bmatrix} 1 & 0 \\ f_0 & 0 \end{bmatrix}\Delta.$$

In these right ideals K, closed right ideals are:

$$0, \begin{bmatrix} 1+f_0 & 0 \\ 0 & 0 \end{bmatrix}\Delta, \begin{bmatrix} f_0 & 0 \\ 0 & 0 \end{bmatrix}\Delta, \begin{bmatrix} 1 & 0 \\ 0 & 0 \end{bmatrix}\Delta, \begin{bmatrix} f_0 & 0 \\ f_0 & 0 \end{bmatrix}\Delta, \text{ and } \begin{bmatrix} 1 & 0 \\ f_0 & 0 \end{bmatrix}\Delta.$$

Also these are direct summands of Δ_Δ. Thus every closed right ideal K of Δ such that $K \cap e_2\Delta = 0$ is a direct summand of Δ_Δ.

Hence, by Proposition 2.2.17, Δ is right extending. From Theorem 7.2.1, $(V,+,\bullet_2)$, $(V,+,\bullet_3)$, and $(V,+,\bullet_4)$ are also right extending.

Next, we show that Δ is not right quasi-continuous. For this, take

$$v := \begin{bmatrix} f_0 & 0 \\ f_0 & 0 \end{bmatrix} \in \Delta.$$

Then $v^2 = v$, so $v\Delta_\Delta \leq^\oplus \Delta_\Delta$. Note that $e_2\Delta_\Delta \leq^\oplus \Delta_\Delta$ and $v\Delta \cap e_2\Delta = 0$. But $v\Delta \oplus e_2\Delta$ is not a direct summand of Δ_Δ. Hence Δ_Δ does not satisfy (C_3) condition. So Δ is not right quasi-continuous. From Theorem 7.2.1, $(V,+,\bullet_2)$, $(V,+,\bullet_3)$, and $(V,+,\bullet_4)$ are not right quasi-continuous.

(iii) By Theorems 7.1.22 and 7.2.1, $(S,+,\circ_{(1,0)})$ $(\cong (S,+,\circ_{(1,2)}))$ is right self-injective, but $(S,+,\circ_{(0,0)})$ $(\cong (S,+,\circ_{(0,2)}))$ is not right FI-extending.

(iv) Let $\Phi = (U,+,\circledcirc_1)$. Then we see that

$$\Phi_\Phi = \begin{bmatrix} f_0 & 0 \\ 0 & 0 \end{bmatrix}\Phi_\Phi \oplus \begin{bmatrix} 1+f_0 & 0 \\ 0 & 0 \end{bmatrix}\Phi_\Phi \oplus \begin{bmatrix} 0 & 0 \\ 0 & 1 \end{bmatrix}\Phi_\Phi.$$

Also $\begin{bmatrix} f_0 & 0 \\ 0 & 0 \end{bmatrix}\Phi_\Phi$, $\begin{bmatrix} 1+f_0 & 0 \\ 0 & 0 \end{bmatrix}\Phi_\Phi$, and $\begin{bmatrix} 0 & 0 \\ 0 & 1 \end{bmatrix}\Phi_\Phi$ are uniform, hence they are extending. Thus, Φ is right FI-extending by Theorem 2.3.6.

Next, take $K = \begin{bmatrix} 2 & 2 \\ 0 & 2 \end{bmatrix} \Phi$. Then the only possible $e^2 = e \in \Phi$ with $K_\Phi \leq e\Phi_\Phi$ is $e = 1$. Thus, if Φ is right extending, then $K_\Phi \leq^{\text{ess}} \Phi_\Phi$. But this is absurd because $\begin{bmatrix} f_0 & 0 \\ 0 & 0 \end{bmatrix} \Phi \cap K = 0$. Thus, Φ is not right extending, and hence from Theorem 7.2.1, $(U, +, \circledcirc_2)$ also is right FI-extending, but not right extending.

(v) Let $\Gamma = (T, +, \diamond_1)$. As A is self-injective, Γ is right FI-extending by Theorem 5.6.19. Let $e_{ij} \in \Gamma$ be the matrix with 1 in the (i, j)-position and 0 elsewhere. Take $I = (e_{12} + 2e_{22})\Gamma$. Then there is no direct summand K_Γ of Γ_Γ such that $I_\Gamma \leq^{\text{ess}} K_\Gamma$. So Γ is not right extending. By Theorem 7.2.1, also $(T, +, \diamond_2)$ is right FI-extending, but not right extending. □

Exercise 7.2.3

1. Let E be as in introduction of this section. By a similar argument as in the proof of Theorem 7.2.1 showing that E does not have a compatible ring structure, verify that both Y and W have no compatible ring structures.
2. Let V be as in Theorem 7.2.1. Prove explicitly the following facts in Cases 1 and 2 in the proof of 3 of Theorem 7.2.1(i).
   (i) $e_1 V e_1 = \begin{bmatrix} A \oplus \text{Hom}(2A_A, A_A) & 0 \\ 0 & 0 \end{bmatrix}$.
   (ii) $e_1 V e_2 = \begin{bmatrix} 0 & 2A \\ 0 & 0 \end{bmatrix}$ and $e_2 V e_2 = \begin{bmatrix} 0 & 0 \\ 0 & A \end{bmatrix}$.
   (iii) $e_2 V e_1 = \begin{bmatrix} 0 & 0 \\ \text{Hom}(2A_A, A_A) & 0 \end{bmatrix}$ or $e_2 V e_1 = \left\{0, \begin{bmatrix} 2 & 0 \\ f_0 & 0 \end{bmatrix}\right\}$.
   (iv) Show that Case 1 can be subdivided into Subcases 1.1 and 1.2.
3. Let U be as in Theorem 7.2.1. Show explicitly that $(U, +, \circledcirc_1)$ and $(U, +, \circledcirc_2)$ are all possible compatible ring structures on U.
4. ([88, Birkenmeier, Park, and Rizvi]) Let A be an algebra over a commutative ring C such that $c1_A \neq 0$ but $c^2 1_A = 0$ for some $c \in C$. Set $R = \begin{bmatrix} A & cA \\ 0 & A \end{bmatrix}$. Show that $E(R_R)$ does not have a compatible C-algebra structure. Note that Example 7.1.13 is a particular case of this result.

7.3 Osofsky Compatibility

Osofsky compatibility of rings is the focus of this section. We note that if R is a ring such that $E(R_R)$ is a rational extension of R_R (i.e., $E(R_R) = Q(R)$, for example when R is right nonsingular), then $E(R_R)$ has a unique compatible ring structure (see also Proposition 7.1.6) and this ring structure is right self-injective. When $E(R_R)$ is not a rational extension of R_R, we shall see in Examples 7.3.3 and 7.3.5 that it is still possible for $E(R_R)$ to have a right self-injective compatible ring structure. In [327], Osofsky had asked: If $E(R_R)$ has a compatible ring structure, must this ring structure necessarily be right self-injective? This question has been answered in the negative by Camillo, Herzog, and Nielsen and their example will be presented in this section.

For the case when R is a commutative Artinian ring, it is shown that R is right Osofsky compatible if and only if R is self-injective. This result shows some constraints in finding classes of right Osofsky compatible commutative Artinian rings. We shall see that this result does not hold true for noncommutative rings.

Theorem 7.3.14 will provide a class of Artinian right Kasch rings R which are both right and left Osofsky compatible. It will be shown that $E(R_R)$ may have distinct compatible ring structures. Indeed, let A be a commutative local QF-ring with $J(A) \neq 0$ and

$$R = \begin{bmatrix} A & A/J(A) \\ 0 & A/J(A) \end{bmatrix}.$$

Then it will be shown that $E(R_R)$ has at least $|\text{Soc}(A)|^2$ distinct compatible ring structures. We shall also identify all possible right essential overrings of R and their properties when $A = \mathbb{Z}_{p^m}$ with p a prime integer and $m \geq 2$. This also provides, together with results in Sect. 7.2, a motivation for the study of ring hulls which we shall discuss in Chap. 8.

Theorem 7.3.1 *Let R be a ring. If $E(R_R)$ is a rational extension of R_R, then R is right Osofsky compatible and $Q(R) = E(R_R)$. Further, $E(R_R)$ has a unique compatible ring structure, which is right self-injective.*

Proof As $E(R_R)$ is a rational extension of R_R, $E(R_R) = Q(R)$ by Corollary 1.3.15. Thus by Proposition 7.1.6, $Q(R)$ has a unique compatible ring structure. From Theorem 7.1.3, $Q(R)$ is right self-injective. □

We remark that in Theorem 7.3.1, if R is right nonsingular, then $E(R_R)$ is a rational extension of R_R and the compatible ring structure on $E(R_R)$ is unique and regular right self-injective (see Theorem 2.1.31).

Theorem 7.3.2 *Let R be a ring. Then R is right nonsingular with finite right uniform dimension if and only if $Q(R)$ is semisimple Artinian.*

Proof Let R be right nonsingular with $\text{udim}(R_R) = n < \infty$ (recall that udim(−) denotes uniform dimension of a module). By Theorem 2.1.31, $Q(R)$ is regular. Now $\text{udim}(Q(R)_{Q(R)}) \leq \text{udim}(Q(R)_R) = \text{udim}(R_R) < \infty$, so $Q(R)$ is orthogonally finite. By Proposition 1.2.15, $Q(R)$ has a complete set of primitive idempotents, say $\{e_1, \ldots, e_n\}$. So $eQ(R) = e_1 Q(R) \oplus \cdots \oplus e_n Q(R)$ and each $e_i Q(R)_{Q(R)}$ is indecomposable. As $Q(R)$ is regular, each $e_i Q(R)_{Q(R)}$ is a simple module, so $Q(R)$ is semisimple Artinian. The converse is clear. □

In the following example, Osofsky constructs a ring R for which an injective hull $E(R_R)$ of R_R is not a rational extension of R_R, but $E(R_R)$ has a right self-injective compatible ring structure (see [327]). This example and Theorem 7.2.1

provide motivation to investigate right essential overrings which are not right rings of quotients.

Example 7.3.3 Let $R = \mathbb{Z}_2\{X, Y\}/I$, where $\mathbb{Z}_2\{X, Y\}$ is the free algebra over $\mathbb{Z}_2$ with indeterminates X, Y, and I is the ideal of $\mathbb{Z}_2\{X, Y\}$ generated by X^2, Y^2, and YX. Let x and y be the images of X and Y in R, respectively. Then it follows that $R = \{a + bx + cy + dxy \mid a, b, c, d \in \mathbb{Z}_2\}$ and $x^2 = y^2 = yx = 0$.

Now $|R| = 2^4$, $J(R) = \{bx + cy + dxy \mid b, c, d \in \mathbb{Z}_2\}$, and $R/J(R) \cong \mathbb{Z}_2$. So R is local Artinian, thus $J(R)$ is the only maximal right ideal of R and $0 \neq y \in \ell_R(J(R))$. Hence R is right Kasch, so $R = Q(R)$ by Proposition 1.3.18. Also, $\text{Soc}(R_R) = yR_R \oplus xyR_R$ and $yR_R \cong xyR_R$. Further, yR_R and xyR_R are simple R-modules. Osofsky shows that R is right Osofsky compatible for which $E(R_R)$ has a right self-injective compatible ring structure, but $E(R_R)$ is not a rational extension of R_R.

Let R and S be two rings. Say ${}_RW_S$ is an (R, S)-bimodule and V_S is a right S-module. For $f \in \text{Hom}_S({}_RW_S, V_S)$ and $r \in R$, let $fr \in \text{Hom}_S({}_RW_S, V_S)$ defined by $(fr)(w) = f(rw)$, where $w \in W$. Then $\text{Hom}_S({}_RW_S, V_S)$ is a right R-module. We use $[\text{Hom}_S({}_RW_S, V_S)]_R$ to denote this right R-module.

Lemma 7.3.4 *Let ${}_RW_S$ and V_S be as above. Assume that ${}_RW$ is a flat R-module and V_S is an injective S-module. Then $[\text{Hom}_S({}_RW_S, V_S)]_R$ is an injective R-module.*

Proof See [262, Lemma 3.5]. □

Similar to Example 7.3.3, there are other right Osofsky compatible rings R such that $E(R_R)$ is not a rational extension of R_R as shown next.

Example 7.3.5 Let A be a ring and let A satisfy the following properties.

(1) A is local with $J(A) \neq 0$.
(2) A is a finite dimensional algebra over a field K.
(3) $A = K \cdot 1_A \oplus J(A)$ as vector spaces over K.
(4) $\text{Soc}(A)$ is a simple A-module.

For example, let $A = \mathbb{Z}_2[C_2]$, the group algebra of the group $C_2 = \{1, g\}$ of order two over $\mathbb{Z}_2$. Then A is local with $J(A) = \{0, 1 + g\}$. Further, we see that $\text{Soc}(A) = J(A)$, which is a simple A-module. Also $A = \mathbb{Z}_2 \cdot 1_A \oplus J(A)$ as vector spaces over $\mathbb{Z}_2$.

Let m be an integer such that $m > 1$ and let R be the set of all matrices of the form

$$\begin{bmatrix} k \cdot 1_A & 0 & \dots & 0 & a_1 \\ 0 & k \cdot 1_A & \dots & 0 & a_2 \\ \vdots & \vdots & \ddots & \vdots & \vdots \\ 0 & 0 & \dots & k \cdot 1_A & a_{m-1} \\ 0 & 0 & \dots & 0 & k \cdot 1_A + w \end{bmatrix},$$

where $a_1, \dots, a_{m-1} \in A$, $k \in K$, $w \in J(A)$, which is a subring of $\text{Mat}_m(A)$.

We observe that R is local Artinian. Let M_R be the K-dual of ${}_R R$, that is, $M_R = [\text{Hom}_K({}_R R_K, K_K)]_R$. Then M_R is injective by Lemma 7.3.4. The K-dual M_R of ${}_R R$ is indecomposable because ${}_R R$ is indecomposable (see [262, p. 92]). Since R_K is finite dimensional over K, M is finite dimensional over K and $\dim_K M = \dim_K R$.

Say $\dim_K A = n$. As vector spaces over K, $\dim_K M = \dim_K R = mn$. Thus, M_R is Artinian, so $\text{Soc}(M_R) \leq^{\text{ess}} M_R$. Therefore, $M_R = E(\text{Soc}(M_R))$. Because M_R is indecomposable, $\text{Soc}(M_R)$ is simple.

Again since R is local, $\text{Soc}(M_R) \cong (R/J(R))_R$. So $M_R \cong E((R/J(R)_R)$.

As R is local Artinian, $\text{Soc}(R_R) \cong (R/J(R))_R^{(\ell)}$ for some $\ell \geq 1$. Also note that $\dim_K \text{Soc}(R_R) = m$ and $\dim_K (R/J(R)) = 1$. Therefore,

$$m = \dim_K \text{Soc}(R_R) = \dim_K (R/J(R))^{(\ell)} = \ell.$$

Thus, $\text{Soc}(R_R) \cong (R/J(R))_R^{(m)}$, so

$$E(R_R) = E((R/J(R))_R)^{(m)} \cong M_R^{(m)}.$$

Thus, $\dim_K E(R_R) = m^2 n = \dim_K \text{Mat}_m(A)$ as vector spaces over K. Since $R_R \leq^{\text{ess}} \text{Mat}_m(A)_R$, $E(R_R) = \text{Mat}_m(A)$. The matrix multiplication on $\text{Mat}_m(A)$ extends the right R-module scalar multiplication of $\text{Mat}_m(A)$ over R. Thus, R is right Osofsky compatible. By Proposition 1.3.18 R is right Kasch, so $R = Q(R)$. Hence $E(R_R)$ is not a rational extension of R_R.

Let R be a right Osofsky compatible ring and E be an injective hull of R_R with a compatible ring structure $(E, +, \circ)$. Osofsky asked in [327]: *Must* $(E, +, \circ)$ *be necessarily right self-injective?* The next example, due to Camillo, Herzog, and Nielsen, gives a negative answer to the question (see [111]).

Example 7.3.6 Let $\mathbb{R}\{X_1, X_2, \ldots\}$ be the free algebra over the field $\mathbb{R}$ with indeterminates $X_1, X_2, \ldots$. Let

$$A = \mathbb{R}\{X_1, X_2, \ldots\}/\langle X_i X_j - \delta_{ij} X_1^2 \mid i, j = 1, 2, \ldots \rangle,$$

where $\langle X_i X_j - \delta_{ij} X_1^2 \mid i, j = 1, 2, \ldots \rangle$ is the ideal of $\mathbb{R}\{X_1, X_2, \ldots\}$ which is generated by $\{X_i X_j - \delta_{ij} X_1^2 \mid i, j = 1, 2, \ldots\}$, and δ_{ij} is the Kronecker delta. We denote by x_i the image of X_i in A. Set $V = \mathbb{R}x_1 \oplus \mathbb{R}x_2 \oplus \cdots$, $P = \mathbb{R}x_1^2$, and let the bilinear form on V be given by $B(x_i, x_j) = \delta_{ij}$. Now we see that $A \cong R$, where

$$R = \left\{ \begin{bmatrix} k & v & p \\ 0 & k & v \\ 0 & 0 & k \end{bmatrix} \mid k \in \mathbb{R}, v \in V, \text{ and } p \in P \right\},$$

where the addition is componentwise and the multiplication is defined by

$$\begin{bmatrix} k_1 & v_1 & p_1 \\ 0 & k_1 & v_1 \\ 0 & 0 & k_1 \end{bmatrix} \begin{bmatrix} k_2 & v_2 & p_2 \\ 0 & k_2 & v_2 \\ 0 & 0 & k_2 \end{bmatrix} = \begin{bmatrix} k_1 k_2 & k_1 v_2 + k_2 v_1 & k_1 p_2 + k_2 p_1 + B(v_1, v_2) x_1^2 \\ 0 & k_1 k_2 & k_1 v_2 + k_2 v_1 \\ 0 & 0 & k_1 k_2 \end{bmatrix},$$

is a commutative local ring. Let $E_R = [\mathrm{Hom}_{\mathbb{R}}(R_{\mathbb{R}}, \mathbb{R}_{\mathbb{R}})]_R$. Then E_R is an injective hull of R_R. Further, E_R has a compatible ring structure, but it is not right self-injective. Also

$$\begin{bmatrix} 0 & 0 & P \\ 0 & 0 & 0 \\ 0 & 0 & 0 \end{bmatrix}$$

is the smallest nonzero ideal of R. Thus R_R is uniform, so it is extending. Hence the compatible ring structure on E_R is right FI-extending (see Theorem 8.1.8(i)).

Following Osofsky's initial work on the right Osofsky compatibility of rings, Lang investigated this notion for commutative Artinian rings. The next three results are due to Lang (see [266]).

Lemma 7.3.7 *Let R be a ring, $E_R = E(R_R)$, and $H = \mathrm{End}(E_R)$. Assume that $E := (E_R, +, \circ)$ is a compatible ring structure on E_R. Then*:

(i) $J(E) = Z(E_R)$.
(ii) $H/J(H) \cong E/J(E)$.
(iii) $E/J(E)$ *is a regular ring*.

Proof (i) and (ii) First, we show that $Z(E_R)$ is an ideal of the ring E. Take $z \in Z(E_R)$. Then $zI = 0$ for some $I_R \leq^{\text{ess}} R_R$. Take $t \in E$. Then we have that $(t \circ z)I = t \circ (zI) = 0$, so $t \circ z \in Z(E_R)$. Let $J = \{r \in R \mid tr \in I\}$. Then J is an essential right ideal of R and $(z \circ t)J = z(tJ) \subseteq zI = 0$. Thus, $Z(E_R)$ is an ideal of E.

Let $\Delta = \{h \in H \mid hI = 0 \text{ for some } I_R \leq^{\text{ess}} R_R\}$. Then

$$\Delta = \{h \in H \mid \mathrm{Ker}(h)_R \leq^{\text{ess}} E_R\}$$

as $R_R \leq^{\text{ess}} E_R$ (see also Lemma 2.1.28(i)).

We define $\theta : H \to E/Z(E_R)$ by $\theta(h) = h(1) + Z(E_R)$. To see that θ is onto, say $x \in E$ and let $I = \{r \in R \mid xr \in R\}$. Then $I_R \leq^{\text{ess}} R_R$. Let $f : I \to E$ be defined by $f(r) = xr$. As E_R is injective, there is an extension $h \in H$ of f. Thus $xr = f(r) = h(r) = h(1)r$ for all $r \in I$. So $(x - h(1))I = 0$, and hence $x - h(1) \in Z(E)$. Thus $x + Z(E_R) = h(1) + Z(E_R)$. Therefore, θ is onto. Clearly, θ is additive.

Note that $\mathrm{Ker}(\theta) = \{h \in H \mid hI = h(1)I = 0 \text{ for some } I_R \leq^{\text{ess}} R_R\} = \Delta$. So the induced map $\overline{\theta} : H/\Delta \to E/Z(E_R)$, defined by $\overline{\theta}(h + \Delta) = h(1) + Z(E_R)$, is an additive isomorphism.

To see that $\overline{\theta}$ is a ring isomorphism, let $h_1 + \Delta,\ h_2 + \Delta \in H/\Delta$ with $h_1, h_2 \in H$. Put $K = \{r \in R \mid h_2(r) \in R\}$. Then $K_R \leq^{\text{ess}} R_R$ as $R_R \leq^{\text{ess}} E_R$. For $a \in K$, we have that

$$\begin{aligned} ((h_1h_2)(1) - h_1(1) \circ h_2(1))a &= h_1(h_2(a)) - h_1(1) \circ h_2(1)a \\ &= h_1(h_2(a)) - h_1(1) \circ h_2(a) = h_1(h_2(a)) - h_1(1)h_2(a) \\ &= h_1(h_2(a)) - h_1(h_2(a)) = 0. \end{aligned}$$

So $((h_1h_2)(1) - h_1(1) \circ h_2(1))K = 0$, and $(h_1h_2)(1) - h_1(1) \circ h_2(1) \in Z(E_R)$. Thus $(h_1h_2)(1) + Z(E_R) = h_1(1) \circ h_2(1) + Z(E_R)$. Therefore, $\overline{\theta}$ is a ring isomorphism.

Since H/Δ is regular by Theorem 2.1.29, $E/Z(E_R)$ is also a regular ring. Hence, $J(E) \subseteq Z(E_R)$. Next, take $z \in Z(E_R)$. Then $r_R(z)_R \leq^{\text{ess}} R_R$. Now we observe that $r_R(z) \cap r_R(1-z) = 0$, so $r_R(1-z) = 0$. As $R_R \leq^{\text{ess}} E_R$, $r_E(1-z) = 0$.

Define $f : E \to (1-z) \circ E$ by $f(t) = (1-z) \circ t$. Then f is an R-isomorphism, hence $(1-z) \circ E$ is an injective R-module. Thus, $(1-z) \circ E$ is an R-direct summand of E_R. If $a \in r_R(z)$, then $za = 0$ and so $a = (1-z)a \in (1-z) \circ E$. Thus $r_R(z) \subseteq (1-z) \circ E$, so $(1-z) \circ E_R \leq^{\text{ess}} E_R$ since $r_R(z)_R \leq^{\text{ess}} R_R \leq^{\text{ess}} E_R$. Therefore $(1-z) \circ E = E$. Hence there is $s \in E$ such that $(1-z) \circ s = 1$, so z is right quasi-regular. Thus $Z(E_R)$ is a right quasi-regular ideal of E, hence it is a quasi-regular ideal (see 1.1.9). So $Z(E_R) \subseteq J(E)$, thus $J(E) = Z(E_R)$. From Theorem 2.1.29, $J(H) = \Delta$, and so $H/J(H) = H/\Delta \cong E/J(E)$.

(iii) As H/Δ is a regular ring by Theorem 2.1.29 and $H/\Delta \cong E/J(E)$ by part (ii), $E/J(E)$ is also a regular ring. □

Proposition 7.3.8 *Let R be a commutative ring with* $\text{Soc}(R)_R \leq^{\text{ess}} R_R$, *and let E be an injective hull of R_R. Assume that $(E, +, \circ)$ is a compatible ring structure. Then:*

(i) $\text{Soc}(R)$ *is a left ideal of the ring E.*
(ii) $\text{Soc}(R) \subseteq \text{Soc}({}_E E)$.
(iii) $\text{Soc}(R) \subseteq \text{Soc}(E_E)$.
(iv) *If $vR \subseteq \text{Soc}(R)$ with vR a minimal ideal of R, then $vR = v \circ E$.*
(v) *If $t\,\text{Soc}(R) = 0$ with $t \in E$, then* $\text{Soc}(R) \circ t = 0$.

Proof (i) Let $x \in E$ and put $x^{-1}\text{Soc}(R) = \{r \in R \mid xr \in \text{Soc}(R)\}$. Then $x^{-1}\text{Soc}(R)_R \leq^{\text{ess}} R_R$ since $\text{Soc}(R)_R \leq^{\text{ess}} R_R$. Thus, $\text{Soc}(R) \subseteq x^{-1}\text{Soc}(R)$ because $\text{Soc}(R)$ is the intersection of all essential ideals of R. Hence, we have that $x\,\text{Soc}(R) \subseteq \text{Soc}(R)$. Therefore, $\text{Soc}(R)$ is a left ideal of E.

(ii) Say V is a homogeneous component of $\text{Soc}(R)$. We prove that V is a left ideal of E. For this, take $t \in E$. If $tV = 0$, then $tV \subseteq V$. So assume that $tV \neq 0$. Then there exists $a \in V$ such that $ta \neq 0$. Write $a = a_1 + \cdots + a_n$, where $a_i R_R$ is simple and $a_i \in V$ for each i. We may assume that there exists m such that $1 \leq m \leq n$ and each $b_i = ta_i \neq 0$ for $1 \leq i \leq m$. By part (i), each $b_i \in \text{Soc}(R)$. Then $a_i R_R \cong b_i R_R$ via left multiplication by t. Hence each $b_i \in V$, so $ta = b_1 + \cdots + b_m \in V$. Thus, V is a left ideal of E.

To see that V is a minimal left ideal of E, take $0 \neq h \in V$. Then $Eh \subseteq V$ since V is a left ideal of E. Next, say $0 \neq k \in V$. Define $\varphi : hR \to kR$ by $\varphi(hr) = kr$ for $r \in R$. Then φ is well-defined. Indeed, say $h = c_1 + \cdots + c_\ell$, where each $c_i R_R$ is simple and $\sum_{i=1}^{\ell} c_i R = \oplus_{i=1}^{\ell} c_i R$. Also $k = d_1 + \cdots + d_m$, where each $d_i R_R$ is simple and $\sum_{i=1}^{m} d_i R = \oplus_{i=1}^{m} d_i R$. Then

$$c_1 R_R \cong \cdots \cong c_\ell R_R \cong d_1 R_R \cong \cdots \cong d_m R_R.$$

Thus $r_R(c_1 R) = \cdots = r_R(c_\ell R) = r_R(d_1 R) = \cdots = r_R(d_m R)$, so it follows that $r_R(c_1) = \cdots = r_R(c_\ell) = r_R(d_1) = \cdots = r_R(d_m)$ since R is commutative.

Now if $hr = 0$ with $r \in R$, then $c_1 r + \cdots + c_\ell r = 0$, so $c_1 r = \cdots = c_\ell r = 0$, so $r \in r_R(c_1) = r_R(d_i)$ for each i, $1 \le i \le m$. Hence $kr = (d_1 + \cdots + d_m)r = 0$, so φ is well-defined. Let $\phi \in \text{End}(E_R)$ which is an extension of φ. Then $k = \varphi(h) = \phi(h) = \phi(1)h \in Eh$. So $V = Eh$. Therefore V is a minimal left ideal of E. Hence $V \subseteq \text{Soc}({}_E E)$. Thus $\text{Soc}(R) \subseteq \text{Soc}({}_E E)$.

(iii) Take $v \in \text{Soc}(R)$ such that vR is a minimal ideal of R. First, we claim that $(v \circ E) \cap \text{Soc}(R) = vR$. Obviously, $vR \subseteq (v \circ E) \cap \text{Soc}(R)$. For the converse containment, we note that there exists an ideal A of R such that $\text{Soc}(R) = vR \oplus A$. Let $\pi : \text{Soc}(R) \to vR$ be the canonical projection. There is $\lambda \in \text{End}(E_R)$ which extends π. Now let $v \circ t \in (v \circ E) \cap \text{Soc}(R)$, where $t \in E$. Then

$$v \circ t = \lambda(v) \circ t = \lambda(1) \circ (v \circ t) = \lambda(1)(v \circ t) = \lambda(v \circ t) = \pi(v \circ t)$$

since $v \circ t \in \text{Soc}(R)$. Thus $v \circ t \in vR$. So $(v \circ E) \cap \text{Soc}(R) \subseteq vR$. Therefore, we obtain $vR = (v \circ E) \cap \text{Soc}(R)$.

We show that $v \circ E$ is a minimal right ideal of E. For this, let $0 \neq k \in v \circ E$. Since $\text{Soc}(R)_R \le^{\text{ess}} R_R$, $(k \circ E) \cap \text{Soc}(R) \neq 0$. Now

$$0 \neq (k \circ E) \cap \text{Soc}(R) \subseteq (v \circ E) \cap \text{Soc}(R) = vR.$$

Thus $(k \circ E) \cap \text{Soc}(R) = vR$ since vR is a minimal ideal of R. So $v \in k \circ E$, hence $v \circ E \subseteq k \circ E$. Therefore $k \circ E = v \circ E$. Thus, $v \circ E$ is a minimal right ideal of E. In particular, $\text{Soc}(R) \subseteq \text{Soc}(E_E)$.

(iv) Let $v \in \text{Soc}(R)$ such that vR is a minimal ideal of R. Then by the proof of part (iii), $vR = (v \circ E) \cap \text{Soc}(R)$.

Say V is a homogeneous component of $\text{Soc}(R)$ such that $v \in V$. Then $\text{Soc}(R) = V \oplus C$ with $C_R \le \text{Soc}(R)_R$. By the proof of part (ii), both V and C are left ideals of E. Thus, $\text{Soc}({}_E E) = B \oplus V \oplus C$ for some ${}_E B \le \text{Soc}({}_E E)$.

Take $t \in E$. Then $v \circ t \in \text{Soc}(R) \circ t \subseteq \text{Soc}({}_E E) \circ t \subseteq \text{Soc}({}_E E)$. Write

$$v \circ t = b + h + c$$

with $b \in B$, $h \in V$, and $c \in C$. We show that $b = 0$. For this, assume on the contrary that $b = v \circ t - (h + c) \neq 0$. Then there exists r in R such that $0 \neq br \in \text{Soc}(R)$ since $\text{Soc}(R)_R \le^{\text{ess}} R_R \le^{\text{ess}} E_R$.

We now show that $(r \circ t - tr)\text{Soc}(R) = 0$. For this, take $p \in \text{Soc}(R)$. Then from part (i), $tp \in t\,\text{Soc}(R) \subseteq E\,\text{Soc}(R) \subseteq \text{Soc}(R)$. Therefore,

$$(r \circ t - tr)p = r \circ (tp) - trp = r \circ (tp) - (tp)r = 0.$$

Hence, $z := r \circ t - tr \in Z(E_R)$. So it follows that

$$\begin{aligned} r \circ b &= r \circ (v \circ t - (h + c)) = (rv) \circ t - r(h + c) \\ &= (vr) \circ t - (h + c)r = v \circ (r \circ t) - (h + c)r \\ &= v \circ (tr + z) - (h + c)r = v \circ (tr) - (h + c)r \\ &= (v \circ t - (h + c))r = br \end{aligned}$$

since $v \circ z \in v \circ Z(E_R) = v \circ J(E) \subseteq \text{Soc}(R) \circ J(E) \subseteq \text{Soc}(E_E) \circ J(E) = 0$ by part (iii) and Lemma 7.3.7. Thus $0 \neq br \in R \circ b \cap \text{Soc}(R) \subseteq B \cap \text{Soc}(R) = 0$, a contradiction. Hence, $b = 0$. So $v \circ t = h + c \in \mathrm{Soc}(R)$. Thus $v \circ E \subseteq \mathrm{Soc}(R)$. Therefore, $vR = (v \circ E) \cap \mathrm{Soc}(R) = v \circ E$.

(v) We observe that $t\,\mathrm{Soc}(R) = 0$ implies that $t \in Z(E_R)$. So $t \in J(E)$ by Lemma 7.3.7. Therefore, $\mathrm{Soc}(R) \circ t \subseteq \mathrm{Soc}(E_E) \circ J(E) = 0$. □

Theorem 7.3.9 *Let R be a commutative ring which satisfies the following conditions.*

(i) $\mathrm{Soc}(R)$ *is the sum of a finite number of minimal ideals.*
(ii) $\mathrm{Soc}(R)_R \leq^{\mathrm{ess}} R_R$.
(iii) *No homogeneous component of* $\mathrm{Soc}(R)$ *is simple.*

Then R is not right Osofsky compatible.

Proof Assume on the contrary that R is right Osofsky compatible. We let $E = E(R_R)$ and $(E, +, \circ)$ be a compatible ring structure on E. Write $\mathrm{Soc}(R) = \oplus_{k=1}^{n} V_k$, where the V_k are the homogeneous components and $V_k = \oplus v_{ki} R$, a finite direct sum of at least two minimal ideals of R.

If $v_{kj} a = 0$ with $a \in R$, then $v_{kj} Ra = 0$. Hence, $v_{ki} Ra = 0$ because $v_{ki} R_R \cong v_{kj} R_R$. So $v_{ki} a = 0$. Consider the maps $f_{ij}^k : \mathrm{Soc}(R)_R \to \mathrm{Soc}(R)_R$ defined by $f_{ij}^k(v_{kj}) = v_{ki}$ and $f_{ij}^k(v_{t\ell}) = 0$, for $(t, \ell) \neq (k, j)$. Let $\lambda \in \mathrm{End}(E_R)$ be an extension of f_{ij}^k. Put $u_{ij}^k = \lambda(1) \in E$. Then $u_{ij}^k v_{kj} = v_{ki}$ and $u_{ij}^k v_{t\ell} = 0$ for $(t, \ell) \neq (k, j)$.

Let $j \neq t$. Then $(u_{jj}^k - u_{jt}^k \circ u_{tj}^k)\mathrm{Soc}(R) = 0$ and $(u_{jt}^k \circ u_{jt}^k)\mathrm{Soc}(R) = 0$. Thus $u_{jj}^k - u_{jt}^k \circ u_{tj}^k \in Z(E_R)$ and $u_{jt}^k \circ u_{jt}^k \in Z(E_R)$. Take $v \in \mathrm{Soc}(R)$ such that vR is a minimal ideal of R. Consider $v \circ u_{jt}^k$ for any k, j, t with $j \neq t$. Then $v \circ u_{jt}^k \in v \circ E = vR$ by Proposition 7.3.8(iv). Thus, $v \circ u_{jt}^k = vr$ for some $r \in R$. We note that from Proposition 7.3.8(v), $\mathrm{Soc}(R) \circ (u_{jt}^k \circ u_{jt}^k) = 0$ since $(u_{jt}^k \circ u_{jt}^k)\mathrm{Soc}(R) = 0$. Therefore

$$0 = v \circ (u_{jt}^k \circ u_{jt}^k) = (v \circ u_{jt}^k) \circ u_{jt}^k = (vr) \circ u_{jt}^k = (rv) \circ u_{jt}^k$$
$$= r \circ (v \circ u_{jt}^k) = r(vr) = vr^2.$$

Since vR is a minimal ideal of R, $\ell_R(vR)$ is a prime ideal of R. It follows that $r \in \ell_R(vR)$ as $r^2 \in \ell_R(vR)$. So $v \circ u_{jt}^k = vr = 0$ for k, j, t with $j \neq t$.

Since $Z(E_R) = J(E)$ by Lemma 7.3.7(i), $u_{jj}^k - u_{jt}^k \circ u_{tj}^k \in J(E)$ for $j \neq t$. Thus, $\mathrm{Soc}(E_E) \circ (u_{jj}^k - u_{jt}^k \circ u_{tj}^k) = 0$. Hence, $\mathrm{Soc}(R) \circ (u_{jj}^k - u_{jt}^k \circ u_{tj}^k) = 0$ because $\mathrm{Soc}(R) \subseteq \mathrm{Soc}(E_E)$ by Proposition 7.3.8(iii). For j, we can take t such that $t \neq j$. Then

$$0 = v \circ (u_{jj}^k - u_{jt}^k \circ u_{tj}^k) = v \circ u_{jj}^k - v \circ u_{jt}^k \circ u_{tj}^k = v \circ u_{jj}^k$$

for all j and k because $v \circ u_{jt}^k = vr = 0$. Thus, $v \circ u_{jj}^k = 0$ for all j and k. But $(\sum_{j,k} u_{jj}^k - 1)\mathrm{Soc}(R) = 0$. Hence, $\mathrm{Soc}(R) \circ (\sum_{j,k} u_{jj}^k - 1) = 0$ by Proposition 7.3.8(v). Thus $v \circ \sum_{j,k} u_{jj}^k = v$. So $v = 0$ as $v \circ u_{jj}^k = 0$ for all j and k. This is a contradiction. Therefore, R is not right Osofsky compatible. □

Proposition 7.3.10 *Let $R = A \oplus B$ (ring direct sum). Then R is right Osofsky compatible if and only if A and B are right Osofsky compatible.*

Proof Let $E = E_R$ be an injective hull of R_R and $(E, +, \circ)$ be a compatible ring structure. Note that $E_R = E(A_R) \oplus E(B_R)$. Also, $E_1 := E(A_R) = E(A_A)$ and $E_2 := E(B_R) = E(B_B)$.

For $x, y \in E_1$, say $(x, 0) \circ (y, 0) = (f(x, y), g(x, y))$ for some $f(x, y) \in E_1$ and $g(x, y) \in E_2$. Consider any $(r_1, r_2) \in R = A \oplus B$. Then

$$[(x, 0) \circ (y, 0)] \circ (r_1, r_2) = (f(x, y), g(x, y)) \circ (r_1, r_2) = (f(x, y)r_1, g(x, y)r_2).$$

Also, $(x, 0) \circ [(y, 0) \circ (r_1, r_2)] = (x, 0) \circ (yr_1, 0) = (f(x, yr_1), g(x, yr_1))$. Hence $g(x, yr_1) = g(x, y)r_2$. By taking $r_1 = 1_A$ and $r_2 = 0$, we get that $g(x, y) = 0$.

From the preceding argument, $(x, 0) \circ (y, 0) = (f(x, y), 0)$ for $x, y \in E_1$. Let $x \circ_1 y = f(x, y)$ for $x, y \in E_1$. For $x \in E_1$ and $a \in A$, $(x, 0) \circ (a, 0) = (xa, 0)$, and so $x \circ_1 a = xa$. Thus, $(E_1, +, \circ_1)$ is a compatible ring structure on E_1 for which the multiplication $\circ_1$ extends the A-module scalar multiplication. Thus, A is right Osofsky compatible. Similarly, B is right Osofsky compatible. The converse can be checked routinely. □

The next result for commutative QF-rings is well known.

Theorem 7.3.11 *For a commutative ring R, the following are equivalent.*

(i) *R is QF.*
(ii) *R is Artinian and every homogeneous component of* Soc(R) *is simple.*
(iii) *$R \cong R_1 \oplus \cdots \oplus R_k$ for some positive integer k, where each R_i is a local Artinian ring with a simple socle.*

Proof See [262, Theorem 15.27] for the proof. □

Now we shall see that a commutative Artinian ring R is right Osofsky compatible precisely when R is self-injective (see [266]).

Theorem 7.3.12 *Let R be a commutative Artinian ring. Then the following are equivalent.*

(i) *R is right Osofsky compatible.*
(ii) *Every homogeneous component of* Soc(R) *is simple.*
(iii) $R = E(R_R)$.

Proof (i)⇒(ii) Let $\{e_1, \ldots, e_n\}$ be a complete set of primitive idempotents of R. Then $R = A_1 \oplus \cdots \oplus A_n$ (ring direct sum), where each $A_i = e_i R$ and local. So every simple A_i-module is isomorphic to $A_i/J(A_i)$, thus each Soc(A_i) has only one homogeneous component, say V_i. From Proposition 7.3.10, each A_i is right Osofsky compatible. Hence by Theorem 7.3.9, each V_i is simple.

(ii)⇒(iii) We see that R is QF from Theorem 7.3.11. Thus $R = E(R_R)$. (iii)⇒(i) is evident. □

There exists a commutative ring which is not right Osofsky compatible as shown in the next example.

Example 7.3.13 (i) For a field K, let $T = K[x]/x^4K[x]$. Say $\overline{x}$ is the image of x in T. Then $T = K + K\overline{x} + K\overline{x}^2 + K\overline{x}^3$.

Let $R = K + K\overline{x}^2 + K\overline{x}^3$, a subring of T. Then R is not self-injective. In fact, note that $K\overline{x}^2$ is an ideal of R. Let $f : K\overline{x}^2 \to R$ defined by $f(a\overline{x}^2) = a\overline{x}^3$, where $a \in K$. Then f is an R-homomorphism. But there is no $\alpha \in R$ such that $\alpha\, a\overline{x}^2 = a\overline{x}^3$ for all $a \in K$. Hence R is not self-injective by Baer's Criterion. So R cannot be right Osofsky compatible by Theorem 7.3.12. Note that R is Artinian.

(ii) Let $\Lambda = \text{End}(\mathbb{Z}_{p^\infty})$, where $\mathbb{Z}_{p^\infty}$ is the Prüfer p-group and p is a prime integer. Let $R = \Lambda \oplus \mathbb{Z}_{p^\infty} \oplus \mathbb{Z}_{p^\infty}$, where the addition is componentwise and the multiplication is defined by:

$$(\lambda,\, m_1,\, n_1)(\mu,\, m_2,\, n_2) = (\lambda\mu,\ \lambda(m_2) + \mu(m_1),\ \lambda(n_2) + \mu(n_1)),$$

for $(\lambda,\, m_1,\, n_1),\ (\mu,\, m_2,\, n_2) \in R$. Then we see that

$$R \cong S := \left\{ \begin{bmatrix} \lambda & m & 0 & 0 \\ 0 & \lambda & 0 & 0 \\ 0 & 0 & \lambda & n \\ 0 & 0 & 0 & \lambda \end{bmatrix} \mid \lambda \in \Lambda \text{ and } m, n \in \mathbb{Z}_{p^\infty} \right\}$$

with the addition componentwise and the multiplication is defined by

$$\begin{bmatrix} \lambda & m_1 & 0 & 0 \\ 0 & \lambda & 0 & 0 \\ 0 & 0 & \lambda & n_1 \\ 0 & 0 & 0 & \lambda \end{bmatrix} \begin{bmatrix} \mu & m_2 & 0 & 0 \\ 0 & \mu & 0 & 0 \\ 0 & 0 & \mu & n_2 \\ 0 & 0 & 0 & \mu \end{bmatrix} = \begin{bmatrix} \lambda\mu & \lambda(m_2) + \mu(m_1) & 0 & 0 \\ 0 & \lambda\mu & 0 & 0 \\ 0 & 0 & \lambda\mu & \lambda(n_2) + \mu(n_1) \\ 0 & 0 & 0 & \lambda\mu \end{bmatrix}.$$

The ring S is commutative since Λ is commutative (in fact, Λ is the ring of p-adic integers). Set

$$V = \begin{bmatrix} 0 & \mathbb{Z}_p & 0 & 0 \\ 0 & 0 & 0 & 0 \\ 0 & 0 & 0 & 0 \\ 0 & 0 & 0 & 0 \end{bmatrix} \text{ and } W = \begin{bmatrix} 0 & 0 & 0 & 0 \\ 0 & 0 & 0 & 0 \\ 0 & 0 & 0 & \mathbb{Z}_p \\ 0 & 0 & 0 & 0 \end{bmatrix}.$$

Then V and W are the only minimal ideals of S. Further,

$$\text{Soc}(S) = V \oplus W \text{ and } V_S \cong W_S,$$

thus $V \oplus W$ is the only homogeneous component of $\text{Soc}(S)$. Moreover, we observe that $\text{Soc}(S)_S \leq^{\text{ess}} S_S$. By Theorem 7.3.9, S is not right Osofsky compatible. Hence, R is not right Osofsky compatible.

We now consider another interesting class of right Osofsky compatible rings. For this, the following preparation is needed.

Let A be a ring and $J = J(A) \neq 0$. Put

$$R = \begin{bmatrix} A & A/J \\ 0 & A/J \end{bmatrix}.$$

Consider a subset $\mathfrak{A} = \{(a, -\overline{a}) \mid a \in A\} \subseteq A \times (A/J)$, where $\overline{a}$ is the image of a in A/J. The addition on $\mathfrak{A}$ is componentwise. Define $(a, -\overline{a})\, b = (ab, -\overline{ab})$ for $(a, -\overline{a}) \in \mathfrak{A}$ and $b \in A$. Then $\mathfrak{A}$ is a right A-module and $\mathfrak{A}_A \cong A_A$ via corresponding $(a, -\overline{a}) \to a$. Let

$$\mathbb{A} = \begin{bmatrix} \mathfrak{A} & 0 \\ 0 & 0 \end{bmatrix} = \left\{ \begin{bmatrix} (a, -\overline{a}) & 0 \\ 0 & 0 \end{bmatrix} \mid a \in A \right\}.$$

Then $\mathbb{A}$ is a right R-module under the componentwise addition and the R-module scalar multiplication is defined by

$$\begin{bmatrix} (a, -\overline{a}) & 0 \\ 0 & 0 \end{bmatrix} \begin{bmatrix} x & \overline{y} \\ 0 & \overline{z} \end{bmatrix} = \begin{bmatrix} (ax, -\overline{ax}) & 0 \\ 0 & 0 \end{bmatrix}$$

for $\begin{bmatrix} (a, -\overline{a}) & 0 \\ 0 & 0 \end{bmatrix} \in \mathbb{A}$ and $\begin{bmatrix} x & \overline{y} \\ 0 & \overline{z} \end{bmatrix} \in R$, where $\overline{y}$ and $\overline{z}$ are the images of y and z in A/J, respectively.

The next theorem provides a class of Artinian right Kasch rings R which are both right and left Osofsky compatible using the preceding preparation. Further, it shows that $E(R_R)$ may have distinct compatible ring structures when $E(R_R)$ is not a rational extension of R_R (cf. Theorem 7.3.1).

Theorem 7.3.14 *Let A be a local commutative QF-ring with $J \neq 0$. Put*

$$R = \begin{bmatrix} A & A/J \\ 0 & A/J \end{bmatrix}.$$

Then we have the following.

(i) $R = Q(R)$.

(ii) $\mathbb{A}_R$ *is an injective hull of* $\mathcal{S}_R = \left\{ \begin{bmatrix} (a, -\overline{a}) & 0 \\ 0 & 0 \end{bmatrix} \mid a \in \mathrm{Soc}(A) \right\}$.

(iii) $E_R = \mathbb{A}_R \oplus \begin{bmatrix} A/J & A/J \\ A/J & A/J \end{bmatrix}_R$ *is an injective hull of* R_R. *Also,* $\begin{bmatrix} a & \overline{b} \\ 0 & \overline{d} \end{bmatrix} \in R$ *can be identified with* $\begin{bmatrix} (a, -\overline{a}) & 0 \\ 0 & 0 \end{bmatrix} + \begin{bmatrix} \overline{a} & \overline{b} \\ \overline{0} & \overline{d} \end{bmatrix} \in E_R$.

(iv) E_R *has a ring structure as a QF-ring that is compatible with its* R*-module structure.*

(v) *There exist* $|\mathrm{Soc}(A)|^2$ *distinct compatible ring structures on* E_R *such that they are QF-rings and each is isomorphic to the ring structure in* (iv).

(vi) *R is right and left Osofsky compatible.*

Proof (i) We see that R is right Kasch, so $R = Q(R)$ by Proposition 1.3.18.

(ii) From Lemma 7.3.4, $[\mathrm{Hom}_A({}_RR_A, A_A)]_R$ is injective. Let

$$V = \begin{bmatrix} A & \overline{0} \\ 0 & \overline{0} \end{bmatrix} \text{ and } W = \begin{bmatrix} 0 & A/J \\ 0 & A/J \end{bmatrix}.$$

Then both V and W are (R, A)-bimodules. Note that ${}_RR_A = {}_RV_A \oplus {}_RW_A$, so V and W are projective left R-modules. By Lemma 7.3.4, $[\mathrm{Hom}_A({}_RV_A, A_A)]_R$ and $[\mathrm{Hom}_A({}_RW_A, A_A)]_R$ are injective right R-modules.

We claim that $\mathbb{A}_R = E(\mathcal{S}_R)$. For this, first we see that $\mathcal{S}_R \leq^{\mathrm{ess}} \mathbb{A}_R$ because $\mathrm{Soc}(A)_A \leq^{\mathrm{ess}} A_A$. Note that $[\mathrm{Hom}_A({}_RV_A, A_A)]_R$ is an injective R-module, $[\mathrm{Hom}_A({}_RV_A, \mathfrak{A}_A)]_R$ is an injective R-module because $\mathfrak{A}_A \cong A_A$ (recall that $\mathfrak{A} = \{(a, -\overline{a}) \mid a \in A\}$).

The map $\theta : [\mathrm{Hom}_A({}_RV_A, \mathfrak{A}_A)]_R \to \mathbb{A}_R$ defined by

$$\theta(\phi) = \begin{bmatrix} (a, -\overline{a}) & 0 \\ 0 & 0 \end{bmatrix},$$

for $\phi \in [\mathrm{Hom}_A({}_RV_A, \mathfrak{A}_A)]_R$ with $\phi \begin{bmatrix} 1 & \overline{0} \\ 0 & \overline{0} \end{bmatrix} = (a, -\overline{a}) \in \mathfrak{A}$, is an R-isomorphism. So $\mathbb{A}_R$ is an injective R-module. Therefore, $\mathbb{A}_R = E(\mathcal{S}_R)$.

(iii) We first show that $[\mathrm{Hom}_A({}_RW_A, A_A)]_R \cong \begin{bmatrix} 0 & 0 \\ A/J & A/J \end{bmatrix}_R$. In fact, take $0 \neq s \in \mathrm{Soc}(A)$. Then $\mathrm{Soc}(A) = sA$ as $\mathrm{Soc}(A)$ is the smallest nonzero ideal of A. Let $\varphi \in \mathrm{Hom}(W_A, A_A)$ with $\varphi \begin{bmatrix} 0 & \overline{1} \\ 0 & \overline{0} \end{bmatrix} = a \in A$ and $\varphi \begin{bmatrix} 0 & \overline{0} \\ 0 & \overline{1} \end{bmatrix} = b \in A$. Then for $x \in A$, $\varphi \begin{bmatrix} 0 & \overline{x} \\ 0 & \overline{0} \end{bmatrix} = ax$ and $\varphi \begin{bmatrix} 0 & \overline{0} \\ 0 & \overline{x} \end{bmatrix} = bx$. If $x \in J$, then $ax = 0$ and $bx = 0$. Therefore, we obtain $a, b \in \mathrm{Soc}(A)$, so $a = sa_0$ and $b = sb_0$ for some $a_0 \in A$ and $b_0 \in A$.

Let $f : \mathrm{Hom}_A({}_RW_A, A_A) \to \begin{bmatrix} 0 & 0 \\ A/J & A/J \end{bmatrix}$ defined by

$$f(\varphi) = \begin{bmatrix} 0 & 0 \\ \overline{a_0} & \overline{b_0} \end{bmatrix}.$$

Then it is routine to check that f is an R-isomorphism. So $\begin{bmatrix} 0 & 0 \\ A/J & A/J \end{bmatrix}_R$ is injective since $[\mathrm{Hom}_A({}_RW_A, A_A)]_R$ is injective. We observe that

$$U := \begin{bmatrix} A/J & A/J \\ A/J & A/J \end{bmatrix}_R = \begin{bmatrix} A/J & A/J \\ 0 & 0 \end{bmatrix}_R \oplus \begin{bmatrix} 0 & 0 \\ A/J & A/J \end{bmatrix}_R$$

and

$$\begin{bmatrix} A/J & A/J \\ 0 & 0 \end{bmatrix}_R \cong \begin{bmatrix} 0 & 0 \\ A/J & A/J \end{bmatrix}_R.$$

So U_R is injective. Also, $W_R \leq^{\text{ess}} U_R$, hence $E(W_R) = U_R$. Therefore, $E(R_R) = E(\mathcal{S}_R) \oplus E(W_R) = \mathbb{A}_R \oplus U_R$ since $\mathcal{S}_R \oplus W_R \leq^{\text{ess}} R_R$.

(iv) The map from $\mathfrak{A}$ to A corresponding $(a, -\overline{a})$ to a is one-to-one and onto. Define a multiplication on $\mathfrak{A}$ by $(a, -\overline{a}) \cdot (b, -\overline{b}) = (ab, -\overline{ab})$ so that $(\mathbb{A}, +, \cdot) \cong A$ as rings via the correspondence $\begin{bmatrix} (a, -\overline{a}) & 0 \\ 0 & 0 \end{bmatrix} \to a$.

Now E_R has an obvious ring multiplication making it isomorphic to a ring direct sum of the QF-ring $\mathbb{A}$ and a simple Artinian ring $\text{Mat}_2(A/J)$. Thus the ring structure on E_R is QF. One can easily check that R is a subring of this ring and it is clear that the ring multiplication of E_R extends the R-module scalar multiplication of E_R over R.

(v) Let $(E, +, \cdot)$ be the ring structure given in part (iv). Assume

$$f \in \text{End}(E_R) \text{ such that } f(r) = r$$

for each $r \in R$ (such f can be constructed by an extension to E_R of the identity map of R_R). Then f is an R-isomorphism, and

$$f \begin{bmatrix} \overline{1} & \overline{0} \\ \overline{0} & \overline{0} \end{bmatrix} = \begin{bmatrix} (\mu, \overline{0}) & 0 \\ 0 & 0 \end{bmatrix} + \begin{bmatrix} \overline{1} & \overline{0} \\ \overline{0} & \overline{0} \end{bmatrix} \text{ and } f \begin{bmatrix} \overline{0} & \overline{0} \\ \overline{1} & \overline{0} \end{bmatrix} = \begin{bmatrix} (\nu, \overline{0}) & 0 \\ 0 & 0 \end{bmatrix} + \begin{bmatrix} \overline{0} & \overline{0} \\ \overline{1} & \overline{0} \end{bmatrix}$$

for some $\mu, \nu \in \text{Soc}(A)$. We use $f_{(\mu,\nu)}$ to denote this f.

Next, for $(\mu, \nu) \in \text{Soc}(A) \times \text{Soc}(A)$, let $g : E_R \to E_R$ defined by

$$g\left(\begin{bmatrix} (s, -\overline{s}) & 0 \\ 0 & 0 \end{bmatrix} + \begin{bmatrix} \overline{a} & \overline{b} \\ \overline{c} & \overline{d} \end{bmatrix}\right) = \begin{bmatrix} (s + \mu(a - s) + \nu c, -\overline{s}) & 0 \\ 0 & 0 \end{bmatrix} + \begin{bmatrix} \overline{a} & \overline{b} \\ \overline{c} & \overline{d} \end{bmatrix}$$

for $\begin{bmatrix} (s, -\overline{s}) & 0 \\ 0 & 0 \end{bmatrix} + \begin{bmatrix} \overline{a} & \overline{b} \\ \overline{c} & \overline{d} \end{bmatrix} \in E_R$. Then g is an R-isomorphism. Furthermore, we obtain that $g = f_{(\mu,\nu)}$.

For $x, y \in E$, define

$$x \circ_{(\mu,\nu)} y = f_{(\mu,\nu)}^{-1}[f_{(\mu,\nu)}(x) \cdot f_{(\mu,\nu)}(y)].$$

Then $\cdot = \circ_{(0,0)}$. For $x \in E$ and $r \in R$, $x \circ_{(\mu,\nu)} r = xr$ as $f_{(\mu,\nu)}(r) = r$ for all $r \in R$ and $(E, +, \cdot)$ is a compatible ring structure on E. Also $f_{(\mu,\nu)}$ is a ring isomorphism from $(E, +, \circ_{(\mu,\nu)})$ to $(E, +, \cdot)$.

For $(\mu, \nu), (\gamma, \delta) \in \text{Soc}(A) \times \text{Soc}(A)$, we see that $\circ_{(\mu,\nu)} = \circ_{(\gamma,\delta)}$ if and only if $(\mu, \nu) = (\gamma, \delta)$ if and only if $\mu = \gamma$ and $\nu = \delta$. Thus, E has $|\text{Soc}(A)|^2$ distinct compatible ring structures such that they are QF-rings and are isomorphic to $(E, +, \cdot)$.

(vi) By part (iv), R is right Osofsky compatible. Consider E^T, the transpose of $(E, +, \cdot)$. Then ${}_RR$ embeds in ${}_RE^T$ by

$$\begin{bmatrix} a & \overline{b} \\ 0 & \overline{d} \end{bmatrix} \to \begin{bmatrix} (a, -\overline{a}) & 0 \\ 0 & 0 \end{bmatrix} + \begin{bmatrix} \overline{a} & \overline{0} \\ \overline{b} & \overline{d} \end{bmatrix}.$$

Then ${}_RE^T$ is an injective hull of ${}_RR$. Note that right multiplication on the ring $(E, +, \cdot)$ is the same as left multiplication by E^T. □

Corollary 7.3.15 *Let A be a commutative QF-ring such that $J(A) \neq 0$. Then the ring*

$$R = \begin{bmatrix} A & A/J(A) \\ 0 & A/J(A) \end{bmatrix}$$

has a right injective hull which has a ring structure, compatible with the right R-module structure, as a ring direct sum of QF-rings and that ring structure is not unique. Its left injective hull can be given a ring structure compatible with the left R-module structure.

Proof A commutative QF-ring is a ring direct sum of local rings. Apply Proposition 7.3.10 and Theorem 7.3.14. □

Proposition 7.3.16 *Let A, R, and E_R be as in Theorem 7.3.14. Put*

$$\mathbb{E}_R = \begin{bmatrix} A \oplus A/J(A) & A/J(A) \\ A/J(A) & A/J(A) \end{bmatrix},$$

where the addition is componentwise and the R-module scalar multiplication over R is given by

$$\begin{bmatrix} s+\overline{a} & \overline{b} \\ \overline{c} & \overline{d} \end{bmatrix} \begin{bmatrix} x & \overline{y} \\ 0 & \overline{z} \end{bmatrix} = \begin{bmatrix} sx+\overline{ax} & \overline{sy}+\overline{ay}+\overline{bz} \\ \overline{cx} & \overline{cy}+\overline{dz} \end{bmatrix},$$

where $\overline{a}, \overline{y} \in A/J(A)$, etc. denote the images of $a, y \in A$, etc., respectively. Then $R_R \leq^{\text{ess}} \mathbb{E}_R$ and $\mathbb{E}_R \cong E_R$. So $\mathbb{E}_R$ is an injective hull of R_R.

Proof Obviously, $\mathbb{E}_R$ is a right R-module. Further, we see that $R_R \leq^{\text{ess}} \mathbb{E}_R$. Define $f : \mathbb{E}_R \to E_R$ by

$$f\begin{bmatrix} s+\overline{a} & \overline{b} \\ \overline{c} & \overline{d} \end{bmatrix} = \begin{bmatrix} (s, -\overline{s}) & 0 \\ 0 & 0 \end{bmatrix} + \begin{bmatrix} \overline{s}+\overline{a} & \overline{b} \\ \overline{c} & \overline{d} \end{bmatrix}$$

for $\begin{bmatrix} s+\overline{a} & \overline{b} \\ \overline{c} & \overline{d} \end{bmatrix} \in \mathbb{E}_R$. Then f is an additive group isomorphism. Put

$$\iota(R) = \left\{ \begin{bmatrix} (x, -\overline{x}) & 0 \\ 0 & 0 \end{bmatrix} + \begin{bmatrix} \overline{x} & \overline{y} \\ \overline{0} & \overline{z} \end{bmatrix} \mid x, y, z \in A \right\}.$$

Now $E_{\iota(R)}$ is an injective hull of $\iota(R)_{\iota(R)}$ by Theorem 7.3.14, and the scalar multiplication of $\mathbb{E}_R$ over R corresponds to that of $E_{\iota(R)}$ over $\iota(R)$ via f. So $\mathbb{E}_R$ is an injective hull of R_R. □

For a ring R as in Theorem 7.3.14, we exhibit a QF-ring A for which every injective hull of R_R has exactly $|\mathrm{Soc}(A)|^2$ distinct compatible ring structures as follows.

Theorem 7.3.17 *Assume that $n = p_1^{m_1} \cdots p_k^{m_k}$, where p_i is a distinct prime integer and m_i is an integer such that $m_i \geq 2$ for each i. Let $A = \mathbb{Z}_n$ and let*

$$R = \begin{bmatrix} A & A/J(A) \\ 0 & A/J(A) \end{bmatrix}$$

as in Theorem 7.3.14. Then R is right Osofsky compatible and every injective hull of R_R has exactly $|\mathrm{Soc}(A)|^2 = p_1^2 \cdots p_k^2$ distinct compatible ring structures. These ring structures are isomorphic and QF.

Proof The ring R is right Osofsky compatible by Corollary 7.3.15. We suppose that $n = p^m$, where p is a prime integer and m is an integer such that $m \geq 2$. We show that every injective hull of R_R has exactly p^2 distinct compatible ring structures and these ring structures are isomorphic and QF. By Proposition 7.3.16, $\mathbb{E}_R = \begin{bmatrix} A \oplus A/J(A) & A/J(A) \\ A/J(A) & A/J(A) \end{bmatrix}$ is an injective hull of R_R.

Claim $\mathbb{E} := \mathbb{E}_R$ *has exactly p^2 distinct compatible ring structures.*

Proof of Claim Take $\alpha,\ \beta \in \mathrm{Soc}(A)$. Define $|\mathrm{Soc}(A)|^2$ distinct multiplications $\bullet_{(\alpha,\beta)}$ on $\mathbb{E}$: For $\begin{bmatrix} s_1 + \overline{a_1} & \overline{b_1} \\ \overline{c_1} & \overline{d_1} \end{bmatrix}, \begin{bmatrix} s_2 + \overline{a_2} & \overline{b_2} \\ \overline{c_2} & \overline{d_2} \end{bmatrix} \in \mathbb{E}$, let

$$\begin{bmatrix} s_1 + \overline{a_1} & \overline{b_1} \\ \overline{c_1} & \overline{d_1} \end{bmatrix} \bullet_{(\alpha,\beta)} \begin{bmatrix} s_2 + \overline{a_2} & \overline{b_2} \\ \overline{c_2} & \overline{d_2} \end{bmatrix} = \begin{bmatrix} x & y \\ z & w \end{bmatrix},$$

where

$$\begin{aligned} x &= s_1 s_2 + \beta a_1 a_2 + \alpha c_1 a_2 + (-\alpha) s_1 c_2 + \beta b_1 c_2 + \alpha d_1 c_2 \\ &\quad + \overline{a_1 a_2} + \overline{a_1 s_2} + \overline{s_1 a_2} + \overline{b_1 c_2}, \\ y &= \overline{a_1 b_2} + \overline{s_1 b_2} + \overline{b_1 d_2}, \\ z &= \overline{c_1 s_2} + \overline{c_1 a_2} + \overline{d_1 c_2}, \text{ and } w = \overline{c_1 b_2} + \overline{d_1 d_2}. \end{aligned}$$

Then $(\mathbb{E}, +, \bullet_{(\alpha,\beta)})$ is a compatible ring structure.

Conversely, assume that $\mathbb{E}$ has a compatible ring structure. By Lemma 7.1.7, $1_{\mathbb{E}} = 1_R = \begin{bmatrix} 1 & 0 \\ 0 & \overline{1} \end{bmatrix}$ because $R_R \leq^{\text{ess}} \mathbb{E}_R$. Put $e_1 = \begin{bmatrix} 1 & 0 \\ 0 & 0 \end{bmatrix}$ and $e_2 = \begin{bmatrix} 0 & 0 \\ 0 & \overline{1} \end{bmatrix}$. Then we have that $e_1 + e_2 = 1_{\mathbb{E}}$, $e_1^2 = e_1$, $e_2^2 = e_2$, and $e_1 e_2 = e_2 e_1 = 0$. Thus, it follows that $\mathbb{E} = e_1 \mathbb{E} e_1 + e_1 \mathbb{E} e_2 + e_2 \mathbb{E} e_1 + e_2 \mathbb{E} e_2$.

We observe that $R_R \leq^{\text{ess}} \mathbb{E}_R$ and $e_1 R e_2 = \begin{bmatrix} 0 & A/J(A) \\ 0 & 0 \end{bmatrix} \subseteq e_1 \mathbb{E} e_2$. Also we get that $e_2 R e_2 = \begin{bmatrix} 0 & 0 \\ 0 & A/J(A) \end{bmatrix} \subseteq e_2 \mathbb{E} e_2$. So we have the following:

Subclaim 1 $e_1\mathbb{E}e_2 = \begin{bmatrix} 0 & A/J(A) \\ 0 & 0 \end{bmatrix}$ *and* $e_2\mathbb{E}e_2 = \begin{bmatrix} 0 & 0 \\ 0 & A/J(A) \end{bmatrix}$.

Proof The proof is routine.

Subclaim 2 $e_1\mathbb{E}e_1 = \begin{bmatrix} A \oplus A/J(A) & 0 \\ 0 & 0 \end{bmatrix}$.

Proof Subclaim 1, the fact that $R_R \leq^{\text{ess}} \mathbb{E}_R$, and standard argument yield that $e_1\mathbb{E}e_1 = \begin{bmatrix} A \oplus A/J(A) & 0 \\ 0 & 0 \end{bmatrix}$.

Also by using Subclaims 1 and 2, we can check the following.

Subclaim 3 $e_2\mathbb{E}e_1 = \left\{ \begin{bmatrix} \alpha c & 0 \\ \overline{c} & 0 \end{bmatrix} \mid c \in A \right\}$ *for some* $\alpha \in \text{Soc}(A)$.

Proof Exercise.

Subclaim 4 $\begin{bmatrix} \overline{1} & 0 \\ 0 & 0 \end{bmatrix}\begin{bmatrix} \overline{1} & 0 \\ 0 & 0 \end{bmatrix} = \begin{bmatrix} \beta + \overline{1} & 0 \\ 0 & 0 \end{bmatrix}$ *for some* $\beta \in \text{Soc}(A)$.

Proof Exercise.

We proceed to determine all possible ring multiplications on $\mathbb{E}$ which extend the R-module scalar multiplication of $\mathbb{E}$ over R. By using Subclaim 1 through Subclaim 4 together with the associativity of the multiplication and its distributive laws, we get the following:

1. $\begin{bmatrix} 1 & 0 \\ 0 & 0 \end{bmatrix}\begin{bmatrix} \overline{1} & 0 \\ 0 & 0 \end{bmatrix} = \begin{bmatrix} \overline{1} & 0 \\ 0 & 0 \end{bmatrix}$, 2. $\begin{bmatrix} 0 & \overline{1} \\ 0 & 0 \end{bmatrix}\begin{bmatrix} \overline{1} & 0 \\ 0 & 0 \end{bmatrix} = 0$, 3. $\begin{bmatrix} 0 & 0 \\ \overline{1} & 0 \end{bmatrix}\begin{bmatrix} \overline{1} & 0 \\ 0 & 0 \end{bmatrix} = \begin{bmatrix} \alpha & 0 \\ \overline{1} & 0 \end{bmatrix}$,

4. $\begin{bmatrix} 0 & 0 \\ 0 & \overline{1} \end{bmatrix}\begin{bmatrix} \overline{1} & 0 \\ 0 & 0 \end{bmatrix} = 0$, 5. $\begin{bmatrix} \overline{1} & 0 \\ 0 & 0 \end{bmatrix}\begin{bmatrix} 0 & 0 \\ \overline{1} & 0 \end{bmatrix} = 0$, 6. $\begin{bmatrix} 1 & 0 \\ 0 & 0 \end{bmatrix}\begin{bmatrix} 0 & 0 \\ \overline{1} & 0 \end{bmatrix} = \begin{bmatrix} -\alpha & 0 \\ 0 & 0 \end{bmatrix}$,

7. $\begin{bmatrix} 0 & \overline{1} \\ 0 & 0 \end{bmatrix}\begin{bmatrix} 0 & 0 \\ \overline{1} & 0 \end{bmatrix} = \begin{bmatrix} \beta + \overline{1} & 0 \\ 0 & 0 \end{bmatrix}$, 8. $\begin{bmatrix} 0 & 0 \\ \overline{1} & 0 \end{bmatrix}\begin{bmatrix} 0 & 0 \\ \overline{1} & 0 \end{bmatrix} = 0$, 9. $\begin{bmatrix} 0 & 0 \\ 0 & \overline{1} \end{bmatrix}\begin{bmatrix} 0 & 0 \\ \overline{1} & 0 \end{bmatrix} = \begin{bmatrix} \alpha & 0 \\ \overline{1} & 0 \end{bmatrix}$.

Using 1 through 9 together with Subclaims 1–4, a ring multiplication on $\mathbb{E}$ which extends the R-module scalar multiplication of $\mathbb{E}$ can be defined by

$$\begin{bmatrix} s_1 + \overline{a_1} & \overline{b_1} \\ \overline{c_1} & d_1 \end{bmatrix}\begin{bmatrix} s_2 + \overline{a_2} & \overline{b_2} \\ \overline{c_2} & d_2 \end{bmatrix} = \begin{bmatrix} x & y \\ z & w \end{bmatrix},$$

where

$$\begin{aligned} x = {} & s_1s_2 + \beta a_1a_2 + \alpha c_1a_2 + (-\alpha)s_1c_2 + \beta b_1c_2 + \alpha d_1c_2 \\ & + \overline{a_1a_2} + \overline{a_1s_2} + \overline{s_1a_2} + \overline{b_1c_2}, \\ y = {} & \overline{a_1b_2} + \overline{s_1b_2} + \overline{b_1d_2}, \end{aligned}$$

$$z = \overline{c_1 s_2} + \overline{c_1 a_2} + \overline{d_1 c_2}, \text{ and } w = \overline{c_1 b_2} + \overline{d_1 d_2}.$$

Thus the ring multiplication is exactly $\bullet_{(\alpha,\, \beta)}$, where $\alpha, \beta \in \mathrm{Soc}(A)$. Therefore, there are exactly p^2 distinct compatible ring structures $(\mathbb{E}, +, \bullet_{(\alpha,\, \beta)})$ with $\alpha, \beta \in \mathrm{Soc}(A)$. This completes the proof of Claim at the beginning of the proof of this theorem.

Define $f_{(\alpha,\beta)} : (\mathbb{E}, +, \bullet_{(\alpha,\beta)}) \to (\mathbb{E}, +, \bullet_{(0,0)})$ by

$$f_{(\alpha,\beta)} \begin{bmatrix} s + \overline{a} & \overline{b} \\ \overline{c} & \overline{d} \end{bmatrix} = \begin{bmatrix} s + (-\beta)a + (-\alpha)c + \overline{a} & \overline{b} \\ \overline{c} & \overline{d} \end{bmatrix}.$$

Then $f_{(\alpha,\beta)}$ is a ring isomorphism. Therefore, these $|\mathrm{Soc}(A)|^2$ distinct rings $(\mathbb{E}, +, \bullet_{(\alpha,\beta)})$ are all isomorphic. Let $V = (\mathbb{E}, +, \bullet_{(0,0)})$. Note that $1_V = \begin{bmatrix} 1 & 0 \\ 0 & \overline{1} \end{bmatrix}$. Let $e = \begin{bmatrix} \overline{1} & 0 \\ 0 & \overline{1} \end{bmatrix}$. Then e is a central idempotent of V and $eV = \mathrm{Mat}_2(A/J(A))$. Also $(1_V - e)V \cong A$ as rings. Thus, $V = eV \oplus (1_V - e)V$ is a QF-ring. Hence, all $(\mathbb{E}, +, \bullet_{(\alpha,\beta)})$ are QF. Also $V \cong E_\lambda$, where E_λ is the compatible ring structure given in Theorem 7.3.14(iv).

Suppose that $n = p_1^{m_1} \cdots p_k^{m_k}$, where p_i is a distinct prime integer and m_i is an integer such that $m_i \geq 2$ for each i. Then $A = \oplus_{i=1}^k A_i$ with $A_i = \mathbb{Z}_{p_i^{m_i}}$. By Proposition 7.3.16,

$$\mathbb{E}_i = \begin{bmatrix} A_i \oplus A_i/J(A_i) & A_i/J(A_i) \\ A_i/J(A_i) & A_i/J(A_i) \end{bmatrix}$$

is an injective hull of $R_{i\, R_i}$, where $R_i = \begin{bmatrix} A_i & A_i/J(A_i) \\ 0 & A_i/J(A_i) \end{bmatrix}$, so

$$\mathbb{E} = \begin{bmatrix} A \oplus A/J(A) & A/J(A) \\ A/J(A) & A/J(A) \end{bmatrix} = \oplus_{i=1}^k \mathbb{E}_i$$

is an injective hull of R_R since $R = \oplus_{i=1}^k R_i$.

Assume that $(\mathbb{E}, +, \bullet)$ is a compatible ring structure on $\mathbb{E}$. By Proposition 7.3.10, each $(\mathbb{E}_i, +, \bullet_i)$ is a compatible ring structure induced from the compatible ring structure $(\mathbb{E}, +, \bullet)$.

From the proof of Claim, $\bullet_i = \bullet_{(\alpha_i,\, \beta_i)}$ for some $\alpha_i, \beta_i \in \mathrm{Soc}(A_i)$. Thus, we have that $(\mathbb{E}, +, \bullet) = (\mathbb{E}_1, +, \bullet_{(\alpha_1,\beta_1)}) \oplus \cdots \oplus (\mathbb{E}_k, +, \bullet_{(\alpha_k,\beta_k)})$. Hence,

$$(x_i)_{i=1}^k \bullet (y_i)_{i=1}^k = (x_i \bullet_{(\alpha_i,\beta_i)} y_i)_{i=1}^k \in \mathbb{E}$$

for $(x_i)_{i=1}^k, (y_i)_{i=1}^k \in \mathbb{E} = \oplus_{i=1}^k \mathbb{E}_i$. We put $\bullet = \oplus_{i=1}^k \bullet_{(\alpha_i,\beta_i)}$.

Thus $\{\oplus_{i=1}^k \bullet_{(\alpha_i,\beta_i)} \mid (\alpha_i, \beta_i) \in \mathrm{Soc}(A_i) \times \mathrm{Soc}(A_i) \text{ for } 1 \leq i \leq k\}$ is the set of all compatible ring multiplications on $\mathbb{E}$. Thus, all compatible ring structures of

$\mathbb{E}$ are precisely the ring structures induced from the compatible ring structures of $\mathbb{E}_1, \ldots, \mathbb{E}_k$. Therefore, $\mathbb{E}$ has exactly $p_1^2 \cdots p_k^2$ distinct compatible ring structures which are isomorphic and QF.

Finally, let $E(R_R)$ be an arbitrary injective hull of R_R. Then by the proof of Proposition 7.1.10, $E(R_R)$ also has exactly $|\mathrm{Soc}(A)|^2 = p_1^2 \cdots p_k^2$ distinct compatible ring structures which are isomorphic and QF. □

Let R be the ring as in Theorem 7.3.17. We present all proper right essential overrings of R and their properties as follows.

Example 7.3.18 Let $n = p^m$, where p is a prime integer and m is an integer such that $m \geq 2$. Put $A = \mathbb{Z}_n$. Let

$$R = \begin{bmatrix} A & A/J(A) \\ 0 & A/J(A) \end{bmatrix} \text{ and } \mathbb{E} = \begin{bmatrix} A \oplus A/J(A) & A/J(A) \\ A/J(A) & A/J(A) \end{bmatrix}.$$

By Proposition 7.3.16, $\mathbb{E}_R$ is an injective hull of R_R. Let

$$T = \begin{bmatrix} A \oplus A/J(A) & A/J(A) \\ 0 & A/J(A) \end{bmatrix}.$$

Then we have the following.

(i) $R_R \trianglelefteq T_R \trianglelefteq \mathbb{E}_R$.
(ii) There are exactly p distinct compatible ring structures $(T, +, \diamond_{(0,\beta)})$ with $\beta \in \mathrm{Soc}(A)$, where $\diamond_{(0,\beta)}$ is the restriction of $\bullet_{(0,\beta)}$ of $\mathbb{E}$ (in Theorem 7.3.17) to T. Further, all these compatible ring structures on T are isomorphic.
(iii) All right essential overrings of R are: $(\mathbb{E}, +, \bullet_{(\alpha,\gamma)})$, $(T, +, \diamond_{(0,\beta)})$ and R itself, where $\alpha, \beta, \gamma \in \mathrm{Soc}(A)$.
(iv) $(T, +, \diamond_{(0,\beta)})$ with $\beta \in \mathrm{Soc}(A)$ are precisely all of the right extending (also, right FI-extending) minimal right essential overrings of R.
(v) $(\mathbb{E}, +, \bullet_{(\alpha,\beta)})$, with $\alpha, \beta \in \mathrm{Soc}(A)$, are all of the right self-injective (also, right quasi-continuous and right continuous, respectively) minimal right essential overrings of R.

Proof (i) It is obvious.

(ii) Let $t_1 = \begin{bmatrix} s_1 + \overline{a_1} & \overline{b_1} \\ 0 & \overline{d_1} \end{bmatrix}$, $t_2 = \begin{bmatrix} s_2 + \overline{a_2} & \overline{b_2} \\ 0 & \overline{d_2} \end{bmatrix} \in T$, and $\beta \in \mathrm{Soc}(A)$. Define

$$t_1 \diamond_{(0,\beta)} t_2 = \begin{bmatrix} s_1 s_2 + \beta a_1 a_2 + \overline{s_1 a_2} + \overline{a_1 s_2} + \overline{a_1 a_2} & \overline{s_1 b_2} + \overline{a_1 b_2} + \overline{b_1 d_2} \\ 0 & \overline{d_1 d_2} \end{bmatrix}.$$

Then $(T, +, \diamond_{(0,\beta)})$ is a compatible ring structure, and $\diamond_{(0,\beta)}$ is the restriction of $\bullet_{(0,\beta)}$ to T.

Conversely, assume that T has a compatible ring structure. By Lemma 7.1.7, $1_T = 1_R$ $(= 1_{\mathbb{E}})$ since $R_R \leq^{\text{ess}} T_R$. Let $e_1 = \begin{bmatrix} 1 & 0 \\ 0 & 0 \end{bmatrix}$ and $e_2 = \begin{bmatrix} 0 & 0 \\ 0 & \overline{1} \end{bmatrix}$. Then it follows that $e_1^2 = e_1$, $e_2^2 = e_2$, $e_1e_2 = e_2e_1 = 0$, and $1_T = e_1 + e_2$. Therefore, we obtain that $T = e_1Te_1 + e_1Te_2 + e_2Te_1 + e_2Te_2$. By direct calculation as in the proof of Theorem 7.3.17,

$$e_1Te_2 = \begin{bmatrix} 0 & A/J(A) \\ 0 & 0 \end{bmatrix}, \; e_2Te_2 = \begin{bmatrix} 0 & 0 \\ 0 & A/J(A) \end{bmatrix}, \; e_2Te_1 = 0,$$

and

$$e_1Te_1 = \begin{bmatrix} A \oplus A/J(A) & 0 \\ 0 & 0 \end{bmatrix}.$$

Moreover, $\begin{bmatrix} \overline{1} & 0 \\ 0 & 0 \end{bmatrix} \begin{bmatrix} \overline{1} & 0 \\ 0 & 0 \end{bmatrix} = \begin{bmatrix} \beta + \overline{1} & 0 \\ 0 & 0 \end{bmatrix}$ for some $\beta \in \text{Soc}(A)$.

We see that there is a ring multiplication, say $\diamond_{(0,\beta)}$:

$$t_1 \diamond_{(0,\beta)} t_2 = \begin{bmatrix} s_1s_2 + \beta a_1a_2 + \overline{s_1a_2} + \overline{a_1s_2} + \overline{a_1a_2} & \overline{s_1b_2} + \overline{a_1b_2} + \overline{b_1d_2} \\ 0 & \overline{d_1d_2} \end{bmatrix},$$

for $t_1 = \begin{bmatrix} s_1 + \overline{a_1} & \overline{b_1} \\ 0 & \overline{d_1} \end{bmatrix}$, $t_2 = \begin{bmatrix} s_2 + \overline{a_2} & \overline{b_2} \\ 0 & \overline{d_2} \end{bmatrix} \in T$. So there exist exactly p distinct compatible ring structures $(T, +, \diamond_{(0,\beta)})$, with $\beta \in \text{Soc}(A)$, on T.

Define $g_{(0,\beta)} : (T, +, \diamond_{(0,\beta)}) \to (T, +, \diamond_{(0,0)})$ by

$$g_{(0,\beta)} \begin{bmatrix} s + \overline{a} & \overline{b} \\ 0 & \overline{d} \end{bmatrix} = \begin{bmatrix} s + (-\beta)a + \overline{a} & \overline{b} \\ 0 & \overline{d} \end{bmatrix}.$$

Then $g_{(0,\beta)}$ is a ring isomorphism.

(iii) All intermediate R-modules between R_R and $\mathbb{E}_R$ are:

$$\mathbb{E}_R, \; S = \begin{bmatrix} A & A/J(A) \\ A/J(A) & A/J(A) \end{bmatrix}, \; T = \begin{bmatrix} A \oplus A/J(A) & A/J(A) \\ 0 & A/J(A) \end{bmatrix}, \text{ and } R.$$

Routinely, we see that S cannot have a compatible ring structure. Thus, part (ii) yields part (iii).

(iv) Put $W = (T, +, \diamond_{(0,0)})$ and let $e = \begin{bmatrix} \overline{1} & 0 \\ 0 & \overline{1} \end{bmatrix} \in W$. Then e is a central idempotent of W. The ring $eW = T_2(A/J(A))$ is Baer by Theorem 5.6.2 because $A/J(A)$ is a field. Thus, eW is right extending by Corollary 3.3.3. Also $(1_W - e)W \cong A$ as rings. Thus, $(1_W - e)T$ is right extending. So the ring $W = eW \oplus (1_W - e)W$ (ring direct sum) is right extending.

From the preceding argument, $(T, +, \diamond_{(0,\beta)})$ is right extending for each $\beta \in \text{Soc}(A)$ since $(T, +, \diamond_{(0,\beta)}) \cong W$ by part (ii). So $(T, +, \diamond_{(0,\beta)})$ is right FI-extending for each $\beta \in \text{Soc}(A)$.

On the other hand, by Theorem 5.6.10, R cannot be right FI-extending. Thus, $(T, +, \diamond_{(0,\beta)})$ with $\beta \in \mathrm{Soc}(A)$ are precisely all of the right extending (also, right FI-extending) minimal right essential overrings of R.

(v) We show that $W = (T, +, \diamond_{(0,0)})$ is not right quasi-continuous. For this, let $e = \begin{bmatrix} 0 & 0 \\ 0 & \overline{1} \end{bmatrix}$ and $f = \begin{bmatrix} 0 & 1 \\ 0 & \overline{1} \end{bmatrix}$ in W. Then we see that $e^2 = e$, $f^2 = f$, and $eW \cap fW = 0$. But we note that

$$eW \oplus fW = \begin{bmatrix} 0 & A/J(A) \\ 0 & A/J(A) \end{bmatrix}$$

is not a direct summand of W_W. So W_W does not satisfy (C_3) condition. Thus, W is not right quasi-continuous. By Theorem 7.3.17 and part (iv), $(\mathbb{E}, +, \bullet_{(\alpha,\beta)})$, with $\alpha, \beta \in \mathrm{Soc}(A)$, are all of the right self-injective (also, right quasi-continuous and right continuous, respectively) minimal right essential overrings of R. □

Exercise 7.3.19

1. Assume that R is the ring in Example 7.3.3. Show that

$$E(R_R) \cong [\mathrm{Hom}_K({}_R R_K, K_K)]_R^{(2)},$$

where $K = \mathbb{Z}_2$ by an argument similar to that used in Example 7.3.5. Thus, it follows that $|E(R_R)| = 2^8$.
2. For a field K, let $T = K[x]/x^4K[x]$. Then $T = K + K\overline{x} + K\overline{x}^2 + K\overline{x}^3$, where $\overline{x}$ is the image of x in T. As in Example 7.3.13(i), let $R = K + K\overline{x}^2 + K\overline{x}^3$ which is a subring of T. Prove that $E(R_R) \cong [\mathrm{Hom}_K({}_R R_K, K_K)]_R^{(2)}$.
3. We can construct a ring R for which every injective hull of R_R has infinitely many distinct compatible ring structures and these are isomorphic and QF-rings. Indeed, let F be an infinite field and

$$R = \begin{bmatrix} \Lambda & \Lambda/J(\Lambda) \\ 0 & \Lambda/J(\Lambda) \end{bmatrix},$$

where Λ is the ring in (i) and (ii) below. Show that R is right Osofsky compatible and every injective hull $E(R_R)$ of R_R has $|F|$ distinct compatible ring structures. These compatible ring structures on $E(R_R)$ are isomorphic and QF.
   (i) $\Lambda = F[x]/f(x)F[x]$, where $f(x) \neq 0$ is not square free by an irreducible polynomial.
   (ii) $\Lambda = F[G]$ is the group algebra, where the characteristic of F is $p > 0$, p a prime integer, and G is a finite Abelian group such that $p \mid |G|$.
4. Give proofs of Subclaims 1, 2, 3, and 4 in the proof of Theorem 7.3.17.
5. In the proof of Example 7.3.18(iii), show that S cannot have a compatible ring structure.
6. ([95, Birkenmeier, Osofsky, Park, and Rizvi]) Let R be a ring and $E = E_R$ be an injective hull of R_R with $\iota : R_R \to E_R$ an essential embedding. An additive

group monomorphism $\lambda : E_R \to H := \mathrm{End}(E_R)$ with $\lambda(u) = u_\lambda$ is called a *left multiplication by E* if the following are satisfied:
(1) $\{u_\lambda \mid u \in E\}$ is closed under composition $\circ$.
(2) $\iota(1_R)_\lambda = 1_H$.
(3) $u_\lambda(\iota(1_R)) = u$.
Define multiplication $\cdot : E \times E \to E$ by $u \cdot v = u_\lambda(v)$. Prove the following.
(i) For $u, v \in E$, $(u \cdot v)_\lambda = u_\lambda \circ v_\lambda$.
(ii) The multiplication $\cdot$ is associative and distributive over addition on both sides.
(iii) $u \cdot \iota(r) = ur$ for $u \in E$ and $r \in R$.
(iv) $(E, +, \cdot)$ is a compatible ring structure on E_R.

Historical Notes Lemma 7.1.7 is due to Osofsky [327]. Example 7.1.8 was obtained by Birkenmeier, Park, and Rizvi in [99]. The idea for the proof of Example 7.1.8 is used to find compatible ring structures on essential extensions of a ring R in Sects. 7.2 and 7.3. Theorem 7.1.14 is from [88]. We apply an idea of Sakano in [367] for E_R in Theorem 7.1.14. Lemma 7.1.19 and the fact that U_Γ is injective in Lemma 7.1.20 are corrected versions of Lemma 6, Lemma 7, and the fact that V_Γ is injective in Lemma 8 in [92]. Result 7.1.18, and Results 7.1.21–7.1.25 are from [92], while Theorem 7.1.26 is in [88]. Theorem 7.2.1 and Theorem 7.2.2 are taken from [99]. In Theorem 7.2.1, by using an explicit description of an injective hull E of R_R, where R is the ring of Example 7.1.13, we provide another method which is different from that of Lam [262] and that of Osofsky [327] for showing that R is not right Osofsky compatible.

Theorem 7.3.1 is from [234, 235, 415], and [327]. Theorem 7.3.2 is known as Gabriel's Theorem. Example 7.3.5, from [266], was obtained by Dlab and Ringel. Results 7.3.7–7.3.10 appear in [266]. Proposition 7.3.10 is also in [99] with a different proof from that of [266]. The proof of Proposition 7.3.10 is taken from [99]. For Theorem 7.3.11, see also [294]. Theorem 7.3.12 was obtained by Lang [266], which exhibits some constraints in finding classes of right Osofsky compatible commutative Artinian rings. Example 7.3.13(i) is initially due to Utumi [395] and Example 7.3.13(ii) appears in [266]. Theorem 7.3.14 is due to Birkenmeier, Osofsky, Park, and Rizvi [95]. We provide an alternate proof Theorem 7.3.14(v). Corollary 7.3.15 is from [95]. Theorem 7.3.17 and Example 7.3.18 were obtained by Birkenmeier, Park, and Rizvi, which are unpublished new results.

See [95, 297, 332], and [334] for further materials on right Osofsky compatibility and compatible ring structures. Osofsky has obtained some new results on this notion in a very recent work that is not published yet [332]. Her earlier related works include [325] and [328].

Chapter 8
Ring and Module Hulls

A motivation for the need to study ring and module hulls that are intermediate between a ring R and $Q(R)$ or $E(R)$, and between a module M and $E(M)$, respectively, can be seen from the following examples. Consider

$$R = \begin{bmatrix} \mathbb{Z} & \mathbb{Q} \\ 0 & \mathbb{Z} \end{bmatrix}.$$

The ring R is neither right nor left Noetherian and its prime radical is nonzero. However, $Q(R) = \mathrm{Mat}_2(\mathbb{Q})$ is simple Artinian. Next, take R to be a domain which is not right Ore. Then $Q(R)$ is a simple regular right self-injective ring (see Theorem 2.1.31 and [262, Corollary 13.38′]) which is neither orthogonally finite nor with bounded index (of nilpotency). The disparity between R and $Q(R)$ in the preceding examples limits the transfer of information between R and $Q(R)$.

Although every module has an injective hull, it is generally hard to construct or explicitly describe it. However, certain known subsets of the injective hull or of the endomorphism ring of the injective hull of a given ring (or module) can be used to generate an overring (or an overmodule) in conjunction with the base ring (or module) to serve as a hull of the ring (module) with some desirable properties. For example, since $Q(R)$ can be constructed for a ring R by Utumi's method (see Theorem 1.3.13), $\mathcal{B}(Q(R))$ can also be determined. Hence, the set of all $f(1)$, where f is a central idempotent in $\mathrm{End}(E(R_R))$ is explicitly described via $\mathcal{B}(Q(R))$ (see Lemma 8.3.10). Therefore, rings or modules generated by such a known subset of the injective hull in conjunction with the base ring or module may provide hulls. Additionally, these hulls may possess properties of interest to us.

These examples and constructions illustrate a need to find overrings of a given ring that have some weaker versions of the properties traditionally associated with right rings of quotients (e.g., semisimple Artinian, or (regular) right self-injective, or right continuous, etc.). These overrings are close enough to the base ring to facilitate an effective exchange of information between the base ring and the overrings. Furthermore, this need is reinforced when one studies the classes of rings for which $R = Q(R)$ (e.g., right Kasch rings). For these classes, the theory of right rings of

G.F. Birkenmeier et al., *Extensions of Rings and Modules*,
DOI 10.1007/978-0-387-92716-9_8,

quotients does not apply as was seen in Chap. 7 (and now in Chap. 8). However, the results presented in Chap. 7 which deal with right essential overrings will still be applicable (as will also be seen for such results from this chapter).

Our goal is to find methods that enable us to describe all right essential overrings of a ring R in a selected class $\mathfrak{K}$ (or essential overmodules of a module M in a selected class $\mathfrak{M}$). For this, our focus is on the study of the following problems:

Problem I. Given a ring R and a class $\mathfrak{K}$ of rings, determine what information transfers between R and its right essential overrings in $\mathfrak{K}$.

Problem II. Assume that a ring R and a class of rings $\mathfrak{K}$ are given.

(i) Determine conditions to ensure the existence of right rings of quotients and that of right essential overrings of R, which are, in some sense, "minimal" with respect to belonging to the class $\mathfrak{K}$.

(ii) Characterize the right rings of quotients and the right essential overrings of R which are in the class $\mathfrak{K}$ possibly by using the "minimal" ones obtained in (i).

Problem III. Given classes of rings $\mathfrak{A}$ and $\mathfrak{B}$, determine those rings $R \in \mathfrak{B}$ such that $Q(R) \in \mathfrak{A}$.

Problem IV. Given a ring R and a class $\mathfrak{K}$ of rings, let $X(R)$ denote some standard type of extension of R (e.g., $X(R) = R[x]$ or $X(R) = \text{Mat}_n(R)$, etc.) and let $H(R)$ denote a right essential overring of R which is "minimal" with respect to belonging to the class $\mathfrak{K}$. Determine when $H(X(R))$ is comparable to $X(H(R))$.

Problems I and II will be discussed in Sects. 8.1–8.3, while Problems III and IV will be studied in Sects. 9.1 and 9.3, respectively. We shall see that the right essential overrings which are minimal with respect to belonging to a specific class of rings are important tools in these investigations. To accommodate various notions of minimality, three basic notions of hulls are included in our discussion (see Definition 8.2.1). Using these notions, we establish the existence and uniqueness of the FI-extending ring hull for a semiprime ring (which, in this case, coincides with the quasi-Baer ring hull). In another basic type of a ring hull, we shall use R and certain subsets of $E(R_R)$ to generate a right essential overring S, so that S is in $\mathfrak{K}$ in some minimal fashion (see Definition 8.2.8). This construction leads to the concept of a pseudo ring hull. Moreover, we show that there is an effective transfer of information between the aforementioned hulls and the base ring. The results we present in this chapter will be applied to the study of boundedly centrally closed C^*-algebras later in Chap. 10.

We will conclude the chapter with a discussion on module hulls in Sect. 8.4. In particular, we will discuss quasi-injective, continuous and quasi-continuous hulls of a module. Conditions for a continuous hull to exist will be shown. We will see that every finitely generated projective module over a semiprime ring has an FI-extending hull. Moreover, it will be shown that the extending and FI-extending properties transfer from a module M to its rational hull.

For the convenience of the reader, **Con**, **qCon**, **E**, and **FI** are used respectively to denote the class of right continuous rings (modules), the class of right quasi-continuous rings (modules), the class of right extending rings (modules), and the class of right FI-extending rings (modules) according to the context. Further, we let **B** and **qB** denote the class of Baer rings and the class of quasi-Baer rings, respectively.

In this chapter, in general, all rings are assumed to have an identity element. However, in Definition 8.1.5, Definition 8.2.1, and Sect. 8.3, we do not require that rings must have an identity element.

8.1 Background and Preliminaries

This section is devoted to background information and preliminary results. Various properties are presented which transfer from a ring to its right rings of quotients or to its right essential overrings.

Definition 8.1.1 A ring R is said to be *right essentially Baer* (resp., *right essentially quasi-Baer*) if the right annihilator of any nonempty subset (resp., ideal) of R is essential in a right ideal generated by an idempotent. Let **eB** (resp., **eqB**) denote the class of right essentially Baer (resp., right essentially quasi-Baer) rings.

It can be seen that **eB** properly contains **E** and **B**, while **eqB** properly contains **FI** and **qB**. If $S = A \oplus \mathbb{Z}_4$, where A is a domain which is not right Ore, then S is neither right extending nor Baer. But S is right essentially Baer. Next let R be the ring as in Example 7.1.13. Then the ring R is neither right FI-extending nor quasi-Baer. But R is right essentially quasi-Baer (see Exercise 8.1.10.1). In Theorem 3.2.37, we have seen that when a ring R is semiprime, R is quasi-Baer if and only if R is right essentially quasi-Baer. The next result shows that replacing semiprime with nonsingularity also yields this equivalence.

Proposition 8.1.2 *Assume that R is a right nonsingular ring.*
(i) *If $R \in$ **eB**, then $R \in$ **B**.*
(ii) *If $R \in$ **eqB**, then $R \in$ **qB**.*

Proof (i) Assume that R is right essentially Baer. Say $\emptyset \neq X \subseteq R$. Then $r_R(X)_R$ is essential in eR_R with $e^2 = e \in R$. As in the proof of Theorem 3.2.38, we obtain that $\ell_R(r_R(X)) = \ell_R(eR) = R(1-e)$, so $r_R(X) = r_R(\ell_R(r_R(X))) = eR$. Thus R is Baer.

(ii) Say R is right essentially quasi-Baer. Take X to be an ideal of R and follow the proof of part (i). □

Lemma 8.1.3 *Let T be a right ring of quotients of R. Then:*

(i) *For right ideals I and J of T, if $I_T \leq^{\text{ess}} J_T$, then $I_R \leq^{\text{ess}} J_R$.*
(ii) *If $A_R \trianglelefteq T_R$, then $A_R \leq^{\text{ess}} TAT_R$.*

Proof (i) Let $0 \neq y \in J$. Then there is $t \in T$ with $0 \neq yt \in I$. As $R_R \leq^{\text{den}} T_R$, there is $r \in R$ satisfying $tr \in R$ and $ytr \neq 0$. Now $ytr \in I$. So $I_R \leq^{\text{ess}} J_R$.

(ii) By Proposition 2.1.32, $\text{End}(T_R) = \text{End}(T_T) \cong T$. Thus $TA \subseteq A$ as $A_R \trianglelefteq T_R$. Let $0 \neq y \in TAT = AT$. Then $y = a_1t_1 + \cdots + a_nt_n$ where $a_i \in A$, $t_i \in T$ for each i, $1 \leq i \leq n$. Since $R_R \leq^{\text{den}} T_R$, there is $r_1 \in R$ with $t_1r_1 \in R$ and $yr_1 \neq 0$. Again there is $r_2 \in R$ with $t_2 r_1 r_2 \in R$ and $y r_1 r_2 \neq 0$. Continuing this process, there is $r \in R$ with $0 \neq yr \in A$. So $A_R \leq^{\text{ess}} TAT_R$. □

Proposition 8.1.4 *Let T be a right ring of quotients of R. Then:*

(i) *T_T is FI-extending if and only if T_R is FI-extending.*
(ii) *T_T is extending if and only if T_R is extending.*

Proof (i) Let T_T be FI-extending. Say $A_R \trianglelefteq T_R$. Then $A_R \leq^{\text{ess}} TAT_R$ by Lemma 8.1.3(ii). There exists $e^2 = e \in T$ satisfying $TAT_T \leq^{\text{ess}} eT_T$. Thus $TAT_R \leq^{\text{ess}} eT_R$ by Lemma 8.1.3(i), so $A_R \leq^{\text{ess}} eT_R$. Hence, T_R is FI-extending.

Conversely, let T_R be FI-extending. Then $\text{End}(T_R) = \text{End}(T_T)$ by Proposition 2.1.32. Take $B \trianglelefteq T$. Then $B_R \trianglelefteq T_R$ because $\text{End}(T_R) = \text{End}(T_T) \cong T$. So there exists $e^2 = e \in \text{End}(T_R) = \text{End}(T_T)$ such that $B_R \leq^{\text{ess}} eT_R$. Hence, $B_T \leq^{\text{ess}} e(1)T_T$ and $e(1)^2 = e(1) \in T$. Therefore, T_T is FI-extending.

(ii) The proof is similar to that of part (i). □

The condition that T is a right ring of quotients of R in Proposition 8.1.4 cannot be replaced by the condition that T is a right essential overring of R (see Exercise 8.1.10.2). The concept of a **D-E** class is introduced in the next definition. Such a class has the advantage that its members have an abundance of idempotents for their "designated" right ideals.

Definition 8.1.5 Let $\mathfrak{K}$ be a class of rings not necessarily with identity and P be a property of right ideals. We say that $\mathfrak{K}$ is a *class determined by P* if:

(i) there exists an assignment $\mathfrak{D}_{\mathfrak{K}}$ on the class of all rings such that $\mathfrak{D}_{\mathfrak{K}}(R)$ is a set of right ideals of a ring R.

(ii) each element of $\mathfrak{D}_{\mathfrak{K}}(R)$ has the property P if and only if $R \in \mathfrak{K}$.

If $\mathfrak{K}$ is such a class where P is the property that a right ideal is essential in a right ideal generated by an idempotent, then we say that $\mathfrak{K}$ is a ***D-E*** *class* and use $\mathfrak{C}$ to denote a **D-E** class. Thus, a **D-E** class exhibits the extending property with respect to a designated set of right ideals of a ring in $\mathfrak{C}$. We note that any **D-E** class always contains the class of right extending rings.

Some examples illustrating Definition 8.1.5 are as follows.

(1) $\mathfrak{K}$ is the class of semisimple Artinian rings, $\mathfrak{D}_{\mathfrak{K}}(R) = \{I \mid I_R \leq R_R\}$, and P is the property that every right ideal is a direct summand.
(2) $\mathfrak{K}$ is the class of right Noetherian rings, $\mathfrak{D}_{\mathfrak{K}}(R) = \{I \mid I_R \leq R_R\}$, and P is the property that every right ideal is finitely generated.
(3) $\mathfrak{K}$ is the class of regular rings, $\mathfrak{D}_{\mathfrak{K}}(R) = \{aR \mid a \in R\}$, and P is the property that every right ideal is generated by an idempotent.
(4) $\mathfrak{K}$ is the class of biregular rings, $\mathfrak{D}_{\mathfrak{K}}(R) = \{RaR \mid a \in R\}$, and P is the property that every ideal is generated by a central idempotent.
(5) $\mathfrak{K}$ is the class of right Rickart rings, $\mathfrak{D}_{\mathfrak{K}}(R) = \{aR \mid a \in R\}$, and P is the property that every right ideal is projective.

(6) $\mathfrak{K} = \mathbf{B}$, $\mathfrak{D}_{\mathbf{B}}(R) = \{r_R(X) \mid \emptyset \neq X \subseteq R\}$, and P is the property that every right ideal is generated by an idempotent.
(7) $\mathfrak{K} = \mathbf{qB}$, $\mathfrak{D}_{\mathbf{qB}}(R) = \{r_R(I) \mid I \trianglelefteq R\}$, and P is the property that every right ideal is generated by an idempotent.
(8) $\mathfrak{C} = \mathbf{E}$ and $\mathfrak{D}_{\mathbf{E}}(R) = \{I \mid I_R \leq R_R\}$.
(9) $\mathfrak{C} = \mathbf{eB}$ and $\mathfrak{D}_{\mathbf{eB}}(R) = \{r_R(X) \mid \emptyset \neq X \subseteq R\}$.
(10) $\mathfrak{C} = \mathbf{eqB}$ and $\mathfrak{D}_{\mathbf{eqB}}(R) = \{r_R(I) \mid I \trianglelefteq R\}$.
(11) $\mathfrak{C} = \mathbf{FI}$ and $\mathfrak{D}_{\mathbf{FI}}(R) = \{I \mid I \trianglelefteq R\}$.
(12) $\mathfrak{C} = \mathbf{pFI}$ (**pFI** is the class of right principally FI-extending rings), and $\mathfrak{D}_{\mathbf{pFI}}(R) = \{RaR \mid a \in R\}$.

We observe that the same class $\mathfrak{K}$ of rings can be determined by more than one $\mathfrak{D}_{\mathfrak{K}}$ and P. For example, with the class **E** we can also use the set of closed right ideals of R for $\mathfrak{D}_{\mathbf{E}}(R)$ and take P to be the property that every right ideal is either essential in a right ideal generated by an idempotent, or P to be the property that every right ideal generated by an idempotent. Also we note that the class of right Rickart rings can also be characterized by $\mathfrak{D}_{\mathfrak{K}}(R) = \{r_R(a) \mid a \in R\}$, and P is the property that every right ideal is generated by an idempotent.

Lemma 8.1.6 (i) *Assume that T is a right ring of quotients of R. If $J_R \leq T_R$, then $\ell_R(J) = \ell_R(J \cap R)$.*

(ii) *Assume that T is a left ring of quotients of R. If ${}_RJ \leq {}_RT$, then $r_R(J) = r_R(J \cap R)$.*

Proof (i) Clearly, $\ell_R(J) \subseteq \ell_R(J \cap R)$. Let $a \in \ell_R(J \cap R)$ and suppose that there is $y \in J$ such that $ay \neq 0$. Since $R_R \leq^{\text{den}} T_R$, there is $r \in R$ such that $yr \in J \cap R$ and $ayr \neq 0$, a contradiction. Thus, $\ell_R(J) = \ell_R(J \cap R)$.

(ii) The proof is similar to that of part (i). □

We say that an overring T of a ring R is a *right intrinsic* (*ideal intrinsic*) *extension* of R if every nonzero right ideal (ideal) of T has a nonzero intersection with R. Note that if T is a right essential overring of R, then T is a right intrinsic extension of R. See [162] and [64] for more details on right intrinsic extensions.

Proposition 8.1.7 (i) *Let $\mathfrak{C}$ be a* **D-E** *class of rings, and let T be a right intrinsic extension of a ring R. Assume that for each $J \in \mathfrak{D}_{\mathfrak{C}}(T)$ there exists $e^2 = e \in R$ such that $eJ \subseteq J$ and $(J \cap R)_R \leq^{\text{ess}} eR_R$. Then $T \in \mathfrak{C}$.*

(ii) *Let $\mathfrak{C}$ be a* **D-E** *class of rings, and T be a right ring of quotients of R. Assume that $R \in \mathfrak{C}$. If $J \in \mathfrak{D}_{\mathfrak{C}}(T)$ implies $J \cap R \in \mathfrak{D}_{\mathfrak{C}}(R)$, then $T \in \mathfrak{C}$.*

Proof (i) We first note that $J = eJ \oplus (1-e)J$. Suppose that $(1-e)J \neq 0$. Then $0 \neq (1-e)J \cap R \subseteq J \cap R \subseteq eR$, a contradiction. Hence, $J = eJ \subseteq eT$. To show that $J_T \leq^{\text{ess}} eT_T$, we take $0 \neq ev \in eT$ with $v \in T$. Then $evT \cap R \neq 0$, hence $0 \neq evu \in R$ for some $u \in T$, and so $0 \neq evu \in eR$. Thus, there is $r \in R$ such that $0 \neq euvr \in J \cap R \subseteq J$. So $J_T \leq^{\text{ess}} eT_T$, therefore $T \in \mathfrak{C}$.

(ii) Let $J \in \mathfrak{D}_{\mathfrak{C}}(T)$. Since $J \cap R \in \mathfrak{D}_{\mathfrak{C}}(R)$ by assumption, there exists $e^2 = e \in R$ with $(J \cap R)_R \leq^{\text{ess}} eR_R$. Because $1 - e \in \ell_R(J \cap R)$, $(1-e)J = 0$ by Lemma 8.1.6. Hence, $J_R \leq eT_R$ and $eR_R \leq^{\text{ess}} eT_R$, so $J_R \leq^{\text{ess}} eT_R$. Thus, $J_T \leq^{\text{ess}} eT_T$ and therefore $T \in \mathfrak{C}$. □

A ring R is called *right finitely Σ-extending* if $R_R^{(n)}$ is extending for each positive integer n (cf. Exercise 6.1.18.1). A ring R is said to be *right uniform-extending* if every uniform right ideal of R is essential in a direct summand of R_R. The following result demonstrates that the right extending property transfers to right rings of quotients, while the right FI-extending property transfers to right intrinsic extensions.

Theorem 8.1.8 (i) *Assume that T is a right intrinsic extension of a ring R. If R_R is FI-extending, then so is T_T.*

(ii) *Let T be an ideal intrinsic extension of a ring R such that $\mathcal{B}(R) \subseteq \mathcal{B}(T)$. If R is semiprime and R is* (*right*) *FI-extending, then T is semiprime and* (*right*) *FI-extending.*

(iii) *Assume that T is a right ring of quotients of a ring R. If R_R is extending, then so is T_T.*

(iv) *Assume that T is a right ring of quotients of a ring R. If R_R is finitely Σ-extending, then so is T_T.*

(v) *Assume that T is a right ring of quotients of a ring R. If R_R is uniform-extending, then so is T_T.*

Proof (i) Let $J \trianglelefteq T$. Then $J \in \mathfrak{D}_{\mathbf{FI}}(T)$ and $J \cap R \in \mathfrak{D}_{\mathbf{FI}}(R)$. Because R_R is FI-extending, $(J \cap R)_R \leq^{\text{ess}} eR_R$ with $e^2 = e \in R$. From Proposition 8.1.7(i), T_T is FI-extending.

(ii) Clearly, T is semiprime. Let $0 \neq I \trianglelefteq T$. Then $(I \cap R)_R \leq^{\text{ess}} eR_R$ for some $e \in \mathcal{B}(R) \subseteq \mathcal{B}(T)$ by Theorem 3.2.37 and assumption. Similar to the proof of Proposition 8.1.7(i), $(1 - e)I = 0$, so $I = eI \subseteq eT$. We show that $I_T \leq^{\text{ess}} eT_T$. For this, we prove that $I_{eTe} \leq^{\text{ess}} eTe_{eTe}$. Say V is a nonzero ideal of eTe. Then V is an ideal of T, so $V \cap R \neq 0$. Hence $0 \neq V \cap R \subseteq eT \cap R = eR$, and thus $0 \neq (V \cap R) \cap (I \cap R) \subseteq V \cap I$ because $(I \cap R)_R \leq^{\text{ess}} eR_R$. Therefore, $I_{eTe} \leq^{\text{ess}} eTe_{eTe}$. As $e \in \mathcal{B}(T)$, $I_T \leq^{\text{ess}} eT_T$. So T is (right) FI-extending.

(iii) The proof follows from Proposition 8.1.7(ii) since the class $\mathbf{E}$ of right extending rings is a **D-E** class and $\mathfrak{D}_{\mathbf{E}}(R)$ is the set of all right ideals of R.

(iv) Let T be a right ring of quotients of a ring R and assume that R_R is finitely Σ-extending. Note that $\text{Mat}_n(R)$ is a right extending ring for every positive integer n (Exercise 6.1.18.1). So $\text{Mat}_n(T)$ is a right extending ring by part (iii) as $\text{Mat}_n(T)$ is a right ring of quotients of $\text{Mat}_n(R)$. Thus, T_T is a finitely Σ-extending.

(v) Let T be a right ring of quotients of R and assume that R_R is a uniform-extending. Say J is a uniform right ideal of T. Let $I = J \cap R$, and take nonzero elements x and y in I. Then $xT \cap yT \neq 0$. Say $xs = yt \neq 0$ with $s, t \in T$. As $R_R \leq^{\text{den}} T_R$, there is $r \in R$ such that $sr \in R$ and $xsr = ytr \neq 0$. Again since $R_R \leq^{\text{den}} T_R$, there exists $a \in R$ with $tra \in R$ and $ytra \neq 0$. So $sra \in R$, $tra \in R$, and $0 \neq xsra = ytra \in xR \cap yR$. Thus, I is a uniform right ideal of R. Hence the proof follows directly from Proposition 8.1.7(ii). □

Theorem 8.1.9 (i) *Assume that T is a right and left essential overring of a ring R. If $R \in$ **qB**, then $T \in$ **qB**.*

(ii) *Assume that T is a right essential overring of a ring R which is also a left ring of quotients of R. If $R \in$ **eqB**, then $T \in$ **eqB**.*

(iii) *Assume that T is a right essential overring of a ring R which is also a left ring of quotients of R. If $R \in$ **B**, then $T \in$ **B**.*

(iv) *Assume that T is a right and left ring of quotients of a ring R. If $R \in$ **eB**, then $T \in$ **eB**.*

Proof (i) Let R be quasi-Baer. Say $J \trianglelefteq T$ and let $I = J \cap R$. There exists $e^2 = e \in R$ with $r_R(I) = eR$. Let $t \in r_T(J)$.

If $(1-e)t \neq 0$, then there is $r \in R$ with $0 \neq (1-e)tr \in R$ as $R_R \leq^{\text{ess}} T_R$. We see that $I(1-e)tr \subseteq Itr \subseteq Jtr = 0$. Hence $(1-e)tr \in r_R(I) = eR$, a contradiction. Therefore, $(1-e)r_T(J) = 0$. Thus $r_T(J) \subseteq eT$. To show that $eT \subseteq r_T(J)$, assume on the contrary that there is $y \in J$ such that $ye \neq 0$. As T is a left essential overring of R, there is $s \in R$ with $0 \neq sye \in R$. Hence, $sye \in J \cap R = I$.

But $sye \in Ie = 0$, a contradiction. Thus, $Je = 0$ and so $r_R(J) = eT$. Therefore, T is quasi-Baer.

(ii) Assume that R is right essentially quasi-Baer. Say $J \trianglelefteq T$ and $I = J \cap R$. There exists $e^2 = e \in R$ such that $r_R(I)_R \leq^{\text{ess}} eR_R$. As in the proof of part (i), we obtain $r_T(J) \subseteq eT$. By Lemma 8.1.6(ii), $r_R(J) = r_R(I)$. Thus we have that $r_R(J)_R \leq^{\text{ess}} eR_R$. Since $r_T(J) \subseteq eT$, $r_T(J)_R \leq^{\text{ess}} eT_R$. Thus $r_T(J)_T \leq^{\text{ess}} eT_T$, so T is right essentially quasi-Baer.

(iii) Let R be Baer. Take $\emptyset \neq X \subseteq T$ and $J = TX$. Then $r_T(X) = r_T(J)$. We now set $I = J \cap R$. Then there exists $e^2 = e \in R$ such that $r_R(I) = eR$. First to show that $r_T(J) \subseteq eT$, suppose that there is $t \in r_T(J)$ with $(1-e)t \neq 0$.

Since $R_R \leq^{\text{ess}} T_R$, there is $r \in R$ with $0 \neq (1-e)tr \in R$. So

$$I(1-e)tr = Itr = 0,$$

hence $0 \neq (1-e)tr \in r_R(I) = eR$, a contradiction. Thus $r_T(J) \subseteq eT$.

If $ye \neq 0$ for some $y \in J$, then there is $s \in R$ with $sy \in R$ and $sye \neq 0$ as ${}_RR$ is dense in ${}_RT$, So $sy \in I$. Hence $0 \neq sye \in Ie = 0$, a contradiction. Thus $ye = 0$ for all $y \in J$, hence $e \in r_T(J)$. Therefore, $eT \subseteq r_T(J)$ and thus $r_T(X) = r_T(J) = eT$. So T is Baer.

(iv) Assume that R is right essentially Baer. Let $\emptyset \neq X \subseteq T$ and $J = TX$. Then $r_T(X) = r_T(J)$. Take $I = J \cap R$. There exists $e^2 = e \in R$ such that $r_R(I)_R$ is essential in eR_R.

We show that $r_T(J)_T \leq^{\text{ess}} eT_T$. For this, say $t \in r_T(J)$. If $(1-e)t \neq 0$, then since $R_R \leq^{\text{den}} T_R$, there exists $r \in R$ with $tr \in R$ and $(1-e)tr \neq 0$. But because $Itr \subseteq Jtr = 0$, $tr \in r_R(I)$. Hence

$$(1-e)tr \in (1-e)r_R(I) \subseteq (1-e)eR = 0,$$

a contradiction. So $r_T(J) \subseteq eT$. To see that $r_T(J)_T \leq^{\text{ess}} eT_T$, use the corresponding part of the proof in part (ii). Therefore T is right essentially Baer. □

As an application of Theorems 8.1.8 and 8.1.9, note that (by direct computation) $T_n(R)_{T_n(R)} \leq^{\text{den}} \text{Mat}_n(R)_{T_n(R)}$ (see Exercise 8.1.10.5). Hence for various conditions in Theorems 8.1.8 and 8.1.9, if the condition holds for $T_n(R)$, then it holds for $\text{Mat}_n(R)$. Proposition 8.1.7, Theorems 8.1.8, and 8.1.9 show that if R is a ring which belongs to a certain class (of rings) and S is a right essential overring of R in that class, then every other right essential overring of R which contains S as a subring, also belongs to that certain class, under some conditions. These results provide information related to Problem I.

Exercise 8.1.10

1. ([89, Birkenmeier, Park, and Rizvi]) Show that the ring R as in Example 7.1.13 is neither quasi-Baer nor right FI-extending, but R is right essentially quasi-Baer.
2. ([89, Birkenmeier, Park, and Rizvi]) For a field K, as in Example 7.3.13(i), let $T = K[x]/x^4K[x]$ and $\overline{x}$ be the image of x in T. Put $T = K + K\overline{x} + K\overline{x}^2 + K\overline{x}^3$ and $R = K + K\overline{x}^2 + K\overline{x}^3$ which is a subring of T. Then T_T is injective. Also T is a right essential overring of R. Prove that T_R is not FI-extending. (Hint: check with $\overline{x}^3R_R \trianglelefteq T_R$.)
3. ([89, Birkenmeier, Park, and Rizvi]) Show that if a ring R is Abelian and right extending, then so is $Q(R)$.
4. ([89, Birkenmeier, Park, and Rizvi]) Let T be a right and left ring of quotients of R. Show that if R is right semihereditary and $\text{Mat}_n(R)$ is orthogonally finite for every positive integer n, then T is right and left semihereditary.
5. Let R be a ring and n a positive integer. Show that $\text{Mat}_n(R)$ is a right ring of quotients of $T_n(R)$. Hence if P is a property that transfers from a ring to its right rings of quotients, then P transfers from $T_n(R)$ to $\text{Mat}_n(R)$ (see Theorems 8.1.8, 8.1.9, [4, Theorem 1 and Corollary 2], and [67, Theorem 3.5 and Corollary 3.6]).

8.2 Ring Hulls and Pseudo Ring Hulls

Motivated by the results of Sect. 8.1 and Chap. 7, we shall introduce and develop ring hull concepts in this section. These enable us to study Problem II mentioned in introduction of this chapter. After illustrating the ring hull notions via examples, we shall discuss some technical machinery which enables us to verify the existence of hulls for various **D-E** classes.

As a standing assumption in our considerations on hulls, for a given ring R, all right essential overrings of R are assumed to be contained as right R-modules in a fixed injective hull $E(R_R)$ of R_R and all right rings of quotients of R are assumed to be subrings of a fixed maximal right ring of quotients $Q(R)$ of R.

We begin with the following definition on various ring hulls.

Definition 8.2.1 Let R be a ring with $\ell_R(R) = 0$, but not necessarily with an identity element. Let $\mathfrak{K}$ denote a class of rings.

(i) The smallest right ring of quotients T of a ring R which belongs to $\mathfrak{K}$ is called the $\mathfrak{K}$ *absolute to* $Q(R)$ *right ring hull* of R (when it exists). We denote $T = \widehat{Q}_{\mathfrak{K}}(R)$.

(ii) The smallest right essential overring S of a ring R which belongs to $\mathfrak{K}$ is called the $\mathfrak{K}$ *absolute right ring hull* of R (when it exists). We denote $S = Q_{\mathfrak{K}}(R)$.

(iii) A minimal right essential overring of a ring R which belongs to $\mathfrak{K}$ is called a $\mathfrak{K}$ *right ring hull* of R (when it exists).

We remark that if R is a ring (not necessarily with identity), then any right R-module M_R has an injective hull $E(M_R)$ (see [153, Theorem 9, p. 19]). Further, if $Z(R_R) = 0$ for such a ring, then $Q(R) = E(R_R)$ (see [153, p. 69]). Next, we note that when $Q(R) = E(R_R)$, $\widehat{Q}_{\mathfrak{K}}(R) = Q_{\mathfrak{K}}(R)$. In particular, from Theorem 2.1.25, $Q_{\mathbf{qCon}}(R)$ exists whenever $Q(R) = E(R_R)$ (e.g., $Z(R_R) = 0$).

Since we are mostly dealing with the right-sided notions, we will drop the word "right" (from the preceding definition) in the future to make it easier on the reader. Thus, when the context is clear, we will use "$\mathfrak{K}$ absolute to $Q(R)$ ring hull" of R instead of "$\mathfrak{K}$ absolute to $Q(R)$ *right* ring hull" of R, etc.

The next example, taken from Theorems 7.2.1, 7.2.2, and their proofs, illustrates some examples of ring hulls defined in Definition 8.2.1.

Example 8.2.2 Let R, V, S, U, and T be as in Theorem 7.2.1. Then:

(i) All right FI-extending ring hulls of R are precisely: $(S, +, \circ_{(1,0)})$, $(S, +, \circ_{(1,2)})$, $(U, +, \circledcirc_1)$, $(U, +, \circledcirc_2)$, $(T, +, \diamond_1)$, and $(T, +, \diamond_2)$.

(ii) All right extending ring hulls of R are precisely: $(V, +, \bullet_1)$, $(V, +, \bullet_2)$, $(V, +, \bullet_3)$, $(V, +, \bullet_4)$, $(S, +, \circ_{(1,0)})$, and $(S, +, \circ_{(1,2)})$.

(iii) All right quasi-continuous ring hulls of R are precisely: $(S, +, \circ_{(1,0)})$ and $(S, +, \circ_{(1,2)})$.

(iv) All right continuous ring hulls of R are precisely: $(S, +, \circ_{(1,0)})$ and $(S, +, \circ_{(1,2)})$.

(v) All right self-injective ring hulls of R are precisely: $(S, +, \circ_{(1,0)})$ and $(S, +, \circ_{(1,2)})$.

The following example also illustrates Definition 8.2.1. In fact, it exhibits a ring R which has several isomorphic right FI-extending ring hulls, but R does not have a quasi-Baer ring hull.

Example 8.2.3 Let A, R, $\mathbb{E} = \mathbb{E}_R$, and T be as in Example 7.3.18. Then from Theorem 7.3.17, $\mathbb{E}$ has exactly p^2 compatible ring structures $(\mathbb{E}, +, \bullet_{(\alpha,\beta)})$, where $\alpha, \beta \in \mathrm{Soc}(A)$. These ring structures on $\mathbb{E}_R$ are isomorphic and they are QF. Also by Example 7.3.18, on T there are exactly p distinct compatible ring structures $(T, +, \diamond_{(0,\beta)})$ where $\beta \in \mathrm{Soc}(A)$ and $\diamond_{(0,\beta)}$ is the restriction of $\bullet_{(\alpha,\beta)}$ to T. Further, all compatible ring structures $(T, +, \diamond_{(0,\beta)})$, $\beta \in \mathrm{Soc}(A)$, on T are isomorphic. The rings $(T, +, \diamond_{(0,\beta)})$ are right FI-extending ring hulls of R by Example 7.3.18. Say $I = \begin{bmatrix} J(A) & 0 \\ 0 & 0 \end{bmatrix}$. Then I is a right ideal of each of R, $(T, +, \diamond_{(0,0)})$, and $(\mathbb{E}, +, \bullet_{(0,0)})$, respectively. We see that $r_R(I)$ is not generated by an idempotent of R, so R is not quasi-Baer. Also the right annihilator of I in $(T, +, \diamond_{(0,0)})$ (resp., $(\mathbb{E}, +, \bullet_{(0,0)})$) is not generated by an idempotent in $(T, +, \diamond_{(0,0)})$ (resp., $(\mathbb{E}, +, \bullet_{(0,0)})$). Thus, neither $(T, +, \diamond_{(0,0)})$ nor $(\mathbb{E}, +, \bullet_{(0,0)})$ is quasi-Baer. So R does not have a quasi-Baer ring hull.

Recall that $\mathbf{I}(R)$ and $\mathcal{B}(R)$ denote the set of all idempotents and the set of all central idempotents of a ring R, respectively. Let R be a ring. Then $R\mathcal{B}(Q(R))$, the subring of $Q(R)$ generated by R and $\mathcal{B}(Q(R))$, has been called the *idempotent closure* of R by Beidar and Wisbauer [42, p. 65]. In the following result, the Baer ring hull $Q_{\mathbf{B}}(R)$ is $R\mathcal{B}(Q(R))$ for a commutative semiprime ring R, is due to Mewborn [298].

Theorem 8.2.4 *Assume that R is a commutative semiprime ring. Then $Q_{\mathbf{B}}(R) = Q_{\mathbf{E}}(R) = Q_{\mathbf{qCon}}(R) = R\mathcal{B}(Q(R))$.*

Proof Say A is a commutative semiprime ring. Then A is reduced, so it is nonsingular by Theorem 1.2.20(ii). From Corollary 3.3.3, A is Baer if and only if A is extending. As A is commutative, A satisfies (C_3) condition. Thus, A is extending if and only if A is quasi-continuous. For the proof it is enough to show that $Q_{\mathbf{qCon}}(R) = R\mathcal{B}(Q(R))$. From Corollary 1.3.15, Theorem 2.1.25, and Proposition 2.1.32, $R\mathcal{B}(Q(R))$ is a quasi-continuous ring. Next, say S is a quasi-continuous (right) ring of quotients of R. Then again by Corollary 1.3.15, Theorem 2.1.25, and Proposition 2.1.32, $\mathcal{B}(Q(R)) \subseteq S$ as $Q(S) = Q(R)$. Thus $R\mathcal{B}(Q(R)) \subseteq S$. So $Q_{\mathbf{qCon}}(R) = R\mathcal{B}(Q(R))$. □

Theorem 8.2.5 *Assume that R is a regular right self-injective ring. Then $R = A \oplus B$ (ring direct sum), where A is a strongly regular ring and B is a ring generated by idempotents.*

Proof See [397, Theorems 2 and 4]. □

Theorem 8.2.6 *Let R be a right nonsingular ring and S the intersection of all right continuous right rings of quotients of R. Then $Q_{\mathbf{Con}}(R) = S$ and S is regular.*

Proof By Theorem 2.1.31 and Theorem 8.2.5, $Q(R) = A \oplus B$ (ring direct sum), where A is strongly regular and B is a ring generated by idempotents. Let T be a right continuous right ring of quotients of R. Since $Z(T_T) = 0$, T is regular by Corollary 2.1.30. Put $A = eQ(R)$ with $e \in \mathcal{B}(Q(R))$. From Theorem 2.1.25, $e \in T$ and $B \subseteq T$. So $T = (eQ(R) \cap T) \oplus B = eT \oplus B$ (ring direct sum).

Let $\{T_\alpha \mid \alpha \in \Lambda\}$ be the set of all right continuous right rings of quotients of R. Then $\cap T_\alpha = [\cap(eT_\alpha)] \oplus B$. In fact, note that $[\cap(eT_\alpha)] \oplus B \subseteq T_\alpha$ for each α as $T_\alpha = eT_\alpha \oplus B$. So $[\cap(eT_\alpha)] \oplus B \subseteq \cap T_\alpha$. Next, say $x \in \cap T_\alpha$ and $\beta \in \Lambda$. Then $x \in T_\beta = eT_\beta \oplus B$, hence $x = y + b$ with $y \in eT_\beta$ and $b \in B$. So $y = ey$ and $y = x - b \in (\cap T_\alpha) + B = \cap T_\alpha$ as $B \subseteq T_\alpha$ for every α. Hence, $y \in T_\alpha$

for every α, so $y = ey \in eT_\alpha$ for every α. Thus, $y \in \cap(eT_\alpha)$, and therefore $x = y + b \in [\cap(eT_\alpha)] \oplus B$. Hence, $\cap T_\alpha = [\cap(eT_\alpha)] \oplus B$.

Say $a \in \cap(eT_\alpha)$. There is a unique element $b \in eQ(R)$ with $a = aba$ and $b = bab$ as $eQ(R)$ is a strongly regular ring (see [264, Exercise 3, p. 36]). Also since each eT_α is strongly regular, there exists $b_\alpha \in eT_\alpha \subseteq eQ(R)$ such that $a = ab_\alpha a$ and $b_\alpha = b_\alpha ab_\alpha$, for each α. By the uniqueness of b, $b = b_\alpha \in eT_\alpha$ for each α. Hence, $b \in \cap(eT_\alpha)$, so $\cap(eT_\alpha)$ is a strongly regular ring.

As B is regular, $\cap T_\alpha = [\cap(eT_\alpha)] \oplus B$ is regular. From $\mathbf{I}(\cap T_\alpha) = \mathbf{I}(Q(R))$, $\cap T_\alpha$ is right quasi-continuous by Theorem 2.1.25. So, $\cap T_\alpha$ is right continuous. □

A ring is called *right duo* if every right ideal is an ideal. The next result shows the existence of the right duo absolute ring hull for a right Ore domain.

Proposition 8.2.7 *If R is a right Ore domain, then R has a right duo absolute ring hull.*

Proof Clearly, $Q(R)$ is right duo. Let S be the intersection of all right duo right rings of quotients of R. Let T and U be right duo right rings of quotients of R. Say $s, x \in S$ with $x \neq 0$. Then there are $t \in T$ and $u \in U$ with $sx = xt = xu$. Hence $x(t-u) = 0$, so $t = u$ and t (or u) $\in T \cap U$. As T and U are arbitrary right duo right rings of quotients of R, $t \in S$ and so $sx = xt \in xS$. Hence, S is the right duo absolute ring hull of R. □

Theorem 8.2.4 and the construction of $Q_{\mathbf{qCon}}(R)$ by Theorem 2.1.25 suggest how to design a "hull" of R by adjoining a certain subset of $E(R_R)$ to R. This leads to the notion of a pseudo ring hull which we define next. To define pseudo ring hulls in Definition 8.2.8, for a **D-E** class $\mathfrak{C}$, we *fix* $\mathfrak{D}_{\mathfrak{C}}(R)$ for the class $\mathfrak{C}$ (e.g., for $\mathbf{E}$, we fix $\mathfrak{D}_{\mathbf{E}} = \{I \mid I_R \leq R_R\}$ rather than $\{J \mid J_R \text{ is closed in } R_R\}$). Define

$$\delta_{\mathfrak{C}}(R) = \{e \in \mathbf{I}(\mathrm{End}(E(R_R)) \mid I_R \leq^{\mathrm{ess}} eE(R_R) \text{ for some } I \in \mathfrak{D}_{\mathfrak{C}}(R)\}$$

and $\delta_{\mathfrak{C}}(R)(1) = \{e(1) \mid e \in \delta_{\mathfrak{C}}(R)\}$. For example,

$$\delta_{\mathbf{FI}}(R) = \{e \in \mathbf{I}(\mathrm{End}(E(R_R)) \mid I_R \leq^{\mathrm{ess}} eE(R_R) \text{ for some } I \trianglelefteq R\}$$

because $\mathfrak{D}_{\mathbf{FI}}(R)$ is the set of all ideals of R.

We next generate a right essential overring in a class $\mathfrak{C}$ from a base ring R and $\delta_{\mathfrak{C}}(R)$. By using an equivalence relation, say ρ on $\delta_{\mathfrak{C}}(R)$, we reduce the size of the subset of idempotents needed to generate a right essential overring of R in $\mathfrak{C}$. For this, we consider $\delta_{\mathfrak{C}}^{\rho}(R)$, which is a set of representatives of all equivalence classes of ρ, and let $\delta_{\mathfrak{C}}^{\rho}(R)(1) = \{h(1) \in E(R_R) \mid h \in \delta_{\mathfrak{C}}^{\rho}(R)\}$.

Recall that $\langle X \rangle_A$ denotes the subring of a ring A generated by a subset X of A (see 1.1.2).

Definition 8.2.8 Let S be a right essential overring of R.

(i) If $\delta_{\mathfrak{C}}(R)(1) \subseteq S$ and $\langle R \cup \delta_{\mathfrak{C}}(R)(1)\rangle_S \in \mathfrak{C}$, then we put

$$\langle R \cup \delta_{\mathfrak{C}}(R)(1)\rangle_S = R(\mathfrak{C}, S).$$

If $S = R(\mathfrak{C}, S)$, then S is called a $\mathfrak{C}$ *pseudo right ring hull of R*.

(ii) If $\delta^{\rho}_{\mathfrak{C}}(R)(1) \subseteq S$ and $\langle R \cup \delta^{\rho}_{\mathfrak{C}}(R)(1)\rangle_S \in \mathfrak{C}$, then we put

$$\langle R \cup \delta^{\rho}_{\mathfrak{C}}(R)(1)\rangle_S = R(\mathfrak{C}, \rho, S).$$

If $S = R(\mathfrak{C}, \rho, S)$, then S is called a $\mathfrak{C}$ ρ *pseudo right ring hull of R*.

If $\delta_{\mathfrak{C}}(R)(1) \subseteq Q(R)$ and S is a right essential overring of R such that $R(\mathfrak{C}, S)$ exists, then $R(\mathfrak{C}, S) = R(\mathfrak{C}, Q(R))$ from Proposition 7.1.11.

For example, assume that $Q(R) = E(R_R)$. Then $Q_{\mathbf{qCon}}(R)$ exists, and we see that $Q_{\mathbf{qCon}}(R) = R(\mathbf{qCon}, Q(R))$.

As we are usually using the right-sided notions, we will drop the word "right" in the preceding definition. Thus we will call "$\mathfrak{C}$ pseudo *right* ring hull of R" just "$\mathfrak{C}$ pseudo ring hull of R", etc.

The next examples illustrate Definitions 8.2.1 and 8.2.8. They show that neither $\mathfrak{C}$ ring hulls nor $\mathfrak{C}$ ρ pseudo ring hulls are unique.

Example 8.2.9 In this example, we see that the intersection of all right FI-extending ring hulls is not necessarily a right FI-extending absolute ring hull. Further, it is shown that a right FI-extending ring hull may not be unique even up to isomorphism (cf. Example 8.2.3). Let F be a field and as in Example 3.2.39, we put

$$R = \left\{ \begin{bmatrix} a & 0 & x \\ 0 & a & y \\ 0 & 0 & c \end{bmatrix} \mid a, c, x, y \in F \right\} \cong \begin{bmatrix} F & F \oplus F \\ 0 & F \end{bmatrix}.$$

Then by [262, Example 13.26(5)], R is right nonsingular and $Q(R) = \mathrm{Mat}_3(F)$.

(i) Let $H_1 = \left\{ \begin{bmatrix} a & 0 & x \\ 0 & b & y \\ 0 & 0 & c \end{bmatrix} \mid a, b, c, x, y \in F \right\} \cong \begin{bmatrix} F \oplus F & F \oplus F \\ 0 & F \end{bmatrix}$, and let

$$H_2 = \left\{ \begin{bmatrix} a+b & a & x \\ 0 & b & y \\ 0 & 0 & c \end{bmatrix} \mid a, b, c, x, y \in F \right\}.$$

Note that H_1 and H_2 are subrings of $\mathrm{Mat}_3(F)$. Define $\phi : H_1 \to H_2$ by

$$\phi \begin{bmatrix} a & 0 & x \\ 0 & b & y \\ 0 & 0 & c \end{bmatrix} = \begin{bmatrix} a & a-b & x-y \\ 0 & b & y \\ 0 & 0 & c \end{bmatrix}.$$

Then ϕ is a ring isomorphism. The ring R is not right FI-extending (see Example 3.2.39), but H_1 is right FI-extending by Corollary 5.6.11. Thus H_2 is right FI-extending because $H_1 \cong H_2$.

Let $F = \mathbb{Z}_2$. Then there is no proper intermediate ring between R and H_1, also between R and H_2. Thus, H_1 and H_2 are right FI-extending ring hulls of R. Since $H_1 \cap H_2 = R$, the intersection of right FI-extending ring hulls is not a right FI-extending absolute ring hull.

(ii) Assume that $F = \mathbb{Z}_2$. Consider

$$H_3 = \left\{ \begin{bmatrix} a+b & b & x \\ b & a & y \\ 0 & 0 & c \end{bmatrix} \mid a, b, c, x, y \in F \right\}.$$

The ring H_3 is right FI-extending from Corollary 5.6.11. Also H_3 is a right FI-extending ring hull of R because there is no proper intermediate ring between R and H_3. Further, $\mathrm{Tdim}(H_1) = 3$, but $\mathrm{Tdim}(H_3) = 2$. Thus $H_3 \not\cong H_1$.

(iii) From Theorem 5.6.5 (see also Example 3.2.39), $R = Q_{\mathbf{qB}}(R)$. Also we see that $R(\mathbf{FI}, Q(R)) = \begin{bmatrix} F & F & F \\ F & F & F \\ 0 & 0 & F \end{bmatrix} \neq \mathrm{Mat}_3(F) = Q_{\mathbf{qCon}}(R) = Q_{\mathbf{Con}}(R)$.

Example 8.2.10 There is a right nonsingular ring which has an infinite number of right FI-extending ρ pseudo ring hulls. Furthermore, none of these pseudo ring hulls is a right FI-extending ring hull, for some equivalence relation ρ on $\delta_{\mathbf{FI}}(R)$. Take $R = T_2(\mathbb{Z})$. Then R is right FI-extending from Theorem 5.6.19. Say e_{ij} is the matrix in R with 1 in (i, j)-position and 0 elsewhere. We note that $\{0, 1_R\} \cup \{e_{11} + qe_{12} \mid q \in \mathbb{Q}\} \subseteq \delta_{\mathbf{FI}}(R)$. Define an equivalence relation ρ on $\delta_{\mathbf{FI}}(R)$ such that: $e \rho f$ if and only if $e = fe$ and $f = ef$. Then each $\delta_{\mathbf{FI}}^{\rho}(R)$ contains $\{0, 1_R, e_{11} + qe_{12}\}$, where $q \in \mathbb{Q}$ is fixed.

Suppose that $q \notin \mathbb{Z}$. Then $\langle R \cup \delta_{\mathbf{FI}}^{\rho}(R)(1)\rangle_{Q(R)}$, the subring of $Q(R)$ generated by $R \cup \delta_{\mathbf{FI}}^{\rho}(R)(1)$, is a right FI-extending ρ pseudo ring hull of R because by Theorem 8.1.8(i) $\langle R \cup \delta_{\mathbf{FI}}^{\rho}(R)(1)\rangle_{Q(R)}$ is right FI-extending. Therefore, we obtain that $R(\mathbf{FI}, \rho, Q(R)) = \langle R \cup \delta_{\mathbf{FI}}^{\rho}(R)(1)\rangle_{Q(R)}$. But $\langle R \cup \delta_{\mathbf{FI}}^{\rho}(R)(1)\rangle_{Q(R)}$ is not a right FI-extending ring hull of R as $\langle R \cup \delta_{\mathbf{FI}}^{\rho}(R)(1)\rangle_{Q(R)} \neq R = Q_{\mathbf{FI}}(R)$.

We introduce two new equivalence relations which will be helpful.

Definition 8.2.11 (i) We define an equivalence relation α on $\delta_{\mathfrak{C}}(R)$ by $e \alpha f$ if $e = fe$ and $f = ef$.

(ii) We define an equivalence relation β on $\delta_{\mathfrak{C}}(R)$ by $e \beta f$ if there exists $I_R \leq R_R$ such that $I_R \leq^{\mathrm{ess}} eE(R_R)$ and $I_R \leq^{\mathrm{ess}} fE(R_R)$.

The equivalence relation α was used as ρ in Example 8.2.10. Note that for e, f in $\delta_{\mathfrak{C}}(R)$, $e \alpha f$ implies $e \beta f$. If $Z(R_R) = 0$, then $\alpha = \beta$.

Lemma 8.2.12 *Let R be a ring and $H = \mathrm{End}(E(R_R))$.*

(i) *If T is a right essential overring of R, then for $e \in \mathbf{I}(T)$, there exists $c \in \mathbf{I}(H)$ such that $c|_T \in \mathrm{End}(T_T)$ and $c(1) = e$.*

(ii) *For $b \in \mathbf{I}(H)$, if $b(1) \in Q(R)$, then $b(1) \in \mathbf{I}(Q(R))$.*

Proof (i) Note that $E(T_R) = E(eT_R) \oplus E((1-e)T_R)$. Let c be the canonical projection from $E(T_R)$ onto $E(eT_R)$. Then $c(t) = c(et) + c((1-e)t) = et$ for $t \in T$. Hence $c(1) = e$. If $s \in T$, then $c(ts) = ets = c(t)s$. So $c|_T \in \mathrm{End}(T_T)$.

(ii) As $E(R_R)$ is an $(H, Q(R))$-bimodule, each element of H is a $Q(R)$-homomorphism. So if $b(1) \in Q(R)$, then $b(1) = b(b(1)) = b(1b(1)) = b(1)b(1)$, thus $b(1) \in \mathbf{I}(Q(R))$. □

Proposition 8.2.13 *Let $\mathfrak{C}$ be a* **D-E** *class of rings, and let T be a right ring of quotients of R, δ be some $\delta_{\mathfrak{C}}^{\alpha}(R)$ such that $\delta(1) \subseteq T$. Take $S = \langle R \cup \delta(1) \rangle_T$. Suppose that for each $J \in \mathfrak{D}_{\mathfrak{C}}(S)$ there is $I \in \mathfrak{D}_{\mathfrak{C}}(R)$ with $I_R \leq^{\mathrm{ess}} J_R$. Then $S = R(\mathfrak{C}, \alpha, T)$, which is a $\mathfrak{C}$ α pseudo ring hull of R.*

Proof Since $\delta(1) \subseteq Q(R)$, $\delta(1) \subseteq \mathbf{I}(S)$ by Lemma 8.2.12(ii). To show that $S = R(\mathfrak{C}, \alpha, T)$, we only need to see that $S \in \mathfrak{C}$. For this, let $J \in \mathfrak{D}_{\mathfrak{C}}(S)$. By assumption, there exists $I \in \mathfrak{D}_{\mathfrak{C}}(R)$ satisfying $I_R \leq^{\mathrm{ess}} J_R$. Therefore we have that $I_R \leq^{\mathrm{ess}} J_R \leq^{\mathrm{ess}} E(J_R) = eE(R_R)$ for some $e \in \mathbf{I}(H)$, where $H = \mathrm{End}(E(R_R))$. Hence, $e \in \delta_{\mathfrak{C}}(R)$, so there exists $f \in \delta$ satisfying $eE(R_R) = fE(R_R)$. Thus we get $J_R \leq^{\mathrm{ess}} fE(R_R)$ and so $J_R \leq^{\mathrm{ess}} fS_R$.

Note that $f \in \mathrm{End}(E_R) = \mathrm{End}(E_{Q(R)})$ by the proof of Theorem 2.1.31, where $E = E(R_R)$. So $J_R \leq^{\mathrm{ess}} fS_R = f(1)S_R$ because S is a subring of $Q(R)$. Hence, $J_S \leq^{\mathrm{ess}} f(1)S_S$ and $f(1)^2 = f(1) \in S$, so $S \in \mathfrak{C}$. □

Proposition 8.2.14 *Let $\mathfrak{C}$ be a* **D-E** *class of rings, and let T be a right essential overring of R. Assume that for each $I \in \mathfrak{D}_{\mathfrak{C}}(R)$ there exists $e \in \mathbf{I}(T)$ satisfying $I_R \leq^{\mathrm{ess}} eT_R$. Then there exists $\delta_{\mathfrak{C}}^{\beta}(R)$ such that, for each $c \in \delta_{\mathfrak{C}}^{\beta}(R)$, $c|_T \in \mathrm{End}(T_T)$ and $c(1) \in \mathbf{I}(T)$.*

Proof Let $b \in \delta_{\mathfrak{C}}(R)$. Then there is $I \in \mathfrak{D}_{\mathfrak{C}}(R)$ with $I_R \leq^{\mathrm{ess}} bE(R_R)$. By assumption, $I_R \leq^{\mathrm{ess}} eT_R$ for some $e \in \mathbf{I}(T)$. From Lemma 8.2.12(i), there is $c^2 = c$ in $\mathrm{End}(E(R_R))$ such that $c|_T \in \mathrm{End}(T_T)$ and $c(1) = e$.

We note that $I_R \leq^{\mathrm{ess}} eT_R = c(1)T_R = cT_R$, so $I_R \leq^{\mathrm{ess}} cE(R_R)$. Thus, $b \beta c$. □

The next result will be used to find right extending right rings of quotients of certain rings in Sect. 9.1.

Theorem 8.2.15 *Let R be a ring such that $\alpha = \beta$ (e.g., $Z(R_R) = 0$), and let T be a right ring of quotients of R. Then the following are equivalent.*

(i) *T is right extending.*
(ii) *There exists a right extending α pseudo ring hull $R(\mathbf{E}, \alpha, Q(R))$ and it is a subring of T.*

Proof (i)⇒(ii) Assume that T is right extending. To apply Proposition 8.2.14, let $I \in \mathfrak{D}_{\mathbf{E}}(R)$, that is, $I_R \leq R_R$. By the proof of Lemma 8.1.3(ii), $I_R \leq^{\mathrm{ess}} IT_R$. Take $J = IT$. Since T is right extending, there is $e \in \mathbf{I}(T)$ with $J_T \leq^{\mathrm{ess}} eT_T$. Thus $J_R \leq^{\mathrm{ess}} eT_R$ by Lemma 8.1.3(i), so $I_R \leq^{\mathrm{ess}} J_R \leq^{\mathrm{ess}} eT_R$.

By Proposition 8.2.14, there is $\delta_{\mathbf{E}}^{\beta}(R)$ with $c|_T \in \mathrm{End}(T_T)$ and $c(1) \in \mathbf{I}(T)$ for each $c \in \delta_{\mathbf{E}}^{\beta}(R)$. Take $S = \langle R \cup \delta_{\mathbf{E}}^{\beta}(R)(1)\rangle_T = \langle R \cup \delta_{\mathbf{E}}^{\beta}(R)(1)\rangle_{Q(R)}$. Now for each $K_S \le S_S$, $(K \cap R)_R \le R_R$ and $(K \cap R)_R \le^{\text{ess}} K_R$. Since $Z(R_R) = 0$, $\alpha = \beta$ (Exercise 8.2.16.3), and hence $S = \langle R \cup \delta_{\mathbf{E}}^{\alpha}(R)(1)\rangle_{Q(R)} = R(\mathbf{E}, \alpha, Q(R))$ by Proposition 8.2.13. Clearly S is a subring of T.

(ii)⇒(i) The proof follows from Theorem 8.1.8(iii). □

Exercise 8.2.16

1. ([89, Birkenmeier, Park, and Rizvi]) Assume that A is a commutative local QF-ring such that $J(A) \neq 0$. In this case, we take $S_0 = \begin{bmatrix} A & J(A) \\ 0 & A \end{bmatrix}$, $S_1 = T_2(A)$, $S_2 = \begin{bmatrix} A & A \\ J(A) & A \end{bmatrix}$, and $S_3 = \mathrm{Mat}_2(A)$. Prove that the following hold true.
   (i) $S_0 \subseteq S_1 \subseteq S_2 \subseteq S_3$ is a chain of subrings of S_3 where S_i, $1 \le i \le 3$ is a right essential overring of its predecessor.
   (ii) $S_{0S_0} \le^{\text{ess}} S_{1S_0}$, $S_{1S_1} \le^{\text{den}} S_{3S_1}$, but S_{0S_0} is not essential in S_{2S_0}.
   (iii) S_1 is a right FI-extending ring hull of S_0, $S_2 = Q_{\mathbf{E}}(S_1)$, and also $S_3 = Q_{\mathbf{SI}}(S_1) = Q_{\mathbf{SI}}(S_2)$, where **SI** is the class of right self-injective rings.
2. ([89, Birkenmeier, Park, and Rizvi]) Assume that $\mathfrak{U}$ denotes the class of rings, $\{R \mid R \cap \mathbf{U}(Q(R)) = \mathbf{U}(R)\}$, where $\mathbf{U}(-)$ is the set of invertible elements of a ring. We let $R_1 = \langle R \cup \{q \in \mathbf{U}(Q(R)) \mid q^{-1} \in R\}\rangle_{Q(R)}$. Let i and j be ordinal numbers. When $j = i + 1$, put
$$R_j = \langle R_i \cup \{q \in \mathbf{U}(Q(R)) \mid q^{-1} \in R_i\}\rangle_{Q(R)}.$$
If j is a limit ordinal, let $R_j = \cup_{i<j} R_i$. Prove the following.
   (i) $\widehat{Q}_{\mathfrak{U}}(R)$ exists and $\widehat{Q}_{\mathfrak{U}}(R) = R_j$ for any j with $|j| > |Q(R)|$.
   (ii) If R is a right Ore ring, then $\widehat{Q}_{\mathfrak{U}}(R) = Q^r_{c\ell}(R)$. Thus $\widehat{Q}_{\mathfrak{U}}(R)$ is a ring hull that coincides with $Q^r_{c\ell}(R)$ when R is right Ore.
3. Let α and β be as in Definition 8.2.11. Show that $\alpha = \beta$ if $Z(R_R) = 0$.
4. ([89, Birkenmeier, Park, and Rizvi]) Let R be the ring in Example 8.2.9. Show that $\cap_{\alpha} R(\mathbf{FI}, \alpha, Q(R)) = T_3(F)$.
5. ([89, Birkenmeier, Park, and Rizvi]) Let T be a right ring of quotients of a ring R and assume that $IT \trianglelefteq T$ for any $I \trianglelefteq R$. Prove that $T \in \mathbf{FI}$ if and only if there exists an $R(\mathbf{FI}, \beta, Q(R))$ which is a subring of T.
6. ([89, Birkenmeier, Park, and Rizvi]) Assume that W is a local ring and V is a subring of W with $J(W) \subseteq V$. Let $R = \begin{bmatrix} V & W \\ 0 & W \end{bmatrix}$, $S = \begin{bmatrix} V & W \\ J(W) & W \end{bmatrix}$, and $T = \mathrm{Mat}_2(W)$. Prove the following.
   (i) For each $e \in \mathbf{I}(T)$, there exists $f \in \mathbf{I}(S)$ such that $e \,\alpha\, f$.
   (ii) $S \in \mathbf{E}$ if and only if $T \in \mathbf{E}$ if and only if $S = R(\mathbf{E}, \rho, T)$ for some ρ.
   (iii) If W is right self-injective, then $S = R(\mathbf{E}, \alpha, T)$.
   (iv) If W is right self-injective, then $Q_{\mathbf{qCon}}(R) = R(\mathbf{E}, T) = T$.
   (v) $R \in \mathbf{FI}$ if and only if $W \in \mathbf{FI}$.

7. ([89, Birkenmeier, Park, and Rizvi]) Let W be a local ring and V be a subring of W. Take $R = \begin{bmatrix} V & W \\ 0 & W \end{bmatrix}$. Show that the following are equivalent.
   (i) R is right extending.
   (ii) $T_2(W)$ is right extending.
   (iii) W is a division ring.
8. ([89, Birkenmeier, Park, and Rizvi]) Assume that A is a right FI-extending ring and $W = \oplus_{i=1}^{n} A_i$, where $A_i = A$ for each i. Let D be the set of all $(a_1, \ldots, a_n) \in W$ such that $a_i = a \in A$ for all $i = 1, \ldots, n$. Say S is a subring of W containing D. Prove that the ring $H = \begin{bmatrix} W & W \\ 0 & A \end{bmatrix}$ is a right FI-extending ring hull of the ring $R = \begin{bmatrix} S & W \\ 0 & A \end{bmatrix}$.
9. ([89, Birkenmeier, Park, and Rizvi]) Assume that R is a ring such that $Q(R)$ is Abelian. Prove the following.
   (i) $\widehat{Q}_{\mathbf{E}}(R) = \widehat{Q}_{\mathbf{qCon}}(R) = R\mathcal{B}(Q(R))$ if and only if $Q(R)$ is right extending.
   (ii) Let R be a right Ore ring such that $r_R(x) = 0$ implies $\ell_R(x) = 0$ for $x \in R$ and $Z(R_R)$ has finite right uniform dimension. Then $Q(R)$ is right extending if and only if $\widehat{Q}_{\mathbf{Con}}(R)$ exists and $\widehat{Q}_{\mathbf{Con}}(R) = H_1 \oplus H_2$ (ring direct sum), where H_1 is a right continuous strongly regular ring and H_2 is a direct sum of right continuous local rings.
10. ([89, Birkenmeier, Park, and Rizvi]) Let R be a commutative ring. Prove the following.
   (i) If R or $Q^r_{c\ell}(R)$ is extending, then $\widehat{Q}_{\mathbf{Con}}(R) = Q^r_{c\ell}(R)$.
   (ii) If $Z(R_R) = 0$, then $\widehat{Q}_{\mathrm{Con}}(R)$ is the intersection of all regular right rings of quotients T of R such that $\mathcal{B}(Q(R)) \subseteq T$.

8.3 Idempotent Closure Classes and Ring Hulls

This section is mainly devoted to discussions and study of Problems I and II mentioned in the introduction of this chapter. As $E(R_R)$ is extending, for each right ideal I of R there exists $e^2 = e \in \mathrm{End}(E(R_R))$ such that $I_R \leq^{\mathrm{ess}} eE(R_R)$. Furthermore, in many cases $Q(R) = E(R_R)$ (e.g., when $Z(R_R) = 0$). So one may expect that $Q(R)$ would satisfy the extending property for a certain subset of the set of right ideals of R.

We let $\mathfrak{D}_{\mathbf{IC}}(R) = \{I \trianglelefteq R \mid I \cap \ell_R(I) = 0 \text{ and } \ell_R(I) \cap \ell_R(\ell_R(I)) = 0\}$. In Theorem 8.3.8, we show that $I \in \mathfrak{D}_{\mathbf{IC}}(R)$ if and only if there exists e in $\mathcal{B}(Q(R))$ such that $I_R \leq^{\mathrm{den}} eQ(R)_R$. This result motivates the definition of the idempotent closure class of rings, we shall consider in this section, denoted by **IC**. This class of rings is a **D-E** class for which $\widehat{Q}_{\mathbf{IC}}(R) = \langle R \cup \mathcal{B}(Q(R)) \rangle_{Q(R)}$ (see Theorem 8.3.11). Thus this hull exists for every ring (not necessarily with identity) for which $Q(R)$ exists (i.e., when $\ell_R(R) = 0$). The set $\mathfrak{D}_{\mathbf{IC}}(R)$ forms a sublattice of the lattice of ideals of R and is quite large, in general. In fact, if R is semiprime, then

$\mathfrak{D}_{\mathbf{IC}}(R)$ is the full lattice of ideals of R. From this if R is a semiprime ring, then $\widehat{Q}_{\mathbf{FI}}(R) = \widehat{Q}_{\mathbf{qB}}(R) = \widehat{Q}_{\mathbf{eqB}}(R) = \langle R \cup \mathcal{B}(Q(R)) \rangle_{Q(R)}$. Further, if R is a semiprime ring with identity, then $\widehat{Q}_{\mathbf{FI}}(R) = R(\mathbf{FI}, Q(R))$ and $\widehat{Q}_{\mathbf{eqB}}(R) = R(\mathbf{eqB}, Q(R))$ (see Theorem 8.3.17).

This result demonstrates that the semiprime condition of a ring R overcomes the somewhat chaotic situation we encountered in Examples 8.2.2, 8.2.3, 8.2.9, and 8.2.10 by providing a unique ring hull which agrees with its pseudo ring hulls. Next we consider the transfer of algebraic information between R and $\langle R \cup \mathcal{B}(Q(R)) \rangle_{Q(R)}$ in terms of prime ideals, various radicals, regularity conditions, and so on (see Problem I). We shall see that for a semiprime ring R with identity, $\widehat{Q}_{\mathbf{pqB}}(R)$, $\widehat{Q}_{\mathbf{pFI}}(R)$, and $\widehat{Q}_{\mathbf{fgFI}}(R)$ all exist and are equal to each other. Also the transfer of algebraic information between R and these various hulls will also be discussed. Finally, we shall apply these results to obtain a proper generalization of Rowen's well-known result: Let R be a semiprime PI-ring. Then $\mathrm{Cen}(R) \cap I \neq 0$ for any $0 \neq I \trianglelefteq R$ (Theorem 3.2.16).

Throughout this section, R does not necessarily have an identity unless mentioned otherwise. However, we assume that $\ell_R(R) = 0$ to guarantee the existence of $Q(R)$ which has an identity (see [395]).

Definition 8.3.1 (i) Let R be a ring. We recall that $\mathfrak{D}_{\mathbf{IC}}(R)$ is the set of all ideals of R such that $I \cap \ell_R(I) = 0$ and $\ell_R(I) \cap \ell_R(\ell_R(I)) = 0$.

(ii) A ring R is called an **IC**-*ring* if for each $I \in \mathfrak{D}_{\mathbf{IC}}(R)$ there exists $e^2 = e \in R$ such that $I_R \leq^{\mathrm{ess}} eR_R$. The class of **IC**-rings is denoted by **IC** and is called the *idempotent closure class*. Thus, **IC** is a **D-E** class.

If a ring R with identity is right FI-extending, then $R \in \mathbf{IC}$. The set $\mathfrak{D}_{\mathbf{IC}}(R)$ was studied by Johnson [236] and denoted by $\mathfrak{F}'(R)$. While Propositions 8.3.2 and 8.3.3 provide examples of $\mathfrak{D}_{\mathbf{IC}}(R)$ when R is right nonsingular or semiprime, we shall see from Theorem 8.3.8 that $R \cap eQ(R) \in \mathfrak{D}_{\mathbf{IC}}(R)$ for any $e \in \mathcal{B}(Q(R))$. Also Theorem 8.3.11(ii) characterizes the **IC** class of rings.

Proposition 8.3.2 *If $Z(R_R) = 0$, then $\mathfrak{D}_{\mathbf{IC}}(R) = \{I \trianglelefteq R \mid I \cap \ell_R(I) = 0\}$.*

Proof Assume that $I \trianglelefteq R$ such that $I \cap \ell_R(I) = 0$. Say J_R is a complement of I_R in R_R. Then $(I \oplus J)_R \leq^{\mathrm{ess}} R_R$. Now $JI \subseteq J \cap I = 0$, thus $J \subseteq \ell_R(I)$. Therefore $(I \oplus \ell_R(I))_R \leq^{\mathrm{ess}} R_R$. If $x(I \oplus \ell_R(I)) = 0$, then $x = 0$ because $Z(R_R) = 0$. Hence $\ell_R(I \oplus \ell_R(I)) = \ell_R(I) \cap \ell_R(\ell_R(I)) = 0$. □

Proposition 8.3.3 (i) *A ring R is semiprime if and only if $\mathfrak{D}_{\mathbf{IC}}(R)$ is precisely the set of all ideals of R.*

(ii) *If $e \in \mathbf{S}_\ell(R)$, then $eR \in \mathfrak{D}_{\mathbf{IC}}(R)$ if and only if $e \in \mathcal{B}(R)$.*

(iii) *Let P be a prime ideal of R. Then $P \in \mathfrak{D}_{\mathbf{IC}}(R)$ if and only if $P \cap \ell_R(P) = 0$.*

(iv) *Let P be a prime ideal of R and $P \in \mathfrak{D}_{\mathbf{IC}}(R)$. If $I \trianglelefteq R$ such that $P \subseteq I$, then $I \in \mathfrak{D}_{\mathbf{IC}}(R)$.*

(v) *If $I \trianglelefteq R$ such that $\ell_R(I) \cap P(R) = 0$, then $I \in \mathfrak{D}_{\mathbf{IC}}(R)$.*

(vi) *If $Z(R_R) = 0$ and $I \trianglelefteq R$ such that $I \cap P(R) = 0$, then $I \in \mathfrak{D}_{\mathbf{IC}}(R)$.*

Proof (i) Assume that R is a semiprime ring. Let $I \trianglelefteq R$. Since R is semiprime and $(I \cap \ell_R(I))^2 = 0$, $I \cap \ell_R(I) = 0$. Now $\ell_R(I) \cap \ell_R(\ell_R(I)) = 0$ because $\ell_R(I) \trianglelefteq R$. So $\mathfrak{D}_{\mathbf{IC}}(R)$ is the set of all ideals of R. Conversely, assume that $\mathfrak{D}_{\mathbf{IC}}(R)$ is the set of all ideals of R. Let $I \trianglelefteq R$ with $I^2 = 0$. Then $I \subseteq \ell_R(I)$. As $I \in \mathfrak{D}_{\mathbf{IC}}(R)$, $I \cap \ell_R(I) = 0$ and so $I = 0$. Hence, R is semiprime.

(ii)–(vi) Exercise. □

Let R be a ring and $I \trianglelefteq R$ with $I \cap \ell_R(I) = 0$. As $I\ell_R(I) \subseteq I \cap \ell_R(I) = 0$, so $I \subseteq \ell_R(\ell_R(I))$. The next lemma will be used in the sequel. We note that every ideal in a semiprime ring satisfies all of these conditions.

Lemma 8.3.4 *Assume that $I \trianglelefteq R$ with $I \cap \ell_R(I) = 0$. Then the following are equivalent.*

(i) $\ell_R(I) \cap \ell_R(\ell_R(I)) = 0$.
(ii) $\ell_R(I \oplus \ell_R(I)) = 0$.
(iii) $(I \oplus \ell_R(I))_R \leq^{\text{den}} R_R$.
(iv) $I_R \leq^{\text{den}} \ell_R(\ell_R(I))_R$.
(v) $I_R \leq^{\text{ess}} \ell_R(\ell_R(I))_R$.

Proof Exercise. □

Proposition 8.3.5 *Let R be a ring. Then $\mathfrak{D}_{\mathbf{IC}}(R)$ is the set of all ideals I of R such that there exists an ideal J of R with $I \cap J = 0$ and $(I \oplus J)_R \leq^{\text{den}} R_R$.*

Proof Let $\mathfrak{D}_1$ be the set of all ideals I of R such that there is an ideal J of R satisfying $I \cap J = 0$ and $(I \oplus J)_R \leq^{\text{den}} R_R$. Then we show that $\mathfrak{D}_{\mathbf{IC}}(R) = \mathfrak{D}_1$. Take $I \in \mathfrak{D}_{\mathbf{IC}}(R)$ and $J = \ell_R(I)$. Then $I \cap J = 0$. Also, $\ell_R(I \oplus J) = \ell_R(I) \cap \ell_R(J) = \ell_R \cap \ell_R(\ell_R(I)) = 0$ as $I \in \mathfrak{D}_{\mathbf{IC}}(R)$. By Lemma 8.3.4 or Proposition 1.3.11(iv), $(I \oplus J)_R \leq^{\text{den}} R_R$. Thus $I \in \mathfrak{D}_1$, and so $\mathfrak{D}_{\mathbf{IC}}(R) \subseteq \mathfrak{D}_1$.

Next, we take $I \in \mathfrak{D}_1$. Then there exists $J \trianglelefteq R$ satisfying that $I \cap J = 0$ and $(I \oplus J)_R \leq^{\text{den}} R_R$. We note that $J \subseteq \ell_R(I)$, $I \subseteq \ell_R(J)$, and by Proposition 1.3.11(iv) $\ell_R(I \oplus J) = \ell_R(I) \cap \ell_R(J) = 0$. Thus $I \cap \ell_R(I) = 0$. Since $J \subseteq \ell_R(I)$, $\ell_R(\ell_R(I)) \subseteq \ell_R(J)$. Hence $\ell_R(I) \cap \ell_R(\ell_R(I)) \subseteq \ell_R(I) \cap \ell_R(J) = 0$, and thus $I \in \mathfrak{D}_{\mathbf{IC}}(R)$. Therefore $\mathfrak{D}_1 \subseteq \mathfrak{D}_{\mathbf{IC}}(R)$. Whence $\mathfrak{D}_{\mathbf{IC}}(R) = \mathfrak{D}_1$. □

We note that $\mathfrak{D}_{\mathbf{IC}}(R)$ contains all ideals of R which are dense in R_R as right R-modules from Proposition 1.3.11(iv). Also if a ring R is semiprime, then by Proposition 8.3.3(i), $\mathfrak{D}_{\mathbf{IC}}(R)$ is precisely the set of all ideals. We provide an example of a nonsemiprime ring R where the cardinality of $\mathfrak{D}_{\mathbf{IC}}(R)$ is greater than or equal to the cardinality of its complement in the set of all ideals of R. Indeed, take $R = T_2(S)$, where S is a right nonsingular prime ring with identity. The set of all ideals of R is $\left\{ \begin{bmatrix} A & B \\ 0 & C \end{bmatrix} \mid A, B, C \trianglelefteq S \text{ with } A, C \subseteq B \right\}$. Since R is right nonsingular, by Proposition 8.3.2

$$\mathfrak{D}_{\mathbf{IC}}(R) = \left\{ \begin{bmatrix} A & B \\ 0 & C \end{bmatrix} \mid A, B, C \trianglelefteq S \text{ with } A, C \subseteq B \text{ and } C \neq 0 \right\} \cup \left\{ \begin{bmatrix} 0 & 0 \\ 0 & 0 \end{bmatrix} \right\}.$$

Hence, we see that the cardinality of $\mathfrak{D}_{\mathbf{IC}}(R)$ is greater than or equal to the cardinality of its complement.

Lemma 8.3.6 *Assume that T is a right ring of quotients of R and let $I \in \mathfrak{D}_{\mathbf{IC}}(T)$. Then $I \cap R \in \mathfrak{D}_{\mathbf{IC}}(R)$.*

Proof Let $I \in \mathfrak{D}_{\mathbf{IC}}(T)$ and put $K = I \cap R$. Then $\ell_R(K) = \ell_R(I)$ from Lemma 8.1.6(i). Hence $K \cap \ell_R(K) = K \cap \ell_R(I) \subseteq I \cap \ell_R(I) \subseteq I \cap \ell_T(I) = 0$.

Say $a \in \ell_R(K \oplus \ell_R(K))$. Then $a \in \ell_R(K) = \ell_R(I)$, so $aI = 0$. We show that $a\,\ell_T(I) = 0$. For this, assume on the contrary that $at \neq 0$ for some $t \in \ell_T(I)$. Then there exists $r \in R$ satisfying $tr \in R$ and $atr \neq 0$ since $R_R \leq^{\text{den}} T_R$. Therefore

$$tr \in R \cap \ell_T(I) = \ell_R(I) = \ell_R(K).$$

Because $a \in \ell_R(K \oplus \ell_R(K))$, $a\,\ell_R(K) = 0$. So $atr = 0$, a contradiction. Hence we get $a\,\ell_T(I) = 0$ and $a \in \ell_T(I) \cap \ell_T(\ell_T(I)) = 0$. So $\ell_R(K \oplus \ell_R(K)) = 0$, as a consequence $K \in \mathfrak{D}_{\mathbf{IC}}(R)$. □

Lemma 8.3.7 *Let I and J be ideals of R.*

(i) *If $I \in \mathfrak{D}_{\mathbf{IC}}(R)$ and $I_R \leq^{\text{ess}} J_R$, then $I_R \leq^{\text{den}} J_R$ and $J \in \mathfrak{D}_{\mathbf{IC}}(R)$.*

(ii) *If $I_R \leq^{\text{den}} J_R$ and $J \in \mathfrak{D}_{\mathbf{IC}}(R)$, then $I \in \mathfrak{D}_{\mathbf{IC}}(R)$.*

(iii) *If $I \cap J = 0$ and $I \oplus J \in \mathfrak{D}_{\mathbf{IC}}(R)$, then $I \in \mathfrak{D}_{\mathbf{IC}}(R)$ and $J \in \mathfrak{D}_{\mathbf{IC}}(R)$.*

(iv) *$I \in \mathfrak{D}_{\mathbf{IC}}(R)$ if and only if $\ell_R(I) \in \mathfrak{D}_{\mathbf{IC}}(R)$ and $I \cap \ell_R(I) = 0$.*

Proof (i) Assume that $I \in \mathfrak{D}_{\mathbf{IC}}(R)$ and $I_R \leq^{\text{ess}} J_R$. From Proposition 8.3.5, there exists $K \trianglelefteq R$ such that $(I \oplus K)_R \leq^{\text{den}} R_R$. By the modular law,

$$(J \cap (I \oplus K))_R = (I \oplus (J \cap K))_R \leq^{\text{den}} J_R.$$

As $I_R \leq^{\text{ess}} J_R$ and $I \cap (J \cap K) = 0$, $J \cap K = 0$, so $I_R \leq^{\text{den}} J_R$. We show that $\ell_R(I) = \ell_R(J)$. For this, it suffices to see that $\ell_R(I) \subseteq \ell_R(J)$. Assume on the contrary that there is $x \in \ell_R(I)$ but $xJ \neq 0$. There is $y \in J$ with $xy \neq 0$. Since $I_R \leq^{\text{den}} J_R$, there is $r \in R$ such that $yr \in I$ and $xyr \neq 0$, which is a contradiction since $xI = 0$. Thus $\ell_R(I) = \ell_R(J)$.

So $I \cap \ell_R(J) = I \cap \ell_R(I) = 0$ and $J \cap \ell_R(J) = 0$. Now $I \oplus \ell_R(I) \subseteq J \oplus \ell_R(J)$. Thus, $(J \oplus \ell_R(J))_R \leq^{\text{den}} R_R$ as $(I \oplus \ell_R(I))_R \leq^{\text{den}} R_R$. Hence, $J \in \mathfrak{D}_{\mathbf{IC}}(R)$.

(ii) Let $J \in \mathfrak{D}_{\mathbf{IC}}(R)$ and $I_R \leq^{\text{den}} J_R$. Then $\ell_R(I) = \ell_R(J)$ by the proof of part (i). From Lemma 8.3.4, $J_R \leq^{\text{ess}} \ell_R(\ell_R(J))_R = \ell_R(\ell_R(I))_R$. Therefore, it follows that $I_R \leq^{\text{ess}} \ell_R(\ell_R(I))_R$. Hence, we obtain $\ell_R(I) \cap \ell_R(\ell_R(I)) = 0$ from Lemma 8.3.4 because $I \cap \ell_R(I) \subseteq J \cap \ell_R(J) = 0$. Therefore, $I \in \mathfrak{D}_{\mathbf{IC}}(R)$.

(iii) Suppose that $I \oplus J \in \mathfrak{D}_{\mathbf{IC}}(R)$. From Proposition 8.3.5, there is $V \trianglelefteq R$ with $((I \oplus J) \oplus V)_R \leq^{\text{den}} R_R$. Therefore, $I \in \mathfrak{D}_{\mathbf{IC}}(R)$ and $J \in \mathfrak{D}_{\mathbf{IC}}(R)$ again by Proposition 8.3.5.

(iv) Say $I \in \mathfrak{D}_{\mathbf{IC}}(R)$. Then $I \cap \ell_R(I) = 0$ and $\ell_R(I) \cap \ell_R(\ell_R(I)) = 0$. Since $I \subseteq \ell_R(\ell_R(I))$, we have that $I \oplus \ell_R(I) \subseteq \ell_R(\ell_R(I)) \oplus \ell_R(I)$. As a consequence $\ell_R[\ell_R(I) \oplus \ell_R(\ell_R(I))] \subseteq \ell_R(I \oplus \ell_R(I)) = 0$. So $\ell_R(I) \in \mathfrak{D}_{\mathbf{IC}}(R)$.

Conversely, $\ell_R(I) \in \mathfrak{D}_{\mathbf{IC}}(R)$ implies $\ell_R(I) \cap \ell_R(\ell_R(I)) = 0$. Therefore, $I \in \mathfrak{D}_{\mathbf{IC}}(R)$ because $I \cap \ell_R(I) = 0$ by assumption. □

Let R be a ring (not necessarily with identity) with $\ell_R(R) = 0$. Say U is a subring of R such that $U_U \leq^{\text{den}} R_U$ (i.e., for $x, y \in R$ with $y \neq 0$, there exists $u \in U$ satisfying $xu \in U$ and $yu \neq 0$). Then $\ell_U(U) = 0$. Indeed, let $x \in \ell_U(U)$. If $xr \neq 0$ for some $r \in R$, then there exists $u \in U$ such that $ru \in U$ and $xru \neq 0$, a contradiction. So $x \in \ell_R(R) = 0$, and hence $\ell_U(U) = 0$. Thus, $Q(U)$ exists. Therefore, $Q(U) = Q(R)$ as R is a right ring of quotients of U.

The following result characterizes the ideals of R which are dense as right R-modules in some ring direct summands of $Q(R)$ as precisely the elements of $\mathfrak{D}_{\mathbf{IC}}(R)$.

Theorem 8.3.8 *Assume that R is a ring and $I \trianglelefteq R$. Then the following are equivalent.*

(i) $I \in \mathfrak{D}_{\mathbf{IC}}(R)$.
(ii) *There exists $e \in \mathcal{B}(Q(R))$ such that $Q(I) = eQ(R)$.*
(iii) $I_R \leq^{\text{den}} eQ(R)_R$ *for some (unique) $e \in \mathcal{B}(Q(R))$.*

Proof (i)⇒(ii) Put $J = \ell_R(I)$. Since $I \in \mathfrak{D}_{\mathrm{IC}}(R)$, $\ell_R(I \oplus J) = 0$ and hence $\ell_{I \oplus J}(I \oplus J) = 0$. Therefore $\ell_I(I) = 0$ and $\ell_J(J) = 0$, hence $Q(I)$ and $Q(J)$ exist. Put $U = I \oplus J$. For $U_U \leq^{\text{den}} R_U$, take $x, y \in R$ with $y \neq 0$. As $U_R \leq^{\text{den}} R_R$, there exists $r \in R$ such that $xr \in U$ and $yr \neq 0$. Again since $U_R \leq^{\text{den}} R_R$, there exists $a \in R$ satisfying that $ra \in U$ and $yra \neq 0$. Because $ra \in U$ and $xra \in U$, we see that $U_U \leq^{\text{den}} R_U$. So, $Q(R) = Q(U) = Q(I \oplus J) = Q(I) \oplus Q(J)$ by [395, (2.1)]. Consequently, $Q(I) = eQ(R)$ for some $e \in \mathcal{B}(Q(R))$.

(ii)⇒(iii) Say $Q(I) = eQ(R)$ for some $e \in \mathcal{B}(Q(R))$. Take $eq_1, eq_2 \in eQ(R)$ with $q_1, q_2 \in Q(R)$ and $eq_2 \neq 0$. As $I_I \leq^{\text{den}} Q(I)_I$, there exists $a \in I$ such that $eq_1a \in I$ and $eq_2a \neq 0$. Since $a \in R$, $I_R \leq^{\text{den}} eQ(R)_R$. If $f \in \mathcal{B}(Q(R))$ satisfying $I_R \leq^{\text{den}} fQ(R)_R$, then $e = f$ as $e \in \mathcal{B}(Q(R))$.

(iii)⇒(i) Let $I_R \leq^{\text{den}} eQ(R)_R$ for some $e \in \mathcal{B}(Q(R))$. Then we have that $I_R \leq^{\text{den}} (eQ(R) \cap R)_R$. Now Lemma 8.3.6 yields that $eQ(R) \cap R \in \mathfrak{D}_{\mathbf{IC}}(R)$ because $eQ(R) \in \mathfrak{D}_{\mathbf{IC}}(Q(R))$. From Lemma 8.3.7(ii), $I \in \mathfrak{D}_{\mathbf{IC}}(R)$. □

We note that if $I \in \mathfrak{D}_{\mathbf{IC}}(R)$, then from Lemma 8.3.4, Lemma 8.3.7(i), and Theorem 8.3.8, there exists $e \in \mathcal{B}(Q(R))$ such that $\ell_R(\ell_R(I))_R \leq^{\text{den}} eQ(R)_R$. Further, $\ell_R(\ell_R(I)) = eQ(R) \cap R$ and $\ell_R(\ell_R(I))$ is the unique closure of I_R in R_R (see Exercise 8.3.58.5).

Corollary 8.3.9 *Assume that* $I \in \mathfrak{D}_{\mathbf{IC}}(R)$ *and* T *is a right ring of quotients of* R. *Then* $(I) \in \mathfrak{D}_{\mathbf{IC}}(T)$ *and* $I_R \leq^{\text{den}} (I)_R$, *where* (I) *is the ideal of* T *generated by* I.

Proof There exists $e \in \mathcal{B}(Q(R))$ with $I_R \leq^{\text{den}} eQ(R)$ from Theorem 8.3.8. Hence, $I_R \leq (I)_R \leq eQ(R)$ as $I = eI$. Therefore $(I)_R \leq^{\text{den}} eQ(R)_R$, and thus we see that $(I)_T \leq^{\text{den}} eQ(R)_T$. Because $Q(R) = Q(T)$, $(I)_T \leq^{\text{den}} eQ(T)_T$. Thus from Theorem 8.3.8, $(I) \in \mathfrak{D}_{\mathbf{IC}}(T)$. □

Say $A \in \mathfrak{D}_{\mathbf{IC}}(Q(R))$. Then $A_{Q(R)} \leq^{\text{den}} eQ(R)_{Q(R)}$ for some $e \in \mathcal{B}(Q(R))$ by Theorem 8.3.8. Thereby $Q(R)$ is an **IC**-ring and this suggests that there may be a smallest right ring of quotients of R which is an **IC**-ring. So one may naturally ask: *Does* $\widehat{Q}_{\mathbf{IC}}(R)$ *exist for every ring* R *when* $\ell_R(R) = 0$? For this question, we need the following lemma.

Lemma 8.3.10 *Assume that* R *is a ring with identity and* $b \in \mathcal{B}(Q(R))$. *Then there exists* $\lambda \in \mathcal{B}(\text{End}(E(R_R)))$ *such that* $b = \lambda(1)$.

Proof Note that $E(R_R)$ is an $(\text{End}(E(R_R)), Q(R))$-bimodule. Define a map

$$\lambda : E(R_R) \to E(R_R) \text{ by } \lambda(x) = xb$$

for $x \in E(R_R)$. Then $\lambda \in \text{End}(E(R_R))$ and $\lambda^2 = \lambda$ because $b \in \mathcal{B}(Q(R))$. Next, say $\varphi \in \text{End}(E(R_R))$. For $x \in E(R_R)$,

$$(\lambda\varphi)(x) = \varphi(x)b = \varphi(xb),$$

since $\text{End}(E(R_R)) = \text{End}(E(R_R)_{Q(R)})$ (see the proof of Theorem 2.1.31). Further, $(\varphi\lambda)(x) = \varphi(xb)$. So $\lambda\varphi(x) = \varphi\lambda(x)$ for all $x \in E(R_R)$, thus $\lambda\varphi = \varphi\lambda$. Hence $\lambda \in \mathcal{B}(\text{End}(E(R_R)))$ and $b = \lambda(1)$. □

Our next result shows that $\widehat{Q}_{\mathbf{IC}}(R)$ exists for all rings R with $\ell_R(R) = 0$ and it can be used to characterize **IC** right rings of quotients of R. When R is a ring with $\ell_R(R) = 0$, we recall from 1.1.2 that $\langle R \cup \mathcal{B}(Q(R)) \rangle_{Q(R)}$ denotes the subring of $Q(R)$ generated by $R \cup \mathcal{B}(Q(R))$. Observe that if R has identity, then we see that $\langle R \cup \mathcal{B}(Q(R)) \rangle_{Q(R)} = R\mathcal{B}(Q(R))$.

Theorem 8.3.11 *Assume that* R *is a ring.*

(i) *Let* T *be a right ring of quotients of* R. *Then* $T \in \mathbf{IC}$ *if and only if* $\mathcal{B}(Q(R)) \subseteq T$.

(ii) $R \in \mathbf{IC}$ *if and only if* $\mathcal{B}(Q(R)) \subseteq R$. *Hence,* **IC**-*rings have identity.*

(iii) $\widehat{Q}_{\mathbf{IC}}(R) = \langle R \cup \mathcal{B}(Q(R)) \rangle_{Q(R)}$.

(iv) *If* R *has identity, then* $\widehat{Q}_{\mathbf{IC}}(R) = R(\mathbf{IC}, Q(R))$.

Proof (i) Say $T \in \mathbf{IC}$. Take $c \in \mathcal{B}(Q(R))$ and we let $I = R \cap cQ(R)$. Then $I_R \leq^{\text{ess}} cQ(R)_R$. We note that $cQ(R) \in \mathfrak{D}_{\mathbf{IC}}(Q(R))$. From Lemma 8.3.6, $Y := cQ(R) \cap T \in \mathfrak{D}_{\mathbf{IC}}(T)$ and $I_R \leq^{\text{ess}} Y_R$. Since $Y \in \mathfrak{D}_{\mathbf{IC}}(T)$ and $T \in \mathbf{IC}$, $Y_T \leq^{\text{ess}} eT_T$ for some $e \in \mathbf{I}(T)$. Thus, $Y_R \leq^{\text{ess}} eT_R$ by Lemma 8.1.3(i). Now $c = e \in T$, as

$I_R \leq^{\text{ess}} Y_R \leq^{\text{ess}} eT_R \leq^{\text{ess}} eQ(R)_R$ and $I_R \leq^{\text{ess}} Y_R \leq^{\text{ess}} cQ(R)_R$. So $\mathcal{B}(Q(R)) \subseteq T$. Conversely, let $\mathcal{B}(Q(R)) \subseteq T$. Take $I \in \mathfrak{D}_{\mathbf{IC}}(T)$. As $Q(R) = Q(T)$, Theorem 8.3.8 yields that there is $e \in \mathcal{B}(Q(T)) \subseteq T$ such that $I_T \leq^{\text{den}} eQ(T)_T$. Hence, we get that $I_T \leq^{\text{den}} eT_T$. Therefore, $T \in \mathbf{IC}$.

(ii) and (iii) These parts follows from part (i) immediately.

(iv) By part (iii) $\widehat{Q}_{\mathbf{IC}}(R) = \langle R \cup \mathcal{B}(Q(R))\rangle_{Q(R)}$. Recall that

$$\delta_{\mathbf{IC}}(R) = \{e^2 = e \in \text{End}(E(R_R)) \mid I_R \leq^{\text{ess}} eE(R_R) \text{ for some } I \in \mathfrak{D}_{\mathbf{IC}}(R)\}$$

and $\delta_{\mathbf{IC}}(R)(1) = \{e(1) \mid e \in \delta_{\mathbf{IC}}(R)\}$.

We prove that $\mathcal{B}(Q(R)) = \delta_{\mathbf{IC}}(R)(1)$. For this, say $c \in \mathcal{B}(Q(R))$. Then it follows that $R \cap cQ(R) \trianglelefteq R$ and $(R \cap cQ(R))_R \leq^{\text{ess}} cQ(R)_R$. From Lemma 8.3.6, we get $R \cap cQ(R) \in \mathfrak{D}_{\mathbf{IC}}(R)$. Also, there exists $\lambda^2 = \lambda \in \mathcal{B}(\text{End}(E(R_R)))$ such that $c = \lambda(1)$ by Lemma 8.3.10.

We note that $(R \cap cQ(R))_R \leq^{\text{ess}} \lambda(1)Q(R)_R = \lambda Q(R)_R \leq^{\text{ess}} \lambda E(R_R)$ because $\lambda \in \text{End}(E(R_R)_{Q(R)})$. Thus $\lambda \in \delta_{\mathbf{IC}}(R)$, so $c = \lambda(1) \in \delta_{\mathbf{IC}}(R)(1)$. As a consequence, $\mathcal{B}(Q(R)) \subseteq \delta_{\mathbf{IC}}(R)(1)$.

Next, say $h \in \delta_{\mathbf{IC}}(R)$. Then there is $I \in \mathfrak{D}_{\mathbf{IC}}(R)$ with $I_R \leq^{\text{ess}} hE(R_R)$. By Theorem 8.3.8, $I_R \leq^{\text{ess}} bQ(R)_R$ for some $b \in \mathcal{B}(Q(R))$. From Lemma 8.3.10, there exists $\gamma \in \mathcal{B}(\text{End}(E(R_R)))$ such that $b = \gamma(1)$. Sometimes we will use E_R for $E(R_R)$.

Observe that $I_R \leq^{\text{ess}} bQ(R)_R = \gamma(1)Q(R)_R = \gamma Q(R)_R \leq^{\text{ess}} \gamma E(R_R)$. So $hE(R_R) = \gamma E(R_R)$ because $\gamma \in \mathcal{B}(\text{End}(E_R))$. Therefore $h(1) = \gamma(x)$ for some $x \in E(R_R)$, and hence $\gamma h(1) = h(1)$. Also $\gamma(1) = h(y)$ with $y \in E(R_R)$. As a consequence, $h\gamma(1) = \gamma(1)$, so $h(1) = \gamma h(1) = h\gamma(1) = \gamma(1) = b$. Hence, it follows that $\delta_{\mathbf{IC}}(R)(1) \subseteq \mathcal{B}(Q(R))$. Therefore, $\mathcal{B}(Q(R)) = \delta_{\mathbf{IC}}(R)(1)$.

Now $\langle R \cup \delta_{\mathbf{IC}}(R)(1)\rangle_{Q(R)} = \langle R \cup \mathcal{B}(Q(R))\rangle_{Q(R)}$. By the definition of pseudo ring hulls, $R(\mathbf{IC}, Q(R)) = \langle R \cup \mathcal{B}(Q(R))\rangle_{Q(R)}$ since $\langle R \cup \mathcal{B}(Q(R))\rangle_{Q(R)}$ is an $\mathbf{IC}$-ring by part (iii). □

From Theorems 8.3.8 and 8.3.11, we see that any intermediate ring T between $\langle R \cup \mathcal{B}(Q(R))\rangle_{Q(R)}$ and $Q(R)$ satisfies that for every $I \in \mathfrak{D}_{\mathbf{IC}}(R)$, there exists $e^2 = e \in T$ such that $I_R \leq^{\text{ess}} eT_R$. Furthermore, we see that for every $J \in \mathfrak{D}_{\mathbf{IC}}(T)$, $J_T \leq^{\text{ess}} fT_T$ for some $f^2 = f \in T$.

Corollary 8.3.12 *Let R be an $\mathbf{IC}$-ring with $Z(R_R) = 0$. Then $R = R_1 \oplus R_2$ (ring direct sum), where R_1 is a semiprime FI-extending ring and $P(R)$ is ideal essential in R_2.*

Proof Exercise. □

The following result is on the lattice properties of $\mathfrak{D}_{\mathbf{IC}}(R)$ as suggested by earlier results.

Theorem 8.3.13 (i) *$\mathfrak{D}_{\mathbf{IC}}(R)$ is a sublattice of the lattice of ideals of R.*

(ii) *If $\mathfrak{D}_{\mathbf{IC}}(R)$ is a complete sublattice of the lattice of ideals of R, then $\mathcal{B}(Q(R))$ is a complete Boolean algebra.*

(iii) *Let* $R \in \mathbf{IC}$ *such that* $\mathfrak{D}_{\mathbf{IC}}(R) = \{I \trianglelefteq R \mid I \cap \ell_R(I) = 0\}$. *Then* $\mathfrak{D}_{\mathbf{IC}}(R)$ *is a complete sublattice of the lattice of ideals of* R.

(iv) *If* R *is right and left FI-extending, then* $\mathfrak{D}_{\mathbf{IC}}(R)$ *is a complete sublattice of the lattice of ideals of* R.

Proof (i) Assume that $I, J \in \mathfrak{D}_{\mathbf{IC}}(R)$. By Theorem 8.3.8 there are unique c_1, c_2 in $\mathcal{B}(Q(R))$ such that $I_R \leq^{\text{den}} c_1 Q(R)_R$ and $J_R \leq^{\text{den}} c_2 Q(R)_R$. Therefore

$$(I \cap J)_R \leq^{\text{den}} c_1 Q(R)_R \cap c_2 Q(R)_R = c_1 c_2 Q(R)_R \text{ and } c_1 c_2 \in \mathcal{B}(Q(R)).$$

By Theorem 8.3.8, $I \cap J \in \mathfrak{D}_{\mathbf{IC}}(R)$.

Let $c = c_1 + c_2 - c_1 c_2$. Then $(I + J)_R \leq (c_1 Q(R) + c_2 Q(R))_R = c Q(R)_R$ and $c \in \mathcal{B}(Q(R))$. Take $K = R \cap \ell_{cQ(R)}(I + J)$. Then $K \subseteq \ell_R(I) \cap \ell_R(J)$. As $I_R \leq^{\text{den}} (R \cap c_1 Q(R))_R$ and $J_R \leq^{\text{den}} (R \cap c_2 Q(R))_R$, it follows that $\ell_R(I) = \ell_R(R \cap c_1 Q(R)) = \ell_R(c_1 Q(R)) = R \cap (1 - c_1) Q(R)$ by Lemma 8.1.6(i) and the proof of Lemma 8.3.7(i). Also $\ell_R(J) = R \cap (1 - c_2) Q(R)$ similarly.

Since $K \subseteq \ell_R(I) \cap \ell_R(J) = R \cap (1 - c_1) Q(R) \cap (1 - c_2) Q(R)$, it follows that $K c_1 = 0$ and $K c_2 = 0$. So $Kc = 0$. But we see that $Kc = K$ because

$$K = R \cap \ell_{cQ(R)}(I + J) \subseteq c Q(R),$$

so $\ell_{cQ(R) \cap R}(I + J) = K = 0$.

Now since $I + J \trianglelefteq c Q(R) \cap R$, $(I + J)_{cQ(R) \cap R} \leq^{\text{den}} (c Q(R) \cap R)_{cQ(R) \cap R}$ from Proposition 1.3.11(iv), and hence $(I + J)_R \leq^{\text{den}} (R \cap c Q(R))_R$. Thus it follows that $(I + J)_R \leq^{\text{den}} c Q(R)_R$. By Theorem 8.3.8, $I + J \in \mathfrak{D}_{\mathbf{IC}}(R)$. Hence $\mathfrak{D}_{\mathbf{IC}}(R)$ is a sublattice of the lattice of ideals of R.

(ii) Let $\{e_i \mid i \in \Lambda\} \subseteq \mathcal{B}(Q(R))$. Then $I_i := e_i Q(R) \cap R \in \mathfrak{D}_{\mathbf{IC}}(R)$ for all $i \in \Lambda$ from Lemma 8.3.6. Put $I = \sum_{i \in \Lambda} I_i$. Then $I \in \mathfrak{D}_{\mathbf{IC}}(R)$ by assumption. From Theorem 8.3.8, there is $e \in \mathcal{B}(Q(R))$ with $I_R \leq^{\text{den}} e Q(R)_R$.

For each $i \in \Lambda$, $I_{iR} \leq^{\text{ess}} e_i Q(R)_R$. Because $I_{iR} \leq I_R \leq^{\text{ess}} e Q(R)_R$, we have that $I_{iR} \leq^{\text{ess}} e e_i Q(R)_R$. Thus, $e_i = e e_i$, so $e_i \leq e$ for all $i \in \Lambda$.

We claim that $e = \sup\{e_i \mid i \in \Lambda\}$. For this, say $f \in \mathcal{B}(Q(R))$ such that $e_i = f e_i$ (i.e., $e_i \leq f$) for all $i \in \Lambda$. By Lemma 8.3.6, $f Q(R) \cap R \in \mathfrak{D}_{\mathbf{IC}}(R)$. Since $I_i = e_i Q(R) \cap R \subseteq f Q(R) \cap R$ for all i, $I \subseteq f Q(R) \cap R \subseteq f Q(R)$. As $I_R \leq^{\text{ess}} e Q(R)_R$, $I_R \leq^{\text{ess}} (e Q(R) \cap f Q(R))_R = ef Q(R)_R \leq^{\text{ess}} e Q(R)_R$, so $ef Q(R) = e Q(R)$. Hence $e = ef = fe$ (i.e., $e \leq f$), so $e = \sup\{e_i \mid i \in \Lambda\}$. Therefore, $\mathcal{B}(Q(R))$ is a complete Boolean algebra.

(iii) Assume that $\{I_i \mid i \in \Lambda\} \subseteq \mathfrak{D}_{\mathbf{IC}}(R)$. Then from Theorem 8.3.8, there exists $\{e_i \mid i \in \Lambda\} \subseteq \mathcal{B}(Q(R))$ with $I_{iR} \leq^{\text{den}} e_i Q(R)_R$ for each $i \in \Lambda$.

Assume that F is a finite nonempty subset of Λ. First, say $F = \{1, 2\}$. Then $I_{1R} \leq^{\text{den}} e_1 Q(R)_R$ and $I_{2R} \leq^{\text{den}} e_2 Q(R)_R$. From the proof of part (i), $(I_1 + I_2)_R \leq^{\text{den}} e Q(R)_R$, where $e = e_1 + e_2 - e_1 e_2$. Inductively, we can see that $\sum_{i \in F} I_{iR} \leq^{\text{den}} \sum_{i \in F} e_i Q(R)_R$. Next, we show that

$$\sum_{i \in \Lambda} I_{iR} \leq^{\text{den}} \sum_{i \in \Lambda} e_i Q(R)_R.$$

For this, let $x, y \in \sum_{i\in\Lambda} e_i Q(R)$ with $y \neq 0$. Then there is a nonempty finite subset F of Λ with $x, y \in \sum_{i\in F} e_i Q(R)$. As $\sum_{i\in F} I_{iR} \leq^{\text{den}} \sum_{i\in F} e_i Q(R)_R$ by the preceding argument, there is $r \in R$ with $xr \in \sum_{i\in F} I_{iR} \leq \sum_{i\in\Lambda} I_{iR}$ and $yr \neq 0$. Therefore, $\sum_{i\in\Lambda} I_{iR} \leq^{\text{den}} \sum_{i\in\Lambda} e_i Q(R)_R$.

From Theorem 8.3.11(ii), $\mathcal{B}(Q(R)) \subseteq R$, hence $e_i \in \mathcal{B}(R)$ for each $i \in \Lambda$. To see that $(\sum_{i\in\Lambda} e_i R) \cap \ell_R(\sum_{i\in\Lambda} e_i R) = (\sum_{i\in\Lambda} e_i R) \cap (\cap_{i\in\Lambda}(1-e_i)R) = 0$, it is enough to prove that

$$(\sum_{i\in F} e_i R) \cap (\cap_{i\in F}(1-e_i)R) = 0$$

for any nonempty finite subset F of Λ. If $F = \{1\}$, then we are done. Say $F = \{1, 2\}$. Then

$$(e_1 R + e_2 R) \cap ((1-e_1)R \cap (1-e_2)R) = (e_1 R + e_2 R) \cap (1-e_1)(1-e_2)R = 0.$$

So $(\sum_{i\in F} e_i R) \cap \ell_R(\sum_{i\in F} e_i R) = (\sum_{i\in F} e_i R) \cap (\cap_{i\in F}(1-e_i)R) = 0$ inductively. Thus, with the hypothesis $\mathfrak{D}_{\mathbf{IC}}(R) = \{I \trianglelefteq R \mid I \cap \ell_R(I) = 0\}$, it follows that $\sum_{i\in\Lambda} e_i R \in \mathfrak{D}_{\mathbf{IC}}(R)$. By Lemma 8.3.7(ii), $\sum_{i\in\Lambda} I_i \in \mathfrak{D}_{\mathbf{IC}}(R)$.

(iv) Let R be right and left FI-extending. Then R is an **IC**-ring, so $\mathcal{B}(Q(R)) \subseteq R$ by Theorem 8.3.11(ii). Let $\{I_i \mid i \in \Lambda\} \subseteq \mathfrak{D}_{\mathbf{IC}}(R)$. From Theorem 8.3.8, there exists a set $\{e_i \mid i \in \Lambda\} \subseteq \mathcal{B}(Q(R))$ with $I_{iR} \leq^{\text{den}} e_i Q(R)_R$ for each $i \in \Lambda$.

Now $(\sum_{i\in\Lambda} e_i R) \cap \ell_R(\sum_{i\in\Lambda} e_i R) = (\sum_{i\in\Lambda} e_i R) \cap r_R(\sum_{i\in\Lambda} e_i R) = 0$ by the preceding argument. From Theorem 2.3.15, there exists $c \in \mathcal{B}(R)$ such that $\ell_R(\sum_{i\in\Lambda} e_i R) = (1-c)R$. We recall that $\sum_{i\in\Lambda} I_{iR} \leq^{\text{den}} \sum_{i\in\Lambda} e_i R_R$ from the proof of part (iii). Therefore, the proof of Lemma 8.3.7(i) yields that

$$\ell_R(\sum_{i\in\Lambda} I_i) = \ell_R(\sum_{i\in\Lambda} e_i R) = (1-c)R$$

from the proof of Lemma 8.3.7(i). So $\ell_R(\sum_{i\in\Lambda} I_i) \in \mathfrak{D}_{\mathbf{IC}}(R)$. Also

$$\begin{aligned}(\sum_{i\in\Lambda} I_i) \cap \ell_R(\sum_{i\in\Lambda} I_i) &= (\sum_{i\in\Lambda} I_i) \cap (1-c)R \subseteq r_R(\ell_R(\sum_{i\in\Lambda} I_i)) \cap (1-c)R \\ &= cR \cap (1-c)R = 0.\end{aligned}$$

From Lemma 8.3.7(iv), $\sum_{i\in\Lambda} I_i \in \mathfrak{D}_{\mathbf{IC}}(R)$. Hence, $\mathfrak{D}_{\mathbf{IC}}(R)$ is a complete sublattice of the lattice of ideals of R. □

Corollary 8.3.14 *If $Q(R)$ is semiprime, then $\mathcal{B}(Q(R))$ is a complete Boolean algebra.*

Proof By Theorem 8.3.11(ii), $Q(R)$ is an **IC**-ring. As $Q(R)$ is semiprime, $Q(R)$ is right FI-extending from Proposition 8.3.3(i). Thus by Theorem 3.2.37, $Q(R)$ is also left FI-extending. So Theorem 8.3.13(ii) and (iv) yield that $\mathcal{B}(Q(R))$ is a complete Boolean algebra. □

Corollary 8.3.15 *If R is a right nonsingular* **IC**-*ring, then $\mathfrak{D}_{\mathbf{IC}}(R)$ is a complete sublattice of the lattice of ideals of R.*

Proof The proof follows from Proposition 8.3.2 and Theorem 8.3.13(iii). □

Proposition 8.3.16 *Assume that R is a semiprime ring. Then, for any ideal I of R, $r_{Q(R)}(Q(R)IQ(R)) = r_{Q(R)}(I)$.*

Proof Let $I \trianglelefteq R$. Clearly, $r_{Q(R)}(IQ(R)) \subseteq r_{Q(R)}(I)$. Let $\alpha \in r_{Q(R)}(I)$ and $\sum x_i q_i \in IQ(R)$ with $x_i \in I$ and $q_i \in Q(R)$. Assume that $(\sum x_i q_i)\alpha \neq 0$. Since $R_R \leq^{\text{den}} Q(R)_R$, there exists $r_1 \in R$ with $\alpha r_1 \in R$ and $(\sum x_i q_i)\alpha r_1 \neq 0$. Thus, $\alpha r_1 \in R \cap r_{Q(R)}(I) = r_R(I) = \ell_R(I)$ because R is semiprime. Also there is $r_2 \in R$ with $0 \neq (\sum x_i q_i)\alpha r_1 r_2 \in R$ since $R_R \leq^{\text{ess}} Q(R)_R$.

Let $y = (\sum x_i q_i)\alpha r_1 r_2$. As $\alpha r_1 \in \ell_R(I)$, $\alpha r_1 r_2 \in \ell_R(I)$ and so $\alpha r_1 r_2 I = 0$. Hence $yRI = (\sum x_i q_i)\alpha r_1 r_2 RI \subseteq (\sum x_i q_i)\alpha r_1 r_2 I = 0$. Further, note that $yR = (\sum x_i q_i \alpha r_1 r_2)R \subseteq IQ(R)$. So $(yR)^2 = (yR)(yR) \subseteq yRIQ(R) = 0$, which is a contradiction because R is semiprime. Therefore $\alpha \in r_{Q(R)}(IQ(R))$, and thus $r_{Q(R)}(I) = r_{Q(R)}(IQ(R)) = r_{Q(R)}(Q(R)IQ(R))$. □

The next result demonstrates the existence and uniqueness of the quasi-Baer and the right FI-extending ring hulls of a semiprime ring. It extends Mewborn's result (Theorem 8.2.4) as a commutative quasi-Baer ring is Baer.

Theorem 8.3.17 *Let R be a semiprime ring. Then:*

(i) $\widehat{Q}_{\mathbf{qB}}(R) = \widehat{Q}_{\mathbf{FI}}(R) = \widehat{Q}_{\mathbf{eqB}}(R) = \langle R \cup \mathcal{B}(Q(R))\rangle_{Q(R)}$.
(ii) *If R has identity, then $\widehat{Q}_{\mathbf{FI}}(R) = R(\mathbf{FI}, Q(R))$.*
(iii) *If R has identity, then $\widehat{Q}_{\mathbf{eqB}}(R) = R(\mathbf{eqB}, Q(R))$.*

Proof (i) Note that $\widehat{Q}_{\mathbf{FI}}(R) = \langle R \cup \mathcal{B}(Q(R))\rangle_{Q(R)}$ by Proposition 8.3.3(i) and Theorem 8.3.11(iii). From Theorem 3.2.37, $\widehat{Q}_{\mathbf{qB}}(R) = \widehat{Q}_{\mathbf{eqB}}(R) = \widehat{Q}_{\mathbf{FI}}(R)$.

(ii) This part follows from Proposition 8.3.3(i) and Theorem 8.3.11(iv).

(iii) To prove that $R(\mathbf{eqB}, Q(R)) = \langle R \cup \mathcal{B}(Q(R))\rangle_{Q(R)}$, we claim that $\mathcal{B}(Q(R)) = \delta_{\mathbf{eqB}}(R)(1)$. For this, let $a \in \mathcal{B}(Q(R))$ and $I = R \cap (1-a)Q(R)$. Then $I_R \leq^{\text{ess}} (1-a)Q(R)_R$, and so $Q(R)IQ(R)_R \leq^{\text{ess}} (1-a)Q(R)_R$. Thus $Q(R)IQ(R)_{Q(R)} \leq^{\text{ess}} (1-a)Q(R)_{Q(R)}$.

By Theorem 8.3.11(ii), $Q(R)$ is an **IC**-ring. As $Q(R)$ is semiprime, $Q(R)$ is a right FI-extending ring from Proposition 8.3.3(i). By Theorem 3.2.37, $Q(R)$ is quasi-Baer. So there is $k \in \mathcal{B}(Q(R))$ with $r_{Q(R)}(Q(R)IQ(R)) = kQ(R)$ by Proposition 1.2.6(ii). Now $Q(R)IQ(R)_{Q(R)} \leq^{\text{ess}} (1-k)Q(R)_{Q(R)}$ by Lemma 2.1.13. Thus $1-a = 1-k$, so $a = k$.

From Lemma 8.3.10, there is $\mu^2 = \mu \in \text{End}(E(R_R))$ such that $a = \mu(1)$. By Proposition 8.3.16, $r_{Q(R)}(I) = r_{Q(R)}(Q(R)IQ(R)) = kQ(R)$. Hence

$$\begin{aligned} r_R(I)_R &= (r_{Q(R)}(I) \cap R)_R = (kQ(R) \cap R)_R \\ &\leq^{\text{ess}} kQ(R)_R = aQ(R)_R = \mu(1)Q(R)_R = \mu Q(R)_R \\ &\leq^{\text{ess}} \mu E(R_R) \end{aligned}$$

because $\mu \in \text{End}(E(R_R)) = \text{End}(E(R_R)_{Q(R)})$. Thus $\mu \in \delta_{\mathbf{eqB}}(R)$, and therefore $a = \mu(1) \in \delta_{\mathbf{eqB}}(R)(1)$. Hence $\mathcal{B}(Q(R)) \subseteq \delta_{\mathbf{eqB}}(R)(1)$.

To show that $\delta_{\mathbf{eqB}}(R)(1) \subseteq \mathcal{B}(Q(R))$, let $\nu \in \delta_{\mathbf{eqB}}(R)$. Then there is $J \trianglelefteq R$ with $r_R(J)_R \leq^{\text{ess}} \nu E(R_R)$. By Proposition 8.3.3(i) and Theorem 8.3.8, $r_R(J)_R \leq^{\text{ess}} dQ(R)_R$ for some $d \in \mathcal{B}(Q(R))$. From Lemma 8.3.10, there exists ϕ in $\mathcal{B}(\text{End}(E(R_R)))$ such that $d = \phi(1)$. Thus $\nu(1) = \phi(1) = d \in \mathcal{B}(Q(R))$ as in the proof of Theorem 8.3.11(iv). Hence, $\delta_{\mathbf{eqB}}(R)(1) \subseteq \mathcal{B}(Q(R))$. Therefore, $\mathcal{B}(Q(R)) = \delta_{\mathbf{eqB}}(R)(1)$. So $\langle R \cup \mathcal{B}(Q(R))\rangle_{Q(R)} = \langle R \cup \delta_{\mathbf{eqB}}(R)(1)\rangle_{Q(R)}$.

Consequently, $\langle R \cup \delta_{\mathbf{eqB}}(R)(1)\rangle_{Q(R)} = R(\mathbf{eqB}, Q(R))$ from the definition of pseudo ring hulls, since $\langle R \cup \mathcal{B}(Q(R))\rangle_{Q(R)}$ is right essentially quasi-Baer by part (i). Hence, $\widehat{Q}_{\mathbf{eqB}}(R) = \langle R \cup \mathcal{B}(Q(R))\rangle_{Q(R)} = R(\mathbf{eqB}, Q(R))$. □

We note that from Theorems 3.2.37 and 8.3.17 when R is a semiprime ring, $\langle R \cup \mathcal{B}(Q(R))\rangle_{Q(R)}$ is also the strongly FI-extending absolute to $Q(R)$ ring hull of R. The following example shows that the semiprimeness of R in Theorem 8.3.17 is not a superfluous condition.

Example 8.3.18 There is a right nonsingular ring R which is not semiprime and $\langle R \cup \mathcal{B}(Q(R))\rangle_{Q(R)} \neq \widehat{Q}_{\mathbf{qB}}(R)$. Let F be a field, and put

$$R = \begin{bmatrix} F & F & F \\ 0 & F & 0 \\ 0 & 0 & F \end{bmatrix}.$$

Observe that $\langle R \cup \mathcal{B}(Q(R))\rangle_{Q(R)} = R\mathcal{B}(Q(R))$ since R has an identity. Also we see that R is quasi-Baer by Corollary 5.4.2 or Theorem 5.6.5. Therefore $\widehat{Q}_{\mathbf{qB}}(R) = R$. As R is right Artinian, $\text{Soc}(R_R) \leq^{\text{ess}} R_R$. Since $\text{Soc}(R_R)$ is the intersection of all essential right ideals of R, $\text{Soc}(R_R)$ is the smallest essential right ideal of R. Also as R is right nonsingular, $\text{Soc}(R_R)$ is the smallest dense right ideal of R from Proposition 1.3.14. If $q \in Q(R)$, then $q\text{Soc}(R_R) \subseteq R$, and so $q\text{Soc}(R_R) \subseteq \text{Soc}(R_R)$. By Proposition 1.3.11(ii), $\ell_{Q(R)}(\text{Soc}(R_R)) = 0$. Hence, $Q(R) \cong \text{End}(\text{Soc}(R_R))$. As $\text{Soc}(R_R) = \ell_R(J(R))$, $\text{Soc}(R_R) = M_R \oplus N_R$, where

$$M = \begin{bmatrix} 0 & F & 0 \\ 0 & F & 0 \\ 0 & 0 & 0 \end{bmatrix} \text{ and } N = \begin{bmatrix} 0 & 0 & F \\ 0 & 0 & 0 \\ 0 & 0 & F \end{bmatrix}.$$

So $Q(R) \cong \text{End}(M_R \oplus N_R)$. In this case, by straightforward computation,

$$Q(R) \cong \text{End}(M_R) \oplus \text{End}(N_R) = \text{End}(M_F) \oplus \text{End}(N_F) \cong \text{Mat}_2(F) \oplus \text{Mat}_2(F).$$

Now $|\mathcal{B}(R)| = 2$. But $|\mathcal{B}(Q(R))| = 4$. Thus, $R = \widehat{Q}_{\mathbf{qB}}(R) \neq R\mathcal{B}(Q(R))$.

Since idempotents as well as various properties lift modulo the prime radical, Theorem 8.3.17 provides an effective mechanism for transferring information between an arbitrary ring R and $\widehat{Q}_{\mathbf{qB}}(R/P(R))$ (or $\widehat{Q}_{\mathbf{FI}}(R/P(R))$) via

$$R \xrightarrow{\mu} R/P(R) \xrightarrow{\iota} \widehat{Q}_{\mathbf{qB}}(R/P(R)),$$

where μ is the natural homomorphism and ι is the inclusion.

Corollary 8.3.19 *Let T be a semiprime right ring of quotients of a ring R. Then T is quasi-Baer (and right FI-extending) if and only if $\mathcal{B}(Q(R)) \subseteq T$.*

Proof Proposition 8.3.3(i), Theorems 3.2.37 and 8.3.17 yield the result. □

It is worth noting that if we modify the ring R in Example 8.2.9 and instead of a field take F to be a commutative domain which is not a field, then R is neither semiprime nor right FI-extending. Now, $T = \text{Mat}_3(F)$ is a semiprime quasi-Baer (and right FI-extending) right ring of quotients of R such that $\mathcal{B}(Q(R)) \subseteq T$. But observe that $T \neq Q(R) = \text{Mat}_3(K)$, where K is the field of fractions of F. If R is a semiprime ring, $Q^s(R)$, $Q^m(R)$, and $Q(R)$ are all semiprime rings. Also, they contain $\mathcal{B}(Q(R))$. If R is a semiprime ring with identity, then the central closure of R and the normal closure of R are semiprime and contain $\mathcal{B}(Q(R))$. So Theorem 8.3.17 or Corollary 8.3.19 yields the following consequence.

Corollary 8.3.20 (i) *If R is a semiprime ring, then $Q^s(R)$, $Q^m(R)$, and $Q(R)$ are quasi-Baer and right FI-extending.*

(ii) *If R is a semiprime ring with identity, then the central closure and the normal closure are quasi-Baer and right FI-extending.*

There is a semiprime ring R for which neither $Q^m(R)$ nor $Q^s(R)$ is Baer. In fact, there is a simple ring R which is not a domain and 0, 1 are its only idempotents (see Example 3.2.7(ii)). Then $Q^m(R) = R$ and $Q^s(R) = R$. So neither $Q^m(R)$ nor $Q^s(R)$ is Baer.

Corollary 8.3.21 *Let R be a right Osofsky compatible ring with identity. If R has a right FI-extending right essential overring which is a subring of $E(R_R)$, then $E(R_R)$ is right FI-extending. In particular, if $Q(R)$ is semiprime, then $E(R_R)$ is right FI-extending.*

Proof Let S be a right FI-extending right essential overring of R which is a subring of the ring $E(R_R)$. Then $E(R_R)$ is a right essential overring of S. Thus $E(R_R)$ is a right FI-extending ring by Theorem 8.1.8(i). If $Q(R)$ is semiprime, then from Corollary 8.3.20(i), $Q(R)$ is right FI-extending. By Proposition 7.1.11, $Q(R)$ is a subring of $E(R_R)$, so $E(R_R)$ is a right essential overring of $Q(R)$. Hence, Theorem 8.1.8(i) yields that $E(R_R)$ is a right FI-extending ring. □

We remark that the ring R in Example 7.3.6 is right FI-extending and right Osofsky compatible, so $E(R_R)$ is right FI-extending by Corollary 8.3.21.

A ring R is said to have *no nonzero n-torsion* (n is a positive integer) if $na = 0$ with $a \in R$ implies $a = 0$.

Theorem 8.3.22 *Let $R[G]$ be the group ring of a group G over a ring R with identity. Then $R[G]$ is semiprime if and only if R is semiprime and R has no $|N|$-torsion for any finite normal subgroup N of G.*

Proof See [264, Proposition 8, p. 162] or [341, Theorem 2.13, p. 131]. □

The next corollary is obtained from Theorems 8.3.22 and 8.3.17. It is of interest to compare this result with Theorem 6.3.10(ii).

Corollary 8.3.23 *Assume that $R[G]$ is the semiprime group ring of a group G over a ring R with identity. If $R[G]$ is quasi-Baer, then $|N|^{-1} \in R$ for any finite normal subgroup N of G.*

Proof Let N be a finite normal subgroup of G. Because $R[G]$ is semiprime, R has no $|N|$-torsion by Theorem 8.3.22. Let $e = |N|^{-1} \sum_{g \in N} g$. Then

$$e \in Q^m(R)[G] \subseteq Q^m(R[G]) \subseteq Q(R[G])$$

(see the proof of Theorem 9.3.1(i)). Further, we see that $e \in \mathcal{B}(Q(R[G]))$. From Theorem 8.3.17, $e \in R[G]$ since $R[G]$ is quasi-Baer. So $|N|^{-1} \in R$. □

The next example illustrates the existence of a right nonsingular ring R which is not semiprime such that $\mathcal{B}(Q(R)) \subseteq R$, but R is not quasi-Baer.

Example 8.3.24 For a field F, as in Example 3.2.9, let

$$R = \begin{bmatrix} F\mathbf{1} & \mathrm{Mat}_2(F) & \mathrm{Mat}_2(F) \\ 0 & F\mathbf{1} & \mathrm{Mat}_2(F) \\ 0 & 0 & F\mathbf{1} \end{bmatrix}$$

be a subring of $T_3(\mathrm{Mat}_2(F))$, where $\mathbf{1}$ is the identity matrix in $\mathrm{Mat}_2(F)$. Then we see that R is right nonsingular. However, $\langle R \cup \mathcal{B}(Q(R))\rangle_{Q(R)} (= R)$ is not quasi-Baer (see Example 3.2.9).

In contrast to Examples 8.3.18 and 8.3.24, there exists a nonsemiprime ring R for which Theorem 8.3.17(ii) holds true as in the next example.

Example 8.3.25 Let A be a QF-ring with $J(A) \neq 0$. Assume that A is right strongly FI-extending, and A has nontrivial central idempotents, while the subring of A generated by 1_A contains no nontrivial idempotents (e.g., $A = \mathbb{Q} \oplus \mathrm{Mat}_2(\mathbb{Z}_4)$). Let $\mathbf{1}$ denote the identity of $\prod_{i=1}^{\infty} A_i$, where $A_i = A$. Take R to be the subring of $\prod_{i=1}^{\infty} A_i$ generated by $\mathbf{1}$ and $\oplus_{i=1}^{\infty} A_i$. We note that $Q(R) = \prod_{i=1}^{\infty} A_i = E(R_R)$ and $\langle R \cup \mathcal{B}(Q(R))\rangle_{Q(R)} = R\mathcal{B}(Q(R))$.

In this case, we have the following:

(i) R is not right FI-extending and $R\mathcal{B}(Q(R))$ is not quasi-Baer.
(ii) $Q_{\mathbf{FI}}(R) = R(\mathbf{FI}, Q(R)) = R\mathcal{B}(Q(R))$.
(iii) R has no right and left essential overring which is quasi-Baer.

Let k be a nontrivial central idempotent of A. Let ι_i denote the i-th canonical injection, respectively of the direct product. Let K be the ideal of R generated by $\{\iota_i(k) \mid 1 \le i < \infty\}$. Then there exists no $b^2 = b \in R$ such that $K_R \le^{\mathrm{ess}} bR_R$. So R is not right FI-extending.

We claim that $R\mathcal{B}(Q(R))$ is not quasi-Baer. For this, first we observe that $\mathbf{S}_\ell(Q(R)) = \mathcal{B}(Q(R))$ as $\mathbf{S}_\ell(A_i) = \mathcal{B}(A_i)$ for each i by [262, Exercise 16, p. 421]. Suppose that $Q(R)$ is quasi-Baer. Take $q \in Q(R)$ such that $qQ(R)q = 0$. Now we note that $r_{Q(R)}(qQ(R)) = \alpha Q(R)$ such that $\alpha \in \mathbf{S}_\ell(Q(R)) = \mathcal{B}(Q(R))$. Since $q \in r_{Q(R)}(qQ(R))$, $q = \alpha q = q\alpha = 0$. Therefore $Q(R)$ is semiprime, a contradiction. So $Q(R)$ is not quasi-Baer.

Because A is QF, $Q(R) = Q^\ell(R) = E(R_R) = E(_RR)$. Therefore the ring $R\mathcal{B}(Q(R))$ is not quasi-Baer by Theorem 8.1.9(i). Further, R has no right and left essential overring which is quasi-Baer from Theorem 8.1.9(i).

We prove that $\delta_{\mathbf{FI}}(R)(\mathbf{1}) = \mathcal{B}(Q(R))$. For this, let $f \in \delta_{\mathbf{FI}}(R)$. Then there exists $I \trianglelefteq R$ such that $I_R \leq^{\text{ess}} fE(R_R) = fQ(R)_R = f(\mathbf{1})Q_R$, because $\text{End}(E(R_R)) = \text{End}(Q(R)_R) = \text{End}(Q(R)_{Q(R)})$.

Furthermore, we note that $f(\mathbf{1})^2 = f(\mathbf{1})f(\mathbf{1}) = f(\mathbf{1}f(\mathbf{1})) = f(f(\mathbf{1})) = f(\mathbf{1})$.

Let π_i be the canonical projection of the direct product. Then $\pi_i(I) \trianglelefteq A_i$. By [262, Exercise 16, p. 421], there is $e_i \in \mathcal{B}(A_i)$ such that $\pi_i(I)_{A_i} \leq^{\text{ess}} e_iA_{i\,A_i}$, because A_i is right strongly FI-extending by assumption. Let $e \in Q(R)$ such that $\pi_i(e) = e_i$ for all i. Then we see that $I_R \leq^{\text{ess}} eQ(R)_R$ and $e \in \mathcal{B}(Q(R))$. So $f(\mathbf{1}) = e$. Thus, $\delta_{\mathbf{FI}}(R)(\mathbf{1}) \subseteq \mathcal{B}(Q(R))$.

Next, say $b \in \mathcal{B}(Q(R))$. Then $(bR \cap R)_R \leq^{\text{ess}} bR_R \leq^{\text{ess}} bQ(R)_R$. There exists $\lambda \in \mathcal{B}(\text{End}(E(R_R)))$ such that $b = \lambda(\mathbf{1})$ from Lemma 8.3.10, and hence $bQ(R)_R = \lambda(\mathbf{1})Q(R)_R = \lambda Q(R)_R$. So $\lambda \in \delta_{\mathbf{FI}}(R)$ and $b = \lambda(\mathbf{1}) \in \delta_{\mathbf{FI}}(R)(\mathbf{1})$, thus $\mathcal{B}(Q(R)) \subseteq \delta_{\mathbf{FI}}(R)(\mathbf{1})$. Hence $\mathcal{B}(Q(R)) = \delta_{\mathbf{FI}}(R)(\mathbf{1})$. Therefore we have that $S := \langle R \cup \delta_{\mathbf{FI}}(R)(\mathbf{1})\rangle_{Q(R)} = R\mathcal{B}(Q(R))$.

To show that $S = R(\mathbf{FI}, Q(R))$, let $J \trianglelefteq R\mathcal{B}(Q(R))$. First, we note that $\text{End}(E(R_R)) = \text{End}(Q(R)_R) = \text{End}(Q(R)_{Q(R)}) \cong Q(R)$. Thus, it follows that $(J \cap R)_R \leq^{\text{ess}} J_R \leq^{\text{ess}} E(J_R) = hQ(R)_R$ with $h^2 = h \in Q(R)$. Since $J \cap R \trianglelefteq R$, there is $g \in \mathcal{B}(Q(R))$ with $(J \cap R)_R \leq^{\text{ess}} gQ(R)_R$ from the preceding argument. Hence $h = g$, and thus $J_R \leq^{\text{ess}} gQ(R)_R$. Therefore, $J = Jg \subseteq R\mathcal{B}(Q(R))$. Hence, we have that $J_R \leq^{\text{ess}} gR\mathcal{B}(Q(R))_R$, and thus $J_{Q(R)} \leq^{\text{ess}} gR\mathcal{B}(Q(R))_{Q(R)}$. Whence $R\mathcal{B}(Q(R))$ is right FI-extending, so $S = R(\mathbf{FI}, Q(R))$.

Next, we show that $S = Q_{\mathbf{FI}}(R)$. Let T be a right FI-extending right ring of quotients of R. Take $c \in \mathcal{B}(Q(R))$. Then $cQ(R) \cap T \trianglelefteq T$. Since T is right FI-extending, there is $s^2 = s \in T$ such that $(cQ(R) \cap T)_T \leq^{\text{ess}} sT_T$.

Therefore $(cQ(R) \cap T)_R \leq^{\text{ess}} sT_R$ from Lemma 8.1.3(i), and hence it follows that $(cQ(R) \cap T)_R \leq^{\text{ess}} sQ(R)_R$, thus $(cQ(R) \cap R)_R \leq^{\text{ess}} sQ(R)_R$. Also we see that $(cQ(R) \cap R)_R \leq^{\text{ess}} cQ(R)_R$. So $c = s \in T$. Thus $\mathcal{B}(Q(R)) \subseteq T$, and hence S is a subring of T. Therefore, $S = Q_{\mathbf{FI}}(R)$.

Now from Theorems 8.3.11 and 8.3.17, we see that $\langle R \cup \mathcal{B}(Q(R))\rangle_{Q(R)}$ is a ring hull for the **IC** class, as well as a ring hull for a semiprime ring R in the **qB** and **FI** classes. This motivates our interest in the transfer of information between R and the ring $\langle R \cup \mathcal{B}(Q(R))\rangle_{Q(R)}$.

Let S be an overring of a ring R. We consider the following properties between prime ideals of R and S (see [248, p. 28]).

(1) *Lying over* (LO). For any prime ideal P of R, there exists a prime ideal Q of S such that $P = Q \cap R$.

(2) *Going up* (GU). Given prime ideals $P_1 \subseteq P_2$ of R and Q_1 of S with $P_1 = Q_1 \cap R$, there exists a prime ideal Q_2 of S with $Q_1 \subseteq Q_2$ and $P_2 = Q_2 \cap R$.

(3) *Incomparable* (INC). Two different prime ideals of S with the same contraction in R are not comparable.

Lemma 8.3.26 *Let R be a subring of a ring T and $\emptyset \neq \mathbb{E} \subseteq \mathbf{S}_\ell(T) \cup \mathbf{S}_r(T)$. Assume that S is the subring of T generated by R and $\mathbb{E}$.*

(i) If K is a prime ideal of S, then $R/(K \cap R) \cong S/K$.

(ii) LO, GU, and INC hold between R and S. In particular, LO, GU, and INC hold between R and $\langle R \cup \mathcal{B}(Q(R))\rangle_{Q(R)}$.

Proof (i) Let $\overline{S} = S/K$. Assume that $e \in \mathbb{E}$ such that $e \notin K$. Then $e \in \mathbf{S}_\ell(T)$ or $e \in \mathbf{S}_r(T)$. First, we show that $\overline{e} = e + K \in S/K$ is an identity of S/K. Without loss of generality, assume that $e \in \mathbf{S}_\ell(T)$. Then $\overline{0} \neq \overline{e} \in \mathbf{S}_\ell(\overline{S})$, so $\overline{S} = \overline{e}\,\overline{S} \oplus r_{\overline{S}}(\overline{e})$. As $\overline{e} \in \mathbf{S}_\ell(\overline{S})$, $(r_{\overline{S}}(\overline{e}))(\overline{e}\,\overline{S}) = \overline{0}$. Thus, $r_{\overline{S}}(\overline{e}) = \overline{0}$ because $\overline{S}$ is a prime ring. So $\overline{e}$ is a left identity for $\overline{S}$. Also, $\overline{S} = \overline{S}\,\overline{e} \oplus \ell_{\overline{S}}(\overline{e})$. As $\overline{e} \in \mathbf{S}_\ell(\overline{S})$, $(\ell_{\overline{S}}(\overline{e}))(\overline{S}\,\overline{e}) = \overline{0}$. Thus, $\ell_{\overline{S}}(\overline{e}) = \overline{0}$ since $\overline{S}$ is a prime ring. So $\overline{S} = \overline{S}\,\overline{e}$. Therefore, $\overline{e}$ is an identity element for $\overline{S}$. A similar argument works if $e \in \mathbf{S}_r(T)$.

From the preceding argument, for $f \in \mathbb{E}$, either $f + K = \overline{0}$ or $f + K$ is an identity of S/K. We define $\varphi : R \to S/K$ by $\varphi(r) = r + K$. Because S is generated by R and $\mathbb{E}$, φ is a ring epimorphism. Also $\mathrm{Ker}(\varphi) = K \cap R$. Thus, $R/(K \cap R) \cong S/K$.

(ii) (LO) Assume that P is a prime ideal of R. By Zorn's lemma, there exists an ideal K of S maximal with respect to $K \cap R \subseteq P$. Then K is a prime ideal of S. By (i), $R/(K \cap R) \cong S/K$. Since $P/(K \cap R)$ is a prime ideal of $R/(K \cap R)$ $(\cong S/K)$, there is a prime ideal K_0 of S with $K \subseteq K_0$, so K_0/K is a prime ideal of S/K, and $K_0/K = \overline{\varphi}(P/(K \cap R))$, where $\overline{\varphi}$ is the isomorphism from $R/(K \cap R)$ to S/K induced from φ in the proof of part (i). Therefore $K_0 = P + K$, hence we obtain that $K_0 \cap R = P + (K \cap R) = P$. Therefore, LO holds.

(GU) Suppose that $P_1 \subseteq P_2$ are prime ideals of R and K_1 is a prime ideal of S such that $K_1 \cap R = P_1$. Then by part (i), $R/P_1 \cong S/K_1$. By the same argument for LO, there is a prime ideal K_2 of S such that $K_1 \subseteq K_2$ and $K_2 \cap R = P_2$. Thus GU holds.

(INC) Suppose that K_1, K_2 are prime ideals of S and P is a prime ideal of R such that $K_1 \cap R = K_2 \cap R = P$. Assume that $K_1 \subseteq K_2$.

First, we show that $K_2/K_1 = \{r + K_1 \mid r \in K_2 \cap R\}$. For this, we observe that $S/K_1 = \{a + K_1 \mid a \in R\}$ by the argument in the proof of part (i). Let $r \in K_2 \cap R$. Then $r + K_1 \in K_2/K_1$, so $\{r + K_1 \mid r \in K_2 \cap R\} \subseteq K_2/K_1$.

Let $k_2 + K_1 \in K_2/K_1$. Then $k_2 + K_1 \in S/K_1$, so $k_2 + K_1 = a + K_1$ for some a in R. Thus, $k_2 = a + k_1$ for some $k_1 \in K_1$, hence

$$a = k_2 - k_1 \in K_2 + K_1 = K_2.$$

Therefore $a \in K_2 \cap R$. Thus $k_2 + K_1 \in \{r + K_1 \mid r \in K_2 \cap R\}$, so we have that

$$K_2/K_1 = \{r + K_1 \mid r \in K_2 \cap R\}.$$

As $P = K_1 \cap R = K_2 \cap R$, we see that $K_2/K_1 = 0$. Hence $K_2 = K_1$. □

The next theorem, due to Fisher and Snider [170], is a characterization of regular rings.

Theorem 8.3.27 *A ring R is regular if and only if the following hold:*

(i) *R is semiprime.*
(ii) *The union of any chain of semiprime ideals of R is semiprime.*
(iii) *Every prime factor ring of R is regular.*

Proof See [170, Theorem 1.1] or [183, Theorem 1.17]. □

A class ϱ of rings (not necessarily satisfying $\ell_R(R)=0$) is called a *special class* if ϱ is a class of prime rings that is hereditary (i.e., closed with respect to ideals) and closed with respect to ideal essential extensions. That is, if I is in ϱ and $I \trianglelefteq R$ that is ideal essential in R, then R is in ϱ (see [176, p. 80]). Let ϱ be a special class of rings. The *special radical* $\varrho(R)$ for a ring R is the intersection of all ideals I of R such that R/I is a ring in the special class ϱ. Note that the class of special radicals includes most well-known radicals (e.g., the prime radical, the Jacobson radical, the Brown-McCoy radical, the nil radical, and the generalized nil radical, etc.). See [139] and [176] for more details.

For a ring R with identity, the *classical Krull dimension* $\text{kdim}(R)$ is the supremum of all lengths of chains of prime ideals of R. We show that various types of information transfer between a ring R and $\langle R \cup \mathcal{B}(Q(R))\rangle_{Q(R)}$. The transference of information in Lemma 8.3.26 and Theorem 8.3.28 is used to study $\widehat{Q}_{\mathbf{qB}}(R)$ (or $\widehat{Q}_{\mathbf{FI}}(R)$) when R is a semiprime ring.

Theorem 8.3.28 *Let R be a subring of a ring T and $\emptyset \neq \mathbb{E} \subseteq \mathbf{S}_\ell(T) \cup \mathbf{S}_r(T)$. Assume that S is the subring of T generated by R and $\mathbb{E}$. Then:*

(i) *$\varrho(R) = \varrho(S) \cap R$, where ϱ is a special radical. In particular, we have that $\varrho(R) = \varrho(\langle R \cup \mathcal{B}(Q(R))\rangle_{Q(R)}) \cap R$.*
(ii) *R is strongly π-regular if and only if S is strongly π-regular. Hence, R is strongly π-regular if and only if $\langle R \cup \mathcal{B}(Q(R))\rangle_{Q(R)}$ is strongly π-regular.*
(iii) *If S is regular, then so is R.*
(iv) *If the ring R has identity, then* $\text{kdim}(R) = \text{kdim}(S)$. *Thus, we have that* $\text{kdim}(R) = \text{kdim}(R\mathcal{B}(Q(R)))$.

Proof (i) Let K be a prime ideal of S such that S/K is in the special class of ρ. From Lemma 8.3.26, $R/(K \cap R)$ is in the special class of ϱ. Therefore $\varrho(R) \subseteq \varrho(S) \cap R$. As in the proof of LO in Lemma 8.3.26, $\varrho(S) \cap R \subseteq \varrho(R)$.

(ii) This part is a consequence of Lemma 8.3.26 and Theorem 1.2.18 (note that Theorem 1.2.18 holds for rings not necessarily with an identity).

(iii) Since S is regular, R is semiprime by part (i). Let $I_1 \subseteq I_2 \subseteq \dots$ be a chain of semiprime ideals of R. Let $\mathbf{U}_k$ be the set of all prime ideals of R containing I_k, for $k = 1, 2, \dots$. Then I_k is the intersection of all prime ideals in $\mathbf{U}_k$. By LO in Lemma 8.3.26, for each $P \in \mathbf{U}_1$, there exists a prime ideal K of S such that

$P = K \cap R$. Let $\mathbf{V}_1$ be the set of all prime ideals K of S such that $K \cap R \in \mathbf{U}_1$, and let J_1 be the intersection of all prime ideals K in $\mathbf{V}_1$. Then $J_1 \cap R = I_1$ by using Lemma 8.3.26.

Next, consider $\mathbf{U}_2$. Then $\mathbf{U}_2 \subseteq \mathbf{U}_1$ since $I_1 \subseteq I_2$. Let $\mathbf{V}_2$ be the set of prime ideals K such that $K \cap R \in \mathbf{U}_2$. Let J_2 be the intersection of all prime ideals in $\mathbf{V}_2$. Because $\mathbf{U}_2 \subseteq \mathbf{U}_1$, $\mathbf{V}_2 \subseteq \mathbf{V}_1$ and so $J_1 \subseteq J_2$. Again by Lemma 8.3.26, $J_2 \cap R = I_2$. Continuing this process, there exists a chain of semiprime ideals $J_1 \subseteq J_2 \subseteq \dots$, of S with $J_n \cap R = I_n$ for each n. So $(\cup J_n) \cap R = \cup I_n$.

Note that $S/(\cup J_n)$ is semiprime by Theorem 8.3.27. Since $\cup J_n$ is a semiprime ideal of S, $\cup J_n = \cap K_\alpha$ for some prime ideals K_α of S. Then each $K_\alpha \cap R$ is a prime ideal of R by Lemma 8.3.26(i). So $\cup I_n = (\cup J_n) \cap R = (\cap K_\alpha) \cap R = \cap (K_\alpha \cap R)$ is a semiprime ideal of R.

Finally, say P is a prime ideal of R. By LO in Lemma 8.3.26, there is a prime ideal K of S with $P = K \cap S$ and $R/P \cong S/K$. Since S/K is regular, so is R/P. By Theorem 8.3.27, the ring R is regular.

(iv) The proof follows immediately from Lemma 8.3.26. □

Lemma 8.3.29 *Assume that T is an overring with identity, of a ring R and $\{f_1, \dots, f_n\} \subseteq \mathcal{B}(T)$. Then there exists a set of orthogonal idempotents $\{e_1, \dots, e_m\} \subseteq \mathcal{B}(T)$ such that $\sum_{i=1}^{n} f_i R \subseteq \sum_{i=1}^{m} e_i R$.*

Proof We use induction on n. If $n = 1$, then we are done by taking $e_1 = f_1$. Assume that $n \geq 2$ and the lemma is true for $n = k - 1$, and let $n = k$.

By induction, there exists a set of orthogonal idempotents $\{e_1, \dots, e_\ell\} \subseteq \mathcal{B}(T)$ such that $\sum_{i=1}^{k-1} f_i R \subseteq \sum_{i=1}^{\ell} e_i R$. Hence,

$$\sum_{i=1}^{k} f_i R = \sum_{i=1}^{k-1} f_i R + f_k R \subseteq \sum_{i=1}^{\ell} e_i R + f_k R$$

$$\subseteq f_k(1 - \sum_{i=1}^{\ell} e_i)R \oplus (\oplus_{i=1}^{\ell}(1 - f_k)e_i R) \oplus (\oplus_{i=1}^{\ell} f_k e_i R).$$

This yields the result. □

Corollary 8.3.30 *For a ring R with identity, the following are equivalent.*

(i) *R is regular.*
(ii) *$R\mathcal{B}(Q(R))$ is regular.*
(iii) *R is semiprime and $\widehat{Q}_{\mathbf{qB}}(R)$ is regular.*

Proof Assume that R is regular. Take $q \in R\mathcal{B}(Q(R))$. From Lemma 8.3.29, $q = a_1 e_1 + \cdots + a_m e_m \in R\mathcal{B}(Q(R))$, where $a_i \in R$, $e_i \in \mathcal{B}(Q(R))$, and e_i are orthogonal. Since R is regular, there is $b_i \in R$ with $a_i = a_i b_i a_i$ for each i. Let

$p = b_1e_1 + \cdots + b_me_m \in R\mathcal{B}(Q(R))$. Then $q = qpq$, so $R\mathcal{B}(Q(R))$ is regular. The rest of the proof follows by an easy application of Theorem 8.3.28(iii) and the fact that $\widehat{Q}_{\mathbf{qB}}(R) = R\mathcal{B}(Q(R))$ from Theorem 8.3.17 when R is semiprime. □

Lemma 8.3.26, Theorem 8.3.28, and Corollary 8.3.30 show the transference of some properties between R and $\widehat{Q}_{\mathbf{qB}}(R)$. Our next example indicates that in general these properties do not transfer between R and its right rings of quotients which properly contain $\widehat{Q}_{\mathbf{qB}}(R)$, in general.

Example 8.3.31 Let $R = \mathbb{Z}[C_2]$ be the group ring of the group $C_2 = \{1, g\}$ over the ring $\mathbb{Z}$. Then $\mathbb{Z}[C_2]$ is semiprime and $Q(\mathbb{Z}[C_2]) = \mathbb{Q}[C_2]$.

Note that $\mathcal{B}(\mathbb{Q}[C_2]) = \{0, 1, (1/2)(1+g), (1/2)(1-g)\}$. Thus, using Theorem 8.3.17, $\widehat{Q}_{\mathbf{qB}}(\mathbb{Z}[C_2]) = \{(a + c/2 + d/2) + (b + c/2 - d/2)g \mid a, b, c, d \in \mathbb{Z}\}$. Therefore

$$\mathbb{Z}[C_2] \subsetneq \widehat{Q}_{\mathbf{qB}}(\mathbb{Z}[C_2]) \subsetneq \mathbb{Z}[1/2][C_2] \subsetneq \mathbb{Q}[C_2],$$

where $\mathbb{Z}[1/2]$ is the subring of $\mathbb{Q}$ generated by $\mathbb{Z}$ and $1/2$.

Note that $\mathbb{Z}[C_2]/2\mathbb{Z}[C_2] \cong \mathbb{Z}_2[C_2]$, and $\mathbb{Z}_2[C_2]$ is a local ring. Thus there exists a prime ideal P (in fact, a maximal ideal) of $\mathbb{Z}[C_2]$ containing $2\mathbb{Z}[C_2]$. Also we note that $P \cap \mathbb{Z} = 2\mathbb{Z}$. Assume on the contrary that LO holds between $\mathbb{Z}[C_2]$ and $\mathbb{Z}[1/2][C_2]$. Then there exists a prime ideal K of $\mathbb{Z}[1/2][C_2]$ with $K \cap \mathbb{Z}[C_2] = P$. Now put $K_0 = K \cap \mathbb{Z}[1/2]$.

We see that $K_0 \cap \mathbb{Z} = K \cap \mathbb{Z}[1/2] \cap \mathbb{Z} = K \cap \mathbb{Z} = K \cap \mathbb{Z}[C_2] \cap \mathbb{Z} = P \cap \mathbb{Z} = 2\mathbb{Z}$. Thus $2 \in K_0$. But because K_0 is an ideal of $\mathbb{Z}[1/2]$, $1 = 2 \cdot (1/2) \in K_0$, hence $K = \mathbb{Z}[1/2][C_2]$, a contradiction. Thus, LO does not hold between $\mathbb{Z}[C_2]$ and $\mathbb{Z}[1/2][C_2]$.

Theorem 8.3.32 *Let R be a semiprime ring with identity. Then R has index of nilpotency at most n if and only if $Q_{\mathbf{qB}}(R)$ has index of nilpotency at most n. In particular, if R is reduced, then $Q_{\mathbf{qB}}(R) = Q_{\mathbf{B}}(R)$ and it is reduced.*

Proof Let R have index of nilpotency at most n. By Theorem 1.2.20(ii), R is right nonsingular. Hence $E(R_R) = Q(R)$ from Corollary 1.3.15. Therefore, we see that $\widehat{Q}_{\mathbf{qB}}(R) = Q_{\mathbf{qB}}(R)$. Now say $q \in Q_{\mathbf{qB}}(R)$. Then Lemma 8.3.29 yields that

$$q = a_1e_1 + \cdots + a_te_t,$$

where $a_i \in R$, $e_i \in \mathcal{B}(Q(R))$, and e_i are orthogonal.

Suppose that $q^k = 0$. We show that $q^n = 0$. If $k \le n$, then we are done. So assume that $k > n$. In this case, $q^k = a_1^k e_1 + \cdots + a_t^k e_t = 0$. Thus $a_i^k e_i = 0$ for all i. Note that $\mathcal{B}(Q(R)) = \mathcal{B}(Q^m(R))$ (recall that $Q^m(R)$ denotes the Martindale right ring of quotients of R). Hence, there is $I_i \trianglelefteq R$ with $\ell_R(I_i) = 0$ and $e_i I_i \subseteq R$. Therefore, $a_i^k e_i I_i = 0$ and $e_i I_i \subseteq r_R(a_i^k)$. Since R has index of nilpotency at most n, by Theorem 1.2.20(i) $r_R(a_i^k) = r_R(a_i^n)$, so $e_i I_i \subseteq r_R(a_i^n)$. Thus $a_i^n e_i I_i = 0$.

As $\ell_R(I_i) = 0$, $\ell_{Q(R)}(I_i) = 0$. Hence $a_i^n e_i = 0$ for each i. So

$$q^n = (a_1e_1 + \cdots + a_te_t)^n = a_1^n e_1 + \cdots + a_t^n e_t = 0.$$

Thus $Q_{\mathbf{qB}}(R)$ has index of nilpotency at most n. The converse is clear.

If R is reduced (so $Z(R_R)=0$), then $Q_{\mathbf{qB}}(R)$ is a reduced quasi-Baer ring by the preceding argument, so it is a Baer ring (see Exercise 3.2.44.10(i)). Say T is a right ring of quotients of R and T is Baer. Then T is quasi-Baer. Hence, $Q_{\mathbf{qB}}(R)\subseteq T$ by Theorem 8.3.17. Therefore, $Q_{\mathbf{qB}}(R)=Q_{\mathbf{B}}(R)$. □

Recall that a ring R is called strongly regular if R is regular and reduced (see 1.1.12). Corollary 8.3.30 and Theorem 8.3.32 yield the next result.

Corollary 8.3.33 *A ring R with identity is strongly regular if and only if $R\mathcal{B}(Q(R))$ is strongly regular.*

If R is a domain with identity which is not right Ore, then $R=Q_{\mathbf{qB}}(R)$ has index of nilpotency 1, but $Q(R)$ does not have bounded index of nilpotency. So we cannot replace $Q_{\mathbf{qB}}(R)$ with $Q(R)$ in Theorem 8.3.32.

By Theorem 8.3.32, a reduced ring with identity always has a Baer absolute ring hull. However a Baer absolute ring hull does not exist even for prime PI-rings with index of nilpotency 2, as shown in the next example.

Example 8.3.34 Let $R=\mathrm{Mat}_k(F[x,y])$, where F is a field and k is an integer such that $k\geq 2$. Then R is a prime PI-ring with index of nilpotency k. (In particular, if $k=2$, then R has index of nilpotency 2.) The ring R has the following properties. We note that $Q(R)=E(R_R)$, hence $\widehat{Q}_{\mathfrak{K}}(R)=Q_{\mathfrak{K}}(R)$ for any class $\mathfrak{K}$ of rings.

(i) The Baer absolute ring hull $Q_{\mathbf{B}}(R)$ does not exist.
(ii) The right extending absolute ring hull $Q_{\mathbf{E}}(R)$ does not exist.

As R is a prime ring, $R=Q_{\mathbf{qB}}(R)=Q_{\mathbf{FI}}(R)$. We claim that $Q_{\mathbf{B}}(R)$ does not exist (the same argument shows that $Q_{\mathbf{E}}(R)$ does not exist). Assume on the contrary that $Q_{\mathbf{B}}(R)$ exists. Note that $F(x)[y]$ and $F(y)[x]$ are Prüfer domains. So $\mathrm{Mat}_k(F(x)[y])$ and $\mathrm{Mat}_k(F(y)[x])$ are Baer rings by Theorem 6.1.4 (and right extending rings by Theorem 6.1.4). Since $Q(R)=\mathrm{Mat}_k(F(x,y))$,

$$Q_{\mathbf{B}}(R)\subseteq \mathrm{Mat}_k(F(x)[y])\cap \mathrm{Mat}_k(F(y)[x])=\mathrm{Mat}_k(F(x)[y]\cap F(y)[x]).$$

To see that $F(x)[y]\cap F(y)[x]=F[x,y]$, let

$$\begin{aligned}\gamma(x,y)&=f_0(x)/g_0(x)+(f_1(x)/g_1(x))y+\cdots+(f_m(x)/g_m(x))y^m\\&=h_0(y)/k_0(y)+(h_1(y)/k_1(y))x+\cdots+(h_n(y)/k_n(y))x^n\end{aligned}$$

be in $F(x)[y]\cap F(y)[x]$, where $f_i(x),\ g_i(x)\in F[x]$, $h_j(y),\ k_j(y)\in F[y]$, and $g_i(x)\neq 0$, $k_j(y)\neq 0$ for $i=0,1,\ldots,m$, $j=0,1,\ldots,n$. Let $\overline{F}$ be the algebraic closure of F. If $\deg(g_0(x))\geq 1$, then there exists $\alpha\in\overline{F}$ such that $g_0(\alpha)=0$. Therefore $\gamma(\alpha,y)$ cannot be defined. On the other hand, we observe that

$$\gamma(\alpha,y)=h_0(y)/k_0(y)+(h_1(y)/k_1(y))\alpha+\cdots+(h_n(y)/k_n(y))\alpha^n,$$

a contradiction. Thus $g_0(x)\in F$. Similarly, $g_1(x),\ldots,g_m(x)\in F$.

Hence $\gamma(x,y)\in F[x,y]$. Therefore $F(x)[y]\cap F(y)[x]=F[x,y]$, and so

$$Q_{\mathbf{B}}(R)=\mathrm{Mat}_k(F(x)[y]\cap F(y)[x])=\mathrm{Mat}_k(F[x,y]).$$

Thus $\text{Mat}_k(F[x,y])$ is a Baer ring, a contradiction because the commutative domain $F[x,y]$ is not Prüfer (see Theorem 6.1.4).

A ring R with identity is called *right Utumi* [382, p. 252] if it is both right nonsingular and right cononsingular. In the proof of Theorem 3.3.1 or by Lemma 4.1.16, every right extending ring is right cononsingular.

Proposition 8.3.35 *Let R be a reduced ring with identity. Then R is right Utumi if and only if $Q(R)$ is strongly regular.*

Proof See [382, Proposition 5.2, p. 254]. □

Proposition 8.3.36 *A reduced ring R with identity is right Utumi if and only if $Q_{\mathbf{qCon}}(R) = Q_{\mathbf{E}}(R) = R\mathcal{B}(Q(R))$.*

Proof Assume that R is right Utumi. Because R is reduced, $Z(R_R) = 0$ and from Theorem 8.3.32 $R\mathcal{B}(Q(R)) = Q_{\mathbf{qB}}(R) = Q_{\mathbf{B}}(R)$. Also, we observe that $Q(R) = Q(R\mathcal{B}(Q(R)))$ is strongly regular from Proposition 8.3.35. So $R\mathcal{B}(Q(R))$ is right Utumi, since $R\mathcal{B}(Q(R))$ is reduced by Theorem 8.3.32. Hence, $R\mathcal{B}(Q(R))$ is right cononsingular. As $R\mathcal{B}(Q(R))$ is Baer, $R\mathcal{B}(Q(R))$ is right extending by Theorem 3.3.1.

From Theorem 8.3.17, $R\mathcal{B}(Q(R)) = Q_{\mathbf{FI}}(R)$. If S is a right extending right ring of quotients of R, then S is right FI-extending, and hence $R\mathcal{B}(Q(R)) \subseteq S$. Thus, $R\mathcal{B}(Q(R)) = Q_{\mathbf{E}}(R)$. As $Q(R)$ is strongly regular, $\mathbf{I}(Q(R)) = \mathcal{B}(Q(R))$.

By Corollary 1.3.15, Theorem 2.1.25, and Proposition 2.1.32, $R\mathcal{B}(Q(R))$ is a right quasi-continuous ring. Let T be a right quasi-continuous right ring of quotients of R. Then again from Corollary 1.3.15, Theorem 2.1.25, and Proposition 2.1.32, $\mathcal{B}(Q(R)) = \mathcal{B}(Q(T)) \subseteq T$ as $Q(R) = Q(T)$. Thus $R\mathcal{B}(Q(R)) \subseteq T$, and hence $Q_{\mathbf{qCon}}(R) = R\mathcal{B}(Q(R))$. So $R\mathcal{B}(Q(R)) = Q_{\mathbf{E}}(R) = Q_{\mathbf{qCon}}(R)$.

Conversely, if $R\mathcal{B}(Q(R)) = Q_{\mathbf{E}}(R)$, then $R\mathcal{B}(Q(R))$ is right cononsingular by Theorem 3.3.1. Hence, $R\mathcal{B}(Q(R))$ is right Utumi. Further, $R\mathcal{B}(Q(R))$ is reduced by Theorem 8.3.32, so $Q(R) = Q(R\mathcal{B}(Q(R)))$ is strongly regular and thus R is right Utumi from Proposition 8.3.35. □

There exists a nonreduced right Utumi ring R for which the equalities

$$Q_{\mathbf{qCon}}(R) = Q_{\mathbf{E}}(R) \text{ and } Q_{\mathbf{qCon}}(R) = R\mathcal{B}(Q(R))$$

in Proposition 8.3.36 do not hold true, as the next example shows.

Example 8.3.37 Let $R = \text{Mat}_k(F[x])$, where F is a field and k is an integer such that $k > 1$. Then R is right Utumi by Proposition 3.3.2. Note that

$$E(R_R) = Q(R) = \text{Mat}_k(F(x)),$$

where $F(x)$ is the field of fractions of $F[x]$.

There is $e^2 = e \in Q(R)$ such that $e \notin R$. By Theorem 2.1.25, R is not right quasi-continuous. Now $\mathbf{R}\mathbf{B}(Q(R)) = R \neq Q_{\mathbf{qCon}}(R)$. From Theorem 6.1.4, R is right extending, so $R = Q_{\mathbf{E}}(R)$. Thus $Q_{\mathbf{E}}(R) \neq Q_{\mathbf{qCon}}(R)$.

For a semiprime ring R with identity, the notions of (right) FI-extending and quasi-Baer coincide by Theorem 3.2.37. Theorem 8.3.17 shows that the quasi-Baer ring hull of a semiprime ring exists and is precisely the same as its right FI-extending ring hull.

In view of this result, it is natural to ask: *Whether the right principally quasi-Baer ring hull and the right principally FI-extending ring hull exist for a semiprime ring and if so, are they equal?* In Theorem 8.3.39, an affirmative answer to these questions will be provided.

Burgess and Raphael [108] study ring extensions of regular rings with bounded index (of nilpotency). In particular, for a regular ring R with bounded index (of nilpotency), they obtain a unique closely related smallest overring, $R^{\#}$, which is "almost biregular" (see [108, p. 76 and Theorem 1.7]). Theorem 8.3.39 shows that their ring $R^{\#}$ is exactly the right principally FI-extending pseudo ring hull of a regular ring R with bounded index (of nilpotency). When R is commutative semiprime, the "weak Baer envelope" defined by Dobbs and Picavet in [141] is exactly the right p.q.-Baer ring hull $\widehat{Q}_{\mathbf{pqB}}(R)$ obtained in Theorem 8.3.39.

We use **pFI** and **fgFI** to denote the class of right principally FI-extending rings and the class of right finitely generated FI-extending rings, respectively (see Proposition 3.2.41 for **pFI** and **fgFI**). The following definition is useful for studying p.q.-Baer ring hulls.

Definition 8.3.38 For a ring R with identity, define

$$\mathcal{B}_p(Q(R)) = \{c \in \mathcal{B}(Q(R)) \mid \text{ there is } x \in R \text{ with } RxR_R \leq^{\text{ess}} cQ(R)_R\}.$$

The next Theorem 8.3.39 unifies the result by Burgess and Raphael [108] and that of Dobbs and Picavet [141].

Theorem 8.3.39 *Let R be a semiprime ring with identity. Then:*

(i) $\widehat{Q}_{\mathbf{pFI}}(R) = \langle R \cup \mathcal{B}_p(Q(R))\rangle_{Q(R)} = R(\mathbf{pFI}, Q(R))$.
(ii) $\widehat{Q}_{\mathbf{pqB}}(R) = \langle R \cup \mathcal{B}_p(Q(R))\rangle_{Q(R)}$.
(iii) $\widehat{Q}_{\mathbf{fgFI}}(R) = \langle R \cup \mathcal{B}_p(Q(R))\rangle_{Q(R)}$.

Proof (i) Using a proof similar to that of Theorem 8.3.11(iv), we obtain that $\delta_{\mathbf{pFI}}(R)(1) = \mathcal{B}_p(Q(R))$. Let $S = \langle R \cup \delta_{\mathbf{pFI}}(R)(1)\rangle_{Q(R)}$. Then we have that $S = \langle R \cup \mathcal{B}_p(Q(R))\rangle_{Q(R)}$. We show that S is right principally FI-extending. For this, take $0 \neq s \in S$. From Lemma 8.3.29, $s = \sum_{i=1}^{n} r_i b_i$, where each $r_i \in R$ and the b_i are orthogonal idempotents in $\mathcal{B}(S)$. From Proposition 8.3.3(i) and Theorem 8.3.8, we see that there is $c_i \in \mathcal{B}(Q(R))$ with $Rr_iR_R \leq^{\text{ess}} c_iQ(R)_R$ for each i. So each $c_i \in \mathcal{B}_p(Q(R))$. Hence, $s = \sum_{i=1}^{n} r_i b_i = \sum_{i=1}^{n} r_i c_i b_i$. Put $e_i = c_i b_i$ for each i. Then $s = \sum_{i=1}^{n} r_i e_i$. We note that the e_i are orthogonal idempotents in $\mathcal{B}(S)$.

Put $D = \oplus_{i=1}^{n} e_i S$. To see that $SsS_S \leq^{\text{ess}} D_S$, say $0 \neq y \in D$. Then there exist $y_i \in S$ for $1 \leq i \leq n$ so that $y = \sum_{i=1}^{n} e_i y_i$. In this case, there exists $e_j y_j \neq 0$ for some j, $1 \le j \le n$, and $v \in R$ such that $0 \ne e_j y_j v \in R$. Because

$$ye_j v = e_j y_j v = c_j b_j y_j v \in c_j R \text{ and } Rr_j R_R \le^{\text{ess}} c_j R_R,$$

there is $w \in R$ with $0 \ne ye_j vw \in Rr_j R$.

So $0 \ne y(e_j vw) = e_j y_j vw \in Rr_j e_j R = Rse_j R \subseteq SsS$ as $se_j = r_j e_j$. Hence $SsS_S \le^{\text{ess}} D_S$. Let $f = \sum_{i=1}^n e_i \in \mathcal{B}(S)$. Then S is right principally FI-extending since $SsS_S \le^{\text{ess}} D_S = \oplus_{i=1}^n e_i S_S = fS_S$. Therefore, $S = R(\mathbf{pFI}, Q(R))$.

Assume that T is a right ring of quotients of R and T is right principally FI-extending. Say $e \in \mathcal{B}_p(Q(R))$. Then there is $x \in R$ with $RxR_R \le^{\text{ess}} eQ(R)_R$. Note that $TxT = T(RxR)T \subseteq T(eQ(R))T = eQ(R)$, so $TxT_R \le^{\text{ess}} eQ(R)_R$. Hence $TxT_T \le^{\text{ess}} eQ(R)_T$. Since T is right principally FI-extending, there exists $c^2 = c \in T$ such that $TxT_T \le^{\text{ess}} cT_T \le^{\text{ess}} cQ(R)_T$. Thus $e = c$ because $e \in \mathcal{B}(Q(R))$. Hence, $e \in T$ for each $e \in \mathcal{B}_p(Q(R))$. So S is a subring of T. Thus, $S = \widehat{Q}_{\mathbf{pFI}}(R)$ and $\widehat{Q}_{\mathbf{pFI}}(R) = \langle R \cup \mathcal{B}_p(Q(R)) \rangle_{Q(R)} = R(\mathbf{pFI}, Q(R))$.

Parts (ii) and (iii) follow from part (i) and Proposition 3.2.41. □

Corollary 8.3.40 *Let R be a semiprime ring with identity. Then R is right p.q.-Baer if and only if $\mathcal{B}_p(Q(R)) \subseteq R$.*

Corollary 8.3.41 *Let R be a semiprime ring with identity.*

(i) *If K is a prime ideal of $\widehat{Q}_{\mathbf{pqB}}(R)$, then $\widehat{Q}_{\mathbf{pqB}}(R)/K \cong R/(K \cap R)$.*
(ii) *LO, GU, and INC hold between R and $\widehat{Q}_{\mathbf{pqB}}(R)$.*

Proof Theorem 8.3.39 and Lemma 8.3.26 yield the result. □

Corollary 8.3.42 *Let R be a semiprime ring with identity. Then:*

(i) *$\varrho(R) = \varrho(\widehat{Q}_{\mathbf{pqB}}(R)) \cap R$, where $\varrho(-)$ is a special radical of a ring.*
(ii) *R is strongly π-regular if and only if $\widehat{Q}_{\mathbf{pqB}}(R)$ is strongly π-regular.*
(iii) $\text{kdim}(R) = \text{kdim}(\widehat{Q}_{\mathbf{pqB}}(R))$.

Proof The proof follows from Theorems 8.3.28 and 8.3.39. □

Corollary 8.3.43 *Let R be a semiprime ring with identity. Then:*

(i) *R is regular if and only if $\widehat{Q}_{\mathbf{pqB}}(R)$ is regular.*
(ii) *R has index of nilpotency at most n if and only if $\widehat{Q}_{\mathbf{pqB}}(R)$ has index of nilpotency at most n.*
(iii) *R is strongly regular if and only if $\widehat{Q}_{\mathbf{pqB}}(R)$ is strongly regular.*

Proof Put $S = \widehat{Q}_{\mathbf{pqB}}(R)$. Then S is semiprime and $\widehat{Q}_{\mathbf{qB}}(S) = \widehat{Q}_{\mathbf{qB}}(R)$ by Theorem 8.3.17.

(i) If R is regular, then $\widehat{Q}_{\mathbf{qB}}(S)$ is regular by Corollary 8.3.30. Since S is semiprime, again by Corollary 8.3.30 S is regular. Conversely, if S is regular, then from Corollary 8.3.30 $\widehat{Q}_{\mathbf{qB}}(S) = \widehat{Q}_{\mathbf{qB}}(R)$ is regular, so R is regular.

(ii) and (iii) The proof follows immediately from Theorem 8.3.32, Corollary 8.3.33, and the argument used for the proof of part (i). □

Theorem 8.3.44 *Let R be a reduced ring with identity. Then the p.q.-Baer absolute ring hull $Q_{\mathbf{pqB}}(R)$ is the Rickart absolute ring hull of R.*

Proof Because R is reduced, $Z(R_R) = 0$. Hence, Corollary 1.3.15 yields that $Q(R) = E(R_R)$. By Theorem 8.3.39, $S := Q_{\mathbf{pqB}}(R)$ exists. From Corollary 8.3.43, S is reduced and so S is Rickart (see Exercise 3.2.44.10(ii)).

Let T be a (right) Rickart right ring of quotients of R. Take $e \in \mathcal{B}_p(Q(R))$. Then $e \in S$ and there exists $x \in R$ such that $RxR_R \leq^{\text{ess}} eQ(R)_R$. Hence $SxS_S \leq^{\text{ess}} eS_S$. As S is right nonsingular, $SxS_S \leq^{\text{den}} eS_S$ by Proposition 1.3.14, as a consequence $\ell_S(SxS) = \ell_S(eS) = S(1-e)$ from the proof of Lemma 8.3.7(i). Since S is semiprime, $r_S(SxS) = \ell_S(SxS)$. So $r_S(SxS) = S(1-e) = (1-e)S$. Further, as S is reduced, $r_S(x) = r_S(SxS) = (1-e)S$.

Because T is right Rickart, $r_T(x) = fT$ for some $f^2 = f \in T$. Observe that $r_R(x) = (1-e)S \cap R$ and $r_R(x) = r_T(x) \cap R$. Therefore, we have that

$$r_R(x)_R \leq^{\text{ess}} (1-e)S_R \leq^{\text{ess}} (1-e)Q(R)_R \text{ and } r_R(x)_R \leq^{\text{ess}} fT_R \leq^{\text{ess}} fQ(R)_R.$$

Thus $1-e = f$ as $1-e$ is central in $Q(R)$. Hence $e = 1-f \in T$, so $\mathcal{B}_p(Q(R)) \subseteq T$. From Theorem 8.3.39, $S \subseteq T$. Whence $Q_{\mathbf{pqB}}(R)$ is the Rickart absolute ring hull of R. □

When R is a semiprime ring with identity, $\widehat{Q}_{\mathbf{pqB}}(R) \subseteq \widehat{Q}_{\mathbf{qB}}(R)$. However, in the following example, we see that there exists a semiprime ring R with identity such that $\widehat{Q}_{\mathbf{pqB}}(R) \subsetneq \widehat{Q}_{\mathbf{qB}}(R)$.

Example 8.3.45 Let R be the ring as in Example 4.5.5. Then R is (right) p.q.-Baer, so $R = \widehat{Q}_{\mathbf{pqB}}(R)$. But R is not quasi-Baer. By Theorem 8.3.17,

$$\widehat{Q}_{\mathbf{qB}}(R) = R\mathcal{B}(Q(R)), \text{ therefore } \widehat{Q}_{\mathbf{qB}}(R) = Q(R) = \prod_{n=1}^{\infty} F_n,$$

where $F_n = \mathbb{Z}_2$ for $n = 1, 2, \ldots$. Thus, $\widehat{Q}_{\mathbf{pqB}}(R) \subsetneq \widehat{Q}_{\mathbf{qB}}(R)$ (further, we observe that $\widehat{Q}_{\mathbf{qB}}(R) = Q_{\mathbf{qB}}(R)$ and $\widehat{Q}_{\mathbf{pqB}}(R) = Q_{\mathbf{pqB}}(R)$ as R is right nonsingular).

In Theorem 8.3.47, we will see that there is a connection between the right FI-extending ring hulls of semiprime homomorphic images of R and the right FI-extending right rings of quotients of R. For this, we need the next lemma.

Lemma 8.3.46 *Assume that I is a proper ideal of a ring R with identity such that I is a complement of a right ideal of R. If $P(R) \subseteq I$, then R/I is a semiprime ring.*

Proof Let J be a right ideal of R such that I is a complement of J. First we show that $(I \oplus J)/I$ is essential in R/I as a right R/I-module. To see this, assume on the contrary that there exists a nonzero right R/I-submodule K/I of R/I such that

$[(I \oplus J)/I] \cap (K/I) = 0$. There is $y \in K$ with $y \notin I$. Then $(I + yR) \cap J \neq 0$. So there exist $c \in I$, $r \in R$, and $0 \neq x \in J$ such that $c + yr = x$. Then

$$yr = -c + x \in (I \oplus J) \cap K \subseteq I.$$

Hence $x \in I \cap J = 0$, a contradiction. So $(I \oplus J)/I$ is essential in R/I as a right R/I-module.

Next, let $0 \neq B/I \trianglelefteq R/I$ such that $(B/I)^2 = 0$. Then $B^2 \subseteq I$. Note that

$$(B/I) \cap [(I \oplus J)/I] \neq 0$$

because $(I \oplus J)/I$ is essential in R/I as a right R/I-module.

From the modular law, $B \cap (I \oplus J) = I \oplus (B \cap J)$. As $B \cap (I \oplus J) \nsubseteq I$, $I \oplus (B \cap J) \nsubseteq I$, and thus $B \cap J \neq 0$. But $(B \cap J)^2 \subseteq I \cap J = 0$ as $B^2 \subseteq I$. Hence $B \cap J \subseteq J \cap P(R) \subseteq J \cap I = 0$, which is a contradiction. Therefore, R/I is a semiprime ring. □

Theorem 8.3.47 *Assume that R is a ring with identity which is either semiprime or $Q(R) = E(R_R)$. Let I be a proper ideal of R such that I_R is closed in R_R. Then:*

(i) *There exists $e \in \mathbf{I}(Q(R))$ such that $I = (1 - e)Q(R) \cap R$.*
(ii) *$eR = eRe$ and $R(1 - e) = (1 - e)R(1 - e)$.*
(iii) *R/I is ring isomorphic to eRe.*
(iv) *If R is semiprime, then $eQ(R)e \subseteq Q(eRe)$.*
(v) *If $E(R_R) = Q(R)$, then $E(eRe_{eRe}) = eQ(R)e$ and $eQ(R)e = Q(eRe)$.*
(vi) *If $P(R) \subseteq I$, then R/I is semiprime and $\widehat{Q}_{\mathbf{FI}}(R/I) \cong \widehat{Q}_{\mathbf{FI}}(eRe)$.*
(vii) *Suppose that R is semiprime (resp., right nonsingular and semiprime). Then $\widehat{Q}_{\mathbf{FI}}(R/I) \cong e\widehat{Q}_{\mathbf{FI}}(R)e$ (resp., $Q_{\mathbf{FI}}(R/I) \cong eQ_{\mathbf{FI}}(R)e$).*

Proof (i) If R is semiprime, use Proposition 8.3.3(i) and Theorem 8.3.8. In this case, we observe that $e \in \mathcal{B}(Q(R))$. If $Q(R) = E(R_R)$, then the proof is routine.

(ii) If R is semiprime, the proof of this part is clear since $e \in \mathcal{B}(Q(R))$. For $Q(R) = E(R_R)$, let $r \in R$ with $er(1 - e) \neq 0$. Since R_R is dense in $Q(R)_R$, there exists $s \in R$ such that $(1 - e)s \in R$ and $er(1 - e)s \neq 0$. Then

$$(1 - e)s \in R \cap (1 - e)Q(R) = I.$$

Hence $0 \neq er(1 - e)s \in eI = 0$, a contradiction. So $eR(1 - e) = 0$. Consequently, $eR = eRe$ and $R(1 - e) = (1 - e)R(1 - e)$.

(iii) Define $f : R/I \to eRe$ by $f(r + I) = er$. As $eI = 0$, f is well defined. Clearly, f is a ring epimorphism. If $x + I \in \text{Ker}(f)$, then $x \in (1 - e)Q(R) \cap R$. By part (i), $x \in I$. Hence $\text{Ker}(f) = 0$. Thus, f is a ring isomorphism.

(iv) As $e \in \mathcal{B}(Q(R))$, $eRe_{eRe} \leq^{\text{den}} eQ(R)e_{eRe}$. So $eQ(R)e \subseteq Q(eRe)$.

(v) Let K be a right ideal of eRe and let $g : K \to eQ(R)e$ be an eRe-homomorphism. From part (ii) K, eRe, and $eQ(R)e$ are right R-modules, and g is an R-homomorphism. As $eQ(R)e \subseteq eQ(R)$ and $eQ(R)$ is the injective hull of eR_R, g can be extended to an R-homomorphism $\overline{g} : eR \to eQ(R)$. Now $\overline{g}$ can be extended to an R-homomorphism $\widetilde{g} : eQ(R) \to eQ(R)$. Therefore, $\widetilde{g}$ is a $Q(R)$-homomorphism as in the proof of Proposition 2.1.32.

As $eR = eRe$, $\overline{g}(eR) = \widetilde{g}(eRe) = \widetilde{g}(eRe)e = \widetilde{g}(eR)e \subseteq eQ(R)e$. By Baer's Criterion, $eQ(R)e$ is an injective right eRe-module. Further, we observe that $eRe_{eRe} \leq^{\text{den}} eQ(R)e_{eRe}$. Hence, $eQ(R)e$ is the injective hull of eRe as a right eRe-module and $eQ(R)e = Q(eRe)$.

(vi) Note that a closed right ideal of R is a complement of some right ideal of R (see Exercise 2.1.37.3). Hence this part is a consequence of part (iii), Lemma 8.3.46, and Theorem 8.3.17.

(vii) Let R be semiprime. Then $1-e \in \mathcal{B}(Q(R))$ by Proposition 8.3.3(i) and Theorem 8.3.8, so $e \in \mathcal{B}(Q(R))$. Hence $\mathcal{B}(eQ(R)e) = e\mathcal{B}(Q(R))e$. Thus we have that $\widehat{Q}_{\mathbf{FI}}(R/I) \cong \langle eRe \cup \mathcal{B}(eQ(R)e)\rangle_{eQ(R)e} = eR\mathcal{B}(Q(R))e = e\widehat{Q}_{\mathbf{FI}}(R)e$ from Theorem 8.3.17. If additionally $Z(R_R) = 0$, then eR_R is nonsingular, so $(R/I)_R$ is right nonsingular since $(R/I)_R \cong eR_R$ by modifying the proof of part (iii). Thus, R/I is a right nonsingular ring by [180, Proposition 1.28] and so eRe is a right nonsingular ring. The result follows from the fact that for any right nonsingular ring T, $\widehat{Q}_{\mathbf{FI}}(T) = Q_{\mathbf{FI}}(T)$ since $Q(T) = E(T_T)$. □

Corollary 8.3.48 *Let R be a semiprime ring with identity, S a ring with identity, and $\theta : R \to S$ a ring epimorphism such that* $\text{Ker}(\theta)$ *is a nonessential ideal of R. Then there exists a nonzero ring homomorphism $h : S \to \widehat{Q}_{\mathbf{FI}}(R)$.*

Proof Let $K = \text{Ker}(\theta)$ and $I = \ell_R(\ell_R(K))$. Then $K \in \mathfrak{D}_{\mathbf{IC}}(R)$ by Proposition 8.3.3(i) since R is semiprime. So I is the unique closure of K_R in R_R (see Exercise 8.3.58.5(i)). From Theorem 8.3.47(i), there exists $e \in \mathcal{B}(Q(R))$ such that $I = (1-e)Q(R) \cap R$. As K is not essential and R is semiprime, $\ell_R(K) \neq 0$ by Proposition 1.3.16, so $I \neq R$. We have the following sequence of ring homomorphisms $S \stackrel{\alpha}{\to} R/K \stackrel{\beta}{\to} R/I \stackrel{\lambda}{\to} \widehat{Q}_{\mathbf{FI}}(R/I) \stackrel{\delta}{\to} e\widehat{Q}_{\mathbf{FI}}(R)e \stackrel{\iota}{\to} \widehat{Q}_{\mathbf{FI}}(R)$, using Theorem 8.3.47, where α and δ are ring isomorphisms, β is a ring epimorphism, and λ and ι are inclusions. Take $h = \iota\delta\lambda\beta\alpha$. □

Proposition 8.3.49 *Let $I \in \mathfrak{D}_{\mathbf{IC}}(R)$. Then* $\text{Cen}(I) = I \cap \text{Cen}(R)$.

Proof Let $I \in \mathfrak{D}_{\mathbf{IC}}(R)$. Then $Q(I) = eQ(R)$ with $e \in \mathcal{B}(Q(R))$ by Theorem 8.3.8. So $\text{Cen}(I) \subseteq \text{Cen}(Q(I)) = \text{Cen}(eQ(R)) \subseteq \text{Cen}(Q(R))$. Therefore we have that $\text{Cen}(I) = I \cap \text{Cen}(R)$. □

A nonempty subset M of a ring R is called an *m-system* if $0 \notin M$ and for any $a, b \in M$ there exists $x \in R$ such that $axb \in M$ (see [296]). We note that an ideal P of a ring R maximal with respect to $P \cap M = \emptyset$, where M is an m-system, is always a prime ideal.

Theorem 8.3.50 *Let R be a semiprime ring with a descending chain of essential ideals $K_1 \supseteq K_2 \supseteq \dots$ such that $\bigcap_{i\geq 1} K_i = 0$. Then R has a prime ideal P such that $K_i \nsubseteq P$ for all $i \geq 1$.*

Proof We use the condition on $\{K_i\}_{i=1}^{\infty}$ to find a properly descending subsequence $\{L_i\}_{i=1}^{\infty}$ and nonzero elements $\{a_i\}$, $\{x_i\}$ such that $a_{i+1} = a_i x_i a_i$, $a_{i+1} \in L_i$ and $a_{i+1} \notin L_{i+1}$ for $i \geq 1$.

Let $L_1 = K_1$ and choose $0 \neq a_1 \in L_1$. Then we show that $a_1K_2a_1 \neq 0$. For this, assume on the contrary that $a_1K_2a_1 = 0$. Then $(K_2a_1K_2)(K_2a_1K_2) = 0$, so $K_2a_1K_2 = 0$ because R is semiprime. Now $\ell_R(K_2) = r_R(K_2) = 0$ since K_2 is essential in R, and hence $K_2a_1 = 0$. Again since $r_R(K_2) = 0$, $a_1 = 0$, a contradiction. Thus, $a_1K_2a_1 \neq 0$. From $\cap_{i\geq 1}K_i = 0$, there exists K_j with j minimal, such that $a_1K_2a_1 \not\subseteq K_j$, and hence there is $x_1 \in K_2$ such that $a_1x_1a_1 \notin K_j$. Let $L_2 = K_j$ and $a_2 = a_1x_1a_1$; then $a_2 \in L_1$ and $a_2 \notin L_2$.

Next, $a_2L_2a_2 \neq 0$ by the preceding argument. Choose L_3 such that $a_2L_2a_2 \not\subseteq L_3$. So there is $x_2 \in L_2$ with $a_3 := a_2x_2a_2 \notin L_3$. Note that $a_3 \in L_2$. Continue this procedure to get L_{i+1} and $a_{i+1} = a_ix_ia_i \in L_i$ but $a_{i+1} \notin L_{i+1}$ as needed. The sequence $\{a_i\}$ constitutes an m-system. In fact, let $a_\ell, a_n \in \{a_i\}$. If $\ell = n$, then $a_\ell x_n a_n = a_{n+1}$. So without of loss of generality, we may assume that $n > \ell$. Then $a_{n+1} = a_\ell[(x_\ell a_\ell)(x_{\ell+1}a_{\ell+1})\cdots(x_{n-1}a_{n-1})x_n]a_n$. Hence, an ideal P maximal with respect to $\{a_i\} \cap P = \emptyset$ is a prime ideal. By construction, $K_i \not\subseteq P$ for all $i \geq 1$. □

Lemma 8.3.51 *Let R be a semiprime ring and $I \trianglelefteq R$. Then:*

(i) *$\ell_R(I)$ is a semiprime ideal of R.*
(ii) *$(I \oplus \ell_R(I))/\ell_R(I)$ is an essential ideal of $R/\ell_R(I)$.*

Proof (i) To show that $\ell_R(I)$ is a semiprime ideal, let $a \in R$ such that $aRa \subseteq \ell_R(I)$. Then $aRaI = 0$, so $(aI)R(aI) = 0$. Thus, $aI = 0$ because R is semiprime. Hence, $a \in \ell_R(I)$, so $\ell_R(I)$ is a semiprime ideal.

(ii) Let $S = R/\ell_R(I)$. By part (i), S is a semiprime ring. To show that $V := (I \oplus \ell_R(I))/\ell_R(I)$ is essential in S, it suffices to see that $\ell_S(V) = 0$ by Proposition 1.3.16. Say $a + \ell_R(I) \in \ell_S(V)$, where $a \in R$. Then $aI \subseteq \ell_R(I)$, so $aI^2 = 0$. Hence, $(aI)^2 = 0$. Thus, $aI = 0$ because R is semiprime. Therefore, $a \in \ell_R(I)$, hence $a + \ell_R(I) = 0$. □

The following theorem is well known (see [366, Remark 1.2.14, Theorems 1.4.1 and 1.6.27]).

Theorem 8.3.52 *Let R be a semiprime PI-ring. Then R satisfies a standard identity $f_n(x_1, \ldots, x_n) = \sum_{\sigma\in S_n} \operatorname{sgn}(\sigma)x_{\sigma(1)}\cdots x_{\sigma(n)}$, where S_n is the symmetric group of degree n and $\operatorname{sgn}(\sigma)$ is the signature of $\sigma \in S_n$. Further, R satisfies $f_m(x_1, \ldots, x_m)$ for $m \geq n$.*

An ideal I of a ring is called a *PI-ideal* if I is a PI-ring as a ring by itself.

Theorem 8.3.53 *Let R be a semiprime ring such that R/P is a PI-ring for each prime ideal P of R. Then R contains a nonzero PI-ideal, and the sum of all PI-ideals of R is an essential ideal of R.*

Proof Put $\mathcal{F}_n = \{P \mid P$ is a prime ideal and R/P satisfies $f_n(x_1, \ldots, x_n)\}$ for $n \geq 2$, and let $K_n = \cap_{P\in\mathcal{F}_n} P$. Since $\mathcal{F}_2 \subseteq \mathcal{F}_3 \subseteq \ldots$ from Theorem 8.3.52, the sequence of ideals $\{K_j\}$ is a descending sequence of semiprime ideals with $\cap_{i\geq 2}K_i = 0$ since R is semiprime and $\cup_{i\geq 2}\mathcal{F}_i$ is the set of all prime ideals. We note that R/K_n embeds in $\prod_{P\in\mathcal{F}_n} R/P$, hence it satisfies a PI. If each K_i is essential, Theorem 8.3.50 yields a prime ideal P which contains none of the K_i. However $P \in \mathcal{F}_m$ for some $m \geq 2$ and so $K_m \subseteq P$, a contradiction. Thus there exists some K_n which is not essential. Hence, $\ell_R(K_n) \neq 0$ by Proposition 1.3.16. As R is semiprime, $\ell_R(K_n) \cap K_n = 0$ and so $\ell_R(K_n)$ embeds in R/K_n. Therefore, $\ell_R(K_n)$ is an PI-ideal.

Let S be the sum of all PI-ideals of R and let $A = \ell_R(S)$. Then $B := \ell_R(A)$ is a semiprime ideal by Lemma 8.3.51(i) and $A \cap B = 0$. Since all prime factor rings of R are PI-rings, all prime factor rings of the semiprime ring R/B are PI-rings. If $B = R$, then $R = \ell_R(A)$, so $A = 0$ because R is semiprime. Thus $\ell_R(S) = 0$, hence by Proposition 1.3.16, S is essential in R.

Next, we assume that $B \neq R$. Then R/B contains a nonzero PI-ideal by the previous argument. To see that S is an essential ideal of R, we need to show that $A = 0$ from Proposition 1.3.16. If $A \neq 0$, then $(A+B)/B$ is essential in R/B by Lemma 8.3.51(ii). So $(A+B)/B$ contains a nonzero PI-ideal, say V/B of R/B. Put

$$K = \{a \in A \mid a + B \in V/B\}.$$

Then $K \trianglelefteq R$ and $K \cong V/B$ as rings since $A \cap B = 0$. So K is a nonzero PI-ideal of R and $K \subseteq A$. Hence $S \cap A \neq 0$, which is a contradiction because $A = \ell_R(S)$. So $A = 0$. Therefore, S is essential in R. □

The next lemma, known as Andrunakievic's lemma, is useful for studying the relationship between the ideal structure of a given ideal of a ring R and that of R (see [9, Lemma 4]).

Lemma 8.3.54 *Let R be a ring and $V \trianglelefteq R$. Assume that $I \trianglelefteq V$ and W is the ideal of R generated by I. Then $W^3 \subseteq I$.*

Proof Since $V \trianglelefteq R$ and $I \trianglelefteq V$, we get $W = I + IR + RI + RIR$. Therefore it follows that $W^3 \subseteq VWV = V(I + IR + RI + RIR)V \subseteq I$. □

Proposition 8.3.55 *Let R be a ring and $V \trianglelefteq R$.*

(i) *If R is a semiprime ring, then V is a semiprime ring.*

(ii) *If R is a prime ring, then V is a prime ring.*

Proof (i) To show that V is a semiprime ring, let $I \trianglelefteq V$ with $I^2 = 0$. Say W is the ideal of R generated by I. By Lemma 8.3.54, $W^3 \subseteq I$. So $W^6 \subseteq I^2 = 0$. As R is semiprime, $W = 0$ and so $I = 0$. Hence, V is a semiprime ring.

(ii) Similarly, we see that V is a prime ring if R is a prime ring. □

Every semiprime PI-ring satisfies the hypothesis of our next result. Example 8.3.57 illustrates that Theorem 8.3.56 is a proper generalization of Theorem 3.2.16.

Theorem 8.3.56 *Let R be a semiprime ring with R/P a PI-ring for each prime ideal P of R. If $0 \neq I \trianglelefteq R$, then $I \cap \mathrm{Cen}(R) \neq 0$.*

Proof From Theorem 8.3.53, there exists $V \trianglelefteq R$ such that

$$V_R \leq^{\mathrm{ess}} R_R \text{ and } V = \sum_{\lambda \in \Lambda} V_\lambda,$$

where each V_λ is a nonzero PI-ideal. If $I \cap V_\lambda = 0$ for all $\lambda \in \Lambda$, then $IV = 0$, and hence $I \cap V = 0$, contrary to $V_R \leq^{\mathrm{ess}} R_R$.

So there is $\beta \in \Lambda$ with $0 \neq I \cap V_\beta \trianglelefteq V_\beta$. By Theorem 3.2.16 and Proposition 8.3.55, $I \cap \mathrm{Cen}(V_\beta) = I \cap V_\beta \cap \mathrm{Cen}(V_\beta) \neq 0$ since V_β is a semiprime PI-ring. Propositions 8.3.3(i) and 8.3.49 yield that $\mathrm{Cen}(V_\beta) = V_\beta \cap \mathrm{Cen}(R)$. As a consequence, $I \cap \mathrm{Cen}(V_\beta) = I \cap V_\beta \cap \mathrm{Cen}(R) \neq 0$. Therefore, $I \cap \mathrm{Cen}(R) \neq 0$. □

Example 8.3.57 There is a semiprime ring R which does not satisfy a PI, but R/P is a PI-ring for every prime ideal P of R. For a field F, let

$$R = \{(A_n)_{n=1}^{\infty} \in \prod_{n=1}^{\infty} \mathrm{Mat}_n(F) \mid A_n \text{ is a scalar matrix eventually}\},$$

which is a subring of $\prod_{n=1}^{\infty} \mathrm{Mat}_n(F)$. Then R is a semiprime ring which does not satisfy a PI. Let P be a prime ideal of R.

Case 1. Assume that the k-th component of all elements of P is zero for some k. Let $e_k = (0, 0, \ldots, 0, 1, 0, \ldots)$, where 1 is in the k-th component. Take $x \in R$ such that x has zero in its k-th component. Then $e_k R x = 0$ and so $x \in P$. Therefore $P = \{(A_n)_{n=1}^{\infty} \in R \mid A_k = 0\}$. Hence $R/P \cong \mathrm{Mat}_k(F)$.

Case 2. Assume that for any k, there is an element of P with a nonzero entry in its k-th component. Then for any k, there is $0 \neq \alpha \in \mathrm{Mat}_k(F)$ such that $\mu_k := (0, 0, \ldots, 0, \alpha, 0, \ldots) \in P$, where α is in the k-th component. Thus $R\mu_k R \subseteq P$, so $\oplus_{k=1}^{\infty} \mathrm{Mat}_k(F) \subseteq P$. As $R/\oplus_{k=1}^{\infty} \mathrm{Mat}_k(F)$ is commutative, and R/P is a ring homomorphic image of $R/\oplus_{k=1}^{\infty} \mathrm{Mat}_k(F)$, R/P is commutative.

By Cases 1 and 2, R/P is a PI-ring for every prime ideal P of R.

Exercise 8.3.58

1. Finish the proof of Proposition 8.3.3 and prove Lemma 8.3.4.
2. Let $I \in \mathfrak{D}_{\mathbf{IC}}(R)$. Prove the following.
   (i) $\ell_R(I) \subseteq r_R(I)$.
   (ii) $\ell_R(I) = r_R(I)$ if and only if $r_R(I) \cap I = 0$.
3. Assume that R is a ring.
   (i) Show that $\mathfrak{D}_{\mathbf{IC}}(R)$ contains no nonzero nilpotent ideals of R.
   (ii) Find an example of a right nonsingular quasi-Baer ring R such that $0 \neq P(R) \in \mathfrak{D}_{\mathbf{IC}}(R)$ (see [232]).
4. Let R be a ring. Show that the following are equivalent.
   (i) $R \in \mathbf{IC}$.

(ii) For each $K \in \mathfrak{D}_{\mathbf{IC}}(R)$ with K_R closed in R_R, there exists $e^2 = e \in R$ such that $K = eR$.
(iii) For each $K \in \mathfrak{D}_{\mathbf{IC}}(R)$ with K_R closed in R_R, there is $c \in \mathcal{B}(R)$ satisfying $K = cR$.

5. Let R be a ring with identity and $I \in \mathfrak{D}_{\mathbf{IC}}(R)$. Prove the following.
   (i) There exists $e \in \mathcal{B}(Q(R))$ such that $\ell_R(\ell_R(I)) = eQ(R) \cap R$ and $\ell_R(\ell_R(I))$ is the unique closure of I_R in R_R.
   (ii) Let $K = \ell_R(\ell_R(I))$. Then $R/K \cong (1-e)R(1-e)$ as rings.
6. Prove Corollary 8.3.12.
7. Show that in Lemma 8.3.26 and in Theorem 8.3.28, the set $\mathbb{E}$ can be a set of idempotents each taken from some set of left or right triangulating idempotents (see [97, Example 2.3]).
8. ([42, Beidar and Wisbauer]) Show that a ring R with identity is biregular if and only if R is semiprime and $R\mathcal{B}(Q(R))$ is biregular.
9. Let R be a ring (not necessarily with identity) and $S = \langle R \cup 1_{Q(R)} \rangle_{Q(R)}$. Show that $Q(R) = Q(S) \subseteq E(S_S) \subseteq E(S_R) = E(R_R)$.

8.4 Module Hulls

It is well known that for every module M, there always exists a unique (up to isomorphism) minimal injective extension (overmodule) which is called its injective hull and is denoted by $E(M)$. While the injective hull has been studied and used extensively, in some instances it is difficult for a fruitful transfer of information to take place between M and $E(M)$. For example, take M to be the $\mathbb{Z}$-module $\mathbb{Z}_p \oplus \mathbb{Z}_{p^3}$, where p is a prime integer. Then $H = \mathbb{Z}_{p^2} \oplus \mathbb{Z}_{p^3}$ is an extending hull of M. We observe that both M and H are finite, but $E(M)$ is infinite.

The studies on module hulls have been rather limited. In this section, we discuss module hulls satisfying some generalizations of injectivity. One may expect that such minimal overmodules will allow for a rich transfer of information similar to the case of rings. This is because each of these hulls, with more general properties than injectivity, sits in between M and a fixed injective hull $E(M)$ of M; and hence it generally lies closer to the module M than $E(M)$.

Definition 8.4.1 Let M be a module. We fix an injective hull $E(M)$ of M. Let $\mathfrak{M}$ be a class of modules. We call, when it exists, a module $H_{\mathfrak{M}}(M)$ the $\mathfrak{M}$ *hull* of M if $H_{\mathfrak{M}}(M)$ is the smallest extension of M in $E(M)$ that belongs to $\mathfrak{M}$ (i.e., $H_{\mathfrak{M}}(M)$ is the $\mathfrak{M}$ absolute hull of M).

We begin this section with a description of a quasi-injective hull of a module M (i.e., $H_{\mathbf{qI}}(M)$, where $\mathbf{qI}$ is the class of quasi-injective modules). We recall that an R-module M is quasi-injective if and only if $f(M) \subseteq M$, for all $f \in \text{End}(E(M))$ (see Theorem 2.1.9). The next result about the existence of quasi-injective hulls is due to Johnson and Wong [238].

Theorem 8.4.2 *Let M be a right R-module and let $S = \mathrm{End}(E(M))$. Then SM is the quasi-injective hull of M.*

Proof We put $U = SM$. Then $M \le U \le E(M)$ and $E(U) = E(M)$. Now take $\phi \in \mathrm{End}(E(U)) = \mathrm{End}(E(M))$. Then $\phi(U) \subseteq U$. By Theorem 2.1.9, U is quasi-injective. Next we assume that $M \le N \le E(M)$ and N is quasi-injective. Then $\varphi(N) \subseteq N$ for any $\varphi \in \mathrm{End}(E(N)) = \mathrm{End}(E(M))$ by Theorem 2.1.9. Thus, $SN \subseteq N$ and so $SM \subseteq SN \subseteq N$. Therefore SM is the quasi-injective hull of M (i.e., $SM = H_{\mathbf{qI}}(M)$). □

The following result for the existence of the quasi-continuous hull of a module is obtained by Goel and Jain [177].

Theorem 8.4.3 *Let M be a right R-module and $S = \mathrm{End}(E(M))$. Let Ω be the subring of S generated by the set of all idempotents of S. Then ΩM is the quasi-continuous hull of M.*

Proof As $E(\Omega M) = E(M)$, Ω is also the subring of $\mathrm{End}(E(\Omega M))$ generated by the set of all idempotents. As $\Omega(\Omega M) = \Omega M$, ΩM is quasi-continuous by Theorem 2.1.25. Say $M \le N \le E(M)$ and N is quasi-continuous. Then $E(N) = E(M)$, so Ω is the subring of $\mathrm{End}(E(N))$ generated by the set of all idempotents. From Theorem 2.1.25, $\Omega N \subseteq N$. Thus, $\Omega M \subseteq \Omega N \subseteq N$. So ΩM is the quasi-continuous hull of M (i.e., $\Omega M = H_{\mathbf{qCon}}(M)$). □

In contrast to Theorems 8.4.2 and 8.4.3, for the case of continuous hulls, there exists a nonsingular uniform cyclic module over a noncommutative ring which does not have an absolute continuous hull as follows.

Example 8.4.4 Let V be a vector space over a field F with basis elements v_m, w_k $(m, k = 0, 1, 2, \ldots)$. We denote by V_n the subspace generated by the v_m $(m \ge n)$ and all the w_k. Also we denote by W_n the subspace generated by the w_k $(k \ge n)$. We write $\mathcal{S}$ for the shift operator such that $\mathcal{S}(w_k) = w_{k+1}$ and $\mathcal{S}(v_i) = 0$ for all k, i. Let R be the set of all $\rho \in \mathrm{End}_F(V)$ with $\rho(v_m) \in V_m$, $\rho(w_0) \in W_0$ and $\rho(w_k) = \mathcal{S}^k \rho(w_0)$, for $m, k = 0, 1, 2, \ldots$.

Note that $\tau\rho(w_k) = \mathcal{S}^k \tau\rho(w_0)$, for $\rho, \tau \in R$, and so $\tau\rho \in R$. Thus, it is routine to check that R is a subring of $\mathrm{End}_F(V)$. Further, we see that $V_n = Rv_n$, $W_n = Rw_n$, and $V_{n+1} \subseteq V_n$ for all n. (When $f \in R$ and $v \in V$, we also use fv for the image $f(v)$ of v under f.)

Consider the left R-module $M = W_0$. First, we show that $M = Rw_0$ is uniform. For this, take $fw_0 \ne 0$, $gw_0 \ne 0$ in M, where $f, g \in R$. We need to find $h_1, h_2 \in R$ such that $h_1 f w_0 = h_2 g w_0 \ne 0$. Let

$$f w_0 = b_0 w_0 + b_1 w_1 + \cdots + b_m w_m \in Rw_0$$

and

$$g w_0 = c_0 w_0 + c_1 w_1 + \cdots + c_m w_m \in Rw_0,$$

where $b_i, c_j \in F$, $i, j = 0, 1, \ldots, m$, and some terms of b_i and c_j may be zero.

Put $h_1w_0 = x_0w_0 + x_1w_1 + \cdots + x_\ell w_\ell$ and $h_2w_0 = y_0w_0 + y_1w_1 + \cdots + y_\ell w_\ell$, where $x_i, y_i \in F$, $i = 0, 1, \ldots, \ell$ (also some terms of x_i and y_j may be zero). Since $h_1(w_k) = \mathcal{S}^k h_1(w_0)$ and $h_2(w_k) = \mathcal{S}^k h_2(w_0)$ for $k = 0, 1, 2 \ldots$, we need to find such $x_i, y_i \in F$, $0 \le i \le \ell$ so that $h_1 f w_0 = h_2 g w_0 \ne 0$ from the following equations:

$$b_0x_0 = c_0y_0,\ b_0x_1 + b_1x_0 = c_0y_1 + c_1y_0,$$

$$b_0x_2 + b_1x_1 + b_2x_0 = c_0y_2 + c_1y_1 + c_2y_0,$$

$$b_0x_3 + b_1x_2 + b_2x_1 + b_3x_0 = c_0y_3 + c_2y_1 + c_2y_1 + c_3y_0,$$

and so on. Now say $\alpha(t) = b_0 + \cdots + b_mt^m \ne 0$ and $\beta(t) = c_0 + \cdots + c_mt^m \ne 0$ in the polynomial ring $F[t]$. Then $\alpha(t)F[t] \cap \beta(t)F[t] \ne 0$. We may note that finding such $x_0, x_1 \ldots, x_\ell, y_0, y_1 \ldots, y_\ell$ in F above is the same job for finding $x_0, x_1 \ldots, x_\ell, y_0, y_1 \ldots, y_\ell$ such that

$$\alpha(t)(x_0 + x_1t + \cdots + x_\ell t^\ell) = \beta(t)(y_0 + y_1t + \cdots + y_\ell t^\ell) \ne 0$$

in the polynomial ring $F[t]$. Observing that $0 \ne \alpha(t)\beta(t) \in \alpha(t)F[t] \cap \beta(t)F[t]$, take $h_1w_0 = c_0w_0 + c_1w_1 + \cdots + c_mw_m$ by putting $\ell = m$, $x_i = c_i$ for $0 \le i \le m$, and $h_2w_0 = b_0w_0 + b_1w_1 + \cdots + b_mw_m$ by putting $\ell = m$, $y_i = b_i$ for $0 \le i \le m$. As $\alpha(t)\beta(t) \ne 0$, $0 \ne h_1 f w_0 = h_2 g w_0 \in Rfw_0 \cap Rgw_0$. So M is uniform.

Next, we show that each V_n is an essential extension of M (hence each V_n is uniform). Indeed, let $0 \ne \mu v_n \in Rv_n = V_n$, where $\mu \in R$. Say

$$\mu v_n = a_{n+k}v_{n+k} + \cdots + a_{n+k+\ell}v_{n+k+\ell} + b_sw_s + \cdots + b_{s+m}w_{k+m}.$$

If $a_{n+k} = \cdots = a_{n+k+\ell} = 0$, then $\mu v_n \in W_0$. Otherwise, we assume that $a_{n+k} \ne 0$. Let $\omega \in R$ such that $\omega(v_{n+k}) = w_0$ and $\omega(v_i) = 0$ for $i \ne n+k$ and $\omega(w_j) = 0$ for all j. Then $0 \ne \omega\mu v_n = a_{n+k}w_0 \in W_0$. Thus $M = W_0$ is essential in V_n. Since M is uniform, V_n is also uniform for all n.

We prove that ${}_RM$ is nonsingular. For this, assume that $u \in Z({}_RM)$ and let $K = \{\alpha \in R \mid \alpha u = 0\}$. Then K is an essential left ideal of R. So $K \cap R\mathcal{S}^2 \ne 0$. Thus there exists $\rho \in R$ such that $\rho\mathcal{S}^2 \ne 0$ and $\rho\mathcal{S}^2(u) = 0$. Say

$$u = a_kw_k + a_{k+1}w_{k+1} + \cdots + a_nw_n \ \text{ with } \ a_k, a_{k+1}, \ldots, a_n \in F.$$

Assume on the contrary that $u \ne 0$. Then we may suppose that $a_k \ne 0$. Because $\rho(w_n) = \mathcal{S}^n\rho(w_0)$ for $n = 0, 1, 2, \ldots$,

$$\begin{aligned} 0 = \rho\mathcal{S}^2(u) &= a_k\rho\mathcal{S}^2(w_k) + a_{k+1}\rho\mathcal{S}^2(w_{k+1}) + \cdots + a_n\rho\mathcal{S}^2(w_n) \\ &= a_k\mathcal{S}^{k+2}\rho(w_0) + a_{k+1}\mathcal{S}^{k+3}\rho(w_0) + \cdots + a_n\mathcal{S}^{n+2}\rho(w_0). \end{aligned}$$

Here we put $\rho(w_0) = b_\ell w_\ell + b_{\ell+1}w_{\ell+1} + \cdots + b_tw_t$. If $\rho(w_0) = 0$, then we see that $\rho\mathcal{S}^2(w_0) = \rho(w_2) = \mathcal{S}^2\rho(w_0) = 0$. Also, $\rho\mathcal{S}^2(w_m) = 0$ for all $m = 1, 2, \ldots$, and

$\rho\mathcal{S}^2(v_i) = 0$ for all $i = 0, 1, \ldots$. Thus $\rho\mathcal{S}^2 = 0$, a contradiction. Hence $\rho(w_0) \neq 0$, and so we may assume that $b_\ell \neq 0$. Note that

$$\mathcal{S}^{k+2}\rho(w_0) = b_\ell w_{\ell+k+2} + b_{\ell+1}w_{\ell+k+3} + \cdots + b_t w_{t+k+2},$$

$$\mathcal{S}^{k+3}\rho(w_0) = b_\ell w_{\ell+k+3} + b_{\ell+1}w_{\ell+k+4} + \cdots + b_t w_{t+k+3},$$

and so on. Thus $0 = \rho\mathcal{S}^2(u) = a_k b_\ell w_{\ell+k+2} + (a_k b_{\ell+1} + a_{k+1}b_\ell)w_{\ell+k+3} + \cdots$, and hence $a_k b_\ell = 0$, which is a contradiction because $a_k \neq 0$ and $b_\ell \neq 0$. Therefore $u = 0$, and so M is nonsingular.

We show now that V_n is continuous. Note that V_n is uniform. So clearly, V_n has (C_1) condition. Thus, to show that V_n is continuous, it suffices to prove that every R-monomorphism of V_n is onto for V_n to satisfy (C_2) condition.

Let $\varphi : V_n \to V_n$ be an R-monomorphism. We put

$$\varphi(v_n) = \rho v_n \in Rv_n = V_n, \quad \text{where} \quad \rho \in R.$$

We claim that $\rho v_n \notin V_{n+1}$. For this, assume on the contrary that $\rho v_n \in V_{n+1}$. Now we let $\lambda \in R$ such that $\lambda v_n = v_n$, $\lambda v_k = 0$ for $k \neq n$, and $\lambda w_m = 0$ for all m. Then $\varphi(\lambda v_n) = \lambda(\rho v_n) = 0$ since $\rho(v_n) \in V_{n+1}$. But $\lambda v_n = v_n \neq 0$. Thus φ is not one-to-one, a contradiction. Therefore $\rho v_n \notin V_{n+1}$.

As $\rho v_n \in V_n$, write

$$\rho v_n = a_n v_n + a_{n+1}v_{n+1} + \cdots + a_{n+\ell}v_{n+\ell} + b_0 w_0 + \cdots + b_h w_h,$$

where $a_n, a_{n+1}, \ldots, a_{n+\ell}, b_0, b_1, \ldots, b_h \in F$, and $a_n \neq 0$.

Take $\nu \in R$ such that $\nu v_n = a_n^{-1} v_n$, $\nu v_k = 0$ for $k \neq n$ and $\nu w_m = 0$ for all m. Then we see that $v_n = \nu\rho v_n \in R\rho v_n$. So $Rv_n \subseteq R\rho v_n$, hence $V_n = Rv_n = R\rho v_n$. Thus $\varphi(Rv_n) = R\varphi(v_n) = R\rho v_n = V_n$, so φ is onto. Therefore each V_n is continuous.

Finally, note that the uniform nonsingular module $M = Rw_0$ is not continuous, since the shifting operator $\mathcal{S}$ provides an R-monomorphism which is not onto. Hence, M does not have a continuous hull (in $E(M) = E(V)$), because such a hull would have to be contained in each V_n, and hence in $M = \cap_n V_n$.

Despite Example 8.4.4, we will show that continuous hulls do exist for certain classes of modules over a commutative ring as shown in the next several results. We start with a lemma.

Lemma 8.4.5 *Assume that R is a commutative ring and M is a nonsingular cyclic R-module. Let $E = E(M_R)$ and T be a subring of* End(E_R). *Then*:

(i) $E\,r_R(M) = 0$.
(ii) *There exists a smallest continuous module V such that $M \leq V \leq E$ and $TV \subseteq V$.*

Proof Let $I = r_R(M) \trianglelefteq R$. Put $\overline{R} = R/I$. Then $M \cong \overline{R}_R$.

(i) Note that E_R is nonsingular because M_R is nonsingular. Let $x \in E(\overline{R}_R)$. Then there is an essential ideal L of R with $xL \subseteq R/I$. Hence $(xI)L = xLI = 0$, so $xI \subseteq Z(E_R) = 0$. Thus, $xI = 0$. Therefore, $EI = 0$.

(ii) *Step 1.* By part (i), E has an $\overline{R}$-module structure induced from the R-module E_R. To see that E is the injective hull of the $\overline{R}$-module M, note that E is an essential extension of M as an $\overline{R}$-module. Let K/I be an ideal of R/I and $\alpha \in \mathrm{Hom}((K/I)_{\overline{R}}, E_{\overline{R}})$. Then $\alpha \in \mathrm{Hom}((K/I)_R, E_R)$ and so there exists an extension $\beta \in \mathrm{Hom}((R/I)_R, E_R)$ of α. We see that $\beta \in \mathrm{Hom}((R/I)_{\overline{R}}, E_{\overline{R}})$. Hence E is an injective $\overline{R}$-module. Therefore, E is an injective hull of M as an $\overline{R}$-module. Further, M is nonsingular as an $\overline{R}$-module by routine arguments.

By Theorem 2.1.31, $E = Q(\overline{R})$, which is a commutative regular ring. Also from Proposition 2.1.32, $E = \mathrm{End}(E_E) = \mathrm{End}(E_{\overline{R}})$ $(= \mathrm{End}(E_R))$. Thus T is a subring of E. Also $\overline{R}$ is a subring of E.

Let P be the subring of E generated by all idempotents of E. We claim that any regular subring A of E satisfying $\overline{R}P \subseteq A$ is continuous as an R-module (or equivalently, as an $\overline{R}$-module).

First, by Theorem 2.1.25 or Theorem 8.4.3, A is a quasi-continuous R-module because $PA = A$. We show that A_R has (C_2) condition. For this, let $A = A_1 \oplus A_2$, which is an R-module decomposition, and let $\varphi : A_1 \to N$ be an R-isomorphism, where $N_R \leq A_R$. Note that $\mathrm{Hom}_R(A, A_1) = \mathrm{Hom}_{\overline{R}}(A, A_1)$. Further, from the proof of Proposition 2.1.32, $\mathrm{Hom}_{\overline{R}}(A, A_1) = \mathrm{Hom}_A(A, A_1)$, because A is a ring of quotients of $\overline{R}$. Thus $\mathrm{Hom}_R(A, A_1) = \mathrm{Hom}_A(A, A_1)$.

We let $\pi_1 : A \to A_1$ be the canonical projection of R-modules. Then we see that π_1 is an A-homomorphism. Therefore $A_1 = \pi_1(A) = \pi_1(1)A$. Similarly, we observe that $\varphi \in \mathrm{Hom}_R(A_1, N) \subseteq \mathrm{Hom}_R(A_1, A) = \mathrm{Hom}_A(A_1, A)$.

So we have that $N = \varphi(A_1) = \varphi(\pi_1(1)A) = \varphi\pi_1(1)A$ is a principal (right) ideal of A. Hence $N_A \leq^{\oplus} A_A$ because A is a regular ring, and so $N_R \leq^{\oplus} A_R$. Thus A_R satisfies (C_2) condition. Therefore, A_R is a continuous module.

Let V be the intersection of all regular subrings V_i of E with $\overline{R}PT \subseteq V_i$. Then as in the proof of Theorem 8.2.6, V is a regular ring. Also $\overline{R}PT \subseteq V$. Thus by the preceding consideration, V_R is continuous. Clearly, $\overline{R} \subseteq V \subseteq E$. Moreover, we obtain $TV \subseteq V$ since $T \subseteq \overline{R}PT \subseteq V$.

Step 2. Let Y be a continuous R-module such that $\overline{R}_R \leq Y_R \leq E_R$ and $TY \subseteq Y$. Put $B = \{b \in E \mid bY \subseteq Y\}$. Then B is a subring of E. Further, $\overline{R} \subseteq B$ and $T \subseteq B$. Since Y is a continuous R-module and $E(Y_R) = E$, $PY = Y$ by Theorem 2.1.25 or Theorem 8.4.3 (recall that P is the subring of E generated by the set of all idempotents of E). So $P \subseteq B$. Thus $\overline{R}PT \subseteq B \subseteq E$.

We claim that B is regular. For this, take $b \in B$. Since E is commutative regular, there exists $c \in E$ such that $b = bcb$ and $c = cbc$ (see [264, Exercise 3, p. 36]). Note that $(cb)^2 = cb \in E$ and so $cb \in P$. Hence, $cbY \subseteq Y$ and $cbY_R \leq^{\oplus} Y_R$. Define

$$\phi : bY \to cbY \text{ by } \phi(by) = cby,$$

where $y \in Y$. Then ϕ is an R-isomorphism because $b = bcb$. Hence by (C_2) condition of Y, there is $g^2 = g \in \mathrm{End}(Y_R)$ such that $bY = gY$. Also there exists $f \in E$, which is an extension of g. Then we have that $bY = gY = fY$ and $(f - f^2)(Y) = (g - g^2)(Y) = 0$.

We show that $(f - f^2)(E) = 0$. Assume on the contrary that there exists $x \in E$ such that $(f - f^2)(x) \neq 0$. Since $Y_R \leq^{\text{ess}} E_R$ and E_R is nonsingular, $Y_R \leq^{\text{den}} E_R$ by Proposition 1.3.14. Thus there exists $r \in R$ such that $xr \in Y$ and $(f - f^2)(x)r \neq 0$. Therefore, $0 \neq (f - f^2)(x)r = (f - f^2)(xr)$, which is a contradiction because $xr \in Y$ and $(f - f^2)(Y) = 0$. Hence $(f - f^2)(E) = 0$, so

$$f^2 = f \in P \text{ and } bY = gY = fY \subseteq Y$$

as $b \in B$. Thus $(1 - f)Y \subseteq Y$ and $Y = fY \oplus (1 - f)Y$. Therefore

$$cY = cfY \oplus c(1 - f)Y = cbY \oplus cbc(1 - f)Y = cbY \oplus c^2(1 - f)bY.$$

As $bY = fY$, $c^2(1 - f)bY = c^2(1 - f)fY = 0$, and hence $cY = cbY \subseteq PY = Y$. Thus $c \in B$, and so B is a regular ring. As $\overline{R}\,PT \subseteq B$ and B is a regular ring, $V \subseteq B$ by the definition of V. So $V = V\overline{R} \subseteq B\overline{R} \subseteq BY \subseteq Y$. □

We remark that, if R is a commutative semiprime ring, then by Lemma 8.4.5 and Theorem 8.4.6 the continuous hull of R_R is the intersection of all intermediate continuous regular rings between R and $Q(R)$. Thus, the continuous hull of R_R is exactly the continuous absolute ring hull $Q_{\mathbf{Con}}(R)$ of R (see Theorem 8.2.6).

Theorem 8.4.6 *Every nonsingular cyclic module over a commutative ring has a continuous hull (which is a regular ring).*

Proof Assume that M be a nonsingular cyclic module over a commutative ring R and $I = r_R(M)$. Put $\overline{R} = R/I$. Then $M \cong \overline{R}_R$. Let $E = E(M_R)$.

From Lemma 8.4.5(i), $EI = 0$. Thus, $T := R/I$ can be considered as a subring of $\text{End}_R(E)$. By Lemma 8.4.5(ii), there exists a smallest continuous module V such that $M \leq V \leq E$ and $TV \subseteq V$. So V is a continuous hull of M. □

The next example shows that quasi-continuous hulls (even for commutative semiprime rings) are distinct from continuous hulls which are, in turn, distinct from (quasi-)injective hulls.

Example 8.4.7 Let $F_n = \mathbb{R}$ for $n = 1, 2, \ldots$ and R the subring of $\prod_{n=1}^{\infty} F_n$ generated by $\oplus_{n=1}^{\infty} F_n$ and $1_{\prod_{n=1}^{\infty} F_n}$. Then $E(R_R) = Q(R) = \prod_{n=1}^{\infty} F_n$. In this case, we see that

$$U = \{(a_n)_{n=1}^{\infty} \in \prod_{n=1}^{\infty} F_n \mid a_n \in \mathbb{Z} \text{ eventually}\}$$

is the quasi-continuous hull of R_R (see Theorem 8.4.3). By Lemma 8.4.5,

$$V = \{(a_n)_{n=1}^{\infty} \in \prod_{n=1}^{\infty} F_n \mid a_n \in \mathbb{Q} \text{ eventually}\}$$

is the continuous hull of R_R because V is the smallest continuous regular ring between R and $Q(R)$ (therefore V is the intersection of all intermediate continuous regular rings between R and $Q(R)$).

Consider an arbitrary cyclic R-module $M = \overline{R} = R/r_R(M)$ over a commutative ring R. We fix the following notations: $E = E(\overline{R}_R)$, $E = E_1 \oplus E_2$, where $E_1 = Z_2(E)$ (note that since E_R is injective, $Z_2(E) \leq^{\oplus} E$ by Proposition 2.3.10). Write $1_{\overline{R}} = e_1 + e_2$ (where $e_1 \in E_1$, and $e_2 \in E_2$) be the corresponding decomposition. Then $E_1 = E(e_1 R)$ and $E_2 = E(e_2 R)$.

Proposition 8.4.8 *Let M be a cyclic module over a commutative ring R and let $I = r_R(e_2 R)$. Then the following conditions are equivalent.*

(i) *$e_1 R + \ell_{E_1}(I)$ has a continuous hull.*
(ii) *M has a continuous hull.*

Proof Note that $e_2 R_R$ is a nonsingular cyclic R-module. Say $\pi_2 : E \to E_2$ is the canonical projection onto E_2. Let T be the subring of $\mathrm{End}_R(E_2)$ generated by the set $\{\pi_2 \pi|_{E_2}\}$, where $\pi^2 = \pi \in \mathrm{End}_R(E)$. By Lemma 8.4.5(i), $E_2 I = 0$. Also from Lemma 8.4.5(ii), there exists a smallest continuous module V_2 with $e_2 R \leq V_2 \leq E_2$ and $T V_2 \subseteq V_2$.

(i)⇒(ii) Assume that there exists a continuous hull V_1 of $e_1 R + \ell_{E_1}(I)$. We claim that $V = V_1 \oplus V_2$ is continuous. For this, first we prove that V is quasi-continuous. Let $\pi^2 = \pi \in \mathrm{End}_R(E)$. Then $\pi|_{E_1} \in \mathrm{End}_R(E_1)$ because $E_1 = Z_2(E) \trianglelefteq E$. Therefore, $\pi(V_1) = \pi|_{E_1}(V_1) \subseteq V_1$ by Theorem 2.1.25 since V_1 is continuous. Let $\pi_1 : E \to E_1$ be the canonical projection onto E_1 and put $\phi = \pi_1 \pi|_{E_2}$. Then $\phi \in \mathrm{Hom}_R(E_2, E_1)$. Also, $\phi(V_2) I = \phi(V_2 I) \subseteq \phi(E_2 I) = 0$, so $\phi(V_2) \subseteq \ell_{E_1}(I) \subseteq V_1$. Hence, $\pi_1 \pi(V_2) \subseteq V_1$.

Next $\pi_2 \pi|_{E_2} \in T$, and hence $\pi_2 \pi(V_2) \subseteq T V_2 \subseteq V_2$. Therefore, we have that $\pi(V) = \pi(V_1) + \pi(V_2) = \pi(V_1) + \pi_1 \pi(V_2) + \pi_2 \pi(V_2) \subseteq V_1 + V_2 = V$. Thus V is quasi-continuous by Theorem 2.1.25.

By Lemma 2.2.4, V_1 and V_2 are relatively injective. Since V_1 and V_2 are continuous, $V = V_1 \oplus V_2$ is continuous by Theorem 2.2.16. Next, we show that $V = V_1 \oplus V_2$ is a continuous hull of $M = \overline{R}_R$. For this, say Y is a continuous module such that $\overline{R} \leq Y \leq E = E_1 \oplus E_2$. Then since $E(Y) = E$, $Y = Y_1 \oplus Y_2$ from Theorem 2.1.25, where $Y_1 = Y \cap E_1$ and $Y_2 = Y \cap E_2$. Observe that $e_1 = \pi_1(1_{\overline{R}}) \in \pi_1(Y) = Y_1$ and $e_2 = \pi_2(1_{\overline{R}}) \in \pi_2(Y) = Y_2$. So $e_1 R \subseteq Y_1$ and $e_2 R \subseteq Y_2$. Since Y is continuous, $\pi(Y) \subseteq Y$ by Theorem 2.1.25 and so $\pi_2 \pi(Y_2) \subseteq \pi_2 \pi(Y) \subseteq \pi_2(Y) = Y_2$. Hence, $T Y_2 \subseteq Y_2$. Note that Y_2 is continuous by Theorem 2.2.16. Thus $V_2 \subseteq Y_2$ since V_2 is the smallest continuous module such that $e_2 R \leq V_2 \leq E_2$ and $T V_2 \subseteq V_2$.

To show that $\ell_{E_1}(I) \subseteq Y_1$ so that $e_1 R + \ell_{E_1}(I) \subseteq Y_1$, take $a \in \ell_{E_1}(I)$. Then the map $f : e_2 R \to aR$ defined by $f(e_2 r) = ar$ for $r \in R$ is an R-homomorphism. Thus, there is $\varphi \in \mathrm{Hom}_R(E_2, E_1)$ with $\varphi|_{e_2 R} = f$. Note that $E_1 = E(Y_1)$ and $E_2 = E(Y_2)$. Since $Y = Y_1 \oplus Y_2$ is continuous, Y_1 is Y_2-injective by Lemma 2.2.4. Thus, $\varphi(Y_2) \subseteq Y_1$ from Theorem 2.1.2. Whence $a = f(e_2) = \varphi(e_2) \in \varphi(Y_2) \subseteq Y_1$.

Therefore, $\ell_{E_1}(I) \subseteq Y_1$, so $e_1 R + \ell_{E_1}(I) \subseteq Y_1$. Hence $V_1 \subseteq Y_1$ because Y_1 is continuous by Theorem 2.2.16. This yields that $V = V_1 \oplus V_2 \subseteq Y_1 \oplus Y_2 = Y$. Therefore V is a continuous hull of $\overline{R}_R$.

(ii)$\Rightarrow$(i) Assume that there exists a continuous hull W of $\overline{R}_R$. Then as in the argument used in the proof of (i)$\Rightarrow$(ii), we have that

$$W = W_1 \oplus W_2,\ e_1 R + \ell_{E_1}(I) \subseteq W_1 \subseteq E_1,\ e_2 R \subseteq W_2 \subseteq E_2,$$

and $T W_2 \subseteq W_2$.

Let $e_1 R + \ell_{E_1}(I) \le U \le E_1$ with U a continuous module. We see that $U \oplus W_2$ is quasi-continuous exactly as in the proof of (i)$\Rightarrow$(ii) for showing that $V = V_1 \oplus V_2$ is quasi-continuous. Thus U and W_2 are relatively injective by Lemma 2.2.4. So $U \oplus W_2$ is continuous by Theorem 2.2.16 as both U and W_2 are continuous. Hence, $W = W_1 \oplus W_2 \le U \oplus W_2$. Therefore, $W_1 \le U$. Thus W_1 is a continuous hull of $e_1 R + \ell_{E_1}(I)$. □

An element $a \in R$ is said to *act regularly* on an R-module M, if $ma = 0$ implies $m = 0$ for $m \in M$. Motivated by the condition in Proposition 8.4.8, we now obtain the following result.

Lemma 8.4.9 *Let E be an indecomposable injective module over a commutative ring R. Assume that $f \in E$ and $I \trianglelefteq R$. Then $fR + \ell_E(I)$ has a continuous hull.*

Proof Let C be the multiplicatively closed set of those elements of R which act regularly on E, and let RC^{-1} be the corresponding right ring of fractions of R (see Proposition 5.5.4). For $c \in C$, we see that $E \cong Ec \le E$. Since E is indecomposable and injective, $E = Ec$. Take $y \in E$ and $rc^{-1} \in RC^{-1}$, where $r \in R$ and $c \in C$. From $E = Ec$, there exists uniquely $y_1 \in E$ such that $y = y_1 c$. Define $yrc^{-1} = y_1 r$. Then E becomes an RC^{-1}-module.

Say V is a continuous R-submodule of E. Then each $c \in C$ defines an R-monomorphism $V \to V$. Thus $V \cong Vc \le V$. Since V is continuous and uniform, $Vc = V$. As in the previous argument, V becomes an RC^{-1}-module.

Let $A = \ell_E(I)$. To see that A is an RC^{-1}-module, we first prove that A is quasi-injective. For this, take $h \in \mathrm{End}(E)$ and let $x \in A$. Then $xI = 0$ and thus $h(x)I = h(xI) = 0$. Therefore, $h(x) \in A$. Thus, $A \trianglelefteq E$. If $A = 0$, then A is quasi-injective. Suppose that $A \neq 0$. As E is indecomposable injective, $E = E(A)$ and so A is quasi-injective by Theorem 2.1.9. Thus A is an RC^{-1}-module by the preceding argument.

We show that $fRC^{-1} + A$ is a continuous R-module. If $f \in A$, then we obtain $fRC^{-1} + A = A$, and therefore $fRC^{-1} + A$ is a continuous R-module. Next, assume that $f \notin A$. We let $\varphi : fRC^{-1} + A \to fRC^{-1} + A$ be an R-monomorphism. Then φ can be extended to an isomorphism $\overline{\varphi}$ of E because $fRC^{-1} + A$ is essential in E, and E is indecomposable and injective.

Write $\varphi(f) = ft + a$, where $t \in RC^{-1}$ and $a \in A$. We note that $\varphi(f) \notin A$. For, if $\varphi(f) = \overline{\varphi}(f) \in A$, then $f \in \overline{\varphi}^{-1}(A) \subseteq A$ as $A \trianglelefteq E$, which is a contradiction. Hence $ft \neq 0$, so $t \neq 0$.

Put $t = rc^{-1}$ with $r \in R$ and $c \in C$. We show that t is invertible in RC^{-1}. Let $\mu \in \mathrm{End}(E)$ such that $\mu(y) = yt$, where $y \in E$. If μ is one-to-one, then r acts regularly on E, thus $r \in C$. Therefore, $t = rc^{-1}$ is invertible in RC^{-1}.

Assume that μ is not one-to-one. Then $\mu \in J(\mathrm{End}(E))$ as $\mathrm{End}(E)$ is a local ring. Thus, $\overline{\varphi} - \mu$ is an isomorphism because $\overline{\varphi}$ is an isomorphism. Put $\psi = \overline{\varphi} - \mu$. By Theorem 2.1.9, $\psi(A) \subseteq A$ because A is quasi-injective.

Next, for $w \in A$, there exists $v \in E$ such that $\psi(v) = w$ as ψ is an isomorphism. Whence $\psi(vI) = \psi(v)I = wI = 0$. Hence $vI = 0$, so $v \in A$. Thus $w \in \psi(A)$. As a consequence, $A = \psi(A) = (\overline{\varphi} - \mu)(A)$.

In particular, $a = (\overline{\varphi} - \mu)(b)$ with $b \in A$. So $\varphi(b) - bt = a$. Let $f' = f - b$. Then $fR + A = f'R + A$. As $f \notin A$, $f' \neq 0$. Recall that $\varphi(f) = ft + a$. Therefore, $\varphi(f') = \varphi(f - b) = \varphi(f) - \varphi(b) = (ft + a) - (a + bt) = ft - bt = f't$. Take $0 \neq x \in E$. Since E is indecomposable injective and $f' \neq 0$, $f'R$ is essential in E. So there exist $r, r' \in R$ with $xr = f'r' \neq 0$.

If $xt = 0$, then $\varphi(f'r') = \varphi(f')r' = f'tr' = xtr = 0$. Hence $f'r' = 0$ as φ is a monomorphism, a contradiction. Thus $xt \neq 0$, so t acts regularly on E. Hence $t \in C$, and thus t is invertible in RC^{-1}.

From $\varphi(f) = ft + a$, $\varphi(f) - a = ft$. Therefore

$$f = (\varphi(f) - a)t^{-1} = \varphi(f)t^{-1} - at^{-1} \in \varphi(f)RC^{-1} + A$$

because A is an RC^{-1}-module. Hence $fRC^{-1} + A \subseteq \varphi(f)RC^{-1} + A$. As $\overline{\varphi}$ is an isomorphism, $A = \overline{\varphi}(A)$ by the preceding argument. Hence $A = \varphi(A)$.

Note that $\varphi \in \mathrm{End}_{RC^{-1}}(fRC^{-1} + A)$. Indeed, for $\alpha \in fRC^{-1} + A$ and $c \in C$, $\varphi(\alpha c^{-1})c = \varphi(\alpha c^{-1}c) = \varphi(\alpha)$ and so $\varphi(\alpha c^{-1}) = \varphi(\alpha)c^{-1}$. Thus we have that $fRC^{-1} + A \subseteq \varphi(f)RC^{-1} + A \subseteq \varphi(f)RC^{-1} + \varphi(A) = \varphi(fRC^{-1} + A)$. Hence φ is onto. From this fact, every R-monomorphism from $fRC^{-1} + A$ to $fRC^{-1} + A$ is onto. Therefore, $fRC^{-1} + A$ is a continuous R-module because $fRC^{-1} + A$ is uniform.

Finally, assume that N is a continuous R-module with $fR + A \subseteq N \subseteq E$. By the preceding argument, N is an RC^{-1}-module (also note that A is an RC^{-1}-module). Thus, $fRC^{-1} + A \subseteq N$. So $fRC^{-1} + A$ is a continuous hull of $fR + A$. □

The following result is an extension of Theorem 8.4.6.

Theorem 8.4.10 *Let R be a commutative ring. Then every cyclic module M with $Z(M)$ uniform, has a continuous hull.*

Proof Let $E = E(M)$. Then $E = E_1 \oplus E_2$, where $E_1 = Z_2(E)$. We observe that $E_1 = E(Z_2(M)) = E(Z(M))$ as $Z(M)$ is essential in $Z_2(M)$. Since $Z(M)$ is uniform, E_1 is indecomposable injective. Let $I = r_R(e_2R)$. By Lemma 8.4.9, $e_1R + \ell_{E_1}(I)$ has a continuous hull. Hence, Proposition 8.4.8 yields that M has a continuous hull. □

When M is a uniform cyclic module over a commutative ring, M has a continuous hull by Theorem 8.4.10. This continuous hull is described explicitly in the next theorem.

Theorem 8.4.11 *Let R be a commutative ring, and $M = fR$ a uniform cyclic R-module. Then $MC^{-1} = fRC^{-1}$ is a continuous hull of M, where C is the multiplicatively closed set of those elements of R which act regularly on M.*

Proof Take $I = R$ in Lemma 8.4.9. Then $\ell_E(I) = 0$. By the proof of Lemma 8.4.9, $MC^{-1} = fRC^{-1}$ is a continuous hull of M. □

The following is an example of a continuous hull of a uniform cyclic module over a commutative ring, which is distinct from its quasi-continuous and injective hulls.

Example 8.4.12 Consider the ring

$$A = \{ \sum_{i\in[0,\infty)} \alpha_i x^i \mid \alpha_i \in \mathbb{Z} \text{ and } \alpha_i = 0 \text{ for all but finitely many } i\}.$$

Let $R = A/I$, where I is the ideal of A generated by x. Then R_R is uniform and nonsingular. Thus $Q(R) = E(R_R)$ by Corollary 1.3.15, and $Q(R)$ is regular by Theorem 2.1.31. Since R_R is uniform, $Q(R)$ has only 0 and 1 as its idempotents (hence $Q(R)$ is a field). So the quasi-continuous hull of R_R is R_R itself by Theorem 8.4.3 or Theorem 2.1.25. Next, consider

$$B = \{ \sum_{i\in[0,\infty)} \alpha_i x^i \mid \alpha_i \in \mathbb{Q} \text{ and } \alpha_i = 0 \text{ for all but finitely many } i\}.$$

Take $Q = B/K$, where K is the ideal of B generated by x. Let C be the set of all non zero-divisors of R. Then $Q = RC^{-1}$, which becomes the classical ring of quotients of R. By Theorem 8.4.11, Q_R is the continuous hull of R_R.

We claim that Q_R is not injective. For this, consider the ideal $\cup_{n=1}^{\infty} x^{1/n}R$ of R and the map $\phi : \cup_{n=1}^{\infty} x^{1/n}R \to Q$, where $\phi|_{x^{1/n}R} = \phi_n$ is given by the multiplication by $1 + x^{1/2} + \cdots + x^{(n-2)/(n-1)} + x^{(n-1)/n}$. Then ϕ is well-defined since $\phi_{n+1}|_{x^{1/n}R} = \phi_n$. Also, ϕ is an R-homomorphism. However, there is no element $q \in Q$, for which $\phi(x) = qx$ for all x in $\cup_{n=1}^{\infty} x^{1/n}R$. Since, in that case, q would have to be an infinite sum, and such q does not lie in Q. Consequently, ϕ cannot be extended to R. Thus, Q_R is not injective.

In the next example, we exhibit a free module of finite rank over a commutative domain, which does not have an extending hull.

Example 8.4.13 Let $R = \mathbb{Z}[x, y]$, the polynomial ring. Put $M = R \oplus R$. Then the R-module M is not extending by Theorem 6.1.4 and Exercise 6.1.18.1 because the commutative domain R is not Prüfer. Let $F = \mathbb{Q}(x, y)$, the field of fractions of R. Note that $E(M) = F \oplus F$.

Let $U = F \oplus R$ and $S = \text{End}(U_R)$. As $\text{Hom}(F_R, R_R) = 0$,

$$S = \begin{bmatrix} \text{End}(F_R) & \text{Hom}(R_R, F_R) \\ 0 & \text{End}(R_R) \end{bmatrix}.$$

By Theorem 4.2.18, U_R is a Baer module. We claim that U_R is a $\mathcal{K}$-cononsingular. For this, say $N_R \le U_R$ such that $\ell_S(N) = 0$. If $N \subseteq F \oplus 0$, then $\ell_S(N) \neq 0$. Also, if $N \subseteq 0 \oplus R$, then $\ell_S(N) \neq 0$. Thus, there are $0 \neq q_0 \in F$ and $0 \neq r_0 \in R$ such that $\alpha := \begin{bmatrix} q_0 \\ r_0 \end{bmatrix} \in N$. Let $f \in \text{Hom}(R_R, F_R)$ defined by $f(r) = (-q_0/r_0)r$ for $r \in R$. Put

$$\varphi = \begin{bmatrix} 1 & f \\ 0 & 0 \end{bmatrix} \in S.$$

Then $\varphi(\alpha) = 0$, and so $\ell_S(\alpha) \neq 0$. If $N = \alpha R$, then $\ell_S(N) = \ell_S(\alpha R) \neq 0$, a contradiction. Therefore, $\alpha R \subsetneq N$. Assume that $\alpha R \cap \beta R \neq 0$ for each $\beta \in N \setminus \alpha R$. Then there are $a, b \in R$ with $\alpha a = \beta b \neq 0$. For $s \in S$, note that $s\alpha = 0$ if and only if $s\alpha a = 0$ if and only if $s\beta b = 0$ if and only if $s\beta = 0$. Thus $\ell_S(\alpha) = \ell_S(\beta)$ for all $\beta \in N \setminus \alpha R$. Take $0 \neq s_0 \in \ell_S(\alpha)$. Then $s_0 \in \ell_S(N)$, which contradicts $\ell_S(N) = 0$. Thus, there exists $\beta \in N \setminus \alpha R$ such that $\alpha R \cap \beta R = 0$.

So $\alpha F \cap \beta F = 0$, hence α and β are linearly independent vectors in the vector space $F \oplus F$ over F. Thus, $\alpha F \oplus \beta F = F \oplus F$. Therefore, we have that $(\alpha R \oplus \beta R)_R \le^{\text{ess}} (\alpha F \oplus \beta F)_R = (F \oplus F)_R$. So $N_R \le^{\text{ess}} (F \oplus R)_R$ because $(\alpha R \oplus \beta R)_R \le N_R \le (F \oplus R)_R \le (F \oplus F)_R$. Hence, U_R is $\mathcal{K}$-cononsingular.

By Theorem 4.1.15, U_R is extending. Similarly, $W_R = (R \oplus F)_R$ is extending. Because $U \cap W = M$ and M is not extending, M cannot have an extending hull.

We use **SFI** to denote the class of strongly FI-extending right modules (or the class of right strongly FI-extending rings according to the context). In contrast to Example 8.4.13, we show that over a semiprime ring R, every finitely generated projective module P_R has the FI-extending module hull $H_{\mathbf{FI}}(P_R)$ (see Definition 8.4.1). This module hull $H_{\mathbf{FI}}(P_R)$ is explicitly described in Theorem 8.4.15. As a consequence, it will be seen that a finitely generated projective module P_R over a semiprime ring R is FI-extending if and only if it is a quasi-Baer module if and only if $\text{End}(P_R)$ is a quasi-Baer ring. This result will also be applied to C^*-algebras in Chap. 10.

Lemma 8.4.14 *Assume that M_R is an FI-extending module. Then $fM \subseteq M$ for any $f \in \mathcal{B}(\text{End}(E(M_R)))$.*

Proof Say $f \in \mathcal{B}(\text{End}(E(M_R)))$. Then $fE(M_R) \cap M \trianglelefteq M$. Because M is FI-extending, there exists $g^2 = g \in \text{End}(M_R)$ satisfying

$$fE(M_R) \cap M \le^{\text{ess}} gM \le^{\text{ess}} \overline{g}E(M_R),$$

where $\overline{g}$ is the canonical projection from $E(M_R) = E(gM_R) \oplus E((1-g)M_R)$ to $E(gM_R)$. Now we note that $fE(M_R) \cap M_R \le^{\text{ess}} fE(M_R)$. Thus $f = \overline{g}$ as f is central in $\text{End}(E(M_R))$. So $fM = \overline{g}M = gM \subseteq M$. □

We observe that Lemma 8.4.14 shows connections to Theorem 2.1.25 (and also Lemma 9.3.12). The next result shows and explicitly describes the unique (up to isomorphism) FI-extending hull for every finitely generated projective module over a semiprime ring.

Theorem 8.4.15 *Every finitely generated projective module P_R over a semiprime ring R has the FI-extending hull $H_{\mathbf{FI}}(P_R)$. Indeed,*

$$H_{\mathbf{FI}}(P_R) \cong e(\oplus^n \widehat{Q}_{\mathbf{FI}}(R)_R),$$

where $P \cong e(\oplus^n R_R)$ for some positive integer n and $e^2 = e \in \mathrm{End}(\oplus^n R_R)$.

Proof Step 1. $\widehat{Q}_{\mathbf{FI}}(R)_R$ is strongly FI-extending. From Theorems 3.2.37 and 8.3.17, $\widehat{Q}_{\mathbf{FI}}(R) = \widehat{Q}_{\mathbf{qB}}(R) = R\mathcal{B}(Q(R))$ is quasi-Baer, right strongly FI-extending, and semiprime. To show that $\widehat{Q}_{\mathbf{FI}}(R)_R$ is strongly FI-extending, take $U_R \trianglelefteq \widehat{Q}_{\mathbf{FI}}(R)_R$. Then by Lemma 8.1.3(ii), $U_R \leq^{\text{ess}} \widehat{Q}_{\mathbf{FI}}(R)U\widehat{Q}_{\mathbf{FI}}(R)_R$. Theorem 3.2.37 yields that $\widehat{Q}_{\mathbf{FI}}(R)U\widehat{Q}_{\mathbf{FI}}(R)_{\widehat{Q}_{\mathbf{FI}}(R)} \leq^{\text{ess}} h\widehat{Q}_{\mathbf{FI}}(R)_{\widehat{Q}_{\mathbf{FI}}(R)}$ for some $h \in \mathcal{B}(\widehat{Q}_{\mathbf{FI}}(R))$. By Lemma 8.1.3(i), $\widehat{Q}_{\mathbf{FI}}(R)U\widehat{Q}_{\mathbf{FI}}(R)_R \leq^{\text{ess}} h\widehat{Q}_{\mathbf{FI}}(R)_R$.

Now $\mathrm{End}(\widehat{Q}_{\mathbf{FI}}(R)_R) = \mathrm{End}(\widehat{Q}_{\mathbf{FI}}(R)_{\widehat{Q}_{\mathbf{FI}}(R)}) \cong \widehat{Q}_{\mathbf{FI}}(R)$ from Proposition 2.1.32. Therefore, $\lambda(h\widehat{Q}_{\mathbf{FI}}(R)) = h(\lambda\widehat{Q}_{\mathbf{FI}}(R))$ for any $\lambda \in \mathrm{End}(\widehat{Q}_{\mathbf{FI}}(R)_R)$. Thus $h\widehat{Q}_{\mathbf{FI}}(R)_R \trianglelefteq \widehat{Q}_{\mathbf{FI}}(R)_R$, so $\widehat{Q}_{\mathbf{FI}}(R)_R$ is strongly FI-extending because $U_R \leq^{\text{ess}} h\widehat{Q}_{\mathbf{FI}}(R)_R$.

Step 2. $H_{\mathbf{FI}}(\oplus^n R_R) = \oplus^n \widehat{Q}_{\mathbf{FI}}(R)_R$. Note that $\widehat{Q}_{\mathbf{FI}}(R)_R$ is FI-extending by Step 1, so $\oplus^n \widehat{Q}_{\mathbf{FI}}(R)_R$ is FI-extending by Theorem 2.3.5. Suppose that N_R is FI-extending such that $\oplus^n R_R \leq N_R \leq E(\oplus^n R_R) = \oplus^n E(R_R)$.

Take $f \in \mathcal{B}(Q(R))$. Then $f = \lambda(1)$ for some $\lambda \in \mathcal{B}(\mathrm{End}(E(R_R)))$ from Lemma 8.3.10. Let $\lambda\mathbf{1}$, which is the $n \times n$ diagonal matrix with λ on the diagonal, where $\mathbf{1}$ is the identity matrix in $\mathrm{End}(\oplus^n E(R_R)) = \mathrm{Mat}_n(\mathrm{End}(E(R_R)))$. Then because $\lambda\mathbf{1} \in \mathcal{B}(\mathrm{End}(\oplus^n E(R_R)))$, $\lambda\mathbf{1}N \subseteq N$ by Lemma 8.4.14, and so

$$\lambda\mathbf{1}\begin{bmatrix} R \\ \vdots \\ R \end{bmatrix} = \begin{bmatrix} fR \\ \vdots \\ fR \end{bmatrix} \subseteq N, \text{ where } \begin{bmatrix} R \\ \vdots \\ R \end{bmatrix} = \oplus^n R_R.$$

As $\widehat{Q}_{\mathbf{FI}}(R) = R\mathcal{B}(Q(R))$ by Theorem 8.3.17, we have that $\oplus^n \widehat{Q}_{\mathbf{FI}}(R)_R \leq N_R$, hence $H_{\mathbf{FI}}(\oplus^n R_R) = \oplus^n \widehat{Q}_{\mathbf{FI}}(R)_R$.

Step 3. $H_{\mathbf{FI}}(e(\oplus^n R_R)) = e(\oplus^n \widehat{Q}_{\mathbf{FI}}(R)_R)$. For this, we first observe that $\oplus^n \widehat{Q}_{\mathbf{FI}}(R)_R = e(\oplus^n \widehat{Q}_{\mathbf{FI}}(R)_R) \oplus (1-e)(\oplus^n \widehat{Q}_{\mathbf{FI}}(R)_R)$. As $\widehat{Q}_{\mathbf{FI}}(R)_R$ is strongly FI-extending by Step 1, $\oplus^n \widehat{Q}_{\mathbf{FI}}(R)_R$ is strongly FI-extending by Theorem 2.3.23. So $e(\oplus^n \widehat{Q}_{\mathbf{FI}}(R)_R)$ is strongly FI-extending from Theorem 2.3.19.

Let V_R be FI-extending such that $e(\oplus^n R_R) \leq V_R \leq E(e(\oplus^n R_R))$. Then

$$\begin{aligned}\oplus^n R_R &= e(\oplus^n R_R) \oplus (1-e)(\oplus^n R_R) \leq V_R \oplus (1-e)(\oplus^n R_R) \\ &\leq V_R \oplus E[(1-e)(\oplus^n R_R)].\end{aligned}$$

Since V_R is FI-extending and $E[(1-e)(\oplus^n R_R)]$ is injective, Theorem 2.3.5 yields that $V_R \oplus E[(1-e)(\oplus^n R_R)]$ is FI-extending. Therefore by Step 2,

$$H_{\mathbf{FI}}(\oplus^n R_R) = \oplus^n \widehat{Q}_{\mathbf{FI}}(R)_R \leq V_R \oplus E[(1-e)(\oplus^n R_R)].$$

To prove that $e(\oplus^n \widehat{Q}_{\mathbf{FI}}(R)_R) \leq V_R$, we take

$$e\alpha \in e(\oplus^n \widehat{Q}_{\mathbf{FI}}(R)_R), \quad \text{where } \alpha \in \oplus^n \widehat{Q}_{\mathbf{FI}}(R)_R.$$

Since $e(\oplus^n \widehat{Q}_{\mathbf{FI}}(R)_R) \leq V_R \oplus E[(1-e)(\oplus^n R_R)]$, $e\alpha = v + y$ for some $v \in V$ and $y \in E[(1-e)(\oplus^n R_R)]$. Thus,

$$e\alpha - v = y \in [e(\oplus^n \widehat{Q}_{\mathbf{FI}}(R)_R) + V] \cap E[(1-e)(\oplus^n R_R)].$$

Since $e(\oplus^n R_R) \leq^{\text{ess}} e(\oplus^n \widehat{Q}_{\mathbf{FI}}(R)_R)$, $E[e(\oplus^n \widehat{Q}_{\mathbf{FI}}(R)_R)] = E[e(\oplus^n R_R)]$. So $[e(\oplus^n \widehat{Q}_{\mathbf{FI}}(R)_R) + V] \cap E((1-e)(\oplus^n R_R)] \leq E[e(\oplus^n R_R)] \cap E[(1-e)(\oplus^n R_R)]$. Hence, $e\alpha - v = y = 0$, so $e\alpha = v \in V$. Therefore, $e(\oplus^n \widehat{Q}_{\mathbf{FI}}(R)_R) \leq V_R$. Consequently, $H_{\mathbf{FI}}(e(\oplus^n R_R)) = e(\oplus^n \widehat{Q}_{\mathbf{FI}}(R)_R)$.

Step 4. $H_{\mathbf{FI}}(P_R) \cong e(\oplus^n \widehat{Q}_{\mathbf{FI}}(R)_R)$. Let $\sigma : P_R \to e(\oplus^n R_R)$ be an isomorphism. Then σ can be extended to an isomorphism $\overline{\sigma} : E(P_R) \to E(e(\oplus^n R_R))$. We see that $H_{\mathbf{FI}}(P_R) = \overline{\sigma}^{-1}(e(\oplus^n \widehat{Q}_{\mathbf{FI}}(R)_R)) \cong e(\oplus^n \widehat{Q}_{\mathbf{FI}}(R)_R)$. □

Remark 8.4.16 By the proof of Theorem 8.4.15, the strongly FI-extending hull and the FI-extending hull of a finitely generated projective module P_R coincide when R is semiprime.

If R is not semiprime, the above remark does not hold. For example, let $R = \mathbb{Z}_3[S_3]$, the group algebra of S_3 over the field $\mathbb{Z}_3$, where S_3 is the symmetric group on $\{1, 2, 3\}$. By Example 2.3.18, R_R is not strongly FI-extending. Thus $H_{\mathbf{SFI}}(R_R)$ does not exist because R_R is injective.

The existence of an FI-extending hull of a module is not always guaranteed, even in the presence of nonsingularity, as the next example shows.

Example 8.4.17 Let R be the ring of Example 8.2.9. Then $H_{\mathbf{FI}}(R_R)$ does not exist. Indeed, let H_1 and H_2 be rings as in Example 8.2.9, which are right FI-extending rings. Since H_1 and H_2 are right rings of quotients of R, H_1 and H_2 are FI-extending right R-modules by Proposition 8.1.4(i). Suppose $H_{\mathbf{FI}}(R_R)$ exists. Then it follows that $H_{\mathbf{FI}}(R_R) \subseteq H_1 \cap H_2 = R$, so $H_{\mathbf{FI}}(R_R) = R_R$. But, R_R is not FI-extending, a contradiction.

Corollary 8.4.18 *Assume that R is a semiprime ring and P_R is a finitely generated projective module. Then $\widehat{Q}_{\mathbf{FI}}(\mathrm{End}(P_R)) \cong \mathrm{End}(H_{\mathbf{FI}}(P_R))$.*

Proof Since $P_R \cong e(\oplus^n R_R)$ with $e^2 = e \in \mathrm{Mat}_n(R)$, $\mathrm{End}(P_R) \cong e\mathrm{Mat}_n(R)e$. Also by Theorem 8.4.15, $H_{\mathbf{FI}}(P_R) \cong e(\oplus^n \widehat{Q}_{\mathbf{FI}}(R))$. Thus it follows that

$$\mathrm{End}(H_{\mathbf{FI}}(P_R)) \cong e\mathrm{Mat}_n(\mathrm{End}(\widehat{Q}_{\mathbf{FI}}(R)_R)e.$$

Now $\mathrm{End}(\widehat{Q}_{\mathbf{FI}}(R)_R) \cong \widehat{Q}_{\mathbf{FI}}(R)$ by Proposition 2.1.32.

Hence $\mathrm{End}(H_{\mathbf{FI}}(P_R)) \cong e\mathrm{Mat}_n(\mathrm{End}_R(\widehat{Q}_{\mathbf{FI}}(R)_R))e \cong e\mathrm{Mat}_n(\widehat{Q}_{\mathbf{FI}}(R))e$. Next, we observe that $\widehat{Q}_{\mathbf{FI}}(e\mathrm{Mat}_n(R)e) = e\widehat{Q}_{\mathbf{FI}}(\mathrm{Mat}_n(R))e$ since $\mathrm{Mat}_n(R)$ is semiprime and $0 \neq e^2 = e \in \mathrm{Mat}_n(R)$ (see Theorem 3.2.37 and Lemma 9.3.9).

So $\mathrm{End}(H_{\mathbf{FI}}(P_R)) \cong e\widehat{Q}_{\mathbf{FI}}(\mathrm{Mat}_n(R))e = \widehat{Q}_{\mathbf{FI}}(e\mathrm{Mat}_n(R)e) \cong \widehat{Q}_{\mathbf{FI}}(\mathrm{End}(P_R))$. □

When P_R is a progenerator, we have the following.

Corollary 8.4.19 *Let R be a semiprime ring. If P_R is a progenerator of the category Mod-R of right R-modules, then $H_{\mathbf{FI}}(P_R)_{\widehat{Q}_{\mathbf{FI}}(R)}$ is a progenerator of the category Mod-$\widehat{Q}_{\mathbf{FI}}(R)$ of right $\widehat{Q}_{\mathbf{FI}}(R)$-modules.*

Proof Assume that P_R is a progenerator for Mod-R. Let $P_R \cong e(\oplus^n R_R)$ with $e^2 = e \in \mathrm{Mat}_n(R)$ and let $S = \mathrm{End}(P_R)$. Then R is Morita equivalent to S and

$$S \cong e\mathrm{Mat}_n(R)e \ \text{ with } \ \mathrm{Mat}_n(R)e\mathrm{Mat}_n(R) = \mathrm{Mat}_n(R).$$

Now $\mathrm{Mat}_n(\widehat{Q}_{\mathbf{FI}}(R))e\mathrm{Mat}_n(\widehat{Q}_{\mathbf{FI}}(R)) = \mathrm{Mat}_n(R\mathcal{B}(Q(R))) = \mathrm{Mat}_n(\widehat{Q}_{\mathbf{FI}}(R))$ by observing that $\widehat{Q}_{\mathbf{FI}}(R) = R\mathcal{B}(Q(R))$ from Theorem 8.3.17.

Since $H_{\mathbf{FI}}(P_R) \cong e(\oplus^n \widehat{Q}_{\mathbf{FI}}(R))$, $\mathrm{End}(H_{\mathbf{FI}}(P_R)_{\widehat{Q}_{\mathbf{FI}}(R)}) \cong e\mathrm{Mat}_n(\widehat{Q}_{\mathbf{FI}}(R))e$. Thus, we get that $H_{\mathbf{FI}}(P_R)_{\widehat{Q}_{\mathbf{FI}}(R)}$ is a progenerator of the category Mod-$\widehat{Q}_{\mathbf{FI}}(R)$ of right $\widehat{Q}_{\mathbf{FI}}(R)$-modules. □

A connection between FI-extending modules and quasi-Baer modules can be seen in the next result.

Theorem 8.4.20 *Assume that P_R is a finitely generated projective module over a semiprime ring R. Then the following are equivalent.*

(i) *P_R is (strongly) FI-extending.*
(ii) *P_R is a quasi Baer module.*
(iii) *$\mathrm{End}(P_R)$ is a quasi-Baer ring.*
(iv) *$\mathrm{End}(P_R)$ is a right FI-extending ring.*

Proof Let $P_R \cong e(\oplus^n R_R)$, where $e^2 = e \in \mathrm{End}(\oplus^n R_R) \cong \mathrm{Mat}_n(R)$ and n is a positive integer.

(i)⇒(ii) If P_R is FI-extending, then $P_R = H_{\mathbf{FI}}(P_R) \cong e(\oplus^n \widehat{Q}_{\mathbf{qB}}(R)_R)$ by Theorems 3.2.37, 8.3.17, and 8.4.15. Note that $\mathrm{End}(\widehat{Q}_{\mathbf{qB}}(R)_R) \cong \widehat{Q}_{\mathbf{qB}}(R)$ from Proposition 2.1.32. By Theorems 3.2.37, 8.3.17, and Proposition 8.1.4(i), $\widehat{Q}_{\mathbf{qB}}(R)_R$ is FI-extending. Next, we show that $\widehat{Q}_{\mathbf{qB}}(R)_R$ is quasi-Baer. For this, take $N_R \trianglelefteq \widehat{Q}_{\mathbf{qB}}(R)_R$. As $\mathrm{End}(\widehat{Q}_{\mathbf{qB}}(R)_R) \cong \widehat{Q}_{\mathbf{qB}}(R)$, N is a left ideal of $\widehat{Q}_{\mathbf{qB}}(R)$. Thus $\ell_{\widehat{Q}_{\mathbf{qB}}(R)}(N) = \widehat{Q}_{\mathbf{qB}}(R)g$ for some $g^2 = g \in \widehat{Q}_{\mathbf{qB}}(R)$. So $\widehat{Q}_{\mathbf{qB}}(R)_R$ is a quasi-Baer module. By Theorem 4.6.15 $\oplus^n \widehat{Q}_{\mathbf{qB}}(R)_R$ is a quasi-Baer module. Hence $e(\oplus^n \widehat{Q}_{\mathbf{qB}}(R)_R)$ is a quasi-Baer module by Theorem 4.6.14. So P_R is quasi-Baer.

(ii)⇒(iii) It follows from Theorem 4.6.16.

(iii)⇒(i) Let $\mathrm{End}(P_R)$ be quasi-Baer. Because $\mathrm{End}(P_R) \cong e\mathrm{Mat}_n(R)e$, $e\mathrm{Mat}_n(R)e = \widehat{Q}_{\mathbf{qB}}(e\mathrm{Mat}_n(R)e) = e\widehat{Q}_{\mathbf{qB}}(\mathrm{Mat}_n(R))e = e\mathrm{Mat}_n(\widehat{Q}_{\mathbf{qB}}(R))e$ (see Proposition 9.3.7 and Lemma 9.3.9). Next, let $f \in \mathcal{B}(Q(R))$. Then we have that $f\mathbf{1} \in \mathcal{B}(\mathrm{Mat}_n(Q(R)))$, where $\mathbf{1}$ is the identity matrix of $\mathrm{Mat}_n(R)$. Thus

$$e(f\mathbf{1})e \in e\mathrm{Mat}_n(\widehat{Q}_{\mathbf{qB}}(R))e = e\mathrm{Mat}_n(R)e.$$

Take $e(f\mathbf{1})e = [\alpha_{ij}] \in e\mathrm{Mat}_n(R)e$. Then

$$e\begin{bmatrix} fR \\ \vdots \\ fR \end{bmatrix} = e(f\mathbf{1})e\begin{bmatrix} R \\ \vdots \\ R \end{bmatrix} = e[\alpha_{ij}]e\begin{bmatrix} R \\ \vdots \\ R \end{bmatrix} \subseteq e\begin{bmatrix} R \\ \vdots \\ R \end{bmatrix}.$$

So $e(\oplus^n \widehat{Q}_{\mathbf{qB}}(R)_R) = e(\oplus^n R_R)$ because $\widehat{Q}_{\mathbf{qB}}(R) = R\mathcal{B}(Q(R))$ by Theorem 8.3.17. From Theorems 8.4.15 and 8.3.17, $H_{\mathbf{FI}}(e(\oplus^n R_R)) = e(\oplus^n R_R)$ since $\widehat{Q}_{\mathbf{qB}}(R) = \widehat{Q}_{\mathbf{FI}}(R)$, and so $e(\oplus^n R_R)$ is (strongly) FI-extending. Therefore, P_R is (strongly) FI-extending.

(iii)⇔(iv) Since $\mathrm{End}(P_R)$ is semiprime, Theorem 3.2.37 yields the equivalence. □

We observe that the rational hull $\widetilde{E}(M)$ of a module M is an $\mathfrak{M}$ hull of M, where $\mathfrak{M}$ is the class of rationally complete modules (see Definition 8.4.1 and [262, p. 277]). Consider $M = \mathbb{Z}_p \oplus \mathbb{Z}_{p^3}$ and $N = \mathbb{Z}_p \oplus p\mathbb{Z}_{p^3}$, where p is a prime integer. Then $N_{\mathbb{Z}} \leq^{\mathrm{ess}} M_{\mathbb{Z}}$ and $N_{\mathbb{Z}}$ is extending (by direct calculation or [301, p. 19]). But recall from Example 2.2.1(ii) that $M_{\mathbb{Z}}$ is not extending. So the extending property does not, in general, transfer to essential extensions of modules. However, Theorem 8.1.8 motivates one to ask: *Does the (FI-)extending property transfer to rational extensions in modules?* Our next result shows this to be the case for rational hulls.

Theorem 8.4.21 *Let M be an (FI-)extending module. Then $\widetilde{E}(M)$ is an (FI-) extending module.*

Proof First, we assume that M is extending. Let $K \leq \widetilde{E}(M)$ and $N = K \cap M$. Then $N \leq^{\mathrm{ess}} eM$ for some $e^2 = e \in \mathrm{End}(M)$. By Proposition 1.3.6 and [262, Theorem 8.24], there exists $f \in \mathrm{End}(\widetilde{E}(M))$ such that $f|_M = e$. As $E(M)$ is injective, there is $g \in \mathrm{End}(E(M))$ satisfying $g|_{\widetilde{E}(M)} = f$.

Let $m \in M$. Then $(g^2 - g)(m) = (e^2 - e)(m) = 0$. From the definition of $\widetilde{E}(M)$ (see the definition of $\widetilde{E}(M)$ after Proposition 1.3.6), $(g^2 - g)(y) = 0$ for all y in $\widetilde{E}(M)$. Hence $f^2 = f$. Assume that there exists $k \in K$ such that $f(k) - k \neq 0$. As $M \leq^{\mathrm{den}} \widetilde{E}(M)$, there exists $r \in R$ satisfying $kr \in M$ and $(f(k) - k)r \neq 0$. Then $kr \in N$, so $(f(k) - k)r = f(kr) - kr = e(kr) - kr = 0$, a contradiction. Hence, $K \leq f\widetilde{E}(M)$. Let $0 \neq f(v) \in f\widetilde{E}(M)$ with $v \in \widetilde{E}(M)$. Then there is $s \in R$ such that $vs \in M$ and $f(v)s \neq 0$. Now we see that $0 \neq f(v)s = f(vs) = e(vs) \in M$. So $0 \neq f(v)st \in N \leq K$ for some $t \in R$. Therefore, $K \leq^{\mathrm{ess}} f\widetilde{E}(M)$, so $\widetilde{E}(M)$ is extending.

Next, assume that M is FI-extending and that $K \trianglelefteq \widetilde{E}(M)$. Put $N = K \cap M$. We claim that $N \trianglelefteq M$. For this, take $h \in \mathrm{End}(M)$. From Proposition 1.3.6 and [262, Theorem 8.24], there exists $f \in \mathrm{End}(\widetilde{E}(M))$ such that $f|_M = h$.

So $h(N) = f(N) \subseteq K \cap M = N$. Thus, $N \trianglelefteq M$. From the proof similar to the case when M is extending, we obtain that $\widetilde{E}(M)$ is FI-extending. □

For an example illustrating Theorem 8.4.21, consider $M = \mathbb{Z} \oplus \mathbb{Z}_p$, where p is a prime integer (see [262, Example 8.21]). By Theorem 2.3.5, $M_{\mathbb{Z}}$ is FI-extending, but not extending (see [301, p. 19]). Now $\widetilde{E}(M_{\mathbb{Z}}) = \mathbb{Z}_P \oplus \mathbb{Z}_p$ is FI-extending from Theorem 8.4.21 or Theorem 2.3.5, but not extending (see [301, p. 19]), where $\mathbb{Z}_P$ is the localization of $\mathbb{Z}$ at $P = p\mathbb{Z}$.

Exercise 8.4.22

1. Let R be the ring in Example 8.4.12. Prove that R_R is uniform and nonsingular.
2. ([98, Birkenmeier, Park, and Rizvi]) Assume that R is a semiprime ring and P_R is a finitely generated projective module. Show that
   (i) $\mathrm{Rad}(H_{\mathbf{FI}}(P_R)_{\widehat{Q}_{\mathbf{FI}}(R)}) \cap P = \mathrm{Rad}(P_R)$.
   (ii) $H_{\mathbf{FI}}(P_R) \cong P \otimes_R \widehat{Q}_{\mathbf{FI}}(R)$ as $\widehat{Q}_{\mathbf{FI}}(R)$-modules.
   (iii) $H_{\mathbf{FI}}(P_R)$ is also a finitely generated projective $\widehat{Q}_{\mathbf{FI}}(R)$-module.
3. Let M be a bounded Abelian group. Prove that $M_{\mathbb{Z}}$ has an extending hull. (Hint: see [172, p. 88] and [301, p. 19].)
4. Let M be a continuous module. Show that $\widetilde{E}(M)$ is quasi-continuous.

Historical Notes Results of Sect. 8.1 are obtained by Birkenmeier, Park, and Rizvi in [89]. The concept of a $\mathfrak{K}$ absolute ring hull in Definition 8.2.1 was already implicit in the paper [307] by Müller and Rizvi from their definition of a type III continuous module hull (see also Definition 8.4.1). Theorem 8.2.6 from [89], is an adaptation of [354, Theorem 4.25]. Other results of Sect. 8.2 appear in [89].

Many results in Sect. 8.3, which were originally stated and proved for a ring with identity, have been extended to rings R with $\ell_R(R) = 0$. Definition 8.3.1 was provided in [96]. Proposition 8.3.2 is due to Johnson [236]. Results 8.3.3–8.3.8 are due to Birkenmeier, Park, and Rizvi in [96]. Theorem 8.3.8(ii) is an unpublished new characterization. Theorem 8.3.11(i), (ii), and (iii) appear in [96]. Corollary 8.3.12 is an unpublished new result. Also Theorem 8.3.13(i), (ii), and (iv) were shown in [96]. Proposition 8.3.16 and Theorem 8.3.17 are due to Birkenmeier, Park, and Rizvi [97]. In [163], Ferrero has shown that $Q^s(R)$ is quasi-Baer for any semiprime ring R. Example 8.3.18 is taken from [262, Example 13.26(4)]. Results 8.3.20, 8.3.21 and 8.3.23 appear in [97].

Theorem 8.3.22 is due to Passman [340] and Connell [131]. Example 8.3.25 appears in [102]. Lemma 8.3.26 and Theorem 8.3.28 are obtained in [97]. In [42], it is shown that LO does hold between R and $R\mathcal{B}(Q(R))$. Lemma 8.3.29 is from [322]. Beidar and Wisbauer [42] show that R is biregular if and only if R is semiprime and $R\mathcal{B}(Q(R))$ is biregular (see Exercise 8.3.58.8). Also, they show that R is regular and biregular if and only if $R\mathcal{B}(Q(R))$ is regular and biregular [42]. Corollary 8.3.30 from [97], complements their results.

Results 8.3.31–8.3.37 appear in [97]. Let R be a semiprime PI-ring. Then so is $Q(R)$ by a result of Martindale [292]. Also by a result of Fisher [168], a semiprime PI-ring R is right nonsingular. Thus $Q(R)$ is a regular right self-injective PI-ring from Theorem 2.1.31. So $Q(R)$ has bounded index (of nilpotency) (see [221, Corollary, p. 226]). Therefore, any semiprime PI-ring R has bounded index (of nilpotency). Also any semiprime right Goldie ring has bounded index (of nilpotency).

Results in [160] show that a semiprime right FPF ring has bounded index (of nilpotency).

For a commutative semiprime ring R, Storrer [386] called the intersection of all regular rings of $Q(R)$ containing R the epimorphic hull of R. By showing this intersection was regular, he showed that every commutative semiprime ring has a smallest regular ring of quotients. The existence of Baer ring hulls shown in [298] for the case of commutative semiprime rings (see also Theorem 8.2.4) and in [208] for the case of reduced Utumi rings, now follow directly from Proposition 8.3.36 (see [323] for the existence of Baer ring hulls of commutative regular rings by a sheaf theoretic method). Results 8.3.39–8.3.44 appear in [101]. Theorem 8.3.44 shows that when R is a commutative semiprime ring, $Q_{\mathbf{pqB}}(R)$ is related to the Baer extension considered in [254]. Lemma 8.3.46 and Theorem 8.3.47 appear in [94]. Theorem 8.3.50, Theorem 8.3.53, and Example 8.3.57 were obtained by Armendariz, Birkenmeier, and Park [29], while Proposition 8.3.49 and Theorem 8.3.56 appear in [96].

Results 8.4.4–8.4.12 are taken from [307], while Results 8.4.14–8.4.20 appear in [98]. Theorem 8.4.20 is a module theoretic version of Theorem 3.2.37 for a finitely generated projective module over a semiprime ring. The proof of Theorem 8.4.21 when M is extending corrects the proof of [1, Theorem 5.3]. We include some more related references such as [43, 86, 87, 90, 133, 143, 146, 197, 225, 257, 258, 337, 351], and [370].

Chapter 9
Hulls of Ring Extensions

Application of results developed in Chap. 8 will be considered in this chapter. Problems II and III (see Introduction of Chap. 8) are studied in Sect. 9.1. Another topic for consideration will be skew group ring extensions of semiprime quasi-Baer rings in Sect. 9.2. The result on skew group ring extensions will be used in the study of boundedly centrally closed C^*-algebras later in Chap. 10. Related to Problem IV (see Introduction of Chap. 8), results on ring hulls of monoid and matrix ring extensions will be included in Sect. 9.3.

9.1 Applications to Certain Matrix Rings

In this section, we apply results on ring hulls and pseudo ring hulls to certain matrix rings. Our focus will be on when certain 2×2 matrix rings are right extending, Baer or are right (semi)hereditary. We include results on the existence (or the lack of existence) of various ring hulls. The first result of this section characterizes a right extending ring whose maximal right ring of quotients is the 2×2 matrix ring over a division ring. Hence this result provides an answer to Problem III (see Introduction of Chap. 8) when $\mathfrak{A}$ is the class of 2×2 matrix rings over division rings and $\mathfrak{B}$ is the class of right extending rings.

Theorem 9.1.1 *Let D be a division ring and assume that T is a ring such that $Q(T) = \mathrm{Mat}_2(D)$ (resp., $Q(T) = Q^\ell(T) = \mathrm{Mat}_2(D)$). Then T is right extending (resp., T is Baer) if and only if the following are satisfied.*

(i) *There exist $v, w \in D$ such that $\begin{bmatrix} 1 & v \\ 0 & 0 \end{bmatrix} \in T$ and $\begin{bmatrix} 0 & 0 \\ w & 1 \end{bmatrix} \in T$.*

(ii) *For each $0 \neq d \in D$ at least one of the following holds.*

(1) $\begin{bmatrix} 0 & d \\ 0 & 1 \end{bmatrix} \in T$.

(2) $\begin{bmatrix} 1 & 0 \\ d^{-1} & 0 \end{bmatrix} \in T$.

G.F. Birkenmeier et al., *Extensions of Rings and Modules*,
DOI 10.1007/978-0-387-92716-9_9,

(3) *there is* $a \in D$ *with* $a - a^2 \neq 0$ *and* $\begin{bmatrix} a & (1-a)d \\ d^{-1}a & d^{-1}(1-a)d \end{bmatrix} \in T$.

Proof Routine calculations show that any nontrivial idempotent of $Q(T)$ has one of the following forms where $a, b, f \in D$ with $a - a^2 \neq 0$ and $b \neq 0$:

$$\begin{bmatrix} 1 & f \\ 0 & 0 \end{bmatrix}, \begin{bmatrix} 0 & 0 \\ f & 1 \end{bmatrix}, \begin{bmatrix} 0 & f \\ 0 & 1 \end{bmatrix}, \begin{bmatrix} 1 & 0 \\ f & 0 \end{bmatrix}, \begin{bmatrix} a & b \\ b^{-1}(1-a)a & b^{-1}(1-a)b \end{bmatrix}.$$

From Definition 8.2.11, we obtain:

- for $f \in D$, $\begin{bmatrix} 1 & f \\ 0 & 0 \end{bmatrix} \alpha \begin{bmatrix} 1 & 0 \\ 0 & 0 \end{bmatrix}$ and $\begin{bmatrix} 0 & 0 \\ f & 1 \end{bmatrix} \alpha \begin{bmatrix} 0 & 0 \\ 0 & 1 \end{bmatrix}$.
- for $0 \neq f, 0 \neq g \in D$, $\begin{bmatrix} 0 & f \\ 0 & 1 \end{bmatrix} \alpha \begin{bmatrix} 1 & 0 \\ g & 0 \end{bmatrix}$ if and only if $g = f^{-1}$.
- for $0 \neq f \in D$, $\begin{bmatrix} 0 & f \\ 0 & 1 \end{bmatrix} \alpha \begin{bmatrix} a & b \\ b^{-1}(1-a)a & b^{-1}(1-a)b \end{bmatrix}$ if and only if $b = (1-a)f$.

Let U be the set of matrices $K \in \mathrm{Mat}_2(D)$ such that for each $0 \neq d \in D$, K has exactly one of the following forms:

$$\begin{bmatrix} 0 & d \\ 0 & 1 \end{bmatrix}, \begin{bmatrix} 1 & 0 \\ d^{-1} & 0 \end{bmatrix}, \text{ or } \begin{bmatrix} a & (1-a)d \\ d^{-1}a & d^{-1}(1-a)d \end{bmatrix}$$

for some $a \in D$ with $a - a^2 \neq 0$. Now for some $v, w \in D$, let

$$Y = U \cup \left\{ \begin{bmatrix} 0 & 0 \\ 0 & 0 \end{bmatrix}, \begin{bmatrix} 1 & 0 \\ 0 & 1 \end{bmatrix} \right\} \cup \left\{ \begin{bmatrix} 1 & v \\ 0 & 0 \end{bmatrix}, \begin{bmatrix} 0 & 0 \\ w & 1 \end{bmatrix} \right\}.$$

Then $Y = \delta^{\alpha}_{\mathbf{E}}(T)(1)$. Since $Z(T_T) = 0$, the result is now a direct consequence of Proposition 8.2.13, Theorem 8.2.15, and Corollary 3.3.3, where R in Theorem 8.2.15 coincides with T in the present result. □

Using Theorem 9.1.1, next we provide an elementwise characterization of a Prüfer domain. The fact that any right ring of quotients of a Prüfer domain, is also a Prüfer domain, follows from Corollary 9.1.2.

Corollary 9.1.2 *Let A be a commutative domain with F as its field of fractions. Then the following are equivalent.*

(i) *A is a Prüfer domain.*
(ii) *For each $0 \neq d \in F$ with $d \notin A$ and $d^{-1} \notin A$, there exists $a \in A$ such that $d^{-1}a \in A$ and $(1-a)d \in A$.*

Proof The proof follows from Theorems 6.1.4 and 9.1.1, where $\mathrm{Mat}_2(A) = T$ in Theorem 9.1.1. □

Corollary 9.1.3 *Let A be a right Ore domain with $D = Q^r_{c\ell}(A)$. Then $T = \begin{bmatrix} A & D \\ 0 & A \end{bmatrix}$ is a right extending ring hull of $T_2(A)$.*

Proof By Theorem 9.1.1, T is right extending. Let $e_{ij} \in Q(T_2(A)) = \mathrm{Mat}_2(D)$ be the matrix with 1 in the (i, j)-position and 0 elsewhere. Assume that S is a right extending intermediate ring between $T_2(A)$ and T. For each $0 \neq d \in D$, it follows that $de_{12} + e_{22} \in S$ by Theorem 9.1.1, so $de_{12} \in S$. So $T = S$. Hence, T is a right extending ring hull of $T_2(A)$. □

By Theorem 5.6.2, if A is a commutative domain with F as its field of fractions and $A \neq F$, then $T_n(A)$ $(n > 1)$ is not Baer. However, by Theorem 8.1.8(iii) and Corollary 3.3.3, any right ring of quotients of $T_n(A)$ which contains $T_n(F)$ is Baer. This result motivates the question: *Assume that A is a commutative domain. For a class of rings $\mathfrak{C}$, can we find $\mathfrak{C}$ ring hulls or $\mathfrak{C}$ ρ pseudo ring hulls for $T_n(A)$? Further, can these hulls be used to describe all $\mathfrak{C}$ right rings of quotients of $T_n(A)$ when $\mathfrak{C}$ is related to the class of Baer rings?* In the next results, we study this question when A is a commutative ring.

For a commutative PID A, we use $\gcd(a, b)$ to denote the greatest common divisor of $a, b \in A$.

Lemma 9.1.4 *Let A be a commutative PID and T a right ring of quotients of $T_2(A)$ such that $\begin{bmatrix} A & A \\ aA & A \end{bmatrix} \subseteq T$ for some $0 \neq a \in A$.*

(i) *If $\begin{bmatrix} 0 & a^{-1} \\ 0 & 0 \end{bmatrix} \in T$, then T is right extending.*

(ii) *If $a = p_1^{k_1} \cdots p_m^{k_m}$ where each p_i is a distinct prime, each k_i is a positive integer, and*

$$\begin{bmatrix} 0 & (p_1^{k_1-1} \cdots p_m^{k_m-1})^{-1} \\ 0 & 0 \end{bmatrix} \in T,$$

then T is right extending.

Proof Let c, $d \in A$ such that $c \neq 0$ and $d \neq 0$. Assume that

$$\begin{bmatrix} 0 & cd^{-1} \\ 0 & 0 \end{bmatrix} \notin T \text{ and } \begin{bmatrix} 0 & 0 \\ dc^{-1} & 0 \end{bmatrix} \notin T.$$

Say $\gcd(c, d) = z \in A$. Then $c = c_1 z$, $d = d_1 z$, and $\gcd(c_1, d_1) = 1$ for some c_1, $d_1 \in A$. By noting that $cd^{-1} = c_1 d_1^{-1}$ and $dc^{-1} = d_1 c_1^{-1}$, we may assume that $\gcd(c, d) = 1$.

Put $g = \gcd(a, d)$. Then $a = sg$, $d = tg$, and $\gcd(s, t) = 1$ with $s, t \in A$.

(i) We note that $\gcd(c, d) = 1$ and $d = tg$, and so $\gcd(c, t) = 1$. Because $\gcd(s, t) = 1 = \gcd(c, t)$, then $\gcd(t, cs) = 1$. Hence, there are $x, y \in A$ with $1 = csx + ty$. Take $b = csx$.

If $b = 0$, then $1 = ty$, thus $t^{-1} = y \in A$. So $cd^{-1} = c(tg)^{-1} = cyg^{-1}$. Hence, $cd^{-1} \in g^{-1}A = a^{-1}sA \subseteq a^{-1}A$, and so $\begin{bmatrix} 0 & cd^{-1} \\ 0 & 0 \end{bmatrix} \in \begin{bmatrix} 0 & a^{-1} \\ 0 & 0 \end{bmatrix} T_2(A) \subseteq T$, which is a contradiction. So $b \neq 0$.

If $b = 1$, then $c^{-1} = sx$ and $dc^{-1} = dsx = tgsx = tax \in aA$, so we have that $\begin{bmatrix} 0 & 0 \\ dc^{-1} & 0 \end{bmatrix} \in \begin{bmatrix} 0 & 0 \\ aA & 0 \end{bmatrix} \subseteq T$, also a contradiction. Thus, $b - b^2 \neq 0$.

We observe that $dc^{-1}b = dc^{-1}csx = dsx = tgsx = tax \in aA$, and therefore $(1-b)cd^{-1} = (ty)c(tg)^{-1} = g^{-1}cy \in g^{-1}A = a^{-1}sA \subseteq a^{-1}A$. Therefore, we see that $\begin{bmatrix} b & (1-b)cd^{-1} \\ dc^{-1}b & 1-b \end{bmatrix} \in T$. From Theorem 9.1.1, T is right extending.

(ii) We put $d = p_1^{h_1} \cdots p_m^{h_m} q$, where each h_i is a nonnegative integer and $\gcd(p_i, q) = 1$ for each i.

Case 1. Assume that whenever $h_i \neq 0$, then $h_i \geq k_i$. We claim that $\gcd(d, s) = 1$. First, if $h_1 = \cdots = h_m = 0$, then $g = \gcd(d, a) = 1$ because $a = p_1^{k_1} \cdots p_m^{k_m}$. So $s = sg = a$. Thus, $\gcd(d, s) = \gcd(d, a) = 1$.

Next, suppose that not all h_i are zero. If $\gcd(d, s) \neq 1$, then there is j in $\{1, \ldots, m\}$ with $p_j \mid d$ and $p_j \mid s$ because $a = p_1^{k_1} \cdots p_m^{k_m}$, $s \mid a$, and $s \neq 1$. So $h_j \geq 1$ and $s = s_1 p_j$ with $s_1 \in A$. Since $a = sg$ and $d = tg$, we have that $at = ds$. As $h_j \geq k_j$ by assumption and $at = ds$,

$$p_1^{k_1} \cdots p_{j-1}^{k_{j-1}} p_{j+1}^{k_{j+1}} \cdots p_m^{k_m} t = p_1^{h_1} \cdots p_{j-1}^{h_{j-1}} p_j^{h_j - k_j} p_{j+1}^{h_{j+1}} \cdots p_m^{h_m} q p_j s_1.$$

So $p_j \mid t$, a contradiction to $\gcd(s, t) = 1$. Thus $\gcd(d, s) = 1$.

As $\gcd(c, d) = 1$ and $\gcd(s, d) = 1$, $\gcd(cs, d) = 1$. Thus, there exist x, y in A with $csx + dy = 1$. Let $b = csx$. If $b = 0$, then $dy = 1$, hence $d^{-1} = y \in A$. So $cd^{-1} = cy \in A$, a contradiction. Thus, $b \neq 0$. If $b = 1$, then $1 = csx$, so $c^{-1} = sx$. Thus $dc^{-1} = dsx = tgsx = tax \in aA$, a contradiction. So $b \neq 1$, and hence $b - b^2 \neq 0$. Note that $dc^{-1}b = dc^{-1}csx = dsx = tgsx = tax \in aA$ therefore $(1-b)cd^{-1} = dycd^{-1} = yc \in A$. Thus,

$$\begin{bmatrix} b & (1-b)cd^{-1} \\ dc^{-1}b & 1-b \end{bmatrix} \in \begin{bmatrix} A & A \\ aA & A \end{bmatrix} \subseteq T.$$

Case 2. Assume that there exists $\ell \in \{1, \ldots, m\}$ with $0 \neq h_\ell < k_\ell$. Let $I = \{i \in \{1, \ldots, m\} \mid 0 \leq h_i < k_i\}$. We may note that $I \neq \emptyset$ because $h_\ell \in I$. Take $J = \{1, \ldots, m\} \setminus I$.

Subcase 2.1. $J \neq \emptyset$. Put $v = |I|$ and $w = |J|$. Denote $I = \{i_1, \ldots, i_v\}$ and $J = \{j_1, \ldots, j_w\}$. Note that $J = \{j \in \{1, 2, \ldots, m\} \mid h_j \geq k_j\}$.

Put $t_{j_1} = h_{j_1} - k_{j_1} + 1, \ldots, t_{j_w} = h_{j_w} - k_{j_w} + 1$, and $\mu = p_{i_1}^{k_{i_1}} \cdots p_{i_v}^{k_{i_v}}$. Then $\gcd(\mu, q) = 1$. As $\gcd(c, d) = 1$, $\gcd(c, q) = 1$. So $\gcd(c\mu, q) = 1$. Let $\xi = p_{j_1}^{t_{j_1}} \cdots p_{j_w}^{t_{j_w}}$. If $\gcd(c, \xi) \neq 1$, then some $p_{j_\ell} \mid c$, where $1 \leq \ell \leq w$. Now because $h_{j_\ell} \geq k_{j_\ell} > 0$, $p_{j_\ell} \mid d$ which contradicts $\gcd(c, d) = 1$. So $\gcd(c, \xi) = 1$. Obviously, $\gcd(\mu, \xi) = 1$. Thus $\gcd(c\mu, \xi) = 1$. Therefore, $\gcd(c\mu, \xi q) = 1$. So there exist $\alpha, \beta \in A$ such that $\alpha c\mu + \beta\xi q = 1$, that is,

$$\alpha c p_{i_1}^{k_{i_1}} \cdots p_{i_v}^{k_{i_v}} + \beta p_{j_1}^{t_{j_1}} \cdots p_{j_w}^{t_{j_w}} q = 1.$$

Take $b = \alpha c p_{i_1}^{k_{i_1}} \cdots p_{i_v}^{k_{i_v}}$. Then $bc^{-1}d = \alpha p_{i_1}^{k_{i_1}} \cdots p_{i_v}^{k_{i_v}} d \in aA$ and

$$\begin{aligned}(1-b)cd^{-1} &= \beta p_{j_1}^{t_{j_1}} \cdots p_{j_w}^{t_{j_w}} qc(p_1^{h_1} \cdots p_m^{h_m} q)^{-1} \\ &= \beta c p_{j_1}^{-k_{j_1}+1} \cdots p_{j_w}^{-k_{j_w}+1} p_{i_1}^{-h_{i_1}} \cdots p_{i_v}^{-h_{i_v}} \\ &\in (p_1^{k_1-1} \cdots p_m^{k_m-1})^{-1} A\end{aligned}$$

because $-h_i \geq -k_i + 1$ for each $i \in I$. Thus, $\begin{bmatrix} b & (1-b)cd^{-1} \\ dc^{-1}b & 1-b \end{bmatrix} \in T$. Also, observe that $b - b^2 \neq 0$ since $b = \alpha c p_{i_1}^{k_{i_1}} \cdots p_{i_v}^{k_{i_v}}$.

Subcase 2.2. $J = \emptyset$. Then $k_i > h_i$ for all i. Note that $\gcd(c, q) = 1$ (since $\gcd(c, d) = 1$) and $\gcd(p_i, q) = 1$ for each i, $\gcd(cp_1^{k_1} \cdots p_m^{k_m}, q) = 1$. Thus there exist $\alpha, \beta \in A$ such that $\alpha c p_1^{k_1} \cdots p_m^{k_m} + \beta q = 1$.

Now we put $b = \alpha c p_1^{k_1} \cdots p_m^{k_m}$. Then $b \in A$ and $b \neq 1$. We also see that

$$bc^{-1}d = \alpha p_1^{k_1} \cdots p_m^{k_m} d = \alpha a d \in aA.$$

On the other hand, since $1 - b = \beta q$, we obtain that

$$(1-b)cd^{-1} = \beta q c d^{-1} = \beta c p_1^{-h_1} \cdots p_m^{-h_m} \in (p_1^{k_1-1} \cdots p_m^{k_m-1})^{-1} A.$$

If $b = 0$, then $cd^{-1} = (1-b)cd^{-1} \in (p_1^{k_1-1} \cdots p_m^{k_m-1})^{-1} A$. So $\begin{bmatrix} 0 & cd^{-1} \\ 0 & 0 \end{bmatrix} \in T$, a contradiction. Thus, $b - b^2 \neq 0$ and $\begin{bmatrix} b & (1-b)cd^{-1} \\ dc^{-1}b & 1-b \end{bmatrix} \in T$.

From Cases 1 and 2, T is right extending by Theorem 9.1.1. □

Proposition 9.1.5 *Let A be a commutative PID with F as its field of fractions, $A \neq F$, and T be a right ring of quotients of $T_2(A)$. If any one of the following conditions holds, then T is right extending and Baer.*

(i) $\begin{bmatrix} A & F \\ 0 & A \end{bmatrix}$ *is a subring of T.*

(ii) $\begin{bmatrix} A & a^{-1}A \\ aA & A \end{bmatrix}$ *is a subring of T for some $0 \neq a \in A$.*

(iii) $\begin{bmatrix} A & (p_1^{k_1-1} \cdots p_m^{k_m-1})^{-1} A \\ aA & A \end{bmatrix}$ *is a subring of T for some $0 \neq a \in A$, where $a = p_1^{k_1} \cdots p_m^{k_m}$, each p_i is a distinct prime, and each k_i is a positive integer.*

Proof It follows from Theorem 9.1.1, Lemma 9.1.4, and Corollary 3.3.3. □

Proposition 9.1.6 *Let R be an overring of a commutative Noetherian ring A such that $A \subseteq \text{Cen}(R)$. Assume that R is a finitely generated as an A-module. Then R is* *a right hereditary ring if and only if $R_P = R \otimes_A A_P$ is a right hereditary ring for every maximal ideal P of A, where A_P is the localization of A at P.*

Proof See [352, Corollary 3.24]. □

A commutative domain A is said to be *Dedekind* if it is a hereditary ring. The following proposition is [295, Example 5.5.11(i)].

Proposition 9.1.7 *Let A be a Dedekind domain (which is not a field), with a maximal ideal M. Then the ring $R = \begin{bmatrix} A & A \\ M & A \end{bmatrix}$ is a hereditary Noetherian prime ring.*

Lemma 9.1.8 *Let A be a commutative PID with F as its field of fractions, $A \neq F$, and*

$$V = \begin{bmatrix} A & (p_1^{k_1-1} \cdots p_m^{k_m-1})^{-1} A \\ p_1^{k_1} \cdots p_m^{k_m} A & A \end{bmatrix},$$

where each p_i is a distinct prime of A and each k_i is a positive integer. Then V is a right hereditary ring.

Proof Let $W = \begin{bmatrix} A & A \\ p_1 \cdots p_m A & A \end{bmatrix}$. Define $\sigma : V \to W$ by

$$\sigma \begin{bmatrix} a & (p_1^{k_1-1} \cdots p_m^{k_m-1})^{-1} b \\ p_1^{k_1} \cdots p_m^{k_m} c & d \end{bmatrix} = \begin{bmatrix} a & b \\ p_1 \cdots p_m c & d \end{bmatrix},$$

where $a, b, c, d \in A$. Then σ is a ring isomorphism. Thus to show that V is right hereditary, we need to prove that W is right hereditary. For this, first note that W is a finitely generated as a right A-module, so W is a Noetherian ring. Let $P = pA$ be a maximal ideal of A, where p is a prime. If $p \notin \{p_1, p_2, \ldots, p_m\}$, then

$$W_P = \begin{bmatrix} A_P & A_P \\ p_1 p_2 \cdots p_m A_P & A_P \end{bmatrix} = \begin{bmatrix} A_P & A_P \\ A_P & A_P \end{bmatrix}$$

is right hereditary. Next, suppose that $p \in \{p_1, p_2, \ldots, p_m\}$. Say $p = p_1$, so $P = p_1 A$. Thus,

$$W_P = \begin{bmatrix} A_P & A_P \\ pA_P & A_P \end{bmatrix}$$

is right hereditary by Proposition 9.1.7 since A_P is a Dedekind domain and pA_P is a maximal ideal of A_P. So W_P is right hereditary for any maximal ideal P of A. Thus, W is right hereditary by Proposition 9.1.6. □

Lemma 9.1.9 *Let A be a commutative PID and T a right ring of quotients of $T_2(A)$ such that $T \cap \mathrm{Mat}_2(A) = \begin{bmatrix} A & A \\ aA & A \end{bmatrix}$, where $0 \neq a = p_1^{k_1} \cdots p_m^{k_m}$, each p_i is a dis-*

tinct prime in A, and each k_i is a positive integer. If T is right extending, then $\begin{bmatrix} A & (p_1^{k_1-1}\cdots p_m^{k_m-1})^{-1}A \\ aA & A \end{bmatrix}$ *is a subring of T.*

Proof Let T be right extending and F be the field of fractions of A. Put $d = p_1^{k_1-1}\cdots p_m^{k_m-1} \in A$. If $\begin{bmatrix} 0 & 0 \\ d & 0 \end{bmatrix} \in T$, then $d \in aA$, so $p_1 \cdots p_m c = 1$ for some $c \in A$, a contradiction. Hence, $\begin{bmatrix} 0 & 0 \\ d & 0 \end{bmatrix} \notin T$. By Theorem 9.1.1, either $\begin{bmatrix} 0 & d^{-1} \\ 0 & 0 \end{bmatrix} \in T$ or there is $b \in F$ such that $b - b^2 \neq 0$ and $\begin{bmatrix} b & (1-b)d^{-1} \\ db & 1-b \end{bmatrix} \in T$.

Suppose that there is $b \in F$ with $b - b^2 \neq 0$ and $\begin{bmatrix} b & (1-b)d^{-1} \\ db & 1-b \end{bmatrix} \in T$. There are $x, y \in A$ with $b = xy^{-1}$ and $\gcd(x, y) = 1$. Hence $\begin{bmatrix} x & (y-x)d^{-1} \\ dx & y-x \end{bmatrix} \in T$. Since $\gcd(x, y) = 1$, there are $v, w \in A$ such that $xv + yw = 1$.

We observe that

$$\begin{bmatrix} wx & (wy-wx)d^{-1} \\ wdx & wy-wx \end{bmatrix} \in T \text{ and } \begin{bmatrix} x & (y-x)d^{-1} \\ dx & y-x \end{bmatrix}\begin{bmatrix} v & 0 \\ 0 & 0 \end{bmatrix} = \begin{bmatrix} vx & 0 \\ vdx & 0 \end{bmatrix} \in T.$$

Since $wy = 1 - vx$, it follows that

$$\begin{bmatrix} wx & (wy-wx)d^{-1} \\ wdx & wy-wx \end{bmatrix} + \begin{bmatrix} vx & 0 \\ vdx & 0 \end{bmatrix} = \begin{bmatrix} (v+w)x & (1-(v+w)x)d^{-1} \\ (v+w)xd & 1-(v+w)x \end{bmatrix} \in T.$$

As $\begin{bmatrix} 0 & 0 \\ xd & 0 \end{bmatrix} = \begin{bmatrix} x & (y-x)d^{-1} \\ xd & y-x \end{bmatrix}\begin{bmatrix} 1 & 0 \\ 0 & 0 \end{bmatrix} - \begin{bmatrix} x & 0 \\ 0 & 0 \end{bmatrix} \in T$, so $\begin{bmatrix} 0 & 0 \\ xd & 0 \end{bmatrix} \in T \cap \mathrm{Mat}_2(A)$ and hence $xd \in aA$.

If x is a unit of A, then $d \in aA$, which is impossible as we did see. Thus x is not a unit of A. Hence $1 - (v+w)x \neq 0$. If $(v+w)x = 0$, then $\begin{bmatrix} 0 & d^{-1} \\ 0 & 0 \end{bmatrix} \in T$. Assume that $(v+w)x \neq 0$ and put $g = (v+w)x$. Then $\begin{bmatrix} g & (1-g)d^{-1} \\ gd & 1-g \end{bmatrix} \in T$. Now $g \in A$ and $g - g^2 \neq 0$. Also $gd = (v+w)xd \in aA$ as $xd \in aA$. Thus $p_i \mid g$ for each $i = 1, \ldots, m$, so $\gcd(1-g, d) = 1$. Hence, there exist $\pi, \sigma \in A$ with $(1-g)\sigma + d\pi = 1$, thus $(1-g)\sigma d^{-1} + \pi = d^{-1}$. Since $gd \in aA$ and $g \in A$,

$$\begin{bmatrix} 0 & (1-g)d^{-1} \\ 0 & 0 \end{bmatrix} = \begin{bmatrix} g & (1-g)d^{-1} \\ gd & g \end{bmatrix} - \begin{bmatrix} g & 0 \\ gd & g \end{bmatrix} \in T.$$

Thus $\begin{bmatrix} 0 & d^{-1} \\ 0 & 0 \end{bmatrix} = \begin{bmatrix} 0 & (1-g)\sigma d^{-1} + \pi \\ 0 & 0 \end{bmatrix} \in T$. So in all cases, $\begin{bmatrix} 0 & d^{-1} \\ 0 & 0 \end{bmatrix} \in T$. As a consequence, $\begin{bmatrix} A & d^{-1}A \\ aA & A \end{bmatrix} = \begin{bmatrix} A & (p_1^{k_1-1}\cdots p_m^{k_m-1})^{-1}A \\ aA & A \end{bmatrix}$ is a subring of T. □

Remark 9.1.10 Recall that a commutative domain A is called Bezout if every finitely generated ideal is principal. Thus for any $a, b \in A$, $\gcd(a, b)$ exists. So Lemma 9.1.4, Proposition 9.1.5, and Lemma 9.1.9 can be extended to the case when A is a commutative Bezout domain. For more details, see [89]. Note that from [248, p. 72] the class of Bezout domains includes the ring of entire functions and the ring of algebraic integers.

The next result shows how both Definitions 8.2.1 and 8.2.8 can be used to characterize certain right rings of quotients of a ring in a **D-E** class (see Problem II in Introduction of Chap. 8).

Theorem 9.1.11 *Let A be a commutative PID with F as its field of fractions and $A \neq F$.*

(i) *Let T be a right ring of quotients of $T_2(A)$. Then T is right extending if and only if either the ring $U := \begin{bmatrix} A & F \\ 0 & A \end{bmatrix}$ is a subring of T, or $\mathrm{Mat}_2(A)$ is a subring of T, or the ring $V := \begin{bmatrix} A & (p_1^{k_1-1} \cdots p_m^{k_m-1})^{-1}A \\ aA & A \end{bmatrix}$ is a subring of T for some $0 \neq a = p_1^{k_1} \cdots p_m^{k_m}$, where each p_i is a distinct prime of A and each k_i is a positive integer.*

(ii) *U is the only right extending ring hull of $T_2(A)$.*

(iii) *$T_2(A)$ has no right extending absolute ring hull.*

(iv) *In (i)–(iii) we can replace "right extending" with "Baer", "right Rickart", or "right semihereditary".*

Proof (i) Let T be right extending. We see that $T \cap \mathrm{Mat}_2(A) = \begin{bmatrix} A & A \\ I & A \end{bmatrix}$, where $I \trianglelefteq A$. First, assume that $I = 0$. Let $0 \neq d \in F$. By using Theorem 9.1.1, we show that $\begin{bmatrix} 0 & d \\ 0 & 1 \end{bmatrix} \in T$. If $\begin{bmatrix} 1 & 0 \\ d^{-1} & 0 \end{bmatrix} \in T$, then $\begin{bmatrix} 0 & 0 \\ d^{-1} & 0 \end{bmatrix} \in T$. Put $d^{-1} = xy^{-1}$ for some $x, y \in A$. Then $\begin{bmatrix} 0 & 0 \\ x & 0 \end{bmatrix} = \begin{bmatrix} 0 & 0 \\ d^{-1} & 0 \end{bmatrix}\begin{bmatrix} y & 0 \\ 0 & 0 \end{bmatrix} \in T \cap \mathrm{Mat}_2(A)$, so $0 \neq x \in I$, a contradiction. If there exists $a \in F$ with $a - a^2 \neq 0$ and $\begin{bmatrix} a & (1-a)d \\ d^{-1}a & 1-a \end{bmatrix} \in T$, then $\begin{bmatrix} 0 & 0 \\ 0 & 1 \end{bmatrix}\begin{bmatrix} a & (1-a)d \\ d^{-1}a & 1-a \end{bmatrix}\begin{bmatrix} 1 & 0 \\ 0 & 0 \end{bmatrix} = \begin{bmatrix} 0 & 0 \\ d^{-1}a & 0 \end{bmatrix} \in T$.

Write $a = st^{-1}$ with $s, t \in A$. Then we see that $sx \neq 0$. Furthermore,

$$\begin{bmatrix} 0 & 0 \\ sx & 0 \end{bmatrix} = \begin{bmatrix} 0 & 0 \\ d^{-1}a & 0 \end{bmatrix}\begin{bmatrix} ty & 0 \\ 0 & 0 \end{bmatrix} \in T \cap \mathrm{Mat}_2(A),$$

which is a contradiction because $I = 0$. As T is right extending, $\begin{bmatrix} 0 & d \\ 0 & 1 \end{bmatrix} \in T$ by Theorem 9.1.1, so $\begin{bmatrix} 0 & d \\ 0 & 0 \end{bmatrix} \in T$. Therefore $\begin{bmatrix} 0 & F \\ 0 & 0 \end{bmatrix} \subseteq T$, and hence U is a subring of T.

Next, if $I = A$, then $\mathrm{Mat}_2(A)$ is a subring of T. Finally, if I is a nonzero proper ideal of A, then $I = aA$, where $a \in A$ is not invertible. By Lemma 9.1.9, V is a subring of T. The converse follows from Proposition 9.1.5.

(ii) We see that U is a right extending ring hull of $T_2(A)$ by Corollary 9.1.3. Let p be a prime in A. Then $H := \begin{bmatrix} A & A \\ pA & A \end{bmatrix}$ is right extending by part (i), and $T_2(A) \subseteq H \subseteq \mathrm{Mat}_2(A)$. Thus, $\mathrm{Mat}_2(A)$ is not a right extending hull of $T_2(A)$.

Now assume that there is another distinct right extending ring hull H of $T_2(A)$ other than U or $\mathrm{Mat}_2(A)$. By part (i) there is $0 \neq a = p_1^{k_1} \cdots p_m^{k_m} \in A$ such that each p_i is a distinct prime of A, each k_i is a positive integer, and $H = V$. Since A is a commutative PID and $A \neq F$, A has infinitely many primes. Let p be a prime in A such that $\gcd(p, p_1 \cdots p_m) = 1$. Take

$$H_1 = \begin{bmatrix} A & (p_1^{k_1-1} \cdots p_m^{k_m-1})^{-1}A \\ paA & A \end{bmatrix}.$$

Then, by part (i), H_1 is a right extending right ring of quotients of $T_2(A)$ such that H_1 is a proper subring of H, a contradiction.

(iii) If $T_2(A)$ has a right extending absolute ring hull S, then by part (ii) we obtain $S \subseteq U \cap \mathrm{Mat}_2(A) = T_2(A)$. So $S = T_2(A)$, thus $T_2(A)$ is right extending. By Corollary 3.3.3, $T_2(A)$ is Baer, which contradicts Theorem 5.6.2.

(iv) We show that in part (i), "right extending" can be replaced by "Baer", "right Rickart" or "right semihereditary". Indeed, Corollary 3.3.3 and Theorem 3.1.25 yield that right extending can be replaced by Baer or right Rickart.

To see that "right extending" can be replaced by "right semihereditary", we first note that a right semihereditary ring is right Rickart. Let T be right ring of quotients of $T_2(A)$. If T is right semihereditary, then it has either U, or $\mathrm{Mat}_2(A)$, or V as a subring.

Conversely, assume that either U, or $\mathrm{Mat}_2(A)$, or V is a subring of T. We first claim that the ring U is right semihereditary. For this, note that U is Baer by Proposition 9.1.5. Thus U is right Rickart. Suppose that I is a finitely generated right ideal of U generated by $\begin{bmatrix} a_i & q_i \\ 0 & b_i \end{bmatrix}$, where $i = 1, 2, \ldots, k$.

Case 1. If there exists i with $a_i \neq 0$, then there are $a, b \in A$ such that $a \neq 0$, $aA = a_1A + \cdots + a_kA$, and $bA = b_1A + \cdots + b_kA$. Moreover,

$$I = \begin{bmatrix} a_1 & q_1 \\ 0 & b_1 \end{bmatrix} U + \cdots + \begin{bmatrix} a_k & q_k \\ 0 & b_k \end{bmatrix} U = \begin{bmatrix} aA & F \\ 0 & bA \end{bmatrix} = \begin{bmatrix} a & 0 \\ 0 & b \end{bmatrix} U.$$

Now since U is right Rickart, I is projective as a right U-module.

Case 2. $a_i = 0$ for all i. Then

$$I = \left\{ \begin{bmatrix} 0 & q_1r_1 \\ 0 & b_1r_1 \end{bmatrix} + \cdots + \begin{bmatrix} 0 & q_kr_k \\ 0 & b_kr_k \end{bmatrix} \mid r_1, \ldots, r_k \in A \right\}.$$

Since A is a commutative PID and I_A is a finitely generated torsion-free A-module, there exist $\begin{bmatrix} 0 & s_j \\ 0 & t_j \end{bmatrix} \in I$ with $j = 1, \ldots, \ell$ such that

$$I = \begin{bmatrix} 0 & s_1 \\ 0 & t_1 \end{bmatrix} A \oplus \cdots \oplus \begin{bmatrix} 0 & s_\ell \\ 0 & t_\ell \end{bmatrix} A,$$

where the scalar multiplication is $\begin{bmatrix} 0 & s_j \\ 0 & t_j \end{bmatrix} r = \begin{bmatrix} 0 & s_j r \\ 0 & t_j r \end{bmatrix}$ for $r \in A$.

So $I = \begin{bmatrix} 0 & s_1 \\ 0 & t_1 \end{bmatrix} U \oplus \cdots \oplus \begin{bmatrix} 0 & s_\ell \\ 0 & t_\ell \end{bmatrix} U$. As U is right Rickart, each $\begin{bmatrix} 0 & s_j \\ 0 & t_j \end{bmatrix} U$ is projective, so I is a projective right ideal. Hence U is right semihereditary.

By Theorem 3.1.29, $\mathrm{Mat}_n(U)$ is right Rickart for any positive integer n. As $\mathrm{Mat}_n(U)$ is orthogonally finite, $\mathrm{Mat}_n(U)$ is Baer by Theorem 3.1.25.

If U is a subring of T, then $\mathrm{Mat}_n(T)$ is Baer for each n by Theorem 8.1.9(iii) ($\mathrm{Mat}_n(T)$ is also a left ring of quotients of $\mathrm{Mat}_n(U)$ as T is a left ring of quotients of U). Thus, T is right semihereditary by Theorem 3.1.29.

Next, assume that $\mathrm{Mat}_2(A)$ is a subring of T. By Theorem 6.1.4, $\mathrm{Mat}_{2n}(A)$ is Baer for each positive integer n. We note that $\mathrm{Mat}_n(T)$ is a right ring of quotients of $\mathrm{Mat}_{2n}(A)$ as T is a right ring of quotients of $\mathrm{Mat}_2(A)$. So $\mathrm{Mat}_n(T)$ is Baer by Theorem 8.1.9(iii) as $\mathrm{Mat}_n(T)$ is also a left ring of quotients of $\mathrm{Mat}_{2n}(A)$. Thus, Theorem 3.1.29 yields that T is right semihereditary.

Finally, by Lemma 9.1.8, V is right hereditary. If V is a subring of T, then we see that T is right semihereditary as in the preceding argument for the case when U is a subring of T.

Similarly, parts (ii) and (iii) also hold when "right extending" is replaced by "Baer","right Rickart", or "right semihereditary". □

We remark that U in Theorem 9.1.11, is a right extending α pseudo ring hull of $T_2(A)$, whereas $Q(T_2(A)) = R(\mathbf{E}, Q(T_2(A)))$. Moreover, if $\{p_1, p_2, \ldots\}$ is an infinite set of distinct primes of A and $V_i = \begin{bmatrix} A & A \\ p_1 \cdots p_i A & A \end{bmatrix}$, then we see that $V_1 \supseteq V_2 \supseteq \ldots$ is an infinite descending chain of right extending rings. Thus no V_i is a right extending ring hull of $T_2(A)$.

Exercise 9.1.12

1. ([89, Birkenmeier, Park, and Rizvi]) Let A be a commutative PID with F as its field of fractions, $A \neq F$, and let T be a right ring of quotients of $R = T_2(A)$. Take

$$S = \begin{bmatrix} A & F \\ 0 & F \end{bmatrix} \text{ and } V = \begin{bmatrix} A & (p_1^{k_1-1} \cdots p_m^{k_m-1})^{-1} A \\ p_1^{k_1} \cdots p_m^{k_m} A & A \end{bmatrix},$$

where each p_i is a distinct prime of A and each k_i is a positive integer. Prove the following.

(i) If T is right hereditary, then either S or V is a subring of T.
(ii) S is the unique right hereditary ring hull of R, but R has no right hereditary absolute ring hull.

9.2 Skew Group Ring Extensions

We discuss the quasi-Baer property for certain skew group rings and fixed rings under a finite group action as an application of results developed in previous sections. Results of this section will be applied to C^*-algebras in Chap. 10.

For a ring R, we let $\text{Aut}(R)$ denote the group of ring automorphisms of R. Let G be a subgroup of $\text{Aut}(R)$. For $r \in R$ and $g \in G$, we let r^g denote the image of r under g. We use R^G to denote the *fixed ring* of R under G, that is, $R^G = \{r \in R \mid r^g = r \text{ for every } g \in G\}$. The *skew group ring*, $R * G$, is defined to be $R * G = \bigoplus_{g \in G} Rg$ with addition given componentwise and multiplication defined as follows: if $a, b \in R$ and $g, h \in G$, then $(ag)(bh) = ab^{g^{-1}}gh \in Rgh$.

Example 9.2.1 There exist a ring R and a finite group G of ring automorphisms of R such that R is quasi-Baer but neither $R * G$ nor R^G is quasi-Baer. Let $R = T_2(F)$, where F is a field of characteristic 2. Then R is (quasi-)Baer by Theorem 5.6.2. Say $e_{ij} \in R$ is the matrix with 1 in the (i, j)-position and 0 elsewhere.

Let $g \in \text{Aut}(R)$ be the conjugation by the element $e_{11} + e_{12} + e_{22}$. Then $g^2 = 1$. Let $G = \{1, g\}$. Then $r_{R*G}((1 + g)(R * G))$ cannot be generated by an idempotent. Thus, $R * G$ is not quasi-Baer. Next, we see that

$$R^G = \left\{ \begin{bmatrix} a & b \\ 0 & a \end{bmatrix} \in R \mid a, b \in F \right\}.$$

So the only idempotents of R^G are 0 and 1, thus R^G is semicentral reduced. If R^G is quasi-Baer, then R^G is a prime ring by Proposition 3.2.5, a contradiction. Thus, R^G is not quasi-Baer.

Definition 9.2.2 Let R be a semiprime ring. For $g \in \text{Aut}(R)$, let

$$\phi_g = \{x \in Q^m(R) \mid xr^g = rx \text{ for each } r \in R\}.$$

We say that g is *X-outer* if $\phi_g = 0$. A subgroup G of $\text{Aut}(R)$ is called *X-outer* on R if every $1 \neq g \in G$ is X-outer.

Assume that R is a semiprime ring. For $g \in \text{Aut}(R)$, let

$$\Phi_g = \{x \in Q(R) \mid xr^g = rx \text{ for each } r \in R\}.$$

Let $g \in \text{Aut}(R)$. We claim that $\Phi_g = \phi_g$. Obviously $\phi_g \subseteq \Phi_g$. Conversely, if $x \in \Phi_g$, then $x \in Q(R)$ and $xR = Rx$. There exists $I_R \leq^{\text{den}} R_R$ such that $xI \subseteq R$. Hence,

$xRI = R(xI) \subseteq R$, $RI \trianglelefteq R$, and $(RI)_R \leq^{\text{den}} R_R$. Therefore, $x \in Q^m(R)$, and thus $x \in \phi_g$. So $\Phi_g = \phi_g$.

If $g \in \text{Aut}(R)$, then g can be extended to a ring automorphism of $Q(R)$. If g is X-outer on R, then g is X-outer on $Q(R)$ by the preceding argument. A group G of ring automorphisms can be considered as that of $Q(R)$. If G is X-outer on R, then G is X-outer on $Q(R)$.

Lemma 9.2.3 *Let R be a semiprime ring and G a group of ring automorphisms of R.*

(i) *If G is X-outer, then every nonzero ideal of $R * G$ intersects R nontrivially. Hence, $R * G$ is semiprime. Additionally, if G is finite, then R^G is semiprime.*

(ii) *If G is finite and R has no nonzero $|G|$-torsion, then $R * G$ and R^G are semiprime.*

Proof The proof follows from [50], [169, Corollary 3, Theorem 7, and Corollary 8], and [302, Theorems 2.1 and 3.1]. □

We observe that if R is a semiprime ring and G is a group of X-outer ring automorphisms of R, then $R * G$ is an ideal intrinsic extension of R by Lemma 9.2.3(i).

Lemma 9.2.4 *Let R be a semiprime ring and G a group of X-outer ring automorphisms of R. Then* $\text{Cen}(R * G) = \text{Cen}(R)^G$.

Proof The proof is straightforward. □

Let R be a ring and G be a group of ring automorphisms of R. We say that a right ideal I of R is *G-invariant* if $I^g \subseteq I$ for every $g \in G$, where $I^g = \{a^g \mid a \in I\}$. Also, we say that R is *G-quasi-Baer* if the right annihilator of every G-invariant ideal is generated by an idempotent as a right ideal.

The condition for rings being G-quasi-Baer is left-right symmetric. In fact, suppose that R is G-quasi-Baer. Say I is a G-invariant ideal of R. Then $\ell_R(I)$ is also a G-invariant ideal, thus $r_R(\ell_R(I)) = eR$ for some $e^2 = e \in R$.

So $\ell_R(I) = \ell_R[r_R(\ell_R(I))] = \ell_R(eR) = R(1-e)$. Obviously if R is quasi-Baer, then R is G-quasi-Baer. But there exist a ring R and a finite group G of X-outer automorphisms of R such that R is G-quasi-Baer, but not quasi-Baer (see Exercise 9.2.15.3).

Theorem 9.2.5 *Let R be a semiprime ring and G be a group of X-outer ring automorphisms of R. Then R is G-quasi-Baer if and only if $R * G$ is quasi-Baer.*

Proof Assume that R is G-quasi-Baer. Say $I \trianglelefteq R * G$. Take $a \in I \cap R$. Then $a^g = g^{-1}ag \in I$ for any $g \in G$. So $I \cap R$ is G-invariant. Thus $\ell_R(I \cap R)$ is G-invariant, hence $r_R(\ell_R(I \cap R))$ is also G-invariant. As R is semiprime G-quasi-Baer, there is $e \in \mathcal{B}(R)$ with $r_R(\ell_R(I \cap R)) = eR$ by Propositions 1.2.2 and 1.2.6(ii). As $r_R(\ell_R(I \cap R))$ is G-invariant, $eR = e^g R$, and so $e = e^g$ for all

$g \in G$. Hence $e \in \mathcal{B}(R)^G$, therefore $e \in \mathrm{Cen}(R * G)$. Further, by Lemma 2.1.13, $(I \cap R)_R \leq^{\text{ess}} eR_R$.

Assume that $(1-e)I \neq 0$. Then $(1-e)I \cap R \neq 0$ by Lemma 9.2.3(i). Thus we get that $0 \neq (1-e)I \cap R \subseteq I \cap R \subseteq eR$, a contradiction. So $I = eI \subseteq e(R * G)$.

As $(I \cap R)_R \leq^{\text{ess}} eR_R$, $I_R \leq^{\text{ess}} e(R * G)_R$. Hence $I_{R*G} \leq^{\text{ess}} e(R * G)_{R*G}$, and therefore $R * G$ is right FI-extending. As $R * G$ is semiprime by Lemma 9.2.3(i), Theorem 3.2.37 yields that $R * G$ is quasi-Baer.

Conversely, assume that $R * G$ is quasi-Baer and J is a G-invariant ideal of R. Then $J * G \trianglelefteq R * G$. Note that $R * G$ is semiprime by Lemma 9.2.3(i). So there exists $e \in \mathcal{B}(R * G)$ such that $r_{R*G}(J * G) = e(R * G)$ by Propositions 1.2.2 and 1.2.6(ii). By Lemma 9.2.4, $e \in \mathrm{Cen}(R)$. Thus $eR \subseteq r_R(J)$. Let $a \in r_R(J)$. Then $Ja = 0$, so $0 = J^g a^g = Ja^g$ for all $g \in G$. Hence $(J * G)a = 0$, and therefore $a \in r_{R*G}(J * G) = e(R * G)$. Thus, $a = ea \in eR$. Thus $r_R(J) = eR$, and so R is G-quasi-Baer. □

In Theorem 9.2.5, the group G need not be finite. When G is a finite group of X-outer ring automorphisms of R, equivalent conditions for the skew group ring $R*G$ to be quasi-Baer are investigated further in Theorem 9.2.10. The next example illustrates Theorem 9.2.5.

Example 9.2.6 Let A be a semiprime quasi-Baer ring, $R = A[x_1, x_2, \dots]$, the polynomial ring with commuting indeterminates $x_1, x_2, \dots$, and G be the group of all permutations on $\{x_1, x_2, \dots\}$ acting on R. Then R is semiprime. By Theorem 6.2.4, R is quasi-Baer. We show that G is X-outer. Let $1 \neq g \in G$. Then there exist x_i and x_j with $i \neq j$ such that $x_i^g = x_j$. Without loss of generality, we assume that $x_i = x_1$ and $x_j = x_2$.

Take $q \in \Phi_g$. Then $qx_1^g = x_1 q$, so $qx_2 = x_1 q$. Because x_1 and x_2 are in $\mathrm{Cen}(R)$, x_1 and x_2 are in $\mathrm{Cen}(Q(R))$. Therefore $q(x_1 - x_2) = 0$. We show that

$$\ell_R(\{x_1 - x_2\}) = 0.$$

Say $0 \neq f \in \ell_R(\{x_1 - x_2\})$. Then $fx_1 = fx_2$. We put $f = hx_1^m x_2^n$ with $h \in R$ and m, n positive integers such that neither x_1 nor x_2 divides h.

From $fx_1 = fx_2$, $hx_1^m x_2^n x_1 = hx_1^m x_2^n x_2$, so $hx_1 = hx_2$. Whence both x_1 and x_2 divide h, a contradiction. Thus $\ell_R(\{x_1 - x_2\}) = 0$, so $r_R(\{x_1 - x_2\}) = 0$. Hence we get that $r_{Q(R)}(\{x_1 - x_2\}) = 0$. Thus $q = 0$, so $\Phi_g = 0$. Hence G is X-outer. So, the skew group ring $R * G$ is semiprime and quasi-Baer by Theorem 9.2.5.

Lemma 9.2.7 *Let R be a ring and G a finite group of ring automorphisms of R. Then $Q(R) * G$ is a maximal right ring of quotients of $R * G$.*

Proof See [285] and [335]. □

Assume that G is a finite group of ring automorphisms of a ring R. Then for $a \in R$, let $\mathrm{tr}(a) = \sum_{g \in G} a^g$, which is called the *trace* of a. Also for a right ideal

I of R, the right ideal $\mathrm{tr}(I) = \{\mathrm{tr}(a) \mid a \in I\}$ of R^G is called the *trace* of I. Say $G = \{g_1, \ldots, g_n\}$. We put $t = g_1 + \cdots + g_n \in R * G$.

When $a \in R$ and $g, h \in G$, we denote $(a^g)^h = a^{gh}$. Suppose that $r \in R$ and $\alpha = a_1 g_1 + \cdots + a_n g_n \in R * G$ with $a_i \in R$. Define

$$r \cdot \alpha = r^{g_1} a_1^{g_1} + \cdots + r^{g_n} a_n^{g_n}.$$

Then R is a right $R * G$-module. Moreover, ${}_{R^G}R_{R*G}$ is an $(R^G, R * G)$-bimodule. Consider the following pairings

$$(\ , \ ) : R \otimes_{R*G} Rt \to R^G \text{ and } [\ , \ ] : Rt \otimes_{R^G} R \to R * G$$

defined by $(a, bt) = \mathrm{tr}(ab)$ and $[at, b] = atb$ for $a, b \in R$. Then

$$(a, bt)c = a \cdot [bt, c] \text{ and } [at, b]ct = at(b, ct)$$

for all $a, b, c \in R$. Also $(\ , \ )$ is an (R^G, R^G)-bimodule homomorphism and $[\ , \ ]$ is an $(R * G, R * G)$-bimodule homomorphism. In this case, we can verify that $(R * G, {}_{R^G}R_{R*G}, {}_{R*G}Rt_{R^G}, R^G)$ is a Morita context with the given pairings (see [129] for more details). The next result is of interest in its own right.

Proposition 9.2.8 *Let R be a semiprime ring and G a finite group of X-outer ring automorphisms of R. Then* $\mathrm{Cen}(Q(R)^G) = \mathrm{Cen}(Q(R))^G$.

Proof Note that $Q(R)$ is semiprime, and G is also X-outer on $Q(R)$. So we may assume that $R = Q(R)$ and claim that $\mathrm{Cen}(R^G) = \mathrm{Cen}(R)^G$. Define

$$\theta : R * G \to \mathrm{End}(Rt_{R^G})$$

by $\theta(x)(rt) = xrt$ for $x \in R * G$ and $r \in R$. Now we claim that θ is a ring isomorphism. First, θ is a ring homomorphism because Rt is a left ideal of $R * G$. Now $\mathrm{Ker}\,(\theta) = \ell_{R*G}(RtR)$. Since $\ell_{R*G}(RtR) \cap R = 0$, $\ell_{R*G}(RtR) = 0$ by Lemma 9.2.3(i). Thus, θ is one-to-one.

To show that θ is onto, let $f \in \mathrm{End}\,(Rt_{R^G})$. Define

$$\lambda : Rt \times R \to R * G$$

by $\lambda(at, b) = f(at)b$ with $a, b \in R$. Then λ is biadditive. Moreover, for $r \in R^G$, $\lambda(atr, b) = \lambda(art, b) = f(art)b = f(atr)b = f(at)rb = \lambda(at, rb)$. Therefore, there exists an additive group homomorphism $\alpha : Rt \otimes_{R^G} R \to R * G$ satisfying $\alpha(a_1 t \otimes b_1 + \cdots + a_k t \otimes b_k) = f(a_1 t)b_1 + \cdots + f(a_k t)b_k$. We can check that $\alpha \in \mathrm{Hom}(Rt \otimes_{R^G} R_{R*G}, R * G_{R*G})$.

Now we define $\overline{f} : RtR \to R * G$ by $\overline{f}(\sum_{i=1}^k a_i t b_i) = \sum_{i=1}^k f(a_i t) b_i$ for $a_i, b_i \in R$, $i = 1, \ldots, k$. To show that $\overline{f}$ is well-defined, suppose that $\sum_{i=1}^m c_i t d_i = \sum_{j=1}^n u_j t v_j$ with $c_i, d_i, u_j, v_j \in R$ for $i = 1, \ldots, m$, $j = 1, \ldots, n$. We can show that $[\alpha(\sum_{i=1}^m c_i t \otimes d_i + \sum_{j=1}^n (-u_j) t \otimes v_j)](RtR) = 0$ by using the Morita context $(R * G, {}_{R^G}R_{R*G}, {}_{R*G}Rt_{R^G}, R^G)$ (Exercise 9.2.15.2).

Since $\ell_{R*G}(RtR) = 0$, we have that

$$0 = \alpha\left(\sum_{i=1}^{m} c_i t \otimes d_i + \sum_{j=1}^{n} (-u_j)t \otimes v_j\right) = \sum_{i=1}^{m} f(c_i t)d_i + \sum_{j=1}^{n} f(-u_j t)v_j.$$

Hence $\sum_{i=1}^{m} f(c_i t)d_i = \sum_{j=1}^{n} f(u_j t)v_j$. So $\overline{f}$ is well-defined. Further, we can show that $\overline{f} \in \text{Hom}(RtR_{R*G}, R*G_{R*G})$.

As $R = Q(R)$, $R*G$ is rationally complete by Lemma 9.2.7 (recall that a ring A is said to be rationally complete if $A = Q(A)$). Also RtR is a dense right ideal of $R*G$ by Proposition 1.3.11(iv) because RtR is an ideal of $R*G$ with $\ell_{R*G}(RtR) = 0$. Therefore, from Proposition 1.3.12, there exists

$$q \in Q(R*G) = R*G \text{ such that } \overline{f} = q_\ell|_{RtR},$$

where q_ℓ is the left multiplication by q.

Now $\theta(q)(rt) = qrt = q_\ell(rt) = \overline{f}(rt) = f(rt)$ for $r \in R$. Thus $\theta(q) = f$, so θ is onto. Therefore, θ is a ring isomorphism.

Now we show that $\text{Cen}(R)^G = \text{Cen}(R^G)$. Clearly, $\text{Cen}(R)^G \subseteq \text{Cen}(R^G)$. Next, let $a \in \text{Cen}(R^G)$. Define $f_a : Rt \to Rt$ by $f_a(rt) = rat$ for $r \in R$. If $rt = 0$, then $rat = rta = 0$, so f_a is well-defined. Take $b \in R^G$. Then

$$f_a(rtb) = f_a(rbt) = rbat = rabt = ratb = f_a(rt)b.$$

So $f_a \in \text{End}(Rt_{R^G})$ because f_a is additive. To see that $f_a \in \text{Cen}(\text{End}(Rt_{R^G}))$, take $g \in \text{End}(Rt_{R^G})$. For $rt \in Rt$ with $r \in R$, we put $g(rt) = s_r t$ with $s_r \in R$. Then $(gf_a)(rt) = g(f_a(rt)) = g(rat) = g(rta) = g(rt)a = s_r at$. Also $(f_a g)(rt) = f_a(g(rt)) = f_a(s_r t) = s_r at$. So $gf_a = f_a g$. Hence $f_a \in \text{Cen}(\text{End}(Rt_{R^G}))$.

Since $\text{Cen}(R*G) \cong \text{Cen}(\text{End}(Rt_{R^G}))$ via θ, there is $q_0 \in \text{Cen}(R*G)$ with $\theta(q_0) = f_a$. So $\theta(q_0)(rt) = f_a(rt)$, thus $q_0 rt = rat$ for $r \in R$. By Lemma 9.2.4, $\text{Cen}(R*G) = \text{Cen}(R)^G$. So $q_0 \in \text{Cen}(R)^G$. Take $r = 1$ in $q_0 rt = rat$. Then $q_0 t = at$. Hence, $a = q_0$ because $q_0 \in \text{Cen}(R)^G \subseteq R$. Thus $a \in \text{Cen}(R)^G$. Consequently, $\text{Cen}(R^G) = \text{Cen}(R)^G$. □

Lemma 9.2.9 *Let R be a semiprime ring and G a finite group of X-outer ring automorphisms of R.*

(i) *For $q \in Q(R^G)$, assume that $J_{R^G} \leq^{\text{den}} R^G_{R^G}$ such that $qJ \subseteq R^G$. Then $JR_R \leq^{\text{den}} R_R$ and the map $\widetilde{q} : JR \to R$ defined by $\widetilde{q}\,(\sum a_i r_i) = \sum q(a_i)r_i$, with $a_i \in J$ and $r_i \in R$, is an R-homomorphism. Moreover, $\widetilde{q} \in Q(R)^G$.*

(ii) *The map $\sigma : Q(R^G) \to Q(R)^G$ by $\sigma(q) = \widetilde{q}$ is a ring isomorphism.*

(iii) *Assume that I is a right ideal of R. Then $I_R \leq^{\text{den}} R_R$ if and only if $\text{tr}(I)_{R^G} \leq^{\text{den}} R^G_{R^G}$.*

Proof See [335] for the proof. □

Despite Example 9.2.1, we obtain the next result for the quasi-Baer property of $R*G$ and R^G when R is semiprime, and G is finite and X-outer.

Theorem 9.2.10 *Assume that R is a semiprime ring and G is a finite group of X-outer ring automorphisms of R. Then the following are equivalent.*

(i) $R * G$ *is quasi-Baer.*
(ii) R *is* G*-quasi-Baer.*
(iii) R^G *is quasi-Baer.*

Proof By Lemma 9.2.3(i), $R * G$ is semiprime. We note that $Q(R) * G$ is a maximal right ring of quotients of $R * G$ by Lemma 9.2.7.

(i)$\Leftrightarrow$(ii) This equivalence is Theorem 9.2.5 for the case when G is finite.

(ii)$\Rightarrow$(iii) Let R be G-quasi-Baer. By Theorem 9.2.5, $R * G$ is quasi-Baer. From Lemma 9.2.3(i), $R * G$ and R^G are semiprime. Thus to show that R^G is quasi-Baer, by Theorem 8.3.17 it suffices to see that $\mathcal{B}(Q(R^G)) \subseteq R^G$. Note that G is also X-outer on $Q(R)$, so by Lemma 9.2.4 and Proposition 9.2.8,

$$\mathcal{B}(Q(R)^G) \subseteq \mathrm{Cen}(Q(R)^G) = \mathrm{Cen}(Q(R))^G = \mathrm{Cen}(Q(R) * G).$$

So $\mathcal{B}(Q(R)^G) \subseteq \mathcal{B}(Q(R) * G) \subseteq R * G$ by Theorem 8.3.17 as $R * G$ is quasi-Baer. Hence $\mathcal{B}(Q(R)^G) \subseteq R$, and thus $\mathcal{B}(Q(R)^G) \subseteq R^G$.

Say $e \in \mathcal{B}(Q(R^G))$. From Lemma 9.2.9(ii), $\widetilde{e} \in \mathcal{B}(Q(R)^G) \subseteq R^G$. Let J be a dense right ideal of R^G with $eJ \subseteq R^G$. By Lemma 9.2.9(i), $(e - \widetilde{e})J = 0$. We note that $\widetilde{e} \in R^G \subseteq Q(R^G)$, so $e - \widetilde{e} = 0$ from Proposition 1.3.11(ii). Consequently, we obtain $e = \widetilde{e} \in R^G$. Hence $\mathcal{B}(Q(R^G)) \subseteq R^G$. Therefore, R^G is quasi-Baer by Theorem 8.3.17.

(iii)$\Rightarrow$(i) Assume that R^G is quasi-Baer. Let $e \in \mathcal{B}(Q(R) * G)$. Then by Lemma 9.2.4, $e \in \mathcal{B}(Q(R))^G$ since G is X-outer on $Q(R)$. Thus $R \cap eR$ is a G-invariant ideal of R. Therefore, $r_R(R \cap eR)$ is also a G-invariant ideal of R. So $\mathrm{tr}(R \cap eR) \subseteq R \cap eR$ and $\mathrm{tr}(r_R(R \cap eR)) \subseteq r_R(R \cap eR)$. As R is semiprime, $(R \cap eR) \oplus r_R(R \cap eR)$ is a dense right ideal of R.

We prove that $e(r_R(R \cap eR)) = 0$. Indeed, let $x \in r_R(R \cap eR)$ (since R is semiprime, $r_R(R \cap eR) = \ell_R(R \cap eR)$). If $ex \neq 0$, then there exists $b \in R$ such that $0 \neq exb \in R \cap eR$ because $(R \cap eR)_R \leq^{\mathrm{ess}} eR_R$. So

$$exb(R \cap eR) \subseteq ex(R \cap eR) = 0,$$

hence $0 \neq exb \in (R \cap eR) \cap r_R(R \cap eR) = 0$, a contradiction. Therefore, we obtain that $e(r_R(R \cap eR)) = 0$. Thus,

$$e[(R \cap eR) \oplus r_R(R \cap eR)] = e(R \cap eR) = R \cap eR \subseteq R.$$

By Lemma 9.2.9(iii), $\mathrm{tr}[(R \cap eR) \oplus r_R(R \cap eR)] = \mathrm{tr}(R \cap eR) \oplus \mathrm{tr}(r_R(R \cap eR))$ is a dense right ideal of R^G. We put $I = (R \cap eR) \oplus r_R(R \cap eR)$. Because I is G-invariant, $\mathrm{tr}(I)R \subseteq I$. By the preceding argument $e(r_R(R \cap eR)) = 0$, so we have that $e[\mathrm{tr}(r_R(R \cap eR))] \subseteq e(r_R(R \cap eR)) = 0$. Thus, it follows that $e\,\mathrm{tr}(I) = e(\mathrm{tr}(R \cap eR)) \subseteq e(R \cap eR) \subseteq R$. So $e\,\mathrm{tr}(I) \subseteq R \cap Q(R)^G = R^G$.

Let e_0 be the restriction of e to $\mathrm{tr}(I)$. Then $e_0\mathrm{tr}(I) \subseteq R^G$. We note that $\mathrm{tr}(I)R_R \leq^{\mathrm{den}} R_R$ by Lemma 9.2.9(i) because $\mathrm{tr}(I)$ is a dense right ideal of R^G. Further, $(e - \widetilde{e_0})\mathrm{tr}(I)R = 0$, and so $e = \widetilde{e_0}$ from Proposition 1.3.11(ii). Now Lemma 9.2.9(ii) yields that $e_0 \in \mathcal{B}(Q(R^G))$. Therefore, we can see that

$$\mathrm{tr}(R \cap eR) = e\,\mathrm{tr}(R \cap eR) = e_0\mathrm{tr}(R \cap eR) \subseteq e_0R^G.$$

We claim that $\mathrm{tr}(R \cap eR)_{R^G} \leq^{\mathrm{ess}} e_0 R^G_{R^G}$. For this, take $0 \neq e_0 a \in e_0 R^G$ with $a \in R^G$. As $R^G_{R^G} \leq^{\mathrm{ess}} Q(R^G)_{R^G}$ and $e_0 \in \mathcal{B}(Q(R^G))$, there is $c \in R^G$ with $0 \neq e_0 ac \in R^G$. Since $[\mathrm{tr}(R \cap eR) \oplus \mathrm{tr}(r_R(R \cap eR))]_{R^G} \leq^{\mathrm{ess}} R^G_{R^G}$, there is $r \in R^G$ such that $0 \neq e_0 acr \in \mathrm{tr}(R \cap eR) \oplus \mathrm{tr}(r_R(R \cap eR))$. Recall that $e r_R(R \cap eR) = 0$, so

$$e_0\, \mathrm{tr}(r_R(R \cap eR)) = e\, \mathrm{tr}(r_R(R \cap eR)) \subseteq e r_R(R \cap eR) = 0.$$

Thus, $0 \neq e_0 acr \in e_0\, \mathrm{tr}(R \cap eR) = e\mathrm{tr}(R \cap eR) = \mathrm{tr}(R \cap eR)$ and $cr \in R^G$. Hence, $\mathrm{tr}(R \cap eR)_{R^G} \leq^{\mathrm{ess}} e_0 R^G_{R^G}$. We note that R^G is semiprime from Lemma 9.2.3(i), $\mathrm{tr}(R \cap eR) \trianglelefteq R^G$, and $e_0 \in \mathcal{B}(Q(R^G))$.

As R^G is quasi-Baer, there is $f \in \mathbf{S}_\ell(R^G)$ with $r_{R^G}(\mathrm{tr}(R \cap eR)) = fR^G$. By Proposition 1.2.6(ii) $f \in \mathcal{B}(R^G)$ since R^G is semiprime. By Lemma 2.1.13, $\mathrm{tr}(R \cap eR)_{R^G} \leq^{\mathrm{ess}} (1-f)R^G_{R^G}$. Thus, $\mathrm{tr}(R \cap eR)_{R^G} \leq^{\mathrm{ess}} (1-f)Q(R^G)_{R^G}$. Also since $\mathrm{tr}(R \cap eR)_{R^G} \leq^{\mathrm{ess}} e_0 R^G_{R^G}$, $\mathrm{tr}(R \cap eR)_{R^G} \leq^{\mathrm{ess}} e_0 Q(R^G)_{R^G}$. Now we note that $e_0, 1-f \in \mathcal{B}(Q(R^G))$, so $e_0 = 1-f$. Therefore, $e_0 \in R^G$ and so $e_0 R^G \subseteq R^G$.

Because $e = \widetilde{e_0}$, $eR = \widetilde{e_0}(R^G R) = (e_0 R^G)R \subseteq R^G R = R$. Hence $e \in R$, so $\mathcal{B}(Q(R) * G) \subseteq R \subseteq R * G$. Since $R * G$ is semiprime by Lemma 9.2.3(i), $R * G$ is quasi-Baer from Theorem 8.3.17. □

Let R be a semiprime ring and G be a group of X-outer ring automorphisms of R. Then G is also a group of X-outer ring automorphisms on the semiprime ring $\widehat{Q}_{\mathbf{qB}}(R)$.

Corollary 9.2.11 *Let R be a semiprime ring and G a group of X-outer ring automorphisms of R. Then $\widehat{Q}_{\mathbf{qB}}(R) * G$ is quasi-Baer. Additionally, if G is finite, then $\widehat{Q}_{\mathbf{qB}}(R)^G$ is quasi-Baer.*

Proof The proof follows from Theorems 9.2.5 and 9.2.10. □

Corollary 9.2.12 *Let R be a semiprime ring and G a group of X-outer ring automorphisms of R. If R is right FI-extending, then $R * G$ is right FI-extending. Additionally, if G is finite, then R^G is right FI-extending.*

Proof The proof follows from Lemma 9.2.3(i), Theorem 3.2.37, Theorem 9.2.5, and Theorem 9.2.10. □

Theorem 9.2.10 does not hold true if quasi-Baer is replaced by Baer. Also Corollary 9.2.12 does not hold true if right FI-extending is replaced by right extending (see Exercise 9.2.15.4). The next corollary also follows from Theorem 9.2.10 because every reduced quasi-Baer ring is Baer.

Corollary 9.2.13 *Let R be a reduced ring with a finite group G of X-outer ring automorphisms of R. Then R is G-quasi-Baer if and only if R^G is Baer.*

From Lemma 9.2.3 and Theorem 9.2.10, one may raise the following question: *Assume that R is a semiprime quasi-Baer ring and G is a finite group of ring automorphisms of R such that R has no nonzero $|G|$-torsion. Then is $R * G$ quasi-Baer?* The next example answers this question in the negative.

Example 9.2.14 For a commutative domain A with no nonzero 2-torsion, let

$$R = A \oplus A \oplus \mathbb{Z}$$

and $g \in \text{Aut}(R)$ defined by $g[(a, b, n)] = (b, a, n)$ for $a, b \in A$ and $n \in \mathbb{Z}$. We now put $G = \{1, g\}$, $S = A \oplus A$, and $h = g|_S$. Then $h \in \text{Aut}(S)$.

Let $H = \{1, h\}$. In this case, $R * G \cong (S * H) \oplus \mathbb{Z}[G]$, where $\mathbb{Z}[G]$ is the group ring of G over $\mathbb{Z}$. If $R * G$ is quasi-Baer, then by Proposition 3.2.8 $\mathbb{Z}[G]$ is quasi-Baer, which is a contradiction by Example 6.3.11 (see also Example 3.1.6). So R is a semiprime quasi-Baer ring with no nonzero $|G|$-torsion, but the ring $R * G$ is not quasi-Baer.

Exercise 9.2.15

1. Prove Lemma 9.2.4.
2. Assume that $\alpha : Rt \otimes_{R^G} R \to R * G$ is the map such that
$$\alpha(a_1 t \otimes b_1 + \cdots + a_k t \otimes b_k) = f(a_1 t)b_1 + \cdots + f(a_k t)b_k,$$
which is in the proof of Proposition 9.2.8.
   (i) Prove that $\alpha \in \text{Hom}(Rt \otimes_{R^G} R_{R*G}, R * G_{R*G})$.
   (ii) Show that if $\sum_{i=1}^m c_i t d_i = \sum_{j=1}^n u_j t v_j$ where $c_i, d_i, u_j, v_j \in R$, then $[\alpha(\sum_{i=1}^m c_i t \otimes d_i + \sum_{j=1}^n (-u_j)t \otimes v_j)](RtR) = 0$.
3. ([233, Jin, Doh, and Park]) Assume that A is a commutative domain which is not a field and A has no 2-torsion (e.g., $A = \mathbb{Z}$). Take a nonzero proper ideal I of A. Let $R = \{(a, b) \in A \oplus A \mid a - b \in I\}$, which is a subring of $A \oplus A$. Define $g \in \text{Aut}(R)$ by $g(a, b) = (b, a)$ for $(a, b) \in R$. Then $g^2 = 1$. Let $G = \{1, g\}$. Show that the following hold:
   (i) G is X-outer.
   (ii) $R * G$ is quasi-Baer and R is G-quasi-Baer.
   (iii) R is not quasi-Baer.
   (iv) $\widehat{Q}_{\mathbf{qB}}(R * G) \neq \widehat{Q}_{\mathbf{qB}}(R) * G$.
4. ([233, Jin, Doh, and Park]) Let $R = \mathbb{Z}[x, y]$. Define $g \in \text{Aut}(R)$ by $g(a(x, y)) = a(y, x)$ for $a(x, y) \in R$. Let $G = \{1, g\}$. Show that the following hold true.
   (i) R is Baer and extending, and G is X-outer.
   (ii) $R * G$ is neither Baer nor right extending.

9.3 Hulls of Monoid and Matrix Ring Extensions

Let $H_{\mathfrak{K}}(R)$ denote a ring hull of R with respect to a class $\mathfrak{K}$ of rings and $X(-)$ denote a ring extension. Then it is natural to ask if $H_{\mathfrak{K}}(X(R))$ has any relation with $X(H_{\mathfrak{K}}(R))$. This question for $\mathfrak{K}$ right ring hulls is considered in this section when $\mathfrak{K}$ is one of the classes of rings such as $\mathfrak{K} = \mathbf{qB}$, $\mathbf{FI}$, $\mathbf{pqB}$, $\mathbf{pFI}$, and $\mathbf{fgFI}$

(see Problem IV in Introduction of Chap. 8). In particular, our primary focus is on the quasi-Baer and the right FI-extending ring hulls of various ring extensions of a ring R.

The extensions considered here, include monoid ring extensions, full and triangular matrix ring extensions, and infinite matrix ring extensions. As an application, we see that for semiprime rings R and S, if R and S are Morita equivalent, then so are $\widehat{Q}_{\mathbf{qB}}(R)$ and $\widehat{Q}_{\mathbf{qB}}(S)$.

Theorem 9.3.1 *Let $R[G]$ be a semiprime monoid ring of a monoid G over a ring R. Then:*

(i) $\widehat{Q}_{\mathbf{qB}}(R)[G] \subseteq \widehat{Q}_{\mathbf{qB}}(R[G])$ *and* $\widehat{Q}_{\mathbf{pqB}}(R)[G] \subseteq \widehat{Q}_{\mathbf{pqB}}(R[G])$.

(ii) *If G is a u.p.-monoid, then we have that* $\widehat{Q}_{\mathbf{qB}}(R[G]) = \widehat{Q}_{\mathbf{qB}}(R)[G]$ *and* $\widehat{Q}_{\mathbf{pqB}}(R[G]) = \widehat{Q}_{\mathbf{pqB}}(R)[G]$.

Proof (i) Note that R is semiprime. Say $q \in Q^m(R)$. Then there exists $I \trianglelefteq R$ such that $\ell_R(I) = 0$ and $qI \subseteq R$. Then we see that

$$I[G] \trianglelefteq R[G] \text{ and } \ell_{R[G]}(I[G]) \subseteq \ell_{R[G]}(I) = 0.$$

Further, $qI[G] \subseteq R[G]$. Because $R[G]$ is semiprime by assumption, we obtain that $q \in Q^m(R[G])$. Hence $Q^m(R) \subseteq Q^m(R[G])$. Therefore, $Q^m(R)[G] \subseteq Q^m(R[G])$.

Let $c \in \mathcal{B}(Q^m(R))$. Then $c \in Q^m(R)[G] \subseteq Q^m(R[G])$ and $cr = rc$ for any $r \in R$, therefore $c\beta = \beta c$ for any $\beta \in R[G]$. So $c \in \mathcal{B}(Q^m(R[G]))$. We note that $\text{Cen}(Q(R)) = \text{Cen}(Q^m(R))$, and so $\mathcal{B}(Q(R)) = \mathcal{B}(Q^m(R))$. Thus

$$\mathcal{B}(Q(R)) = \mathcal{B}(Q^m(R)) \subseteq \mathcal{B}(Q^m(R[G])) = \mathcal{B}(Q(R[G])).$$

From Theorem 8.3.17, $\widehat{Q}_{\mathbf{qB}}(R)[G] \subseteq \widehat{Q}_{\mathbf{qB}}(R[G])$.

We show that $\mathcal{B}_p(Q(R)) \subseteq \mathcal{B}_p(Q(R[G]))$. For this, let $e \in \mathcal{B}_p(Q(R))$. Then there exists $a \in R$ such that $RaR_R \leq^{\text{ess}} eQ(R)_R$, and so $RaR_R \leq^{\text{ess}} eR_R$. Hence, $(RaR)[G]_R \leq^{\text{ess}} eR[G]_R$. Therefore $(RaR)[G]_{R[G]} \leq^{\text{ess}} eR[G]_{R[G]}$.

On the other hand, $e \in \mathcal{B}_p(Q(R)) \subseteq \mathcal{B}(Q(R)) \subseteq \mathcal{B}(Q(R[G]))$ by the preceding argument. So $e \in \mathcal{B}_p(Q(R[G]))$ because $(RaR)[G] = R[G]aR[G]$. Thus, $\mathcal{B}_p(Q(R)) \subseteq \mathcal{B}_p(Q(R[G]))$. Therefore, $\widehat{Q}_{\mathbf{pqB}}(R)[G] \subseteq \widehat{Q}_{\mathbf{pqB}}(R[G])$ from Theorem 8.3.39.

(ii) This is a consequence of part (i) and Theorem 6.2.3. □

Goel and Jain [177] have posed the question: *If G is an infinite cyclic group and R is a prime right quasi-continuous ring, is $R[G]$ right quasi-continuous*? While the general question remains open (as of the writing of this book), Corollary 9.3.2 provides an affirmative answer to the question for the case when R is a commutative semiprime quasi-continuous ring.

Corollary 9.3.2 *Let $R[G]$ be the group ring of a torsion-free Abelian group G over a commutative semiprime quasi-continuous ring R. Then $R[G]$ is quasi-continuous.*

Proof From Theorem 8.3.22, $R[G]$ is semiprime. Since G is a commutative u.p.-monoid, by Theorems 9.3.1 and 3.2.37, $R[G]$ is extending. As $R[G]$ is commutative, it satisfies the (C_3) condition. So $R[G]$ is quasi-continuous. □

Corollary 9.3.3 *Let R be a semiprime ring and let X be a nonempty set of not necessarily commuting indeterminates. Then*:

(i) $\widehat{Q}_{\mathbf{qB}}(R[x,x^{-1}]) = \widehat{Q}_{\mathbf{qB}}(R)[x,x^{-1}]$.
(ii) $\widehat{Q}_{\mathbf{pqB}}(R[x,x^{-1}]) = \widehat{Q}_{\mathbf{pqB}}(R)[x,x^{-1}]$.
(iii) $\widehat{Q}_{\mathbf{qB}}(R[X]) = \widehat{Q}_{\mathbf{qB}}(R)[X]$.
(iv) $\widehat{Q}_{\mathbf{pqB}}(R[X]) = \widehat{Q}_{\mathbf{pqB}}(R)[X]$.

Proof (i) and (ii) Note that $R[x,x^{-1}] \cong R[C_\infty]$, which is semiprime ($C_\infty$ denotes the infinite cyclic group). Therefore, $\widehat{Q}_{\mathbf{qB}}(R[x,x^{-1}]) = \widehat{Q}_{\mathbf{qB}}(R)[x,x^{-1}]$ and $\widehat{Q}_{\mathbf{pqB}}(R[x,x^{-1}]) = \widehat{Q}_{\mathbf{pqB}}(R)[x,x^{-1}]$ from Theorem 9.3.1.

(iii) Since R is semiprime, so is $R[X]$. Now $\widehat{Q}_{\mathbf{qB}}(R[X]) = \widehat{Q}_{\mathbf{qB}}(R)[X]$ follows from Theorem 9.3.1.

(iv) The proof follows from Theorem 9.3.1. □

Example 9.3.4 (i) Let $\mathbb{Z}[C_2]$ be the group ring of the group $C_2 = \{1, g\}$ over $\mathbb{Z}$. Then $\mathbb{Z}[C_2]$ is semiprime. Moreover, $\widehat{Q}_{\mathbf{qB}}(\mathbb{Z})[C_2] = \mathbb{Z}[C_2] \subsetneq \widehat{Q}_{\mathbf{qB}}(\mathbb{Z}[C_2])$ (see Example 8.3.31). Hence, the u.p.-monoid condition is not superfluous in Theorem 9.3.1(ii).

(ii) Let F be a field. Then $F[x] = Q(F)[x] \neq Q(F[x]) = F(x)$, so $Q(-)$ cannot replace $\widehat{Q}_{\mathbf{qB}}(-)$ in Theorem 9.3.1(ii).

(iii) There is a semiprime ring R with $\widehat{Q}_{\mathbf{pqB}}(R[[x]]) \neq \widehat{Q}_{\mathbf{pqB}}(R)[[x]]$. In Example 6.2.9, there is a commutative regular ring R (hence right p.q.-Baer), but the ring $R[[x]]$ is not right p.q.-Baer. Whence $\widehat{Q}_{\mathbf{pqB}}(R) = R$ and so $\widehat{Q}_{\mathbf{pqB}}(R)[[x]] = R[[x]]$. Because $R[[x]]$ is not right p.q.-Baer, it follows that $\widehat{Q}_{\mathbf{pqB}}(R[[x]]) \neq \widehat{Q}_{\mathbf{pqB}}(R)[[x]]$.

Proposition 9.3.5 *Let $\mathfrak{K}$ be a class of rings such that $\Omega \in \mathfrak{K}$ if and only if* $\mathrm{Mat}_n(\Omega) \in \mathfrak{K}$ *for any positive integer n, and let $H_{\mathfrak{K}}(-)$ denote any of the ring hulls given in Definition* 8.2.1 *for the class $\mathfrak{K}$. Then for a ring R, $H_{\mathfrak{K}}(R)$ exists if and only if $H_{\mathfrak{K}}(\mathrm{Mat}_n(R))$ exists for any positive integer n. In this case,* $H_{\mathfrak{K}}(\mathrm{Mat}_n(R)) = \mathrm{Mat}_n(H_{\mathfrak{K}}(R))$.

Proof We prove the case when $H_{\mathfrak{K}}(R) = Q_{\mathfrak{K}}(R)$. The other cases can be shown similarly. Assume that $Q_{\mathfrak{K}}(R)$ exists. By hypothesis, $\mathrm{Mat}_n(Q_{\mathfrak{K}}(R))$ is in $\mathfrak{K}$. Let T be a right essential overring of $\mathrm{Mat}_n(R)$. Then T has a set of $n \times n$ matrix units (see 1.1.16). So $T = \mathrm{Mat}_n(V)$ for some ring V. Now V is a right essential overring of R. By assumption if $T \in \mathfrak{K}$, then $V \in \mathfrak{K}$. Hence $Q_{\mathfrak{K}}(R)$ is a subring of V, so $\mathrm{Mat}_n(Q_{\mathfrak{K}}(R))$ is a subring of T. Thus $\mathrm{Mat}_n(Q_{\mathfrak{K}}(R)) = Q_{\mathfrak{K}}(\mathrm{Mat}_n(R))$. So if $Q_{\mathfrak{K}}(R)$ exists, then $Q_{\mathfrak{K}}(\mathrm{Mat}_n(R))$ exists.

If $Q_{\mathfrak{K}}(\mathrm{Mat}_n(R))$ exists, then the preceding argument yields that there is a right essential overring S of R with $Q_{\mathfrak{K}}(\mathrm{Mat}_n(R)) = \mathrm{Mat}_n(S)$. By hypothesis, $S \in \mathfrak{K}$. If

W is a right essential overring of R and $W \in \mathfrak{K}$, then $Q_{\mathfrak{K}}(\mathrm{Mat}_n(R))$ is a subring of $\mathrm{Mat}_n(W)$. Thus, S is a subring of W, so $S = Q_{\mathfrak{K}}(R)$. □

Lemma 9.3.6 *Let $\delta \subseteq \mathcal{B}(Q(R))$ and $\Delta = \{c\mathbf{1} \mid c \in \delta\}$, where $\mathbf{1}$ is the identity matrix of $\mathrm{Mat}_n(R)$. Then:*

(i) $\mathrm{Mat}_n(\langle R \cup \delta\rangle_{Q(R)}) = \langle \mathrm{Mat}_n(R) \cup \Delta\rangle_{Q(\mathrm{Mat}_n(R))}$.
(ii) $\mathrm{Mat}_n(Q(R)) = Q(\mathrm{Mat}_n(R)) = Q(T_n(R))$.
(iii) $T_n(\langle R \cup \delta\rangle_{Q(R)}) = \langle T_n(R) \cup \Delta\rangle_{Q(\mathrm{Mat}_n(R))}$.

Proof The proof is routine. For (ii), see also Exercise 8.1.10.5. □

Proposition 9.3.7 *Let R be a ring and n a positive integer. Then:*

(i) $\widehat{Q}_{\mathbf{IC}}(\mathrm{Mat}_n(R)) = \mathrm{Mat}_n(\widehat{Q}_{\mathbf{IC}}(R)) = \mathrm{Mat}_n(R\mathcal{B}(Q(R)))$.
(ii) *If R is a semiprime ring, then:*
  (1) $\widehat{Q}_{\mathfrak{K}}(\mathrm{Mat}_n(R)) = \mathrm{Mat}_n(\widehat{Q}_{\mathfrak{K}}(R)) = \mathrm{Mat}_n(R\mathcal{B}(Q(R)))$, *where* $\mathfrak{K} = \mathbf{qB}$ *or* $\mathbf{FI}$.
  (2) $\widehat{Q}_{\mathfrak{K}}(\mathrm{Mat}_n(R)) = \mathrm{Mat}_n(\widehat{Q}_{\mathfrak{K}}(R)) = \mathrm{Mat}_n(\langle R \cup \mathcal{B}_p(Q(R))\rangle_{Q(R)})$, *where* $\mathfrak{K} = \mathbf{pqB}$, $\mathbf{pFI}$, *or* $\mathbf{fgFI}$.

Proof (i) This part follows from Theorem 8.3.11 and Lemma 9.3.6.

(ii)(1) Theorems 3.2.12, 3.2.37, 8.3.17, and Lemma 9.3.6 yield the result.

(ii)(2) Note that $\mathrm{Mat}_n(\widehat{Q}_{\mathbf{pFI}}(R))$ is right p.q.-Baer from Theorem 3.2.36 and Proposition 3.2.41. Let $e \in \mathcal{B}_p(Q(R))$. Then there exists $x \in R$ such that RxR_R is essential in $eQ(R)_R$. Let $\alpha \in \mathrm{Mat}_n(R)$ with x in $(1, 1)$-position and 0 elsewhere. Then $\mathrm{Mat}_n(R)\alpha\mathrm{Mat}_n(R)_{\mathrm{Mat}_n(R)} \leq^{\mathrm{ess}} (e\mathbf{1})\mathrm{Mat}_n(Q(R))_{\mathrm{Mat}_n(R)}$, where $\mathbf{1}$ is the identity matrix of $\mathrm{Mat}_n(R)$. Hence $e\mathbf{1} \in \mathcal{B}_p(Q(\mathrm{Mat}_n(R)))$ from Lemma 9.3.6(ii). From Theorem 8.3.39, $\mathrm{Mat}_n(\widehat{Q}_{\mathbf{pFI}}(R)) \subseteq \widehat{Q}_{\mathbf{pFI}}(\mathrm{Mat}_n(R))$.

Since $\widehat{Q}_{\mathbf{pFI}}(R)$ is right p.q.-Baer by Proposition 3.2.41, $\mathrm{Mat}_n(\widehat{Q}_{\mathbf{pFI}}(R))$ is right p.q.-Baer from Theorem 3.2.36. Thus, $\mathrm{Mat}_n(\widehat{Q}_{\mathbf{pFI}}(R))$ is right principally FI-extending by Proposition 3.2.41. Hence, $\widehat{Q}_{\mathbf{pFI}}(\mathrm{Mat}_n(R)) \subseteq \mathrm{Mat}_n(\widehat{Q}_{\mathbf{pFI}}(R))$, and so $\widehat{Q}_{\mathbf{pFI}}(\mathrm{Mat}_n(R)) = \mathrm{Mat}_n(\widehat{Q}_{\mathbf{pFI}}(R))$. From Theorem 8.3.39 and Proposition 3.2.41, $\widehat{Q}_{\mathfrak{K}}(\mathrm{Mat}_n(R)) = \mathrm{Mat}_n(\widehat{Q}_{\mathfrak{K}}(R)) = \mathrm{Mat}_n(\langle R \cup \mathcal{B}_p(Q(R))\rangle_{Q(R)})$, for $\mathfrak{K} = \mathbf{pqB}$, $\mathbf{pFI}$, or $\mathbf{fgFI}$. □

The next example shows that Theorem 9.3.1(ii) and Proposition 9.3.7 do not hold true if $\mathfrak{K} = \mathbf{qCon}$.

Example 9.3.8 In general, we have that $\widehat{Q}_{\mathbf{qCon}}(R[x]) \neq \widehat{Q}_{\mathbf{qCon}}(R)[x]$ and $\widehat{Q}_{\mathbf{qCon}}(\mathrm{Mat}_n(R)) \neq \mathrm{Mat}_n(\widehat{Q}_{\mathbf{qCon}}(R))$, where n is an integer such that $n > 1$. In Example 8.3.37, $\mathrm{Mat}_n(F[x]) = \mathrm{Mat}_n(F)[x]$, where F is a field, is not right quasi-continuous. We take $R = \mathrm{Mat}_n(F)$. Then $\widehat{Q}_{\mathbf{qCon}}(R[x]) \neq \widehat{Q}_{\mathbf{qCon}}(R)[x]$. Next, let $S = F[x]$. Then $\widehat{Q}_{\mathbf{qCon}}(\mathrm{Mat}_n(S)) \neq \mathrm{Mat}_n(\widehat{Q}_{\mathbf{qCon}}(S))$.

Lemma 9.3.9 *Let R be a semiprime ring. Then $\widehat{Q}_{\mathbf{qB}}(eRe) = e\widehat{Q}_{\mathbf{qB}}(R)e$ for any nonzero idempotent $e \in R$.*

Proof Take $f \in \mathcal{B}(Q^m(R))$. Note that $\mathcal{B}(Q^m(R)) = \mathcal{B}(Q(R))$. There is $I \trianglelefteq R$ such that $\ell_R(I) = 0$ and $fI \subseteq R$. Let $0 \neq ete \in eRe$ with $t \in R$. Then $eteI \neq 0$ because $\ell_R(I) = 0$. As R is semiprime, $(eteI)^2 \neq 0$. So $0 \neq eteIe \subseteq eIe$, hence $eIe_{eRe} \leq^{\text{ess}} eRe_{eRe}$. Since $fI \subseteq R$, $efeeIe \subseteq eRe$ and $eIe \trianglelefteq eRe$. Further, eRe is semiprime. So $\ell_{eRe}(eIe) = 0$ by Proposition 1.3.16. Hence, $efe \in Q^m(eRe)$ and $(efe)^2 = efe$. For $eae \in eRe$, $efeeae = eaeefe$, therefore $efe \in \mathcal{B}(Q^m(eRe))$ and $e\mathcal{B}(Q^m(R))e \subseteq \mathcal{B}(Q^m(eRe))$. From Theorem 8.3.17,

$$\begin{aligned} e\widehat{Q}_{\mathbf{qB}}(R)e &= eR\mathcal{B}(Q^m(R))e = eRee\mathcal{B}(Q^m(R))e \\ &\subseteq eRe\mathcal{B}(Q^m(eRe)) = \widehat{Q}_{\mathbf{qB}}(eRe). \end{aligned}$$

Hence, $eRe \subseteq e\widehat{Q}_{\mathbf{qB}}(R)e \subseteq \widehat{Q}_{\mathbf{qB}}(eRe) \subseteq Q(eRe)$. As $\widehat{Q}_{\mathbf{qB}}(R)$ is quasi-Baer, $e\widehat{Q}_{\mathbf{qB}}(R)e$ is quasi-Baer from Theorem 3.2.10. So $\widehat{Q}_{\mathbf{qB}}(eRe) \subseteq e\widehat{Q}_{\mathbf{qB}}(R)e$ and therefore $\widehat{Q}_{\mathbf{qB}}(eRe) = e\widehat{Q}_{\mathbf{qB}}(R)e$. □

The following natural result establishes a Morita equivalence of quasi-Baer ring hulls between two semiprime rings which are Morita equivalent.

Theorem 9.3.10 *Let R be a semiprime ring. If R and a ring S are Morita equivalent, then $\widehat{Q}_{\mathbf{qB}}(R)$ and $\widehat{Q}_{\mathbf{qB}}(S)$ are Morita equivalent.*

Proof Since R and S are Morita equivalent, S is semiprime. Also there exist a positive integer n and $e^2 = e \in \text{Mat}_n(R)$ such that $S = e\text{Mat}_n(R)e$ and $\text{Mat}_n(R)e\text{Mat}_n(R) = \text{Mat}_n(R)$. Thus, by Lemma 9.3.9 and Proposition 9.3.7, $\widehat{Q}_{\mathbf{qB}}(S) = \widehat{Q}_{\mathbf{qB}}(e\text{Mat}_n(R)e) = e\widehat{Q}_{\mathbf{qB}}(\text{Mat}_n(R))e = e\text{Mat}_n(\widehat{Q}_{\mathbf{qB}}(R))e$. Also,

$$\begin{aligned} &\text{Mat}_n(\widehat{Q}_{\mathbf{qB}}(R))e\text{Mat}_n(\widehat{Q}_{\mathbf{qB}}(R)) = \text{Mat}_n(R)e\text{Mat}_n(R)\mathcal{B}(\text{Mat}_n(Q(R))) \\ &= \text{Mat}_n(R)\mathcal{B}(Q(R))\mathbf{1} = \text{Mat}_n(R\mathcal{B}(Q(R))) = \text{Mat}_n(\widehat{Q}_{\mathbf{qB}}(R)), \end{aligned}$$

where $\mathbf{1}$ is the identity matrix of $\text{Mat}_n(R)$. Note that $e \in \text{Mat}_n(\widehat{Q}_{\mathbf{qB}}(R))$. Therefore, $\widehat{Q}_{\mathbf{qB}}(R)$ is Morita equivalent to $\widehat{Q}_{\mathbf{qB}}(S)$. □

Remark 9.3.11 The conclusion of Theorem 9.3.10 does not hold true for the case of Baer ring hulls of two Morita equivalent semiprime rings. In other words, $\widehat{Q}_{\mathbf{B}}(R)$ and $\widehat{Q}_{\mathbf{B}}(S)$ cannot replace $\widehat{Q}_{\mathbf{qB}}(R)$ and $\widehat{Q}_{\mathbf{qB}}(S)$, respectively in Theorem 9.3.10. For example, we take $R = F[x, y]$ with F a field, and take $S = \text{Mat}_n(R)$ with $n > 1$. Since R is a Baer ring, $\widehat{Q}_{\mathbf{B}}(R) = R$, however $\widehat{Q}_{\mathbf{B}}(S)$ does not even exist (see Example 8.3.34).

Lemma 9.3.12 *Let R be a right FI-extending ring. Then $\mathcal{B}(T) \subseteq R$ for any right essential overring T of R.*

Proof Assume that $e \in \mathcal{B}(T)$. Then $eT \cap R \trianglelefteq R$, hence there is $f^2 = f \in R$ such that $(eT \cap R)_R \leq^{\text{ess}} fR_R$. So $(eT \cap R)_R \leq^{\text{ess}} fT_R$. Also, $(eT \cap R)_R \leq^{\text{ess}} eT_R$. Since $e \in \mathcal{B}(T)$, $e = f \in R$. Therefore, $\mathcal{B}(T) \subseteq R$. □

When R is a semiprime ring, $\widehat{Q}_{\mathbf{qB}}(R) = \widehat{Q}_{\mathbf{FI}}(R) = R\mathcal{B}(Q(R))$ by Theorem 8.3.17. The following result provides a class of nonsemiprime rings for which such equality also holds true.

Theorem 9.3.13 *Let R be a ring and n a positive integer. Then*:

(i) $\widehat{Q}_{\mathbf{IC}}(T_n(R)) = T_n(R)\mathcal{B}(Q(T_n(R))) = T_n(R\mathcal{B}(Q(R))) = T_n(\widehat{Q}_{\mathbf{IC}}(R))$.

(ii) *If R is semiprime, then*:

(1) $\widehat{Q}_{\mathbf{qB}}(T_n(R)) = T_n(\widehat{Q}_{\mathbf{qB}}(R)) = T_n(R)\mathcal{B}(Q(T_n(R)))$.

(2) $\widehat{Q}_{\mathbf{FI}}(T_n(R)) = T_n(\widehat{Q}_{\mathbf{FI}}(R)) = T_n(R)\mathcal{B}(Q(T_n(R)))$.

(3) $\widehat{Q}_{\mathbf{pqB}}(T_n(R)) = T_n(\widehat{Q}_{\mathbf{pqB}}(R))$.

Proof (i) This part follows from Theorem 8.3.11 and Lemma 9.3.6.

(ii)(1) Put $T = T_n(R)$. From Theorems 5.6.7 and 8.3.17, $T_n(\widehat{Q}_{\mathbf{qB}}(R))$ is quasi-Baer and $T_n(\widehat{Q}_{\mathbf{qB}}(R)) = T_n(R\mathcal{B}(Q(R)))$.

Let S be a quasi-Baer right ring of quotients of T. We show that S contains all $n \times n$ constant diagonal matrices whose diagonal entries are from $\mathcal{B}(Q(R))$. For this, take $e \in \mathcal{B}(Q(R))$ and put $I = R \cap (1-e)Q(R)$. Then $Q(R)IQ(R)_R \leq^{\text{ess}} (1-e)Q(R)_R$, so $Q(R)IQ(R)_{Q(R)} \leq^{\text{ess}} (1-e)Q(R)_{Q(R)}$. By Proposition 8.3.16, $r_{Q(R)}(I) = r_{Q(R)}(Q(R)IQ(R))$. From Corollary 8.3.19, $Q(R)$ is quasi-Baer. Hence from Propositions 1.2.2 and 1.2.6(ii), there exists $f^2 = f \in \mathcal{B}(Q(R))$ such that $r_{Q(R)}(Q(R)IQ(R)) = fQ(R)$. From Lemma 2.1.13, $Q(R)IQ(R)_{Q(R)} \leq^{\text{ess}} (1-f)Q(R)_{Q(R)}$. Therefore, $1-e = 1-f$ and so $e = f$. Hence $r_{Q(R)}(I) = r_{Q(R)}(Q(R)IQ(R)) = fQ(R) = eQ(R)$.

Let K be the $n \times n$ matrix with I in the $(1,1)$-position and 0 elsewhere. Thus TKT is the $n \times n$ matrix with I throughout the top row and 0 elsewhere. Also $Q(T)KQ(T) = \text{Mat}_n(Q(R)IQ(R))$ because $Q(T) = \text{Mat}_n(Q(R))$ from Lemma 9.3.6.

Now we have that $TKT \subseteq SKS \subseteq Q(T)KQ(T)$, $r_{Q(R)}(I) = eQ(R)$, and $r_{Q(R)}(Q(R)IQ(R)) = eQ(R)$. We let $g = e\mathbf{1}$, where $\mathbf{1}$ is the identity matrix of $\text{Mat}_n(R)$. So g is the diagonal matrix in $Q(T) = \text{Mat}_n(Q(R))$ with e on the diagonal. Then

$$gQ(T) = r_{Q(T)}(Q(T)KQ(T)) \subseteq r_{Q(T)}(SKS) \subseteq r_{Q(T)}(TKT) = gQ(T).$$

Because S is quasi-Baer, there exists $c^2 = c \in S$ with $r_S(SKS) = cS$. Also $r_S(SKS) = S \cap gQ(T)$ since $r_{Q(T)}(SKS) = gQ(T)$. Hence $cQ(T) \subseteq gQ(T)$.

Further, $cS_T \leq^{\text{ess}} gQ(T)_T$. Thus $cQ(T)_{Q(T)} \leq^{\text{ess}} gQ(T)_{Q(T)}$, and hence from the modular law, $cQ(T) = gQ(T)$.

Now $g = c \in S$ because g is central in $Q(T)$. So S contains all $n \times n$ constant diagonal matrices whose diagonal entries are from $\mathcal{B}(Q(R))$. By Theorem 8.3.17, $T_n(\widehat{Q}_{\mathbf{qB}}(R)) \subseteq S$. Thus

$$\widehat{Q}_{\mathbf{qB}}(T) = T_n(\widehat{Q}_{\mathbf{qB}}(R)) = T_n(R\mathcal{B}(Q(R))) = T_n(R)\mathcal{B}(Q(T_n(R))).$$

(ii)(2) As R is semiprime, $\widehat{Q}_{\mathbf{FI}}(R) = R\mathcal{B}(Q(R))$ by Theorem 8.3.17. From Theorem 5.6.19, $T_n(R\mathcal{B}(Q(R)))$ is right FI-extending. Let S be a right FI-extending right ring of quotients of $T_n(R)$. Lemmas 9.3.6 and 9.3.12 yield that

$$\mathcal{B}(Q(T_n(R))) = \mathcal{B}(\mathrm{Mat}_n(Q(R))) = \mathcal{B}(Q(R))\mathbf{1} \subseteq S,$$

where $\mathbf{1}$ is the identity matrix of $\mathrm{Mat}_n(R)$. So

$$T_n(R\mathcal{B}(Q(R))) \subseteq S \text{ and } \widehat{Q}_{\mathbf{FI}}(T_n(R)) = T_n(R\mathcal{B}(Q(R))).$$

Therefore $\widehat{Q}_{\mathbf{FI}}(T_n(R)) = T_n(\widehat{Q}_{\mathbf{FI}}(R)) = T_n(R)\mathcal{B}(Q(T_n(R)))$.

(ii)(3) Put $T = T_n(R)$. First, we note that $T_n(\widehat{Q}_{\mathbf{pqB}}(R))$ is a right p.q.-Baer ring by Proposition 5.6.8. Let S be a right p.q.-Baer right ring of quotients of T. Say $e \in \mathcal{B}_p(Q(R))$. There exists $x \in R$ with $RxR_R \leq^{\text{ess}} eQ(R)_R$. Hence $Q(R)xQ(R)_{Q(R)} \leq^{\text{ess}} eQ(R)_{Q(R)}$. Because $e \in \mathcal{B}(Q(R))$, we see that $eQ(R)xQ(R)e_{eQ(R)e} \leq^{\text{ess}} eQ(R)e_{eQ(R)e}$. As $eQ(R)e$ is semiprime, by Proposition 1.3.16, $r_{eQ(R)e}(eQ(R)xQ(R)e) = 0$.

Because e is central, it follows that $r_{Q(R)}(Q(R)xQ(R))e = 0$. Therefore we have that $r_{Q(R)}(Q(R)xQ(R)) \subseteq (1-e)Q(R)$. Since $Q(R)xQ(R) \subseteq eQ(R)$,

$$(1-e)Q(R) \subseteq r_{Q(R)}(Q(R)xQ(R)).$$

Therefore, $r_{Q(R)}(Q(R)xQ(R)) = (1-e)Q(R)$. Now from Proposition 8.3.16, we obtain $r_{Q(R)}(RxR) = r_{Q(R)}(Q(R)xQ(R)) = (1-e)Q(R)$.

Let $\sigma \in T = T_n(R)$ be the $n \times n$ matrix with x in the $(1,1)$-position and 0 elsewhere. Thus $T\sigma T$ is the $n \times n$ matrix with RxR throughout the top row and zero elsewhere. Further, we see that $Q(T)\sigma Q(T) = \mathrm{Mat}_n(Q(R)xQ(R))$ because $Q(T) = \mathrm{Mat}_n(Q(R))$. Observe that $T\sigma T \subseteq S\sigma S \subseteq Q(T)\sigma Q(T)$ and $r_{Q(R)}(RxR) = (1-e)Q(R)$.

We take $h = (1-e)\mathbf{1}$, where $\mathbf{1}$ is the identity matrix of $\mathrm{Mat}_n(R)$. Then

$$hQ(T) = r_{Q(T)}(Q(T)\sigma Q(T)) \subseteq r_{Q(T)}(S\sigma S) \subseteq r_{Q(T)}(T\sigma T) = hQ(T).$$

There is $c^2 = c \in S$ with $cS = r_S(S\sigma S) = S \cap r_{Q(T)}(S\sigma S) = S \cap hQ(T)$, since S is right p.q.-Baer. Thus, $cQ(T) \subseteq hQ(T)$.

Further, we see that, as in the proof of part (i), $cQ(T)_{Q(T)} \leq^{\text{ess}} hQ(T)_{Q(T)}$. Because h is central in $Q(T)$, $h = c \in S$. Therefore, S contains all $n \times n$ constant diagonal matrices whose diagonal entries are from $\mathcal{B}_p(Q(R))$. Thus, $T_n(\widehat{Q}_{\mathbf{pqB}}(R)) \subseteq S$ by Theorem 8.3.39(ii). So $\widehat{Q}_{\mathbf{pqB}}(T) = T_n(\widehat{Q}_{\mathbf{pqB}}(R))$. □

For a nonsemiprime version of Theorem 9.3.13(ii)(2), see Exercise 9.3.17.2.

Theorem 9.3.14 *Let R be a semiprime ring. Then:*

(i) $\widehat{Q}_{\mathbf{qB}}(\mathrm{CFM}_\Gamma(R)) \subseteq \mathrm{CFM}_\Gamma(\widehat{Q}_{\mathbf{qB}}(R))$.
(ii) $\widehat{Q}_{\mathbf{qB}}(\mathrm{RFM}_\Gamma(R)) \subseteq \mathrm{RFM}_\Gamma(\widehat{Q}_{\mathbf{qB}}(R))$.
(iii) $\widehat{Q}_{\mathbf{qB}}(\mathrm{CRFM}_\Gamma(R)) \subseteq \mathrm{CRFM}_\Gamma(\widehat{Q}_{\mathbf{qB}}(R))$.

Proof Since R is semiprime, so are $\mathrm{CFM}_\Gamma(R)$, $\mathrm{RFM}_\Gamma(R)$ and $\mathrm{CRFM}_\Gamma(R)$. Let $e \in \mathcal{B}(Q(R))$. Then $e \in \mathcal{B}(Q^m(R))$, so there exists $J \trianglelefteq R$ with $\ell_R(J) = 0$ and $eJ \subseteq R$. Hence $\mathrm{CFM}_\Gamma(J) \trianglelefteq \mathrm{CFM}_\Gamma(R)$, $\ell_{\mathrm{CFM}_\Gamma(R)}(\mathrm{CFM}_\Gamma(J)) = 0$, and also $(e\mathbf{1})\mathrm{CFM}_\Gamma(J) \subseteq \mathrm{CFM}_\Gamma(R)$, where $\mathbf{1}$ is the identity matrix in $\mathrm{CFM}_\Gamma(R)$. Hence $e\mathbf{1} \in Q^m(\mathrm{CFM}_\Gamma(R))$, so $e\mathbf{1} \in \mathcal{B}(Q^m(\mathrm{CFM}_\Gamma(R)))$. Thus

$$\begin{aligned}\mathrm{CFM}_\Gamma(\widehat{Q}_{\mathbf{qB}}(R)) = \mathrm{CFM}_\Gamma(R\mathcal{B}(Q(R))) &\subseteq \mathrm{CFM}_\Gamma(R)\mathcal{B}(Q(R))\mathbf{1}\\ &\subseteq Q^m(\mathrm{CFM}_\Gamma(R)) \subseteq Q(\mathrm{CFM}_\Gamma(R)).\end{aligned}$$

From Theorem 6.1.16, $\mathrm{CFM}_\Gamma(\widehat{Q}_{\mathbf{qB}}(R))$ is quasi-Baer. Therefore, we have that $\widehat{Q}_{\mathbf{qB}}(\mathrm{CFM}_\Gamma(R)) \subseteq \mathrm{CFM}_\Gamma(\widehat{Q}_{\mathbf{qB}}(R))$.

Similarly, $\widehat{Q}_{\mathbf{qB}}(\mathrm{RFM}_\Gamma(R)) \subseteq \mathrm{RFM}_\Gamma(\widehat{Q}_{\mathbf{qB}}(R))$ by the preceding argument and Theorem 6.1.16. Also, we see that $\mathrm{CRFM}_\Gamma(\widehat{Q}_{\mathbf{qB}}(R))$ is also quasi-Baer by Theorem 6.1.16. Moreover, $\mathrm{CRFM}_\Gamma(\widehat{Q}_{\mathbf{qB}}(R)) \subseteq Q(\mathrm{CRFM}_\Gamma(R))$ from the preceding argument. So $\widehat{Q}_{\mathbf{qB}}(\mathrm{CRFM}_\Gamma(R)) \subseteq \mathrm{CRFM}_\Gamma(\widehat{Q}_{\mathbf{qB}}(R))$. □

In view of Proposition 9.3.7 and Theorem 9.3.14, one may expect that some of the following may hold true:

$$\widehat{Q}_{\mathbf{qB}}(\mathrm{CFM}_\Gamma(R)) = \mathrm{CFM}_\Gamma(\widehat{Q}_{\mathbf{qB}}(R)),\ \widehat{Q}_{\mathbf{qB}}(\mathrm{RFM}_\Gamma(R)) = \mathrm{RFM}_\Gamma(\widehat{Q}_{\mathbf{qB}}(R)),$$

or

$$\widehat{Q}_{\mathbf{qB}}(\mathrm{CRFM}_\Gamma(R)) = \mathrm{CRFM}_\Gamma(\widehat{Q}_{\mathbf{qB}}(R)).$$

However, the next example shows that there exists a commutative regular ring R and a nonempty ordered set Γ such that none of these equalities holds.

Example 9.3.15 There exist a commutative regular ring R and a nonempty ordered set Γ such that:

(i) $\widehat{Q}_{\mathbf{qB}}(\mathrm{CFM}_\Gamma(R)) \subsetneq \mathrm{CFM}_\Gamma(\widehat{Q}_{\mathbf{qB}}(R))$.
(ii) $\widehat{Q}_{\mathbf{qB}}(\mathrm{RFM}_\Gamma(R)) \subsetneq \mathrm{RFM}_\Gamma(\widehat{Q}_{\mathbf{qB}}(R))$.
(iii) $\widehat{Q}_{\mathbf{qB}}(\mathrm{CRFM}_\Gamma(R)) \subsetneq \mathrm{CRFM}_\Gamma(\widehat{Q}_{\mathbf{qB}}(R))$.

Let F be a field. Take a set Λ with $|\Lambda| = |F|\aleph_0$, and let $F_i = F$ for all $i \in \Lambda$. Put

$$R = \{(\gamma_i)_{i\in\Lambda} \in \prod_{i\in\Lambda} F_i \mid \gamma_i \text{ is constant for all but finitely many } i\},$$

a subring of $\prod_{i\in\Lambda} F_i$. Then $Q(R) = \prod_{i\in\Lambda} F_i$ and R is a commutative regular ring. Take $\Gamma = \widehat{Q}_{\mathbf{qB}}(R) = R\mathcal{B}(Q(R))$ as a set.

We observe that $\widehat{Q}_{\mathbf{qB}}(\mathrm{CFM}_\Gamma(R)) \subseteq \mathrm{CFM}_\Gamma(\widehat{Q}_{\mathbf{qB}}(R)) \subseteq Q(\mathrm{CFM}_\Gamma(R))$ by Theorem 9.3.14 and its proof. Because $\mathrm{CFM}_\Gamma(R)$ is semiprime, Theorem 8.3.17 yields that $\widehat{Q}_{\mathbf{qB}}(\mathrm{CFM}_\Gamma(R)) = \mathrm{CFM}_\Gamma(R)\mathcal{B}(Q(\mathrm{CFM}_\Gamma(R)))$. Therefore, $\mathcal{B}(Q(\mathrm{CFM}_\Gamma(R))) \subseteq \widehat{Q}_{\mathbf{qB}}(\mathrm{CFM}_\Gamma(R))$. Hence,

$$\mathcal{B}(Q(\mathrm{CFM}_\Gamma(R))) \subseteq \mathcal{B}(\widehat{Q}_{\mathbf{qB}}(\mathrm{CFM}_\Gamma(R))).$$

Thus $\mathcal{B}(Q(\mathrm{CFM}_\Gamma(R))) = \mathcal{B}(\widehat{Q}_{\mathbf{qB}}(\mathrm{CFM}_\Gamma(R)))$.

Assume on the contrary that $\widehat{Q}_{\mathbf{qB}}(\mathrm{CFM}_\Gamma(R)) = \mathrm{CFM}_\Gamma(\widehat{Q}_{\mathbf{qB}}(R))$. Then $\mathcal{B}(Q(\mathrm{CFM}_\Gamma(R))) = \mathcal{B}(\widehat{Q}_{\mathbf{qB}}(\mathrm{CFM}_\Gamma(R))) = \mathcal{B}(\mathrm{CFM}_\Gamma(\widehat{Q}_{\mathbf{qB}}(R)))$.

We let $\mu \in \mathrm{CFM}_\Gamma(\widehat{Q}_{\mathbf{qB}}(R))$ be a diagonal matrix whose diagonal entries are all distinct elements of $\widehat{Q}_{\mathbf{qB}}(R)$. Then $\mu \in \widehat{Q}_{\mathbf{qB}}(\mathrm{CFM}_\Gamma(R))$ by assumption. By Theorem 8.3.17, $\widehat{Q}_{\mathbf{qB}}(\mathrm{CFM}_\Gamma(R)) = \mathrm{CFM}_\Gamma(R)\mathcal{B}(\mathrm{CFM}_\Gamma(\widehat{Q}_{\mathbf{qB}}(R)))$, since $\mathcal{B}(Q(\mathrm{CFM}_\Gamma(R))) = \mathcal{B}(\mathrm{CFM}_\Gamma(\widehat{Q}_{\mathbf{qB}}(R)))$. Let $\mathbf{1}$ be the identity matrix in $\mathrm{CFM}_\Gamma(R)$. Then there are $\theta_1, \ldots, \theta_n \in \mathrm{CFM}_\Gamma(R)$ and $f_1, \ldots, f_n \in \widehat{Q}_{\mathbf{qB}}(R)$, where $f_1\mathbf{1}, \ldots, f_n\mathbf{1} \in \mathcal{B}(\mathrm{CFM}_\Gamma(\widehat{Q}_{\mathbf{qB}}(R)))$ are orthogonal by Lemma 8.3.29 (note that $f_i \in \mathcal{B}(\widehat{Q}_{\mathbf{qB}}(R))$ for all i) and $\mu = \theta_1 f_1\mathbf{1} + \cdots + \theta_n f_n\mathbf{1}$.

Hence for each entry of the diagonal of μ, or equivalently, each element of $R\mathcal{B}(Q(R))$, say a, there exist diagonal entries $\theta_i(a)$ of θ_i for $i = 1, \ldots, n$ such that $a = \theta_1(a) f_1 + \cdots + \theta_n(a) f_n$. Thus $R\mathcal{B}(Q(R)) \subseteq \sum_{i=1}^n Rf_i \subseteq R\mathcal{B}(Q(R))$, so $R\mathcal{B}(Q(R)) = Rf_1 + \cdots + Rf_n$. Hence $|R\mathcal{B}(Q(R))| = |R|$.

Assume that $|F|$ is finite or countably infinite. Then we see that $|R| = \aleph_0$, but $|R| = |R\mathcal{B}(Q(R))| \geq |\mathcal{B}(Q(R))| = 2^{\aleph_0}$ as $|\Lambda| = \aleph_0$ and $Q(R) = \prod_{i\in\Lambda} F_i$, a contradiction.

If $|F|$ is uncountably infinite, then $|R| = |\Lambda|$. But in this case, we note that $|R| = |R\mathcal{B}(Q(R))| \geq |\mathcal{B}(Q(R))| = 2^{|\Lambda|}$, also a contradiction. Therefore, $\widehat{Q}_{\mathbf{qB}}(\mathrm{CFM}_\Gamma(R)) \subsetneq \mathrm{CFM}_\Gamma(\widehat{Q}_{\mathbf{qB}}(R))$.

Similarly, we can verify that $\widehat{Q}_{\mathbf{qB}}(\mathrm{RFM}_\Gamma(R)) \subsetneq \mathrm{RFM}_\Gamma(\widehat{Q}_{\mathbf{qB}}(R))$ and $\widehat{Q}_{\mathbf{qB}}(\mathrm{CRFM}_\Gamma(R)) \subsetneq \mathrm{CRFM}_\Gamma(\widehat{Q}_{\mathbf{qB}}(R))$.

Proposition 9.3.7 and Theorem 9.3.14 motivate the following questions: (1) *Is the right p.q.-Baer property preserved under the various infinite matrix ring extensions?* (2) *Does $\widehat{Q}_{\mathbf{pqB}}(R)$ of a ring R have a behavior similar to that of $\widehat{Q}_{\mathbf{qB}}(R)$ for the various infinite matrix ring extensions?* The next example provides negative answers to both of these questions.

Example 9.3.16 Let F be a field and $F_n = F$ for $n = 1, 2\ldots$. Put

$$R = \left\{ (q_n)_{n=1}^{\infty} \in \prod_{n=1}^{\infty} F_n \mid q_n \text{ is constant eventually} \right\},$$

which is a subring of $\prod_{n=1}^{\infty} F_n$. Then R is a commutative p.q.-Baer ring.

Put $S = \mathrm{CFM}_\Gamma(R)$, where $\Gamma = \{1, 2, \ldots\}$. We now take

$$a_1 = (0, 1, 0, 0, \ldots),\ a_2 = (0, 1, 0, 1, 0, 0, \ldots),\ a_3 = (0, 1, 0, 1, 0, 1, 0, 0, \ldots),$$

and so on, in R. Let x be the element in S with a_n in the (n, n)-position for $n = 1, 2, \ldots$ and 0 elsewhere.

Take $e = (q_n)_{n=1}^{\infty} \in Q(R) = \prod_{n=1}^{\infty} F_n$ such that

$$q_{2n} = 1 \text{ and } q_{2n-1} = 0 \text{ for } n = 1, 2, \ldots.$$

Then $e^2 = e \in \mathcal{B}(Q(R))$, hence $e\mathbf{1} \in \mathrm{CFM}_\Gamma(\widehat{Q}_{\mathbf{qB}}(R))$ as $\widehat{Q}_{\mathbf{qB}}(R) = R\mathcal{B}(Q(R))$ from Theorem 8.3.17, where $\mathbf{1}$ is the identity matrix in S. From the proof of Theorem 9.3.14, $\mathrm{CFM}_\Gamma(\widehat{Q}_{\mathbf{qB}}(R)) \subseteq Q(S)$. Therefore, we obtain that $e\mathbf{1} \in \mathcal{B}(Q(S))$ because $e\mathbf{1} \in \mathcal{B}(\mathrm{CFM}_\Gamma(\widehat{Q}_{\mathbf{qB}}(R)))$.

By direct computation $SxS_S \leq^{\text{ess}} (e\mathbf{1})S_S$ (Exercise 9.3.17.3). Hence, $e\mathbf{1} \in \mathcal{B}_p(Q(S))$. But $e\mathbf{1} \notin S$ since $e \notin R$. Note that S is a semiprime ring as R is semiprime. Therefore, the ring S is not right p.q.-Baer by Theorem 8.3.39(ii). Furthermore, because R is p.q.-Baer, $\widehat{Q}_{\mathbf{pqB}}(R) = R$. Thus

$$\widehat{Q}_{\mathbf{pqB}}(\text{CFM}_\Gamma(R)) \nsubseteq \text{CFM}_\Gamma(\widehat{Q}_{\mathbf{pqB}}(R)),$$

and hence $S = \text{CFM}_\Gamma(\widehat{Q}_{\mathbf{pqB}}(R))$ is not right p.q.-Baer by Theorem 8.3.39(ii).

To show that $\widehat{Q}_{\mathbf{pqB}}(\text{CRFM}_\Gamma(R)) \nsubseteq \text{CRFM}_\Gamma(\widehat{Q}_{\mathbf{pqB}}(R))$, let x and e be as in the case of the column finite matrix ring. Then, by the same method, we can show that $e\mathbf{1} \in \mathcal{B}_p(Q(\text{CRFM}_\Gamma(R)))$, but $e\mathbf{1} \notin \text{CRFM}_\Gamma(R)$. Therefore by Theorem 8.3.39(ii), $\text{CRFM}_\Gamma(R)$ $(= \text{CRFM}_\Gamma(\widehat{Q}_{\mathbf{pqB}}(R)))$ is not right p.q.-Baer. Also $\widehat{Q}_{\mathbf{pqB}}(\text{CRFM}_\Gamma(R)) \nsubseteq \text{CRFM}_\Gamma(\widehat{Q}_{\mathbf{pqB}}(R))$.

Finally for $\widehat{Q}_{\mathbf{pqB}}(\text{RFM}_\Gamma(R)) \nsubseteq \text{RFM}_\Gamma(\widehat{Q}_{\mathbf{pqB}}(R))$, let $U = \text{RFM}_\Gamma(R)$ and x, e be as before. Then ${}_UUxU \leq^{\text{ess}} {}_U(e\mathbf{1})U$, where $\mathbf{1}$ is the identity matrix in U (Exercise 9.3.17.4). Note that $e\mathbf{1}$ is a central idempotent. So we have that ${}_{(e\mathbf{1})U(e\mathbf{1})}UxU \leq^{\text{ess}} {}_{(e\mathbf{1})U(e\mathbf{1})}(e\mathbf{1})U(e\mathbf{1})$. As UxU is an ideal of the semiprime ring $(e\mathbf{1})U(e\mathbf{1})$, $r_{(e\mathbf{1})U(e\mathbf{1})}(UxU) = \ell_{(e\mathbf{1})U(e\mathbf{1})}(UxU) = 0$, so

$$UxU_{(e\mathbf{1})U(e\mathbf{1})} \leq^{\text{ess}} (e\mathbf{1})U(e\mathbf{1})_{(e\mathbf{1})U(e\mathbf{1})}$$

from Proposition 1.3.16. Therefore $UxU_U \leq^{\text{ess}} (e\mathbf{1})U_U$.

As $e \in \mathcal{B}(Q(R)) = \mathcal{B}(Q^m(R))$, there is $J \trianglelefteq R$ such that $\ell_R(J) = 0$ and $eJ \subseteq R$. Hence $\text{RFM}_\Gamma(J) \trianglelefteq U$, $\ell_U(\text{RFM}_\Gamma(J)) = 0$, and $(e\mathbf{1})\text{RFM}_\Gamma(J) \subseteq U$. Therefore $e\mathbf{1} \in Q^m(U)$. Hence $e\mathbf{1} \in \mathcal{B}(Q^m(U))$ and $e\mathbf{1} \in \mathcal{B}(Q(U))$, so $e\mathbf{1} \in \mathcal{B}_p(Q(U))$. But we note that $e\mathbf{1} \notin U$ because $e \notin R$. Thus $U = \text{RFM}_\Gamma(R)$ $(= \text{RFM}_\Gamma(\widehat{Q}_{\mathbf{pqB}}(R))$ is not right p.q.-Baer by Theorem 8.3.39(ii). Also,

$$\widehat{Q}_{\mathbf{pqB}}(\text{RFM}_\Gamma(R)) \nsubseteq \text{RFM}_\Gamma(\widehat{Q}_{\mathbf{pqB}}(R)).$$

Exercise 9.3.17

1. ([100, Birkenmeier, Park, and Rizvi]) Let $\mathfrak{K}$ denote a class of rings and S be a right essential overring of R. The smallest intermediate ring V between R and S which belongs to $\mathfrak{K}$ is called the $\mathfrak{K}$ *absolute to S ring hull* of R (when it exists). We denote $V = Q^S_{\mathfrak{K}}(R)$. Show that $Q^S_{\mathbf{qB}}(R)$ exists if and only if $Q^{T_n(S)}_{\mathbf{qB}}(T_n(R))$ exists for all positive integers n. In this case, $Q^{T_n(S)}_{\mathbf{qB}}(T_n(R)) = T_n(Q^S_{\mathbf{qB}}(R))$ for all positive integers n.
2. ([100, Birkenmeier, Park, and Rizvi]) Assume that S is a right ring of quotients of R. Prove that the following are equivalent.
   (i) $Q^S_{\mathbf{FI}}(R)$ exists.
   (ii) $Q^{T_n(S)}_{\mathbf{FI}}(T_n(R))$ exists for all positive integers n.
   (iii) $Q^{T_n(S)}_{\mathbf{FI}}(T_n(R))$ exists for some positive integer k.
   In this case, $Q^{T_n(S)}_{\mathbf{FI}}(T_n(R)) = T_n(Q^S_{\mathbf{FI}}(R))$ for all positive integers n.
3. Let $S = \text{CFM}_\Gamma(R)$, $x \in S$, and $e \in Q(R)$ as in Example 9.3.16. Show that $SxS_S \leq^{\text{ess}} (e\mathbf{1})S_S$, where $\mathbf{1}$ is the identity matrix of S.

4. Let $U = \mathrm{RFM}_\Gamma(R)$, $x \in U$, and $e \in Q(R)$ as in Example 9.3.16. Prove that ${}_U UxU \leq^{\text{ess}} {}_U(e\mathbf{1})U$, where $\mathbf{1}$ is the identity matrix of U.

Historical Notes Most results of Sect. 9.1 are in [89]. The proof of Theorem 9.1.11(i) corrects the proof of [89, Corollary 3.9(i)]. X-outer automorphisms initially were considered by Kharchenko [249] in the study of group actions on rings. For more details on X-outer ring automorphisms of a ring, see [249] and [169]. Theorem 9.2.5 is also an unpublished new result. Major results of Sect. 9.2 including Theorem 9.2.10 are in [233]. See also [339] for skew group rings.

There is a flaw in the proof of $\widehat{Q}_{\mathbf{qB}}(R[[X]]) = \widehat{Q}_{\mathbf{qB}}(R)[[X]]$ in [100]. Lemma 9.3.12 is from [58], while Example 9.3.16 appears in [101]. All major results of Sect. 9.3 appear in [100]. Related references include [305] and [413].

Chapter 10
Applications to Rings of Quotients and C*-Algebras

We shall now present necessary and sufficient conditions on a ring R for which $Q(R)$ can be decomposed into a direct product of indecomposable rings or into a direct product of prime rings. This will be done by using the idempotent closure class we discussed in Chap. 8 and a dimension on bimodules which will be introduced in Sect. 10.1. An application of these results helps provide a structure theorem for the quasi-Baer ring hull of a semiprime ring having only finitely many minimal prime ideals.

An important focus in this chapter is to showcase some of the applications of algebraic techniques developed in earlier chapters (as well as new results of this chapter) to Functional Analysis. These applications will include obtaining results on C^*-algebras, AW^*-algebras and skew group C^*-algebras. More specifically, we shall see applications to boundedly centrally closed C^*-algebras, local multipliers of C^*-algebras, extended centroids of C^*-algebras, $A * G$ where A is a unital C^*-algebra and G is a finite group of X-outer $*$-automorphisms of A, and on C^*-algebras with a polynomial identity.

10.1 The Structure of Rings of Quotients

The structure of rings and of their rings of quotients, especially the structure of the maximal right ring of quotients $Q(R)$ of a ring R, has been of interest for a long time. The main topic of this section is the characterization theorem of $Q(R)$ as a direct product of prime rings (or indecomposable rings).

We first recall Theorem 7.3.2 which characterizes a ring R such that $Q(R)$ is semisimple Artinian. Further, Goodearl showed that a regular right self-injective ring R is isomorphic to a direct product of prime rings if and only if every nonzero ideal of R contains a minimal nonzero ideal (see [183, Corollary 12.24] and [179]). Let R and S be rings and ${}_SM_R$ be an (S, R)-bimodule. The two-sided uniform dimension $\text{udim}({}_SM_R)$ is the supremum of the set of positive integers n for which M contains a direct sum of n nonzero (S, R)-subbimodules. A closely related invariant, $d(M) = d({}_SM_R)$, is defined by taking the supremum of the set of positive

G.F. Birkenmeier et al., *Extensions of Rings and Modules*,
DOI 10.1007/978-0-387-92716-9_10,

integers n for which there exists a direct sum of nonzero (S, R)-subbimodules $N := M_1 \oplus \cdots \oplus M_n$ such that $N_R \leq^{\text{ess}} M_R$ (see [226]). For $I \trianglelefteq R$, $d({}_R I_R)$ is denoted by $d(I)$. Related to Goodearl's result, the following result was obtained by Jain, Lam, and Leroy in [226] as follows:

Theorem 10.1.1 *Let R be a right nonsingular ring. Then $Q(R)$ is a direct product of prime rings if and only if there exist ideals I_i $(i \in \Lambda)$ of R such that $d(I_i) = 1$ for all i and $(\bigoplus_{i\in\Lambda} I_i)_R \leq^{\text{ess}} R_R$.*

In Theorems 10.1.10, 10.1.12, and 10.1.13, we discuss the structure of $Q(R)$ as a direct product of prime rings (or indecomposable rings). These results generalize Theorem 10.1.1 by removing the right nonsingularity condition of R. We see that this leads to a structure theorem for the quasi-Baer ring hull $\widehat{Q}_{\mathbf{qB}}(R)$ when R is semiprime. If R is a semiprime ring, then R has exactly n minimal prime ideals $P_1, \ldots, P_n$ if and only if $\widehat{Q}_{\mathbf{qB}}(R) \cong R/P_1 \oplus \cdots \oplus R/P_n$ (Theorem 10.1.20). Further, the results of this section have useful applications to the study of C^*-algebras in Sect. 10.3.

Definition 10.1.2 Let R and S be rings and M an (S, R)-bimodule.

(i) We let $\mathfrak{D}_{\mathbf{IC}}(M)$ be the set of all subbimodules ${}_S N_R$ of ${}_S M_R$ such that there exists ${}_S L_R \leq {}_S M_R$ with $N \cap L = 0$ and $(N \oplus L)_R \leq^{\text{den}} M_R$.

(ii) We call *jdim*(M) the *Johnson dimension* of M, where jdim(M) denotes the supremum of the set of positive integers n for which there is a direct sum of n nonzero (S, R)-subbimodules, with $(M_1 \oplus \cdots \oplus M_n)_R \leq^{\text{den}} M_R$.

For $I \trianglelefteq R$, $\mathfrak{D}_{\mathbf{IC}}(I)$ and jdim(I) are defined by considering I as an (R, R)-bimodule. In particular, we consider $M = R$ as an (R, R)-bimodule. Then by Proposition 8.3.5, $\mathfrak{D}_{\mathbf{IC}}(M)$ is exactly $\mathfrak{D}_{\mathbf{IC}}(R)$ of Definition 8.3.1(i). The following relation compares jdim(M) with other dimensions:

$$\text{jdim}(M) \leq d(M) \leq \text{udim}({}_S M_R) \leq \text{udim}(M_R).$$

Proposition 10.1.3 *Let $I, J \trianglelefteq R$ with $I \subseteq J$. Then:*

(i) *If $I \in \mathfrak{D}_{\mathbf{IC}}(R)$, then $I \in \mathfrak{D}_{\mathbf{IC}}(J)$.*
(ii) *If $I \in \mathfrak{D}_{\mathbf{IC}}(J)$ and $J \in \mathfrak{D}_{\mathbf{IC}}(R)$, then $I \in \mathfrak{D}_{\mathbf{IC}}(R)$.*

Proof (i) As $I \in \mathfrak{D}_{\mathbf{IC}}(R)$, there is $Y \trianglelefteq R$ such that $(I \oplus Y)_R \leq^{\text{den}} R_R$ by Proposition 8.3.5. So $(J \cap (I \oplus Y))_R = (I \oplus (J \cap Y))_R \leq^{\text{den}} J_R$ from the modular law. By Definition 10.1.2, $I \in \mathfrak{D}_{\mathbf{IC}}(J)$.

(ii) There exists $K \trianglelefteq R$ such that $I \cap K = 0$ and $(I \oplus K)_R \leq^{\text{den}} J_R$. So we get that $\ell_R(I \oplus K) = \ell_R(J)$ from the proof of Lemma 8.3.7(i). By Proposition 8.3.5, there is $V \trianglelefteq R$ with $J \cap V = 0$ and $(J \oplus V)_R \leq^{\text{den}} R_R$ because $J \in \mathfrak{D}_{\mathbf{IC}}(R)$. Therefore $\ell_R(J \oplus V) = 0$ by Proposition 1.3.11(iv), and so

$$\ell_R((I \oplus K) \oplus V) = \ell_R(I \oplus K) \cap \ell_R(V) = \ell_R(J) \cap \ell_R(V) = \ell_R(J \oplus V) = 0.$$

Thus $((I \oplus K) \oplus V)_R \leq^{\text{den}} R_R$ from Proposition 1.3.11(iv). Hence $I \in \mathfrak{D}_{\text{IC}}(R)$ by Proposition 8.3.5. □

Proposition 10.1.4 *Assume that R is a semiprime ring and $I \trianglelefteq R$. Then* $\text{jdim}(I) = d(I) = \text{udim}({}_R I_R)$. *In particular,* $\text{jdim}(R) = d(R) = \text{udim}({}_R R_R)$.

Proof Let $U \trianglelefteq R$ such that $U \subseteq I$. We claim that ${}_R U_R \leq^{\text{ess}} {}_R I_R$ if and only if $U_R \leq^{\text{ess}} I_R$. Clearly, $U_R \leq^{\text{ess}} I_R$ implies that ${}_R U_R \leq^{\text{ess}} {}_R I_R$. Next suppose that ${}_R U_R \leq^{\text{ess}} {}_R I_R$. Take $0 \neq K \trianglelefteq I$ and let W be the ideal of R generated by K. Then by Lemma 8.3.54, $W^3 \subseteq K$. As $K \neq 0$ and R is semiprime, $W^3 \neq 0$ and further ${}_R W^3_R \leq {}_R I_R$. So $W^3 \cap U \neq 0$ because ${}_R U_R \leq^{\text{ess}} {}_R I_R$. Thus $K \cap U \neq 0$, and hence ${}_I U_I \leq^{\text{ess}} {}_I I_I$. Since I is a semiprime ring from Proposition 8.3.55, $U_I \leq^{\text{ess}} I_I$ by modification of Proposition 1.3.16, and thus $U_R \leq^{\text{ess}} I_R$. Therefore $d(I) = \text{udim}({}_R I_R)$.

Next, we show that $U_R \leq^{\text{ess}} I_R$ if and only if $U_R \leq^{\text{den}} I_R$. For this, we observe that $U_R \leq^{\text{den}} I_R$ implies $U_R \leq^{\text{ess}} I_R$. For the converse, suppose that $U_R \leq^{\text{ess}} I_R$. As R is semiprime, $\ell_R(U) \cap U = 0$ and so $\ell_I(U) = \ell_R(U) \cap I = 0$. Take $x, y \in I$ with $y \neq 0$. Then $yU \neq 0$, hence there exists $u \in U \subseteq R$ such that $yu \neq 0$. In this case, $xu \in U \subseteq I$. Hence $U_R \leq^{\text{den}} I_R$. Therefore $\text{jdim}(I) = d(I)$. Consequently, $\text{jdim}(I) = d(I) = \text{udim}({}_R I_R)$. □

If ${}_S M_R$ is an (S, R)-bimodule and $Z(M_R) = 0$, then $\text{jdim}(M) = d(M)$ by Proposition 1.3.14.

Proposition 10.1.5 (i) *A ring R is prime if and only if R is semiprime and* $\text{jdim}(R) = 1$.

(ii) *Let R be a right Kasch ring. Then R is an indecomposable ring if and only if* $\text{jdim}(R) = 1$.

Proof (i) Let R be a prime ring. If $0 \neq I \trianglelefteq R$, then $\ell_R(I) = 0$. So $I_R \leq^{\text{den}} R_R$ by Proposition 1.3.11(iv). Thus $\text{jdim}(R) = 1$. Conversely, assume that R is semiprime and $\text{jdim}(R) = 1$. If R is not prime, then there is $0 \neq J \trianglelefteq R$ such that $\ell_R(J) \neq 0$. Since R is semiprime, $J \in \mathfrak{D}_{\text{IC}}(R)$ by Proposition 8.3.3(i), and hence we obtain that $(J \oplus \ell_R(J))_R \leq^{\text{den}} R_R$. Hence $\text{jdim}(R) \geq 2$, a contradiction.

(ii) If R is right Kasch, then R itself is the only dense right ideal of R by Proposition 1.3.18. Therefore, the result is a direct consequence of Definition 10.1.2(ii). □

The next example shows that $\text{jdim}(-)$ and $d(-)$ are distinct.

Example 10.1.6 (i) Let $T = K[x]/x^4K[x]$, where K is a field. Let $\overline{x}$ be the image of x in T and $R = K + K\overline{x}^2 + K\overline{x}^3$, which is a subring of T. Then $\text{jdim}(R) = 1$ from Proposition 10.1.5(ii) as R is an indecomposable right Kasch ring. However, $d(R) = \text{udim}(R_R) = 2$.

(ii) Assume that $R = \begin{bmatrix} \mathbb{Z}_4 & 2\mathbb{Z}_4 \\ 0 & \mathbb{Z}_4 \end{bmatrix}$. Then $\mathrm{jdim}(R) = 1$ from Proposition 10.1.5(ii) because R is an indecomposable right Kasch ring. However, we see that $d(R) = \mathrm{udim}({}_R R_R) = \mathrm{udim}(R_R) = 3$.

Proposition 10.1.7 *Let R be a ring and let $I \in \mathfrak{D}_{\mathbf{IC}}(R)$ with $\mathrm{jdim}(I) < \infty$. Then $\mathrm{jdim}(I)$ is the supremum of the set of positive integers k for which there exist nonzero $I_i \in \mathfrak{D}_{\mathbf{IC}}(I)$, $i = 1, \ldots, k$ such that $(I_1 \oplus \cdots \oplus I_k)_R \leq I_R$.*

Proof Let $\mathrm{jdim}(I) = n < \infty$. Then there exist $0 \neq V_j \trianglelefteq R$, $j = 1, \ldots, n$, with $(\oplus_{j=1}^n V_j)_R \leq^{\mathrm{den}} I_R$. By Definition 10.1.2(i), each $V_j \in \mathfrak{D}_{\mathbf{IC}}(I)$. Suppose that there exist $0 \neq I_i \in \mathfrak{D}_{\mathbf{IC}}(I)$, $1 \leq i \leq k$ such that $(I_1 \oplus \cdots \oplus I_k)_R \leq I_R$. We claim that $k \leq n$. For this, put $K = \oplus_{i=1}^k I_i$. From Proposition 10.1.3(ii), each $I_i \in \mathfrak{D}_{\mathbf{IC}}(R)$ because $I_i \in \mathfrak{D}_{\mathbf{IC}}(I)$ and $I \in \mathfrak{D}_{\mathbf{IC}}(R)$.

We see that $\ell_R(K \oplus \ell_R(K)) = \ell_R(K) \cap \ell_R(\ell_R(K)) = 0$ as $K \in \mathfrak{D}_{\mathbf{IC}}(R)$ by Theorem 8.3.13(i). From Proposition 1.3.11(iv), $(K \oplus \ell_R(K))_R \leq^{\mathrm{den}} R_R$. By the modular law, $((K \oplus \ell_R(K)) \cap I)_R = (K \oplus (\ell_R(K) \cap I))_R \leq^{\mathrm{den}} I_R$. Whence $(I_1 \oplus \cdots \oplus I_k \oplus (\ell_R(K) \cap I))_R \leq^{\mathrm{den}} I_R$, so $k \leq \mathrm{jdim}(I) = n$. Thus $\mathrm{jdim}(I)$ is the supremum as desired. □

Let $0 \neq g \in \mathcal{B}(R)$ such that $g = g_1 + \cdots + g_t$, where $\{g_i \mid 1 \leq i \leq t\}$ is a set of orthogonal centrally primitive idempotents of R. Recall from Exercise 5.2.21.1 that t is uniquely determined. *We let* $t = \mathbf{n}(gR)$. The following result is related to Theorem 8.3.8.

Theorem 10.1.8 *Let $I \in \mathfrak{D}_{\mathbf{IC}}(R)$. Then $\mathrm{jdim}(I) = n < \infty$ if and only if there is $e \in \mathcal{B}(Q(R))$ such that $I_R \leq^{\mathrm{den}} eQ(R)_R$ and $\mathbf{n}(eQ(R)) = n$.*

Proof Assume that $\mathrm{jdim}(I) = n < \infty$. Then by Definition 10.1.2, there are $0 \neq I_k \trianglelefteq R$, $1 \leq k \leq n$ such that $(\oplus_{k=1}^n I_k)_R \leq^{\mathrm{den}} I_R$. By Lemma 8.3.7(ii) and (iii), each $I_k \in \mathfrak{D}_{\mathbf{IC}}(R)$, so each $I_k \in \mathfrak{D}_{\mathbf{IC}}(I)$ by Proposition 10.1.3(i).

Let $\mathrm{jdim}(I_1) \geq 2$. Then there exist nonzero $A, B \in \mathfrak{D}_{\mathbf{IC}}(I_1)$ satisfying

$$(A \oplus B)_R \leq^{\mathrm{den}} I_{1R}.$$

As $I_1 \in \mathfrak{D}_{\mathbf{IC}}(R)$, $A \oplus B \in \mathfrak{D}_{\mathbf{IC}}(R)$ and so $A, B \in \mathfrak{D}_{\mathbf{IC}}(R)$ by Lemma 8.3.7(ii) and (iii). Hence $A, B \in \mathfrak{D}_{\mathbf{IC}}(I)$ by Proposition 10.1.3(i). Therefore

$$(A \oplus B \oplus I_2 \oplus \cdots \oplus I_n)_R \leq I_R \text{ and } A,\ B,\ I_2,\ \ldots,\ I_n \in \mathfrak{D}_{\mathbf{IC}}(I),$$

a contradiction by Proposition 10.1.7 because $\mathrm{jdim}(I) = n$. So $\mathrm{jdim}(I_1) = 1$. Similarly, $\mathrm{jdim}(I_k) = 1$ for each k.

By Theorem 8.3.8, there exists $f_k \in \mathcal{B}(Q(R))$ with $I_{kR} \leq^{\mathrm{den}} f_k Q(R)_R$ for each $k = 1, \ldots, n$. In this case, each f_k is centrally primitive. Indeed, suppose that f_k is not centrally primitive for some k. Then there are nonzero h_1, $h_2 \in \mathcal{B}(Q(R))$ such that $f_k Q(R) = h_1 Q(R) \oplus h_2 Q(R)$.

Let $J_i = I_k \cap h_i Q(R)$ for $i = 1, 2$. Then $J_{iR} \leq^{\text{den}} h_i Q(R)_R$ for $i = 1, 2$ as $I_{kR} \leq^{\text{den}} f_k Q(R)_R$. By Theorem 8.3.8, $J_i \in \mathfrak{D}_{\mathbf{IC}}(R)$. So $0 \neq J_i \in \mathfrak{D}_{\mathbf{IC}}(I_k)$ by Proposition 10.1.3(i). Now $(J_1 \oplus J_2)_R \leq I_{kR}$. From Proposition 10.1.7, we have a contradiction because $\text{jdim}(I_k) = 1$. Thus, each f_k is centrally primitive.

Note that $(\oplus_{k=1}^n I_k)_R \leq^{\text{ess}} \oplus_{k=1}^n f_k Q(R)_R = (f_1 + \cdots + f_n) Q(R)_R$. Put

$$f = f_1 + \cdots + f_n.$$

Then $\mathbf{n}(f Q(R)) = n$ since $f_1, \ldots, f_n$ are orthogonal centrally primitive idempotents. By Theorem 8.3.8, there exists $e \in \mathcal{B}(Q(R))$ with $I_R \leq^{\text{den}} e Q(R)_R$. Thus,

$$(\oplus_{k=1}^n I_k)_R \leq^{\text{ess}} I_R \leq^{\text{ess}} e Q(R)_R \text{ and } (\oplus_{k=1}^n I_k)_R \leq^{\text{ess}} f Q(R)_R.$$

Hence $f = e$, so $\mathbf{n}(e Q(R)) = \mathbf{n}(f Q(R)) = n$.

Conversely, assume that there is $e \in \mathcal{B}(Q(R))$ such that $I_R \leq^{\text{den}} e Q(R)_R$ and $\mathbf{n}(e Q(R)) = n$. Then $e = e_1 + \cdots + e_n$, where $e_1, \ldots, e_n$ are orthogonal centrally primitive idempotents in $Q(R)$. Because $(I \cap e_k Q(R))_R \leq^{\text{den}} e_k Q(R)_R$ for each k, from Theorem 8.3.8 each $I \cap e_k Q(R) \in \mathfrak{D}_{\mathbf{IC}}(R)$.

Hence, each $I \cap e_k Q(R) \in \mathfrak{D}_{\mathbf{IC}}(I)$ by Proposition 10.1.3(i). So $\text{jdim}(I) \geq n$ by Proposition 10.1.7 as $[\oplus_{k=1}^n (I \cap e_k Q(R))]_R \leq I_R$ and $I \cap e_k Q(R) \neq 0$ for each k.

Assume on the contrary that $\text{jdim}(I) > n$. By Proposition 10.1.7, there are $0 \neq V_k \in \mathfrak{D}_{\mathbf{IC}}(I), k = 1, \ldots, m$ such that $m > n$ and $(\oplus_{k=1}^m V_k)_R \leq I_R$. Note that from Proposition 10.1.3(ii), each $V_k \in \mathfrak{D}_{\mathbf{IC}}(R)$ as $V_k \in \mathfrak{D}_{\mathbf{IC}}(I)$ and $I \in \mathfrak{D}_{\mathbf{IC}}(R)$. Thus from Theorem 8.3.8, there are $g_k \in \mathcal{B}(Q(R))$ such that $V_{kR} \leq^{\text{den}} g_k Q(R)_R$, $k = 1, \ldots, m$. Put $g = g_1 + \cdots + g_m$. Then

$$(\oplus_{k=1}^m V_k)_R \leq^{\text{ess}} g Q(R)_R \text{ and } (\oplus_{k=1}^m V_k)_R \leq I_R \leq e Q(R)_R.$$

Therefore, $(\oplus_{k=1}^m V_k)_R \leq^{\text{ess}} g Q(R)_R \cap e Q(R)_R = g e Q(R)_R \leq^{\text{ess}} g Q(R)_R$. Thus $ge = g$, and hence $g Q(R) \subseteq e Q(R)$. So $m = \mathbf{n}(g Q(R)) \leq \mathbf{n}(e Q(R)) = n$, a contradiction. Therefore, $\text{jdim}(I) = n$. □

The next formula is obtained from Theorem 10.1.8 as follows.

Theorem 10.1.9 *Assume that* R *is a ring and let* $I, J \in \mathfrak{D}_{\mathbf{IC}}(R)$. *Then* $\text{jdim}(I) + \text{jdim}(J) = \text{jdim}(I + J) + \text{jdim}(I \cap J)$.

Proof By Theorem 8.3.8, there are $e, f \in \mathcal{B}(Q(R))$ with $I_R \leq^{\text{den}} e Q(R)_R$ and $J_R \leq^{\text{den}} f Q(R)_R$. So $(I + J)_R \leq^{\text{den}} g Q(R)_R$, where $g = e + f - ef$, from the proof of Theorem 8.3.13(i). Also $(I \cap J)_R \leq^{\text{den}} ef Q(R)_R$.

Let $\text{jdim}(I) = m < \infty$ and $\text{jdim}(J) = n < \infty$. From Theorem 10.1.8,

$$\text{jdim}(I) = \mathbf{n}(e Q(R)) \quad \text{and} \quad \text{jdim}(J) = \mathbf{n}(f Q(R)).$$

By the modular law, now we see that

$$e Q(R) = ef Q(R) \oplus h_1 Q(R) \text{ and } f Q(R) = ef Q(R) \oplus h_2 Q(R)$$

for some $h_1, h_2 \in \mathcal{B}(Q(R))$. As $\mathbf{n}(eQ(R))$ and $\mathbf{n}(fQ(R))$ are finite,

$$\mathbf{n}(efQ(R)),\ \mathbf{n}(h_1Q(R)),\ \text{and}\ \mathbf{n}(h_2Q(R))$$

are also finite. Hence $\text{jdim}(I \cap J) = \mathbf{n}(efQ(R))$ by Theorem 10.1.8.

Further, $h_1Q(R) \cap h_2Q(R) \subseteq eQ(R) \cap fQ(R) = efQ(R)$, thus we have that $h_1Q(R) \cap h_2Q(R) \subseteq efQ(R) \cap h_1Q(R) = 0$. Also note that $efh_1 = 0$ and $efh_2 = 0$, thus $ef(h_1 + h_2 - h_1h_2) = 0$. So

$$efQ(R) \cap (h_1Q(R) \oplus h_2Q(R)) = efQ(R) \cap (h_1 + h_2 - h_1h_2)Q(R) = 0.$$

Hence, $gQ(R) = eQ(R) + fQ(R) = efQ(R) \oplus h_1Q(R) \oplus h_2Q(R)$. Note that $\mathbf{n}(efQ(R))$, $\mathbf{n}(h_1Q(R))$, and $\mathbf{n}(h_2Q(R))$ are finite. Hence, $\mathbf{n}(gQ(R))$ is finite and $\mathbf{n}(gQ(R)) = \mathbf{n}(efQ(R)) + \mathbf{n}(h_1Q(R)) + \mathbf{n}(h_2Q(R))$. So by Theorem 10.1.8, $\text{jdim}(I + J) = \mathbf{n}(gQ(R))$. Thus $\text{jdim}(I) + \text{jdim}(J) = \text{jdim}(I + J) + \text{jdim}(I \cap J)$.

Next, assume that either $\text{jdim}(I)$ or $\text{jdim}(J)$ is infinite. Say $\text{jdim}(I)$ is infinite. Suppose that $\text{jdim}(I + J) = \ell < \infty$. As in the preceding argument, there exist $e, f \in \mathcal{B}(Q(R))$ such that

$$I_R \leq^{\text{den}} eQ(R)_R,\ J_R \leq^{\text{den}} fQ(R)_R,\ \text{and}\ (I + J)_R \leq^{\text{den}} gQ(R)_R,$$

where $g = e + f - ef$. From Theorem 10.1.8, there exist orthogonal centrally primitive idempotents $b_1, \dots, b_\ell$ in $Q(R)$ such that

$$gQ(R)_R = b_1Q(R)_R \oplus \cdots \oplus b_\ell Q(R)_R = (b_1 + \cdots + b_\ell)Q(R)_R.$$

Therefore $eQ(R)_R \leq (b_1 + \cdots + b_\ell)Q(R)_R$.

As each b_i is centrally primitive, $e = e(b_1 + \cdots + b_\ell) = b_{i_1} + \cdots + b_{i_k}$, where $\{b_{i_1}, \dots, b_{i_k}\} = \{eb_j \mid eb_j \neq 0, 1 \leq j \leq \ell\}$. Therefore, $\mathbf{n}(eQ(R)) \leq \ell$. By Theorem 10.1.8 $\text{jdim}(I) \leq \ell$, a contradiction. So $\text{jdim}(I + J)$ is infinite. □

Theorem 10.1.10 *Assume that T is a right ring of quotients of a ring R such that $\mathcal{B}(Q(R)) \subseteq T$. Let $0 \neq I \in \mathfrak{D}_{\text{IC}}(R)$ and Λ be an index set. Then:*

(i) *There exists $e \in \mathcal{B}(Q(R))$ such that $I_R \leq^{\text{den}} eT_R$.*

(ii) *If $eT = \prod_{i \in \Lambda} Q_i$, then $I_i := Q_i \cap I$ $(i \in \Lambda)$ are ideals of R such that $(\bigoplus_{i \in \Lambda} I_i)_R \leq^{\text{den}} I_R$.*

(iii) *Assume that A_i $(i \in \Lambda)$ is a set of nonzero right ideals of R such that $A_iA_j = 0$ whenever $i \neq j$, for $i, j \in \Lambda$, and $(\sum_{i \in \Lambda} A_i)_R \leq^{\text{den}} I_R$. Then:*
   (1) *$\sum_{i \in \Lambda} A_i = \bigoplus_{i \in \Lambda} A_i$ (the internal direct sum of the A_i).*
   (2) *$Q(A_i) = Q(RA_iR) = e_iQ(R)$ for some $e_i \in \mathcal{B}(Q(R))$, for each $i \in \Lambda$.*
   (3) *if Λ is finite or $T = Q(R)$, then $eT \cong \prod_{i \in \Lambda} T \cap Q(A_i)$.*

(iv) *Assume that Λ is finite or $T = Q(R)$. Then eT is a direct product of $|\Lambda|$ indecomposable rings if and only if there exist nonzero ideals I_i of R $(i \in \Lambda)$ such that $I_i \subseteq I$, $\text{jdim}(I_i) = 1$, $\sum_{i \in \Lambda} I_i = \bigoplus_{i \in \Lambda} I_i$, and $(\bigoplus_{i \in \Lambda} I_i)_R \leq^{\text{den}} I_R$.*

(v) *Assume that Λ is finite or $T = Q(R)$. Then eT is a direct product of $|\Lambda|$ prime rings if and only if there exist nonzero ideals I_i of R $(i \in \Lambda)$ such that $I_i \subseteq I$, $\text{jdim}(I_i) = 1$, $\sum_{i \in \Lambda} I_i = \bigoplus_{i \in \Lambda} I_i$, $(\bigoplus_{i \in \Lambda} I_i)_R \leq^{\text{den}} I_R$, and eT is semiprime.*

Proof (i) The proof follows as a consequence of Theorem 8.3.8.

(ii) Obviously, $I_i \trianglelefteq R$. Also we note that $(\oplus_{i\in\Lambda} Q_i)_T \leq^{\text{den}} eT_T$ because $(\oplus_{i\in\Lambda} Q_i)_{eT} \leq^{\text{den}} eT_{eT}$. We claim that $(\oplus_{i\in\Lambda} Q_i)_R \leq^{\text{den}} eT_R$. For this, take ex, $ey \in eT$ with $x, y \in T$ and $ey \neq 0$. Then, since $(\oplus_{i\in\Lambda} Q_i)_T \leq^{\text{den}} eT_T$, there is $t \in T$ with $ext \in \oplus_{i\in\Lambda} Q_i$ and $eyt \neq 0$. As $R_R \leq^{\text{den}} T_R$ and $0 \neq eyt \in T$, there is $r \in R$ such that $tr \in R$ and $eytr \neq 0$. Here $extr \in \oplus_{i\in\Lambda} Q_i$ because $ext \in \oplus_{i\in\Lambda} Q_i$. Hence, $(\oplus_{i\in\Lambda} Q_i)_R \leq^{\text{den}} eT_R$.

Next to show that $(\oplus_{i\in\Lambda} I_i)_R \leq^{\text{den}} I_R$, let $u, v \in I$ with $v \neq 0$. Then there is $r \in R$ with $ur \in \oplus_{i\in\Lambda} Q_i$ and $vr \neq 0$ since $(\oplus_{i\in\Lambda} Q_i)_R \leq^{\text{den}} eT_R$. So

$$ur = u_{k_1} + u_{k_2} + \cdots + u_{k_n},$$

where $u_{k_1} \in Q_{k_1}, u_{k_2} \in Q_{k_2}, \ldots$, and $u_{k_n} \in Q_{k_n}$. Because $I_R \leq^{\text{den}} eT_R$, there exists $r_1 \in R$ with $u_{k_1} r_1 \in Q_{k_1} \cap I = I_{k_1}$ and $vrr_1 \neq 0$. Then there is $r_2 \in R$ with $u_{k_2} r_1 r_2 \in I_{k_2}$ and $vrr_1 r_2 \neq 0$. By this process, we obtain $a = r_1 r_2 \cdots r_n$ such that $ura \in \oplus_{i\in\Lambda} I_i$ and $vra \neq 0$. Therefore $(\oplus_{i\in\Lambda} I_i)_R \leq^{\text{den}} I_R$.

(iii)(1) Because $(\sum_{i\in\Lambda} A_i)_R \leq^{\text{den}} I_R$, $(\sum_{\in\Lambda} RA_i R)_R \leq^{\text{den}} I_R$. Hence, $\sum_{i\in\Lambda} RA_i R \in \mathfrak{D}_{\mathbf{IC}}(R)$ by Lemma 8.3.7(ii) as $I \in \mathfrak{D}_{\mathbf{IC}}(R)$. Let $k \in \Lambda$ and take $x \in RA_k R \cap (\sum_{i\neq k} RA_i R)$. Then $x(\sum_{i\neq k} RA_i R) = 0$ and $x(RA_k R) = 0$. So $x(\sum_{i\in\Lambda} RA_i R) = 0$, and thus $x \in \ell_R(\sum_{i\in\Lambda} RA_i R) \cap \sum_{i\in\Lambda} RA_i R = 0$. Hence, $\sum_{i\in\Lambda} RA_i R = \oplus_{i\in\Lambda} RA_i R$. So $\sum_{i\in\Lambda} A_i = \oplus_{i\in\Lambda} A_i$.

(2) We show that $A_{kR} \leq^{\text{den}} RA_k R_R$ for each $k \in \Lambda$. Let $x, y \in RA_k R$ with $y \neq 0$. Since $(\sum_{i\in\Lambda} A_i)_R \leq^{\text{den}} I_R$, there exists $r \in R$ such that

$$xr \in \sum_{i\in\Lambda} A_i = \oplus_{i\in\Lambda} A_i \text{ and } yr \neq 0.$$

We note that $xr \in RA_k R$. As $\sum_{i\in\Lambda} RA_i R = \oplus_{\in\Lambda} RA_i R$, $(\oplus_{i\in\Lambda} A_i) \cap RA_k R = A_k$ and so $xr \in A_k$. Hence, $A_{kR} \leq^{\text{den}} RA_k R_R$.

Next, we claim that $\ell_R(A_k) = \ell_R(RA_k R)$ for each $k \in \Lambda$. Obviously, we first see that $\ell_R(RA_k R) \subseteq \ell_R(A_k)$. Assume on the contrary that there exists $a \in \ell_R(A_k)$, but $a \notin \ell(RA_k R)$. Then $ab \neq 0$ for some $b \in RA_k R$. Since $0 \neq ab \in RA_k R$ and $A_{kR} \leq^{\text{den}} RA_k R_R$, there exists $r \in R$ such that $br \in A_k$ and $abr \neq 0$.

But because $a \in \ell_R(A_k)$ and $br \in A_k$, $abr = 0$, a contradiction. As a consequence, $\ell_R(A_k) = \ell_R(RA_k R)$ for each $k \in \Lambda$.

Further, we claim that

$$A_{kA_k} \leq^{\text{den}} RA_k R_{A_k}$$

for every $k \in \Lambda$. We take $x, y \in RA_k R$ and $y \neq 0$. Since $A_{kR} \leq^{\text{den}} RA_k R_R$, there exists $r \in R$ such that $xr \in A_k$ and $yr \neq 0$. Now if $yrA_k = 0$, then $0 \neq yr \in \ell_R(A_k) = \ell_R(RA_k R)$. Because $\oplus_{i\in\Lambda} RA_k R = \sum_{i\in\Lambda} RA_i R \in \mathfrak{D}_{\mathbf{IC}}(R)$ by the preceding argument, $RA_k R \in \mathfrak{D}_{\mathbf{IC}}(R)$ from Lemma 8.3.7(iii) and so $\ell_R(RA_k R) \cap RA_k R = 0$. But $0 \neq yr \in \ell_R(RA_k R) \cap RA_k R$, a contradiction. Therefore, $yrA_k \neq 0$. Hence, $yra \neq 0$ for some $a \in A_k$. Moreover, $xra \in A_k$ since $xr \in A_k$.

So we have that $ra, yra \in RA_k R$ and $yra \neq 0$. As $A_{kR} \leq^{\text{den}} RA_k R_R$, there exists $b \in R$ such that $rab \in A_k$ and $yrab \neq 0$. Thus, $x(rab) \in A_k$ (because

$xr \in A_k$) and $y(rab) \neq 0$ with $rab \in A_k$. Therefore, $A_{k A_k} \leq^{\text{den}} RA_kR_{A_k}$ for each $k \in \Lambda$. So $Q(A_k) = Q(RA_kR)$. By Theorem 8.3.8, there exists $e_k \in \mathcal{B}(Q(R))$ with $Q(RA_kR) = e_kQ(R)$ since $RA_kR \in \mathfrak{D}_{\mathbf{IC}}(R)$. So $Q(A_k) = Q(RA_kR) = e_kQ(R)$ for each $k \in \Lambda$.

(3) Say $j \neq k$. Then $RA_jR \cap RA_kR = 0$, so $Q(RA_jR) \cap Q(RA_kR) = 0$. Thus $e_jQ(R) \cap e_kQ(R) = 0$, and hence $e_je_k = 0$ for $j \neq k$. As $\mathcal{B}(Q(R)) \subseteq T$, $H_i := T \cap Q(A_i) = T \cap Q(RA_iR) = T \cap e_iQ(R) = e_iT$ is a ring with identity e_i, for each $i \in \Lambda$. Note that $Q(RA_iR) \cap (1-e)T = 0$ as $RA_iR \cap (1-e)T = 0$. Thus $e_iT \cap (1-e)T = 0$ since $e_iT \subseteq Q(RA_iR)$. So $e_i(1-e) = 0$, hence $e_i = e_ie$ for all $i \in \Lambda$. Thus $e_iQ(R) \subseteq eQ(R)$ for all $i \in \Lambda$.

Define $h : eQ(R) \to \prod_{i\in\Lambda} e_iQ(R)$ by $h(eq) = (e_iq)_{i\in\Lambda}$. Then h is a ring homomorphism. We can check that $\ell_{eQ(R)}(\oplus_{i\in\Lambda}e_iQ(R)) = 0$ from Proposition 1.3.11(iv), because

$$\oplus_{i\in\Lambda}A_i \subseteq \oplus_{i\in\Lambda}e_iQ(R) \text{ and } (\oplus_{i\in\Lambda}A_i)_R \leq^{\text{den}} I_R \leq^{\text{den}} eQ(R)_R$$

imply that $(\oplus_{i\in\Lambda}e_iQ(R))_{eQ(R)} \leq^{\text{den}} eQ(R)_{eQ(R)}$. Take $eq \in \text{Ker}(h)$. Then we see that

$$\begin{aligned} eq &\in \cap_{i\in\Lambda}r_{eQ(R)}(e_i) = \cap_{i\in\Lambda}\ell_{eQ(R)}(e_i) = \cap_{i\in\Lambda}\ell_{eQ(R)}(e_iQ(R)) \\ &= \ell_{eQ(R)}(\oplus_{i\in\Lambda}e_iQ(R)) = 0, \end{aligned}$$

thus $\text{Ker}(h) = 0$. Therefore h is a ring monomorphism, and hence the restriction, $h|_{eT} : eT \to \prod_{i\in\Lambda} e_iT$ is a ring monomorphism.

If Λ is finite and $(e_it_i)_{i\in\Lambda} \in \prod_{i\in\Lambda} e_iT$, let $t = \sum_{i\in\Lambda} e_it_i$. Then we see that $h(et) = (e_it_i)_{i\in\Lambda}$. So $h|_{eT}$ is a ring isomorphism. Therefore, we obtain that $eT \cong \prod_{i\in\Lambda} e_iT = \prod_{i\in\Lambda} H_i$. Let $T = Q(R)$. We claim that

$$h(eQ(R))_{h(eQ(R))} \leq^{\text{den}} (\prod_{i\in\Lambda} e_iQ(R))_{h(eQ(R))}.$$

Say $(x_i)_{\in\Lambda}, (y_i)_{i\in\Lambda} \in \prod_{i\in\Lambda} e_iQ(R)$ with $(y_i)_{i\in\Lambda} \neq 0$. Then there is $j \in \Lambda$ with $y_j \neq 0$, so $(x_i)_{i\in\Lambda}h(ee_j) = (x_i)_{i\in\Lambda}(e_ie_j)_{i\in\Lambda} = h(ex_j) \in h(eQ(R))$ and $(y_i)_{i\in\Lambda}h(ee_j) = (y_i)_{i\in\Lambda}(e_ie_j)_{i\in\Lambda} \neq 0$. Hence $Q(h(eQ(R))) = Q(\prod_{i\in\Lambda} e_iQ(R))$. Because $h(eQ(R)) \cong eQ(R)$ and $Q(eQ(R)) = eQ(R)$, we have that

$$h(eQ(R)) = Q(h(eQ(R))) = Q(\prod_{i\in\Lambda} e_iQ(R)) = \prod_{i\in\Lambda} Q(e_iQ(R)) = \prod_{i\in\Lambda} e_iQ(R).$$

So h is onto, hence h is a ring isomorphism.

(iv) Let Λ be finite or $T = Q(R)$. Assume that eT is a direct product of $|\Lambda|$ indecomposable rings. Let $eT = \prod_{i\in\Lambda} e_iT$ with e_iT indecomposable and $e_i \in \mathcal{B}(e_iT)$ for each i. Then

$$e_i \in \text{Cen}(eT) = e\text{Cen}(T) \subseteq e\text{Cen}(Q(R)).$$

Hence, $e_i \in \mathcal{B}(Q(R))$ for each i. Put $I_i = e_iT \cap I$. As $e_iT_R \leq^{\text{den}} e_iQ(R)_R$ and $I_R \leq^{\text{den}} eQ(R)_R$, we have that $I_{i\,R} \leq^{\text{den}} e_iQ(R)_R$ for each i. Because each e_iT

is indecomposable and $\mathcal{B}(Q(R)) \subseteq T$, each e_i is centrally primitive. By Theorem 10.1.8, $\text{jdim}(I_i) = 1$ for each i. Also by part (ii), $(\oplus_{i\in\Lambda} I_i)_R \leq^{\text{den}} I_R$.

Conversely by the proof of part (iii), $Q(I_i) = e_i Q(R)$ with

$$e_i \in \mathcal{B}(Q(R)),\ eT \cong \prod_{i\in\Lambda} e_i T, \quad \text{and} \quad e_i T = e_i Q(R) \cap T.$$

Because $I_{iR} \leq^{\text{den}} e_i Q(R)_R$ and $\text{jdim}(I_i) = 1$, e_i is centrally primitive by Theorem 10.1.8. Therefore, each $e_i T$ is indecomposable.

(v) The necessity follows similarly as in the proof of part (iv). Conversely, by part (iv), $eT \cong \prod_{i\in\Lambda} e_i T$, where $e_i T$ is indecomposable. Because eT is semiprime, so is each $e_i T$. We may observe that $e_i T_{e_i T} \leq^{\text{den}} e_i Q(R)_{e_i T}$ since $e_i T_T \leq^{\text{den}} e_i Q(R)_T$. Also we note that $e_i Q(R) = Q(e_i T)$, so $\text{jdim}(e_i T) = 1$ by Theorem 10.1.8. From Proposition 10.1.5(i), each $e_i T$ is a prime ring. □

Let A is a ring. For $e, f \in \mathcal{B}(A)$, $e \leq f$ means that $ef = e$. Then we see that $e \leq f$ if and only if $eR \subseteq fR$ if and only if $Re \subseteq Rf$.

Corollary 10.1.11 *Let $I \in \mathfrak{D}_{\text{IC}}(R)$ with $I_R \leq^{\text{den}} eQ(R)_R$, where $e \in \mathcal{B}(Q(R))$, such that* $\text{jdim}(I) = \infty$ *and $eQ(R)$ is a semiprime ring. Then $eQ(R)$ is an infinite direct product of nonzero rings.*

Proof As $e \in \mathcal{B}(Q(R))$ and $Q(eQ(R)) = eQ(R)$ is semiprime, $\mathcal{B}(eQ(R))$ is a complete Boolean algebra by Corollary 8.3.14. There are infinitely many orthogonal idempotents in $\mathcal{B}(eQ(R))$ since otherwise there will be a set of primitive idempotents, say $g_1, \dots, g_n$ in $\mathcal{B}(eQ(R))$ such that $e = g_1 + \cdots + g_n$ from Proposition 1.2.15. As $I_R \leq^{\text{den}} eQ(R)_R$, Theorem 10.1.8 yields that $\text{jdim}\,(I) = n$, a contradiction. So we get an infinite set of nonzero orthogonal idempotents $e_1, e_2, \dots$ in $\mathcal{B}(eQ(R))$. Since $\mathcal{B}(eQ(R))$ is a complete Boolean algebra, there exists an idempotent $f \in \mathcal{B}(eQ(R))$ such that $f = \sup\{e_i\}_{i=1}^{\infty}$.

Let $e_0 = e - f \in \mathcal{B}(eQ(R))$. Then $e_0, e_1, e_2, \dots$ are orthogonal. We show that $e = \sup\{e_0, e_1, e_2, \dots\}$. For this, note that $e_i \leq e$ as $e_i \in \mathcal{B}(eQ(R))$ for $i = 0, 1, 2, \dots$. Therefore $k := \sup\{e_0, e_1, e_2, \dots\} \leq e$. Furthermore, $f \leq e$ because $f = \sup\{e_1, e_2, \dots\}$. Then $e_0 \leq k$ and $f \leq k$, so $e_0 R + fR \subseteq kR$. Now

$$e_0 R + fR = (e - f)R + fR = (e - f + f - (e - f)f)R = eR,$$

and therefore $eR \subseteq kR$. Whence $e = k$ and thus $e = \sup\{e_0, e_1, e_2, \dots\}$.

Define $h : eQ(R) \to \prod_{i=0}^{\infty} e_i Q(R)$ by $h(eq) = (e_i q)_{i=0}^{\infty}$. Then h is a ring homomorphism. We note that

$$\text{Ker}(h) = \cap_{i=0}^{\infty} r_{eQ(R)}(e_i eQ(R)) = r_{eQ(R)}\Big(\sum_{i=0}^{\infty} e_i eQ(R)\Big) = g(eQ(R))$$

for some $g \in \mathcal{B}(eQ(R)) \subseteq \mathcal{B}(Q(R))$ because $Q(eQ(R)) = eQ(R)$ is semiprime quasi-Baer (see Theorem 8.3.17 and Proposition 1.2.6(ii)). First, we show that $g = \inf\{e - e_i\}_{i=0}^{\infty}$. For this, note that $0 = h(ge) = (e_i ge)_{i=0}^{\infty} = (ge_i)_{i=0}^{\infty}$, so $g(e - e_i) = ge - ge_i = ge = g$. Therefore, $g \leq e - e_i$ for all i. Let $u \in \mathcal{B}(eQ(R))$

such that $u \leq e - e_i$ for all i. Then $u = (e - e_i)u = eu - e_iu = u - e_iu$, hence $e_iu = 0$ for all i. So $u \in \cap_{i=0}^{\infty} r_{eQ(R)}(e_ieQ(R)) = g(eQ(R))$. Thus $u \leq g$, so $g = \inf\{e - e_i\}_{i=0}^{\infty}$, and hence $g = \inf\{e - e_i\}_{i=0}^{\infty} = e - \sup\{e_i\}_{i=0}^{\infty} = e - e = 0$. Therefore $\mathrm{Ker}(h) = g(eQ(R)) = gQ(R) = 0$, and thus h is one-to-one. Next as in the proof of Theorem 10.1.10, we see that h is onto. Hence, $eQ(R)$ is an infinite direct product of nonzero rings. □

For the important case when $I = R$ in Theorem 10.1.10(iv) and (v), the next result describes the structure of $Q(R)$. The following theorem generalizes and extends Theorem 10.1.1 by removing the right nonsingularity condition.

Theorem 10.1.12 (i) *$Q(R)$ is a direct product of indecomposable rings if and only if there are ideals $\{I_i \mid i \in \Lambda\}$ of R such that* $\mathrm{jdim}(I_i) = 1$ *for all $i \in \Lambda$,*

$$\sum_{i\in\Lambda} I_i = \bigoplus_{i\in\Lambda} I_i, \quad and \quad \Big(\bigoplus_{i\in\Lambda} I_i\Big)_R \leq^{\mathrm{den}} R_R.$$

(ii) *$Q(R)$ is a direct product of prime rings if and only if there exist ideals of the ring R, $\{I_i \mid i \in \Lambda\}$ such that* $\mathrm{jdim}(I_i) = 1$ *for all $i \in \Lambda$,*

$$\sum_{i\in\Lambda} I_i = \bigoplus_{i\in\Lambda} I_i, \ \Big(\bigoplus_{i\in\Lambda} I_i\Big)_R \leq^{\mathrm{den}} R_R, \ \ and \ \ Q(R) \ \ is\ semiprime.$$

Proof Put $e = 1$ and $I = R$ in Theorem 10.1.10(iv) and (v). Then the proof follows immediately. □

Theorem 10.1.13 (i) *Let T be a right ring of quotients of a ring R such that $\mathcal{B}(Q(R)) \subseteq T$, and n be a positive integer. Then T is a direct product of n indecomposable rings if and only if* $\mathrm{jdim}(R) = n$.

(ii) *Let T be a right ring of quotients of a ring R with $\mathcal{B}(Q(R)) \subseteq T$, and n be a positive integer. Then T is a direct product of n prime rings if and only if* $\mathrm{jdim}(R) = n$ *and T is a semiprime ring.*

Proof Take $e = 1$, $I = R$, and $\Lambda = \{1, \ldots, n\}$ in Theorem 10.1.10(iv) and (v). Then the proof is obvious. □

The following example illustrates that Theorem 10.1.12 properly generalizes Theorem 10.1.1 for the case of rings R with $Z(R_R) \neq 0$.

Example 10.1.14 Let Δ be a prime ring such that $Z(\Delta_\Delta) \neq 0$ (see Example 3.2.7(i)) and $R = T_n(\Delta)$, where $n \geq 1$. Then $Q(R)$ is a prime ring, but R is not right nonsingular.

The center $\mathrm{Cen}(Q(R))$ of $Q(R)$ is called the *extended centroid* of R. If R is semiprime (not necessarily with identity), then $\ell_R(R) = 0$. In this case, $Q^m(R)$ and $Q^s(R)$ of R also can be defined as in Definition 1.3.17. We note that $\mathrm{Cen}(Q(R)) = \mathrm{Cen}(Q^m(R)) = \mathrm{Cen}(Q^s(R))$. The next well known result, due to Amitsur [7, Theorem 5], is necessary for the proof of Theorem 10.1.17.

Theorem 10.1.15 *Assume that R is a semiprime ring (not necessarily with identity). Then:*

(i) $\mathrm{Cen}(Q(R))$ *is a regular ring.*
(ii) *If R is a prime ring, then* $\mathrm{Cen}(Q(R))$ *is a field.*

Proof See [262, Proposition 14.20 and Corollary 14.22]. □

The converse of Theorem 10.1.15(ii) also holds true by Amitsur [7, Theorem 5]. See also the next remark.

Remark 10.1.16 Let R be a semiprime ring (not necessarily with identity), and let $S = \{r + n1_{Q(R)} \mid r \in R \text{ and } n \in \mathbb{Z}\}$. We observe that S is semiprime and $Q(R) = Q(S)$ because $R_R \leq^{\text{den}} S_R$. If $\mathrm{Cen}(Q(R))$ is a field, then $\mathcal{B}(Q(R))$ is $\{0, 1\}$, and so $\widehat{Q}_{\mathbf{qB}}(S) = S$ from Theorem 8.3.17. Thus S is quasi-Baer. As S is semiprime, $\mathbf{S}_\ell(S) = \mathcal{B}(S)$ by Proposition 1.2.6(ii). Hence $\mathbf{S}_\ell(S) = \{0, 1\}$, so S is semicentral reduced. Thus, S is a prime ring by Proposition 3.2.5. Proposition 8.3.55(ii) yields that R is a prime ring because $R \trianglelefteq S$.

Let A be an algebra (not necessarily with identity) over a commutative ring C with identity satisfying $\ell_A(A) = 0$. Define

$$A^1 = \{a + c1_{Q(A)} \mid a \in A \text{ and } c \in C\},$$

which is a subring of $Q(A)$. Then A^1 is an algebra over C. Note that $A \trianglelefteq A^1$ and $Q(A) = Q(A^1)$ as was noted before. A characterization for $Q(A)$ to be a direct product of prime rings is given as follows.

Theorem 10.1.17 *Let A be a ring (not necessarily with identity) which is an algebra over a commutative ring C with identity and $\ell_A(A) = 0$. Assume that $Q(A)$ is a semiprime ring and Λ is an index set. Then the following are equivalent.*

(i) $Q(A) = \prod_{i \in \Lambda} Q_i$, *where each Q_i is a prime ring.*
(ii) $\mathrm{Cen}(Q(A)) = \prod_{i \in \Lambda} F_i$, *where each F_i is a field.*
(iii) *There exist nonzero ideals I_i $(i \in \Lambda)$ of A^1 such that* $\mathrm{jdim}(I_i) = 1$ *for each $i \in \Lambda$, $\sum_{i \in \Lambda} I_i = \bigoplus_{i \in \Lambda} I_i$, and $\ell_{A^1}(\bigoplus_{i \in \Lambda} I_i) = 0$.*
(iv) *There exist nonzero ideals V_i $(i \in \Lambda)$ of A such that for each $i \in \Lambda$,* $\mathrm{jdim}(V_i) = 1$, *where for* $\mathrm{jdim}(V_i)$, *V_i is considered as an (A^1, A^1)-bimodule, $CV_i \subseteq V_i$, $\sum_{i \in \Lambda} V_i = \bigoplus_{i \in \Lambda} V_i$, and $\ell_A(\bigoplus_{i \in \Lambda} V_i) = 0$.*
(v) *There is a set of orthogonal primitive idempotents $\{e_i \mid i \in \Lambda\}$ in $\mathcal{B}(Q(A))$ with supremum* 1.

If A is semiprime, then the above conditions are equivalent to the following.

(vi) *For each $i \in \Lambda$, there is a prime ring T_i such that A is a subring of $\prod_{i \in \Lambda} T_i$ and $A_A \leq^{\text{den}} (\prod_{i \in \Lambda} T_i)_A$.*

Proof (i)⇒(ii) Since $\text{Cen}(Q(A)) = \prod_{i\in\Lambda} \text{Cen}(Q_i)$, $\text{Cen}(Q(A))$ is regular by Theorem 10.1.15. So each commutative domain $\text{Cen}(Q_i)$ is a field.

(ii)⇒(i) Assume that e_i is the identity of F_i. Then because $e_i \in \mathcal{B}(Q(A))$, $\text{Cen}(e_i Q(A)) = e_i\text{Cen}(Q(A)) = F_i$. So $e_i Q(A)$ is prime by Remark 10.1.16 as $e_i Q(A)$ is semiprime.

(i)⇔(iii) Theorem 10.1.12(ii) yields the equivalence.

(iii)⇒(iv) Put $V_i = I_i \cap A$ for $i \in \Lambda$. Then $V_i \trianglelefteq A$ and $CV_i \subseteq V_i$. Also note that $V_i \trianglelefteq A^1$ because $I_i \trianglelefteq A^1$ and $A \trianglelefteq A^1$. Let $a, b \in I_i$ with $b \neq 0$. Since $A_A \leq^{\text{den}} A^1{}_A$, there exists $r \in A$ such that $ar \in A$ and $br \neq 0$. Therefore, $ar \in A \cap I_i = V_i$, so $V_{i\,A^1} \leq^{\text{den}} I_{i\,A^1}$.

As in the proof of Lemma 8.3.7(i), $\ell_{A^1}(V_i) = \ell_{A^1}(I_i)$ for each i. So we have that $\ell_{A^1}(\oplus_{i\in\Lambda} V_i) = \cap_{i\in\Lambda}\ell_{A^1}(V_i) = \cap_{i\in\Lambda}\ell_{A^1}(I_i) = \ell_{A^1}(\oplus_{i\in\Lambda} I_i) = 0$. Therefore, $\ell_A(\oplus_{i\in\Lambda} V_i) = 0$.

Assume on the contrary that $\text{jdim}(V_i) \neq 1$. For each $0 \neq K \in \mathfrak{D}_{\mathbf{IC}}(V_i)$, if $K_{A^1} \leq^{\text{den}} V_{i\,A^1}$, then $\text{jdim}(V_i) = 1$. So there is $0 \neq U \in \mathfrak{D}_{\mathbf{IC}}(V_i)$ such that U_{A^1} is not dense in $V_{i\,A^1}$. By Definition 10.1.2, there exists a nonzero (A^1, A^1)-subbimodule W of V_i with $U \cap W = 0$ and $(U \oplus W)_{A^1} \leq^{\text{den}} V_{i\,A^1}$. Thus $W \in \mathfrak{D}_{\mathbf{IC}}(V_i)$ again by Definition 10.1.2. On the other hand, we note that $\ell_{A^1}(V_i \oplus (\oplus_{j\neq i} V_j)) = 0$, so $(V_i \oplus (\oplus_{j\neq i} V_j))_{A^1} \leq^{\text{den}} A^1{}_{A^1}$ from Proposition 1.3.11(iv). Hence $V_i \in \mathfrak{D}_{\mathbf{IC}}(A^1)$, therefore $U \in \mathfrak{D}_{\mathbf{IC}}(A^1)$ and $W \in \mathfrak{D}_{\mathbf{IC}}(A^1)$ from Lemma 8.3.7(ii) and (iii). Because $U \subseteq I_i$ and $W \subseteq I_i$, $U \in \mathfrak{D}_{\mathbf{IC}}(I_i)$ and $W \in \mathfrak{D}_{\mathbf{IC}}(I_i)$ by Proposition 10.1.3(i). This is a contradiction by Proposition 10.1.7 because $\text{jdim}\,(I_i) = 1$. Thus, $\text{jdim}(V_i) = 1$ for each i.

(iv)⇒(iii) As $CV_i \subseteq V_i$ for all i, each V_i is an ideal of A^1 and so the implication follows immediately.

(ii)⇒(v) It is straightforward.

(v)⇒(iii) By Corollary 8.3.19, $Q(A)$ is quasi-Baer since $Q(A)$ is semiprime. Hence each $e_i Q(A)$ is semiprime and quasi-Baer by Theorem 3.2.10. Note that $e_i Q(A)$ is indecomposable (as a ring), $e_i Q(A)$ is semicentral reduced by Proposition 1.2.6(ii). From Proposition 3.2.5, $e_i Q(A)$ is prime.

Let $I_i = e_i Q(A) \cap A^1 \trianglelefteq A^1$. Then $\sum_{i\in\Lambda} I_i = \oplus_{i\in\Lambda} I_i$. From Theorem 8.3.8, $I_i \in \mathfrak{D}_{\mathbf{IC}}(A^1)$ as $I_{i\,A^1} \leq^{\text{den}} e_i Q(A^1)_{A^1}$. Also by Theorem 10.1.8, $\text{jdim}(I_i) = 1$. From Lemma 8.1.6(i), $\ell_{A^1}(I_i) = \ell_{A^1}(e_i Q(A))$ for each i. Hence,

$$\begin{aligned}\ell_{A^1}(\oplus_{i\in\Lambda} I_i) &= \cap_{i\in\Lambda}\ell_{A^1}(I_i) = \cap_{i\in\Lambda}\ell_{A^1}(e_i Q(A))\\ &\subseteq \cap_{i\in\Lambda}\ell_{Q(A)}(e_i Q(A)) = \cap_{i\in\Lambda}(1-e_i)Q(A)\\ &= r_{Q(A)}(\sum_{i\in\Lambda} e_i Q(A)) = hQ(A)\end{aligned}$$

for some $h \in \mathcal{B}(Q(A))$ by Corollary 8.3.19 and Proposition 1.2.6(ii) as $Q(A)$ is semiprime quasi-Baer. So $h = \inf\{1 - e_i\}_{i\in\Lambda}$. Indeed, say $k = \inf\{1 - e_i\}_{i\in\Lambda}$. As $hQ(A) \subseteq (1 - e_i)Q(A)$ for each i, $h \leq k$. Since $k \leq 1 - e_i$ for each i,

$$kQ(A) \subseteq \cap_{i\in\Lambda}(1 - e_i)Q(A) = hQ(A),$$

$k \leq h$. Therefore, $h = k$. Now because $1 = \sup\{e_i \mid i \in \Lambda\}$, it follows that $\ell_{A^1}(\oplus_{i\in\Lambda} I_i) \subseteq hQ(A) = (\inf\{1 - e_i\}_{i\in\Lambda})Q(A) = (1 - \sup\{e_i\}_{i\in\Lambda})Q(A) = 0$.

Finally, we assume additionally that A is a semiprime ring.

(i)$\Rightarrow$(vi) There is a set of orthogonal idempotents $\{e_i \mid i \in \Lambda\} \subseteq \mathcal{B}(Q(A))$ such that $Q(A) = \prod_{i\in\Lambda} e_i Q(A)$, where each $e_i Q(A)$ is a prime ring. Thus for $a \in A$, $a = (e_i a)_{i\in\Lambda} \in \prod_{i\in\Lambda} e_i Q(A)$. Hence, A is a subring of $\prod_{i\in\Lambda} e_i A$. Now we get that $A_A \leq^{\text{den}} (\prod_{i\in\Lambda} e_i A)_A$ since $A_A \leq^{\text{den}} Q(A)_A = (\prod_{i\in\Lambda} e_i Q(A))_A$.

To show that each $e_i A$ is a prime ring, first we claim that each $e_i A$ is a semiprime ring. Let $0 \neq K \trianglelefteq e_i A$ and take $0 \neq x \in K$. Say $x = e_i a$ with $a \in A$.

Since $A_A \leq^{\text{den}} Q(A)_A$, there is $b \in A$ with $e_i b \in A$ and $xb \neq 0$. Thus

$$xb = e_i ab = (e_i a)(e_i b) = x(e_i b) \in K$$

and $xb = e_i ab = a(e_i b) \in A$ as $e_i b \in A$. Hence $0 \neq xb \in K \cap A$, so $K \cap A \neq 0$.

Further, take $y \in K \cap A$ and $r \in A$. Then $yr \in A$. Say $y = e_i \alpha$ with $\alpha \in A$. Then $yr = e_i \alpha r = (e_i \alpha)(e_i r) = y(e_i r) \in K$. Thus $yr \in K \cap A$, so $K \cap A$ is a right ideal of A. Because A is semiprime, $(K \cap A)^2 \neq 0$, and so $K^2 \neq 0$. Thus, $e_i A$ is semiprime.

Next, we prove that $e_i A$ is prime. We see that $e_i A_{e_i A} \leq^{\text{den}} e_i Q(A)_{e_i A}$ as $A_A \leq^{\text{den}} Q(A)_A$. So $Q(e_i A) = Q(e_i Q(A)) = e_i Q(A)$. Because $e_i Q(A)$ is prime, $\text{Cen}(Q(e_i Q(A))) = \text{Cen}(Q(e_i A))$ is a field by Theorem 10.1.15. Therefore, each $e_i A$ is prime from Remark 10.1.16 (because $e_i A$ is semiprime). Take $T_i = e_i A$ for each $i \in \Lambda$.

(vi)$\Rightarrow$(i) We note that $\prod_{i\in\Lambda} T_i$ is a right ring of quotients of A. Therefore, $Q(A) = Q(\prod_{i\in\Lambda} T_i) = \prod_{i\in\Lambda} Q(T_i)$. Since T_i is prime, so is $Q(T_i)$. □

In view of Theorem 10.1.17, one might conjecture that if A is semiprime with $\mathcal{B}(Q(A)) \subseteq A$ (i.e., A is a semiprime quasi-Baer ring by Theorem 8.3.17), and A satisfies any one of the conditions (i)–(vi) of Theorem 10.1.17, then A itself must be a direct product of prime rings. However this conjecture is not true, in general, as the next example shows.

Example 10.1.18 Let F be a field with a proper subfield K, set $F_n = F$ for $n = 1, 2, \ldots$, and let $A = \{(x_n)_{n=1}^{\infty} \in \prod_{n=1}^{\infty} F_n \mid x_n \in K \text{ eventually}\}$, a subring of $\prod_{n=1}^{\infty} F_n$. Then A is commutative regular, $Q(A) = \prod_{n=1}^{\infty} F_n$, and $\mathcal{B}(Q(A)) \subseteq A$. If A is a direct product of prime rings, then A is a direct product of fields because A is commutative regular. So A is self-injective, a contradiction.

An ideal I of a ring R is called a *uniform ideal* if $\text{udim}({}_R I_R) = 1$. Thus a nonzero ideal I of R is uniform if $J \cap K \neq 0$ for any nonzero ideals J and K of R with $J \subseteq I$ and $K \subseteq I$. Recall that an ideal V of a semiprime ring is said to be an annihilator ideal if $V = r_R(W)$ for some $W \trianglelefteq R$. Thus, V is an annihilator ideal if and only if $V = r_R(\ell_R(V))$.

The next result shows that some of the well known finiteness conditions on a ring yield that it has only finitely many minimal prime ideals.

Theorem 10.1.19 *Assume that A is a semiprime ring (not necessarily with identity) which is an algebra over a commutative ring C with identity. Say n is a positive* *integer and let* $A^1 = \{a + c1_{Q(A)} \mid a \in A$ *and* $c \in C\}$. *Then the following are equivalent.*

(i) $\mathrm{udim}({}_{A^1}A^1{}_{A^1}) = n$.
(ii) $\mathrm{udim}({}_{A^1}A_{A^1}) = n$.
(iii) A^1 *has exactly n minimal prime ideals.*
(iv) A *has exactly n minimal prime ideals.*
(v) $\mathrm{Cen}(Q(A))$ *has a complete set of primitive idempotents with n elements.*

Proof Put $R = A^1$. As A is semiprime, R is semiprime. For an ideal K of R (resp., A), we note that $r_R(K) = \ell_R(K)$ (resp., $r_A(K) = \ell_A(K)$) since R and A are semiprime. So without any ambiguity, in this proof, we use $\mathrm{Ann}_R(K)$ for $r_R(K)$ or $\ell_R(K)$ (resp., $\mathrm{Ann}_A(K)$ for $r_A(K)$ or $\ell_A(K)$).

(i)⇔(v) By Proposition 10.1.4, $\mathrm{jdim}(I) = \mathrm{udim}({}_RI_R)$ for $I \trianglelefteq R$. Therefore (i)⇔(v) follows from Proposition 1.3.16 and (iii)⇔(v) in Theorem 10.1.17.

(i)⇒(iii) Let $\mathrm{udim}({}_RR_R) = n$. There exist uniform ideals U_i, $1 \le i \le n$ of R with ${}_R(U_1 \oplus \cdots \oplus U_n)_R \le^{\text{ess}} {}_RR_R$. Put $P_i = \mathrm{Ann}_R(U_i)$. We claim that each P_i is a maximal annihilator ideal. For this, assume that $P_i \subseteq V$ and V is an annihilator ideal. If $V \cap U_i = 0$, then $VU_i = 0$ and so $V \subseteq \mathrm{Ann}_R(U_i) = P_i$. Thus, $V = P_i$. Next, if $V \cap U_i \neq 0$, then ${}_R(V \cap U_i)_R \le^{\text{ess}} {}_RU_{iR}$ since ${}_RU_{iR}$ is uniform. Therefore ${}_R((V \cap U_i) \oplus \mathrm{Ann}_R(U_i))_R \le^{\text{ess}} {}_R(U_i \oplus \mathrm{Ann}_R(U_i))_R \le^{\text{ess}} {}_RR_R$.

Note that ${}_R((V \cap U_i) \oplus \mathrm{Ann}_R(U_i))_R \le {}_RV_R$ as $\mathrm{Ann}_R(U_i) = P_i \subseteq V$. Hence ${}_RV_R \le^{\text{ess}} {}_RR_R$, so $\mathrm{Ann}_R(V) = 0$ by Proposition 1.3.16. As V is an annihilator ideal, $V = \mathrm{Ann}_R(\mathrm{Ann}_R(V)) = R$. Therefore, each $P_i = \mathrm{Ann}_R(U_i)$ is a maximal annihilator ideal.

To see that each P_i is a minimal prime ideal, note that $P_i = \mathrm{Ann}_R(U_i) \neq R$. Now say $I, J \trianglelefteq R$ such that $IJ \subseteq P_i$. Assume that $J \not\subseteq P_i$. Then we see that $0 \neq JU_i \subseteq U_i$, hence $\mathrm{Ann}_R(U_i) \subseteq \mathrm{Ann}_R(JU_i) \neq R$. As $P_i = \mathrm{Ann}_R(U_i)$ is a maximal annihilator ideal, $\mathrm{Ann}_R(JU_i) = \mathrm{Ann}_R(U_i)$. Now $IJU_i = 0$, and hence $IU_i = 0$. Therefore $I \subseteq \mathrm{Ann}_R(U_i) = P_i$, so P_i is a prime ideal. Next, assume that P is a prime ideal of R such that $P \subsetneq P_i$. From the fact that $0 = U_iP_i \subseteq P$, $U_i \subseteq P \subseteq P_i$. Thus, $U_i^2 \subseteq U_iP_i = 0$. As R is semiprime, $U_i = 0$, a contradiction. Hence, each P_i is a minimal prime ideal of R.

Further, all P_i are distinct. Indeed, assume that $P_i = P_j$, where $i \neq j$. Then as $U_i \cap U_j = 0$, $U_iU_j = 0$, so $U_i \subseteq \mathrm{Ann}_R(U_j) = P_j = P_i$ and hence $U_i^2 \subseteq U_iP_i = 0$. So $U_i = 0$, a contradiction.

Finally, $P_1 \cap \cdots \cap P_n = \mathrm{Ann}_R(U_1 \oplus \cdots \oplus U_n) = 0$ from Proposition 1.3.16. If P is a minimal prime ideal of R, then $P = P_k$ for some k, $1 \le k \le n$ because $0 = P_1 \cap \cdots \cap P_n \subseteq P$. Therefore, R has exactly n minimal prime ideals, which are precisely $P_1, \ldots, P_n$.

(iii)⇒(i) To show that $\mathrm{udim}({}_RR_R) = n$, let $\{P_1, \ldots, P_n\}$ be the set of all minimal prime ideals of R. If $n = 1$, then $P_1 = 0$ and hence R is a prime ring. Thus, $\mathrm{udim}({}_RR_R) = 1$. So we assume that $n \ge 2$.

Put $U_i = \mathrm{Ann}_R(P_i)$ for $i = 1, \ldots, n$. We show that each U_i is a uniform ideal. Since $P_1 \cap P_2 \cap \cdots \cap P_n = 0$, $P_1P_2 \cdots P_n = 0$, and so $P_2 \cdots P_n \subseteq \mathrm{Ann}_R(P_1) = U_1$. If $U_1 = 0$, then $P_2 \cdots P_n = 0$ and hence $P_2 \cdots P_n \subseteq P_1$.

Thus $P_k = P_1$ for some $k \neq 1$. Therefore $U_1 \neq 0$. Say A and B are nonzero ideals of R such that $A, B \subseteq U_1$. If $A \cap B = 0$, then $AB = 0$. Hence $A \subseteq P_1$ or $B \subseteq P_1$. If $A \subseteq P_1$, then $A \subseteq P_1 \cap U_1 = 0$, so $A = 0$. If $B \subseteq P_1$, then $B = 0$ as $B \subseteq P_1 \cap U_1 = 0$. Thus U_1 is a uniform ideal. Similarly, all U_i are uniform ideals.

Next, we claim that $\mathrm{Ann}_R(\mathrm{Ann}_R(P_i)) = P_i$. Since $U_i = \mathrm{Ann}_R(P_i)$ is a uniform ideal, $\mathrm{Ann}_R(U_i) = \mathrm{Ann}_R(\mathrm{Ann}_R(P_i))$ is a minimal prime ideal by the argument used in the proof of (i)$\Rightarrow$(iii). But since $P_i \subseteq \mathrm{Ann}_R(\mathrm{Ann}_R(P_i))$, we have that $P_i = \mathrm{Ann}_R(\mathrm{Ann}_R(P_i))$.

We show that $U_i \cap U_j = 0$ for $i \neq j$. For this, suppose that $U_i \cap U_j \neq 0$ for some $i \neq j$. Then $U_i \cap U_j$ is also a uniform ideal because U_i is a uniform ideal. So as in the proof of (i)$\Rightarrow$(iii), $\mathrm{Ann}_R(U_i \cap U_j)$ is a minimal prime ideal. But because

$$P_i = \mathrm{Ann}_R(\mathrm{Ann}_R(P_i)) = \mathrm{Ann}_R(U_i) \subseteq \mathrm{Ann}_R(U_i \cap U_j),$$

$P_i = \mathrm{Ann}_R(U_i \cap U_j)$. Similarly, $P_j = \mathrm{Ann}_R(U_i \cap U_j)$. Hence $P_i = P_j$, a contradiction. Thus $U_i \cap U_j = 0$ for $i \neq j$. So $U_i U_j = 0$ for $i \neq j$. Therefore

$$\sum_{i=1}^{n} U_i = \oplus_{i=1}^{n} U_i.$$

Now $\mathrm{Ann}_R(\oplus_{i=1}^{n} U_i) = \cap_{i=1}^{n} \mathrm{Ann}_R(U_i) = \cap_{i=1}^{n} P_i = 0$, so ${}_R(U_1 \oplus \cdots \oplus U_n)_R$ is essential in ${}_R R_R$ by Proposition 1.3.16. Thus $\mathrm{udim}({}_R R_R) = n$.

(i)$\Leftrightarrow$(ii) From Proposition 1.3.16, ${}_R A_R \leq^{\mathrm{ess}} {}_R R_R$. Therefore, we have that $\mathrm{udim}({}_R R_R) = \mathrm{udim}({}_R A_R)$.

(ii)$\Leftrightarrow$(iv) We first show that $\mathrm{udim}({}_A A_A) = \mathrm{udim}({}_R A_R)$. Then it follows that $\mathrm{udim}({}_A A_A) = \mathrm{udim}({}_R R_R)$ as $\mathrm{udim}({}_R R_R) = \mathrm{udim}({}_R A_R)$ from (i)$\Leftrightarrow$(ii).

Let $\{W_i\}_{i \in \Lambda}$ be nonzero ideals of R with ${}_R(\oplus_{i \in \Lambda} W_i)_R \leq {}_R R_R$. We now take $U_i = W_i \cap A$. Then $U_i \neq 0$ since $A_R \leq^{\mathrm{ess}} R_R$. As ${}_A(\oplus_{i \in \Lambda} U_i)_A \leq {}_A A_A$, we have that $\mathrm{udim}({}_R A_R) = \mathrm{udim}({}_R R_R) \leq \mathrm{udim}({}_A A_A)$.

Next, let $\{V_i\}_{i \in \Omega}$ be nonzero ideals of A such that ${}_A(\oplus_{i \in \Omega} V_i)_A \leq {}_A A_A$. We show that $V_i R V_j = 0$ for $i \neq j$. For this, say $r = a + c 1_{Q(A)} \in R$ with $a \in A$ and $c \in C$. Then it follows that

$$V_i r V_j = V_i (a + c 1_{Q(A)}) V_j \subseteq V_i a V_j + V_i c 1_{Q(A)} V_j \subseteq V_i V_j + c V_i V_j = 0$$

as $V_i V_j = 0$. So $V_i R V_j = 0$. Put $W_i = R V_i R$ for each i. Then we see that each W_i is an ideal of R and $W_i W_j = 0$ for $i \neq j$ since $V_i R V_j = 0$. Therefore,

$$\sum_{i \in \Omega} W_i = \oplus_{i \in \Omega} W_i \quad \text{and} \quad {}_R(\oplus_{i \in \Omega} W_i)_R \leq {}_R R_R.$$

So $\mathrm{udim}({}_A A_A) \leq \mathrm{udim}({}_R R_R) = \mathrm{udim}({}_R A_R)$. Thus $\mathrm{udim}({}_R A_R) = \mathrm{udim}({}_A A_A)$.

As $\mathrm{udim}({}_R R_R) = \mathrm{udim}({}_A A_A)$, the proof of (ii)$\Leftrightarrow$(iv) is similar to that of (i)$\Leftrightarrow$(iii). □

The structure theorem for $\widehat{Q}_{\mathbf{qB}}(R)$ when R is a semiprime ring with only finitely many minimal prime ideals is obtained as follows. It will be used in Sects. 10.2 and 10.3 (see Theorems 10.2.21 and 10.3.41).

We recall that MinSpec $(-)$ denotes the set of all minimal prime ideals of a ring.

Theorem 10.1.20 *Let R be a ring. Then the following are equivalent.*

(i) *R is semiprime and has exactly n minimal prime ideals.*
(ii) *$\widehat{Q}_{\mathbf{qB}}(R) = R\mathcal{B}(Q(R))$ is a direct sum of n prime rings.*
(iii) *$\widehat{Q}_{\mathbf{qB}}(R) = R\mathcal{B}(Q(R)) \cong R/P_1 \oplus \cdots \oplus R/P_n$, where each P_i is a minimal prime ideal of R.*

Proof (i)$\Rightarrow$(ii) Let R be semiprime with exactly n minimal prime ideals $P_1, \ldots, P_n$. From Theorem 8.3.17, $\widehat{Q}_{\mathbf{qB}}(R) = R\mathcal{B}(Q(R))$. Proposition 10.1.4 and Theorem 10.1.19 yield that $\operatorname{jdim}(R) = d(R) = \operatorname{udim}({}_R R_R) = n$ since R is a semiprime ring. Because $R\mathcal{B}(Q(R))$ is semiprime, we see from Theorem 10.1.13(ii) that $R\mathcal{B}(Q(R)) = \oplus_{i=1}^n S_i$ with each S_i a prime ring.

(ii)$\Rightarrow$(i) Say $R\mathcal{B}(Q(R)) = \oplus_{i=1}^n S_i$, where each S_i is a prime ring. Then $R\mathcal{B}(Q(R))$ is semiprime and has exactly n minimal prime ideals K_i, where $K_i = \oplus_{j\neq i} S_j$. From Theorem 10.1.19, $\operatorname{Cen}(Q(R)) = \operatorname{Cen}(Q(R\mathcal{B}(Q(R))))$ has a complete set of primitive idempotents with n elements. Since R is semiprime, Theorem 10.1.19 yields that R has also exactly n minimal prime ideals.

(ii)$\Rightarrow$(iii) Assume that $R\mathcal{B}(Q(R)) = \oplus_{i=1}^n S_i$, where each S_i is a prime ring. So $R\mathcal{B}(Q(R))$ has exactly n minimal prime ideals K_i, where $K_i = \oplus_{j\neq i} S_j$. Note that $|\operatorname{MinSpec}(R)| = n = |\operatorname{MinSpec}(R\mathcal{B}(Q(R)))|$ from (i)$\Leftrightarrow$(ii). For each $P_i \in \operatorname{MinSpec}(R)$, $1 \le i \le n$, we can choose $K \in \operatorname{MinSpec}(R\mathcal{B}(Q(R)))$ such that $P_i = K \cap R$ from Lemma 8.3.26(ii). Define

$$\lambda : \operatorname{MinSpec}(R) \to \operatorname{MinSpec}(R\mathcal{B}(Q(R)))$$

by $\lambda(P_i) = K$. As $|\operatorname{MinSpec}(R)| = n = |\operatorname{MinSpec}(R\mathcal{B}(Q(R)))|$, λ is a one-to-one correspondence. So, without loss of generality, we may put $K = K_i$. Observe that $R\mathcal{B}(Q(R))/K_i \cong R/(K_i \cap R) = R/P_i$ by Lemma 8.3.26(i). Therefore, $R\mathcal{B}(Q(R)) = \oplus_{i=1}^n S_i \cong \oplus_{i=1}^n (R\mathcal{B}(Q(R))/K_i) \cong \oplus_{i=1}^n (R/P_i)$.

(iii)$\Rightarrow$(i) Assume that $R\mathcal{B}(Q(R)) \cong \oplus_{i=1}^n (R/P_i)$, where each P_i is a minimal prime ideal of R. From the proof of (ii)$\Rightarrow$(i), R is semiprime and has exactly n minimal prime ideals which are consequently $\{P_1, \ldots, P_n\}$. □

Corollary 10.1.21 *Let R be a ring with only finitely many minimal prime ideals $P_1, \ldots, P_n$. Then $\widehat{Q}_{\mathbf{qB}}(R/P(R)) \cong R/P_1 \oplus \cdots \oplus R/P_n$.*

We have the next result in which $\widehat{Q}_{\mathbf{pqB}}(R)$ does coincide with $\widehat{Q}_{\mathbf{qB}}(R)$.

Theorem 10.1.22 *Assume that R is a semiprime ring with only finitely many minimal prime ideals, say $P_1, \ldots, P_n$. Then $\widehat{Q}_{\mathbf{pqB}}(R) = \widehat{Q}_{\mathbf{qB}}(R)$ and thus, it follows that $\widehat{Q}_{\mathbf{pqB}}(R) \cong R/P_1 \oplus \cdots \oplus R/P_n$.*

Proof As R has exactly n minimal prime ideals, $\mathrm{Cen}(Q(R))$ has a complete set of primitive idempotents with n elements by Theorem 10.1.19. The extended centroid of R is equal to that of $\widehat{Q}_{\mathbf{pqB}}(R)$. As $\widehat{Q}_{\mathbf{pqB}}(R)$ is semiprime, $\widehat{Q}_{\mathbf{pqB}}(R)$ also has exactly n minimal prime ideals from Theorem 10.1.19. By Proposition 5.4.5 and Theorem 5.4.20, $\widehat{Q}_{\mathbf{pqB}}(R)$ is quasi-Baer, hence $\widehat{Q}_{\mathbf{pqB}}(R) = \widehat{Q}_{\mathbf{qB}}(R)$. So the proof follows from Theorem 10.1.20. □

Exercise 10.1.23

1. Let R and S be two rings, and ${}_SM_R$ be an (S,R)-bimodule. Show that $\mathrm{jdim}(M) \leq d(M) \leq \mathrm{udim}({}_SM_R) \leq \mathrm{udim}(M_R)$.
2. ([96, Birkenmeier, Park, and Rizvi]) Let T be a right ring of quotients of R such that $\mathcal{B}(Q(R)) \subseteq T$. Prove that the following are equivalent.
   (i) T is semiprime.
   (ii) $R \cap TKT \in \mathfrak{D}_{\mathbf{IC}}(R)$ for all $K \trianglelefteq R$.
   (iii) For each $K \trianglelefteq R$, there is $c \in \mathcal{B}(T)$ such that $TKT_R \leq^{\mathrm{den}} cT_R$.
3. ([96, Birkenmeier, Park, and Rizvi]) Assume that R is a ring. Show that $Q(R)$ is a direct product of prime rings if and only if there exist ideals $\{I_i \mid i \in \Lambda\}$ of R such that $(\oplus_{i\in\Lambda} I_i)_R \leq^{\mathrm{den}} R_R$, $\mathrm{jdim}(I_i) = 1$ for all $i \in \Lambda$, and $R \cap Q(R)KQ(R) \in \mathfrak{D}_{\mathbf{IC}}(R)$ for all $K \trianglelefteq R$.
4. ([96, Birkenmeier, Park, and Rizvi]) Let T be a right ring of quotients of a ring R such that $\mathcal{B}(Q(R)) \subseteq T$ and let n be a positive integer. Prove that T is a direct product of n prime rings if and only if $\mathrm{jdim}(R) = n$ and $R \cap TKT \in \mathfrak{D}_{\mathbf{IC}}(R)$ for all $K \lhd R$.
5. ([40, Beidar]) Let R be a semiprime ring. Prove that $\mathrm{Cen}(Q(R))$ is self-injective (hence $\mathrm{Cen}(Q(R))$ is regular self-injective by Theorem 10.1.15).
6. Let R be a semiprime ring and T a right ring of quotients of R such that $\widehat{Q}_{\mathbf{qB}}(R) \subseteq T$. Show that the following are equivalent.
   (i) R has exactly n minimal prime ideals.
   (ii) $\mathrm{Tdim}\,(T) = n$.
   (iii) T is a direct sum of n prime rings.
7. Let $Q(R)$ be semiprime. Show that $\mathrm{jdim}(R) = \mathrm{Tdim}(Q(R))$.
8. Prove that $\mathrm{jdim}(R) = \mathrm{jdim}(T)$ for any right ring of quotients of R.

10.2 Rings with Involution

We study conditions for a $*$-ring to be a Baer $*$-ring or a quasi-Baer $*$-ring. The concept of Baer $*$-rings is naturally motivated in the study of Functional Analysis. For example, every von Neumann algebra is a Baer $*$-algebra. Characterizations of Baer $*$-rings and quasi-Baer $*$-rings in terms of $*$-ideal structures and semiproper involutions are provided.

Using these results, it is shown that the quasi-Baer $*$-ring property can be transferred to polynomial ring and formal power series ring extensions without any additional requirements. Further, the existence of quasi-Baer $*$-ring hulls for rings with a semiproper involution is presented. A nonempty subset X of a $*$-ring R is called *self-adjoint* if $X^* = X$. Self-adjoint ideals of quasi-Baer $*$-rings are studied. A criterion for $\widehat{Q}_{\mathbf{qB}}(R)$ of a semiprime $*$-ring R with only finitely many minimal prime ideals to be a quasi-Baer $*$-ring is discussed.

Definition 10.2.1 Assume that R is a ring. Then a map $* : R \to R$ defined by $x \mapsto x^*$ is called an *involution* if it satisfies the following conditions for all $x, y \in R$: (i) $(x+y)^* = x^* + y^*$. (ii) $(xy)^* = y^* x^*$. (iii) $(x^*)^* = x$.

A ring R with an involution $*$ is called a *$*$-ring*. An idempotent e of a $*$-ring R is called a *projection* if $e^* = e$.

We remark that an involution map is one-to-one and onto.

Definition 10.2.2 (i) A $*$-ring R is called a *Baer $*$-ring* if the right annihilator of every nonempty subset of R is generated by a projection.

(ii) A $*$-ring R is called a *quasi-Baer $*$-ring* if the right annihilator of every ideal of R is generated by a projection.

The next result indicates that the Baer $*$-ring property is left-right symmetric similar to the case of Baer rings.

Proposition 10.2.3 *The following are equivalent for a $*$-ring R.*

(i) *R is a Baer $*$-ring*
(ii) *The left annihilator of every nonempty subset of R is generated by a projection.*

Proof For (i)⇒(ii), let $\emptyset \neq X \subseteq R$. Then $r_R(X^*) = (\ell_R(X))^*$. There exists a projection e of R with $r_R(X^*) = eR$. So $\ell_R(X) = Re$. (ii)⇒(i) follows similarly. □

In the next result for a characterization of a quasi-Baer $*$-ring, we see that the definition of a quasi-Baer $*$-ring is also left-right symmetric.

Proposition 10.2.4 *The following are equivalent for a $*$-ring R.*

(i) *R is a quasi-Baer $*$-ring.*
(ii) *R is a quasi-Baer ring in which every left semicentral idempotent is a projection.*
(iii) *R is a semiprime quasi-Baer ring in which every central idempotent is a projection.*
(iv) *The left annihilator of every ideal is generated by a projection.*

Proof (i)⇒(ii) Let R be a quasi-Baer $*$-ring. Clearly R is a quasi-Baer ring. Take $e \in \mathbf{S}_\ell(R)$. Then by Proposition 1.2.2, $1 - e \in \mathbf{S}_r(R)$ and $R(1-e) \trianglelefteq R$. Further $r_R(R(1-e)) = eR$. Also $r_R(R(1-e)) = fR$ for some projection $f \in R$. Since $fR \trianglelefteq R$, $f \in \mathbf{S}_\ell(R)$ from Proposition 1.2.2. As f is a projection, $(fR)^* = Rf \trianglelefteq R$. Again by Proposition 1.2.2, $f \in \mathbf{S}_r(R)$. So $f \in \mathcal{B}(R)$ from Proposition 1.2.6(i). Since $r_R(R(1-e)) = eR = fR$ and $f \in \mathcal{B}(R)$, $e = f$ and thus e is a projection.

(ii)⇒(iii) As every central idempotent is left semicentral, it suffices to show that R is semiprime. Say $x \in R$ with $xRx = 0$. Then $x \in r_R(xR) = eR$ with $e \in \mathbf{S}_\ell(R)$. So $e^* = e$ by assumption, and hence $(eR)^* = Re^* = Re \trianglelefteq R$. Thus $e \in \mathbf{S}_r(R)$. By Proposition 1.2.6(i), $e \in \mathcal{B}(R)$. So $x = ex = xe = 0$, hence R is semiprime.

(iii)⇒(iv) Let $I \trianglelefteq R$. Since R is quasi-Baer, there exists $e \in \mathbf{S}_r(R)$ with $\ell_R(I) = Re$. So e is central by Proposition 1.2.6(ii) because R is semiprime. Hence, e is a projection by assumption.

(iv)⇒(i) Let $J \trianglelefteq R$. Note that $(r_R(J))^* = \ell_R(J^*)$. There is a projection e with $\ell_R(J^*) = Re$. Hence, $r_R(J) = eR$, so R is a quasi-Baer $*$-ring. □

Corollary 10.2.5 *If R is a prime ring with an involution $*$, then R is a quasi-Baer $*$-ring.*

Proof Proposition 10.2.4 yields the result as a prime ring is quasi-Baer. □

Example 10.2.6 There exists a semiprime quasi-Baer ring R with an involution $*$ which is not a quasi-Baer $*$-ring. Take $R = F \oplus F$, where F is a field and $*$ is defined by $(a, b)^* = (b, a)$ for all $a, b \in F$. Then $(1, 0)$ is a central idempotent, but it is not a projection, so R is not a quasi-Baer $*$-ring by Proposition 10.2.4.

The next result provides a certain class of quasi-Baer $*$-group algebras.

Proposition 10.2.7 *Let $F[G]$ be the group algebra of a group G over a field F. Consider an involution $*$ on $F[G]$ defined by $(\sum a_g g)^* = \sum a_g g^{-1}$, where $a_g \in F$ and $g \in G$. If the following conditions are satisfied, then $F[G]$ is a quasi-Baer $*$-ring.*

(1) *$F[G]$ is semiprime.*
(2) *each annihilator ideal of $F[G]$ is finitely generated.*
(3) *for each finite normal subgroup N of G, $x^2 = 1$ for all $x \in N$.*

Proof From Theorem 6.3.2, $F[G]$ is semiprime quasi-Baer. Say $e = \sum a_s s$ be a nonzero central idempotent of $F[G]$, where $a_s \in F$ and $s \in G$. Let N be the subgroup of G generated by the support of e. Then [341, Theorem 3.8, p. 136] yields that N is a finite normal subgroup of G. Since $s = s^{-1}$, $e = e^*$. By Proposition 10.2.4, $F[G]$ is a quasi-Baer $*$-group algebra. □

Definition 10.2.8 (i) An involution $*$ of a ring R is called a *proper* involution if, for any $a \in R$, $aa^* = 0$ implies $a = 0$.

(ii) We say that an involution $*$ of a ring R is a *semiproper* involution if, for any $a \in R$, $aRa^* = 0$ implies $a = 0$.

We see that a semiproper involution on a reduced ring is a proper involution. Obviously, if $*$ is a proper involution, then $*$ is a semiproper involution. The converse is not true as shown in the following example.

Example 10.2.9 Let n be an integer such that $n > 1$ and p a prime integer with $p \le n$. Consider the prime ring $R = \mathrm{Mat}_n(\mathbb{Z}_p)$ with transpose involution $*$. Let $e_{ij} \in R$ be the matrix with 1 in the (i, j)-position and 0 elsewhere. Say $\alpha = [a_{ij}] \in R$ with $\alpha R \alpha^* = 0$. Then $(e_{ii}\alpha e_{jj})(e_{jj}\alpha^* e_{ii}) = 0$ for all i, j, so $a_{ij}^2 = 0$. Hence $a_{ij} = 0$ for all i, j, thus $\alpha = 0$. So $*$ is a semiproper involution. Next take

$$\beta = e_{11} + e_{12} + \cdots + e_{1p}.$$

Then $\beta\beta^* = 0$, but $\beta \ne 0$. Therefore $*$ is not a proper involution on R.

Lemma 10.2.10 (i) *If R is a Baer $*$-ring, then $*$ is proper.*
(ii) *If R is a quasi-Baer $*$-ring, then $*$ is semiproper.*

Proof (i) Let R be a Baer $*$-ring. Say $aa^* = 0$ with $a \in R$. Since R is a Baer $*$-ring, $r_R(a) = eR$ for some projection e in R, so $ea^* = a^*$ because $a^* \in r_R(a) = eR$. Thus $a = ae = 0$, so $*$ is proper.

(ii) Let R be a quasi-Baer $*$-ring. Say $aRa^* = 0$. Then $(RaR)(RaR)^* = 0$ and hence $(RaR)^* \subseteq r_R(RaR)$. As R is a quasi-Baer $*$-ring, R is semiprime and so there exists a central projection e such that $r_R(RaR) = eR$ by Propositions 10.2.4 and 1.2.6(ii). Thus, $eR = (eR)^* = (r_R(RaR))^* = \ell_R(Ra^*R)$. Hence, we obtain $(RaR)^* \subseteq r_R(RaR) = eR = \ell_R(Ra^*R)$, so $Ra^*R \subseteq \ell_R(Ra^*R)$. Since R is semiprime, $Ra^*R = 0$. Hence, $a^* = 0$, so $a = 0$. Thus, $*$ is semiproper. □

The next example exhibits a Baer ring with a proper involution $*$ which is not a Baer $*$-ring.

Example 10.2.11 Let $R = \mathrm{Mat}_2(\mathbb{Z})$ with the transpose involution $*$. By Theorem 6.1.4, R is a Baer ring. We note that $*$ is a proper involution. Let $e_{ij} \in R$ be the matrix with 1 in the (i, j)-position and 0 elsewhere. Then we see that $r_R(e_{11} + 2e_{12})$ cannot be generated by a projection of R. So R is not a Baer $*$-ring.

Baer $*$-rings are characterized in terms of proper involutions as follows.

Proposition 10.2.12 *The following are equivalent for a $*$-ring R.*

(i) *R is a Baer $*$-ring.*
(ii) *$*$ is a proper involution and the right annihilator of every nonempty self-adjoint subset is generated by a projection.*
(iii) *$*$ is a proper involution and the left annihilator of every nonempty self-adjoint subset is generated by a projection.*

Proof (i)⇒(ii) By Lemma 10.2.10, $*$ is proper. The remainder of the proof follows from the definition of a Baer $*$-ring.

(ii)⇒(i) Let $\emptyset \ne X \subseteq R$. Let $Y = \{x^*x \mid x \in X\}$. Then there is a projection e with $r_R(Y) = eR$. Let $x \in X$. Then $x^*xe = 0$, thus $(xe)^*(xe) = ex^*xe = 0$. Because $*$ is a proper involution, $xe = 0$. So $r_R(Y) = eR \subseteq r_R(X)$.

Also we see that $r_R(X) \subseteq r_R(Y) = eR$. Thus $r_R(X) = eR$, and so R is a Baer $*$-ring.

(i)⇔(iii) The proof is analogous to the proof of (i)⇔(ii). □

Lemma 10.2.13 (i) *If a ring R has a semiproper involution $*$, then R is semiprime and every central idempotent is a projection.*

(ii) *If a ring R has a proper involution $*$, then R is right (and left) nonsingular.*

Proof (i) Let $xRx = 0$ with $x \in R$. Then $xrx^*Rxr^*x^* = (xrx^*)R(xrx^*)^* = 0$ for all $r \in R$. Hence $xRx^* = 0$, so $x = 0$. Thus R is semiprime. Let $e \in \mathcal{B}(R)$. Then $(e^*e - e)R(e^*e - e)^* = R(e^*e - e)(e^*e - e)^* = 0$. Hence we have that $e = e^*e$, so $e^* = e^*e = e$.

(ii) Assume on the contrary that $Z(R_R) \neq 0$. Take $0 \neq x \in Z(R_R)$. Then $r_R(x)_R \leq^{\text{ess}} R_R$. So there exists $s \in R$ with $0 \neq x^*s \in r_R(x)$. Hence $(s^*x)(s^*x)^* = s^*xx^*s = 0$. But as $*$ is proper, $s^*x = 0$ and so $x^*s = 0$, a contradiction. Thus, $Z(R_R) = 0$. Similarly, R is left nonsingular. □

Using previous results, in the following, another characterization of quasi-Baer $*$-rings is given in terms of semiproper involution.

Proposition 10.2.14 *The following are equivalent for a $*$-ring R.*

(i) *R is a quasi-Baer $*$-ring.*
(ii) *$*$ is a semiproper involution and the right annihilator of every self-adjoint ideal is generated by a projection.*
(iii) *$*$ is a semiproper involution and the left annihilator of every self-adjoint ideal is generated by a projection.*
(iv) *R is a quasi-Baer ring and $*$ is a semiproper involution.*

Thereby the center of a quasi-Baer $*$-ring is a Baer $*$-ring.

Proof (i)⇒(iii) Lemma 10.2.10(ii) yields that $*$ is semiproper. The remainder of the proof follows from the definition of a quasi-Baer $*$-ring.

(iii)⇒(i) From Lemma 10.2.13(i), R is semiprime. Take $0 \neq I \trianglelefteq R$. It is clear that $\ell_R(I) \subseteq \ell_R(II^*)$. Say $x \in \ell_R(II^*)$ and $a \in I$. Then we have that $(xa)R(xa)^* = xaRa^*x^* \subseteq xII^* = 0$, hence $xa = 0$. Therefore, $x \in \ell_R(I)$. Thus, $\ell_R(II^*) = \ell_R(I)$. Since II^* is self-adjoint, $\ell_R(I)$ is generated by a projection.

(i)⇔(ii) The proof is analogous to the proof of (i)⇔(iii).

(i)⇒(iv) It follows from Lemma 10.2.10(ii).

(iv)⇒(i) By Lemma 10.2.13(i), R is semiprime and every central idempotent is a projection. Thus from Proposition 10.2.4, R is a quasi-Baer $*$-ring.

Let $C := \text{Cen}(R)$. From Theorem 3.2.13, C is a Baer ring. Also $*$ is a proper involution on C. Let $\emptyset \neq Y \subseteq C$. Then $r_C(Y) = fC$ for some $f^2 = f \in C$. By Lemma 10.2.13(i), f is a projection. So C is a Baer $*$-ring. □

A $*$-ring R is called a *Rickart $*$-ring* if the right annihilator of every element of R is generated by a projection. From the proof of Lemma 10.2.10(i), if R is a Rickart

$*$-ring, then the involution $*$ is a proper involution. As in the case of Baer $*$-rings and quasi-Baer $*$-rings (see Propositions 10.2.3 and 10.2.4), the left-right symmetry property also holds for Rickart $*$-rings.

Proposition 10.2.15 *The following are equivalent for a $*$-ring R.*

(i) *R is a Rickart $*$-ring.*
(ii) *The left annihilator of each element of R is generated by a projection.*

Proof (i)$\Rightarrow$(ii) Take $x \in R$. Then $r_R(x^*) = eR$ for some projection $e \in R$. Then we see that $\ell_R(x) = Re$.

(ii)$\Rightarrow$(i) It follows similarly as in the proof of (i)$\Rightarrow$(ii). □

For projections e and f in a $*$-ring R, we write $e \leq f$ if $e = ef$ (therefore $ef = fe = e$). The relation $e \leq f$ is a partial ordering of projections.

We see that $e \leq f$ if and only if $eR \subseteq fR$ if and only if $Re \subseteq Rf$. Also note that $e = f$ if and only if $eR = fR$ if and only if $Re = Rf$.

Baer $*$-rings and Rickart $*$-rings are connected to each other as follows.

Proposition 10.2.16 *The following are equivalent.*

(i) *R is a Baer $*$-ring.*
(ii) *R is a Rickart $*$-ring whose projections form a complete lattice.*
(iii) *R is a Rickart $*$-ring in which every orthogonal set of projections has a supremum.*

Proof See [45, Proposition 4.1]. □

A $*$-ring R is called *p.q.-Baer $*$-ring* if the right annihilator of every principal ideal of R is generated by a projection. For some examples of p.q.-Baer $*$-rings, see Exercises 10.2.24.4 and 10.2.24.5.

When R is a $*$-ring, the involution $*$ can be naturally extended to an involution on $R[x, x^{-1}]$, $R[[x, x^{-1}]]$, $R[X]$, and $R[[X]]$, where X is a nonempty set of commuting indeterminates. One might expect that $R[x, x^{-1}]$, $R[[x, x^{-1}]]$, $R[X]$, and $R[[X]]$ are Baer $*$-rings when R is a Baer $*$-ring. However, the next example eliminates some of these expectations.

Example 10.2.17 Let $R = \mathrm{Mat}_2(\mathbb{C})$ and $S = R[x]$. Then S is a Baer ring by Theorem 6.1.4. We see that R is a Baer $*$-ring, where $*$ denotes the conjugate transpose involution. It can be checked that the right annihilator

$$r_S\left(\begin{bmatrix} 0 & 2 \\ 0 & 0 \end{bmatrix} + \begin{bmatrix} 1 & 0 \\ 0 & 0 \end{bmatrix} x\right)$$

is not generated by a projection of S. Thus S is not a Baer $*$-ring.

In contrast to Example 10.2.17, $\text{Mat}_2(\mathbb{C})[[x]]$ is a Baer $*$-ring (see Exercise 10.2.24.2). But $\text{Mat}_2(\mathbb{C})[[x]][[y]]$ is not a Baer $*$-ring (see [77, Example 2.3]). Thereby, the Baer $*$-ring property may not transfer to polynomial rings or formal power series rings in general. However, the quasi-Baer $*$-ring property transfers to polynomial or formal power series ring extensions without additional requirements. The following result shows this and also provides examples of quasi-Baer $*$-rings which are not Baer $*$-rings.

Theorem 10.2.18 *Let R be a $*$-ring and X a nonempty set of commuting indeterminates. Then the following are equivalent.*

(i) *R is a quasi-Baer $*$-ring.*
(ii) *$R[X]$ is a quasi-Baer $*$-ring.*
(iii) *$R[[X]]$ is a quasi-Baer $*$-ring.*
(iv) *$R[x, x^{-1}]$ is a quasi-Baer $*$-ring.*
(v) *$R[[x, x^{-1}]]$ is a quasi-Baer $*$-ring.*

Proof We prove the equivalence (i)⇔(v). The other equivalences follows similarly. Let R be a quasi-Baer $*$-ring. Say $T = R[[x, x^{-1}]]$. Since R is semiprime from Proposition 10.2.4, so is T. Let $K \trianglelefteq T$. As T is quasi-Baer by Theorem 6.2.4, $r_T(K) = e(x)T$ for some $e(x) \in \mathbf{S}_\ell(T)$. But since T is semiprime, $e(x)$ is central from Proposition 1.2.6(ii). So $e(x) = e_0$, where e_0 is the constant term of $e(x)$. By Lemma 10.2.13(i), e_0 is a projection in R, so it is a projection on T. Thus, T is a quasi-Baer $*$-ring.

Conversely, assume that $T = R[[x, x^{-1}]]$ is a quasi-Baer $*$-ring. Let I be a right ideal of R. Then $r_T(IT) = e(x)T$ with $e(x) \in \mathbf{S}_\ell(T)$. By Propositions 10.2.4 and 1.2.6(ii), $e(x)$ is central, so $e(x) = e_0 \in R$, where e_0 is the constant term of $e(x)$. So $r_R(I) = r_T(IT) \cap R = e_0T \cap R = e_0R$. By Lemma 10.2.13(i), e_0 is a projection in R. Hence, R is a quasi-Baer $*$-ring. □

In the Baer $*$-ring $\text{Mat}_n(\mathbb{C}[[x]])$, where $*$ is conjugate transpose, we see that every ideal is self-adjoint. However, the next example shows that, in general, not every ideal of a Baer $*$-ring is self-adjoint.

Example 10.2.19 There exists a commutative Baer $*$-ring R which has an ideal I such that I is not self-adjoint. Let $*$ be the involution on $\mathbb{C}[x]$ induced by the conjugate involution on $\mathbb{C}$. Take $\alpha \in \mathbb{C}$ which is not a real. Then we see that the ideal $I = (x - \alpha)\mathbb{C}[x]$ is not self-adjoint.

In Theorem 10.2.20 next, we show that every ideal in a quasi-Baer $*$-ring is an essential extension of a self-adjoint ideal and is also essential in a self-adjoint ideal.

Theorem 10.2.20 *Let R be a quasi-Baer $*$-ring.*

(i) *If I is an ideal of R, then $\ell_R(I + I^*) = \ell_R(I) = \ell_R(II^*) = eR$, where e is a central projection.*

(ii) *If I is an ideal of R, then there is a central projection $f \in R$ such that*

$$II^*_R \leq^{\text{ess}} I_R \leq^{\text{ess}} (I+I^*)_R \leq^{\text{ess}} fR_R,$$

and

$$_RII^* \leq^{\text{ess}} {}_RI \leq^{\text{ess}} {}_R(I+I^*) \leq^{\text{ess}} {}_RfR.$$

Proof (i) We note that R is semiprime from Proposition 10.2.4. Also $*$ is semiproper by Lemma 10.2.10(ii). If $I = 0$, then we are done. So assume that $I \neq 0$. By Propositions 1.2.6(ii) and 10.2.4, there exists a central projection $e \in R$ such that $\ell_R(I + I^*) = eR$. Then by the modular law, we obtain that $\ell_R(I) = eR \oplus J$, where $J = (1-e)R \cap \ell_R(I)$. Thus $J \subseteq \ell_R(I)$, so $JI = 0$ and $IJ = 0$ as R is semiprime. Hence $(J \cap J^*)I = 0$ and also $(J \cap J^*)I^* \subseteq J^*I^* = (IJ)^* = 0$. So it follows that $(J \cap J^*)(I + I^*) = 0$. Thus $J \cap J^* \subseteq eR \cap J = 0$, and so $JJ^* = 0$.

Take $y \in J$. Then $yRy^* \subseteq JJ^* = 0$. Since $*$ is semiproper by Lemma 10.2.10(ii), $y = 0$, so $J = 0$. Thus $\ell_R(I) = eR$ and hence $\ell_R(I + I^*) = \ell_R(I)$. Next, from the proof of (iii)⇒(i) of Proposition 10.2.14, $\ell_R(II^*) = \ell_R(I)$.

(ii) Say $0 \neq a \in I$. Then there exists $x \in R$ with $axa^* \neq 0$ by Lemma 10.2.10(ii). Note that $0 \neq a(xa^*) \in II^*$. So $II^*_R \leq^{\text{ess}} I_R$. As R is semiprime quasi-Baer, by Theorem 3.2.37 there exists $f \in \mathcal{B}(R)$ such that $I_R \leq^{\text{ess}} fR_R$. From Proposition 10.2.4, f is a projection.

Because fR is a self-adjoint ideal of R and $I_R \leq^{\text{ess}} fR_R$, $I^* \subseteq fR$ and $(I + I^*)_R \leq^{\text{ess}} fR_R$. Hence $II^*_R \leq^{\text{ess}} I_R \leq^{\text{ess}} (I + I^*)_R \leq^{\text{ess}} fR_R$. Similarly, $_RII^* \leq^{\text{ess}} {}_RI \leq^{\text{ess}} {}_R(I + I^*) \leq^{\text{ess}} {}_RfR$. □

Let R be a semiprime ring with an involution $*$. It is well known that an involution $*$ of any semiprime ring R extends to $Q^s(R)$ (see Exercise 10.2.24.3). Because $\widehat{Q}_{\mathbf{qB}}(R) = R\mathcal{B}(Q(R)) \subseteq Q^s(R)$ and $\mathcal{B}(Q(R)) = \mathcal{B}(Q^s(R))$, $*$ extends to $\widehat{Q}_{\mathbf{qB}}(R)$. By Corollary 10.2.5, if R is a prime ring with an involution $*$, then R is a quasi-Baer $*$-ring. However, there exists a semiprime quasi-Baer ring with an involution $*$, but it is not a quasi-Baer $*$-ring (see Example 10.2.6). In the following theorem, we discuss a criterion for a semiprime $*$-ring R with only finitely many minimal prime ideals to ensure that $\widehat{Q}_{\mathbf{qB}}(R)$ is a quasi-Baer $*$-ring.

Theorem 10.2.21 *Let R be a semiprime $*$-ring with only finitely many minimal prime ideals. Then the following are equivalent.*

(i) *$\widehat{Q}_{\mathbf{qB}}(R)$ is a quasi-Baer $*$-ring.*
(ii) *Every minimal prime ideal of R is self-adjoint.*

Proof (i)⇒(ii) Let $\widehat{Q}_{\mathbf{qB}}(R)$ be a quasi-Baer $*$-ring. Say $P_1, \ldots, P_n$ are all the minimal prime ideals of R. Then as in the proof of Theorem 10.1.20, there are minimal prime ideals $K_1, \ldots, K_n$ of $\widehat{Q}_{\mathbf{qB}}(R)$ such that $K_i \cap R = P_i$ and $K_i = e_i\widehat{Q}_{\mathbf{qB}}(R)$ with $e_i \in \mathcal{B}(\widehat{Q}_{\mathbf{qB}}(R))$. From Lemma 10.2.4, e_i is a projection, so each K_i is self-adjoint. Hence, $P_i = e_i\widehat{Q}_{\mathbf{qB}}(R) \cap R$ is self-adjoint for each i.

(ii)⇒(i) By Theorem 10.1.20, $\widehat{Q}_{\mathbf{qB}}(R) = \oplus_{i=1}^{n} S_i$, where each S_i is a prime ring. Let $K_i = \oplus_{j\neq i} S_j$ and $P_i = K_i \cap R$ for each i. Then $P_1, \ldots, P_n$ are all the minimal prime ideals of R. As all P_i are self-adjoint by assumption, $K_i^* \cap R = P_i^* = P_i$ for all i. In this case, $K_i, K_i^* \in \text{MinSpec}(\widehat{Q}_{\mathbf{qB}}(R))$ by Lemma 8.3.26, hence $K_i = K_i^*$ for each i by the proof of Theorem 10.1.20 because $K_i \cap R = P_i$ and $K_i^* \cap R = P_i$.

We show that $S_i^* = S_i$ for each i. Indeed, since $S_1 = K_2 \cap K_3 \cap \cdots \cap K_n$, we see that $S_1^* = S_1$. Similarly, $S_i^* = S_i$ for each i. Hence, each S_i is a quasi-Baer $*$-ring by Corollary 10.2.5. Therefore, $\widehat{Q}_{\mathbf{qB}}(R)$ is a quasi-Baer $*$-ring. □

Proposition 10.2.22 *Let R be a $*$-ring and T a right ring of quotients of R such that $*$ extends to T.*

(i) *If $*$ is semiproper on R, then $*$ is semiproper on T.*
(ii) *$*$ is proper on R if and only if $*$ is proper on T.*

Proof (i) If there exists $0 \neq t \in T$ such that $tTt^* = 0$, then there is $x \in R$ with $0 \neq tx \in R$, so $txR(tx)^* = t(xRx^*)t^* = 0$. Hence $tx = 0$, a contradiction, so $*$ is a semiproper involution on T.

(ii) Let $0 \neq t \in T$ such that $tt^* = 0$. Put $s = t^*$. Then $s^*s = 0$. If $s \neq 0$, then there is $x \in R$ such that $0 \neq sx \in R$. So $(sx)^*(sx) = x^*(s^*s)x = 0$, and hence $(sx)^*((sx)^*)^* = 0$. Thus $(sx)^* = 0$, so $sx = 0$, a contradiction. Hence $s = 0$, so $t = 0$. Thus, $*$ is proper on T. The converse is obvious. □

Theorem 10.2.23 *Assume that R is a ring with a semiproper involution $*$ and T is a right ring of quotients of R. If $*$ extends to T, then the following are equivalent.*

(i) *T is a quasi-Baer $*$-ring.*
(ii) *$\widehat{Q}_{\mathbf{qB}}(R)$ is a subring of T.*
(iii) *$\mathcal{B}(Q(R)) \subseteq T$.*

Thus, $Q^s(R)$ is a quasi-Baer $*$-ring. Also $\widehat{Q}_{\mathbf{qB}}(R)$ is the quasi-Baer $*$-ring absolute to $Q(R)$ ring hull of R. If R is reduced, then $Q_{\mathbf{qB}}(R)$ is the Baer $*$-ring absolute ring hull of R.

Proof By Lemma 10.2.13(i), R is semiprime, hence (i)⇒(ii)⇒(iii) are consequences of Theorem 8.3.17. For (iii)⇒(i), note that by Proposition 10.2.22, $*$ is semiproper on T. By Theorem 8.3.17 and Proposition 10.2.14, T is a quasi-Baer $*$-ring. Hence $Q^s(R)$ and $\widehat{Q}_{\mathbf{qB}}(R)$ are quasi-Baer $*$-rings. The rest of the proof follows from the preceding arguments and Theorem 8.3.32. □

Exercise 10.2.24

1. ([77, Birkenmeier, Kim, and Park]) Assume that R is a $*$-ring and n is a positive integer. Define an involution $\star$ on $\text{Mat}_n(R)$ by $[a_{ij}]^\star = [a_{ji}^*]$. Prove that R is quasi-Baer $*$-ring if and only if $\text{Mat}_n(R)$ is a quasi-Baer $\star$-ring.
2. ([77, Birkenmeier, Kim, and Park]) Let F be a field with an involution such that $1 + aa^*$ is invertible for every $a \in F$. Let $*$ be the induced transpose involution

on $R = \mathrm{Mat}_2(F[[x]])$, that is, $[a_{ij}]^* = [a_{ji}^*]$ for $[a_{ij}] \in R$. Show that R is a Baer $*$-ring, but the formal power series ring $R[[y]]$ is not a Baer $*$-ring.
3. Prove that an involution $*$ on a semiprime ring R can be uniquely extended to $Q^s(R)$.
4. Let A be a domain, $A_n = A$ for all $n = 1, 2, \ldots$, and B be the ring of $(a_n)_{n=1}^{\infty} \in \prod_{n=1}^{\infty} A_n$ such that a_n is eventually constant, which is a subring of $\prod_{n=1}^{\infty} A_n$. Take $R = \mathrm{Mat}_n(B)$, where n is an integer such that $n > 1$. Let $*$ be the transpose involution of R. Show that R is a p.q.-Baer $*$-ring that is not quasi-Baer. Also, show that if A is commutative which is not Prüfer, then R is not a Rickart $*$-ring.
5. Let R be a $*$-ring. Prove that R is a p.q.-Baer $*$-ring if and only if R is a right (or left) p.q.-Baer ring and $*$ is semiproper. Hence if R is biregular and $*$ is semiproper, then R is a p.q.-Baer $*$-ring.
6. Let F be a field with $\mathrm{char}(F) \neq 2$, where $\mathrm{char}(F)$ is the characteristic of F, and $G = D_\infty \times C_2$. Prove that $F[G]$ is a quasi-Baer $*$-ring, where $*$ is defined as in Proposition 10.2.7.
7. Show that R is a prime ring with involution $*$ if and only if R is an indecomposable quasi-Baer $*$-ring.

10.3 Boundedly Centrally Closed C^*-Algebras

We consider *C^*-algebras which do not necessarily have identity* (i.e., are not unital C^*-algebras, in general). We note that a boundedly centrally closed C^*-algebra and the bounded central closure of a C^*-algebra are the C^*-algebra analogues of a centrally closed subring and the central closure of a semiprime ring, respectively.

In this section, boundedly centrally closed C^*-algebras are discussed by using results developed in previous chapters and sections. The results on boundedly centrally closed C^*-algebras are applied to study the extended centroid of a C^*-algebra. It is shown that a unital boundedly centrally closed C^*-algebra is precisely a C^*-algebra which is quasi-Baer. Also, boundedly centrally closed intermediate C^*-algebras between A and the local multiplier algebra $M_{\mathrm{loc}}(A)$ are investigated. When $\aleph$ is a cardinal number, the class of C^*-algebras with the extended centroid $\mathbb{C}^{\aleph}$ is characterized. Furthermore, C^*-algebras satisfying a polynomial identity which have only finitely many minimal prime ideals are described.

A *Banach algebra* is a complex normed algebra A which is complete (as a topological space) such that $||ab|| \leq ||a||\,||b||$ for all $a, b \in A$. A *Banach $*$-algebra* is a complex Banach algebra A with an involution $*$ satisfying that $(\lambda a)^* = \overline{\lambda} a^*$ for $\lambda \in \mathbb{C}$ and $a \in A$, where $\overline{\lambda}$ is the conjugate of λ.

Definition 10.3.1 A *C^*-algebra* A is a Banach $*$-algebra with the additional norm condition that $||a^*a|| = ||a||^2$ for all $a \in A$.

We observe that if A is a C^*-algebra, then

$$||a||^2 = ||a^*a|| \leq ||a^*||\,||a||$$

for all $a \in A$. Thus, $||a|| \leq ||a^*|| \leq ||a^{**}|| = ||a||$ for all $a \in A$. So $||a|| = ||a^*||$ for each $a \in A$. An involution $*$ on a C^*-algebra is a proper involution. By applying Lemma 10.2.13 to the case of C^*-algebras (not necessarily unital), we see that every C^*-algebra is semiprime and right (and left) nonsingular.

An operator (linear transformation) T from a normed vector space V (over $\mathbb{C}$) to V is called bounded if there is a real number $k \geq 0$ such that

$$||T(x)|| \leq k||x||$$

for all $x \in V$. The smallest such k is the norm of T, denoted $||T||$. In this case, we note that $||T|| = \sup_{||x||=1} ||T(x)||$. An operator is continuous if and only if it is bounded. The set of all bounded operators from V to V is denoted $B(V)$. Then $B(V)$ is closed under addition and scalar multiplication. Indeed,

$$||S+T|| \leq ||S|| + ||T|| \text{ and } ||\alpha T|| = |\alpha| \; ||T||$$

for $S, T \in B(V)$ and $\alpha \in \mathbb{C}$. Note that $||ST|| \leq ||S|| \, ||T||$ for $S, T \in B(V)$.

Let R and S be $*$-algebras. An algebra isomorphism g from R to S is called a *$*$-isomorphism* if $g(a^*) = g(a)^*$ for all $a \in R$. When $R = S$, such g is called a *$*$-automorphism*.

Example 10.3.2 (i) An inner product space (over $\mathbb{C}$) which is complete with respect to the induced norm is called a Hilbert space. Let H be a Hilbert space. Then each $T \in B(H)$ has an adjoint, denoted T^*, in $B(H)$ satisfying $\langle x, T(y)\rangle = \langle T^*(x), y\rangle$ for all $x, y \in H$, where $\langle \; , \; \rangle$ is the inner product on H. Adjoints have the following properties for all $S, T \in B(H)$ and $\alpha \in \mathbb{C}$:

(1) $(S+T)^* = S^* + T^*$, $(T^*)^* = T$, and $(\alpha T)^* = \overline{\alpha}\, T^*$.
(2) $||T^*|| = ||T||$ and $||T^*T|| = ||T||^2$.

Thus, $B(H)$ is a C^*-algebra. Any C^*-algebra is isometrically $*$-isomorphic to a C^*-subalgebra of $B(H)$ for some Hilbert space H.

(ii) Let X be a locally compact Hausdorff space, and $C_\infty(X)$ be the set of complex valued continuous functions on X vanishing at infinity. The usual pointwise operations and supremum norm are given on $C_\infty(X)$. Define an involution by $f^*(x) = \overline{f(x)}$ for $x \in X$, where $\overline{f(x)}$ is the conjugate of $f(x) \in \mathbb{C}$. Then $C_\infty(X)$ is a commutative C^*-algebra. Also $C_\infty(X)$ is unital if and only if X is compact.

For a subset X of a C^*-algebra, the *norm closure* (i.e., the topological closure) of X is denoted by $\overline{X}$. An ideal I of a C^*-algebra is said to be *norm closed* if $I = \overline{I}$. When I is a norm closed ideal of a C^*-algebra A, it is well known that I is self-adjoint (see [136, Lemma I.5.1]). If I is a norm closed ideal of C^*-algebra A, the involution on A/I is defined as $(a+I)^* = a^* + I$ for $a \in A$; and the norm on A/I is defined as $||a + I|| = \inf_{x \in I} ||a - x||$. Then A/I is a C^*-algebra (see [136, Theorem I.5.4] and [140, Proposition 1.8.2] for further details).

An involution $*$ on a $*$-algebra is called *positive-definite* if for any finite subset $\{x_i\}_{i=1}^n$, the relation $\sum_{i=1}^n x_i^* x_i = 0$ implies that $x_i = 0$ for all i. The involution on a C^*-algebra is positive-definite.

Say R is a semiprime ring with an involution $*$. Then $*$ can be extended to an involution $*$ on $Q^s(R)$ (Exercise 10.2.24.3). If $*$ is positive-definite on R, then the extended involution $*$ on $Q^s(R)$ is also positive-definite. Indeed, say $\sum_{i=1}^n x_i^* x_i = 0$ for $x_i \in Q^s(R)$ and $i = 1, \ldots, n$. Then there exists an essential ideal I of R such that $x_i I + I x_i \subseteq R$ for all i. Therefore, for each $y \in I$, $\sum_{i=1}^n (x_i y)^* (x_i y) = 0$. Thus, $x_i y = 0$ for all i. Hence $x_i I = 0$, and so $x_i = 0$ for all i.

Let Q be a unital complex $*$-algebra for which $*$ is positive-definite. An element $x \in Q$ is said to be *bounded* if $x^* x \leq n1$ for some positive integer n. This is equivalent to the existence of a finite subset $\{y_i\}$ of Q such that

$$x^* x + \sum y_i^* y = n1.$$

The set of all bounded elements of Q is denoted by Q_b. Then Q_b is a $*$-subalgebra of Q.

Let A be a C^*-algebra. We denote by $Q_b(A)$ the set of all bounded elements of $Q^s(A)$. The norm $\|x\|$ of $x \in Q_b(A)$ is defined to be $\sqrt{\lambda}$ for the least nonnegative real number λ such that $x^* x \leq \lambda 1$.

Definition 10.3.3 Let A be a C^*-algebra and let I be a norm closed essential ideal of A. We put $M(I) = \{q \in Q^s(A) \mid qI + Iq \subseteq I\}$.

If I is a norm closed essential ideal of a C^*-algebra A, then I is also a C^*-algebra. Also in this case $M(I)$ is a C^*-algebra (see [17, Proposition 2.1.3] and [109]). Further, $M(I)$ is the largest C^*-algebra in which I is contained as a norm closed essential ideal (see [147]).

Let $\mathbf{I}_{ce}$ denote the set of all norm closed essential ideals of a C^*-algebra A. Then for $I, J \in \mathbf{I}_{ce}$, it is clear that $I \cap J \in \mathbf{I}_{ce}$. Further, if $I \supseteq J$, then $M(I) \subseteq M(J)$. Indeed, let $q \in M(I)$. Then $q \in Q^s(A)$ and $qI + Iq \subseteq I$. So $qJ \subseteq I$ and $Jq \subseteq I$. It is well known that $J^2 = J$ as J is a norm closed ideal (see [104, II.5.1.4(ii)]). Thus $qJ = qJ^2 \subseteq IJ \subseteq J$. Similarly, $Jq \subseteq J$. Hence $qJ + Jq \subseteq J$. Therefore $q \in M(J)$ and so $M(I) \subseteq M(J)$. Hence, $\{M(I)\}_{I \in \mathbf{I}_{ce}}$ is ordered by inverse inclusion.

We let $\text{alg.lim}_{I \in \mathbf{I}_{ce}} M(I)$ denote the algebraic direct limit of $\{M(I)\}_{I \in \mathbf{I}_{ce}}$. For every C^*-algebra A, there exists a unique isometric $*$-isomorphism from $\text{alg.lim}_{I \in \mathbf{I}_{ce}} M(I)$ onto $Q_b(A)$ (see [17, Proposition 2.2.2]).

Henceforth, $\text{alg.lim}_{I \in \mathbf{I}_{ce}} M(I)$ is identified with $Q_b(A)$ and $Q_b(A)$ is called the *bounded symmetric algebra of quotients* of A.

Definition 10.3.4 The (topological) completion $M_{\text{loc}}(A)$ of $Q_b(A)$ is called the *local multiplier algebra* of A. Thus, $M_{\text{loc}}(A)$ is the norm closure of $Q_b(A)$.

The local multiplier algebra $M_{\text{loc}}(A)$ was first used by Elliott in [147] and Pedersen in [343] to show the innerness of certain $*$-automorphisms and derivations. Its properties have been extensively studied in [17].

In this section, when A is a C^*-algebra, we let

$$A^1 = \{a + \lambda 1_{Q(A)} \mid a \in A \text{ and } \lambda \in \mathbb{C}\}.$$

We note that A^1 is a C^*-algebra. If A is unital, then $A^1 = A$. Also observe that A is a norm closed essential ideal of A^1. Therefore, $Q_b(A) = Q_b(A^1)$ and $M_{\text{loc}}(A) = M_{\text{loc}}(A^1)$.

Say A is a C^*-algebra. Projections e and f in A are said to be *(Murray–von Neumann) equivalent*, written $e \sim f$ if there exists $w \in A$ such that $e = w^*w$ and $f = ww^*$. The relation $\sim$ is an equivalence relation in the set of projections of A.

Lemma 10.3.5 (i) *Let e and f be projections in a C^*-algebra such that $||e - f|| < 1$. Then $e \sim f$.*

(ii) *Let A be a C^*-algebra. Then for any projection $e \in M_{\text{loc}}(A)$, there exists a projection $f \in Q_b(A)$ such that $||e - f|| < 1$ (hence $e \sim f$).*

Proof (i) See [136, Proposition IV.1.2].

(ii) Note that $M_{\text{loc}}(A) = \overline{Q_b(A)}$, $Q_b(A) = \text{alg.lim}_{I \in \mathbf{I}_{ce}} M(I)$, and $M(I)$ is a C^*-algebra for $I \in \mathbf{I}_{ce}$. So [136, Lemma III.3.1] yields part (ii). □

Proposition 10.3.6 *Let A be a C^*-algebra. Then:*

(i) *Every central idempotent of A is a projection.*
(ii) *Every idempotent in the extended centroid of a C^*-algebra is a projection, and bounded.*
(iii) $\text{Cen}(M_{\text{loc}}(A))$ *and* $\text{Cen}(Q_b(A))$ *contain the same projections.*

Proof (i) It follows from Lemma 10.2.13(i) since $*$ is a proper involution.

(ii) Let $e \in \mathcal{B}(Q(A))$. Then $e \in \mathcal{B}(Q^s(A))$. Note that the induced involution $*$ on $Q^s(A)$ is positive-definite. As $(e - ee^*)^*(e - ee^*) = 0$, $e - ee^* = 0$ and so $e = ee^*$. Also $e^* = ee^*$. Hence, $e = e^*$. Since $e^*e + (1 - e)^*(1 - e) = 1$, e is a bounded element in $Q^s(A)$, and so $e \in Q_b(A)$.

(iii) Clearly, projections of $\text{Cen}(Q_b(A))$ are in $\text{Cen}(M_{\text{loc}}(A))$. Let e be a projection in $\text{Cen}(M_{\text{loc}}(A))$. Then there exists a projection $f \in Q_b(A)$ such that $||e - f|| < 1$ by Lemma 10.3.5(ii). Thus from Lemma 10.3.5(i), $e \sim f$. So there exists $w \in M_{\text{loc}}(A)$ such that $e = w^*w$ and $f = ww^*$.

Since $e \in \text{Cen}(M_{\text{loc}}(A))$, $fe = ww^*e = wew^* = (ww^*)(ww^*) = f^2 = f$. Thus $f = fe = ef$, and hence $e - f$ is a projection. Therefore,

$$||e - f|| = ||(e - f)(e - f)^*|| = ||e - f||^2,$$

so $||e - f|| = 0$ or $||e - f|| = 1$.

But since $||e - f|| < 1$, $||e - f|| = 0$ and so $e - f = 0$. Thus $e = f \in Q_b(A)$. □

When A is a C^*-algebra, $\widehat{Q}_{\mathbf{qB}}(A^1) = Q_{\mathbf{qB}}(A^1)$ as A^1 is right nonsingular by Lemma 10.2.13(ii). Recall that $Q_{\mathbf{qB}}(A) = \langle A \cup \mathcal{B}(Q(A))\rangle_{Q(A)}$ by Theorem 8.3.17. Since $A^1 = \{a + \lambda 1_{Q(R)} \mid a \in A \text{ and } \lambda \in \mathbb{C}\}$, $Q_{\mathbf{qB}}(A) \subsetneq Q_{\mathbf{qB}}(A^1)$ in general. For example, let A be a nonunital prime C^*-algebra. We assume that $\lambda \in \mathbb{C} \setminus \mathbb{Z}$. Then $\lambda 1_{Q(A)} \in A^1 \subseteq Q_{\mathbf{qB}}(A^1)$. If $\lambda 1_{Q(A)} \in Q_{\mathbf{qB}}(A)$, then $\lambda 1_{Q(A)} = a + n 1_{Q(A)}$ for some

$a \in A$ and $n \in \mathbb{Z}$ because $\mathcal{B}(Q(A)) = \{0, 1_{Q(A)}\}$. So $(\lambda - n)1_{Q(A)} = a$, thus we have that $1_{Q(A)} = (\lambda - n)^{-1}a \in A$, a contradiction.

Theorem 10.3.7 *Let A be a C^*-algebra. Then:*

(i) *$Q_{\mathbf{qB}}(A^1)$ is a $*$-subalgebra of $Q_b(A)$.*
(ii) *$Q_b(A)$ is a quasi-Baer $*$-algebra.*

Proof (i) Note that $Q_b(A) = Q_b(A^1)$. From Theorem 10.2.23, $Q_{\mathbf{qB}}(A^1)$ is a $*$-algebra. By Proposition 10.3.6(ii), $\mathcal{B}(Q(A)) \subseteq \text{Cen}(Q_b(A))$. So $Q_{\mathbf{qB}}(A^1)$ is a $*$-subalgebra of $Q_b(A)$ as $Q_{\mathbf{qB}}(A^1) = A^1\mathcal{B}(Q(A))$ by Theorem 8.3.17.

(ii) Theorems 8.3.17 and 10.2.23 yield that $Q_b(A)$ is a quasi-Baer $*$-algebra since $Q_b(A) \subseteq Q^s(A)$ and $\mathcal{B}(Q(A)) \subseteq Q_b(A)$. □

From Theorem 10.3.7, if A is a C^*-algebra, then $Q_{\mathbf{qB}}(A^1)$ is a $*$-subalgebra of $M_{\text{loc}}(A)$ because $Q_b(A)$ is a $*$-subalgebra of $M_{\text{loc}}(A)$.

Definition 10.3.8 (i) A C^*-algebra is called an *AW^*-algebra* if it is a Baer $*$-ring. For example, the commutative C^*-algebra $C(X)$ of continuous functions on a Stonian space X is an AW^*-algebra. Also, the C^*-algebra $B(H)$ of all bounded operators on a Hilbert space H is an AW^*-algebra.

(ii) In analogy with an AW^*-algebra, we say that a unital C^*-algebra A is a *quasi-AW^*-algebra* if it is also a quasi-Baer $*$-ring. Thus by Proposition 10.2.14, a unital C^*-algebra A is a quasi-AW^*-algebra if A is quasi-Baer.

Lemma 10.3.9 *Let A be a C^*-algebra and J a norm closed ideal of $M_{\text{loc}}(A)$. Then $J = \overline{J \cap Q_b(A)}$.*

Proof See [17, Lemma 1.2.32] for the proof. □

The next result shows that the local multiplier algebra of any C^*-algebra is a quasi-Baer ring.

Theorem 10.3.10 *$M_{\text{loc}}(A)$ is a quasi-AW^*-algebra for any C^*-algebra A.*

Proof Let $M = M_{\text{loc}}(A)$ and $J \trianglelefteq M$. We claim that $r_M(J)$ is generated by an idempotent of M. For this, note that $r_M(J) = r_M(\overline{J})$. Thus, we may assume that J is a norm closed ideal of M. Put $J_0 = J \cap Q_b(A)$. We show that

$$r_M(J) \cap Q_b(A) = r_{Q_b(A)}(J_0).$$

Obviously, $r_M(J) \cap Q_b(A) \subseteq r_{Q_b(A)}(J_0)$. Next, we take $x \in r_{Q_b(A)}(J_0)$. Consequently, $(J \cap Q_b(A))x = 0$, hence $Jx = \overline{(J \cap Q_b(A))}\,x = 0$ by Lemma 10.3.9 as J is norm closed. Thus, $x \in r_M(J) \cap Q_b(A)$. So $r_M(J) \cap Q_b(A) = r_{Q_b(A)}(J_0)$.

On the other hand, from Theorem 10.3.7, $Q_b(A)$ is a quasi-Baer $*$-ring. Hence, there is a projection $e \in Q_b(A)$ with $r_{Q_b(A)}(J_0) = eQ_b(A)$. Now we observe that

$e \in \mathbf{S}_\ell(Q_b(A))$ from Proposition 1.2.2. Because $Q_b(A)$ is semiprime, therefore we get $e \in \mathcal{B}(Q_b(A)) \subseteq \mathcal{B}(M_{\text{loc}}(A))$ from Proposition 1.2.6(ii).

Further, $r_M(J)$ is norm closed. Note that $e \in \mathcal{B}(M_{\text{loc}}(A))$ is a projection by Proposition 10.3.6(i). Also from Lemma 10.3.9, $r_M(J) = \overline{r_M(J) \cap Q_b(A)}$. Thus, $r_M(J) = \overline{r_{Q_b(A)}(J_0)} = \overline{eQ_b(A)} = e\,\overline{Q_b(A)} = eM_{\text{loc}}(A)$. Hence, $M_{\text{loc}}(A)$ is a quasi-Baer ring. Therefore, $M_{\text{loc}}(A)$ is a quasi-AW^*-algebra. □

Corollary 10.3.11 $\text{Cen}(M_{\text{loc}}(A))$ *is an AW^*-algebra for any C^*-algebra A.*

Proof Theorem 10.3.10 and Proposition 10.2.14 yield this result. □

Example 10.3.12 (i) Recall that from Theorem 10.3.10, $M_{\text{loc}}(A)$ is a quasi-AW^*-algebra for any C^*-algebra A.

(ii) Unital boundedly centrally closed C^*-algebras are precisely quasi-AW^*-algebras (see Definition 10.3.19 and Theorem 10.3.20).

(iii) Any unital prime C^*-algebra is a quasi-AW^*-algebra since any prime ring is quasi-Baer by Proposition 3.2.5. There are unital prime C^*-algebras (hence quasi-AW^*-algebras) which are not AW^*-algebras [186, pp. 150–158].

(iv) Recall that a C^*-algebra is said to be projectionless if there is no projection except zero or identity. From [45, Corollary, p. 43] and [246, p. 10], $\mathbb{C}$ is the only prime projectionless AW^*-algebra. In [136, pp. 124–129 and pp. 205–214], various unital prime projectionless C^*-algebras (hence quasi-AW^*-algebras) are provided.

The following result gives a relationship between A and its projections (see [45, Exercise 7, p. 142]).

Proposition 10.3.13 *Let A be an AW^*-algebra and I a norm closed ideal of A. Then I is the norm closure of the linear span of all projections of I.*

The next lemma follows from Propositions 10.3.6 and 10.3.13.

Lemma 10.3.14 *Let A be a C^*-algebra. Then*:

(i) $\mathcal{B}(M_{\text{loc}}(A)) = \mathcal{B}(Q(A)) = \mathcal{B}(Q^s(A)) = \mathcal{B}(Q_b(A))$.
(ii) $\text{Cen}(M_{\text{loc}}(A))$ *is the norm closure of the linear span of $\mathcal{B}(Q(A))$.*

Proof (i) By definition of $Q^s(A)$, $\mathcal{B}(Q(A)) = \mathcal{B}(Q^s(A))$. Proposition 10.3.6(ii) yields that $\mathcal{B}(Q(A)) \subseteq \text{Cen}(Q_b(A)) \subseteq \text{Cen}(M_{\text{loc}}(A))$. Thus we have that $\mathcal{B}(Q(A)) \subseteq \mathcal{B}(Q_b(A)) \subseteq \mathcal{B}(M_{\text{loc}}(A))$. Say $e \in \mathcal{B}(M_{\text{loc}}(A))$. From Proposition 10.3.6(i), e is a projection. By Proposition 10.3.6(iii), $e \in \text{Cen}(Q_b(A))$, thus $e \in \mathcal{B}(Q_b(A))$. Hence $e \in \mathcal{B}(Q(A))$, and so $\mathcal{B}(M_{\text{loc}}(A)) \subseteq \mathcal{B}(Q(A))$. Consequently,

$$\mathcal{B}(M_{\text{loc}}(A)) = \mathcal{B}(Q(A)) = \mathcal{B}(Q^s(A)) = \mathcal{B}(Q_b(A)).$$

(ii) From Proposition 10.3.6(i), each element of $\mathcal{B}(Q(A)) = \mathcal{B}(M_{\text{loc}}(A))$ (see part (i)) is a projection in $\text{Cen}(M_{\text{loc}}(A))$. Since $\text{Cen}(M_{\text{loc}}(A))$ is commutative, any projection in $\text{Cen}(M_{\text{loc}}(A))$ is in $\mathcal{B}(M_{\text{loc}}(A)) = \mathcal{B}(Q(A))$. So the set of all projections in $\mathrm{Cen}(M_{\mathrm{loc}}(A))$ is precisely $\mathcal{B}(Q(A))$. From Corollary 10.3.11, $\mathrm{Cen}(M_{\mathrm{loc}}(A))$ is an AW^*-algebra. Hence by Proposition 10.3.13, $\mathrm{Cen}(M_{\mathrm{loc}}(A))$ is the norm closure of the linear span of $\mathcal{B}(Q(A))$. □

Proposition 10.3.15 *Let A be a unital C^*-algebra and I a proper ideal of A such that I_A is (essentially) closed in A_A. Then:*

(i) *A/I is a C^*-algebra.*
(ii) *$Q_{\mathbf{qB}}(A/I)$ is $*$-isomorphic to $eQ_{\mathbf{qB}}(A)e$ for some $e \in \mathcal{B}(Q(A))$.*

Proof (i) We show first that I is norm closed. It is straightforward to see that $\ell_A(\overline{I}) = \ell_A(I)$. Since A is a semiprime ring, $\ell_A(I) \cap \overline{I} = \ell_A(\overline{I}) \cap \overline{I} = 0$. Let $0 \neq y \in \overline{I}$. Then $yI \neq 0$. Hence, $I_A \leq^{\mathrm{ess}} \overline{I}_A$. Therefore, $I = \overline{I}$ as I_A is a closed submodule of A_A. So I is norm closed. Thus, A/I is a C^*-algebra.

(ii) From Theorem 8.3.47(i), $I = A \cap (1-e)Q(A)$ with $e \in \mathcal{B}(Q(A))$. From Theorem 8.3.47(iii), $f : A/I \to eAe$ defined by $f(a+I) = eae$ for $a \in A$ is an algebra isomorphism. As $e \in \mathcal{B}(M_{\mathrm{loc}}(A))$, e is a projection by Proposition 10.3.6(ii). Therefore, $f((a+I)^*) = f(a^*+I) = ea^*e = (eae)^* = f(a+I)^*$. Thus, f is a $*$-isomorphism. By Theorem 8.3.47(vii), we see that f induces a $*$-isomorphism from $Q_{\mathbf{qB}}(A/I)$ to $eQ_{\mathbf{qB}}(A)e$. □

An element u in a C^*-algebra is said to be *unitary* if $uu^* = u^*u = 1$. The following is a well-known fact.

Theorem 10.3.16 *Each element of a unital C^*-algebra A is a finite linear combination of unitary elements of A.*

Proof See [142, Proposition 22.6] and [240, Theorem 4.1.7]. □

Assume that A is an AW^*-algebra. For projections g and h in A, $g \leq h$ means that $g = hg$. We observe that $g \leq h$ if and only if $g = gh$ if and only if $gA \subseteq hA$ if and only if $Ag \subseteq Ah$. Let $\{e_i \mid i \in \Lambda\}$ and $\{f_i \mid i \in \Lambda\}$ be two sets of orthogonal projections in a C^*-algebra A such that $e_i \sim f_i$ for all $i \in \Lambda$.

Take $e = \sup\{e_i \mid i \in \Lambda\}$ and $f = \sup\{f_i \mid i \in \Lambda\}$ in A. Note that the existence of e and f in A is guaranteed because A is an AW^*-algebra (see [45, Proposition 4.1]). By [45, Theorem 20.1], it is shown that $e \sim f$ as follows.

Theorem 10.3.17 *Let A be an AW^*-algebra. Assume that $\{e_i \mid i \in \Lambda\}$ and $\{f_i \mid i \in \Lambda\}$ are two sets of orthogonal projections in A such that $w_i^* w_i = e_i$ and $w_i w_i^* = f_i$ with $w_i \in A$ (i.e., $e_i \sim f_i$) for each $i \in \Lambda$. Take*

$$e = \sup\{e_i \mid i \in \Lambda\} \quad \text{and} \quad f = \sup\{f_i \mid i \in \Lambda\}.$$

Then there exists $w \in A$ such that $w^*w = e$ and $ww^* = f$ (i.e., $e \sim f$). Further, it follows that $we_i = w_i = f_i w$ for each $i \in \Lambda$.

For the next result, we note that if R is a Baer $*$-ring, then eRe is a Baer $*$-ring for any projection e of R (see [246, p. 30] and [45, Proposition 4.6]).

Theorem 10.3.18 *Let A be an AW^*-algebra and I a norm closed essential ideal of A. Then $M(I) = A$. Therefore, $M_{\text{loc}}(A) = A$.*

Proof Let I be a norm closed essential ideal of A, and let $X = \{e_i \mid i \in \Lambda\}$ be a maximal set of nonzero orthogonal projections in I. The existence of such X is guaranteed by Zorn's lemma and Proposition 10.3.13. As A is an AW^*-algebra, $r_A(X) = gA$ for some projection $g \in A$.

First, we show that $g = 0$. For this, assume on the contrary that $g \neq 0$. As I is an essential ideal of A and A is semiprime, $I_A \leq^{\text{ess}} A_A$ by Proposition 1.3.16. Hence, there exists $k \in A$ such that $0 \neq gk \in I$. From Proposition 8.3.55(i), I is a semiprime ring because $I \trianglelefteq A$ and A is a semiprime ring. Hence, $(gk)I(gk) \neq 0$, so $gIg \neq 0$.

Since g is a projection, gAg is also an AW^*-algebra because gAg is a Baer $*$-ring, and gIg is a nonzero norm closed ideal of gAg. By Proposition 10.3.13, gIg contains a nonzero projection, say h. Since $Xg = 0$ and $h \in gIg \subseteq gA$, $e_i h = 0$ for all $i \in \Lambda$. Thus $0 = (e_i h)^* = h^* e_i^* = he_i$, so $e_i h = 0$ and $he_i = 0$ for all $i \in \Lambda$. If $h \in X$, then $hg = 0$, so $gh = (hg)^* = 0$. As $h \in gIg$, we have that $h = gh = 0$, it is absurd. Therefore $h \notin X$, a contradiction to the maximality of X. Hence $g = 0$, so $r_A(X) = 0$. Because each e_i is a projection, we see that $\ell_A(X) = 0$.

Let $u \in M(I)$ be a unitary element. Then $uu^* = u^*u = 1$. Put $w_i = ue_i$ and $f_i = ue_i u^*$ for each $i \in \Lambda$. Then $w_i = ue_i \in uI \subseteq I$ as $u \in M(I)$ and $I \trianglelefteq M(I)$. Also $w_i^* w_i = e_i$ and $f_i = ue_i u^* = w_i w_i^*$.

Further, we have that $f_i = ue_i u^* \in M(I)IM(I) \subseteq I$, and each f_i is a projection. Therefore $e_i \sim f_i$ for each $i \in \Lambda$. Also $f_i f_j = 0$ for $i \neq j$. Thus $\{f_i \mid i \in \Lambda\}$ is a set of nonzero orthogonal projections in A.

By Theorem 10.3.17, there is $w \in A$ such that $we_i = w_i = f_i w$ for each $i \in \Lambda$. Thus $(u - w)e_i = ue_i - we_i = w_i - w_i = 0$ for all $i \in \Lambda$. As $A \subseteq M(I)$, we see that $w \in A \subseteq M(I)$. So $u - w \in \ell_{M(I)}(X)$. Since $\ell_A(X) = 0$, $\ell_I(X) = 0$. Therefore $\ell_{M(I)}(X) = 0$ because I is essential in $M(I)$. Hence $u - w = 0$, so $u = w \in A$.

Consequently, every unitary element of $M(I)$ is in A. From Theorem 10.3.16, it follows that $M(I) \subseteq A$. Since $A \subseteq M(I)$, $M(I) = A$. Therefore $Q_b(A) = A$, so $M_{\text{loc}}(A) = A$. □

Definition 10.3.19 Assume that A is a C^*-algebra. Then the C^*-subalgebra $\overline{A\text{Cen}(Q_b(A))}$ of $M_{\text{loc}}(A)$ is called the *bounded central closure* of A. We say that A is *boundedly centrally closed* if $A = \overline{A\text{Cen}(Q_b(A))}$.

Boundedly centrally closed algebras are used for studying local multiplier algebras and have been treated extensively in [15, 16], and [17]. The next result shows that a unital boundedly centrally closed C^*-algebra is precisely a C^*-algebra which is quasi-Baer.

Theorem 10.3.20 *The following are equivalent for a unital C^*-algebra A.*

(i) *A is boundedly centrally closed.*
(ii) *A is a quasi-AW^*-algebra.*

Proof (i)⇒(ii) Assume that A is boundedly centrally closed. Then we have that $\overline{A\mathrm{Cen}(Q_b(A))} = A$. Because $\mathcal{B}(Q(A)) = \mathcal{B}(Q_b(A)) \subseteq \mathrm{Cen}(Q_b(A))$ from Lemma 10.3.14(i), $A = \overline{A\mathcal{B}(Q(A))}$. Thus $A = A\mathcal{B}(Q(A))$ and hence Theorem 8.3.17 yields that A is a quasi-Baer ring. So A is a quasi-AW^*-algebra.

(ii)⇒(i) Let A be a quasi-AW^*-algebra. By Lemma 10.3.14(ii), we see that $\mathrm{Cen}(M_{\mathrm{loc}}(A))$ is the norm closure of the linear span of $\mathcal{B}(Q(A))$. Thus, $\mathrm{Cen}(Q_b(A)) \subseteq \mathrm{Cen}(M_{\mathrm{loc}}(A)) \subseteq \overline{A\mathcal{B}(Q(A))} = \overline{Q_{\mathbf{qB}}(A)} = A$ as $Q_{\mathbf{qB}}(A) = A$ from Theorem 8.3.17. Therefore $A = \overline{A\mathrm{Cen}(Q_b(A))}$, and hence A is boundedly centrally closed. □

Corollary 10.3.21 next is a direct consequence of Theorems 10.3.10 and 10.3.20. It exhibits some interesting classes of C^*-algebras which are boundedly centrally closed.

Corollary 10.3.21 (i) *The local multiplier algebra $M_{\mathrm{loc}}(A)$ is boundedly centrally closed for any C^*-algebra A.*
(ii) *Any AW^*-algebra is boundedly centrally closed.*

Proof (i) The proof follows from Theorems 10.3.10 and 10.3.20.

(ii) Let A be an AW^*-algebra. Then A is a Baer ring, so it a quasi-Baer ring. Thus, A is boundedly centrally closed by Theorem 10.3.20. □

From [17, p. 74], every norm closed essential ideal of a boundedly centrally closed C^*-algebra is also boundedly centrally closed. By Proposition 10.2.14, the center of a quasi-Baer $*$-ring is a Baer $*$-ring. Hence Theorem 10.3.20 and Corollary 10.3.21 yield the next corollary.

Corollary 10.3.22 *The center of a unital boundedly centrally closed C^*-algebra is an AW^*-algebra.*

The following result provides another class of boundedly centrally closed C^*-algebras.

Theorem 10.3.23 *Every prime C^*-algebra is boundedly centrally closed.*

Proof Let A be a prime C^*-algebra. Then A^1 is a unital prime C^*-algebra. Hence A^1 is a quasi-Baer ring, so A^1 is boundedly centrally closed by Theorem 10.3.20. Thus $\mathrm{Cen}(A^1)$ is a AW^*-algebra from Corollary 10.3.22. By Proposition 10.3.13, $\mathrm{Cen}(A^1) = \mathbb{C}$ as 1 is the only nonzero projection of $\mathrm{Cen}(A^1)$.

Note that $\mathcal{B}(Q(A)) = \{0, 1\}$ and $\mathrm{Cen}(M_{\mathrm{loc}}(A)) = \mathbb{C}$ by Lemma 10.3.14(ii). So $\mathrm{Cen}(Q_b(A)) = \mathbb{C}$ as $\mathbb{C} = \mathrm{Cen}(A^1) \subseteq \mathrm{Cen}(Q_b(A)) \subseteq \mathrm{Cen}(M_{\mathrm{loc}}(A)) = \mathbb{C}$. Therefore, $A = \overline{A\mathrm{Cen}(Q_b(A))}$. So A is boundedly centrally closed. □

The next result illustrates an interesting relationship between the extended centroid of a C^*-algebra A and the center of $Q_b(A)$.

Theorem 10.3.24 $\mathrm{Cen}(Q(A)) = Q^r_{c\ell}(\mathrm{Cen}(Q_b(A)))$ *for any* C^**-algebra* A.

Proof As A is semiprime, $\mathrm{Cen}(Q(A)) = \mathrm{Cen}(Q^s(A))$ is regular from Theorem 10.1.15. Say $x \in \mathrm{Cen}(Q^s(A))$. We claim that $1 + xx^*$ is invertible in $\mathrm{Cen}(Q^s(A))$. Since $\mathrm{Cen}(Q^s(A))$ is regular, we need to see that $1+xx^*$ is a nonzero-divisor. For this, say $(1+xx^*)y = 0$ with $y \in \mathrm{Cen}(Q^s(A))$. Then $(1+xx^*)yy^* = 0$, so $yy^* + (xy)(xy)^* = 0$.

Recall that the involution $*$ induced on $Q^s(A)$ from A is positive-definite. Thus $y = 0$, and so $1 + xx^*$ is invertible in $\mathrm{Cen}(Q^s(A))$. Put

$$c = x(1 + xx^*)^{-1} \quad \text{and} \quad d = (1 + xx^*)^{-1}.$$

Then $(1 + xx^*)d = 1$ and $(1 + xx^*)d^* = 1$, so $(1 + xx^*)^2 dd^* = 1$. Thus

$$dd^* + (xd)(xd)^* + (xd)(xd)^* + (x^2 d)(x^2 d)^* = 1,$$

equivalently, $d^*d + (xd)^*(xd) + (xd)^*(xd) + (x^2d)^*(x^2d) = 1$. Hence, d and xd are bounded elements of $Q^s(A)$. Therefore, $d \in Q_b(A)$ and $c = xd \in Q_b(A)$. Thus, $c \in \mathrm{Cen}(Q_b(A))$ and $d \in \mathrm{Cen}(Q_b(A))$. So $x = cd^{-1} \in Q^r_{c\ell}(\mathrm{Cen}(Q_b(A)))$. Hence, $\mathrm{Cen}(Q^s(A)) \subseteq Q^r_{c\ell}(\mathrm{Cen}(Q_b(A)))$.

Next, let $ab^{-1} \in Q^r_{c\ell}(\mathrm{Cen}(Q_b(A)))$, where $a, b \in \mathrm{Cen}(Q_b(A))$ and b is a nonzero-divisor. Then $b \in \mathrm{Cen}(Q^s(A))$. We show that b is a nonzero-divisor in $\mathrm{Cen}(Q^s(A))$. Indeed, let $bz = 0$ with $z \in \mathrm{Cen}(Q^s(A)) \subseteq Q^r_{c\ell}(\mathrm{Cen}(Q_b(A)))$. Then $z = vw^{-1}$ for some $v, w \in \mathrm{Cen}(Q_b(A))$. So $bvw^{-1} = bz = 0$, hence $bv = 0$. Thus, $v = 0$ as b is a nonzero-divisor in $\mathrm{Cen}(Q_b(A))$, so $z = 0$.

As $\mathrm{Cen}(Q^s(A))$ is regular from Theorem 10.1.15, $b^{-1} \in \mathrm{Cen}(Q^s(A))$, and hence $ab^{-1} \in \mathrm{Cen}(Q^s(A))$. Therefore, $Q^r_{c\ell}(\mathrm{Cen}(Q_b(A))) \subseteq \mathrm{Cen}(Q^s(A))$. So, $\mathrm{Cen}(Q^s(A)) = Q^r_{c\ell}(\mathrm{Cen}(Q_b(A)))$. □

As in Theorem 10.3.24, the following result also provides a useful relationship between $\mathrm{Cen}(M_{\mathrm{loc}}(A))$ and $\mathrm{Cen}(Q_b(A))$.

Theorem 10.3.25 $\mathrm{Cen}(M_{\mathrm{loc}}(A)) = \overline{\mathrm{Cen}(Q_b(A))}$ *for any* C^**-algebra* A.

Proof As $\mathrm{Cen}(Q_b(A)) \subseteq \mathrm{Cen}(M_{\mathrm{loc}}(A))$, $\overline{\mathrm{Cen}(Q_b(A))} \subseteq \mathrm{Cen}(M_{\mathrm{loc}}(A))$. From Lemma 10.3.14(ii), $\mathrm{Cen}(M_{\mathrm{loc}}(A))$ is the norm closure of the linear span of $\mathcal{B}(Q(A))$. Therefore, $\mathrm{Cen}(M_{\mathrm{loc}}(A)) \subseteq \overline{\mathrm{Cen}(Q_b(A))}$ as $\mathcal{B}(Q(A)) \subseteq \mathrm{Cen}(Q_b(A))$ from Lemma 10.3.14(i). □

Lemma 10.3.26 *Let* A *be a* C^**-algebra. If* $\alpha A \beta = 0$ *for* $\alpha, \beta \in M_{\mathrm{loc}}(A)$, *then* $\alpha M_{\mathrm{loc}}(A) \beta = 0$.

Proof See [17, Proposition 2.3.3(ii)] for the proof. □

Let B be a C^*-algebra and A a C^*-subalgebra of B. Recall that $\mathbf{I}_{ce}$ denotes the set of norm closed essential ideals of A. Then B is said to be an $\mathbf{I}_{ce}$*-enlargement* of A if $B = B_{\mathbf{I}_{ce}}$, where $B_{\mathbf{I}_{ce}}$ is the norm closure of the union of all B_I with $I \in \mathbf{I}_{ce}$, and $B_I = \{x \in B \mid xI + Ix \subseteq I\}$. Further, we say that an $\mathbf{I}_{ce}$-enlargement B is said to be *essential* if, for $I \in \mathbf{I}_{ce}$ and $x \in B$, $Ix = 0$ or $xI = 0$ implies $x = 0$. The local multiplier algebra $M_{\text{loc}}(A)$ is an essential $\mathbf{I}_{ce}$-enlargement of A. For more details, see [17, p. 66].

The next Theorem 10.3.27 will be used in the further study of boundedly centrally closed C^*-algebras.

Theorem 10.3.27 *Let A be a C^*-algebra and B an intermediate C^*-algebra between A and $M_{\text{loc}}(A)$. Then:*

(i) $\text{Cen}(B) \subseteq \text{Cen}(M_{\text{loc}}(A))$.
(ii) *If* $\text{Cen}(Q_b(A)) \subseteq B$, *then* B *is a quasi-AW^*-algebra, and furthermore* $\text{Cen}(B) = \text{Cen}(M_{\text{loc}}(A))$.

Proof (i) Say $x \in \text{Cen}(B)$ and $q \in Q_b(A)$. Let $\mathbf{I}_{ce}$ be the set of all norm closed essential ideals of A. There is $I \in \mathbf{I}_{ce}$ with $qI + Iq \subseteq I$. For any $a \in I$,

$$(xq - qx)a = x(qa) - q(xa) = x(qa) - (qa)x = x(qa) - x(qa) = 0,$$

so $(xq - qx)I = 0$. Thus $xq - qx = 0$ because $M_{\text{loc}}(A)$ is an essential $\mathbf{I}_{ce}$-enlargement of A. Hence $xq = qx$ for all $q \in Q_b(A)$.

Thus, $x \in \text{Cen}(M_{\text{loc}}(A))$ as $M_{\text{loc}}(A) = \overline{Q_b(A)}$. So $\text{Cen}(B) \subseteq \text{Cen}(M_{\text{loc}}(A))$.

(ii) Let J be an ideal of B. Put $M = M_{\text{loc}}(A)$. There exists $e^2 = e \in M$ such that $r_M(MJM) = eM$ by Theorem 10.3.10. Since M is semiprime and $e \in \mathbf{S}_\ell(M)$, $e \in \mathcal{B}(M)$ by Proposition 1.2.6(ii). Thus, $e \in \mathcal{B}(Q_b(A)) \subseteq B$ by Lemma 10.3.14(i). Note that $Je = 0$, thus $eB \subseteq r_B(J)$.

Next, let $b \in r_B(J)$ and take $a \in J$. Then $aBb \subseteq Jb = 0$, hence $aAb = 0$. By Lemma 10.3.26 $aMb = 0$. So $MaMb = 0$ for each $a \in J$, and therefore $MJMb = 0$. Hence $b \in r_M(MJM) = eM$, so $b = eb \in eB$. Thus $r_B(J) \subseteq eB$ and consequently $r_B(J) = eB$ with $e \in B$. Therefore B is a quasi-AW^*-algebra.

From part (i), $\text{Cen}(B) \subseteq \text{Cen}(M_{\text{loc}}(A))$. By assumption and Theorem 10.3.25, $\text{Cen}(M_{\text{loc}}(A)) = \overline{\text{Cen}(Q_b(A))} \subseteq B$. So $\text{Cen}(B) = \text{Cen}(M_{\text{loc}}(A))$. □

Definition 10.3.28 Let A be a C^*-algebra. We call the smallest boundedly centrally closed C^*-subalgebra of $M_{\text{loc}}(A)$ containing A the *boundedly centrally closed hull* of A.

From the next result, when A is a unital C^*-algebra, $\overline{Q_{\mathbf{qB}}(A)}$ is the boundedly centrally closed hull of A. Also this result is a unital C^*-algebra analogue of Theorem 8.3.17.

Theorem 10.3.29 *Let A be a unital C^*-algebra. Then:*

(i) $\overline{Q_{\mathbf{qB}}(A)} = \overline{A\text{Cen}(Q_b(A))}$.
(ii) $\overline{Q_{\mathbf{qB}}(A)}$ *is the boundedly centrally closed hull of A.*

(iii) *Let B be an intermediate C^*-algebra between A and $M_{\rm loc}(A)$. Then B is a quasi-AW^*-algebra if and only if $\mathcal{B}(Q(A)) \subseteq B$.*

Proof (i) We note that $\mathrm{Cen}(M_{\rm loc}(A)) \subseteq \overline{A\mathcal{B}(Q(A))} = \overline{Q_{\mathbf{qB}}(A)}$ from Lemma 10.3.14(ii) and Theorem 8.3.17. Hence

$$\mathrm{Cen}(Q_b(A)) \subseteq \mathrm{Cen}(M_{\rm loc}(A)) \subseteq \overline{Q_{\mathbf{qB}}(A)} \subseteq \overline{A\mathrm{Cen}(Q_b(A))},$$

because $\mathcal{B}(Q(A)) = \mathcal{B}(Q_b(A)) \subseteq \mathrm{Cen}(Q_b(A))$ by Lemma 10.3.14(i). Therefore

$$\overline{Q_{\mathbf{qB}}(A)} = \overline{A\mathrm{Cen}(Q_b(A))}.$$

(ii) By part (i), Theorem 10.3.27(ii), and Theorem 10.3.20, $\overline{Q_{\mathbf{qB}}(A)}$ is boundedly centrally closed. Let B be a boundedly centrally closed intermediate C^*-algebra between A and $M_{\rm loc}(A)$. Then B is quasi-Baer by Theorem 10.3.20.

Say $0 \neq e \in \mathcal{B}(Q(A))$. Then $(eQ_{\mathbf{qB}}(A) \cap A)_A \leq^{\rm ess} eQ_{\mathbf{qB}}(A)_A$. As B is semiprime quasi-Baer, there is $c \in \mathcal{B}(B)$ with $B(eQ_{\mathbf{qB}}(A) \cap A)B_B \leq^{\rm ess} cB_B$ by Theorem 3.2.37. Note that, by Theorem 10.3.27(i), $c \in \mathcal{B}(M_{\rm loc}(A))$. From Lemma 10.3.14(i) and Theorem 8.3.17, $\mathcal{B}(M_{\rm loc}(A)) = \mathcal{B}(Q(A)) \subseteq Q_{\mathbf{qB}}(A)$, so $c \in \mathcal{B}(Q(A)) \subseteq Q_{\mathbf{qB}}(A)$. Hence, $(eQ_{\mathbf{qB}}(A) \cap A)_A \leq ceQ_{\mathbf{qB}}(A)_A \leq eQ_{\mathbf{qB}}(A)_A$. Since $(eQ_{\mathbf{qB}}(A) \cap A)_A \leq^{\rm ess} eQ_{\mathbf{qB}}(A)_A$, $ceQ_{\mathbf{qB}}(A) = eQ_{\mathbf{qB}}(A)$, and so $ce = e$.

We claim that $c(1-e)M_{\rm loc}(A) = 0$. For this, assume on the contrary that $c(1-e)M_{\rm loc}(A) \neq 0$. Note that $c(1-e)M_{\rm loc}(A)$ is a norm closed ideal since $c(1-e) \in \mathcal{B}(M_{\rm loc}(A))$ (so $c(1-e)$ is a projection by Proposition 10.3.6(i)). Hence, from Lemma 10.3.9

$$c(1-e)M_{\rm loc}(A) = \overline{c(1-e)M_{\rm loc}(A) \cap Q_b(A)},$$

so $c(1-e)M_{\rm loc}(A) \cap Q_b(A) \neq 0$. Therefore $c(1-e)M_{\rm loc}(A) \cap A \neq 0$, and hence $0 \neq [c(1-e)M_{\rm loc}(A) \cap B]_B \leq cB_B$. Recall that $B(eQ_{\mathbf{qB}}(A) \cap A)B_B \leq^{\rm ess} cB_B$ from the preceding argument. Therefore,

$$[c(1-e)M_{\rm loc}(A) \cap B] \cap B(eQ_{\mathbf{qB}}(A) \cap A)B \neq 0.$$

Take $y \in [c(1-e)M_{\rm loc}(A) \cap B] \cap B(eQ_{\mathbf{qB}}(A) \cap A)B$. Then $y = (1-e)y$ and $y = ey$ because $e \in \mathcal{B}(Q(A)) \subseteq \mathrm{Cen}(M_{\rm loc}(A))$ (see Lemma 10.3.14(i)). So $y = 0$, a contradiction. Hence, $c(1-e)M_{\rm loc}(A) = 0$. Thus, $c(1-e) = 0$. Thus $e = ce = c \in B$. Therefore $Q_{\mathbf{qB}}(A) = A\mathcal{B}(Q(A)) \subseteq B$, so $\overline{Q_{\mathbf{qB}}(A)} \subseteq B$. Thus, $\overline{Q_{\mathbf{qB}}(A)}$ is the boundedly centrally closed hull of A.

(iii) Let B be a quasi-AW^*-algebra. Then from Theorem 10.3.20, B is boundedly centrally closed. From part (ii), $Q_{\mathbf{qB}}(A) \subseteq B$. So $\mathcal{B}(Q(A)) \subseteq B$ by Theorem 8.3.17. Conversely, if $\mathcal{B}(Q(A)) \subseteq B$, then by Theorem 8.3.17 $Q_{\mathbf{qB}}(A) = A\mathcal{B}(Q(A)) \subseteq B$. Therefore part (i) yields that $\overline{A\mathrm{Cen}(Q_b(A))} \subseteq B$. Now from Theorem 10.3.27(ii), B is a quasi-AW^*-algebra. □

Motivated by Theorem 8.3.17, it is natural to ask: *If T is any semiprime intermediate ring, not necessarily a C^*-algebra, between A and $M_{\rm loc}(A)$, then is the* *condition, $\mathcal{B}(Q(A)) \subseteq T$, necessary and/or sufficient for T to be quasi-Baer?* Our next two results provide partial affirmative answers to this question.

Theorem 10.3.30 *Assume that A is a C^*-algebra. If T is a semiprime quasi-Baer intermediate ring between A and $M_{\text{loc}}(A)$, then $\mathcal{B}(Q(A)) \subseteq T$.*

Proof For $\mathcal{B}(T) \subseteq \mathcal{B}(M_{\text{loc}}(A))$, say $f \in \mathcal{B}(T)$ and $q \in Q_b(A)$. Then there is a norm closed essential ideal I of A with $qI + Iq \subseteq I$. For any $a \in I$,

$$(fq - qf)a = fqa - qfa = qaf - qaf = 0, \text{ hence } (fq - qf)I = 0.$$

As $M_{\text{loc}}(A)$ is an essential $\mathbf{I}_{\text{ce}}$-enlargement of A, $fq - qf = 0$, so $fq = qf$. Hence f commutes with every element of $Q_b(A)$, and thus $f \in \text{Cen}(M_{\text{loc}}(A))$. Therefore, $\mathcal{B}(T) \subseteq \mathcal{B}(M_{\text{loc}}(A))$.

Let $e \in \mathcal{B}(Q(A))$ and $K = T \cap eM_{\text{loc}}(A)$. From Theorem 3.2.37, there exists $c \in \mathcal{B}(T)$ such that $K_T \leq^{\text{ess}} cT_T$. As $\mathcal{B}(M_{\text{loc}}(A)) = \mathcal{B}(Q(A))$ by Lemma 10.3.14(i) and $\mathcal{B}(T) \subseteq \mathcal{B}(M_{\text{loc}}(A))$, we have that $e, c \in \mathcal{B}(M_{\text{loc}}(A))$. As a consequence, we get $K \subseteq eT \cap cT \subseteq ecT$, so $K \subseteq ecM_{\text{loc}}(A) \subseteq eM_{\text{loc}}(A)$. By the modular law,

$$eM_{\text{loc}}(A) = ecM_{\text{loc}}(A) \oplus J,$$

where $J = eM_{\text{loc}}(A) \cap (1 - ec)M_{\text{loc}}(A) = e(1 - ec)M_{\text{loc}}(A)$.

If $J \neq 0$, then $A \cap J \neq 0$ from the proof of Theorem 10.3.29(ii). Therefore,

$$0 \neq A \cap J \subseteq T \cap eM_{\text{loc}}(A) = K \subseteq ecT \subseteq ecM_{\text{loc}}(A),$$

a contradiction. Thus, $J = 0$, so $e = ec$. Hence, $eM_{\text{loc}}(A) \subseteq cM_{\text{loc}}(A)$. Then again by the modular law,

$$cM_{\text{loc}}(A) = eM_{\text{loc}}(A) \oplus W, \quad \text{where} \quad W = cM_{\text{loc}}(A) \cap (1 - e)M_{\text{loc}}(A).$$

Assume that $W \neq 0$. Again by the proof of Theorem 10.3.29(ii), $A \cap W \neq 0$. Hence, $0 \neq T \cap W = T \cap cM_{\text{loc}}(A) \cap (1 - e)M_{\text{loc}}(A) = cT \cap (1 - e)M_{\text{loc}}(A)$. So $cT \cap (1 - e)M_{\text{loc}}(A) \cap K = cT \cap (1 - e)M_{\text{loc}}(A) \cap T \cap eM_{\text{loc}}(A) \neq 0$, as $K_T \leq^{\text{ess}} cT_T$. Thus, we have a contradiction. Hence $W = 0$, so $e = c \in T$. Therefore, $\mathcal{B}(Q(A)) \subseteq T$. □

When A is a C^*-algebra, by Theorem 8.3.17, $Q_{\mathbf{qB}}(A)$ is the smallest (semiprime) quasi-Baer intermediate ring between A and $Q(A)$. Further, from Theorem 10.3.30, $Q_{\mathbf{qB}}(A)$ is also the smallest semiprime quasi-Baer intermediate ring between A and $M_{\text{loc}}(A)$.

Corollary 10.3.31 *Let A be a C^*-algebra and T be an intermediate $*$-algebra between A and $M_{\text{loc}}(A)$. Then T is a quasi-Baer $*$-algebra if and only if $\mathcal{B}(Q(A)) \subseteq T$.*

Proof Let T be a quasi-Baer $*$-algebra. By Proposition 10.2.4, T is semiprime. From Theorem 10.3.30, $\mathcal{B}(Q(A)) \subseteq T$. Conversely, if $\mathcal{B}(Q(A)) \subseteq T$, then A^1 is a subalgebra of T. By Theorem 10.3.29(iii), $\overline{T}$ is a quasi-AW^*-algebra. Let $I \trianglelefteq T$. Then $\ell_{\overline{T}}(\overline{I}) = e\overline{T}$ for some $e \in \mathcal{B}(\overline{T})$ from Proposition 1.2.6(ii). By Theorem 10.3.27(i), $\mathcal{B}(\overline{T}) \subseteq \mathcal{B}(M_{\text{loc}}(A))$.

Because $\mathcal{B}(M_{\text{loc}}(A)) = \mathcal{B}(Q(A))$ from Lemma 10.3.14(i) and $\mathcal{B}(Q(A)) \subseteq T$, $e \in \mathcal{B}(Q(A)) \subseteq T$. But $\ell_{\overline{T}}(\overline{I}) = \ell_{\overline{T}}(I)$. So $\ell_T(I) = \ell_{\overline{T}}(I) \cap T = e\overline{T} \cap T = eT$. Hence, T is quasi-Baer. By Proposition 10.3.6(i), e is a projection and so T is a quasi-Baer $*$-algebra. □

The definitions of C^*-direct product and C^*-direct sum of C^*-algebras are as follows.

Definition 10.3.32 (i) Let $\{A_i\}$ be a set of C^*-algebras. By $\prod_i^{C^*} A_i$, we denote the C^*-algebra $\{(a_i) \in \prod_i A_i \mid \sup_i \|a_i\| < \infty\}$, which is called the C^*-*direct product* of $\{A_i\}$.

(ii) We use $\bigoplus_i^{C^*} A_i$ to denote the C^*-subalgebra of $\prod_i^{C^*} A_i$, consisting of all elements $(a_i) \in \prod_i^{C^*} A_i$ with the set $\{i \mid \|a_i\| > \varepsilon\}$ finite for any given $\varepsilon > 0$. The C^*-algebra $\bigoplus_i^{C^*} A_i$ is called the C^*-*direct sum* of $\{A_i\}$.

The following is a well known fact.

Proposition 10.3.33 (i) *Assume that* $\{A_i \mid i \in \Lambda\}$ *is a set of* C^*-*algebras. Then* $M(\bigoplus_{i\in\Lambda}^{C^*} A_i) = \prod_{i\in\Lambda}^{C^*} M(A_i)$.

(ii) *Assume that* A *is a* C^*-*algebra. We let* $\{e_i \mid i \in \Lambda\}$ *be a set of orthogonal central projections in* $M_{\text{loc}}(A)$ *such that* $\sup\{e_i \mid i \in \Lambda\} = 1$. *Then it follows that* $M_{\text{loc}}(A) = \prod_{i\in\Lambda}^{C^*} e_i M_{\text{loc}}(A)$.

Proof See [17, Lemma 1.2.21 and Lemma 3.3.6]. □

The next example illustrates Theorem 10.3.30.

Example 10.3.34 Let A be the C^*-direct sum of $\aleph_0$ copies of $\mathbb{C}$. Then $M_{\text{loc}}(A)$ is the C^*-direct product of $\aleph_0$ copies of $\mathbb{C}$. Let T be the set of all bounded sequences of complex numbers whose imaginary parts approach zero. Then T is a semiprime quasi-Baer intermediate ring between A and $M_{\text{loc}}(A)$, but T is not a C^*-subalgebra of $M_{\text{loc}}(A)$ since T does not form an algebra over $\mathbb{C}$.

The next result provides a characterization and shows the existence of the boundedly centrally closed hull of A in terms of $\mathcal{B}(Q(A))$.

Theorem 10.3.35 *Let* A *be a* C^*-*algebra and* B *an intermediate* C^*-*algebra between* A *and* $M_{\text{loc}}(A)$. *Then*:

(i) $\mathcal{B}(Q(A)) \subseteq \text{Cen}(Q_b(B)) \subseteq \text{Cen}(M_{\text{loc}}(A))$.

(ii) B *is boundedly centrally closed if and only if* $B = \overline{B\mathcal{B}(Q(A))}$.

(iii) $\overline{A\mathcal{B}(Q(A))}$ *is the boundedly centrally closed hull of* A.

Proof (i) Note that the left (resp., right) multiplication by $1_{M_{\text{loc}}(A)}$ on B is the identity map as a left (resp., right) B-module homomorphism of B. Thus it follows that

$1_{M_{\rm loc}(A)} = 1_{M(B)} (= 1_{M_{\rm loc}(B)} = 1_{Q(B)})$. So

$$B^1 = \{b + \lambda 1_{M_{\rm loc}(A)} \mid b \in B \text{ and } \lambda \in \mathbb{C}\},$$

the subalgebra of $M_{\rm loc}(A)$ generated by B and $1_{M_{\rm loc}(A)}$.

We claim that $Q(B^1\mathcal{B}(Q(A))) = Q(B^1) = Q(B)$. For this, first we prove that $B^1_{B^1} \leq^{\rm ess} B^1\mathcal{B}(Q(A))_{B^1}$. Indeed, take $0 \neq \alpha \in B^1\mathcal{B}(Q(A)) \subseteq M_{\rm loc}(A)$. Then $\alpha = b_1e_1 + \cdots + b_ne_n$, where $b_i \in B^1$ and $e_i \in \mathcal{B}(Q(A)) \subseteq Q_b(A)$ for each i. Since $Q(A) = Q(A^1)$ and $Q_b(A) = Q_b(A^1)$, there exist norm closed essential ideals I_i of A^1 such that $e_iI_i \subseteq A^1$ for each i. Hence, $I := I_1 \cap \cdots \cap I_n$ is a norm closed essential ideal of A^1 and $\alpha I \subseteq b_1A^1 + \cdots + b_nA^1 \subseteq B^1$. As $M_{\rm loc}(A^1) = M_{\rm loc}(A)$ is an essential $\mathbf{I}_{ce}$-enlargement of A^1, $\alpha I \neq 0$. Hence, $B^1_{A^1} \leq^{\rm ess} B^1\mathcal{B}(Q(A))_{A^1}$ and therefore $B^1_{B^1} \leq^{\rm ess} B^1\mathcal{B}(Q(A))_{B^1}$.

Because $Z(B^1{}_{B^1}) = 0$ and $B^1_{B^1} \leq^{\rm ess} B^1\mathcal{B}(Q(A))_{B^1}$, $Z(B^1\mathcal{B}(Q(A))_{B^1}) = 0$ and so $B^1_{B^1} \leq^{\rm den} B^1\mathcal{B}(Q(A))_{B^1}$ from Proposition 1.3.14. Therefore, we have that $Q(B^1\mathcal{B}(Q(A))) = Q(B^1) = Q(B)$.

If $e \in \mathcal{B}(Q(A))$, then $e \in B^1\mathcal{B}(Q(A)) \subseteq Q(B^1\mathcal{B}(Q(A))) = Q(B^1) = Q(B)$. As $e \in \mathcal{B}(M_{\rm loc}(A))$ by Lemma 10.3.14(i), e commutes with every element of B. Further, $e \in Q(B)$, hence $e \in \text{Cen}(Q(B)) = \text{Cen}(Q^s(B))$. By Proposition 10.3.6(ii), e is bounded, so $e \in Q_b(B)$. Thus, $e \in \text{Cen}(Q_b(B))$. Therefore,

$$\mathcal{B}(Q(A)) \subseteq \text{Cen}(Q_b(B)).$$

Note that the involutions on B, B^1, and $B^1\mathcal{B}(Q(A))$, respectively, are restricted from the involution of $M_{\rm loc}(A)$. As $B \subseteq B^1 \subseteq B^1\mathcal{B}(Q(A)) \subseteq Q_b(B^1)$, the involution on $M_{\rm loc}(A)$ and that on $M_{\rm loc}(B)$ agree on $B^1\mathcal{B}(Q(A))$. So the norm defined on $M_{\rm loc}(A)$ and that defined on $M_{\rm loc}(B)$ agree on $B^1\mathcal{B}(Q(A))$. Thus, we see that $\overline{B^1\mathcal{B}(Q(A))}_{M_{\rm loc}(A)} = \overline{B^1\mathcal{B}(Q(A))}_{M_{\rm loc}(B)}$, where $\overline{(-)}_{M_{\rm loc}(A)}$ (resp., $\overline{(-)}_{M_{\rm loc}(B)}$) denotes the norm closure in $M_{\rm loc}(A)$ (resp., $M_{\rm loc}(B)$).

By Theorem 10.3.29(iii), $\overline{B^1\mathcal{B}(Q(A))}_{M_{\rm loc}(A)}$ is boundedly centrally closed. So, from Theorem 10.3.20, $\overline{B^1\mathcal{B}(Q(A))}_{M_{\rm loc}(B)} = \overline{B^1\mathcal{B}(Q(A))}_{M_{\rm loc}(A)}$ is quasi-Baer. Hence $\overline{B^1\mathcal{B}(Q(A))}_{M_{\rm loc}(B)}$ is boundedly centrally closed in $M_{\rm loc}(B)$ by Theorem 10.3.20. Now $\overline{B^1\text{Cen}(Q_b(B))}_{M_{\rm loc}(B)} \subseteq \overline{B^1\mathcal{B}(Q(A))}_{M_{\rm loc}(B)}$ by Theorem 10.3.29(i) and (ii). So

$$\text{Cen}(Q_b(B)) \subseteq \overline{B^1\mathcal{B}(Q(A))}_{M_{\rm loc}(B)} = \overline{B^1\mathcal{B}(Q(A))}_{M_{\rm loc}(A)} \subseteq M_{\rm loc}(A).$$

From the preceding argument, there is no ambiguity to use $\overline{B^1\mathcal{B}(Q(A))}$ for $\overline{B^1\mathcal{B}(Q(A))}_{M_{\rm loc}(A)}$ or $\overline{B^1\mathcal{B}(Q(A))}_{M_{\rm loc}(B)}$. Note that elements of $\text{Cen}(Q_b(B))$ commute with elements of $B^1\mathcal{B}(Q(A))$ since $\mathcal{B}(Q(A)) \subseteq \text{Cen}(Q_b(B))$. Thus, elements of $\text{Cen}(Q_b(B))$ commute with elements of $\overline{B^1\mathcal{B}(Q(A))}$. Therefore,

$$\text{Cen}(Q_b(B)) \subseteq \text{Cen}(\overline{B^1\mathcal{B}(Q(A))})$$

because $\text{Cen}(Q_b(B)) \subseteq \overline{B^1\mathcal{B}(Q(A))}$. Since $\overline{B^1\mathcal{B}(Q(A))}$ is an intermediate C^*-algebra between A and $M_{\text{loc}}(A)$, $\text{Cen}(\overline{B^1\mathcal{B}(Q(A))}) \subseteq \text{Cen}(M_{\text{loc}}(A))$ from Theorem 10.3.27(i). Thus, $\text{Cen}(Q_b(B)) \subseteq \text{Cen}(M_{\text{loc}}(A))$. Consequently, we have that $\mathcal{B}(Q(A)) \subseteq \text{Cen}(Q_b(B)) \subseteq \text{Cen}(M_{\text{loc}}(A))$.

(ii) First, we note that part (i) and Lemma 10.3.14(ii) yield that

$$\overline{B\mathcal{B}(Q(A))} \subseteq \overline{B\text{Cen}(Q_b(B))} \subseteq \overline{B\text{Cen}(M_{\text{loc}}(A))} \subseteq \overline{B\mathcal{B}(Q(A))}.$$

Thus, $\overline{B\mathcal{B}(Q(A))} = \overline{B\text{Cen}(Q_b(B))}$. Assume that B is boundedly centrally closed. Then $B = \overline{B\text{Cen}(Q_b(B))} = \overline{B\mathcal{B}(Q(A))}$. Conversely, if $B = \overline{B\mathcal{B}(Q(A))}$, then we get that $B = \overline{B\text{Cen}(Q_b(B))}$, so B is boundedly centrally closed.

(iii) Clearly, $A \subseteq \overline{A\mathcal{B}(Q(A))}$. Let $U = \overline{A\mathcal{B}(Q(A))}$. Then $\overline{U\mathcal{B}(Q(A))} = U$, so U is boundedly centrally closed by part (ii). Next, let B be a boundedly centrally closed intermediate C^*-algebra between A and $M_{\text{loc}}(A)$. From part (ii), $U \subseteq B$. So $\overline{A\mathcal{B}(Q(A))}$ is the boundedly centrally closed hull of A. □

The next result is of interest in its own right.

Corollary 10.3.36 *Let A be a C^*-algebra and B an intermediate C^*-algebra between A and $M_{\text{loc}}(A)$. Then:*

(i) $\mathcal{B}(Q(B)) = \mathcal{B}(Q(A))$.
(ii) $\text{Cen}(M_{\text{loc}}(B)) = \text{Cen}(M_{\text{loc}}(A))$.

Proof (i) Say $f \in \mathcal{B}(Q(B))$. Then $f \in \text{Cen}(Q_b(B))$ by Proposition 10.3.6(ii), so $f \in \mathcal{B}(M_{\text{loc}}(A)) = \mathcal{B}(Q(A))$ by Theorem 10.3.35(i) and Lemma 10.3.14(i). Conversely, let $e \in \mathcal{B}(Q(A))$. Then $e \in \text{Cen}(Q_b(B))$ by Theorem 10.3.35(i). So $e \in \mathcal{B}(Q(B))$ from Lemma 10.3.14(i).

(ii) As in the proof of Theorem 10.3.35(i), the norm closure of the linear span of $\mathcal{B}(Q(A))$ and that of the linear span of $\mathcal{B}(Q(B))$ is the same in both $M_{\text{loc}}(A)$ and $M_{\text{loc}}(B)$ because $\mathcal{B}(Q(A)) = \mathcal{B}(Q(B))$ by part (i). So we have that $\overline{\mathbb{C}\mathcal{B}(Q(A))}_{M_{\text{loc}}(A)} = \overline{\mathbb{C}\mathcal{B}(Q(A))}_{M_{\text{loc}}(B)} = \overline{\mathbb{C}\mathcal{B}(Q(B))}_{M_{\text{loc}}(B)}$ as in the proof of Theorem 10.3.35(i). Hence $\text{Cen}(M_{\text{loc}}(B)) = \text{Cen}(M_{\text{loc}}(A))$ by Lemma 10.3.14(ii). □

Now we study the extended centroid of a C^*-algebra. For this, a prime C^*-algebra A is characterized via the extended centroid of A as follows.

Lemma 10.3.37 *Let A be a C^*-algebra. Then A is prime if and only if the extended centroid of A is $\mathbb{C}$.*

Proof Let A be prime. By the proof of Theorem 10.3.23, $\text{Cen}(Q_b(A)) = \mathbb{C}$. Since $\text{Cen}(Q(A)) = Q^r_{c\ell}(\text{Cen}(Q_b(A)))$ from Theorem 10.3.24, $\text{Cen}(Q(A)) = \mathbb{C}$. Conversely, if $\text{Cen}(Q(A)) = \mathbb{C}$, then A is prime by Remark 10.1.16 because A is semiprime. □

In Theorem 10.3.38, a C^*-algebra whose extended centroid is a direct product of copies of $\mathbb{C}$ is characterized. Moreover, Theorem 10.3.38 characterizes a C^*-algebra A such that $M_{\text{loc}}(A)$ is a C^*-direct product of prime C^*-algebras.

Theorem 10.3.38 *Let A be a C^*-algebra and let Λ be an index set. Then the following are equivalent.*

(i) *There is a set of uniform ideals $\{U_i \mid i \in \Lambda\}$ of A such that $\mathbb{C}U_i = U_i$ for each i, $\sum_{i\in\Lambda} U_i = \bigoplus_{i\in\Lambda} U_i$, and $\ell_A(\bigoplus_{i\in\Lambda} U_i) = 0$.*
(ii) *The extended centroid of A is $\mathbb{C}^{|\Lambda|}$.*
(iii) *$M_{\text{loc}}(A)$ is a C^*-direct product of $|\Lambda|$ unital prime C^*-algebras.*
(iv) *$\text{Cen}(M_{\text{loc}}(A))$ is a C^*-direct product of $|\Lambda|$ copies of $\mathbb{C}$.*

Proof (i)⇒(ii) From Theorem 10.1.17 and Proposition 10.1.4, there exists a set of nonzero orthogonal idempotents $\{e_i \mid i \in \Lambda\} \subseteq \mathcal{B}(Q(A))$ such that

$$Q(A) = \prod_{i\in\Lambda} e_i Q(A)$$

and each $e_i Q(A)$ is prime. As $e_i \in \mathcal{B}(Q(A)) = \mathcal{B}(M_{\text{loc}}(A))$ (Lemma 10.3.14(i)) and e_i is a projection (Proposition 10.3.6(ii)), $e_i A$ is a C^*-algebra. As in the argument for the proof of (i)⇒(vi) of Theorem 10.1.17, $e_i A$ is prime and $Q(e_i A) = e_i Q(A)$. Thus, $\mathbb{C} = \text{Cen}(Q(e_i A)) = \text{Cen}(e_i Q(A))$ by Lemma 10.3.37. Therefore we get that $\text{Cen}(Q(A)) = \prod_{i\in\Lambda} \text{Cen}(e_i Q(A)) = \prod_{i\in\Lambda} \mathbb{C} = \mathbb{C}^{|\Lambda|}$.

(ii)⇒(i) The proof follows from Theorem 10.1.17 and Proposition 10.1.4.

(i)⇒(iii) As in the proof of (i)⇒(ii), there exists a set of nonzero orthogonal idempotents $\{e_i \mid i \in \Lambda\} \subseteq \mathcal{B}(Q(A))$ such that $Q(A) = \prod_{i\in\Lambda} e_i Q(A)$ and each $e_i Q(A)$ is prime. Hence, $\{e_i \mid i \in \Lambda\} \subseteq \mathcal{B}(M_{\text{loc}}(A)) = \mathcal{B}(Q(A))$. Further, each e_i is a projection by Proposition 10.3.6(ii).

By Theorem 10.1.17, $\sup\{e_i \mid i \in \Lambda\} = 1$. Indeed, we note that A is right nonsingular and thus A^1 is right nonsingular. Hence, $Q(A) = Q(A^1)$ is regular right self-injective from Theorem 2.1.31. So $\mathcal{B}(Q(A))$ is a complete Boolean algebra from Corollary 8.3.14. Say $f \in \mathcal{B}(Q(A))$ such that $f = \sup\{e_i \mid i \in \Lambda\}$. As $e_i = e_i f$ for all i, $1 - f = (e_i(1-f))_{i\in\Lambda} = 0$ in $\prod_{i\in\Lambda} e_i Q(A)$, so $f = 1$. From Proposition 10.3.33(ii), $M_{\text{loc}}(A) = \prod_{i\in\Lambda}^{C^*} e_i M_{\text{loc}}(A)$.

Now $M_{\text{loc}}(A) = e_i M_{\text{loc}}(A) \oplus (1 - e_i) M_{\text{loc}}(A)$, therefore we have that $\mathcal{B}(e_i M_{\text{loc}}(A)) = e_i \mathcal{B}(M_{\text{loc}}(A))$. By Lemma 10.3.14(i), $\mathcal{B}(M_{\text{loc}}(A)) = \mathcal{B}(Q(A))$. Hence, $\mathcal{B}(e_i M_{\text{loc}}(A)) = e_i \mathcal{B}(Q(A)) = \mathcal{B}(e_i Q(A))$.

Note that $e_i Q(A)$ is prime, so $e_i M_{\text{loc}}(A)$ is indecomposable as a ring since $\mathcal{B}(e_i M_{\text{loc}}(A)) = \mathcal{B}(e_i Q(A)) = \{0, e_i\}$. Because $e_i M_{\text{loc}}(A)$ is semiprime, we see that $\mathbf{S}_\ell(e_i M_{\text{loc}}(A)) = \mathcal{B}(e_i M_{\text{loc}}(A)) = \{0, e_i\}$ by Proposition 1.2.6(ii). From Theorem 10.3.10, $M_{\text{loc}}(A)$ is quasi-Baer. Thus Theorem 3.2.10 yields that $e_i M_{\text{loc}}(A)$ is quasi-Baer. As $e_i M_{\text{loc}}(A)$ is semicentral reduced, $e_i M_{\text{loc}}(A)$ is prime by Proposition 3.2.5. Further, as e_i is a projection, $e_i M_{\text{loc}}(A)$ is a prime C^*-algebra for each i.

(iii)⇒(iv) Let $M_{\text{loc}}(A) = \prod_{i\in\Lambda}^{C^*} M_i$, where each M_i is a unital prime C^*-algebra. Then $\text{Cen}(M_{\text{loc}}(A)) = \prod_{i\in\Lambda}^{C^*} \text{Cen}(M_i)$. Each M_i is boundedly centrally closed by

Theorem 10.3.23. Thus $\mathrm{Cen}(M_i)$ is an AW^*-algebra from Corollary 10.3.22. So Proposition 10.3.13 yields that $\mathrm{Cen}(M_i) = \mathbb{C}$ because 1 is the only nonzero projection in $\mathrm{Cen}(M_i)$.

(iv)⇒(i) From the fact that $\mathrm{Cen}(M_{\mathrm{loc}}(A)) = \prod_{i\in\Lambda}^{C^*} \mathbb{C}$, Lemma 10.3.14(i), and modification of the proof of (i)⇒(iii), there is a set of nonzero orthogonal idempotents $\{e_i \mid i \in \Lambda\} \subseteq \mathcal{B}(M_{\mathrm{loc}}(A)) = \mathcal{B}(Q(A))$, and $1 = \sup\{e_i \mid i \in \Lambda\}$. By the proof of Corollary 10.1.11, there exists a ring isomorphism

$$\phi : Q(A) \to \prod_{i\in\Lambda} e_i Q(A),$$

where $\phi(q) = (e_i q)_{i\in\Lambda}$ for $q \in Q(A)$.

As $\mathbb{C} = e_i\mathrm{Cen}(M_{\mathrm{loc}}(A)) = \mathrm{Cen}(e_i M_{\mathrm{loc}}(A))$, $\mathcal{B}(e_i M_{\mathrm{loc}}(A)) = \{0, e_i\}$. Now $\mathcal{B}(e_i Q(A)) = e_i\mathcal{B}(Q(A)) = e_i\mathcal{B}(M_{\mathrm{loc}}(A)) = \mathcal{B}(e_i M_{\mathrm{loc}}(A))$ since $e_i \in \mathcal{B}(Q(A))$ and $\mathcal{B}(Q(A)) = \mathcal{B}(M_{\mathrm{loc}}(A))$ by Lemma 10.3.14(i). Thus, $\mathrm{Cen}(e_i Q(A))$ has 0 and e_i as its only idempotents. So each $\mathrm{Cen}(e_i Q(A)) = \mathrm{Cen}(Q(e_i Q(A)))$ is a field as $e_i Q(A)$ is semiprime and $\mathrm{Cen}(e_i Q(A))$ is regular from Theorem 10.1.15. By Remark 10.1.16, each $e_i Q(A)$ is prime. Let $Q_i = \phi^{-1}(e_i Q(A))$. Then it follows that $Q(A) = \prod_{i\in\Lambda} Q_i$ and each Q_i is prime. So Theorem 10.1.17 and Proposition 10.1.4 yield part (i). □

Theorem 10.3.38 is applied to AW^*-algebras as follows.

Corollary 10.3.39 *Let A be an AW^*-algebra and $\aleph$ a cardinal number. Then the following are equivalent.*

(i) *The extended centroid of A is $\mathbb{C}^{\aleph}$.*
(ii) *A is a C^*-direct product of $\aleph$ prime AW^*-algebras.*
(iii) *$\mathrm{Cen}(A)$ is a C^*-direct product of $\aleph$ copies of $\mathbb{C}$.*

Proof If A is an AW^*-algebra, then $M_{\mathrm{loc}}(A) = A$ by Theorem 10.3.18. Thus, the result follows immediately from Theorem 10.3.38. □

Corollary 10.3.40 *Let A be a C^*-algebra, B an intermediate C^*-algebra between A and $M_{\mathrm{loc}}(A)$, and $\aleph$ a cardinal number. Then the extended centroid of A is $\mathbb{C}^{\aleph}$ if and only if the extended centroid of B is $\mathbb{C}^{\aleph}$.*

Proof Corollary 10.3.36(ii) and Theorem 10.3.38 yield the result. □

Next we consider the case when the index set Λ in Theorem 10.3.38 is finite. In general, $Q_{\mathbf{qB}}(A^1)$ may not be norm complete. However, the following result provides a sufficient condition for $Q_{\mathbf{qB}}(A^1)$ to be a C^*-algebra.

Theorem 10.3.41 *Let A be a C^*-algebra and n a positive integer. Then the following are equivalent.*

(i) *A has exactly n minimal prime ideals.*
(ii) $Q_{\mathbf{qB}}(A^1)$ *is a direct sum of n prime C^*-algebras.*
(iii) *The extended centroid of A is $\mathbb{C}^n$.*
(iv) $M_{\mathrm{loc}}(A)$ *is a direct sum of n prime C^*-algebras.*
(v) $\mathrm{Cen}(M_{\mathrm{loc}}(A)) = \mathbb{C}^n$.

Proof The equivalence of (iii), (iv), and (v) follows from Theorem 10.3.38.

(i)⇒(ii) By Theorem 10.1.19, A^1 has exactly n minimal prime ideals. Let $P_1, \dots, P_n$ be all the minimal prime ideals of A^1. Then by Theorem 10.1.20, $Q_{\mathbf{qB}}(A^1) = S_1 \oplus \cdots \oplus S_n$, where each S_i is a prime ring. Let $K_i = \oplus_{j \neq i} S_j$ for each i. From the proof of Theorem 10.1.20, $\{K_1, \dots, K_n\}$ is the set of all minimal prime ideals of $Q_{\mathbf{qB}}(A^1)$ and

$$Q_{\mathbf{qB}}(A^1) \cong Q_{\mathbf{qB}}(A^1)/K_1 \oplus \cdots \oplus Q_{\mathbf{qB}}(A^1)/K_n,$$

where $K_i \cap A^1 = P_i$ and $K_i = e_i Q_{\mathbf{qB}}(A^1)$ with $e_i \in \mathcal{B}(Q(A)) = \mathcal{B}(Q(A^1))$.

We show that each P_i is norm closed in A^1. For this, assume that $\{x_n\}$ is a sequence in P_i with $\lim_{n\to\infty} x_n = x \in A^1$. Observe from Theorems 8.3.17 and 10.3.7(i), that $e_i \in Q_{\mathbf{qB}}(A^1) \subseteq Q_b(A^1) \subseteq M_{\mathrm{loc}}(A^1)$. Therefore,

$$xe_i = (\lim_{n\to\infty} x_n)e_i = \lim_{n\to\infty} (x_n e_i) = \lim_{n\to\infty} x_n = x$$

because $P_i \subseteq Q_{\mathbf{qB}}(A^1)e_i \subseteq M_{\mathrm{loc}}(A^1)$. Hence $x = xe_i \in Q_{\mathbf{qB}}(A^1)e_i \cap A^1 = P_i$, so P_i is a norm closed ideal of A^1. Thus P_i is self-adjoint and A^1/P_i is a C^*-algebra (since $e_i \in \mathcal{B}(Q(A^1)) = \mathcal{B}(M_{\mathrm{loc}}(A^1))$ is a projection from Lemma 10.3.14(i) and Proposition 10.3.6(ii), we also see that each P_i is self-adjoint).

As $Q_{\mathbf{qB}}(A^1) = \oplus_{i=1}^n S_i$ and $K_i = \oplus_{j\neq i} S_j = e_i Q_{\mathbf{qB}}(A^1)$, we have that

$$S_i = (1 - e_i)Q_{\mathbf{qB}}(A^1).$$

Since e_i is a projection and central, S_i is a $*$-algebra for each i. Also by Theorem 10.3.7(i), $Q_{\mathbf{qB}}(A^1)$ is a $*$-subalgebra of $M_{\mathrm{loc}}(A^1)$.

Now let $\phi : A^1/P_i \to Q_{\mathbf{qB}}(A^1)/K_i$ be defined by $\phi(a + P_i) = a + K_i$, where $a \in A^1$. Then ϕ is an isomorphism from Lemma 8.3.26(i). Further, K_i is self-adjoint as S_i is self-adjoint. So ϕ is a $*$-isomorphism. We define

$$\varphi : Q_{\mathbf{qB}}(A^1)/K_i \to (1 - e_i)Q_{\mathbf{qB}}(A^1) = S_i \text{ by } \varphi(q + K_i) = (1 - e_i)q,$$

where $q \in Q_{\mathbf{qB}}(A^1)$. Then φ is also a $*$-isomorphism. So $\varphi\phi$ is a $*$-isomorphism from A^1/P_i to S_i. Thus S_i is a C^*-subalgebra of $M_{\mathrm{loc}}(A^1)$. Hence,

$$Q_{\mathbf{qB}}(A^1) = \oplus_{i=1}^n S_i$$

is a C^*-subalgebra of $M_{\mathrm{loc}}(A^1)$, where each S_i is a prime C^*-algebra.

(ii)⇒(iii) Assume that $Q_{\mathbf{qB}}(A^1) = \oplus_{i=1}^n S_i$, where each S_i a prime C^*-algebra. Then $\mathrm{Cen}(Q(S_i)) = \mathbb{C}$ from Lemma 10.3.37. Therefore,

$$\mathrm{Cen}(Q(A)) = \mathrm{Cen}(Q(A^1)) = \mathrm{Cen}(Q(Q_{\mathbf{qB}}(A^1))) = \oplus_{i=1}^n \mathrm{Cen}(Q(S_i)) = \mathbb{C}^n.$$

(iii)⇒(i) It follows from Theorem 10.1.19. □

Corollary 10.3.42 *Let A be a C^*-algebra and B an intermediate C^*-algebra between A and $M_{\text{loc}}(A)$. Then for any positive integer n, A has exactly n minimal prime ideals if and only if B has exactly n minimal prime ideals.*

Proof The result follows from Corollary 10.3.36(ii) and Theorem 10.3.41. □

It is well known that $A * G$ and A^G are C^*-algebras when A is a C^*-algebra and G is a finite group of X-outer $*$-automorphisms of A (see [17, p. 140]). Since C^*-algebras are semiprime, we obtain the next result from Theorems 9.2.10 and 10.3.20.

Theorem 10.3.43 *Let A be a unital C^*-algebra and G a finite group of X-outer $*$-automorphisms of A. Then the following are equivalent.*

(i) *$A * G$ is a quasi-AW^*-algebra.*
(ii) *A is G-quasi-Baer.*
(iii) *A^G is a quasi-AW^*-algebra.*

Let A be a quasi-AW^*-algebra and G be a finite group of X-outer $*$-automorphisms of A. Then $A * G$ and A^G are quasi-AW^*-algebras from Theorem 10.3.43. We recall that if G is a group of X-outer $*$-automorphisms of a C^*-algebra A, then G is also a group of X-outer $*$-automorphisms on $Q_b(A)$.

Corollary 10.3.44 (i) *Let A be a unital C^*-algebra and G a finite group of X-outer $*$-automorphisms of A. Then $Q_b(A) * G$ and $Q_b(A)^G$ are quasi-Baer.*

(ii) *Let A be a C^*-algebra. If G is a finite group of X-outer $*$-automorphisms of $M_{\text{loc}}(A)$, then $M_{\text{loc}}(A) * G$ and $M_{\text{loc}}(A)^G$ are quasi-AW^*-algebras.*

Proof (i) By Theorem 10.3.7(ii), $Q_b(A)$ is a quasi-Baer $*$-algebra. Thus from Theorem 9.2.10, $Q_b(A) * G$ and $Q_b(A)^G$ are quasi-Baer.

(ii) By Theorem 10.3.20 and Corollary 10.3.21(i), $M_{\text{loc}}(A)$ is a quasi-AW^*-algebra. So Theorem 10.3.43 yields that $M_{\text{loc}}(A) * G$ and $M_{\text{loc}}(A)^G$ are quasi-AW^*-algebras. □

Corollary 10.3.45 *If A is a unital C^*-algebra with only finitely many minimal prime ideals and G is a finite group of X-outer $*$-automorphisms of A, then $Q_{\mathbf{qB}}(A) * G$ and $Q_{\mathbf{qB}}(A)^G$ are quasi-AW^*-algebras.*

Proof From Theorem 10.3.41, $Q_{\mathbf{qB}}(A)$ is a C^*-algebra. Because $Q_{\mathbf{qB}}(A)$ is quasi-Baer, it is a quasi-AW^*-algebra. We observe that G is also a group of X-outer $*$-automorphisms of $Q_{\mathbf{qB}}(A)$. Therefore Theorem 10.3.43 yields that $Q_{\mathbf{qB}}(A) * G$ and $Q_{\mathbf{qB}}(A)^G$ are quasi-AW^*-algebras. □

Theorem 10.3.46 *Let A be a C^*-algebra and n a positive integer. Assume that any one of conditions* (i) *through* (v) *of Theorem* 10.3.41 *is satisfied. Then every boundedly centrally closed intermediate C^*-algebra between A and $M_{\text{loc}}(A)$ is a direct sum of n prime C^*-algebras.*

Proof Let B be any boundedly centrally closed intermediate C^*-algebra between A and $M_{\text{loc}}(A)$. We prove that B is a direct sum of n prime C^*-algebras.

First suppose that B is unital. From Theorem 10.3.20, B is a quasi-Baer ring. Also $\text{Cen}(M_{\text{loc}}(B)) = \text{Cen}(M_{\text{loc}}(A)) = \mathbb{C}^n$ by Corollary 10.3.36(ii) and Theorem 10.3.41. Thus $B = Q_{\mathbf{qB}}(B)$ is a direct sum of n prime C^*-algebras by Theorem 10.3.41.

Next assume that B is nonunital. Now because B is boundedly centrally closed, $B = \overline{B\mathcal{B}(Q(A))}$ by Theorem 10.3.35(ii). As in the proof of Theorem 10.3.35(i), $B^1 = \{b + \lambda 1_{M_{\text{loc}}(A)} \mid b \in B \text{ and } \lambda \in \mathbb{C}\} \subseteq M_{\text{loc}}(A)$. Now $B\mathcal{B}(Q(A))$ is an ideal of $B^1\mathcal{B}(Q(A))$, and so $B = \overline{B\mathcal{B}(Q(A))} \trianglelefteq \overline{B^1\mathcal{B}(Q(A))}$. Note that the C^*-algebra $\overline{B^1\mathcal{B}(Q(A))}$ is unital and boundedly centrally closed by Theorem 10.3.29(iii), so it is a direct sum of n prime C^*-algebras from the preceding argument. Say

$$\overline{B^1\mathcal{B}(Q(A))} = e_1\,\overline{B^1\mathcal{B}(Q(A))} \oplus \cdots \oplus e_n\,\overline{B^1\mathcal{B}(Q(A))},$$

where each $e_i\,\overline{B^1\mathcal{B}(Q(A))}$ is a prime C^*-algebra and $e_1, \ldots, e_n$ are nonzero orthogonal central idempotents in $\overline{B^1\mathcal{B}(Q(A))}$ with $e_1 + \cdots + e_n = 1$.

Hence $e_i \in \mathcal{B}((M_{\text{loc}}(A)))$ as $\text{Cen}(\overline{B^1\mathcal{B}(Q(A))}) \subseteq \text{Cen}(M_{\text{loc}}(A))$ from Theorem 10.3.27(i) (so $e_i \in \mathcal{B}(Q(A))$ by Lemma 10.3.14(i)). Thus, we have that

$$M_{\text{loc}}(A) = e_1 M_{\text{loc}}(A) \oplus \cdots \oplus e_n M_{\text{loc}}(A).$$

Also note that each e_i is a projection by Proposition 10.3.6(i).

Since each $e_i M_{\text{loc}}(A)$ is a C^*-algebra, it is a norm closed ideal of $M_{\text{loc}}(A)$. Therefore, $e_i M_{\text{loc}}(A) \cap A \neq 0$ from the proof of Theorem 10.3.29(ii). Hence, $0 \neq e_i M_{\text{loc}}(A) \cap B = e_i M_{\text{loc}}(A) \cap B \cap \overline{B^1\mathcal{B}(Q(A))} = e_i\,\overline{B^1\mathcal{B}(Q(A))} \cap B$ as $e_i M_{\text{loc}}(A) \cap \overline{B^1\mathcal{B}(Q(A))} = e_i\,\overline{B^1\mathcal{B}(Q(A))}$. Since $B \trianglelefteq \overline{B^1\mathcal{B}(Q(A))}$,

$$B = (e_1\,\overline{B^1\mathcal{B}(Q(A))} \cap B) \oplus \cdots \oplus (e_n\,\overline{B^1\mathcal{B}(Q(A))} \cap B).$$

We claim that $e_i\,\overline{B^1\mathcal{B}(Q(A))} \cap B = e_i B$. First, we observe that $e_i B \subseteq B$ because $B \trianglelefteq \overline{B^1\mathcal{B}(Q(A))}$ and $e_i \in \mathcal{B}(Q(A)) \subseteq \overline{B^1\mathcal{B}(Q(A))}$. Therefore,

$$e_i B \subseteq e_i\,\overline{B^1\mathcal{B}(Q(A))} \cap B.$$

Clearly, $e_i\,\overline{B^1\mathcal{B}(Q(A))} \cap B \subseteq e_i B$. Thus $e_i\,\overline{B^1\mathcal{B}(Q(A))} \cap B = e_i B$.

Therefore, $B = e_1 B \oplus \cdots \oplus e_n B$ and $e_i B \neq 0$ for each i (by our preceding argument $e_i B = e_i M_{\text{loc}}(A) \cap B \neq 0$). From Proposition 8.3.55(ii), each $e_i B$ is a prime ring because $e_i B \trianglelefteq e_i\,\overline{B^1\mathcal{B}(Q(A))}$ and $e_i\,\overline{B^1\mathcal{B}(Q(A))}$ is a prime ring. Therefore, B is a direct sum of n prime C^*-algebras $e_i B$. □

For other application of Theorem 10.3.41, we start with the next lemma.

Lemma 10.3.47 *Let R be a ring with a finite group G of ring automorphisms of R such that $R * G$ is semiprime.*

(i) *If I is a nonzero G-invariant right ideal of R, then* $\mathrm{tr}(I) \neq 0$.

(ii) $\mathrm{udim}(R_R) < \infty$ *if and only if* $\mathrm{udim}({R^G}_{R^G}) < \infty$, *and in this case we have that* $\mathrm{udim}({R^G}_{R^G}) \leq \mathrm{udim}(R_R) \leq \mathrm{udim}({R^G}_{R^G})\,|G|$.

Proof (i) Recall that

$$\mathrm{tr}(a) = \sum_{g \in G} a^g, \ \mathrm{tr}(I) = \{\mathrm{tr}(a) \mid a \in I\}, \ \text{ and } t = \sum_{g \in G} g \in R * G.$$

Then for $a \in I$, $tat = \mathrm{tr}(a)t$. So, if $\mathrm{tr}(I) = 0$, then $tIt = 0$. As a consequence, $(tI)(tI) = (tIt)I = 0$. Note that tI is a right ideal of $R * G$ and $R * G$ is semiprime, and therefore $tI = 0$. Thus $I = 0$, a contradiction.

(ii) Let $\mathrm{udim}(R_R) = n < \infty$ and $\oplus_{i=1}^k a_i R^G$ (with $a_i \in R^G$) be a direct sum of nonzero right ideals $a_i R^G$ of R^G. We show that $\sum_{i=1}^k a_i R = \oplus_{i=1}^k a_i R$. For this, let $I = a_1 R \cap (a_2 R + \cdots + a_k R)$. Then I is a G-invariant right ideal of R. We see that $\mathrm{tr}(I) \subseteq a_1 \mathrm{tr}(R) \cap (\sum_{i=2}^k a_i \mathrm{tr}(R)) \subseteq a_1 R^G \cap (\sum_{i=2}^k a_i R^G) = 0$. By part (i), $I = 0$. Similarly, $a_2 R \cap (a_1 R + a_3 R + \cdots + a_k R) = 0$, and so on. Consequently, $\sum_{i=1}^k a_i R = \oplus_{i=1}^k a_i R$. Therefore $k \leq n$, and so $\mathrm{udim}({R^G}_{R^G}) \leq n$.

Next, assume that $\mathrm{udim}({R^G}_{R^G}) = m < \infty$. If $I \cap R^G \neq 0$ for all nonzero right ideal I of R, then $\mathrm{udim}(R_R) \leq m$ and hence $\mathrm{udim}(R_R) \leq m|G|$.

Say $I \cap R^G = 0$ for some nonzero right ideal I of R. By Zorn's lemma, there exists a nonzero right ideal K of R maximal with respect to $K \cap R^G = 0$. We claim that $\mathrm{udim}((R/K)_R) \leq m$. For this, we assume that $\oplus_{i=1}^{\ell}(K_i/K)$ is a direct sum of nonzero R-submodules of $(R/K)_R$. As $K_i/K \neq 0$, $K_i \cap R^G \neq 0$ for each i, by the choice of K.

We note that $(K_1 \cap R^G) \cap (\sum_{i=2}^{\ell}(K_i \cap R^G)) \subseteq K \cap R^G = 0$, thus we obtain that $(K_1 \cap R^G) \cap (\sum_{i=2}^{\ell}(K_i \cap R^G)) = 0$. Also, $(K_j \cap R^G) \cap (\sum_{i \neq j}^{\ell}(K_i \cap R^G)) = 0$ for all j. Thus, $\sum_{i=1}^{\ell}(K_i \cap R^G) = \oplus_{i=1}^{\ell}(K_i \cap R^G)$, and so $\ell \leq \mathrm{udim}({R^G}_{R^G}) = m$. Thus, we obtain $\mathrm{udim}((R/K)_R) \leq m$.

Observe that $K^g = \{x^g \mid x \in K\}$ is a nonzero right ideal of R which is also maximal with respect to the property that $K^g \cap R^G = 0$. Similarly, we obtain that $\mathrm{udim}((R/K^g)_R) \leq m$ for all $g \in G$. Because $\cap_{g \in G} K^g$ is a G-invariant right ideal of R and $(\cap_{g \in G} K^g) \cap R^G = 0$, $\mathrm{tr}(\cap_{g \in G} K^g) = 0$. Therefore, $\cap_{g \in G} K^g = 0$ by part (i). Thus, R can be embedded in $\prod_{g \in G}(R/K^g)$ as a right R-module. Consequently, $\mathrm{udim}(R_R) \leq m|G|$. □

Recall that MinSpec(−) denotes the set of all minimal prime ideals of a ring. For an application of Theorem 10.3.41, we discuss the following result which exhibits an interesting relationship between the numbers of minimal prime ideals of A, A^G, and $A * G$ when G is a finite group of X-outer $*$-automorphisms of a unital C^*-algebra A.

Theorem 10.3.48 *Let A be a unital C^*-algebra and G a finite group of X-outer $*$-automorphisms of A. Then the following are equivalent.*

(i) $|\text{MinSpec}(A)| < \infty$.
(ii) $|\text{MinSpec}(A^G)| < \infty$.
(iii) $|\text{MinSpec}(A * G)| < \infty$.

In this case, $|\text{MinSpec}(A * G)| = |\text{MinSpec}(A^G)|$ *and*

$$|\text{MinSpec}(A^G)| \leq |\text{MinSpec}(A)| \leq |\text{MinSpec}(A^G)|\,|G|.$$

Proof We use udim(−) to denote the right uniform dimension of a ring. Since G is X-outer on $Q(A)$, Lemmas 9.2.4, 9.2.7, 9.2.9(ii), and Proposition 9.2.8 yield that

$$\begin{aligned}\text{Cen}(Q(A * G)) &= \text{Cen}(Q(A) * G) = [\text{Cen}(Q(A))]^G = \text{Cen}(Q(A)^G)\\ &\cong \text{Cen}(Q(A^G)).\end{aligned}$$

(i)⇒(ii) and (i)⇒(iii) Let $|\text{MinSpec}(A)| = n < \infty$. Then $\text{Cen}(Q(A)) = \mathbb{C}^n$ by Theorem 10.3.41. Hence, $\text{udim}(\text{Cen}(Q(A))) = n$. We observe that G induces a group H of ring automorphisms of $\text{Cen}(Q(A))$ and H is a homomorphic image of G. Because $|H|$ is invertible in $\text{Cen}(Q(A))$, $\text{Cen}(Q(A)) * H$ is semiprime by Lemma 9.2.3(ii). Now $\text{udim}([\text{Cen}(Q(A))]^H) \leq \text{udim}(\text{Cen}(Q(A)))$ from Lemma 10.3.47. We let $k = \text{udim}([\text{Cen}(Q(A))]^H)$. Then $k \leq n$.

We observe that $[\text{Cen}(Q(A))]^H = [\text{Cen}(Q(A))]^G \cong \text{Cen}(Q(A^G))$. Also note by Lemma 9.2.3(ii) that A^G is semiprime. Hence $\text{Cen}(Q(A^G))$ is regular by Theorem 10.1.15. Therefore, $\text{Cen}(Q(A^G))$ is a direct sum of k fields because $\text{udim}(\text{Cen}(Q(A^G))) = \text{udim}([\text{Cen}(Q(A))]^H) = k < \infty$. So $|\text{MinSpec}(A^G)| = k$ by Theorem 10.1.19. As $\text{Cen}(Q(A * G)) \cong \text{Cen}(Q(A^G))$, $\text{Cen}(Q(A * G))$ is a direct sum of k fields. Thus, $|\text{MinSpec}(A * G)| = |\text{MinSpec}(A^G)| = k \leq n$ from Theorem 10.1.19.

(ii)⇔(iii) This equivalence follows immediately from Theorem 10.1.19 and the fact that $\text{Cen}(Q(A * G)) \cong \text{Cen}(Q(A^G))$.

(ii)⇒(i) Let $|\text{MinSpec}(A^G)| = k < \infty$. Since A^G is a C^*-algebra, we have that $\text{Cen}(Q(A^G)) = \mathbb{C}^k$ from Theorem 10.3.41. As before, G induces a group H of ring automorphisms of $\text{Cen}(Q(A))$ and H is a homomorphic image of G. We note that $[\text{Cen}(Q(A))]^H = [\text{Cen}(Q(A))]^G \cong \text{Cen}(Q(A^G)) = \mathbb{C}^k$, thus $\text{udim}([\text{Cen}(Q(A))]^H) = k$. Since $|H|$ is invertible, $\text{Cen}(Q(A)) * H$ is semiprime by Lemma 9.2.3(ii). Therefore, from Lemma 10.3.47,

$$\text{udim}([\text{Cen}(Q(A))]^H) \leq \text{udim}(\text{Cen}(Q(A))) \leq \text{udim}([\text{Cen}(Q(A))]^H)|H|.$$

Hence, $\text{udim}([\text{Cen}(Q(A))]^G) \leq \text{udim}(\text{Cen}(Q(A))) \leq \text{udim}([\text{Cen}(Q(A))]^G)|G|$ because $|H| \leq |G|$ and $[\text{Cen}(Q(A))]^G = [\text{Cen}(Q(A))]^H$. Now we observe that $\text{udim}([\text{Cen}(Q(A))]^G) = \text{udim}([\text{Cen}(Q(A))]^H) = k$. Therefore,

$$n := \text{udim}(\text{Cen}(Q(A))) \leq k|G|.$$

Since $\text{Cen}(Q(A))$ is regular from Theorem 10.1.15, $\text{Cen}(Q(A))$ is a finite direct sum of n fields. So $|\text{MinSpec}(A)| = n$ by Theorem 10.1.19. In this case,

$$\begin{aligned}|\text{MinSpec}(A * G)| &= \text{udim}(\text{Cen}(Q(A) * G)) = \text{udim}([\text{Cen}(Q(A))]^G) \\ &= \text{udim}(\text{Cen}(Q(A^G))) = |\text{MinSpec}(A^G)|.\end{aligned}$$

Thus, $|\text{MinSpec}(A^G)| \leq |\text{MinSpec}(A)| \leq |\text{MinSpec}(A^G)|\,|G|$. □

Recall that Tdim(−) denotes the triangulating dimension of a ring.

Theorem 10.3.49 *Let A be a quasi-AW^*-algebra and G a finite group of X-outer $*$-automorphisms of A. Then*:

(i) $\text{Tdim}(A * G) = \text{Tdim}(A^G)$.
(ii) $\text{Tdim}(A * G) \leq \text{Tdim}(A) \leq \text{Tdim}(A * G)\,|G|$.
(iii) *If* $\text{Tdim}(A) = n < \infty$, *then there exists a positive integer $k \leq n$ such that both $A * G$ and A^G are direct sums of k prime C^*-algebras.*

Proof (i) From Theorem 10.3.43, $A * G$ and A^G are quasi-AW^*-algebras. So $A * G$ and A^G are quasi-Baer rings. If $\text{Tdim}(A * G) = n < \infty$, then we see that $n = |\text{MinSpec}(A * G)| = |\text{MinSpec}(A^G)| = \text{Tdim}(A^G)$ by Theorems 5.4.20 and 10.3.48. Next, if $\text{Tdim}(A * G) = \infty$, then $|\text{MinSpec}(A * G)| = \infty$ from Theorem 5.4.20. By Theorem 10.3.48, $|\text{MinSpec}(A^G)| = \infty$. Hence, $\text{Tdim}(A^G) = \infty$ from Theorem 5.4.20. Therefore, $\text{Tdim}(A * G) = \text{Tdim}(A^G)$.

(ii) By Theorems 10.3.48 and 5.4.20, if any one of $\text{Tdim}(A)$, $\text{Tdim}(A * G)$, and $\text{Tdim}(A^G)$ is finite, then all are finite. Also

$$\text{Tdim}(A * G) \leq \text{Tdim}(A) \leq \text{Tdim}(A * G)\,|G|.$$

If one of $\text{Tdim}(A)$, $\text{Tdim}(A * G)$, and $\text{Tdim}(A^G)$ is infinite, then we are also done by Theorems 10.3.48 and 5.4.20.

(iii) Assume that $\text{Tdim}(A) = n < \infty$. Then $|\text{MinSpec}(A)| = n$ by Theorem 5.4.20. From Theorem 10.3.48, $|\text{MinSpec}(A * G)| = |\text{MinSpec}(A^G)| = k \leq n$ for some k. Therefore, $A * G$ and A^G are direct sums of k prime C^*-algebras by Theorem 10.3.41. □

There exist a quasi-AW^*-algebra A and a finite group G of X-outer $*$-automorphisms of A such that $\text{Tdim}(A * G) \lneq \text{Tdim}(A)$ as the next example illustrates.

Example 10.3.50 Let $A = \mathbb{C}^n$ with $n \geq 2$ and $*$ be the componentwise conjugate involution. Define $g \in \text{Aut}(A)$ by $g(a_1, a_2, \ldots, a_n) = (a_2, a_3, \ldots, a_n, a_1)$ for $(a_1, a_2, \ldots, a_n) \in A$. Then g is an X-outer $*$-automorphism and $g^n = 1$. Let G be the cyclic group generated by g. Then G is X-outer. By Lemma 9.2.3(i), A^G is semiprime. Note that $\mathbf{S}_\ell(A^G) = \mathcal{B}(A^G) = \{0, 1\}$. Therefore A^G is semicentral reduced, so $\text{Tdim}(A^G) = 1$. By Theorem 10.3.49, $\text{Tdim}(A * G) = \text{Tdim}(A^G) = 1$. But $\text{Tdim}(A) = n \geq 2$.

Proposition 10.3.51 *Let A be a unital C^*-algebra and let G be a finite group of $*$-automorphisms of A. Then the following are equivalent.*

(i) A_{A*G} *is a (strongly) FI-extending module.*
(ii) A_{A*G} *is a quasi-Baer module.*
(iii) A^G *is a quasi-AW^*-algebra.*

Further, if G is X-outer, then the conditions (iv) *and* (v) *are equivalent to the conditions* (i)–(iii).

(iv) $A * G$ *is a quasi-AW^*-algebra.*
(v) A *is G-quasi-Baer.*

Proof Note that $\text{End}(A_{A*G}) \cong A^G$. Because $|G|$ is invertible, we see that $e = |G|^{-1}(\sum_{g\in G} g) \in A * G$ is an idempotent. Define $\theta : A_{A*G} \to e(A * G)_{A*G}$ by $\theta(a) = ea$ for $a \in A$. Then $A_{A*G} \cong e(A * G)_{A*G}$ via θ. Thus A_{A*G} is a finitely generated projective module. Hence, the equivalence of (i)–(iii) follows from Theorems 3.2.37, 8.4.20 and Lemma 9.2.3(ii).

Further, assume that G is X-outer. Then (iii), (iv), and (v) are equivalent by Theorem 10.3.43. □

The following is a well known result which describes finite dimensional C^*-algebras (see [136, Theorem III.1.1]).

Theorem 10.3.52 *Any finite dimensional C^*-algebra is unital and $*$-isomorphic to* $\text{Mat}_{m_1}(\mathbb{C}) \oplus \cdots \oplus \text{Mat}_{m_t}(\mathbb{C})$, *where $m_1, \ldots, m_t$ are positive integers.*

In the next result, we characterize C^*-algebras satisfying a PI with only finitely many minimal prime ideals.

Theorem 10.3.53 *Let A be a C^*-algebra. Then the following are equivalent.*

(i) *A satisfies a PI and has exactly n minimal prime ideals.*
(ii) $A \cong \text{Mat}_{k_1}(\mathbb{C}) \oplus \cdots \oplus \text{Mat}_{k_n}(\mathbb{C})$ *($*$-isomorphic) for some positive integers* $k_1, \ldots, k_n$.

Proof (i)⇒(ii) Assume that A satisfies a polynomial identity. Then since A is semiprime, by Theorem 8.3.52, A satisfies a standard identity

$$f_m(x_1, x_2, \ldots, x_m) = \sum_{\sigma\in S_m} \text{sgn}(\sigma)x_{\sigma(1)}x_{\sigma(2)} \cdots x_{\sigma(m)}.$$

Also for any $k \geq m$, A satisfies $f_k(x_1, x_2, \ldots, x_k)$ by Theorem 8.3.52. Thus we may assume that $m \geq 3$.

Let $g(x_1, \ldots, x_m) = f_k(x_1x_2 - x_2x_1, \ldots, x_1x_m - x_mx_1, x_2x_3 - x_3x_2, \ldots, x_2x_m - x_mx_2, \ldots, x_{m-1}x_m - x_mx_{m-1})$, where $k = (m-1)m/2$. Then the coefficient of one of the monomials in g with maximal degree is 1.

We claim that $A = A^1$. For this, assume on the contrary that A is not unital. Since $A^1/A \cong \mathbb{C}$, $\alpha\beta - \beta\alpha \in A$ for all $\alpha, \beta \in A^1$. Thus A^1 satisfies $g(x_1, \ldots, x_m)$. Hence A^1 is also a PI-ring. By Theorem 10.1.19, A^1 also has exactly n minimal prime ideals. Let $\{P_1, \ldots, P_n\}$ be the set of all minimal prime ideals of A^1. From Theorem 10.1.20 and the proof of Theorem 10.3.41,

$$Q_{\mathbf{qB}}(A^1) \cong A^1/P_1 \oplus \cdots \oplus A^1/P_n$$

and each A^1/P_i is a prime C^*-algebra.

For each i, $Q(A^1/P_i) = (A^1/P_i)\mathrm{Cen}(Q(A^1/P_i)) = A^1/P_i$, and A^1/P_i is finite dimensional over its center $\mathbb{C}$ by Theorem 3.2.19 since A^1/P_i is prime PI. Also there exist a positive integer k_i and a division ring D_i such that $A^1/P_i \cong \mathrm{Mat}_{k_i}(D_i)$ for each i by Theorem 3.2.19. Now $D_i = \mathbb{C}$ because D_i is finite dimensional over its center $\mathbb{C}$, and $\mathbb{C}$ is algebraically closed. Therefore, $A^1/P_i \cong \mathrm{Mat}_{k_i}(\mathbb{C})$ for all i, so each P_i is a maximal ideal. As $\cap_{i=1}^n P_i = 0$,

$$A^1 \cong A^1/P_1 \oplus \cdots \oplus A^1/P_n \cong \mathrm{Mat}_{k_1}(\mathbb{C}) \oplus \cdots \oplus \mathrm{Mat}_{k_n}(\mathbb{C})$$

($*$-isomorphic) by the Chinese Remainder Theorem and Theorem 10.3.52. Since A^1 is finite dimensional, A is also finite dimensional, so A is unital by Theorem 10.3.52, a contradiction. Hence $A = A^1$, so

$$A \cong \mathrm{Mat}_{k_1}(\mathbb{C}) \oplus \cdots \oplus \mathrm{Mat}_{k_n}(\mathbb{C}).$$

(ii)$\Rightarrow$(i) Clearly, A is a PI-ring with exactly n minimal prime ideals. □

Exercise 10.3.54

1. Prove that $M(A)$ and A^1 are C^*-algebras for any C^*-algebra A.
2. ([17, Ara and Mathieu]) Prove that the bounded central closure of a C^*-algebra A is boundedly centrally closed.
3. ([97, Birkenmeier, Park, and Rizvi]) Assume that A is a C^*-algebra and B is an intermediate C^*-algebra between A and $M_{\mathrm{loc}}(A)$. Show the following holds true.
   (i) $\overline{M(B)\mathrm{Cen}(Q_b(M(B)))} = \overline{M(B)\mathcal{B}(Q(A))}$.
   (ii) $M(B)$ is boundedly centrally closed if and only if $\mathcal{B}(Q(A)) \subseteq M(B)$.
4. ([96, Birkenmeier, Park, and Rizvi]) Let A be a C^*-algebra, and let Λ be an index set. Show that the following are equivalent.
   (i) Any one of conditions (i) through (iv) of Theorem 10.3.38.
   (ii) There exists a $*$-monomorphism φ from $M(A)$ to a C^*-direct product M of $|\Lambda|$ unital prime C^*-algebras such that $\varphi(M(A))_{\varphi(A)} \leq^{\mathrm{ess}} M_{\varphi(A)}$.
   (iii) There exists a set of nonzero orthogonal idempotents $\{e_i \mid i \in \Lambda\}$ in $\mathcal{B}(Q(A))$ such that:
   (1) Each e_iA is a prime C^*-algebra.
   (2) For each $a \in A$, a is identified with $(e_ia)_{i\in\Lambda} \in \prod_{i\in\Lambda}^{C^*} e_iA$.

   In this case, prove that $(\oplus_{i\in\Lambda}^{C^*}(e_iA \cap A))_A \leq^{\mathrm{ess}} A_A \leq^{\mathrm{ess}} (\prod_{i\in\Lambda}^{C^*} e_iA)_A$ and each $e_iA \cap A$ is a prime C^*-algebra.

5. Let A be a unital C^*-algebra. Show that the bounded central closure of A contains a nonzero homomorphic image of A/K for every nonessential ideal K of A. (Hint: see Corollary 8.3.48.)
6. ([30, Armendariz, Birkenmeier, and Park]) Let A be a semiprime Banach algebra. Show that if $V \trianglelefteq A$, then $V_A \leq^{\text{ess}} \overline{V}_A$. Hence, every (essentially) closed ideal is norm closed.

Historical Notes When M is an (S, R)-bimodule, $d(M) = d({}_SM_R)$ is defined by Jain, Lam, and Leroy in [226]. Definition 10.1.2 and Results 10.1.3–10.1.9 are taken from [96]. Theorems 10.1.10, 10.1.12, and 10.1.13 were shown by Birkenmeier, Park, and Rizvi [96]. We remark that the statement and the proof of part (iii) of Theorem 10.1.10 correct [96, Theorem 3.9(iii)]. The proof of Corollary 10.1.11 from [96], modifies the proof of a result of Jain, Lam, and Leroy in [226]. Also Example 10.1.14 appears in [96]. Theorem 10.1.17, Example 10.1.18, and Theorem 10.1.20 are in [96]. Theorem 10.1.19 is a well known result. The equivalence of (iv) and (v) of Theorem 10.1.19 was shown by Amitsur in [7]. Theorem 10.1.22 appears in [101].

In [10] and [384], it is shown that if R is a reduced ring, then R/P is a domain for any minimal prime ideal P of R. By Theorem 10.1.20, if R is a reduced ring with only n minimal prime ideals, then $\widehat{Q}_{\mathbf{qB}}(R)$ is a direct sum of n domains.

Proposition 10.2.7 is a new unpublished result. Example 10.2.9 is taken from [246]. Most results of Sect. 10.2 are taken from [65, 77], and [97]. Results of this section are concerned with Baer $*$-rings and quasi-Baer $*$-rings. Some of them are applied in Sect. 10.3. Quasi-Baer $*$-rings in Definition 10.2.2 and semiproper involutions in Definition 10.2.8 were introduced in [65]. Example 10.2.17 and Theorem 10.2.18 appear in [77], while Theorem 10.2.20 is in [65]. Theorems 10.2.21 and 10.2.23 appear in [97]. Further work on Baer $*$-rings appears in [401] and [402].

Proposition 10.3.6(i) and (ii) are in [17, Remark 2.2.9], while Proposition 10.3.6(iii) appears in [14] and is [17, Lemma 3.1.2]. Theorem 10.3.7 appears in [97] which shows an interesting connection between $Q_{\mathbf{qB}}(A^1)$ and $Q_b(A)$ when A is a C^*-algebra. Theorem 10.3.10 and Corollary 10.3.11 appear in [14] and [17, Lemma 3.1.3 and Proposition 3.1.5]. We provide a different proof for Theorem 10.3.10 and Corollary 10.3.11 by using quasi-Baer $*$-rings and semicentral idempotents. Lemma 10.3.14 appears in [97], while Proposition 10.3.15 is in [94]. Theorem 10.3.18 has been shown in [344] and in [15]. However, we give a proof in more detail.

Definition 10.3.19 is taken from [17, Definition 3.2.1]. Theorem 10.3.20 was proved by Birkenmeier, Park, and Rizvi in [97]. Corollary 10.3.21(i) is in [14], while Corollary 10.3.21(ii) appears in [17, Example 3.3.1]. Theorem 10.3.24 follows from [17, Theorem 2.2.8 and Remark 2.2.9]. However, we provide a different proof. Theorem 10.3.25 is due to Ara and Mathieu [14] (also see [17, Theorem 3.1.1]), however we give a different proof by Lemma 10.3.14. Theorem 10.3.27(i) is [17, Lemma 3.2.2(i)], while Theorem 10.3.27(ii) is due to Ara and Mathieu [17, Theorem 3.2.8]. Also we provide a different proof for Theorem 10.3.27(ii) by using semicentral idempotents and the fact that unital boundedly centrally closed C^*-algebras are exactly quasi-AW^*-algebras (see Theorem 10.3.20).

Definition 10.3.28 is indicated in [97]. Theorem 10.3.29, due to Birkenmeier, Park, and Rizvi [97], provides a characterization for an intermediate C^*-algebra between a unital C^*-algebra A and $M_{\text{loc}}(A)$ to be boundedly centrally closed. Theorem 10.3.30, Corollary 10.3.31, Theorem 10.3.35, and Corollary 10.3.36 are due to Birkenmeier, Park, and Rizvi in [97]. Example 10.3.34 appears in [96]. Lemma 10.3.37 appears in [12] and is [17, Proposition 2.2.10], but we provide a different proof. Results 10.3.38–10.3.42 and Theorem 10.3.46 have been shown by Birkenmeier, Park, and Rizvi in [96]. Theorem 10.3.43, and Corollaries 10.3.44, 10.3.45 are in [233]. Lemma 10.3.47 appears in [302].

In [17], the skew group ring $A * G$ and the fixed ring A^G of a C^*-algebra A with a group G of $*$-automorphisms of A have been investigated. Jin, Doh, and Park [233] obtained Theorem 10.3.43, Corollaries 10.3.44, 10.3.45, Theorem 10.3.48, Theorem 10.3.49, and Example 10.3.50. Proposition 10.3.51 illustrates a quasi-Baer module over a unital C^*-algebra and it is taken from [98]. In [17] and [336], C^*-algebras and Banach algebras satisfying a polynomial identity have been studied. Theorem 10.3.53 was obtained by Birkenmeier, Park, and Rizvi in [96]. Additional references on related material include [11, 13, 244, 288, 293], and [394].

Chapter 11
Open Problems and Questions

1. Characterize a ring R for which every (cyclic, finitely generated, projective, etc.) module is FI-extending (see Sect. 2.3, [83, 84], and [85]).

2. Characterize a ring R for which every (cyclic, finitely generated, projective, etc.) module is strongly FI-extending (see Sect. 2.3, [83, 84], and [85]).

3. Characterize the class of rings such that every (finite) direct sum of strongly FI-extending modules is strongly FI-extending.

4. Prove or disprove that a direct summand of an FI-extending module is FI-extending (see Sect. 2.3, [83, 84], and [85]). Also prove or disprove that the right FI-extending property of a ring is Morita invariant.

5. Find necessary and sufficient conditions under which a (finite) direct sum of Baer modules is a Baer module (see Sect. 4.2, [357], and [360]).

6. Find necessary and sufficient conditions for a (finite) direct sum of Rickart modules to be a Rickart module (see Sect. 4.2, [269, 270], and [271]).

7. Obtain a characterization for an arbitrary (finite) direct sum of quasi-Baer modules to be quasi-Baer (see [357], Exercise 4.6.21.2, and Exercise 4.6.21.3).

8. Provide an internal characterization of each type of a Baer module (see Sect. 4.4 and [359]).

9. Provide an internal characterization for each type of $\mathcal{K}$-nonsingular extending module (see Sect. 4.1 and [359]).

10. Let R be a ring with finite triangulating dimension. Find classes of ring extensions of R with finite triangulating dimensions (see Sects. 5.2, 5.4, and 9.3).

11. Let R be a PWP ring. Find classes of ring extensions of R which are PWP rings (see Sect. 5.4 and [70]).

12. Is the property of a ring having finite triangulating dimension Morita invariant?

13. Let R be a quasi-Baer ring such that $R \cong \Gamma(\mathrm{Spec}(R), \mathfrak{K}(R))$. Prove or disprove that R is semiprime (see Theorem 5.5.14 for a partial answer).

14. Characterize all Baer, all right Rickart, all quasi-Baer, all right p.q.-Baer, and all right FI-extending group algebras (cf. Theorem 6.3.2, Corollary 6.3.3, Example 6.3.6, [39], and Proposition 10.2.7).

G.F. Birkenmeier et al., *Extensions of Rings and Modules*,

DOI 10.1007/978-0-387-92716-9_11,

15. If $E(R_R)$, a fixed injective hull of R_R, has a ring multiplication which extends its R-module scalar multiplication, is every right essential overring S of R (S_R may not be a submodule of $E(R_R)$) isomorphic to a subring of $E(R_R)$? (cf. Chap. 7.)

16. If R is a semiprime ring, must R be right (or left) Osofsky compatible?

17. If a ring R is semiprime and right Osofsky compatible, must $E(R_R)$ be right self-injective?

18. If R is a right Osofsky compatible ring, must $E(R_R)$ be an **IC**-ring? (Note that $Q(R)$ is an **IC**-ring by Theorem 8.3.11.)

19. If R is a right Osofsky compatible ring, then is $Q(E(R_R)) = E(R_R)$?

20. Assume that S is a right essential overring of a ring R. Find meaningful properties P such that whenever R has P (e.g., right extending), then S has P and/or conversely (see Sect. 8.1 and [89]).

21. Determine necessary and sufficient conditions for R to have a maximal generalized right essential overring. (For this, see Sect. 8.2 and [89]. According to [89] an overring S of a ring R is said to be a *generalized right essential overring* of R if there exists a finite chain $R = S_0 \subseteq S_1 \subseteq \cdots \subseteq S_n = S$ of subrings such that S_{i+1} is a right essential overring of S_i for each i. We observe that if S_k is right self-injective, then $S_k = S$ is a maximal generalized right essential overring of R. See also Exercise 8.2.16.1.)

22. For a given class $\mathfrak{M}$ of modules, determine necessary and/or sufficient conditions on a ring R such that $H_{\mathfrak{M}}(R_R) = \widehat{Q}_{\mathfrak{M}}(R)$ (it is shown in Theorem 8.4.15 that if $\mathfrak{M} = \mathbf{FI}$ and R is semiprime, then $H_{\mathbf{FI}}(R_R) = \widehat{Q}_{\mathbf{FI}}(R)$, see also Sect. 8.4 and [98]).

23. If R is a semiprime ring, must $\widehat{Q}_{\mathbf{qB}}(R) = Q_{\mathbf{qB}}(R)$?

24. Let $\mathfrak{K}$ be a class of rings (e.g., (quasi-)Baer, right (FI-)extending, right Rickart, etc.). Characterize the class of rings $\mathfrak{H}$ such that each $R \in \mathfrak{H}$ has a $\mathfrak{K}$ right ring hull. In particular, does every ring have a right FI-extending ring hull?

25. Characterize the class $\mathfrak{K}$ of rings such that each ring in $\mathfrak{K}$ has a right continuous (absolute) ring hull (see Sect. 8.2, Sect. 8.3, and [89]).

26. Let $\mathfrak{M}$ be a class of modules. Characterize the class of modules $\mathfrak{H}$ such that each $M \in \mathfrak{H}$ has an $\mathfrak{M}$ module hull.

27. If R is a regular ring, when do $Q_{\mathbf{B}}(R)$ and/or $Q_{\mathbf{E}}(R)$ exist?

28. Let $\mathfrak{K}$ be a class of rings and $\mathfrak{X}$ a type of ring extension (e.g., matrices, polynomials, essential overrings, etc.). Characterize a class $\mathfrak{H}$ of rings such that if $R \in \mathfrak{H}$ and S is an $\mathfrak{X}$ ring extension of R, then $S \in \mathfrak{K}$ (see Chap. 6, Theorem 8.1.8, and Sect. 9.3).

29. Let M be an R-module. Find conditions on the module M (or the ring R) so that M has (i) a Baer module hull, (ii) a Rickart module hull, (iii) a quasi-Baer module hull, (iv) a continuous hull, (v) an extending module hull, or (vi) an FI-extending module hull. Characterize these module hulls if possible when they exist.

30. Characterize the semiprime quasi-Baer group algebras that are quasi-Baer $*$-algebras (see Proposition 10.2.7).

31. Characterize the Baer $*$-rings (quasi-Baer $*$-rings) in which every ideal is self-adjoint (see Theorems 10.2.20 and 10.2.21).

32. Characterize all unital C^*-algebras A such that $Q_{\mathbf{qB}}(A)$ is a C^*-algebra (see Sect. 10.3).

33. Characterize unital C^*-algebras which are p.q.-Baer (cf. A unital C^*-algebra A is boundedly centrally closed if and only if A is quasi-Baer. See Theorem 10.3.20).

34. If T is any semiprime intermediate ring, not necessarily a C^*-algebra, between A and $M_{\mathrm{loc}}(A)$, then is the condition, $\mathcal{B}(Q(A)) \subseteq T$, necessary and/or sufficient for T to be quasi-Baer? (cf. Theorem 10.3.30 and Corollary 10.3.31).

References

1. Akalan, E., Birkenmeier, G.F., Tercan, A.: Goldie extending modules. Commun. Algebra **37**, 663–683 (2009). Corrigendum **38**, 4747–4748 (2010). Corrigendum **41**, 2005 (2010)
2. Akalan, E., Birkenmeier, G.F., Tercan, A.: A characterization of Goldie extending modules over Dedekind domain. J. Algebra Appl. **10**, 1291–1299 (2011)
3. Akalan, E., Birkenmeier, G.F., Tercan, A.: Characterizations of extending and $\mathcal{G}$-extending generalized triangular matrix rings. Commun. Algebra **40**, 1069–1085 (2012)
4. Akalan, E., Birkenmeier, G.F., Tercan, A.: Goldie extending rings. Commun. Algebra **40**, 423–428 (2012)
5. Alahmadi, A.N., Alkan, M., López-Permouth, S.R.: Poor modules: the opposite of injectivity. Glasg. Math. J. **52**, 7–17 (2010)
6. Albu, T., Wisbauer, R.: Kasch modules. In: Jain, S.K., Rizvi, S.T. (eds.) Advances in Ring Theory. Trends in Math., pp. 1–16. Birkhäuser, Boston (1997)
7. Amitsur, S.A.: On rings of quotients. Symp. Math. **8**, 149–164 (1972)
8. Anderson, F.W., Fuller, K.R.: Rings and Categories of Modules, 2nd edn. Springer, Berlin (1992)
9. Andrunakievic, V.A.: Radicals of associative rings I. Transl. Am. Math. Soc. **52**, 95–128 (1966)
10. Andrunakievic, V.A., Rjabuhin, Ju.M.: Rings without nilpotent elements and completely simple ideals. Sov. Math. Dokl. **9**, 565–568 (1968)
11. Ara, P.: Centers of maximal quotient rings. Arch. Math. **50**, 342–347 (1988)
12. Ara, P.: The extended centroid of C^*-algebras. Arch. Math. **54**, 358–364 (1990)
13. Ara, P.: On the symmetric algebra of quotients of a C^*-algebra. Glasg. Math. J. **32**, 377–379 (1990)
14. Ara, P., Mathieu, M.: A local version of the Dauns-Hofmann theorem. Math. Z. **208**, 349–353 (1991)
15. Ara, P., Mathieu, M.: An application of local multipliers to centralizing mappings of C^*-algebras. Q. J. Math. Oxf. **44**, 129–138 (1993)
16. Ara, P., Mathieu, M.: On the central Haagerup tensor product. Proc. Edinb. Math. Soc. **37**, 161–174 (1993)
17. Ara, P., Mathieu, M.: Local Multipliers of C^*-Algebras. Springer Monographs in Math. Springer, London (2003)
18. Ara, P., Park, J.K.: On continuous semiprimary rings. Commun. Algebra **19**, 1945–1957 (1991)
19. Arens, R., Kaplansky, I.: Topological representation of algebras. Trans. Am. Math. Soc. **63**, 457–481 (1948)
20. Armendariz, E.P.: A note on extensions of Baer and p.p.-rings. J. Aust. Math. Soc. **18**, 470–473 (1974)

G.F. Birkenmeier et al., *Extensions of Rings and Modules*,
DOI 10.1007/978-0-387-92716-9,

21. Armendariz, E.P.: Rings with dcc on essential left ideals. Commun. Algebra **8**, 299–308 (1980)
22. Armendariz, E.P.: On semiprime rings of bounded index. Proc. Am. Math. Soc. **85**, 146–148 (1982)
23. Armendariz, E.P., Berberian, S.K.: Baer rings satisfying $J^2 = J$ for all ideals. Commun. Algebra **17**, 1739–1758 (1989)
24. Armendariz, E.P., Park, J.K.: Compressible matrix rings. Bull. Aust. Math. Soc. **30**, 295–298 (1984)
25. Armendariz, E.P., Park, J.K.: Self-injective rings with restricted chain conditions. Arch. Math. **58**, 24–33 (1992)
26. Armendariz, E.P., Steinberg, S.A.: Regular self-injective rings with a polynomial identity. Trans. Am. Math. Soc. **190**, 417–425 (1974)
27. Armendariz, E.P., Koo, H.K., Park, J.K.: Compressible group algebras. Commun. Algebra **13**, 1763–1777 (1985). Corrigendum **215**, 99–100 (2011)
28. Armendariz, E.P., Koo, H.K., Park, J.K.: Isomorphic Ore extensions. Commun. Algebra **15**, 2633–2652 (1987)
29. Armendariz, E.P., Birkenmeier, G.F., Park, J.K.: Rings containing ideals with bounded index. Commun. Algebra **30**, 787–801 (2002)
30. Armendariz, E.P., Birkenmeier, G.F., Park, J.K.: Ideal intrinsic extensions with connections to PI-rings. J. Pure Appl. Algebra **213**, 1756–1776 (2009). Corrigendum **215**, 99–100 (2011)
31. Aydoğdu, P., López-Permouth, S.R.: An alternative perspective on injectivity of modules. J. Algebra **338**, 207–219 (2011)
32. Azumaya, G.: On maximally central algebras. Nagoya Math. J. **2**, 119–150 (1951)
33. Azumaya, G.: Strongly π-regular rings. J. Fac. Sci. Hokkaido Univ. **13**, 34–39 (1954)
34. Baba, Y., Oshiro, K.: Classical Artinian Rings and Related Topics. World Scientific, Singapore (2009)
35. Baer, R.: Abelian groups that are direct summands of every containing Abelian group. Bull. Am. Math. Soc. **46**, 800–806 (1940)
36. Baer, R.: Linear Algebra and Projective Geometry. Academic Press, New York (1952)
37. Barthwal, S., Jain, S.K., Kanwar, P., López-Permouth, S.R.: Nonsingular semiperfect CS-rings. J. Algebra **203**, 361–373 (1998)
38. Bass, H.: Finitistic dimension and a homological generalization of semiprimary rings. Trans. Am. Math. Soc. **95**, 466–488 (1960)
39. Behn, A.: Polycyclic group rings whose principal ideals are projective. J. Algebra **232**, 697–707 (2000)
40. Beidar, K.I.: Rings with generalized identities I, II. Mosc. Univ. Math. Bull. **32**, 15–20; 27–33 (1977)
41. Beidar, K.I., Jain, S.K.: The structure of right continuous right π-rings. Commun. Algebra **32**, 315–332 (2004)
42. Beidar, K., Wisbauer, R.: Strongly and properly semiprime modules and rings. In: Jain, S.K., Rizvi, S.T. (eds.) Ring Theory. Proc. Ohio State-Denison Conf., pp. 58–94. World Scientific, Singapore (1993)
43. Beidar, K., Wisbauer, R.: Properly semiprime self-pp-modules. Commun. Algebra **23**, 841–861 (1995)
44. Beidar, K.I., Jain, S.K., Kanwar, P.: Nonsingular CS-rings coincide with tight PP rings. J. Algebra **282**, 626–637 (2004)
45. Berberian, S.K.: Baer ∗-Rings. Springer, Berlin (1972)
46. Berberian, S.K.: The center of a corner of a ring. J. Algebra **71**, 515–523 (1981)
47. Berberian, S.K.: Baer Rings and Baer ∗-Rings. University of Texas, Austin (1991)
48. Bergman, G.M.: Hereditary commutative rings and centres of hereditary rings. Proc. Lond. Math. Soc. **23**, 214–236 (1971)
49. Bergman, G.M.: Some examples of non-compressible rings. Commun. Algebra **12**, 1–8 (1984)

50. Bergman, G.M., Issacs, I.M.: Rings with fixed-point-free group actions. Proc. Lond. Math. Soc. **27**, 69–87 (1973)
51. Birkenmeier, G.F.: A Decomposition Theory of Rings. Doctoral Dissertation, University of Wisconsin at Milwaukee (1975)
52. Birkenmeier, G.F.: On the cancellation of quasi-injective modules. Commun. Algebra **4**, 101–109 (1976)
53. Birkenmeier, G.F.: Self-injective rings and the minimal direct summand containing the nilpotents. Commun. Algebra **4**, 705–721 (1976)
54. Birkenmeier, G.F.: Modules which are subisomorphic to injective modules. J. Pure Appl. Algebra **13**, 169–177 (1978)
55. Birkenmeier, G.F.: Indecomposable decompositions and the minimal direct summand containing the nilpotents. Proc. Am. Math. Soc. **73**, 11–14 (1979)
56. Birkenmeier, G.F.: Baer rings and quasi-continuous rings have a MDSN. Pac. J. Math. **97**, 283–292 (1981)
57. Birkenmeier, G.F.: Idempotents and completely semiprime ideals. Commun. Algebra **11**, 567–580 (1983)
58. Birkenmeier, G.F.: A generalization of FPF rings. Commun. Algebra **17**, 855–884 (1989)
59. Birkenmeier, G.F.: Decompositions of Baer-like rings. Acta Math. Hung. **59**, 319–326 (1992)
60. Birkenmeier, G.F.: When does a supernilpotent radical essentially split off? J. Algebra **172**, 49–60 (1995)
61. Birkenmeier, G.F., Huang, F.K.: Annihilator conditions on polynomials. Commun. Algebra **29**, 2097–2112 (2001)
62. Birkenmeier, G.F., Huang, F.K.: Annihilator conditions on formal power series. Algebra Colloq. **9**, 29–37 (2002)
63. Birkenmeier, G.F., Huang, F.K.: Annihilator conditions on polynomials II. Monatshefte Math. **141**, 265–276 (2004)
64. Birkenmeier, G.F., Lennon, M.J.: Extending sets of idempotents to ring extensions. Commun. Algebra (to appear)
65. Birkenmeier, G.F., Park, J.K.: Self-adjoint ideals in Baer $*$-rings. Commun. Algebra **28**, 4259–4268 (2000)
66. Birkenmeier, G.F., Park, J.K.: Triangular matrix representations of ring extensions. J. Algebra **265**, 457–477 (2003)
67. Birkenmeier, G.F., Tercan, A.: When some complement of a submodule is a summand. Commun. Algebra **35**, 597–611 (2007)
68. Birkenmeier, G.F., Kim, J.Y., Park, J.K.: A characterization of minimal prime ideals. Glasg. Math. J. **40**, 223–236 (1998)
69. Birkenmeier, G.F., Kim, J.Y., Park, J.K.: When is the CS condition hereditary? Commun. Algebra **27**, 3875–3885 (1999)
70. Birkenmeier, G.F., Heatherly, H.E., Kim, J.Y., Park, J.K.: Triangular matrix representations. J. Algebra **230**, 558–595 (2000)
71. Birkenmeier, G.F., Kim, J.Y., Park, J.K.: On polynomial extensions of principally quasi-Baer rings. Kyungpook Math. J. **40**, 247–253 (2000)
72. Birkenmeier, G.F., Kim, J.Y., Park, J.K.: A counterexample for CS-rings. Glasg. Math. J. **42**, 263–269 (2000)
73. Birkenmeier, G.F., Kim, J.Y., Park, J.K.: Quasi-Baer ring extensions and biregular rings. Bull. Aust. Math. Soc. **61**, 39–52 (2000)
74. Birkenmeier, G.F., Kim, J.Y., Park, J.K.: A sheaf representation of quasi-Baer rings. J. Pure Appl. Algebra **146**, 209–223 (2000)
75. Birkenmeier, G.F., Kim, J.Y., Park, J.K.: On quasi-Baer rings. In: Huynh, D.V., Jain, S.K., López-Permouth, S.R. (eds.) Algebras and Its Applications. Contemp. Math., vol. 259, pp. 67–92. Amer. Math. Soc., Providence (2000)
76. Birkenmeier, G.F., Kim, J.Y., Park, J.K.: Rings with countably many direct summands. Commun. Algebra **28**, 757–769 (2000)

77. Birkenmeier, G.F., Kim, J.Y., Park, J.K.: Polynomial extensions of Baer and quasi-Baer rings. J. Pure Appl. Algebra **159**, 25–42 (2001)
78. Birkenmeier, G.F., Kim, J.Y., Park, J.K.: Principally quasi-Baer rings. Commun. Algebra **29**, 639–660 (2001). Erratum **30**, 5609 (2002)
79. Birkenmeier, G.F., Kim, J.Y., Park, J.K.: Semicentral reduced algebras. In: Birkenmeier, G.F., Park, J.K., Park, Y.S. (eds.) International Symposium on Ring Theory. Trends in Math., pp. 67–84. Birkhäuser, Boston (2001)
80. Birkenmeier, G.F., Călugăreanu, G., Fuchs, L., Goeters, H.P.: The fully-invariant-extending property for Abelian groups. Commun. Algebra **29**, 673–685 (2001)
81. Birkenmeier, G.F., Heatherly, H.E., Kim, J.Y., Park, J.K.: Algebras generated by semicentral idempotents. Acta Math. Hung. **95**, 101–104 (2002)
82. Birkenmeier, G.F., Kim, J.Y., Park, J.K.: Triangular matrix representations of semiprimary rings. J. Algebra Appl. **2**, 123–131 (2002)
83. Birkenmeier, G.F., Müller, B.J., Rizvi, S.T.: Modules in which every fully invariant submodule is essential in a direct summand. Commun. Algebra **30**, 1395–1415 (2002)
84. Birkenmeier, G.F., Park, J.K., Rizvi, S.T.: Modules with fully invariant submodules essential in fully invariant summands. Commun. Algebra **30**, 1833–1852 (2002)
85. Birkenmeier, G.F., Park, J.K., Rizvi, S.T.: Generalized triangular matrix rings and the fully invariant extending property. Rocky Mt. J. Math. **32**, 1299–1319 (2002)
86. Birkenmeier, G.F., Park, J.K., Rizvi, S.T.: On ring hulls. In: Rizvi, S.T., Zaidi, S.M.A. (eds.) Trends in Theory of Rings and Modules, pp. 17–34. Anamaya Pub., New Delhi (2005)
87. Birkenmeier, G.F., Park, J.K., Rizvi, S.T.: Ring hulls of extension rings. In: Chen, J.L., Ding, N.Q., Marubayashi, H. (eds.) Advances in Ring Theory. Proc. of 4th China-Japan-Korea International Conference, pp. 12–25. World Scientific, Englewood Cliffs (2005)
88. Birkenmeier, G.F., Park, J.K., Rizvi, S.T.: An essential extension with nonisomorphic ring structures. In: Huynh, D.V., Jain, S.K., López-Permouth, S.R. (eds.) with the cooperation of P. Kanwar. Algebra and Its Applications. Contemp. Math., vol. 419, pp. 29–48. Amer. Math. Soc., Providence (2006)
89. Birkenmeier, G.F., Park, J.K., Rizvi, S.T.: Ring hulls and applications. J. Algebra **304**, 633–665 (2006)
90. Birkenmeier, G.F., Park, J.K., Rizvi, S.T.: Ring hulls of generalized triangular matrix rings. Aligarh Bull. Math. **25**, 65–77 (2006)
91. Birkenmeier, G.F., Huynh, D.V., Kim, J.Y., Park, J.K.: Extending the property of a maximal right ideal. Algebra Colloq. **13**, 163–172 (2006)
92. Birkenmeier, G.F., Park, J.K., Rizvi, S.T.: An essential extension with nonisomorphic ring structures II. Commun. Algebra **35**, 3986–4004 (2007)
93. Birkenmeier, G.F., Park, J.K., Rizvi, S.T.: On triangulating dimension of rings. Commun. Algebra **36**, 1520–1526 (2008)
94. Birkenmeier, G.F., Park, J.K., Rizvi, S.T.: Ring hulls of semiprime homomorphic images. In: Brzeziński, T., Gómez-Pardo, J.L., Shestakov, I., Smith, P.F. (eds.) Modules and Comodules. Trends in Math., pp. 101–111. Birkhäuser, Boston (2008)
95. Birkenmeier, G.F., Osofsky, B.L., Park, J.K., Rizvi, S.T.: Injective hulls with distinct ring structures. J. Pure Appl. Algebra **213**, 732–736 (2009)
96. Birkenmeier, G.F., Park, J.K., Rizvi, S.T.: The structure of rings of quotients. J. Algebra **321**, 2545–2566 (2009)
97. Birkenmeier, G.F., Park, J.K., Rizvi, S.T.: Hulls of semiprime rings with applications to C^*-algebras. J. Algebra **322**, 327–352 (2009)
98. Birkenmeier, G.F., Park, J.K., Rizvi, S.T.: Modules with FI-extending hulls. Glasg. Math. J. **51**, 347–357 (2009)
99. Birkenmeier, G.F., Park, J.K., Rizvi, S.T.: An example of Osofsky and essential overrings. In: Dung, N.V., Guerriero, F., Hammoudi, L., Kanwar, P. (eds.) Rings, Modules and Representations. Contemp. Math., vol. 480, pp. 13–33. Amer. Math. Soc., Providence (2009)
100. Birkenmeier, G.F., Park, J.K., Rizvi, S.T.: Hulls of ring extensions. Can. Math. Bull. **53**, 587–601 (2010)

101. Birkenmeier, G.F., Park, J.K., Rizvi, S.T.: Principally quasi-Baer ring hulls. In: Huynh, D.V., López-Permouth, S.R. (eds.) Advances in Ring Theory. Trends in Math., pp. 47–61. Birkhäuser, Boston (2010)
102. Birkenmeier, G.F., Park, J.K., Rizvi, S.T.: A theory of hulls for rings and modules. In: Albu, T., Birkenmeier, G.F., Erdogan, A., Tercan, A. (eds.) Ring and Module Theory. Trends in Math., pp. 27–71. Birkhäuser, Boston (2010)
103. Birkenmeier, G.F., Kim, J.Y., Park, J.K.: The factor ring of a quasi-Baer ring by its prime radical. J. Algebra Appl. **10**, 157–165 (2011)
104. Blackadar, B.: Operator Algebras, Theory of C^*-Algebras and von Neumann Algebras. Encyclopaedia Math. Sciences. Springer, Berlin (2006)
105. Blecher, D.P., Le Merdy, C.: Operator Algebras and Their Modules—An Operator Space Approach. London Math. Soc. Monograph. Clarendon Press, Oxford (2004)
106. Brown, B., McCoy, N.H.: The maximal regular ideal of a ring. Proc. Am. Math. Soc. **1**, 165–171 (1950)
107. Brown, K.A.: The singular ideals of group rings. Q. J. Math. Oxf. **28**, 41–60 (1977)
108. Burgess, W.D., Raphael, R.: On extensions of regular rings of finite index by central elements. In: Jain, S.K., Rizvi, S.T. (eds.) Advances in Ring Theory. Trends in Math., pp. 73–86. Birkhäuser, Boston (1997)
109. Busby, R.C.: Double centralizers and extensions of C^*-algebras. Trans. Am. Math. Soc. **132**, 79–99 (1968)
110. Camillo, V.P., Costa-Cano, F.J., Simon, J.J.: Relating properties of a ring and its ring of row and column finite matrices. J. Algebra **244**, 435–449 (2001)
111. Camillo, V.P., Herzog, I., Nielsen, P.P.: Non-self-injective injective hulls with compatible multiplication. J. Algebra **314**, 471–478 (2007)
112. Castella, D.: Anneaux régulaiers de Baer compressibles. Commun. Algebra **15**, 1621–1635 (1987)
113. Chari, V., Pressley, A.: A Guide to Quantum Groups. Cambridge Univ. Press, Cambridge (1994)
114. Chase, S.U.: Direct products of modules. Trans. Am. Math. Soc. **97**, 457–473 (1960)
115. Chase, S.U.: A generalization of the ring of triangular matrices. Nagoya Math. J. **18**, 13–25 (1961)
116. Chatters, A.W.: The restricted minimum condition in Noetherian hereditary rings. J. Lond. Math. Soc. **4**, 83–87 (1971)
117. Chatters, A.W.: A decomposition theorem for Noetherian hereditary rings. Bull. Lond. Math. Soc. **4**, 125–126 (1972)
118. Chatters, A.W.: Two results on p.p. rings. Commun. Algebra **4**, 881–891 (1976)
119. Chatters, A.W., Hajarnavis, C.R.: Rings in which every complement right ideal is a direct summand. Q. J. Math. Oxf. **28**, 61–80 (1977)
120. Chatters, A.W., Hajarnavis, C.R.: Rings with Chain Conditions. Pitman, Boston (1980)
121. Chatters, A.W., Khuri, S.M.: Endomorphism rings of modules over non-singular CS rings. J. Lond. Math. Soc. **21**, 434–444 (1980)
122. Chen, J., Zhang, X.: On modules over formal triangular matrix rings. East-West J. Math. 69–77 (2001)
123. Chen, J., Li, Y., Zhou, Y.: Morphic group rings. J. Pure Appl. Algebra **205**, 621–639 (2006)
124. Cheng, Y., Huang, F.K.: A note on extensions of principally quasi-Baer rings. Taiwan. J. Math. **12**, 1721–1731 (2008)
125. Clark, J.: A note on the fixed subring of an FPF ring. Bull. Aust. Math. Soc. **40**, 109–111 (1989)
126. Clark, J., Huynh, D.V.: Simple rings with injectivity conditions on one-sided ideals. Bull. Aust. Math. Soc. **76**, 315–320 (2007)
127. Clark, W.E.: Baer rings which arise from certain transitive graphs. Duke Math. J. **33**, 647–656 (1966)
128. Clark, W.E.: Twisted matrix units semigroup algebras. Duke Math. J. **34**, 417–424 (1967)

129. Cohen, M.: A Morita context related to finite automorphism groups of rings. Pac. J. Math. **98**, 37–54 (1982)
130. Cohn, P.M.: Free Rings and Their Relations. Academic Press, London (1971)
131. Connell, I.G.: On the group rings. Can. J. Math. **15**, 650–685 (1963)
132. Contessa, M.: A note on Baer rings. J. Algebra **118**, 20–32 (1988)
133. Curtis, C.W., Reiner, I.: Representation Theory of Finite Groups and Associative Algebras. Interscience, New York (1962)
134. Dauns, J.: Modules and Rings. Cambridge Univ. Press, Cambridge (1994)
135. Dauns, J., Hofmann, K.H.: Representation of Rings by Sections. Memoir, vol. 83. Amer. Math. Soc., Providence (1968)
136. Davidson, K.R.: C^*-Algebra by Example. Fields Institute Monograph, vol. 6. Amer. Math. Soc., Providence (1996)
137. DeMeyer, F., Ingraham, E.: Separable Algebras over Commutative Rings. Lecture Notes in Math., vol. 181. Springer, Berlin (1971)
138. Dischinger, F.: Sur les anneaux fortement π-réguliers. C. R. Acad. Sci. Paris **283**, 571–573 (1976)
139. Divinsky, N.: Rings and Radicals. Univ. Toronto Press, Toronto (1965)
140. Dixmier, J.: C^*-Algebras. North-Holland, Amsterdam (1977)
141. Dobbs, D.E., Picavet, G.: Weak Baer going-down rings. Houst. J. Math. **29**, 559–581 (2003)
142. Doran, R.S., Belfi, V.A.: Characterizations of C^*-Algebras: The Gelfand-Naimark Theorems. Marcel Dekker, New York (1986)
143. Dorsey, T.J., Mesyan, Z.: On minimal extensions of rings. Commun. Algebra **37**, 3463–3486 (2009)
144. Dung, N.V., Smith, P.F.: Rings for which certain modules are CS. J. Pure Appl. Algebra **102**, 273–287 (1995)
145. Dung, N.V., Huynh, D.V., Smith, P.F., Wisbauer, R.: Extending Modules. Longman, Harlow (1994)
146. Eckmann, B., Schopf, A.: Über injektive Moduln. Arch. Math. **4**, 75–78 (1953)
147. Elliott, G.A.: Automorphisms determined by multipliers on ideals of a C^*-algebra. J. Funct. Anal. **23**, 1–10 (1976)
148. Endo, S.: Note on p.p. rings (a supplement to Hattori's paper). Nagoya Math. J. **17**, 167–170 (1960)
149. Er, N.: Rings whose modules are direct sums of extending modules. Proc. Am. Math. Soc. **137**, 2265–2271 (2009)
150. Er, N., López-Permouth, S.R., Sökmez, N.: Rings whose modules have maximal or minimal injectivity domains. J. Algebra **330**, 404–417 (2011)
151. Evans, M.W.: On commutative P.P. rings. Pac. J. Math. **41**, 687–697 (1972)
152. Faith, C.: Rings with ascending condition on annihilators. Nagoya Math. J. **27**, 179–191 (1966)
153. Faith, C.: Lectures on Injective Modules and Quotient Rings. Lecture Notes in Math., vol. 49. Springer, Berlin (1967)
154. Faith, C.: Algebra I: Rings, Modules, and Categories. Springer, Berlin (1973)
155. Faith, C.: Algebra II, Ring Theory. Springer, Berlin (1976)
156. Faith, C.: Semiperfect Prüfer rings and FPF rings. Isr. J. Math. **26**, 166–177 (1977)
157. Faith, C.: Injective quotient rings of commutative rings. In: Module Theory. Lecture Notes in Math., vol. 700, pp. 151–203. Springer, Berlin (1979)
158. Faith, C.: The maximal regular ideal of self-injective and continuous rings splits off. Arch. Math. **44**, 511–521 (1985)
159. Faith, C.: Embedding torsionless modules in projectives. Publ. Mat. **34**, 379–387 (1990)
160. Faith, C., Page, S.: FPF Ring Theory: Faithful Modules and Generators of Mod-R. Cambridge Univ. Press, Cambridge (1984)
161. Faith, C., Utumi, Y.: Quasi-injective modules and their endomorphism rings. Arch. Math. **15**, 166–174 (1964)
162. Faith, C., Utumi, Y.: Intrinsic extensions of rings. Pac. J. Math. **14**, 505–512 (1964)

163. Ferrero, M.: Closed submodules of normalizing bimodules over semiprime rings. Commun. Algebra **29**, 1513–1550 (2001)
164. Fields, K.L.: On the global dimension of residue rings. Pac. J. Math. **32**, 345–349 (1970)
165. Findlay, G.D., Lambek, J.: A generalized ring of quotients I. Can. Math. Bull. **1**, 77–85 (1958)
166. Findlay, G.D., Lambek, J.: A generalized ring of quotients II. Can. Math. Bull. **1**, 155–167 (1958)
167. Fisher, J.W.: On the nilpotency of nil subrings. Can. J. Math. **22**, 1211–1216 (1970)
168. Fisher, J.W.: Structure of semiprime P.I. rings. Proc. Am. Math. Soc. **39**, 465–467 (1973)
169. Fisher, J.W., Montgomery, S.: Semiprime skew group rings. J. Algebra **52**, 241–247 (1978)
170. Fisher, J.W., Snider, R.L.: On the von Neumann regularity of rings with regular prime factor rings. Pac. J. Math. **54**, 135–144 (1974)
171. Fraser, J.A., Nicholson, W.K.: Reduced PP-rings. Math. Jpn. **34**, 715–725 (1989)
172. Fuchs, L.: Infinite Abelian Groups I. Academic Press, New York (1970)
173. Fuchs, L.: The cancellation property for modules. In: Lecture Notes in Math., vol. 700, pp. 191–212. Springer, Berlin (1972)
174. Fuchs, L.: Infinite Abelian Groups II. Academic Press, New York (1973)
175. Garcia, J.L.: Properties of direct summands of modules. Commun. Algebra **17**, 73–92 (1989)
176. Gardner, B.J., Wiegandt, R.: Radical Theory of Rings. Marcel Dekker, New York (2004)
177. Goel, V.K., Jain, S.K.: π-injective modules and rings whose cyclics are π-injective. Commun. Algebra **6**, 59–73 (1978)
178. Goldie, A.W.: Semi-prime rings with maximum condition. Proc. Lond. Math. Soc. **3**, 201–220 (1960)
179. Goodearl, K.R.: Prime ideals in regular self-injective rings. Can. J. Math. **25**, 829–839 (1973)
180. Goodearl, K.R.: Ring Theory: Nonsingular Rings and Modules. Marcel Dekker, New York (1976)
181. Goodearl, K.R.: Direct sum properties of quasi-injective modules. Bull. Am. Math. Soc. **82**, 108–110 (1976)
182. Goodearl, K.R.: Simple Noetherian rings—the Zalesskii-Neroslavskii examples. In: Handelman, D., Lawrence, J. (eds.) Ring Theory, Waterloo, 1978. Lecture Notes in Math., vol. 734, pp. 118–130. Springer, Berlin (1979)
183. Goodearl, K.R.: Von Neumann Regular Rings. Krieger, Malabar (1991)
184. Goodearl, K.R., Boyle, A.K.: Dimension Theory for Nonsingular Injective Modules. Memoirs, vol. 177. Amer. Math. Soc., Providence (1976)
185. Goodearl, K.R., Warfield, R.B. Jr.: An Introduction to Noncommutative Noetherian Rings. Cambridge Univ. Press, Cambridge (1989)
186. Goodearl, K.R., Handelman, D.E., Lawrence, J.W.: Affine Representations of Grothendieck Groups and Applications to Rickart C*-Algebras and $\aleph_0$-Continuous Regular Rings. Memoirs, vol. 234. Amer. Math. Soc., Providence (1980)
187. Gordon, R., Small, L.W.: Piecewise domains. J. Algebra **23**, 553–564 (1972)
188. Groenewald, N.J.: A note on extensions of Baer and p.p.-rings. Publ. Inst. Math. **34**, 71–72 (1983)
189. Haghany, A.: Injectivity conditions over a formal triangular matrix ring. Arch. Math. **78**, 268–274 (2002)
190. Haghany, A., Varadarajan, K.: Study of formal triangular matrix rings. Commun. Algebra **27**, 5507–5525 (1999)
191. Haghany, A., Varadarajan, K.: Study of modules over formal triangular matrix rings. J. Pure Appl. Algebra **147**, 41–58 (2000)
192. Han, J., Hirano, Y., Kim, H.: Semiprime Ore extensions. Commun. Algebra **28**, 3795–3801 (2000)
193. Han, J., Hirano, Y., Kim, H.: Some results on skew polynomial rings over a reduced ring. In: Birkenmeier, G.F., Park, J.K., Park, Y.S. (eds.) International Symposium on Ring Theory. Trends in Math., pp. 123–129. Birkhäuser, Boston (2001)
194. Handelman, D.E.: Prüfer domains and Baer $*$-rings. Arch. Math. **29**, 241–251 (1977)

195. Hannah, J.: Quotient rings of semiprime rings with bounded index. Glasg. Math. J. **23**, 53–64 (1982)
196. Harada, M.: Hereditary semi-primary rings and triangular matrix rings. Nagoya Math. J. **27**, 463–484 (1966)
197. Harada, M.: On the small hulls of a commutative ring. Osaka J. Math. **15**, 679–682 (1978)
198. Harada, M.: On modules with extending property. Osaka J. Math. **19**, 203–215 (1982)
199. Harmanci, A., Smith, P.F., Tercan, A., Tiraş, Y.: Direct sums of CS-modules. Houst. J. Math. **22**, 61–71 (1996)
200. Hattori, A.: A foundation of torsion theory for modules over general rings. Nagoya Math. J. **17**, 147–158 (1960)
201. Hausen, J.: Modules with the summand intersection property. Commun. Algebra **17**, 135–148 (1989)
202. Heatherly, H.E., Tucci, R.P.: Central and semicentral idempotents. Kyungpook Math. J. **40**, 255–258 (2000)
203. Herstein, I.N.: Noncommutative Rings. Carus Math. Monograph, vol. 15. Math. Assoc. Amer., Washington (1968)
204. Hirano, Y.: Regular modules and V-modules. Hiroshima Math. J. **11**, 125–142 (1981)
205. Hirano, Y.: Open problems. In: Birkenmeier, G.F., Park, J.K., Park, Y.S. (eds.) International Symposium on Ring Theory. Trends in Math., pp. 441–446. Birkhäuser, Boston (2001)
206. Hirano, Y.: On ordered monoid rings over a quasi-Baer ring. Commun. Algebra **29**, 2089–2095 (2001)
207. Hirano, Y., Park, J.K.: On self-injective strongly π-regular rings. Commun. Algebra **21**, 85–91 (1993)
208. Hirano, Y., Hongan, M., Ohori, M.: On right P.P. rings. Math. J. Okayama Univ. **24**, 99–109 (1982)
209. Hofmann, K.H.: Representations of algebras by continuous sections. Bull. Am. Math. Soc. **78**, 291–373 (1972)
210. Holland, S.S. Jr.: Remarks on type I Baer and Baer $*$-rings. J. Algebra **27**, 516–522 (1973)
211. Holston, C., Jain, S.K., Leroy, A.: Rings over which cyclics are direct sums of projective and CS or Noetherian. Glasg. Math. J. **52**, 103–110 (2010)
212. Hudson, T.D., Katsoulis, E.G., Larson, D.R.: Extreme points in triangular UHF algebras. Trans. Am. Math. Soc. **349**, 3391–3400 (1997)
213. Huynh, D.V.: The symmetry of the CS condition on one-sided ideals in a prime ring. J. Pure Appl. Algebra **212**, 9–13 (2008)
214. Huynh, D.V., Rizvi, S.T.: On some classes of Artinian rings. J. Algebra **223**, 133–153 (2000)
215. Huynh, D.V., Rizvi, S.T.: Characterizing rings by a direct decomposition property of their modules. J. Aust. Math. Soc. **80**, 359–366 (2006)
216. Huynh, D.V., Rizvi, S.T.: An affirmative answer to a question on Noetherian rings. J. Algebra Appl. **7**, 47–59 (2008)
217. Huynh, D.V., Dung, N.V., Wisbauer, R.: Quasi-injective modules with acc or dcc on essential submodules. Arch. Math. **53**, 252–255 (1989)
218. Huynh, D.V., Smith, P.F., Wisbauer, R.: A note on GV-modules with Krull dimension. Glasg. Math. J. **32**, 389–390 (1990)
219. Huynh, D.V., Rizvi, S.T., Yousif, M.F.: Rings whose finitely generated modules are extending. J. Pure Appl. Algebra **111**, 325–328 (1996)
220. Huynh, D.V., Jain, S.K., López-Permouth, S.R.: Rings characterized by direct sums of CS modules. Commun. Algebra **28**, 4219–4222 (2000)
221. Jacobson, N.: Structure of Rings. Amer. Math. Soc. Colloq. Publ., vol. 37. Amer. Math. Soc., Providence (1964)
222. Jain, S.K., Saleh, H.H.: Rings whose (proper) cyclic modules have cyclic π-injective hulls. Arch. Math. **48**, 109–115 (1987)
223. Jain, S.K., López-Permouth, S.R., Rizvi, S.T.: Continuous rings with acc on essentials are Artinian. Proc. Am. Math. Soc. **108**, 583–586 (1990)

224. Jain, S.K., López-Permouth, S.R., Rizvi, S.T.: A characterization of uniserial rings via continuous and discrete modules. J. Aust. Math. Soc. **50**, 197–203 (1991)
225. Jain, S.K., López-Permouth, S.R., Syed, S.R.: Mutual injective hulls. Can. Math. Bull. **39**, 68–73 (1996)
226. Jain, S.K., Lam, T.Y., Leroy, A.: On uniform dimensions of ideals in right nonsingular rings. J. Pure Appl. Algebra **133**, 117–139 (1998)
227. Jain, S.K., Kanwar, P., Malik, S., Srivastava, J.B.: KD_∞ is a CS-algebra. Proc. Am. Math. Soc. **128**, 397–400 (1999)
228. Jain, S.K., Kanwar, P., López-Permouth, S.R.: Nonsingular semiperfect CS-rings II. Bull. Lond. Math. Soc. **32**, 421–431 (2000)
229. Jategaonkar, A.V.: Endomorphism rings of torsionless modules. Trans. Am. Math. Soc. **161**, 457–466 (1971)
230. Jeremy, L.: Modules et anneaux quasi-continus. Can. Math. Bull. **17**, 217–228 (1974)
231. Jeremy, L.: L'arithmétique de von Neumann pour les facteurs injectifs. J. Algebra **62**, 154–169 (1980)
232. Jin, H.L., Doh, J., Park, J.K.: Quasi-Baer rings with essential prime radicals. Commun. Algebra **34**, 3537–3541 (2006)
233. Jin, H.L., Doh, J., Park, J.K.: Group actions on quasi-Baer rings. Can. Math. Bull. **52**, 564–582 (2009)
234. Johnson, R.E.: The extended centralizer of a ring over a module. Proc. Am. Math. Soc. **2**, 891–895 (1951)
235. Johnson, R.E.: Structure theory of faithful rings I. Trans. Am. Math. Soc. **84**, 508–522 (1957)
236. Johnson, R.E.: Structure theory of faithful rings II, restricted rings. Trans. Am. Math. Soc. **84**, 523–544 (1957)
237. Johnson, R.E.: Quotient rings of rings with zero singular ideal. Pac. J. Math. **11**, 1385–1395 (1961)
238. Johnson, R.E., Wong, E.T.: Quasi-injective modules and irreducible rings. J. Lond. Math. Soc. **36**, 260–268 (1961)
239. Jøndrup, S.: p.p. rings and finitely generated flat ideals. Proc. Am. Math. Soc. **28**, 431–435 (1971)
240. Kadison, R.V., Ringrose, J.R.: Fundamentals of the Theory of Operator Algebras, Vol. I: Elementary Theory. Graduate Studies in Math., vol. 15. Amer. Math. Soc., Providence (1997)
241. Kamal, M.A., Müller, B.J.: Extending modules over commutative domains. Osaka J. Math. **25**, 531–538 (1988)
242. Kamal, M.A., Müller, B.J.: The structure of extending modules over Noetherian rings. Osaka J. Math. **25**, 539–551 (1988)
243. Kaplansky, I.: Topological representation of algebra II. Trans. Am. Math. Soc. **68**, 62–75 (1950)
244. Kaplansky, I.: Projections in Banach algebras. Ann. Math. **53**, 235–249 (1951)
245. Kaplansky, I.: Rings of operators. Univ. Chicago Mimeographed Lecture Notes (Notes by S.K. Berberian, with an Appendix by R. Blattner), Univ. Chicago (1955)
246. Kaplansky, I.: Rings of Operators. Benjamin, New York (1968)
247. Kaplansky, I.: Infinite Abelian Groups. Univ. Michigan Press, Ann Arbor (1969)
248. Kaplansky, I.: Commutative Rings. Univ. Chicago Press, Chicago (1974)
249. Kharchenko, V.K.: Algebras of invariants of free algebras. Algebra Log. **17**, 478–487 (1978). English Transl. 316–321 (1979)
250. Khuri, S.M.: Endomorphism rings and nonsingular modules. Ann. Sci. Math. Qué. **IV**, 145–152 (1980)
251. Khuri, S.M.: Nonsingular retractable modules and their endomorphism rings. Bull. Aust. Math. Soc. **43**, 63–71 (1991)
252. Khuri, S.M.: Endomorphism rings and modules. Manuscript (2004)
253. Kim, J.Y., Park, J.K.: When is a regular ring a semisimple Artinian ring? Math. Jpn. **45**, 311–313 (1997)

254. Kist, J.: Minimal prime ideals in commutative semigroups. Proc. Lond. Math. Soc. **13**, 31–50 (1963)
255. Koh, K.: On functional representations of a ring without nilpotent elements. Can. Math. Bull. **14**, 349–352 (1971)
256. Koh, K.: On a representation of a strongly harmonic ring by sheaves. Pac. J. Math. **41**, 459–468 (1972)
257. Kulikov, L.Ya.: On the theory of Abelian group of arbitrary cardinality. Mat. Sb. **9**, 165–182 (1941)
258. Kulikov, L.Ya.: On the theory of Abelian group of arbitrary cardinality. Mat. Sb. **16**, 129–162 (1945)
259. Lam, T.Y.: A First Course in Noncommutative Rings. Springer, Berlin (1991)
260. Lam, T.Y.: A lifting theorem, and rings with isomorphic matrix rings. In: Chan, K.Y., Liu, M.C. (eds.) Five Decades as a Mathematician and Educator: On the 80th Birthday of Professor Y.C. Wong, pp. 169–186. World Scientific, Singapore (1995)
261. Lam, T.Y.: Modules with Isomorphic Multiples and Rings with Isomorphic Matrix Rings—A Survey. PAM, vol. 736. Univ. California, Berkeley (1998)
262. Lam, T.Y.: Lectures on Modules and Rings. Springer, Berlin (1999)
263. Lambek, J.: On Utumi's ring of quotients. Can. J. Math. **15**, 363–370 (1963)
264. Lambek, J.: Lectures on Rings and Modules. Blaisdell, Waltham (1966)
265. Lambek, J.: On the representations of modules by sheaves of factor modules. Can. Math. Bull. **14**, 359–368 (1971)
266. Lang, N.C.: On ring properties of injective hulls. Can. Math. Bull. **18**, 233–239 (1975)
267. Lawrence, J.: A singular primitive ring. Proc. Am. Math. Soc. **45**, 59–62 (1974)
268. Lee, G.: Theory of Rickart Modules. Doctoral Dissertation, Ohio State Univ. (2010)
269. Lee, G., Rizvi, S.T., Roman, C.: Rickart modules. Commun. Algebra **38**, 4005–4027 (2010)
270. Lee, G., Rizvi, S.T., Roman, C.: Dual Rickart modules. Commun. Algebra **39**, 4036–4058 (2011)
271. Lee, G., Rizvi, S.T., Roman, C.: Direct sums of Rickart modules. J. Algebra **353**, 62–78 (2012)
272. Lee, G., Rizvi, S.T., Roman, C.: When do the direct sums of modules inherit certain properties? In: Kim, J.Y., Huh, C., Kwak, T.K., Lee, Y. (eds.) Proc. 6th China-Japan-Korea International Conf. on Ring Theory, pp. 47–77. World Scientific, Englewood Cliffs (2012)
273. Lee, G., Rizvi, S.T., Roman, C.: Modules whose endomorphism rings are von Neumann regular. Commun. Algebra (to appear)
274. Lee, T.K., Zhou, Y.: Reduced modules. In: Facchini, A., Houston, E., Salce, L. (eds.) Rings, Modules, Algebras, and Abelian Groups. Lecture Notes in Pure and Appl. Math., vol. 236, pp. 365–377. Marcel Dekker, New York (2004)
275. Lee, T.K., Zhou, Y.: Regularity and morphic property of rings. J. Algebra **322**, 1072–1085 (2009)
276. Lee, T.K., Zhou, Y.: A characterization of von Neumann regular rings and applications. Linear Algebra Appl. **433**, 1536–1540 (2010)
277. Lenzing, H.: Halberbliche Endomorphismenringe. Math. Z. **118**, 219–240 (1970)
278. Levitzki, J.: On a problem of A. Kurosh. Bull. Am. Math. Soc. **52**, 1033–1035 (1946)
279. Levy, L.: Torsion-free and divisible modules over non-integral-domains. Can. J. Math. **15**, 132–151 (1963)
280. Li, M.S., Zelmanowitz, J.M.: Artinian rings with restricted primeness conditions. J. Algebra **124**, 139–148 (1989)
281. Liu, Z.: A note on principally quasi-Baer rings. Commun. Algebra **30**, 3885–3890 (2002)
282. Liu, Z.: Baer rings of generalized power series. Glasg. Math. J. **44**, 463–469 (2002)
283. Liu, Z., Zhao, R.: A generalization of pp-rings and p.q.-Baer rings. Glasg. Math. J. **48**, 217–229 (2006)
284. López-Permouth, S.R., Oshiro, K., Rizvi, S.T.: On the relative (quasi-)continuity of modules. Commun. Algebra **26**, 3497–3510 (1998)
285. Louden, K.: Maximal quotient rings of ring extensions. Pac. J. Math. **62**, 489–496 (1976)

286. Mackey, G.W.: On infinite-dimensional linear spaces. Trans. Am. Math. Soc. **57**, 155–207 (1945)
287. Maeda, S.: On a ring whose principal right ideals generated by idempotents form a lattice. J. Sci. Hiroshima Univ. Ser. A **24**, 509–525 (1960)
288. Maeda, S., Holland, S.S. Jr.: Equivalence of projections in Baer $*$-rings. J. Algebra **39**, 150–159 (1976)
289. Mao, L.: Baer endomorphism rings and envelopes. J. Algebra Appl. **9**, 365–381 (2010)
290. Mao, L.: Properties of P-coherent and Baer modules. Period. Math. Hung. **60**, 97–114 (2010)
291. Martindale, W.S. III: Prime rings satisfying a generalized polynomial identity. J. Algebra **12**, 576–584 (1969)
292. Martindale, W.S. III: On semiprime P.I. rings. Proc. Am. Math. Soc. **40**, 365–369 (1973)
293. Mathieu, M.: The local multiplier algebra: blending noncommutative ring theory and functional analysis. In: Brzeziński, T., Gómez-Pardo, J.L., Shestakov, I., Smith, P.F. (eds.) Modules and Comodules. Trends in Math., pp. 301–312. Birkhäuser, Boston (2008)
294. Matlis, E.: Injective modules over Noetherian rings. Pac. J. Math. **8**, 511–528 (1958)
295. McConnell, J.C., Robson, J.C.: Noncommutative Noetherian Rings. Wiley, New York (1987)
296. McCoy, N.H.: The Theory of Rings. Chelsea, New York (1973)
297. Menal, P., Vamos, P.: Pure ring extensions and self FP-injective rings. Proc. Camb. Philos. Soc. **105**, 447–458 (1989)
298. Mewborn, A.C.: Regular rings and Baer rings. Math. Z. **121**, 211–219 (1971)
299. Michler, G.O.: Structure of semi-perfect hereditary Noetherian rings. J. Algebra **13**, 327–344 (1969)
300. Mohamed, S., Bouhy, T.: Continuous modules. Arab. J. Sci. Eng. **2**, 107–122 (1977)
301. Mohamed, S.H., Müller, B.J.: Continuous and Discrete Modules. Cambridge Univ. Press, Cambridge (1990)
302. Montgomery, S.: Outer automorphisms of semi-prime rings. J. Lond. Math. Soc. **18**, 209–220 (1978)
303. Moody, R.V., Pianzola, A.: Lie Algebras with Triangular Decompositions. Wiley, New York (1995)
304. Moussavi, A., Javadi, H.H.S., Hashemi, E.: Generalized quasi-Baer rings. Commun. Algebra **33**, 2115–2129 (2005)
305. Müller, B.J.: The quotient category of a Morita context. J. Algebra **28**, 389–407 (1974)
306. Müller, B.J., Rizvi, S.T.: On the decomposition of continuous modules. Can. Math. Bull. **25**, 296–301 (1982)
307. Müller, B.J., Rizvi, S.T.: On the existence of continuous hulls. Commun. Algebra **10**, 1819–1838 (1982)
308. Müller, B.J., Rizvi, S.T.: On injective and quasi-continuous modules. J. Pure Appl. Algebra **28**, 197–210 (1983)
309. Müller, B.J., Rizvi, S.T.: Direct sums of indecomposable modules. Osaka J. Math. **21**, 365–374 (1984)
310. Müller, B.J., Rizvi, S.T.: Ring decompositions of CS-rings. In: Abstracts for Methods in Module Theory Conference, Colorado Springs, May (1991)
311. Murray, F.J., von Neumann, J.: On rings of operators. Ann. Math. **37**, 116–229 (1936)
312. Nasr-Isfahani, A.R., Moussavi, A.: On Ore extensions of quasi-Baer rings. J. Algebra Appl. **7**, 211–224 (2008)
313. Nasr-Isfahani, A.R., Moussavi, A.: Baer and quasi-Baer differential polynomial rings. Commun. Algebra **36**, 3533–3542 (2008)
314. Nicholson, W.K.: Semiregular modules and rings. Can. J. Math. **28**, 1105–1120 (1976)
315. Nicholson, W.K.: Lifting idempotents and exchange rings. Trans. Am. Math. Soc. **229**, 269–278 (1977)
316. Nicholson, W.K.: On PP-rings. Period. Math. Hung. **27**, 85–88 (1993)
317. Nicholson, W.K., Yousif, M.F.: Quasi-Frobenius Rings. Cambridge Univ. Press, Cambridge (2003)

318. Nicholson, W.K., Park, J.K., Yousif, M.F.: Principally quasi-injective modules. Commun. Algebra **27**, 1683–1693 (1999)
319. Nicholson, W.K., Park, J.K., Yousif, M.F.: Extensions of simple-injective rings. Commun. Algebra **28**, 4665–4675 (2000)
320. Okniński, J.: Semigroup Algebras. Marcel Dekker, New York (1991)
321. Ornstein, D.: Dual vector spaces. Ann. Math. **69**, 520–534 (1959)
322. Oshiro, K.: On torsion free modules over regular rings. Math. J. Okayama Univ. **16**, 107–114 (1973)
323. Oshiro, K.: On torsion free modules over regular rings III. Math. J. Okayama Univ. **18**, 43–56 (1975/76)
324. Oshiro, K., Rizvi, S.T.: The exchange property of quasi-continuous modules with finite exchange property. Osaka J. Math. **33**, 217–234 (1996)
325. Osofsky, B.L.: Homological Properties of Rings and Modules. Doctoral Dissertation, Rutgers Univ. (1964)
326. Osofsky, B.L.: Rings all of whose finitely generated modules are injective. Pac. J. Math. **14**, 645–650 (1964)
327. Osofsky, B.L.: On ring properties of injective hulls. Can. Math. Bull. **7**, 405–413 (1964)
328. Osofsky, B.L.: A generalization of quasi-Frobenius rings. J. Algebra **4**, 373–387 (1966). Erratum **9**, 120 (1968)
329. Osofsky, B.L.: A non-trivial ring with non-rational injective hull. Can. Math. Bull. **10**, 275–282 (1967)
330. Osofsky, B.L.: Endomorphism rings of quasi-injective modules. Can. J. Math. **20**, 895–903 (1968)
331. Osofsky, B.L.: Minimal cogenerators need not be unique. Commun. Algebra **19**, 2071–2080 (1991)
332. Osofsky, B.L.: Compatible ring structures on the injective hulls of finitely embedded rings. Manuscript
333. Osofsky, B.L., Smith, P.F.: Cyclic modules whose quotients have all complement submodules direct summands. J. Algebra **139**, 342–354 (1991)
334. Osofsky, B.L., Park, J.K., Rizvi, S.T.: Properties of injective hulls of a ring having a compatible ring structure. Glasg. Math. J. **52**, 121–138 (2010)
335. Osterburg, J., Park, J.K.: Morita contexts and quotient rings of fixed rings. Houst. J. Math. **10**, 75–80 (1984)
336. Palmer, T.W.: Banach Algebras and the General Theory of $*$-Algebras, Vol. I: Algebras and Banach Algebras. Cambridge Univ. Press, Cambridge (1994)
337. Papp, Z.: On algebraically closed modules. Publ. Math. (Debr.) **6**, 311–327 (1959)
338. Pardo, J.L.G., Asensio, P.L.G.: Every Σ-CS-module has an indecomposable decomposition. Proc. Am. Math. Soc. **129**, 947–954 (2000)
339. Park, J.K.: Artinian Skew Group Rings and Semiprime Twisted Group Rings. Doctoral Dissertation, Univ. Cincinnati (1978)
340. Passman, D.S.: Nil ideals in group rings. Mich. Math. J. **9**, 375–384 (1962)
341. Passman, D.S.: The Algebraic Structure of Group Rings. Wiley, New York (1977)
342. Passman, D.S.: A Course in Ring Theory. Brooks/Cole, Pacific Grove (1991)
343. Pedersen, G.K.: Approximating derivations on ideals of C^*-algebras. Invent. Math. **45**, 299–305 (1978)
344. Pedersen, G.K.: Multipliers of AW^*-algebras. Math. Z. **187**, 23–24 (1984)
345. Pierce, R.S.: Modules over Commutative Regular Rings. Memoirs, vol. 70. Amer. Math. Soc., Providence (1967)
346. Pierce, R.S.: Associative Algebras. Springer, New York (1982)
347. Pollingher, A., Zaks, A.: On Baer and quasi-Baer rings. Duke Math. J. **37**, 127–138 (1970)
348. Posner, E.C.: Prime rings satisfying a polynomial identity. Proc. Am. Math. Soc. **11**, 180–183 (1960)
349. Rangaswamy, K.M.: Representing Baer rings as endomorphism rings. Math. Ann. **190**, 167–176 (1970)

350. Rangaswamy, K.M.: Regular and Baer rings. Proc. Am. Math. Soc. **42**, 354–358 (1974)
351. Raphael, R.M., Woods, R.G.: The epimorphic hull of $C(X)$. Topol. Appl. **105**, 65–88 (2000)
352. Reiner, I.: Maximal Orders. Academic Press, London (1975)
353. Rickart, C.E.: Banach algebras with an adjoint operation. Ann. Math. **47**, 528–550 (1946)
354. Rizvi, S.T.: Contributions to the Theory of Continuous Modules. Doctoral Dissertation, McMaster Univ. (1981)
355. Rizvi, S.T.: Commutative rings for which every continuous module is quasi-injective. Arch. Math. **50**, 435–442 (1988)
356. Rizvi, S.T.: Open problems. In: Birkenmeier, G.F., Park, J.K., Park, Y.S. (eds.) International Symposium on Ring Theory. Trends in Math., pp. 441–446. Birkhäuser, Boston (2001)
357. Rizvi, S.T., Roman, C.S.: Baer and quasi-Baer modules. Commun. Algebra **32**, 103–123 (2004)
358. Rizvi, S.T., Roman, C.S.: Baer property of modules and applications. In: Chen, J.L., Ding, N.Q., Marubayashi, H. (eds.) Advances in Ring Theory. Proc. of 4th China-Japan-Korea International Conference, pp. 225–241. World Scientific, Englewood Cliffs (2005)
359. Rizvi, S.T., Roman, C.S.: On $\mathcal{K}$-nonsingular modules and applications. Commun. Algebra **35**, 2960–2982 (2007)
360. Rizvi, S.T., Roman, C.S.: On direct sums of Baer modules. J. Algebra **321**, 682–696 (2009)
361. Roman, C.S.: Baer and Quasi-Baer Modules. Doctoral Dissertation, Ohio State Univ. (2004)
362. Rotman, J.J.: The Theory of Groups: An Introduction. Allyn and Bacon, Boston (1973)
363. Rotman, J.J.: An Introduction to Homological Algebra. Academic Press, New York (1979)
364. Rowen, L.H.: Some results on the center of a ring with polynomial identity. Bull. Am. Math. Soc. **79**, 219–223 (1973)
365. Rowen, L.H.: Maximal quotients of semiprime PI-algebras. Trans. Am. Math. Soc. **196**, 127–135 (1974)
366. Rowen, L.H.: Polynomial Identities in Ring Theory. Academic Press, New York (1980)
367. Sakano, K.: Maximal quotient rings of generalized matrix rings. Commun. Algebra **12**, 2055–2065 (1984)
368. Shepherdson, J.C.: Inverses and zero divisors in matrix rings. Proc. Lond. Math. Soc. **1**, 71–85 (1951)
369. Shin, G.: Prime ideals and sheaf representation of a pseudo symmetric ring. Trans. Am. Math. Soc. **184**, 43–60 (1973)
370. Shoda, K.: Zur Theorie der algebraischen Erweiterungen. Osaka Math. J. **4**, 133–143 (1952)
371. Simón, J.J.: Finitely generated projective modules over row and column finite matrix rings. J. Algebra **208**, 165–184 (1998)
372. Small, L.W.: Semihereditary rings. Bull. Am. Math. Soc. **73**, 656–658 (1967)
373. Smith, M.K.: Group algebras. J. Algebra **18**, 477–499 (1971)
374. Smith, P.F.: Modules for which every submodule has a unique closure. In: Jain, S.K., Rizvi, T.S. (eds.) Ring Theory, pp. 302–313. World Scientific, Singapore (1993)
375. Smith, P.F., Tercan, A.: Continuous and quasi-continuous modules. Houst. J. Math. **18**, 339–348 (1992)
376. Smith, P.F., Tercan, A.: Generalizations of CS-modules. Commun. Algebra **21**, 1809–1847 (1993)
377. Smith, P.F., Tercan, A.: Direct summands of modules which satisfy (C_{11}). Algebra Colloq. **11**, 231–237 (2004)
378. Smith, S.P.: An example of a ring Morita-equivalent to the Weyl algebra A_1. J. Algebra **73**, 552–555 (1981)
379. Speed, T.P.: A note on commutative Baer rings II. J. Aust. Math. Soc. **14**, 257–263 (1972)
380. Speed, T.P.: A note on commutative Baer rings III. J. Aust. Math. Soc. **15**, 15–21 (1973)
381. Speed, T.P., Evans, M.W.: A note on commutative Baer rings. J. Aust. Math. Soc. **12**, 1–6 (1972)
382. Stenström, B.: Rings of Quotients. Springer, Berlin (1975)
383. Stephenson, B.W., Tsukerman, G.M.: Rings of endomorphisms of projective modules. Sib. Math. J. **11**, 181–184 (1970)

384. Stewart, P.N.: Semi-simple radical classes. Pac. J. Math. **32**, 249–255 (1970)
385. Stone, M.H.: Applications of the theory of Boolean rings to general topology. Trans. Am. Math. Soc. **41**, 375–481 (1937)
386. Storrer, H.: Epimorphismen von kommutativen Ringen. Comment. Math. Helv. **43**, 378–401 (1968)
387. Sun, S.H.: Duality on compact prime ringed spaces. J. Algebra **169**, 805–816 (1994)
388. Sun, S.H.: A unification of some sheaf representations of rings by quotient rings. Commun. Algebra **22**, 687–696 (1994)
389. Suzuki, Y.: On automorphisms of an injective module. Proc. Jpn. Acad. **44**, 120–124 (1968)
390. Takil, F., Tercan, A.: Modules whose submodules are essentially embedded in direct summands. Commun. Algebra **37**, 460–469 (2009)
391. Teply, M.L.: Right hereditary, right perfect rings are semiprimary. In: Jain, S.K., Rizvi, S.T. (eds.) Advances in Ring Theory. Trends in Math., pp. 313–316. Birkhäuser, Boston (1997)
392. Tercan, A.: CS-modules and Generalizations. Doctoral Dissertation, Univ. Glasgow (1992)
393. Tercan, A.: On certain CS-rings. Commun. Algebra **23**, 405–419 (1995)
394. Tomforde, M.: Continuity of ring $*$-homomorphisms between C^*-algebras. N.Y. J. Math. **15**, 161–167 (2009)
395. Utumi, Y.: On quotient rings. Osaka Math. J. **8**, 1–18 (1956)
396. Utumi, Y.: On a theorem on modular lattice. Proc. Jpn. Acad. **35**, 16–21 (1959)
397. Utumi, Y.: On continuous regular rings and semisimple self injective rings. Can. J. Math. **12**, 597–605 (1960)
398. Utumi, Y.: On continuous rings and self injective rings. Trans. Am. Math. Soc. **118**, 158–173 (1965)
399. Utumi, Y.: Self-injective rings. J. Algebra **6**, 56–64 (1967)
400. Vanaja, N., Purav, V.M.: Characterisations of generalised uniserial rings in terms of factor rings. Commun. Algebra **20**, 2253–2270 (1992)
401. Vas, L.: Dimension and torsion theories for a class of Baer $*$-rings. J. Algebra **289**, 614–639 (2005)
402. Vas, L.: Class of Baer $*$-rings defined by a relaxed set of axioms. J. Algebra **297**, 470–473 (2006)
403. von Neumann, J.: Mathematisches Grundlagen der Quantenmechanik. Springer, Berlin (1932)
404. von Neumann, J.: Continuous geometry. Proc. Natl. Acad. Sci. **22**, 92–100 (1936)
405. von Neumann, J.: Examples of continuous geometry. Proc. Natl. Acad. Sci. **22**, 101–108 (1936)
406. von Neumann, J.: On regular rings. Proc. Natl. Acad. Sci. **22**, 707–713 (1936)
407. von Neumann, J.: Continuous Geometries. Princeton Univ. Press, Princeton (1960)
408. Wang, X., Chen, J.: On FI-extending rings and modules. Northeast. Math. J. **24**, 77–84 (2008)
409. Ware, R.: Endomorphism rings of projective modules. Trans. Am. Math. Soc. **155**, 233–256 (1971)
410. Warfield, R.B. Jr.: Exchange rings and decompositions of modules. Math. Ann. **199**, 31–36 (1972)
411. Wilson, G.V.: Modules with the summand intersection property. Commun. Algebra **14**, 21–38 (1986)
412. Wisbauer, R.: Foundations of Module and Ring Theory. Gordon and Breach, Philadelphia (1991)
413. Wisbauer, R.: Modules and Algebras: Bimodule Structure and Group Action on Algebras. Longman, Harlow (1996)
414. Wolfson, K.G.: Baer rings of endomorphisms. Math. Ann. **143**, 19–28 (1961)
415. Wong, E.T., Johnson, R.E.: Self-injective rings. Can. Math. Bull. **2**, 167–173 (1959)
416. Wong, T.-L.: Jordan isomorphisms of triangular rings. Proc. Am. Math. Soc. **133**, 3381–3388 (2005)
417. Xue, W.: On PP-rings. Kobe Math. J. **7**, 77–80 (1990)

418. Yi, Z., Zhou, Y.: Baer and quasi-Baer properties of group rings. J. Aust. Math. Soc. **83**, 285–296 (2007)
419. Yohe, C.R.: Commutative rings whose matrix rings are Baer. Proc. Am. Math. Soc. **22**, 189–191 (1969)
420. Zelmanowitz, J.: Regular modules. Trans. Am. Math. Soc. **163**, 341–355 (1972)
421. Zhang, X., Chen, J.: Properties of modules and rings relative to some matrices. Commun. Algebra **36**, 3682–3707 (2008)
422. Zhou, Y.: A simple proof of a theorem on quasi-Baer rings. Arch. Math. **81**, 253–254 (2003)

Index

G.F. Birkenmeier et al., *Extensions of Rings and Modules*,
DOI 10.1007/978-0-387-92716-9,

Printed in the United States
By Bookmasters